sed on carbon-12 isotope

NAME	SYMBOL	ATOMIC NUMBER	ATOMIC WEIGHT
MERCURY	Hg	80	200.59
MOLYBDENUM	Mo	42	95.94
NEODYMIUM	Nd	60	144.24
NEON	Ne	10	20.179
NEPTUNIUM	Np	93	237.0482
NICKEL	Ni	28	58.71
NIOBIUM	Nb	41	92.9064
NITROGEN	N	7	14.0067
NOBELIUM	No	102	(254)
OSMIUM	Os	76	190.2
OXYGEN	O	8	15.9994
PALLADIUM	Pd	46	106.4
PHOSPHORUS	P	15	30.97376
PLATINUM	Pt	78	195.09
PLUTONIUM	Pu	94	(242)
POLONIUM	Po	84	(210)
POTASSIUM	K	19	39.098
PRASEODYMIUM	Pr	59	140.9077
PROMETHIUM	Pm	61	(145)
PROTACTINIUM	Pa	91	231.0359
RADIUM	Ra	88	226.0254
RADON	Rn	86	(222)
RHENIUM	Re	75	186.2
RHODIUM	Rh	45	102.9055
RUBIDIUM	Rb	37	85.4678
RUTHENIUM	Ru	44	101.07
SAMARIUM	Sm	62	150.4
SCANDIUM	Sc	21	44.9559
SELENIUM	Se	34	78.96
SILICON	Si	14	28.086
SILVER	Ag	47	107.868
SODIUM	Na	11	22.98977
STRONTIUM	Sr	38	87.62
SULFUR	S	16	32.06
TANTALUM	Ta	73	180.9479
TECHNETIUM	Tc	43	98.9062
TELLURIUM	Te	52	127.60
TERBIUM	Tb	65	158.9254
THALLIUM	Tl	81	204.37
THORIUM	Th	90	232.0381
THULIUM	Tm	69	168.9342
TIN	Sn	50	118.69
TITANIUM	Ti	22	47.90
TUNGSTEN	W	74	183.85
URANIUM	U	92	238.029
VANADIUM	V	23	50.9414
XENON	Xe	54	131.30
YTTERBIUM	Yb	70	173.04
YTTRIUM	Y	39	88.9059
ZINC	Zn	30	65.38
ZIRCONIUM	Zr	40	91.22

SEVENTH EDITION

COLLEGE CHEMISTRY

SEVENTH EDITION
COLLEGE CHEMISTRY

G. BROOKS KING
Washington State University

WILLIAM E. CALDWELL
Oregon State University

MAX B. WILLIAMS
Oregon State University

D. Van Nostrand Company

New York   Cincinnati   Toronto   London   Melbourne

The illustration on the cover is a photomicrograph of the surface of a natural diamond from Kimberley, South Africa, before cutting and polishing. The geometric figures are the diamond trigons which represent the way the crystal molecules grew. The photo was taken through an interference microscope which translates very fine gradations of height and depth into color. From Encyclopedia of Minerals by Roberts, Rapp & Weber © 1974 Litton Educational Publishing, Inc. All rights reserved. Published by Van Nostrand Reinhold.

D. Van Nostrand Company Regional Offices:
New York   Cincinnati

D. Van Nostrand Company International Offices:
London   Toronto   Melbourne

Copyright © 1977 by Litton Educational Publishing, Inc.

Library of Congress Catalog Card Number: 77–72074
ISBN: 0–442–24417–7

Published by D. Van Nostrand Company
450 West 33rd Street, New York, N.Y. 10001

10 9 8 7 6 5 4 3 2 1

Preface

The Seventh Edition of *College Chemistry* is a major revision of a book used successfully in one-year introductory courses designed for students who are not majoring in chemistry but who need a comprehensive survey of modern chemical principles and descriptive chemistry. Every effort has been made to produce a book that retains the same readability as the preceding editions. This edition is the combined result of many thousands of hours of class testing and of extensive users' comments.

The authors believe that a balance of theoretical and practical chemistry is essential in the introductory course, which for many students may be their final course in the subject. For this reason, such topics as colloids, geological chemistry, modern organic molecules, nuclear chemistry, and biochemistry are included without being overwhelming in their treatment.

Much attention has been given to teaching students how to write chemical equations properly. Over one hundred worked-out examples and problems are included. The approximately 900 exercises at the ends of chapters are the result of considerable revision and changes. Many short-answer and matching exercises have been included as well as conventional exercises and problems. Suggested supplementary readings are given at the end of each chapter.

SI units are introduced in this Edition, while some of the older accepted units, such as calorie, Å, torr, and liter, have been retained. Footnotes are used when the SI units first appear in order to show the relationship between the SI units and the older ones. The term gram atom has been deleted and the basic SI unit *mole* used throughout the text. Reduction potentials have replaced oxidation potentials to conform to modern usage.

Emphasis has been given to the *Chemical and Engineering News* list of *50 Top Chemicals* produced in the U.S., the least of which has a use of over a billion pounds annually. The authors believe students should have some concept of the uses and importance of our major chemicals, and they are studied in their proper places as the course develops.

The four chapters on organic and biochemistry and the two chapters on nuclear chemistry have been placed last in the book and are written in such a way that instructors may use them wherever they wish in the course. Instructors may wish to omit them entirely, and this may be done without affecting the rest of the material.

The authors gratefully acknowledge the assistance of users of previous editions whose suggestions have been incorporated in this revision. Special thanks and appreciation are given to Professors H. D. Reese and I. J. Tinsley for their encouragement and help. In addition we wish to thank Professors Ruth Bowen, Peter Popovich, and Maurice Stringer for their helpful critiques.

G. B. King
M. B. Williams

Contents

Chapter 1

"in all the operations of art and nature, nothing is created; an equal quantity of matter exists both before and after the experiment. . . ."

Lavoisier

Fundamental Principles

1.1   INTRODUCTION

Chemistry concerns itself with the composition and properties of matter and the principles governing the changes which matter may undergo. In order to maintain a working relationship with other areas of science, chemistry must use the ground rules which all sciences have in common with each other. A fundamental rule that all science follows is that it progresses primarily with information acquired through observation and experiment. An **experiment** is an operation carried out under controlled conditions in order to gain information for some purpose. For instance, determining the melting point of ice by placing a known amount of ice in an insulated container and carefully varying the temperature so that the temperature in the container is known at all times. When the ice begins to melt, the temperature inside the container is noted. An **observation** is essentially the witnessing of an event under controlled conditions. An example would be watching the melting of the ice in the container described above by the heat of the sun.

The **facts** obtained by these methods of science are organized and classified to discover relationships between them. A precise statement which describes a mode of behavior or a pattern of action concerning a group of related facts is a **law.** After facts and laws concerning these facts have been made a part of knowledge, the next step is to use imagination or speculation regarding particular phenomena. The why, how, or what, pertaining to a set may suggest an **hypothesis.** This is just a provisional conjecture or speculation about a possible law of nature. The hypothesis is then tested by experimentation, and more facts are obtained which lead either to a verification of the hypothesis or to its modification or rejection. A **theory** is a verified hypothesis. It consists essentially of a set of assumptions by which factual evidence and laws may be explained. It is not a permanent conclusion; it may be changed and modified as experimental results are obtained which are not explained by the assumptions of the theory. Theories are useful because they suggest further experiments which lead to the discovery of new phenomena. We will see as we proceed further into the book how this happens in practice.

Traditionally, chemistry embraces such a broad area of knowledge that subdivisions of the subject are common. **Organic** chemistry is a study of the compounds of carbon; **inorganic** chemistry deals with the properties and characteristics of all of the other chemical elements; **analytical** chemistry is concerned with the analysis and identification of substances; **physical** chemistry studies the natural laws and principles which govern both physical and chemical changes; **biochemistry** deals with materials found in living organisms. In this text we shall endeavor to survey these different areas of chemistry—not in detail, but sufficiently to introduce the student to the science, and to show him how chemistry applies to almost everything he does.

In recent years there has been a trend to integrate the traditional fields of chemistry into the categories of **structure, synthesis,** and **energetics.** This, consequently, cuts across any boundaries between the various branches of chemistry.

Chemistry may also be characterized as **descriptive chemistry** which involves chemical facts such as properties, preparations, reactions, and uses of the various elements and their compounds; and **theoretical chemistry**, which provides the theories which explain and unify the empirical facts into a system of logical principles. Both are important for a knowledge and understanding of chemistry and will be studied in this course.

1.2   IMPORTANCE OF CHEMISTRY

Chemistry is utilized in many branches of knowledge. It is therefore a required course in a variety of college curricula, including medicine, pharmacy, veterinary science, agriculture, engineering, and home economics. The application of chemical principles has done much to give man control over nature, to alleviate human suffering, and to provide the comforts of modern living. The chemist uses chemical change to produce thousands of modern-day conveniences and health aids. Through chemistry, coal may be converted into coke, coal gas, coal tar, and condensible liquids, each having commercial value. Ammonia and fuel gas may be obtained from coal gas; coal tar may be converted into pharmaceuticals, dyes, plastics, explosives, and numerous other useful products. Some idea of the magnitude of the chemical industry can be gained from Tables 1.1 and 1.2, which show the 50 top chemicals produced in the U.S. and the 50 top chemical producers. The ranking of the top chemicals and their producers changes slightly from year to year, reflecting changes in the economy and needs. Each of these chemicals, however, is tremendously important. Even the lowest one in this list, vinyl acetate, has an annual production of 1,140,000,000 pounds. In the chapters to follow the significance and uses of the most important of the industrial chemicals will be illustrated as they are studied in their proper places.

We turn now to the fundamental principles which are the basic tools for a true understanding of chemistry.

1.3   UNITS OF MEASUREMENT

Metric System

The metric system is now used universally in scientific work. It is based on the decimal system of multiples of ten. Changes are made within a unit simply by moving the decimal point an appropriate number of places. Metric system conventions are established by international agreement and are recommended by the International Committee on Weights and Measures.

Table 1.1

The 50 top chemicals produced in the U.S.

Rank 1975	Rank 1974[a]		Production (Billions of lb) 1975	Production (Billions of lb) 1974	Production (Common units[b]) 1975	Exports (Common units[b]) 1975	Imports (Common units[b]) 1975
1	1	Sulfuric acid	61.18	66.10	30,591 tt	142 tt	303 tt
2	2	Lime	36.27	40.75	18,135 tt[a]	na	0.13 tt
3	4	Ammonia, anhydrous	31.56	31.61	15,781 tt	301 tt	817 tt
4	3	Oxygen, high and low purity	29.06	32.28	351 bcf	142 bcf	na
5	5	Ethylene	19.78	23.89	19,782 mp	3 mp	[c]
6	6	Sodium hydroxide	18.53	22.38	9,265 tt	1,110 tt	108 tt
7	7	Chlorine, gas	18.06	21.51	9,029 tt	17 tt	74 tt
8	8	Nitrogen, high and low purity	17.17	17.68	237 bcf	36 bcf	na
9	12	Phosphoric acid, total	14.32	14.43	7,158 tt	313 tt	0.7 tt
10	10	Sodium carbonate[d]	14.29	15.11	7,146 tt	530 tt	2.4 tt
11	9	Nitric acid	14.15	16.24	7,073 tt	na	4 tt
12	11	Ammonium nitrate[e]	13.93	15.08	6,963 tt	46 tt	3 tt
13	13	Benzene	7.66	10.91	1,045 mg	31 mg[f]	70 mg
14	14	Propylene	7.60	10.46	7,598 mp	17 mp	[c]
15	16	Urea, primary solution	7.39	7.58	7,390 mp	1,114 mp[g]	1308 mp
16	15	Ethylene dichloride	5.94	9.17	5,940 mp	130 mp	na
17	18	Toluene, all grades	5.87	6.68	811 mg	117 mg[f]	21 mg
18	17	Methanol, synthetic	5.18	6.88	5,177 mp	na	111 mp
19	21	Xylene, all grades	5.08	5.78	704 mg	70 mg[h]	36 mg
20	19	Ethylbenzene	4.72	6.05	4,715 mp	84 mp	na
21	25	Terephthalic acid[i]	4.65	4.26	4,645 mp	207 mp	na
22	22	Formaldehyde, 37% by weight	4.62	5.76	4,616 mp	127 mp[j]	0.008 mp
23	20	Styrene	4.40	5.96	4,398 mp	574 mp	7 mp
24	23	Vinyl chloride	4.20	5.62	4,196 mp	415 mp	3 mp
25	24	Hydrochloric acid	3.95	4.90	1,973 tt	18.8 tt[k]	43 tt
26	27	Ethylene oxide	3.94	3.89	3,938 mp	na	4 mp
27	26	Ammonium sulfate	3.84	4.24	1,921 tt	726 tt	218 tt

28	31	Ethylene glycol	3.80	3.34	3,799 mp	97 mp	371 mp
29	29	Carbon dioxide, all forms	2.76	3.53	1,380 tt	0.76 tt	na
29	30	Carbon black	2.76	3.35	2,756 mp	88 mp	33 mp
31	28	Butadiene (1,3-), rubber grade	2.65	3.68	2,645 mp	73 mp	554 mp
32	35	Sodium sulfate[l]	2.53	2.70	1,267 tt	78 tt	276 tt
33	34	p-Xylene	2.43	2.71	2,425 mp	246 mp	na
34	33	Calcium chloride[m]	2.12	2.74	1,062 tt[n]	28 tt	12 tt
34	37	Aluminum sulfate, commercial	2.12	2.57	1,062 tt	48 tt	12 tt
36	36	Acetic acid, synthetic	2.10	2.58	2,097 mp	5 mp	0.6 mp
37	32	Cumene	1.97	2.91	1,973 mp	46 mp	209 mp
38	38	Cyclohexane	1.77	2.35	1,773 mp	283 mp	na
39	39	Phenol, synthetic	1.72	2.30	1,722 mp	51 mp	25 mp
40	40	Acetone	1.59	1.98	1,591 mp	65 mp	3 mp
41	41	Isopropanol	1.57	1.94	1,573 mp	123 mp[o]	p
42	42	Sodium tripolyphosphate	1.55	1.81	777 tt	76 tt	na
43	43	Propylene oxide	1.52	1.76	1,523 mp	na	21 mp
44	44	Acetic anhydride	1.47	1.63	1,472 mp	na	20 mp
45	48	Adipic acid	1.36	1.48	1,359 mp[n]	10 mp	na
46	47	Sodium silicate (water glass)	1.31	1.54	655 tt	na	0.03 tt
47	45	Ethanol, synthetic	1.29	1.62	1,292 mp	95 mp	51 mp
48	49	Acrylonitrile	1.22	1.41	1,215 mp	198 mp	7 mp
49	46	Titanium dioxide	1.21	1.57	604 tt	16 tt	27 tt
50	50	Vinyl acetate	1.14	1.40	1,136 mp	202 mp	3 mp

[a] Revised. [b] tt = thousand tons, bcf = billion cubic feet, mp = million pounds, mg = million gallons. [c] Ethylene, propylene, and butylene are included in one category at 55 million gal. [d] Synthetic and natural. [e] Original solution. [f] Crude and pure. [g] Includes all urea fertilizer material. [h] Crude only. [i] Includes both the acid and the ester without double counting. Acid production was multiplied by 1.16 to convert to equivalent dimethyl terephthalate. [j] Includes small amounts of other aldehydes and ketones. [k] Includes chlorosulfonic acid. [l] High and low purity. [m] Solid and liquid. [n] C&EN estimate. [o] Includes all propanol. [p] Negligible. na = not available. Sources: International Trade Commission, Bureau of Mines, Bureau of the Census.

Table 1.2

The 50 top chemical producers in the U.S.

Rank 1975	1974	Company	Chemical sales 1975 ($ millions)	Change from 1974 (%)	Total sales 1975 ($ millions)	Chemical sales as % of total sales
1	1	Du Pont	$5,550	2%	$ 7,222	76%
2	3	Union Carbide	3,425	5	5,665	60
3	2	Dow Chemical	3,360	−7	4,888	69
4	4	Monsanto	3,054	4	3,625	84
5	5	Exxon	2,594	−7	44,864	6
6	7	W. R. Grace[a]	1,800	2	3,529	51
7	6	Celanese	1,716	−1	1,900	90
8	9	Allied Chemical	1,522	1	2,333	65
9	8	Occidental Petroleum	1,447	−8	5,346	27
10	11	Shell Oil	1,203	6	8,143	15
11	10	Hercules	1,145	−8	1,413	81
12	13	American Cyanamid	1,100	7	1,928	57
13	14	Eastman Kodak	1,059	7	4,959	21
14	15	Rohm & Haas	994	1	1,046	95
15	17	Borden[a]	990	5	3,367	29
16	16	FMC	936	−1	2,292	41
17	21	Stauffer Chemical	926	10	950	98
18	19	Mobil Oil[a]	909	4	20,620	4
19	23	Standard Oil (Ind.)	889	16	9,955	9
20	22	Ethyl Corp.[a]	834	5	1,029	81
21	12	Phillips Petroleum	830	−21	5,134	16
22	18	Gulf Oil	812	−9	14,268	6
23	20	Texaco	800	−5	24,507	3
24	29	PPG Industries	735	22	1,887	39
25	41	International Minerals[b]	719	63	1,303	55
26	27	Diamond Shamrock[a]	697	11	1,129	62
27	32	Air Products[a]	679	24	699	97
28	24	U.S. Steel	656	−11	8,167	8
29	25	Ashland Oil[c]	639	−5	3,637	18
30	28	Standard Oil of California	633	5	16,822	4
31	31	Olin	625	11	1,261	50
32	26	NL Industries[a]	620	−3	1,279	48
33	33	Ciba-Geigy	572	14	833	65
34	35	BASF Wyandotte	560	15	590	95
35	38	Williams Cos.	518	12	882	59
36	30	B. F. Goodrich[a]	517	−13	1,901	27
37	50	American Hoechst	470	60	617	76

Table 1.2 (Continued)

Rank 1975	1974	Company	Chemical sales 1975 ($ millions)	Change from 1974 (%)	Total sales 1975 ($ millions)	Chemical sales as % of total sales
38	37	Atlantic Richfield	444	−4	7,308	6
39	34	Akzona	429	−13	682	63
40	45	Lubrizol	419	11	419	100
41	40	Mobay Chemical	418	8	418	100
42	39	Cities Service[a]	417	2	3,201	13
43	47	Kerr-McGee[a]	414	20	1,799	23
44	35	Reichhold Chemicals	408	−16	408	100
45		Esmark[a,d]	390	30	4,731	8
46	44	National Distillers	382	0	1,267	30
47	49	Pennwalt	363	10	714	51
48	42	Goodyear Tire	350	−12	5,452	6
49	46	Tenneco	349	−5	5,630	6
50		Nalco Chemical	307	14	314	98

[a] Chemical sales include significant nonchemical sales (welding and cryogenic equipment, fabricated plastics products, coatings, metals, minerals, adhesives, and the like). [b] For fiscal year ended June 30. [c] For fiscal year ended Sept. 30. [d] For fiscal year ended Oct. 25. Enterprise Standard Industrial Classifications used above are as follows: 20.1 Meat products; 20.2 Dairy; 20.8 Beverages; 28.1 Industrial chemicals and synthetics; 28.5 Paints and allied products; 28.7 Agricultural chemicals; 28.9 Miscellaneous chemical products; 29.1 Petroleum refining; 30.5 Rubber products; 32.1 Glass products; 33.1 Blast furnaces and steel mills; 38.6 Photographic equipment and supplies.

Reprinted with permission of the copyright owner, THE AMERICAN CHEMICAL SOCIETY, from the May 3rd, 1976 issue of CHEMICAL & ENGINEERING NEWS.

The most recent recommendation of the committee is the adoption of the **International System of Units** (Le Système International d' Unites) officially designated **SI** units. See Figure 1.1, pages 8–9. This system of units is founded on seven **base units** and two **supplementary units.** Various other **derived units** are calculated from these by algebraic combinations. Some of the SI units are listed in the Appendix.

While a number of scientific groups, such as the National Bureau of Standards, have adopted SI units, most chemists have acted cautiously and adopted the units only partially. Many have retained such widely used and commonly accepted non-SI units as the calorie, liter, angstrom, atmosphere, and torr. In this edition of the text we will follow the latter procedure and make only partial use of SI units.

The standard unit of linear measurement is the **meter** (abbreviated, **m,** without the decimal point) which is defined as 1,650,763.73 wavelengths in vacuum of the orange-red line of the spectrum of krypton-86. A meter is equivalent to 39.370 inches. The standard unit of weight measurement is

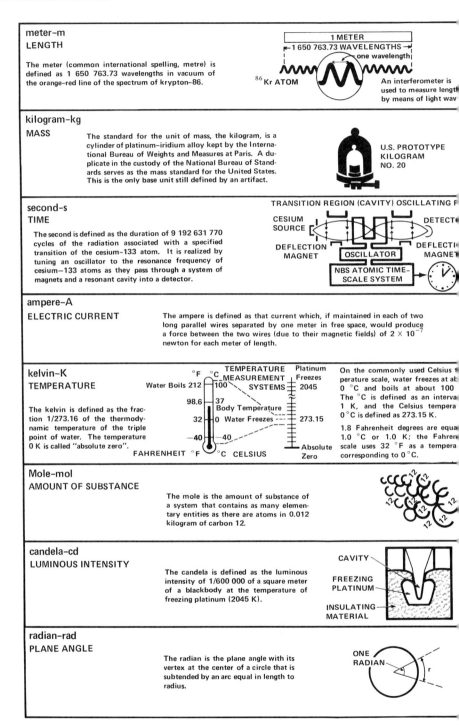

Figure 1.1

The Modernized Metric System — The International System of Units-SI.

The SI unit of volume is the cubic meter (m^3). The liter (0.001 cubic meter), although not an SI unit, is commonly used to measure fluid volume.

The SI unit of area is the square meter (m^2).

The SI unit of force is the newton (N). One newton is the force which, when applied to a 1 kilogram mass, will give the kilogram mass an acceleration of 1 (meter per second) per second.

$$1N = 1kg \cdot m/s^2$$

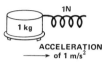

ACCELERATION
of 1 m/s^2

The SI unit for pressure is the pascal (Pa).

$$1Pa = 1N/m^2$$

The SI unit for work and energy of any kind is the joule (J).

$$1J = 1N \cdot m$$

The SI unit for power of any kind is the watt (W).

$$1W = 1J/s$$

The number of periods or cycles per second is called frequency. The SI unit for frequency is the hertz (Hz). One hertz equals one cycle per second.

The SI unit for speed is the meter per second (m/s).

The SI unit for acceleration is the (meter per second) per second (m/s^2).

Standard frequencies and correct time are broadcast from WWV, WWVB, and WWVH, and stations of the U.S. Navy. Many short-wave receivers pick up WWV and WWVH, on frequencies of 2.5, 5, 10, 15, and 20 megahertz.

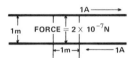

FORCE = 2×10^{-7}N

The SI unit of voltage is the volt (V).

$$1V = 1W/A$$

The SI unit of electric resistance is the ohm (Ω).

$$1\,\Omega = 1V/A$$

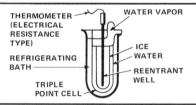

THERMOMETER (ELECTRICAL RESISTANCE TYPE)

WATER VAPOR

ICE WATER

REFRIGERATING BATH

REENTRANT WELL

TRIPLE POINT CELL

The standard temperature at the triple point of water is provided by a special cell, an evacuated glass cylinder containing pure water. When the cell is cooled until a mantle of ice forms around the reentrant well, the temperature at the interface of solid, liquid, and vapor is 273.16 K. Thermometers to be calibrated are placed in the reentrant well.

When the mole is used, the elementary entities must be specified and may be atoms, molecules, ions, electrons, other particles, or specified groups of such particles.

The SI unit of concentration (of amount of substance) is the mole per cubic meter (mol/m^3).

The SI unit of light flux is the lumen (lm). A source having an intensity of 1 candela in all directions radiates a light flux of 4π lumens.

A 100–watt light bulb emits about 1700 lumens

steradian–sr

SOLID ANGLE

The steradian is the solid angle with its vertex at the center of a sphere that is subtended by an area of the spherical surface equal to that of a square with sides equal in length to the radius.

Area
r^2

ONE STERADIAN

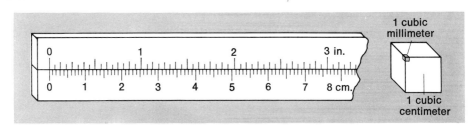

Figure 1.2

Comparison scale of inches and centimeters.

the **kilogram** (abbreviated, **kg**). A kilogram is defined as the weight of a cylinder of platinum-iridium alloy kept by the International Bureau of Weights and Measures in Paris. A duplicate is kept by the National Bureau of Standards in the United States. A kilogram equals 2.2046 pounds. The most commonly used unit of volume, although not an SI unit, is the **liter** (abbreviated l). A liter is the volume of one thousand cubes, each cube measuring one centimeter along each edge. Thus one liter is equal to 1000 cubic centimeters (cc or cm³). The SI unit of volume is the cubic meter (m³). A liter is 0.001 m³. Since the metric system is based on the decimal system of multiples of ten, the above units may be subdivided into smaller

Table 1.3
Metric units

LINEAR		
1 millimeter°	(mm) =	0.001 meter (m)
1 centimeter	(cm) =	0.01 m.
1 decimeter	(dm) =	0.1 m
1 kilometer	(km) =	1000.0 m
1 micron	(μ) =	0.000001 (1×10^{-6})m
1 micrometer	(μm) =	0.000001 (1×10^{-6})m
1 nanometer	(nm) =	0.000000001 (1×10^{-9})m
1 angstrom	(Å) =	0.0000000001 (1×10^{-10})m
1 picometer	(pm) =	0.000000000001 (1×10^{-12})m
WEIGHT		
1 milligram	(mg) =	0.001 gram (g)
1 centigram	(cg) =	0.01 g
1 decigram	(dg) =	0.1 g
1 kilogram	(kg) =	1000.0 g
VOLUME		
1 liter = 1000 milliliters (ml) = 1000 cubic centimeters (cc or cm³) = 0.001 m³		

° The prefix **milli-** stands for one-thousandth, **centi-** for one-hundredth, **deci-** for one-tenth, and **kilo-** for one thousand.

units such as are shown in Table 1.3. This table includes the units commonly used in chemistry.

Weights are usually expressed decimally in grams. Thus 10 grams and
150 milligrams is written as 10.150 g. The relationship between a few units
of measurement in the English and metric systems is shown in Table 1.4.

Table 1.4

English-metric equivalents

1 inch	=	2.540 centimeters
1 meter	=	39.37 inches
1 pound	=	453.6 grams
1 liter	=	1.057 quarts
1 kilogram	=	2.204 pounds

Significant Figures

Even the most careful observations are subject to experimental errors. It
is customary to indicate accuracy by using a certain number of figures or
digits, called **significant figures.** Suppose we weigh an object on an analytical
balance sensitive to one milligram; that is, 0.001 g and find its weight to be
25.382. This result has **five** significant figures and indicates that the accuracy
is within 0.001 g. If the balance is sensitive to only 0.1 g, we express the
result as 25.4 g, which has three significant figures. The result is rounded
off to indicate that only three figures are significant. If the number had been
25.342, it would be rounded off to 25.3, etc. In rounding off a number, the
rule is: if the number following the last significant figure is 5 or greater, the
next higher number is used for the last significant figure; if it is smaller than
5, the last significant figure is retained.

Actually, in using the convention of significant figures, the last figure
is uncertain. When we express the above weight as 25.382 g (five significant
figures), the last digit (2) is uncertain, since our balance is sensitive to only
0.001 g. The rule is to use only the number of digits which are known reliably
(except that the last one is uncertain). With a number less than 1, the significant figures are counted from the first non-zero digit. Thus 0.00746 has three
significant figures, and the number 6 is uncertain.

Very frequently we carry out mathematical operations, such as addition,
subtraction, multiplication, and division, with our experimental data. The
errors involved in the measurement of each quantity we use as well as the
mathematical operation performed, will influence the number of significant
figures of the result. There are a number of methods used to determine how
many significant figures should be retained in the answer of a mathematical
operation. However, the following simple rules, although approximate, will
give adequate results.

For addition or subtraction, the result should be reported to the same number of decimal places as the term in our data with the least number of decimal places. For example, in the addition

$$
\begin{array}{r}
23.156 \\
13.2 \\
62.1834 \\
\hline
98.5394
\end{array}
$$

the answer should be reported as 98.5 because the number 13.2 in the data has only one digit following the decimal point. In adding the numbers 45.315 and 21.3, the result should be expressed as 66.6 and not as 66.615, since the second term is accurate only to the nearest 0.1. It would be pointless to include more digits, since they would not be significant.

Likewise, for multiplication or division the answer should be reported to the same number of significant figures as the term in our data with the lowest number of significant figures. For example, we may calculate the area of a square, the edge of which is 98 mm, the measurement being accurate to 1 mm.

$$ \text{Area} = 98 \text{ mm} \times 98 \text{ mm} = 9604 \text{ mm}^2 $$

Since we know the length of the edge to only two significant figures, our calculation of the area cannot be any more reliable. It is therefore approximately 9600 mm^2. To indicate clearly the reliability of the result, we write it as 9.6×1000 or, better, as 9.6×10^3. If we had measured the length to 0.1 mm, however, and had found it to be 98.0 mm (that is, three significant figures), the result would have been written as 9.60×10^3 mm^2, rounding off the last 4 to the next decimal place. Our result then has three significant figures, which is the same as the number of significant figures in 98.0. We must keep in mind that in a calculation involving several independent measurements, the accuracy of the result is determined by the least reliable measurement, and it is pointless to record the result to more significant figures than are warranted.

EXAMPLE 1.1   How many significant figures are implied in each of the following weights?

(1)   62.3 g   Ans. 3
(2)   24.654 g   Ans. 5
(3)   0.0623 g   Ans. 3
(4)   43.0 g   Ans. 3
(5)   600 g   Ans. 1 to 3, we are not sure

In case (3) the zero to the right of the decimal point is there only to show the decimal point, and is not a significant figure. The weight could be written in scientific notation as 6.23×10^{-2} g or simply as 62.3 mg which clearly have only 3 significant figures.

In case (4) the zero is not necessary to show the decimal point and would only have been placed there by the experimenter to make clear that the weight was good to the nearest 0.1 g.

In case (5) the two zeros to the left of the decimal point are necessary to show where the decimal point is and may or may not be significant—only the experimenter knows. The weight 600 g may be good to the nearest 1 g or 10 g or 100 g. Again the uncertainty could be avoided by the experimenter using scientific notation. 6.00×10^{2} g would mean 3 significant figures; 6.0×10^{2}, 2 significant figures; 6×10^{2}, 1 significant figure.

EXAMPLE 1.2   Add the following volumes and express the answer in liters:

$$4.51 \, 1 \,,\, 2502 \text{ ml} \,,\, 1.55 \text{ dl} \,,\, 50.0 \text{ cc}$$

SOLUTION: Changing all the volumes to liters, we have

$$
\begin{array}{l}
4.51 \ 1 \\
2.502 \ 1 \\
0.155 \ 1 \\
\underline{0.050 \ 1} \\
7.217 \ 1
\end{array}
$$

which should be rounded off to 7.22 l since this answer has the same number of decimal places (2) as the least precise of the data, 4.51 l.

EXAMPLE 1.3   A cube measured 1.22 cm on a side. What is the volume of the cube?

SOLUTION: Volume $= (1.22 \text{ cm})^{3} = 1.815848 \text{ cm}^{3}$, which should be rounded off to 1.82 cm³ since this answer has 3 significant figures, the same as the original data.

Measurement of Temperature

Centigrade (or Celsius) and absolute (or Kelvin) scales of temperature, rather than the Fahrenheit scale are employed in scientific work. The absolute scale was proposed by Lord Kelvin in 1848 and temperatures on this scale are usually designated by the letter K. The Kelvin temperature is one of the basic units in the SI system. The relationship between the three scales to 3 significant figures is shown in Figure 1.3. The horizontal lines A and B denote the boiling point and the freezing point of water. Note that there are

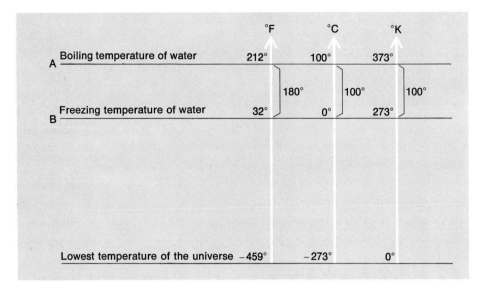

Figure 1.3

Comparison of temperature scales.

100 degrees between the freezing and boiling points of water on *both* the centigrade and the absolute scales of temperature. A reading on the absolute scale, however, will always be 273 (more exactly 273.15) degrees higher than on the centigrade scale, since 273 on the absolute scale corresponds to 0 on the centigrade scale. To convert centigrade temperature to absolute temperature we simply add 273 to the centigrade reading; thus

$$°K = °C + 273.$$

The absolute scale of temperature is based on the fact that when a gas is cooled, it contracts (decreases in volume) $\frac{1}{273}$ of its volume at 0°C for every degree the temperature is lowered. Theoretically, then, a gas would possess no volume at a temperature of −273.15°C. Actually this condition is never realized, since all substances become solids before this temperature is reached. This theoretical temperature is termed **absolute zero,** and is the lowest possible temperature of the universe.

Again referring to Figure 1.3, we note that the freezing point of water on the Fahrenheit scale is 32° and the boiling point 212°. Between the freezing and boiling points there are 180 degrees or divisions (212 − 32); on the centigrade scale there are only 100 such divisions. One division on the centigrade scale, then, is equal to $\frac{180}{100} = 1.8$ or $\frac{9}{5}$ divisions on the Fahrenheit scale; conversely, one Fahrenheit degree is equal to $\frac{5}{9}$ of a centigrade degree. A little study of Figure 1.3 will show that a general formula may be derived for conversion of centigrade to Fahrenheit or vice versa.

$$F = \tfrac{9}{5}C + 32 \quad \text{or} \quad C = \tfrac{5}{9}(F - 32)$$

Density

The density of a substance is its **mass in relation to its volume**—in other words, mass per unit volume. A piece of iron has a greater density than a piece of wood because the mass of a given volume of iron is greater than the mass of the same volume of wood. Expressed mathematically:

$$\text{Density} = \frac{\text{Mass}}{\text{Volume}} \quad \text{or} \quad D = \frac{M}{V}$$

In chemical work, the density of a substance is usually expressed in grams per milliliter (g/ml) or grams per cubic centimeter (g/cm³).

EXAMPLE 1.4   A block of metal measuring 15.0 cm × 80.0 mm × 2.50 dm weighs 33.0 kg. What is the density of the metal?

SOLUTION:

Volume of metal = 15.0 cm × 8.00 cm × 25.0 cm = 3000 cm³

$$\text{Density} = \frac{M}{V} = \frac{33000 \text{ g}}{3000 \text{ cm}^3} = 11.0 \text{ g/cm}^3$$

1.4 MATTER AND ITS CHANGES

Matter and Energy

Chemistry is *concerned with* **matter and the changes which it undergoes. Matter** may be defined as anything which possesses **weight or mass°** and occupies space. Rocks, soil, water, and the air we breathe are examples. Any matter which is homogeneous (uniform throughout), such as sugar, table salt, iron, or baking soda, is called a **substance.**

Matter appears in any one of three physical states: **solid, liquid,** or **gas.** Solids are characterized by rigidity and a definite form; liquids flow; gases diffuse and fill any container in which they may be placed. Most substances can be made to appear in all three of these states—for example, water may exist as a solid (ice), a liquid, or a gas (steam). By adding heat, a liquid may be changed to a gas; by subtracting heat, a liquid may be changed to a solid.

° The mass of a body is the quantity of matter it contains. Mass is determined by the body's inertia—that is, the resistance offered to a change in its motion. **Weight,** on the other hand, is the attraction between the earth and the body. Since the attraction between two bodies is inversely proportional to the square of the distance between the centers of the bodies, the masses of two bodies will be proportional to their weights, as long as the weights are measured at the same distance from the earth's center. Consequently, we may compare masses of objects by weighing them. The attraction of the earth for an object is counterbalanced against arbitarily chosen units of weight. While the mass of an object is constant, its weight varies with its distance from the center of the earth. For our study, however, we shall use the terms "mass" and "weight" interchangeably.

Changes in matter are accompanied by changes in energy. **Energy is the ability to do work** and manifests itself in many forms. The more familiar forms of energy are **heat, mechanical, radiant, electrical,** and **chemical.** Though it may pass from one form to another, **energy may be neither created nor destroyed.** This law of nature is known as the **Law of Conservation of Energy.**

When coal is burned, heat is given up or evolved. This heat, termed heat energy, may be made to do useful work. For example, the heat from burning coal can be used to generate steam, which in turn can operate a steam engine. The moving parts of a steam engine yield mechanical energy, which may be transformed into electrical energy by means of a dynamo. Electrical energy may be converted into light energy, as in a light bulb, or into heat energy, as in an electric toaster.

Another form of energy is potential energy, in which matter has position or other factors that can quickly translate mass into motion. Examples are water behind a dam and a compressed coiled spring.

Chemists are more interested in **chemical energy,** which is "stored energy"—the energy possessed by a substance that allows it to be changed into a new substance. Thus when coal is burned, the chemical energy possessed by the coal is transformed into heat energy and light energy. Various other forms of energy may also be transformed into chemical energy. For example, when plants absorb radiant energy from the sun, changes occur in the plants which result in the production of plant tissue.

The Classification of Matter

On the basis of its chemical composition, matter may be classified as follows:

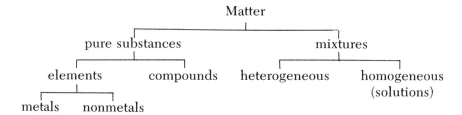

Pure substances are **homogeneous**—perfectly uniform in composition—every part is like every other part. Every minute bit of sugar is like every other small particle of sugar. Salt, soda, aspirin, and gold are further examples of pure or homogeneous substances.

Changes in Matter

We can hardly fail to notice the many changes in matter which are continually taking place about us. Ice melts to liquid water, which in turn may be converted into water vapor by the application of sufficient heat. Liquid water may be changed into ice, or steam may be condensed to a liquid, by a lowering of the temperature. In these processes, although water is changed from one physical state to another, its chemical composition remains the same. Similarly, no change in chemical composition occurs in the magnetization of iron, melting of lead, or stretching of rubber. Changes which take place without modifying the chemical composition of matter are called **physical changes.**

In contrast to physical changes are the processes which take place with a modification of composition. When a piece of iron rusts, a brittle reddish deposit is formed on the surface; the iron is converted into a new substance with a new set of properties. Changes which involve alteration of composition are called **chemical changes.** Decay of animal and vegetable matter, digestion of food, and action of acids on metals are further examples of chemical changes.

Chemical changes are always accompanied by either the liberation or the absorption of heat energy. If heat energy is liberated, the change is said to be **exothermic;** if heat energy is absorbed, the change is **endothermic.**

A common unit of heat energy is the **calorie** (cal) which is defined as the quantity of heat necessary to raise the temperature of one gram of water one degree centigrade.° A **kilocalorie** (kcal) is 1000 calories—the quantity of heat necessary to raise the temperature of one kilogram of water one degree. In the SI system the **Joule** (J) is the unit used for energy. 1 cal = 4.1840 J. A Joule is equivalent to 10^7 ergs.

Properties of Matter

Every substance possesses qualities or characteristics by which we are able to identify it. Sugar is a white solid, dissolves in water, and tastes sweet; iron has a metallic luster, may be magnetized, melts at a certain temperature, and rusts with the formation of a reddish deposit. These qualities or characteristics are called **properties.**

Properties which are associated with physical changes are termed **physical properties.** The melting point (temperature at which a substance changes from the solid state to the liquid state) of a substance is a physical property, since the liquid has the same composition as the solid. Properties which may be perceived by the senses, such as color, odor, and taste, are physical properties. Other common physical properties of a substance are boiling point, density, and solubility in water and other solvents. Less common physical properties include thermal or electrical conductivity, heat of fusion or vaporization, coefficient of thermal expansion, hardness, malleability, ductility, and crystalline form.

Chemical properties are involved only when matter undergoes chemical change. The rusting of iron is a chemical property of iron, since a change in composition occurs and a new substance is formed. Similarly, as coal burns

° Unless otherwise designated, all temperatures will hereafter be given in °C.

it combines with oxygen, and new gaseous substances are formed. Water, soil minerals, and carbon dioxide are converted to many chemical substances by growing plants. The fact that acids will act on most metals is a general chemical property of acids. In chemical change, the substance undergoing change is nearly always associated with at least one other substance; physical change involves only a single substance.

Mixtures

Materials which are variable in composition are called **mixtures.** Most of the materials we encounter in everyday life are of this type. Food products, rocks, soil, wood, and cement are examples. These materials are mixtures of pure substances.

The constituents of a mixture may be present in different proportions. For example, various cements contain variable proportions of calcium and aluminum silicates. Bread is a mixture of ingredients which may be present in different proportions.

The constituents of a mixture retain their identity, since their physical and chemical properties have not been changed by simple mixing—for example, a mixture of powdered iron and sulfur retains the properties of both iron and sulfur. The sulfur, which dissolves in carbon disulfide, may be separated from the iron by extraction with that solvent; the iron, which is attracted to a magnet, may be separated from the sulfur by magnetic attraction.

Compounds and Elements

A **compound** is a pure substance which may be broken down, or decomposed, by chemical change into two or more simpler pure substances. Water is a compound substance, since it may be decomposed into hydrogen and oxygen by an electric current. Sugar may be decomposed by heat into carbon and water. Baking soda, alum, and table salt are also compound substances.

There are a number of substances, however, which have never been decomposed into simpler substances by ordinary chemical means.* These include such familiar substances as iron, copper, gold, sulfur, oxygen, hydrogen, and tin. Such substances are called **elements.** More than one hundred elementary substances have been discovered up to the present time. (A list of these will be found inside the back cover of this text.) Although several of these elements were known in ancient times, they were not then considered elementary substances.

Ten elements make up more than 99% of the earth's crust; oxygen alone makes up nearly 50% of all matter of earth's crust plus hydrosphere. The distribution of elements in the earth's crust is shown in Figure 1.4.

---

* Certain elements (termed "radioactive") spontaneously decompose into simpler elements. See Chapter 35.

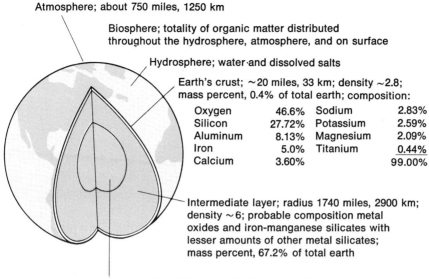

Atmosphere; about 750 miles, 1250 km

Biosphere; totality of organic matter distributed
throughout the hydrosphere, atmosphere, and on surface

Hydrosphere; water and dissolved salts

Earth's crust; ~20 miles, 33 km; density ~2.8;
mass percent, 0.4% of total earth; composition:

Oxygen	46.6%	Sodium	2.83%
Silicon	27.72%	Potassium	2.59%
Aluminum	8.13%	Magnesium	2.09%
Iron	5.0%	Titanium	0.44%
Calcium	3.60%		99.00%

Intermediate layer; radius 1740 miles, 2900 km;
density ~6; probable composition metal
oxides and iron-manganese silicates with
lesser amounts of other metal silicates;
mass percent, 67.2% of total earth

Inner core; radius 2082 miles, 3470 km; density ~11; enormous pressure
lessens distinction between liquid and solid; probable composition
mainly iron with some nickel; mass percent, 32.4% of total earth

Figure 1.4

Structure and composition of the Earth.

Compounds are made of various combinations of two or more elements.
Sugar, for example, is a combination of the elements carbon, hydrogen, and
oxygen; water is composed of the elements hydrogen and oxygen.
Compounds are characterized by their definite or invariable composition.
All samples of a given compound contain the same elements in a fixed proportion. For example, the proportion of hydrogen to oxygen by weight in water
is 1.008:8. Sodium chloride (table salt) contains sodium and chlorine in the
proportion of 22.990 parts by weight of sodium to 35.453 parts by weight of
chlorine. This principle of the constancy of composition is called the **Law of
Definite Composition. (For further discussion of this law, see page 22.)**

Contrasting the Properties of Mixtures and Compounds

The characteristics of mixtures and compounds are summarized as follows:

MIXTURES	COMPOUNDS
1. Heterogeneous or homogeneous.	1. Homogeneous.
2. Constituents retain original identity and properties.	2. Compounds have characteristic properties which are not those of the constituents.

3. Constituents may be present in any proportion by weight.

3. Constituent elements are present in fixed and invariable ratio by weight.

4. Constituents usually may be separated by mechanical means.

4. Not readily broken down into constituent elements.

Metals and Nonmetals

Elements may be conveniently classified as metals and nonmetals. We are probably more familiar with the properties of metals than those of nonmetals. For example, we know that the elements iron, copper, and tin have, in general, properties of ductility (may be drawn into a wire) and malleability (may be hammered or rolled into shape) and a metallic luster. In addition, metals usually have high melting points and high densities, and are good conductors of heat and electricity. Nonmetals, on the other hand, have generally low melting points and low densities, and are brittle, nonductile, and poor conductors of heat and electricity. Typical nonmetals are sulfur, phosphorus, and carbon. It should be pointed out, however, that there is no sharp line of distinction between many metals and nonmetals.

The only metals found as uncombined elements in naturally occurring rocks are gold, silver, copper, platinum, and rare platinum-like metals, plus small, unimportant quantities of mercury, antimony, bismuth, and arsenic. Practically all of our zinc, nickel, copper, lead, and arsenic are mined from deposits in which these elements are chemically combined with the nonmetal sulfur. The main ores of iron and aluminum are oxygen-containing compounds.

There are a number of metallic elements which, though not in common use as metals, are widely used in compounds with nonmetals. For example, most people never see or use metallic sodium, yet the compound sodium chloride (common salt) is an essential food item. The metal calcium has little use, but its compounds are important in bone matter, glass, cement, marble, and innumerable industrial substances.

Some Types of Compounds

Compounds are many and varied, but we shall mention at this point only three important types — salts, acids, and bases. Concepts of chemistry require gradual development; hence, our preliminary statements as to the nature of salts, acids, and bases will be expanded in later discussion. A **salt** is often a compound of a metal with a single nonmetal other than oxygen. Lead sulfide, magnesium bromide, and zinc phosphide are salts. Many salts however, consist of a metal combined with more than one nonmetal (including oxygen)—for example, calcium carbonate, a salt, is composed of calcium, carbon, and oxygen.

Many nonmetals will combine with hydrogen, and most of these compounds are **acids.** Chlorine, bromine, or sulfur, for example, react with hydrogen to form acids.

Compounds of metals with an oxygen-hydrogen group are called *bases*, or sometimes **hydroxide bases,** since the oxygen-hydrogen grouping is termed "hydroxide." Sodium hydroxide and calcium hydroxide are bases.

Salts, acids, and bases comprise the principal classes of compounds which we shall study in this course. They will be considered in detail in Chapter 7.

Solutions

We have all observed that sugar dissolves in water to form a clear solution. The sugar distributes itself uniformly throughout the liquid, so that every part of the solution is exactly like every other part. In other words, a solution is a homogeneous mixture. We know, too, that the amount of sugar which will dissolve in a given weight of water is variable; we may dissolve a teaspoonful or a cup of sugar in a pint of water. Solutions, then, are like compounds in that they are homogeneous, but they are unlike compounds in that the proportion of constituents is variable.

Although the solution of a solid in a liquid is the most common type of solution we shall encounter, many other types exist. We shall study this subject in more detail in Chapter 13.

1.5   LAWS OF CHEMICAL CHANGE

Law of Conservation of Mass

In 1785 the French chemist Lavoisier showed that **there is no detectable**[*] **gain or loss of mass in a chemical change.** In other words, the total weight of the substances entering into a chemical change is equal to the total weight of the substances produced as a result of the change. Since that time much evidence has accumulated in support of this principle, which is now known as the **Law of Conservation of Mass.** The law may be demonstrated, within the limits of the accuracy of our weighing devices, by carefully measuring the masses of the reactants and products of a chemical change.

[*] Einstein and others have demonstrated experimentally that mass and energy are interchangeable under certain conditions according to the Einstein equation $E = mc^2$ (where E = energy, m = mass, and c = velocity of light). A small mass change conceivably accompanies every chemical change, thus accounting for the energy change. However, in ordinary chemical changes the mass loss is so small that it is not detected by our most sensitive balances. The Law of Conservation of Mass thus is a special case of the Law of Conservation of Energy. See Chapter 36 for a more complete discussion of the equivalence of mass and energy.

In certain chemical reactions it may seem that the reactants weigh more than the products. For example, if we burn a piece of wood, ashes are the only visible evidence of any product of the reaction. However, in the burning process the wood combines with oxygen from the air and forms two invisible gases, carbon dioxide and water vapor, which pass into the atmosphere. If the carbon dioxide and water vapor are collected and added to the mineral matter (ashes) which remains, the total weight is found to be equal to the weight of the wood with which we started plus the weight of oxygen which has been taken from the atmosphere. Thus, if we take into account all of the reactants and products, we find no change in total mass during the reaction.

In other cases, the reactants seem to weigh less than the products. Iron rust, for example, weighs more than the iron from which it is formed. However, if we take into account the weight of the oxygen and water which combines with the iron during the rusting process, we find that the weight of iron plus oxygen and water equals the weight of the iron rust (hydrated iron oxide) produced in the chemical change.

Law of Definite Proportions

The law of definite composition which was stated on page 19 is sometimes referred to as the **Law of Definite Proportions.** It may also be stated in the following form: **When elements combine to form a given compound, they do so in a fixed and invariable ratio by weight.** This law may be readily demonstrated in the laboratory by determining the ratio in which elements combine with one another. For example, exactly 55.85 g of iron, no more and no less, will combine with 32.06 g of sulfur to form the compound iron(II) sulfide. No matter what ratio of weights of iron and sulfur we start with, we will find that the ratio 55.85 to 32.06 is always the exact ratio in which iron and sulfur combine to form iron(II) sulfide. Naturally, if a fixed ratio is maintained when these elements react, then the compound formed [iron(II) sulfide] must have a fixed and definite composition.

Law of Multiple Proportions

Although elements always combine in a definite proportion to yield a given compound, they may also combine in a different proportion to yield a different compound. For example, copper and oxygen may combine to form two oxides of copper: copper(I) oxide and copper(II) oxide. Chemical analysis of these compounds shows that in copper(I) oxide 63.55 g of copper combine with 8.000 g of oxygen; in copper(II) oxide 31.78 g of copper combine with 8.000 g of oxygen. Note that the ratio of the weights of copper combining with 8.000 g of oxygen in the two compounds (63.55:31.78) is in small whole numbers — that is, 2 to 1.

As another example, note that the weights of oxygen (8, 16, 24, 32, 40) which combine with a fixed weight of nitrogen (14 g) in the series of compounds below are in the ratio of small whole numbers (1, 2, 3, 4, 5).

In nitrous oxide	14 g nitrogen + 8 g oxygen
In nitric oxide	14 g nitrogen + 16 g oxygen
In nitrogen trioxide	14 g nitrogen + 24 g oxygen
In nitrogen dioxide	14 g nitrogen + 32 g oxygen
In nitrogen pentoxide	14 g nitrogen + 40 g oxygen

The relationship shown here is termed the **Law of Multiple Proportions.** It may be stated: **The weights of an element which combine with a fixed weight of a second element in different compounds of the two elements are in the ratio of small whole numbers.**

There are many other examples of a series of compounds between two elements in which this law holds. We should keep in mind, however, that for any one compound the proportion in which the elements combine is fixed and invariable.

EXAMPLE 1.5   Two different compounds of nitrogen and oxygen yielded the following results on analysis:

Compound (1) 5.60 g of N are combined with 3.20 g of O.
Compound (2) 1.75 g of N are combined with 4.00 g of O.

What are the small whole numbers involved in the law of multiple proportions as applied to this data?

SOLUTION: Fix the weight of one of the elements in the two compounds, say the oxygen, to be the same in both compounds.

$$\text{Compound (1)} \frac{5.60 \text{ g N}}{3.20 \text{ g O}} = 1.75 \text{ g N to } 1.00 \text{ g O}$$

$$\text{Compound (2)} \frac{1.75 \text{ g N}}{4.00 \text{ g O}} = 0.438 \text{ g N to } 1.00 \text{ g O}$$

Now dividing the weight of N in one compound by the weight of N in the other compound, both of which are combined with a fixed weight of oxygen (1.00 g), we get the small whole number ratio of 4 to 1.

$$\frac{1.75}{0.438} = 4$$

Atoms and Molecules

The fact that elements combine in fixed ratios by weight suggests that elements may be made up of particles, or chunks of matter, which act as units in chemical change. In recent years much evidence has accumulated to indicate that such units of matter do exist, and the term **atom** has been used to designate them. An atom may be defined as **the smallest particle of an element which can enter into chemical change.**

Let us imagine subdividing a piece of iron into smaller and smaller particles. Eventually we would obtain a piece of iron which could no longer be divided or broken down into smaller pieces of iron. This ultimate particle would be an atom of iron.

The ultimate particle of a compound substance is termed a **molecule.** For example, in the subdivision of a compound substance, such as sugar, composed of the elements hydrogen, oxygen, and carbon, the molecule is the smallest particle into which it may be divided without losing its identity. If sugar were divided into atoms, it would lose its properties, and its identity as sugar would be destroyed. (Atoms and molecules will be studied in more detail in the next chapter.)

1.6 SYMBOLS AND FORMULAS

The chemist finds it convenient to use abbreviations or **symbols** for the chemical elements. Instead of writing out *hydrogen,* for instance, the element is designated by the symbol H. The symbol of an element is usually derived from the first letter or letters of the English or Latin name of the element. Often only a single capitalized letter is used, such as N for *nitrogen* and P for *phosphorus.* In other cases two letters from the name are used, but only the first is capitalized; examples are He for *helium* and Ne for *neon.* The symbol for *sodium* is Na, which comes from the Latin, *natrium.* Many other symbols are derived from the Latin names of the elements — for example, *copper,* Cu; *iron,* Fe; *Mercury,* Hg; and *silver,* Ag. (A list of the chemical elements and their symbols will be found on the inside front cover of this book.)

To the chemist the symbol is not only an abbreviation for an element; it also stands for one atom of the element. Thus S stands for one atom of sulfur, H for one atom of hydrogen, and O for one atom of oxygen. A number placed in front of the symbol indicates the number of atoms of that element; thus 3 H stands for three atoms of hydrogen, and 10 N stands for ten atoms of nitrogen.

A chemical compound is expressed in abbreviated form as a **formula,** which consists of the proper combination of symbols, representing the elements present in the compound. Thus iron sulfide, composed of the elements iron and sulfur, is represented as FeS, in which Fe is the symbol for iron and S the symbol for sulfur. Sodium chloride, a compound of sodium and chlorine, has the formula NaCl, in which Na is the symbol for sodium and Cl the symbol for chlorine.

A formula, like a symbol, is more than an abbreviation, since it tells at once the relative number of atoms of the constituent elements in the compound. FeS stands for iron(II) sulfide, which contains one atom of iron and one atom of sulfur. H_2O stands for two atoms of H (hydrogen) and one atom of O (oxygen), which compose water. A subscript 2 is placed after the hydrogen to indicate that two atoms of hydrogen are present in the molecule. A molecule of sulfuric acid, which contains two atoms of hydrogen, one atom of sulfur, and four atoms of oxygen, is represented as H_2SO_4; the subscript of the H is 2 to denote two atoms of hydrogen; the subscript of the O is 4 to indicate four atoms of oxygen present in each molecule of the compound. No subscript is necessary if only one atom of an element is present; thus in the formula H_2SO_4 no subscript is placed after S.

The compound sugar contains the elements carbon, hydrogen, and oxygen. Its formula is $C_{12}H_{22}O_{11}$, which tells us that each molecule of sugar contains twelve atoms of carbon, twenty-two atoms of hydrogen, and eleven atoms of oxygen.

Chemical Equations

When iron powder and sulfur powder are heated together, iron combines with sulfur to form the compound iron sulfide. This chemical change might be represented by the word statement:

$$\text{Iron plus sulfur produces iron sulfide.}$$

It is more convenient to represent the chemical change in the form of a **chemical equation** by using the symbols or formulas for the substances taking part in the chemical change:

$$Fe + S = FeS$$

The symbols appearing to the left of the = sign are symbols for the **reactants** — that is, those things entering into the chemical change. The formulas to the right of the = sign represent the **products**, those things produced as a result of the chemical change. The chemist uses the word **reaction** synonymously with **chemical change.**

The above representation for the reaction between iron and sulfur is a true equation: the left side is equal to the right side in that one atom of iron and one atom of sulfur appear on each side. The equation is said to be **balanced** because the same number of atoms of each kind appear on each side. The = sign is often replaced by an arrow, which indicates the direction in which the chemical change is taking place:

$$Fe + S \rightarrow FeS$$

Exercises

1.1. Define or illustrate the following terms or symbols: (a) metric system, (b) SI units, (c) m, (d) kg, (e) l, (f) μm, (g) significant figures, (h) kelvin temperature, (i) density, (j) kinetic energy, (k) matter.

1.2. Explain the meaning of the following: (a) pure substance, (b) mixture, (c) element, (d) nonmetal, (e) calorie, (f) scientific law, (g) atom, (h) molecule, (i) symbol, (j) formula.

1.3. Add the following weights and express the answers decimally in grams.
   (a) 4.50 l, 2500 ml, 0.155 l, 78.0 ml
   (b) 4.70 l, 0.25 l, 3500 ml, 750 ml
   (c) 300 cm³, 10.30 l, 76 ml
   (d) 45.00 g, 2.0000 kg, 375 mg. 22.0 dg, 88 cg

1.4. Add the following units of linear measurement and express the answers in meters.
   (a) 35 cm, 624 mm, 0.200 km, 18.0 dm
   (b) 0.45 m, 365 mm, 1.63 dm, 0.0010 km
   (c) 5.00 km, 25.0 cm, 0.340 m, 44 mm

1.5. Add the following volume units and express the answers in liters.
   (a) 4.50 l, 2500 ml, 0.155 l, 78.0 ml
   (b) 4.70 l, 0.25 l, 3500 ml, 750 ml
   (c) 300 cm³, 10.30 l, 76 ml

1.6. A volume of 54.3 ml = _____l = _____cl (centiliters) = _____kl.

1.7. A length of 0.023 m = _____mm = _____μ = _____nm

1.8. What is the volume in cubic centimeters of a cube which is 3200.0 mm along each edge?

1.9. What volume (in cm³) is occupied by a block of wood: 35.0 m × 20.0 cm × 400 mm? All edges are at 90° to one another.

1.10. A small drop of oil (0.0010 cm³) is spread on the surface of water so that the oil film has an area of 1 sq. m. What is the thickness of the oil film in angstrom units?

1.11. How much area in square centimeters will one liter of paint cover if it is brushed out in a uniform thickness of 100 μ?

1.12. How many significant figures are there in each of the following numbers?
   (a) 136.54, (b) 0.0054, (c) 1.830, (d) 6.02×10^{23}, (e) 1.057, (f) 1,650,763.73, (g) 500.

1.13. Calculate the following, giving the answers to the correct number of significant figures:
   (a) Add: 12.33 m + 11.1000 m + 145.9 m
   (b) Multiply and divide: $\dfrac{(3.333)(1.11)}{5.5}$

1.14. A cube measured 4.40 cm on a side. Calculate, to the correct number of significant figures, the volume of the cube in cm³.

1.15. Fill in the blanks in the following table of temperatures:

	°C	°K	°F
(a)	0	—	—
(b)	—	0	—
(c)	—	—	0
(d)	25	—	—
(e)	—	250	—
(f)	—	—	−20
(g)	−20	—	—
(h)	—	175	—
(i)	—	—	3500

1.16. Normal body temperature is 98.6°F. What is this temperature on a centigrade thermometer?

1.17. Liquid helium boils at 4°K. What is the boiling temperature on the Fahrenheit scale?

1.18. How many calories of heat must be added to 6.0 l of water to raise the temperature of the water from 20° to 100°C?

1.19. A container has the dimensions 3.0 m × 50 cm × 60 mm. If it is filled with water at 32°F, how many kilocalories will be necessary to heat the water to the boiling point (212°F)?

1.20. A 1.98-l sample of metallic osmium, the "heaviest" substance known, weighs 44.5 kg. Calculate the density of osmium in grams per cubic centimeter.

1.21. Calculate the density of a block of wood which weighs 850 kg and has the dimensions 25 cm × 0.10 m × 50.0 m.

1.22. The density of alcohol is 0.8 g/ml. What is the weight of 600 ml of alcohol?

1.23. The density of concentrated sulfuric acid is 1.85 g/ml. What volume of the acid would weigh 65.0 g?

1.24. What is the density of a cube of wood which weighs 850 g and measures 12 cm on a side?

1.25. Calculate the volume in milliliters of 246 g of chloroform. The density of chloroform is 1.50 g/ml.

1.26. Distinguish between (a) matter, (b) mass, and (c) weight.

1.27. Classify the following as pure substances or mixtures: (a) copper, (b) sea water, (c) gasoline, (d) sugar, (e) wood, (f) air, (g) hydrogen gas.

1.28. Classify the following as heterogeneous or homogeneous mixtures: (a) soil, (b) salt water, (c) air, (d) lubricating oil, (e) alcohol solution.

1.29. Classify the following as physical or chemical changes: (a) rusting of iron, (b) molding of clay, (c) melting of silver, (d) digestion of food, (e) setting of mortar, (f) magnetization of iron, (g) rotting of wood, (h) souring of milk, (i) breaking of glass, (j) stretching of rubber, (k) burning of coal.

1.30. Pure water (a) has a density of approximately 1g/ml, (b) does not burn, (c) boils at 100°C, (d) causes iron to rust, (e) in small amounts is colorless, (f) dissolves sugar, (g) reacts with sodium to give hydrogen gas. Which of the above properties are physical and which are chemical?

1.31. It has been determined experimentally that two elements, A and B, react chemically to produce a compound or compounds. Experimental data obtained on combining proportions of the elements are:

	Grams of A	Grams of B	Grams of Compound
Experiment 1	10.02	4.00	14.02
Experiment 2	30.06	12.00	42.06
Experiment 3	5.01	2.00	7.01

Which two laws of chemical change are illustrated by the above data?

1.32. Six samples containing elements A and B are analyzed. Five are pure compounds; the other is a mixture. Results of the analysis are

Sample No.	Grams of A	Grams of B
1	7.0	16
2	14	8.0
3	28	32
4	3.5	16
5	6.5	8.0
6	1.4	3.2

(a) Which two of the samples are the same compound? (b) Which sample is a mixture? (c) Which law of chemical change is applied in obtaining the answer to (a)? (d) What other law is illustrated by these data?

1.33. Three oxides of lead on analysis gave the following data:

	Lead (Percent)	Oxygen (Percent)
First Compound	92.8	7.20
Second Compound	86.6	13.4
Third Compound	90.67	9.33

Does the Law of Multiple Proportions apply here? If so, explain.

1.34. When sugar ($C_{12}H_{22}O_{11}$) is heated, it decomposes as follows:

$$C_{12}H_{22}O_{11} \xrightarrow{\text{heat}} 12\ C + 11\ H_2O$$

Explain the following terms as applied to this reaction: symbol, formula, equation.

Supplementary Readings

A. The following references are general and include material of interest in many chapters of the book.
  1.1. Historical: (a) *Discovery of the Elements*, M. E. Weeks and H. M. Leicester, Chemical Education Publishing, Easton, Pa., 1968. (b) *The Historical Background of Chemistry*, H. M. Leicester, Wiley, New York, 1956. (c) *Centennial American Chemical Society*, 1876 to 1976, Chem. and Eng. News, April 6, 1976. The past 100 years in chemistry. (d) *Alembic Club Reprints*, Williams and Wilkins, Baltimore, Md.
  1.2. Calculations: *Problems For General Chemistry and Qualitative Analysis*, C. J. Nyman and G. B. King, Wiley, New York, 1975 (paper).
  1.3. (a) Readings from Scientific American, *General Chemistry*, J. B. Ifft and John E. Hearst, Freeman, San Francisco, 1974 (paper). (b) Readings From Scientific American, *Chemistry in the Environment*, Carole L. Hamilton, Freeman, San Francisco, 1973 (paper).
B. References relating to this chapter.
  1.1. "International System of Units (SI)," M. A. Paul, *Chemistry*, **45**, 9 (1972).
  1.2. "What is Matter?", Erwin Schroedinger, *Sci. American*, September, 1953; p. 52.
  1.3. "Physical versus Chemical Change," W. J. Gensler, *J. Chem. Educ.*, **47**, 154 (1970).

Chapter 2

In reality there is nothing but atoms
and space—solid and indivisible atoms
in an infinite void.

Democritus, 500 B.C.

Atoms and
Subatomic
Particles

2.1 ATOMIC THEORIES

The laws of chemical change studied in the last chapter lead to the inevitable conclusion that matter is composed of discrete particles which function as units during chemical change. The idea that matter is discontinuous — that it is made up of finite particles and is not all in one piece — was first conceived by the Greek philosophers. Democritus reasoned that matter consists of very small particles, which he called **atoms.** He did not distinguish between "atoms" of compounds and elementary substances, but believed that there were as many kinds of atoms as kinds of substances. It is surprising that the Greek philosophers were able to guess so nearly the truth of the structure of matter, since their conclusions were based on pure speculation, without support of experimental data.

Dalton's Atomic Theory

It was not until 1804 that the atomic theory was developed more fully and accepted by chemists and physicists. In that year John Dalton, an English schoolmaster, reached the following conclusions about the structure of matter:

(a) Substances are composed of tiny, indivisible particles called "atoms."
(b) Atoms of any given substance are identical and have the same weight, size, form, and so on.
(c) The atom is the smallest part of an element which can enter into a chemical change.
(d) Atoms of elements are permanent and cannot be decomposed.
(e) Compounds are formed by the union of two or more atoms of elementary substances.

These conclusions enabled Dalton to explain the laws of chemical change. Since, according to the theory, atoms are permanent and not decomposable, there is no change in mass during chemical reaction. When chemical compounds react with one another, there is simply a rearrangement of the atoms within the compounds. This rearrangement or regrouping produces new substances which have entirely new properties. Since all the atoms present are simply regrouped during a chemical change and the atoms retain their weights, there is no gain or loss of weight (Law of Conservation of Mass). Furthermore, if the atoms possess characteristic weights, the union of two or more atoms would require that combination take place in a fixed ratio by weight (Law of Definite Proportions).

Dalton did not recognize differences between units of elements and units of compounds. Avogadro, an Italian physicist, was the first to differentiate clearly between the two. He defined an atom as the smallest particle of an element which can enter into chemical change. He used the term **molecule** to

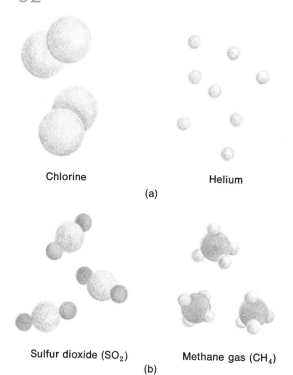

Chlorine   Helium

(a)

Sulfur dioxide (SO₂)   Methane gas (CH₄)

(b)

Figure 2.1

(a) A combination of like atoms results in a molecule of an element (thus Cl_2). Some elements are monoatomic (thus He). (b) A combination of unlike atoms forms a compound.

denote the smallest unit of a substance (element or compound) which could exist and retain all of the properties of the substance. A molecule may be a combination of two or more like or unlike atoms (Figure 2.1).

The difference between molecules of compounds and of elements is a bit confusing. Many elements will exist in the uncombined or free state as single atoms, and in this case the atom and molecule are the same; others may exist in the free state only in pairs or some other number of atoms, e.g. as molecules.

Revisions of Dalton's Theory

Dalton's atomic theory has undergone considerable change in recent years, and we now perceive that there were several errors in the original assumptions. We now know that not all atoms of any given element are identical. For example, two kinds of chlorine atoms are known, and four kinds of lead atoms have been shown to exist. Such atoms are called **isotopes.** (See page 63.)

We now also know that atoms may be disintegrated. Many elements, such as uranium and radium (**radioactive** elements), undergo constant, spontaneous disintegration, their atoms being transformed into atoms of simpler elements. (See Chapter 35.)

Atomic Weights

In dealing with the very small weight of an atom, it is necessary to introduce a new unit of mass – an **atomic mass unit,** designated **u** or **amu** – which is defined as $\frac{1}{12}$ the weight of one atom of the most abundant isotope of carbon° which has been assigned a weight of 12 amu. This atom of carbon is an arbitrary standard but provides a basis for expressing weights of other atoms. For example, a sulfur atom is $2\frac{2}{3}$ as heavy as a carbon atom and will have an atomic weight of $2\frac{2}{3} \times 12 = 32$ amu. An atom of iron is approximately $4\frac{2}{3}$ as heavy as the standard carbon atom; therefore its atomic weight is about 56 amu. Atomic weights have been assigned to the other elements in the same way. (See back inside cover of this text). An atomic weight may also be regarded as simply a number which expresses the relative weight of an atom compared to the standard carbon atom. In the usual practical situations where chemists weigh out definite amounts of chemicals, very large numbers of atoms are always involved. To avoid the repeated use of these large numbers of atoms, it is convenient when making calculations to make use of a unit which stands for a definite number of atoms and is also related to atomic weights. Such a unit, one of the basic units of the SI system, is called a **mole** (abbreviation **mol**). A mole is defined as **the amount of a substance that contains as many elementary entities as there are atoms in exactly 12 g of standard carbon-12 (^{12}C).** This number of atoms has been determined experimentally to be equal to 6.022×10^{23} and is known as **Avogadro's number (N)** in honor of Amadeo Avogadro, an early Italian physicist. According to this definition a mole of magnesium atoms is 6.022×10^{23} Mg atoms and weighs 24.305 (atomic weight of Mg) grams. Likewise a mole of mercury (atomic weight 200.59) contains 6.022×10^{23} Hg atoms and weighs 200.59 g. A mole of atoms then, for an element, has a weight in grams equal to the atomic weight of the element and contains Avogadro's number of atoms. Other terms, **gram atomic weight** or **gram atom,** are often used to mean the same thing as a mole, for an element.

Molecular Weights (Formula Weights)

The smallest particle of a compound which can exist as such has been termed a **molecule.** A molecule of a compound is formed by the union of two or more atoms of the elements of which the compound is composed. The weight of a molecule is the sum of the weights of the constituent atoms. For example a molecule of water is composed of two atoms of hydrogen and one atom of oxygen H_2O. The weight of a molecule of water, or the **molecular weight** of water, is $(2 \times 1 \text{ amu} = 2 \text{ amu}) + (1 \times 16 \text{ amu} = 16 \text{ amu}) = 18$ amu. A molecular

---

° Until recently oxygen, with an assigned weight of 16, was the standard for atomic weights. In 1961 the Council of the International Union of Pure and Applied Chemistry and its Physics counterpart adopted the principal isotope of carbon as a new standard of atomic weights.

weight, like an atomic weight may be regarded as a **relative** weight. For a given compound, the molecular weight is the weight of a molecule of the compound compared with the weight of one atom of carbon. The molecular weight of 18 for water means that one molecule of water is $\frac{18}{12}$ as heavy as one atom of carbon, the standard of comparison. The term **mole** may be applied to **molecules** as well as atoms. A mole of a compound has a weight in grams numerically equal to its molecular weight and contains Avogadro's number of molecules. For example, a mole of sulfuric acid, H_2SO_4 (molecular weight 98.1) weighs 98.1 g and consists of 6.022×10^{23} molecules of H_2SO_4. Note, from the formula H_2SO_4, that one mole of H_2SO_4 contains 2 moles of H atoms, 1 mole of S atoms, and 4 moles of O atoms. The terms **gram molecular weight** or **gram mole** are sometimes used for a mole of molecules.

Actually the term mole is now commonly applied to a number of elementary entities in addition to atoms and molecules, such as ions, electrons, or particles. Care must be taken to specify the entity that the mole refers to. For example, a mole of H atoms weighs 1.0 g and contains **N** (Avogadro's number) **atoms** of H, whereas a mole of H_2 molecules weighs 2.0 g and contains N **molecules** of H_2 (2 N atoms of H). An ionic solid such as $ZnCl_2$ consists of Zn^{2+} and Cl^- **ions** rather than molecules of $ZnCl_2$. For simplicity we will continue using the term molecular weight for these substances, whether molecular or ionic, although some chemists prefer **formula weight** to be used.

A mole of $ZnCl_2$ (molecular weight 136.3) then consists of N formula units of $ZnCl_2$ which contains N Zn^{2+} ions (1 mole of Zn^{2+} ions) and 2 N Cl^- ions (2 moles of Cl^- ions).

EXAMPLE 2.1   Consider arsenous acid, H_3AsO_3. (Atomic weights H = 1.01, As = 74.9, O = 16.0)

(1) What is the molecular weight of H_3AsO_3?
(2) What is the weight in grams of 1 mole of As in 1 mole of H_3AsO_3?
(3) What is the weight of 1 mole of H_3AsO_3?
(4) How many moles is 500 g of H_3AsO_3?
(5) How many grams is 0.050 mole of H_3AsO_3?
(6) How many moles of hydrogen atoms are present in 5 moles of H_3AsO_3?

SOLUTION:
(1) 125.9 amu ($3 \times 1.01 + 74.9 + 3 \times 16.0 = 125.9$)
(2) 74.9 g
(3) 125.9 g
(4) Number of moles of $H_3AsO_3 = \dfrac{500 \text{ g}}{125.9 \text{ g/mole}} = 3.97$ moles
(5) grams of $H_3AsO_3 = (0.050 \text{ mole}) (125.9 \text{ g/mole})$
$= 6.30$ g
(6) Moles of H atoms $= (5 \text{ moles of } H_3AsO_3) (3 \text{ moles H/mole } H_3AsO_3)$
$= 15$ moles of H

2.2   SUBATOMIC PARTICLES OF MATTER

The idea that atoms are indivisible persisted for many years after the postulations of Dalton, and only in the last sixty years or so has evidence been accumulated which indicates that atoms may actually be decomposed and resolved into subatomic particles much smaller than any known atom. Atoms of all the elements are now believed to be composed of three main subatomic particles of matter:

Electrons: units of matter carrying negative charges of electricity.
Protons: units of matter carrying positive charges.
Neutrons: neutral units of matter of about the same mass as the proton.

The Electron

The electron was the first of the subatomic particles to be discovered. Knowledge of the electron began with a study of the discharge of electricity through gases at reduced pressures. For this study we may use a so-called cathode ray tube (Figure 2.2). It consists of a tube of glass in which two metal discs are sealed and a glass side arm through which the tube may be evacuated by a pump. The metal discs act as electrodes between which there will be a discharge of electricity when the electrodes are connected to a source of high potential, such as an induction coil. The nature of the discharge, then, is studied as the pressure is reduced. At atmospheric pressure an irregular

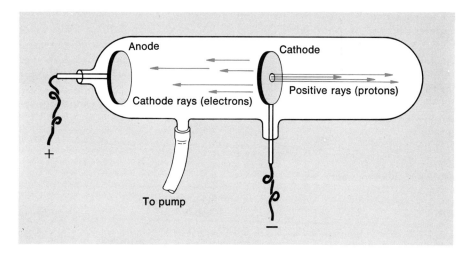

Figure 2.2
Cathode-ray tube. Electrons are emitted from the cathode in straight lines.

violet discharge takes place. As the pressure is reduced, the discharge becomes pink, and finally the whole tube glows. At very low pressures, streaks
of white light seem to emerge at right angles from the negative pole, or
cathode, and cause the appearance of a greenish fluorescence on the glass
walls where the rays strike. (These rays were originally referred to as **cathode
rays.**) An obstacle placed between the cathode and the glass wall casts a
sharp shadow on the wall opposite the cathode, thus indicating that the rays
travel in straight lines. A pinwheel placed in the path of the rays revolves, or
if the rays are allowed to strike a piece of metal foil, the foil becomes hot.
These observations suggested that the rays were material in nature. In 1879
Sir William Crookes showed that the rays were bent in an arc by an electromagnetic field. Subsequently, Sir J. J. Thomson showed that the rays were deflected toward a positively charged plate in an electric field and therefore
must possess a negative charge. These negatively charged particles issuing
from the cathode came to be called **electrons.** By observing the extent or
magnitude of the electron's deflection in magnetic and electric fields, Thomson determined that the mass of a single electron is about $\frac{1}{1837}$ that of the hydrogen atom. Since electrons can be generated by using any cathode or any
gas in the tube, it would seem that electrons are constituents of atoms of all
elements.

The magnitude of the **charge** of an electron was determined by R. A. Millikan. A container was constructed with electrically chargeable plates at top and bottom. Electrons and positively charged particles were produced in the container by a beam of
X rays. Fog droplets were introduced into the container, and the rate of fall of a given
fog droplet was viewed as the plates were electrically charged or discharged. A fog
particle could be accelerated in its fall, suspended in midair, or actually made to rise,
depending on the charge of the plates.

The explanation is that the fog droplet, gathering electrons and thus becoming
negatively charged, would tend to move to the positively charged plate. The number
of electrons adsorbed per droplet would influence the rate of movement to the positive
plate, but all changes in velocity as electrons were adsorbed could be compensated
by appropriate changes in charge between plates. It was noted that all changes in
velocity of the rise or fall of the fog droplets were related to multiples of a certain
unit charge between the plates. This unit charge between the plates must equal the
electron charge and was found to be 1.6022×10^{-19} coulombs.

Table 2.1

Properties of subatomic particles

Particle	Mass	Relative Mass to Carbon-12	Unit Charge°
electron	9.1096×10^{-28} g	0.000549 amu	−1
proton	1.6726×10^{-24} g	1.00728 amu	+1
neutron	1.6749×10^{-24} g	1.00867 amu	0

° 1.6022×10^{-19} coulomb. The coulomb (C) is the SI derived unit of charge.

The actual **mass** of an electron has not been determined directly but may be calculated, as has already been mentioned, from data obtained by measuring the effect of magnetic and electrostatic fields upon a stream of cathode rays. From the deflection of the stream in a magnetic field and its counterbalancing by an electrostatic field, the ratio of charge to mass of the electron is obtained. From this the mass has been calculated to be 9.1096×10^{-28} g (Table 2.1).

The Proton

After electrons were found to be generated in a cathode ray tube, it was natural to ask whether or not positively charged particles were emanating from the region of the positive pole, or anode, of the tube at the same time that electrons were being generated at the cathode. In 1886 Goldstein found that if the cathode is perforated with a cylindrical hole, a beam of rays passes through the cathode into a region behind the cathode and away from the anode. These rays were called **positive rays** or **canal rays.** When these rays were studied in magnetic and electric fields, in much the same manner as the cathode rays were investigated, it was shown that they possess a unit positive charge, or an integral multiple of this charge, and are composed of particles of the residual gas used in the tube. Particles of lowest mass were found when hydrogen was the residual gas. The particles consist of hydrogen atoms with a unit positive charge (Table 2.1); actually, each particle is a hydrogen atom which has lost one electron.

There is strong evidence that these particles are fundamental building units in the structure of atoms of all of the elements. These subatomic particles are called **protons.**

The Neutron

In 1932 Chadwick observed that neutral particles of a mass approximately equal to that of the proton were produced when beryllium was exposed to the radiations from a radioactive substance (Chapter 35). Although it was at first thought that each particle consisted of one proton and one electron, these neutral units of matter, called **neutrons,** are now regarded as a new and fundamentally distinct kind of particle. See Table 2.1.

X RAYS   In 1895 Roentgen discovered a new kind of radiation of very short wave length emanating from a cathode ray tube in which a metallic element was made a target of the cathode rays (Figure 2.3). For lack of a name, the radiations were called **X rays.** The X rays are capable of penetrating wood, paper, and flesh but are stopped by bones and metallic substances. Modern medical practice calls for the use of X-ray photographs in diagnosis of disease, and X rays have been a powerful tool in studies of the structure of matter.

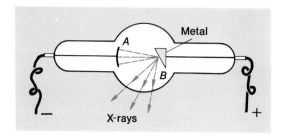

Figure 2.3
X-ray tube. Electrons are emitted and brought to a focus by the concave cathode A. The cathode rays, on striking a metal target B, produce X rays of a frequency characteristic of the metal.

2.3 OTHER SUBATOMIC PARTICLES

In recent years and primarily in the last decade intensive research into the structure of atoms has resulted in a knowledge of subatomic particles that is both unprecedented and exciting. Scores of unsuspected new particles are being found and new properties of old ones discovered. Protons and neutrons —often called **nucleons**—along with electrons are thought to be the major particles in atom building. However, under certain conditions other particles are created, destroyed, or converted to still different particles. And it seems to be a universal principle that every minute particle has its counterpart particle of opposite electrical charge, but of like mass and other physical characteristics. These are known as **matter** and **antimatter.** For example, the positron (or antielectron) has the same mass as the electron but has an opposite electric charge.

At present some 200 subatomic particles have been identified, 34 of which are designated as fundamental (Table 2.2). Eight of these are called **baryons** (the proton, the neutron, and 6 heavier particles referred to as hyperons). Eight are antibaryons; 8 are mesons and antimesons; 8 are leptons and antileptons (among which are the electron and the positron); and finally the photon and the graviton (a gravitational wave theoretically predicted but not yet verified experimentally).

Most of the fundamental particles are extremely unstable with half-lives

Table 2.2
Fundamental particles of matter

Subatomic Particles			Electric Charge	Mass, amu
Baryons	Nucleons	proton (p)	+	1.00728
		neutron (n)	0	1.00867
	Hyperons ($\equiv$, Σ, Λ)		+, 0, −	1.2 to 1.4
Mesons (η, K, π)			+, 0, −	0.15 to 0.59
Electron, positron (e⁻, e⁺)			−, +	0.000549
Other leptons (μ, v)			+, −, 0	0 to 0.11
Photon (γ)			0	0

(Chapter 35) less than a millionth of a second long and as a result change rapidly to other particles. The stable particles consist of the proton, anti-proton, electron, positron, and the particles which move with the speed of light (photon, neutrino, and antineutrino). The neutron is unstable in free space but has a much longer half-life. In the nucleus, the neutron is **stabilized** by strong nuclear forces, particularly pion ($\pi°$) exchange. Just how these fundamental particles and more than 160 other particles which have been discovered fit into the picture of the structure of atoms still has to be determined.

Experimentation has pushed far ahead of theory, and scientists are faced with a composite of facts about subatomic particles for which as yet no comprehensive theory has been advanced. How simple it was to consider that atoms were the ultimate particles of matter instead of having to think about the many subatomic particles discovered in recent research. But how stimulating it is to be living on the frontier of new knowledge!

Historical Development of an Atomic Model

After the discovery of the electron and proton, J. J. Thomson in 1898 suggested an atom model composed of a sphere of positive electricity with imbedded electrons (like raisins in a round loaf of bread). This concept was abandoned in 1911 as a result of experiments carried out by Lord Rutherford at McGill University.

In experimenting with the deflection of alpha particles (rays from radioactive substances) by thin sheets of metals, Lord Rutherford found that most of the particles passed through unaffected. (See Figure 2.4). When a gold foil 0.0004 mm thick was used, only about one alpha particle in 20,000 was deflected at a large angle. Rutherford concluded that a large deflection was the result of a collision between an alpha particle and a portion of a gold atom, in which there was a very high concentration of mass. Since protons are rela-

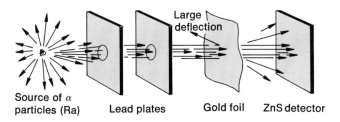

Source of α
particles (Ra)   Lead plates   Gold foil   ZnS detector

Figure 2.4
Rutherford's scattering experiment. Only a few alpha particles are deflected at large angles; most pass through the gold foil unaffected. All of the deflected and undeflected alpha particles are detected by the light flashes produced on a scintillating screen covered with zinc sulfide.

tively heavy compared with electrons, Rutherford proposed that the positive charges or protons are concentrated in a very small space, called the nucleus, of an atom. The region of space outside and surrounding the nucleus was pictured as occupied by electrons.

The discovery of the neutron in 1932 added to the atomic model the concept of neutrons interspaced with protons in the nucleus. Our current model of an atom is a neutron-proton nucleus with some kind of space occupation by electrons about the nucleus. A number of questions may be raised, and for some of these there are as yet no adequate answers. Scientists are not at all sure about the nature of the nucleus. What binding energy holds the protons together, since "like" charges repel? Do neutrons really change into protons and electrons, or protons into neutrons and positrons in certain radioactive changes? What is the part played by mesons, leptons, hyperons, and the nearly 200 subatomic particles that the physicists are now studying? The chemist is most interested in the location and energies of electrons, since the electrons appear to be the most important factors in chemical change. Before we can adequately describe an atom, we need to know more about the properties of the electron.

Atomic Spectra

Much of the information regarding electrons comes from spectral analysis. When white light such as sunlight is passed through a prism, a continuous spectrum of colors ranging from red to violet (as in a rainbow) is obtained. However, light from a flame in which an element (or a vaporizable compound) is being heated does not give such a continuous spectrum; instead, only certain wavelengths of light are emitted which are characteristic of that particular element. If the light is passed through a slit and a prism, these wavelengths of light may be recorded on a photographic plate as a series of lines, called a **line spectrum.** (See Figure 2.5.)

Each of the elements gives a characteristic set of lines which is different from any other element. Some lines are in the visible portion of the spectrum, others in the ultraviolet or infrared. Sodium produces a strong emission of light in the yellow portion of the spectrum, and this is easily visible to the naked eye when a sodium compound is heated in a flame. The yellow flame may serve as a qualitative test for the presence of sodium.

The Bohr Model of the Hydrogen Atom

Light may be regarded as composed of particles called **photons,** which have an energy expressed as $E = h\nu$ where ν is the frequency of the light and h is Planck's constant—a conversion factor for evaluating the frequency in energy units. Each line of a line spectrum of an element such as those shown

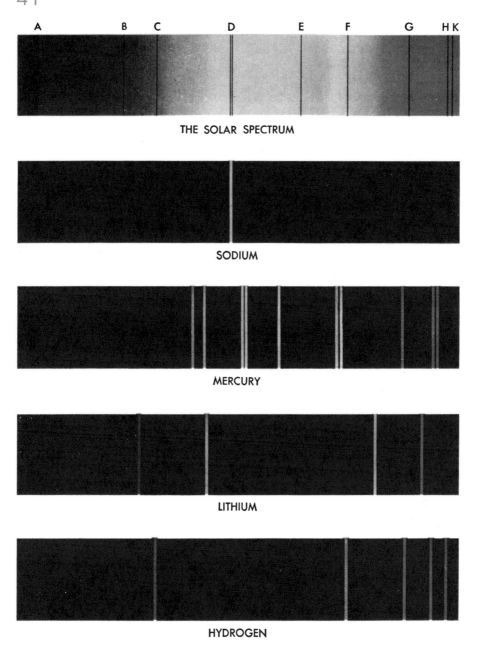

THE SOLAR SPECTRUM

SODIUM

MERCURY

LITHIUM

HYDROGEN

Figure 2.5
The solar spectrum and a few bright-line spectra.

for mercury in Figure 2.5 corresponds to a definite frequency of light, the energy of which can be evaluated with this relationship.

Hydrogen gas subject to high voltage in a discharge tube emits what appears to the eye to be magenta-colored light, but on analysis in a spectroscope it yields a line spectrum which may be evaluated into energy as recorded in Table 2.3.

Table 2.3
**Evaluation of line spectrum of hydrogen*

	Wavelength (angstroms)	Frequency (hertz)□	Energy of Photons (kcal per mole)
VISIBLE	6563	4.57×10^{14}	43.6
	4861	6.17×10^{14}	58.8
	4340	6.91×10^{14}	65.9
	4102	7.31×10^{14}	69.7
ULTRAVIOLET	1216	2.465×10^{15}	234.9
	1026	2.922×10^{15}	278.5
	973	3.081×10^{15}	293.7
	950	3.156×10^{15}	300.8

° The following relationship exists between wave length (λ) and frequency (ν); $\nu = c/\lambda$, where c is the speed of light—2.998×10^{10} cm per sec. The energy in Joules (J) is obtained from $E = h\nu$; h is **Planck's constant** and is equal to 6.626×10^{-34} J sec. To convert energy in Joules per hydrogen atom to kilocalories per mole, one must multiply by 2.390×10^{-4} kcal per J and 6.022×10^{23} atoms per mole (Avogadro's number). For example, we calculate for frequency 7.31×10^{14} Hz:
Energy in Joules = $(6.626 \times 10^{-34}$ J sec$) \times (7.31 \times 10^{14}$ per sec$) = 48.4 \times 10^{-20}$ J
E in kcal per mole = 48.4×10^{-20} J $\times 2.390 \times 10^{-4}$ kcal/J $\times 6.022 \times 10^{23}$ per mole = 69.7 kcal/mole.
□ The hertz (Hz) is a derived SI unit. It is the same as cycles per second, s^{-1}.

In 1912 Niels Bohr interpreted the line spectrum of hydrogen and proposed a model for the hydrogen atom with a proton in the nucleus and an electron in orbital motion about the nucleus. According to Bohr the electron might be found in any one of several definite orbits about the nucleus, depending on the amount of energy absorbed by the electron.

Each of the represented orbits of an electron about a nucleus of a hydrogen atom, according to Bohr, corresponds to a definite amount of energy. In moving from an inner to an outer orbit an electron requires energy; an electron moving from an outer to an inner orbit of lower energy emits energy as photons to produce a line spectrum. A schematic representation of the energy levels is shown in Figure 2.6. The first energy level, which is the normal state for the hydrogen atom, is frequently referred to as the ground state. The fall of an electron to the ground state releases a quantum of energy in the ultraviolet region of the spectrum. For example, an electron dropping from energy level two to one would emit 234.9 kcal of energy per mole. A

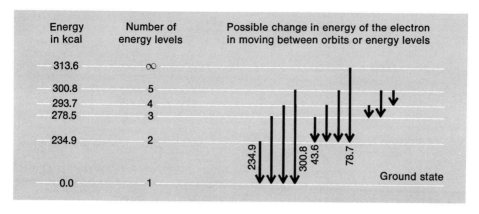

Figure 2.6
Energy levels for hydrogen. Differences in energy values are shown in kcal/mole.

drop between higher levels yields visible line spectra; for example, a drop from energy level three to two yields 43.6 kcal (278.5 − 234.9). A drop from level five to level two yields 65.9 kcal (300.8 − 234.9), and so on.

It was one of the interesting developments of science that the various spectral lines could be related by a simple empirical mathematical relationship. For the hydrogen lines,

$$\nu \text{ (frequency)} = 3.29 \times 10^{15} \text{ cycles/sec} \left(\frac{1}{n_1^{\,2}} - \frac{1}{n_2^{\,2}} \right)$$

where $n_1 = 1, 2, 3. \ldots$ and $n_2 = 2, 3, 4 \ldots$

What is it in the atom that has a *whole number* relationship with light spectra?

In recent years it has been necessary to modify the Bohr model of the atom. While we may attribute definite energy values or levels to the electrons, we cannot be sure of the location of a given electron or its path in its motion about the nucleus. According to the **Uncertainty Principle**, page 99, one cannot simultaneously determine both the position of a very small particle such as an electron and also its momentum (product of mass and velocity). Inherent faults or difficulties in the measuring devices preclude the determination of both these properties at the same moment in time. The energy level of an electron, then, relates to its totality of mass-velocity-probability in a position in space outside the nucleus. Since adequate pictures of atoms cannot be drawn to show the location of electrons, our discussion of atomic structure in the next chapter will be based on a proton-neutron containing nucleus with a **diagrammatic representation** of the well-defined energy levels of electrons somewhere outside the nucleus. A modern electron probability concept of the atom will be given in Chapter 5.

Exercises

2.1. Explain the three laws of chemical change (p. 21) making use of the assumptions in Dalton's atomic theory. Is there anything in these laws which can not be satisfactorily explained by Dalton's theory?

2.2. Distinguish between: atom, element, molecule, and compound.

2.3. Define the terms (a) atomic weight, (b) molecular weight, (c) mole, (d) amu.

2.4. Determine the molecular weights for $BaCO_3$, K_2O, $C_6H_{12}O_6$, $NH_4C_2H_3O_2$.

2.5. How many moles are represented by 250 g of each of the following compounds? (a) CO_2, (b) $CaCO_3$, (c) H_2SO_4, (d) CH_3OH.

2.6. How many moles of C are present in 5 moles of each of the following organic compounds? (a) C_2H_5OH, (b) C_9H_{20}, (c) CH_3CH_2COOH, (d) $C_{17}H_{35}COOH$.

2.7. How many moles of N are in 100 g of (a) NH_3, (b) $(NH_4)_3PO_4$, (c) HNO_3, (d) $CO(NH_2)_2$?

2.8. Consider phosphoric acid, H_3PO_4, and its elements H, P, and O (atomic weights $H = 1.0$, $P = 31$, $O = 16$). *Use proper units in your answers.*
(a) What is the molecular weight of H_3PO_4?
(b) What is the weight of one mole of phosphorus?
(c) What is the weight of one mole of H_3PO_4?
(d) What is the gram atomic weight of oxygen?
(e) What is the gram molecular weight of phosphoric acid?
(f) How many moles is 490 g of H_3PO_4?
(g) How many grams is 0.020 mole of H_3PO_4?
(h) How many moles of oxygen are present in 2 moles of H_3PO_4?

2.9. Calculate the weight in grams of each of the following:
(a) 2 l of water (density 1 g/ml)
(b) 50 moles of $CaCO_3$ (molecular weight 100)
(c) A block of sulfur (density 2.0 g/cm³) which measures 35 cm × 110 mm × 0.05 dm.

2.10. What was the contribution of the following people to the theory of atomic structure? (a) Democritus, (b) Dalton, (c) Crookes, (d) Thompson, (e) Millikan, (f) Goldstein, (g) Chadwick, (h) Rutherford, (i) Bohr.

2.11. Explain the following terms or symbols: (a) nucleon, (b) antimatter, (c) subatomic particle, (d) baryons, (e) π meson, (f) positron, (g) photon.

2.12. List the physical properties of the subatomic particles–electrons, protons, and neutrons, including mass and charge.

2.13. What are X rays? Who discovered them and how are they produced?

2.14. Explain how Rutherford's experiment indicates that atoms possess high concentrations of mass.

2.15. What is a line spectrum of an element? How did Bohr explain the line spectra of hydrogen?

2.16. Explain the following terms concerning the line spectrum of hydrogen in Table 2.3: (a) wavelength, (b) frequency, (c) hertz, (d) photon energy.

2.17. What are the frequency and energy of red light with a wavelength of 7000 Å and of violet light with a wavelength of 4000 Å?

2.18. Explain Figure 2.6.

2.19. Calculate (a) the frequency, ν, and (b) the wavelength, λ, for light emitted when an electron in a hydrogen atom falls from energy level three to the ground state.

Supplementary Readings

2.1. *The Mole Concept in Chemistry*, W. F. Kieffer, Van Nostrand, New York, 1963.
2.2. "Elementary Particles," M. Gell-Mann and E. P. Rosenbaum, *Sci. American*, July, 1957; p. 72.
2.3. "Quarks with Color and Flavor," Sheldon Lee Glashow, *Sci. American*, Oct., 1975; p. 38.

Chapter 3

The Structure and Behavior of Atoms

3.1 ATOMIC STRUCTURE

Dalton believed that atoms were solid, impenetrable particles of uniform density throughout, but scientists today know that even within the most dense of atoms there is a great deal of "free space." According to the modern view, the atom consists of a positively charged compact center, or **nucleus,** of very high density,° surrounded by a negatively charged region of space of relatively great volume but very low density. Except for the hydrogen atom, the nucleus contains both protons and neutrons, but no electrons, while the external structure of the atom consists of electrons only. These electrons are characterized by differing energy relationships and are grouped according to principal energy levels (*K, L, M, N* . . . or 1, 2, 3, 4 . . . starting from the nucleus) and according to sublevels (within the principal energy levels). In some ways this justifies the concept of an "electron cloud" surrounding the nucleus. More particularly, for each electron there is, at any instant, a region (atomic orbital†) within which the electron is most probably located. For such simple atoms as hydrogen and helium, this region of greatest probability—the atomic orbital—is a sphere, larger than but concentric with the nucleus. For more complex atoms, the regions within which the electrons are most probably located involve geometrical interpretations of various types, but because of their general symmetry relative to the nucleus, the over-all result is a sphere of influence. Hence we frequently refer to the (approximate) spherical character of atoms. Since only protons and neutrons are contained in the nucleus, the latter must always bear a net positive charge. All atoms are electrically neutral, so that the number of electrons must equal the number of protons in an atom.

Diagrams of Atoms

It is often convenient to show the component parts of atoms and their relationships to one another in some sort of diagram. For example, the hydrogen atom, which is the simplest of all atoms, may be represented as in Figure 3.1. A single proton, indicated by a + sign, makes up the nucleus and a single electron, denoted by e^- is located in the ground state energy level shown by the shaded portion. This is not intended to be a picture of the atom—it merely shows that a hydrogen atom consists of a single proton in the nucleus and a single electron which moves in an energy level (shaded area) somewhere outside the nucleus.

° The relative size of the nucleus and its density may be visualized from the following approximate figures. An average atom has a radius of about 1×10^{-8} cm, whereas the nucleus has a radius of about 1×10^{-12} cm. In other words, the radius of the nucleus is about $\frac{1}{10000}$ that of the atom. While the density of liquid or solid elements varies from less than 1 to 22.4 g/cm³, the density of the nucleus is estimated at about 10^{14} g/cm³, which would be 10^8 or 100 million tons per cubic centimeter.

† See Chapter 5 for a more detailed discussion of atomic orbitals.

Figure 3.1
The hydrogen atom.

Figure 3.2
The hydrogen ion.

Under certain conditions the electron may be removed from the hydrogen atom, and only the nucleus, which is a single proton, remains. This particle may be represented as H⁺ (Figure 3.2). A particle having a preponderance of either positive or negative charge is termed an **ion**; thus H^+ is a hydrogen ion. It is evident, then, that a hydrogen ion and a proton are the same thing.

The atomic weight of hydrogen is approximately 1 (actually 1.008) and, since the mass of the hydrogen atom is almost entirely contained in the nucleus, it follows that the mass of the nucleus of a hydrogen atom — and therefore of a proton — is very nearly equal to one. Neutrons, which have practically the same mass as protons, therefore must have a weight also very close to unity.

The Structure of Other Atoms

The element of simplest structure after hydrogen is helium, which has an atomic weight of 4. The number 4 indicates that four units of weight are necessary in the nucleus of the atom. These four units of weight must be in protons or neutrons or both. As a matter of fact, the helium nucleus is composed of two protons and two neutrons. Two electrons must be present in the atom to balance the charge of the two protons, and these two electrons are moving within the first energy level (Figure 3.3). The symbol n represents a neutron.

As we progress to the elements of higher atomic weight, the structures become more complex, with additional energy levels in which additional electrons move (Figures 3.4, 3.5, 3.6, 3.7).

We must remember, again, that these diagrams are not intended as pictures of atoms, but are simply a means of showing the component parts in their approximate relative positions. Atoms are three-dimensional rather than two-dimensional, as shown here. Nor do the diagrams give us a true

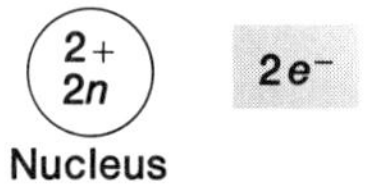

Figure 3.3
The helium atom.

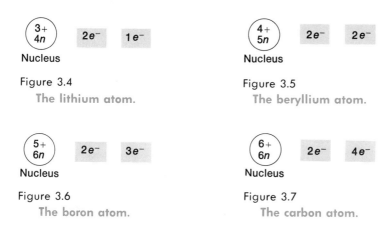

Figure 3.4
The lithium atom.

Figure 3.5
The beryllium atom.

Figure 3.6
The boron atom.

Figure 3.7
The carbon atom.

picture of atomic dimensions. If a nucleus were pictured as about the size of a pea, the first energy level of electrons would be shown at a distance of more than 200 feet away. It is difficult to visualize the immense spaciousness of the atom.

Electrons and Energy Levels

The capacity of the first energy level (also called the K level) of an atom is two electrons, and this number is never exceeded. We may say that two electrons represent a stable or saturated condition for the first energy level of an atom. A maximum of eight electrons may appear in the second energy level (L level); eighteen electrons in the third energy level (M level); and thirty-two electrons in the fourth energy level* (N level). See Figure 3.8.

We shall learn, however, that the **number of electrons in the outermost energy level of an atom may never exceed eight.** The third energy level may have as many as eighteen electrons, providing it is *not* the outer energy level, the fourth energy level may have as many as thirty-two electrons, provided it is *not* the outermost level, and so on. The distribution of electrons in the various energy levels about the nucleus for the elements is shown in Table 3.1.

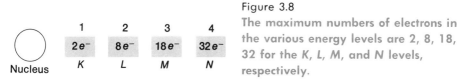

Figure 3.8
The maximum numbers of electrons in the various energy levels are 2, 8, 18, 32 for the K, L, M, and N levels, respectively.

° The maximum number of electrons that may appear in an energy level may be determined from the equation

$$\text{Max. No. of electrons} = 2\ n^2$$

where n is the number of the energy level.

Table 3.1

Distribution of electrons of the elements

Atomic Number (Z)	Element	1 (K) s	2 (L) s p	3 (M) s p d	4 (N) s p d f	5 (O) s p d f	6 (P) s p d	7 (Q) s
1	H	1						
2	He	2						
3	Li	2	1					
4	Be	2	2					
5	B	2	2 1					
6	C	2	2 2					
7	N	2	2 3					
8	O	2	2 4					
9	F	2	2 5					
10	Ne	2	2 6					
11	Na	2	2 6	1				
12	Mg	2	2 6	2				
13	Al	2	2 6	2 1				
14	Si	2	2 6	2 2				
15	P	2	2 6	2 3				
16	S	2	2 6	2 4				
17	Cl	2	2 6	2 5				
18	Ar	2	2 6	2 6				
19	K	2	2 6	2 6	1			
20	Ca	2	2 6	2 6	2			
21	Sc	2	2 6	2 6 1	2			
22	Ti	2	2 6	2 6 2	2			
23	V	2	2 6	2 6 3	2			
24	Cr	2	2 6	2 6 5	1			
25	Mn	2	2 6	2 6 5	2			
26	Fe	2	2 6	2 6 6	2			
27	Co	2	2 6	2 6 7	2			
28	Ni	2	2 6	2 6 8	2			
29	Cu	2	2 6	2 6 10	1			
30	Zn	2	2 6	2 6 10	2			
31	Ga	2	2 6	2 6 10	2 1			
32	Ge	2	2 6	2 6 10	2 2			
33	As	2	2 6	2 6 10	2 3			
34	Se	2	2 6	2 6 10	2 4			
35	Br	2	2 6	2 6 10	2 5			
36	Kr	2	2 6	2 6 10	2 6			
37	Rb	2	2 6	2 6 10	2 6	1		
38	Sr	2	2 6	2 6 10	2 6	2		
39	Y	2	2 6	2 6 10	2 6 1	2		
40	Zr	2	2 6	2 6 10	2 6 2	2		
41	Nb	2	2 6	2 6 10	2 6 4	1		
42	Mo	2	2 6	2 6 10	2 6 5	1		
43	Tc	2	2 6	2 6 10	2 6 6	1		
44	Ru	2	2 6	2 6 10	2 6 7	1		
45	Rh	2	2 6	2 6 10	2 6 8	1		
46	Pd	2	2 6	2 6 10	2 6 10			
47	Ag	2	2 6	2 6 10	2 6 10	1		
48	Cd	2	2 6	2 6 10	2 6 10	2		
49	In	2	2 6	2 6 10	2 6 10	2 1		
50	Sn	2	2 6	2 6 10	2 6 10	2 2		
51	Sb	2	2 6	2 6 10	2 6 10	2 3		
52	Te	2	2 6	2 6 10	2 6 10	2 4		

Table 3.1   (cont'd.)

Atomic Number (Z)	Element	1 (K) s	2 (L) s p	3 (M) s p d	4 (N) s p d f	5 (O) s p d f	6 (P) s p d	7 (Q) s
53	I	2	2 6	2 6 10	2 6 10	2 5		
54	Xe	2	2 6	2 6 10	2 6 10	2 6		
55	Cs	2	2 6	2 6 10	2 6 10	2 6	1	
56	Ba	2	2 6	2 6 10	2 6 10	2 6	2	
57	La	2	2 6	2 6 10	2 6 10	2 6 1	2	
58	Ce	2	2 6	2 6 10	2 6 10 1	2 6 1	2	
59	Pr	2	2 6	2 6 10	2 6 10 3	2 6	2	
60	Nd	2	2 6	2 6 10	2 6 10 4	2 6	2	
61	Pm	2	2 6	2 6 10	2 6 10 5	2 6	2	
62	Sm	2	2 6	2 6 10	2 6 10 6	2 6	2	
63	Eu	2	2 6	2 6 10	2 6 10 7	2 6	2	
64	Gd	2	2 6	2 6 10	2 6 10 7	2 6 1	2	
65	Tb	2	2 6	2 6 10	2 6 10 9	2 6	2	
66	Dy	2	2 6	2 6 10	2 6 10 10	2 6	2	
67	Ho	2	2 6	2 6 10	2 6 10 11	2 6	2	
68	Er	2	2 6	2 6 10	2 6 10 12	2 6	2	
69	Tm	2	2 6	2 6 10	2 6 10 13	2 6	2	
70	Yb	2	2 6	2 6 10	2 6 10 14	2 6	2	
71	Lu	2	2 6	2 6 10	2 6 10 14	2 6 1	2	
72	Hf	2	2 6	2 6 10	2 6 10 14	2 6 2	2	
73	Ta	2	2 6	2 6 10	2 6 10 14	2 6 3	2	
74	W	2	2 6	2 6 10	2 6 10 14	2 6 4	2	
75	Re	2	2 6	2 6 10	2 6 10 14	2 6 5	2	
76	Os	2	2 6	2 6 10	2 6 10 14	2 6 6	2	
77	Ir	2	2 6	2 6 10	2 6 10 14	2 6 7	2	
78	Pt	2	2 6	2 6 10	2 6 10 14	2 6 9	1	
79	Au	2	2 6	2 6 10	2 6 10 14	2 6 10	1	
80	Hg	2	2 6	2 6 10	2 6 10 14	2 6 10	2	
81	Tl	2	2 6	2 6 10	2 6 10 14	2 6 10	2 1	
82	Pb	2	2 6	2 6 10	2 6 10 14	2 6 10	2 2	
83	Bi	2	2 6	2 6 10	2 6 10 14	2 6 10	2 3	
84	Po	2	2 6	2 6 10	2 6 10 14	2 6 10	2 4	
85	At	2	2 6	2 6 10	2 6 10 14	2 6 10	2 5	
86	Rn	2	2 6	2 6 10	2 6 10 14	2 6 10	2 6	
87	Fr	2	2 6	2 6 10	2 6 10 14	2 6 10	2 6	1
88	Ra	2	2 6	2 6 10	2 6 10 14	2 6 10	2 6	2
89	Ac	2	2 6	2 6 10	2 6 10 14	2 6 10	2 6 1	2
90	Th	2	2 6	2 6 10	2 6 10 14	2 6 10	2 6 2	2
91	Pa	2	2 6	2 6 10	2 6 10 14	2 6 10 2	2 6 1	2
92	U	2	2 6	2 6 10	2 6 10 14	2 6 10 3	2 6 1	2
93	Np	2	2 6	2 6 10	2 6 10 14	2 6 10 4	2 6 1	2
94	Pu	2	2 6	2 6 10	2 6 10 14	2 6 10 6	2 6	2
95	Am	2	2 6	2 6 10	2 6 10 14	2 6 10 7	2 6	2
96	Cm	2	2 6	2 6 10	2 6 10 14	2 6 10 7	2 6 1	2
97	Bk	2	2 6	2 6 10	2 6 10 14	2 6 10 9	2 6	2
98	Cf	2	2 6	2 6 10	2 6 10 14	2 6 10 10	2 6	2
99	Es	2	2 6	2 6 10	2 6 10 14	2 6 10 11	2 6	2
100	Fm	2	2 6	2 6 10	2 6 10 14	2 6 10 12	2 6	2
101	Md	2	2 6	2 6 10	2 6 10 14	2 6 10 13	2 6	2
102	No	2	2 6	2 6 10	2 6 10 14	2 6 10 14	2 6	2
103	Lr	2	2 6	2 6 10	2 6 10 14	2 6 10 14	2 6 1	2

Atomic Numbers

It may be noted from Figures 3.4, 3.5, 3.6, and 3.7 that, as we proceed from one element to another, the number of protons in the atom increases by one in each instance. Thus the element lithium has three protons, beryllium has four protons, boron has five protons, and so on. If we arrange the elements according to their increasing atomic weights, we may assign numbers to the chemical elements, starting with hydrogen as one. These numbers will represent the number of protons present in the atom. The number, called the **atomic number** and designated by the letter Z, may be determined experimentally by X-ray studies of the element (see below). Since the number of electrons in a neutral atom must equal the number of protons, the atomic number is also equal to the number of electrons in the atom.

X RAYS AND ATOMIC NUMBERS   Henry G. J. Moseley, a British physicist, discovered that the wave length (or frequency) of X rays produced when different metallic elements were used for the target of high velocity electrons in a cathode ray tube varied over a considerable range. He arranged the metals in order of increasing frequencies and then assigned each an **atomic number** corresponding to its position in the arrangement. Moseley reasoned that the frequency of radiation from the elements depends on the number and arrangement of unit particles in their atoms, and the atomic number was therefore an important consideration in the structure of the atom. We now know that the atomic number of an element corresponds to the number of protons in the atom, as well as to the number of electrons.

Mass Numbers

The mass of an atom is made up primarily of the protons and neutrons present in the nucleus; the contribution of electrons to the mass is so slight that ordinarily it may be neglected. If the mass of each proton and of each neutron is regarded as unity (this is approximately true), the mass of the atom will be an integral or whole number equal to the sum of the numbers of protons and neutrons present. This integral number is called the **mass number** and is designated by the letter A. The constituents of the nucleus, both protons and neutrons, are often referred to as **nucleons.** Hence the mass number of an atom is the number of its nucleons.

These mass numbers are only approximations of the actual comparative masses (atomic weights). In most cases, atomic weights deviate from whole numbers, for reasons we shall discuss in a later section of this chapter. In descriptions of the structures of atoms, the mass number identifies an atom and is used in the diagrammatic representations. We shall use the following notation to show the mass number and atomic number of an element: the mass number will be shown as a superscript and the atomic number as a subscript to the left of the symbol for the element—for example, $^{12}_{6}C$ shows the element carbon to have a mass number of 12 (A = 12) and an atomic number of 6 (Z = 6); $^{238}_{92}U$ means that uranium has a mass number of 238 (A = 238) and atomic number 92 (Z = 92), and so on.

From the mass number and the atomic number of an element, we are able to describe its structure. Since the mass number is equal to the sum of protons and neutrons and the atomic number is equal to the number of protons, the difference between the mass number and atomic number gives us the number of neutrons present in the atom. The number of electrons in the external structure is, of course, equal to the atomic number. Remembering that all of the neutrons and protons appear in the nucleus and all electrons in the external structure, we are in a position to diagram any particular atom, providing we know how the electrons arrange themselves in the various energy levels or energy states.

Valence Electrons

The electrons in the outermost energy level of an atom largely determine the chemical properties of the element. As we have indicated, the maximum number of electrons that can be present in the outermost level is *eight*, which represents a stable configuration. Atoms tend to approach this stable octet configuration by gaining or losing electrons. To illustrate, let us consider the sodium atom, which has a mass number of 23 and an atomic number of 11. Since the atomic number is equal to the number of protons in the atom, and also is equal to the number of electrons, the sodium atom must contain 11 protons and 11 electrons. The difference between the mass number and the atomic number (23 − 11 = 12) gives the number of neutrons. Hence the nucleus will contain 11 protons and 12 neutrons. The first energy level will contain its capacity of two electrons, and the second will have its capacity of eight electrons. This leaves one electron, then, to be put into the third or *M* level, which in this case is the outermost level (Figure 3.9). The sodium atom will tend toward a stable configuration (in which the outer energy level contains eight electrons) by gaining or losing electrons. In order to reach such a stable condition, the sodium atom may give up one electron, in which case the third level disappears and the second level becomes the outermost level. Or it may take up seven electrons to complete the quota of eight in the level which already contains one. This latter process, however, is extremely unlikely; addition of successive electrons would give the atom an increasingly high negative charge, which would repel additional electrons. Certainly it would seem easier for the atom to give up one electron than to take up seven, and as a matter of fact, the sodium atom does readily lose one electron, in which case it becomes a sodium ion, Na^+.

The electrons present in the outermost level of the atom are known as **valence** electrons; they determine the manner in which atoms combine with

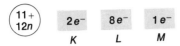

Figure 3.9

The sodium atom.

Figure 3.10

The chlorine atom.

one another. The valence of sodium is +1, since the atom tends to lose one electron and become a sodium ion, Na^+, with a net positive charge of one.

Metallic sodium is a good conductor of electricity. Since sodium has one outer electron which it tends to give up readily, a flow of electricity in sodium is construed to be a movement of this outer electron from one atom to the next. In general, a metal is a good electrical conductor, since the outer or valence electrons are comparatively lightly held by the atomic nucleus and are free to move successively from one atom to another.

Let us examine the structure of the chlorine atom, which has mass number 35 and atomic number 17. The nucleus contains 17 protons (the atomic number) and 18 neutrons (mass number minus atomic number); 7 electrons appear in the outer level (Figure 3.10). From what has been said, we conclude that the chlorine atom should tend to take on one electron to complete its quota of eight in the outer level. This would be much easier than giving up seven electrons. If the chlorine atom gains one electron, it will have a net charge of −1 and may be represented as Cl^-, which is a chloride ion. The valence of chlorine is said to be −1.

As a general rule, we may say that atoms which have fewer than four electrons in the outer configuration tend to give up electrons, while those containing more than four electrons in the outer configuration tend to take up electrons to complete the quota of eight. The number of electrons given up or lost by an atom is its **positive valence;** the number of electrons taken up or gained by an atom is its **negative valence.**

In general we find that those elements which we classify as metals tend to give up electrons rather than take them up; consequently, metals have a positive valence. On the other hand, those elements classified as nonmetals usually take up electrons; consequently, they have a negative valence.

Electronic Symbols

Rather than a complete diagram for an atom, an **electronic symbol** is often used in which only the valence electrons are shown. After all, these are the important ones as far as chemical activity or reaction is concerned. A sodium atom with one valence electron may be shown as Na·, where the dot represents the valence electron and the symbol represents the nucleus and all of the electrons *except* the valence electron. A chlorine atom with seven valence electrons is shown as : C̈l· . The usefulness of electronic symbols will be evident as soon as we start representing the combination of atoms to form molecules.

It will be helpful to know that the **number of valence electrons** used in the electronic symbols of most of the major common elements (**A** family elements) is equal to the element's **group number** in the periodic table (Chapter 4).

Quantum Numbers and Energy Sublevels

We have noted that electrons group themselves into successive energy levels, with maxima of two electrons in the first or *K* level, eight in the second or *L* level, eighteen in the third or *M* level, and thirty-two in the fourth or *N* level. These levels are what we may term the main or principal energy levels which accommodate electrons. Spectral studies indicate that not all electrons within a given principal energy level have exactly the same energies, but that they probably move within sublevels of the main level. In other words, each of the principal energy levels may be resolved into sublevels in which certain electrons are situated. The electrons in the various sublevels are best described by sets of quantum numbers, of which there are four.

°1. The **principal quantum number,** designated by n, corresponds to the main energy level in which the electron moves — that is, 1, 2, 3, or *K, L, M.* Thus a *K* electron has a principal quantum number of 1; for an *L* electron the value of n is 2, and so on.
2. The **azimuthal quantum number,** designated by l, gives a measure of the angular momentum of an electron in its motion about the nucleus. The introduction of this term indicates that the main energy levels of electrons may be considered as being made up of one or more **sublevels.** The term l may have values of 0 to $(n - 1)$. Hence for the first main level where $n = 1$, l can have only a value of 0. For the first level, then, the sublevel and main level would coincide. When $n = 2$, l can have values of 0 and 1. This would correspond to two sublevels within the second main level. When $n = 3$, l may have the values 0, 1, 2, corresponding to 3 sublevels.
3. The **magnetic quantum number,** designated by m, indicates the behavior of the electrons in a magnetic field; m may have values from -1 to $+1$ including 0. Hence for the first main level, where $n = 1$ and $l = 0$, m can have only a value of 0. In the second main level, where $n = 2$, l can have the values 0 and 1. When $l = 0$, $m = 0$, and when $l = 1$, m may be -1, 0, or $+1$.
4. The **spin quantum number,** designated by s, indicates the spin of an electron about its own axis in a clockwise (written $+$) or counterclockwise (written $-$) direction. If we follow the above set of rules for quantum numbers, we may easily calculate the number of electrons in the various sublevels of the atoms: For the first level, where $n = 1$, $l = 0$, $m = 0$, and $s = +$ or $-$. Since, according to **Pauli's exclusion principle,**

° This section in small type may be omitted without loss of continuity.

no more than one electron can have the same values for the four quantum numbers, the first, or K, level can have only two electrons.

$$n = 1, l = 0, m = 0, s = +$$
$$n = 1, l = 0, m = 0, s = -$$

For the second main level, or L level

$$n = 2, l = 0, m = 0, s = +$$
$$s = -$$

2 electrons

$$n = 2, l = 1, m = -1, s = +$$
$$s = -$$
$$m = 0, s = +$$
$$s = -$$
$$m = +1, s = +$$
$$s = -$$

6 electrons

The L level, then, consists of two sublevels, the first of which holds two electrons and the second six electrons, to make a total of eight. Following a similar procedure, it can be shown that the third main level consists of three sublevels, the first with 2 electrons, the second with 6 electrons, and the third with 10 electrons, to make a total of 18. The fourth main level would have 4 sublevels holding 2, 6, 10, and 14 electrons, a total of 32.

Arrangement of Electrons in Sublevels

In using quantum numbers for describing the electrons in the energy levels of atoms, we may visualize the main levels as made up of sublevels arranged in the following way.

The first main energy level (K) can accommodate only two electrons; the second main energy level (L) is composed of two sublevels, one of which accommodates 2 electrons and the other 6. The third main energy level (M) consists of three sublevels, the first with 2 electrons, the second with 6 electrons, and the third with 10 electrons, to make a total of 18 electrons. The fourth main energy level (N) would have four sublevels holding 2, 6, 10, and 14 electrons, respectively, to make a total of 32. See Table 3.1.

The various levels are designated by letters: s for the 2-electron sublevel, p for the 6-electron sublevel, d for the 10, and f for the 14. The maximum number of electrons in the s, p, d, and f sublevels are 2, 6, 10, and 14, as shown in Figure 3.11.*

Notations are convenient in describing the electronic configuration of atoms — that is, the number of electrons in the main levels and the sublevels

---

* The letters s, p, d, f originally were spectroscopic designations given to certain line spectra which correlated with atomic structure. s stood for the **sharp** series of lines and p, d, f for the **principle, diffuse,** and **fundamental** series, respectively.

Figure 3.11

The sublevels of the K, L, M, and N energy levels. Numbers in the colored areas represent electrons.

of these main levels. For example, neon, with 10 electrons, may be designated as $1s^2$; $2s^2$, $2p^6$, which indicates 2 s electrons in the first main energy level, and 2 s and 6 p electrons in the second main energy level. The first number is the principal quantum number and gives the number of the main energy level. 1 corresponds to the K level, 2 to the L level, 3 to the M level, and so on. Iron has 26 electrons, arranged in the following order: $1s^2$; $2s^2$, $2p^6$; $3s^2$, $3p^6$, $3d^6$; $4s^2$. This means 2 s electrons in the K level; 2 s and 6 p electrons in the second level (L); 2 s, 6 p, and 6 d electrons in the third level (M); and 2 s electrons in the fourth level (N).

Modifications of this standard notation are sometimes employed. One may represent the main levels which are completely filled with electrons as K, L, M, N, and so on, followed by the standard notation for the remaining sublevels. For example, iron has the K level filled with 2 s electrons and the L level filled with 2 s and 6 p electrons, but the third and fourth levels are incomplete, and we may write K, L, $3s^2$, $3p^6$, $3d^6$; $4s^2$.

Another method of notation represents the configuration possessed by one of the inert gases (He, Ne, Ar, Kr, and so on) by the symbol for that element followed by the standard notation. For example, Fe may be shown as Ar-$3d^6$; $4s^2$, since Ar has the configuration $1s^2$; $2s^2$, $2p^6$; $3s^2$, $3p^6$.

Energies of Sublevels

The relative energies of the main levels and the sublevels are shown diagrammatically in Figure 3.12. In this arrangement, the first main level (K), which is closest to the nucleus, has the lowest energy electrons, the second level (L) considerably higher energy electrons, and so on. Each main level (except the first) is made up of two or more sublevels—these are the s, p, d, and f sublevels. Within a given main energy level, we note that the s electrons have the lowest energy, p electrons slightly higher energies, and d and f electrons still higher energies. The squares which appear in the diagram are called **orbitals**. An **orbital** may be regarded as an energy state which may accommodate one or two but no more than two electrons. Two electrons, which are just alike except for direction of spin (one spins clockwise, the other counterclockwise) can be present in each orbital. Note that the first main level has only one orbital which may accommodate one or two s electrons, so that its capacity is 2 electrons. The second main level, made up of the s and p sublevels, can accommodate a maximum of 8 electrons—2 in the one s orbital and 6 in the three p orbitals (that is, 2 in each orbital). The third main level has three sets of

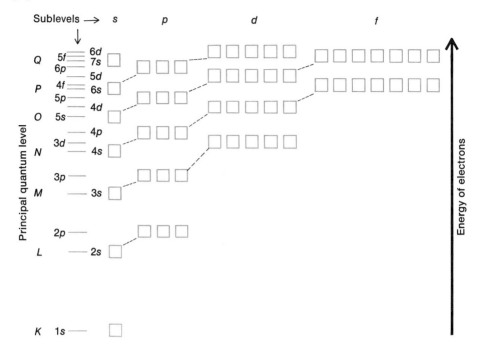

Figure 3.12
Orbitals of atoms are filled with electrons starting with the lowest (K) energy
level.

orbitals, each set representing a different sublevel — one *s* orbital which can
accommodate 2 electrons, three *p* orbitals which can accommodate 6 electrons, and five *d* orbitals which hold 10 electrons, a total of 18 electrons.
The fourth main level is made up of four sublevels of orbitals corresponding
to 2 *s*, 6 *p*, 10 *d*, and 14 *f* electrons — a total of 32 electrons.

Order of Filling Orbitals

As electrons are fitted into the various orbitals shown in Figure 3.12, in general
it can be said that the lower energy levels are filled before the higher levels —
that is, an added electron will go into the orbital of lowest energy. This building process is known as the "aufbau" principle. It follows that the first electrons go into the 1s orbital. When this is filled with 2 electrons, the next 2
electrons go into the 2s orbital, and the next 6 into the 2 *p* orbitals. Following
the filling of the 1s, 2s, and 2p orbitals, added electrons fill the 3s and 3p
orbitals. It will be noted in the diagram that beyond the 3p level of orbitals
there begins to be some overlapping of levels — for example, the 4s level is
slightly lower than the 3d level, the 5s level lower than the 4d, and so on.

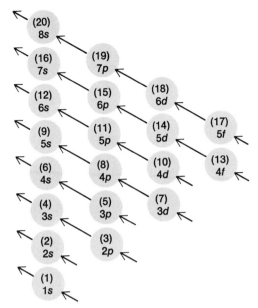

Figure 3.13
Approximate order of filling orbitals.
Starting with the 1s orbital, electrons
enter successive orbitals in the order
indicated by the arrows.

Consequently, in these regions, electrons will enter the next higher main energy level before the lower main level is filled. For example, after the $3p$ orbitals are filled, the next two electrons go into the $4s$ level rather than the $3d$ level, since the energy of the $4s$ level is slightly lower than the $3d$ level. Beyond the $4s$ level, added electrons fill the $3d$ level before going into the $4p$ level, since the $3d$ level is lower in energy than the $4p$ level.

We may summarize the order of filling orbitals to this point as follows: $1s, 2s, 2p, 3s, 3p, 4s, 3d, 4p$. Extending this order according to Figure 3.12 beyond the $4p$ level, electrons would successively enter orbitals as follows: $5s, 4d, 5p, 6s, 4f, 5d, 6p, 7s, 5f$. (See Figure 3.13.)

Note that as we go from bottom to top of the diagram in Figure 3.12, the electron energy levels get closer together. Because there are small energy differences in some orbital levels, certain elements show irregularities in order of orbital filling. For example, in copper, with the electronic configuration $1s^2; 2s^2, 2p^6; 3s^2, 3p^6, 3d^{10}; 4s^1$, the $3d$ level has filled before the $4s$ orbital has been completed. If the order had been followed exactly, copper should have a configuration of $3d^9; 4s^2$. This deviation from predicted configuration is observed in a number of instances where either a half-filled or a filled sublevel seems to give an enhanced stability. In the case of copper, the $3d^{10}; 4s^1$ configuration is more stable than the predicted $3d^9; 4s^2$ because the $3d$ level is filled. Chromium with a predicted configuration of $3d^4; 4s^2$ is actually $3d^5; 4s^1$, presumably because of the half-filled $3d$ level. Other deviations may be noted in Table 3.1.

The aufbau principle is useful in predicting electronic configurations as electrons are added. When electrons are taken away or lost in ionization procedures, they are usually lost in *reverse* order of principal quantum number—that is from the highest level first.

Paired and Unpaired Electrons

As we have seen, each orbital can accommodate a maximum of two electrons. If there are two electrons, they must have opposite spins—that is, one of the two electrons spins in a clockwise direction, and the other in a counterclockwise direction. Two electrons with opposed spins are termed **paired electrons.**

Evidence indicates that as electrons are added to the orbitals in a given energy level, successive electrons will occupy orbitals *singly* until all of the orbitals in that particular energy level are filled. In other words, the electrons tend to remain **unpaired** as long as possible. This is known as **Hund's rule of maximum multiplicity.** In the *s* level, where there is only one orbital, a second electron must enter the orbital which already has one. The two electrons must be paired, of course, and therefore their spins are neutralized. However, in the *p, d,* and *f* levels, which have 3, 5, and 7 orbitals, respectively, single or unpaired electrons occupy the orbitals until each orbital in a given level has one electron. Only after the entire set of orbitals has been occupied by unpaired electrons will added electrons become paired.

Consider nitrogen, with the configuration $1s^2; 2s^2, 2p^3$. We may represent the three *p* electrons as occupying the three orbitals in this manner: [↑] [↑] [↑] , where each of the three orbitals has a single electron. Consequently, the nitrogen atom has three unpaired electrons. The 1*s* and 2*s* levels, of course, contain paired electrons.

Another example is vanadium, V, with the configuration $1s^2; 2s^2, 2p^6; 3s^2, 3p^6, 3d^3; 4s^2$. In the unfilled 3*d* level, the three electrons would be placed [↑] [↑] [↑] [ ] [ ] , not [↑↓] [↑] [ ] [ ] [ ]

From a knowledge of the number of electrons in an unfilled level of orbitals, we may predict the number of unpaired electrons which the atom will possess. For example, Fe has the configuration $1s^2; 2s^2, 2p^6; 3s^2, 3p^6, 3d^6; 4s^2$, and the 3*d* level is incomplete. The expected distribution of electrons in the 3*d* level would be [↑↓] [↑] [↑] [↑] [↑] Thus Fe would possess four unpaired electrons.

Paramagnetism

The unpaired electrons in an atom affect the physical properties of an element, and are important in determining the manner in which an element will combine with other elements. Among other things, unpaired electrons give rise to **paramagnetism,** which is the property of a substance of being weakly attracted to a magnet. A spinning electron acts like a tiny magnet, and unless its magnetism is neutralized by an electron of opposite spin, magnetic properties for the substance result. All substances which contain one or more unpaired electrons are paramagnetic to a degree dependent on the number of unpaired electrons. Materials which have no unpaired electrons tend to migrate out of magnetic fields and are referred to as **diamagnetic. Ferromagnetism,** which is the kind of magnetism we usually hear about in everyday life, is a much stronger form of magnetism exhibited by the elements iron, cobalt, and nickel. These elements are strongly attracted into a magnetic field and remain permanently magnetized after removal from such a field. Ferromagnetism is actually an extreme form of paramagnetism where the combined forces of paramagnetism of a large number of atoms are pooled in a certain definite direction to give a strong magnetic effect. In most cases of paramagnetism the atoms exert their individual magnetic effects in random directions and the reinforcement found in ferromagnetism is absent.

Incomplete Main Levels

In considering energy levels (Figure 3.12), we noted that after the second main level (L level) is filled, electrons do not completely fill the third main level (M) before the fourth main level (N) is started. Thus main energy levels beyond the L level may be incomplete. Table 3.1 shows that between the elements calcium ($_{20}$Ca) and copper ($_{29}$Cu) the third level is not filled to its capacity of 18 electrons. Calcium has the configuration 2, 8, 8, 2, or $1s^2$; $2s^2$, $2p^6$; $3s^2$, $3p^6$; $4s^2$. The next element, scandium ($_{21}$Sc), has the configuration 2, 8, 9, 2, or $1s^2$; $2s^2$, $2p^6$; $3s^2$, $3p^6$, $3d^1$; $4s^2$. The additional electron goes into the 3d level rather than the 4p level, since the 3d level is of somewhat lower energy (see Figure 3.12). Following scandium, successive electrons enter the 3d level until it is filled.

Elements in which the 3d orbitals are being filled are called **transition** elements. Of course, main energy levels beyond the second contain d orbitals, and all of these (into which electrons are being fitted) are included in the category of transition elements. Although these elements will be discussed later (Chapter 28), we may point out now that electron movement between an incomplete level and the outer valence level seems possible. For example, scandium, with the configuration 2, 8, 9, 2, (Figure 3.14), might be expected to exhibit a valence of +2, since it contains two electrons in the outer level (4s). Actually, the common valence of Sc is +3. Presumably the ninth electron (a 3d electron) in the third level, as well as the two electrons (4s electrons) in the outer level, may be lost to give scandium a valence of +3, (Figure 3.15).

Removal of electrons from more than one level gives rise to **variable** valences. For example, iron, with the configuration 2, 8, 14, 2, exhibits valences of +2 and +3. In the first case, two electrons (the 4s electrons)

Figure 3.14
Scandium atom.

Figure 3.15
Scandium ion.

in the fourth level are lost; in the second, an additional electron (a $3d$ electron) is lost from the third level.

Similarities of Structure

Since the outermost configuration of an atom largely determines the chemical properties of the element, we can expect those atoms exhibiting the same outer structure (same number of electrons in outer main energy level) to be quite similar in chemical behavior. We can see that this is indeed so by considering the structures of several elements. Let us first consider the elements lithium, sodium, and potassium (Figure 3.16).

While there is little similarity in the inner structure of these atoms, we note that the outer configuration of each contains one electron. Each of these atoms could easily lose one electron, so that the valence of each would be $+1$. Later we shall find that they are very similar in chemical properties. All are extremely active metals, and this activity is related to the ease with which one electron can be removed from each of their atoms.

In a similar manner, consider the elements fluorine, chlorine, and bromine (Figure 3.17). Because of their similar structures — the outer structure of each atom contains seven electrons — we should expect fluorine, chlorine, and bromine to show similar chemical behavior. Each atom would tend to take on one electron to complete the quota of eight (the valence of each would be -1), and since electrons are gained, we class these elements as nonmetals.

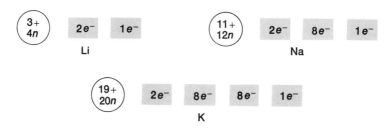

Figure 3.16
The outer structures of atoms of lithium, sodium, and potassium are similar; each atom has one valence electron. (Only main energy levels are shown.)

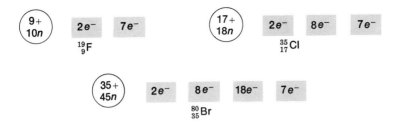

Figure 3.17
Atoms of fluorine, chlorine, and bromine have 7 electrons in the valence level.

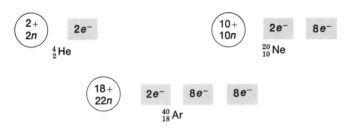

Figure 3.18
Structures of the inert gases helium, neon, and argon.

The elements helium, neon, and argon (Figure 3.18) have stable outer structures. Although helium has only two electrons in its outer energy level, we may recall that two electrons represent a stable configuration for the first energy level in the same way that eight electrons represent a stable configuration for the second. Since these atoms have a completed outer configuration, we should expect these elements to be inactive and to exhibit little tendency to lose or gain electrons. Such is the case—these elements show little tendency to combine with other than extremely active elements (see Chapter 19)

Isotopes

We have pointed out that the mass of an atom is due almost entirely to the protons and neutrons present in the nucleus. Electrons have a negligible weight compared with that of protons and neutrons. Since protons and neutrons have very nearly unit mass, it would seem that the weights of atoms (atomic weights) should be nearly whole numbers, if we neglect the very small mass due to the electrons. An examination of the table of atomic weights (see inside of front cover) shows, however, that many of the elements have atomic weights which deviate considerably from whole numbers. For example, chlorine has an atomic weight of 35.45, iron an atomic weight of 55.85.

Studies have shown that most elements are really mixtures of two or more kinds of the element. Chlorine, for example, has been found to be made up of two kinds of chlorine, one of which has a mass number of 35 and the other of 37. In ordinary chlorine these two kinds are in proportions that give the average atomic weight of 35.45. The two kinds of chlorine have the same chemical properties, but atoms of different mass (Figure 3.19).

We note that the outer structure of the atoms is exactly the same and, since there are seven electrons in the outer configuration of each, the atoms have the same chemical properties. The atomic number is the same for the

Figure 3.19

Isotopes of chlorine. The chlorine on the left has a mass number of 35; the chlorine on the right has a mass number of 37.

Table 3.2

Natural stable isotopes of selected elements*

Element	Atomic Number	Atomic Weight	Mass of Isotopes (mass numbers)
Hydrogen	1	1.0079	1, 2
Helium	2	4.00260	4, 3
Lithium	3	6.941	7, 6
Carbon	6	12.011	12, 13
Nitrogen	7	14.0067	14, 15
Oxygen	8	15.9994	16, 18, 17
Fluorine	9	18.9984	19
Sodium	11	22.9898	23
Magnesium	12	24.305	24, 25, 26
Aluminum	13	26.9815	27
Silicon	14	28.086	28, 29, 30
Sulfur	16	32.06	32, 34, 33
Chlorine	17	35.453	35, 37
Potassium	19	39.098	39, 41, 40
Calcium	20	40.08	40, 44, 42, 43
Chromium	24	51.996	52, 53, 50, 54
Iron	26	55.847	56, 54, 57, 58
Nickel	28	58.70	58, 60, 62, 61, 64
Copper	29	63.546	63, 65
Zinc	30	65.38	64, 66, 68, 67, 70
Bromine	35	79.904	79, 81
Silver	47	107.868	107, 109
Tin	50	118.69	120, 118, 116, 119, 117, 124, 122, 112, 114, 115
Iodine	53	126.9045	127
Mercury	80	200.59	202, 200, 199, 201, 198, 204, 196

° Isotopes are listed according to decreasing abundance; the most abundant isotope is listed first. Atomic weights are based on the most abundant isotope of carbon with an assigned weight of 12.0000.

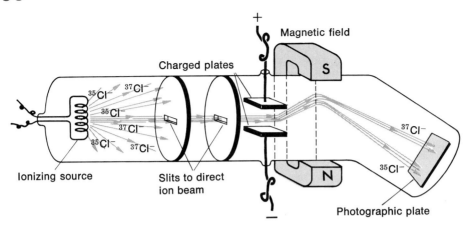

Figure 3.20
Mass spectrograph. Particles having a constant mass-to-charge ratio are focused at the same position on the photographic plate. By calibration of the apparatus, the isotopic masses may be ascertained.

two atoms. The difference appears in the nucleus: the atom of mass number 35 has 18 neutrons, while the atom of mass number 37 has 20 neutrons. Elements with the same atomic number but different mass numbers are termed **isotopes;** thus ordinary chlorine is a mixture of two isotopes.

Isotopes of most of the chemical elements are known. Table 3.2 includes the stable isotopes of several of the elements. We may note that the number of isotopes of the elements varies from two in some of the elements to ten in tin.

MEASURING THE MASS OF ISOTOPES   When gas molecules are bombarded with cathode rays, electrons are dislodged, leaving positively charged particles (ions). A beam of such ionized particles will be deflected on passage through magnetic and electrical fields. The magnitude of the deflection depends on various factors, such as mass, charge, and velocity of the positive ions, and dimensions of the apparatus. By keeping various factors constant, the mass of the positive particles may be determined. Such an arrangement is called a **mass spectrograph** (Figure 3.20) when the masses are recorded on a photograph film and a **mass spectrometer** when an electrical recorder is used (Figure 3.21).

Atomic Weights from the Mass Spectrograph

As pointed out above, isotopic masses may be determined accurately with the mass spectrograph. In the case of an element with only one isotope — such as iodine — the mass determined with the mass spectrograph is its atomic weight. For elements with more than one isotope, it is necessary to know not only the mass of each isotope but also the percentage of each isotope present in the naturally occurring form of the element, since atomic weights as we use them

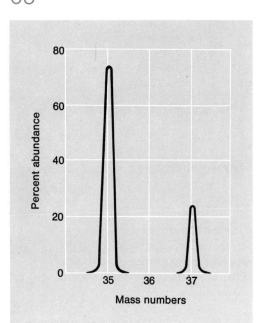

Figure 3.21
Mass spectrum for naturally occurring chlorine.

are based on the average mass of the atoms in the naturally occurring mixture. Fortunately, this information can be derived from mass spectrometer data, and the atomic weight can be calculated. For example, a mass spectrum for chlorine (Figure 3.21) shows that in ordinary chlorine the two isotopes occur in the following proportions:

	% Abundance	Actual Mass
^{35}Cl	75.77	34.969
^{37}Cl	24.23	36.966

The atomic weight of naturally occurring chlorine may then be obtained in the following manner:

$$(0.7577)(34.969) + (0.2423)(36.966) = 35.453$$

Because of its great accuracy, the mass spectrometer method of determining atomic weights is becoming more important than methods based on chemical analysis.

3.2 HOW ATOMS COMBINE

The theory of atomic structure gives us an explanation of the manner in which atoms combine.* The valence electrons in the atom—that is, those electrons

---

* Study of modern theory to account for diverse manners of bonding between atoms, bond energies and distances is deferred to Chapter 5. In this chapter, elementary bonding principles will suffice for usual chemical compounds.

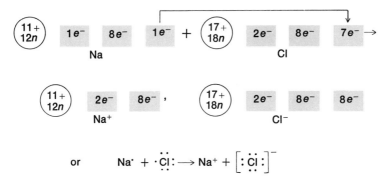

or $\mathrm{Na^{\cdot}} + \cdot\ddot{\underset{\cdot\cdot}{Cl}}: \longrightarrow \mathrm{Na^{+}} + \left[:\ddot{\underset{\cdot\cdot}{Cl}}:\right]^{-}$

Figure 3.22
In the formation of the ionic compound sodium chloride, an electron is transferred from the sodium atom to the chlorine atom.

present in the outermost energy level—seem to be the determinants of chemical combination. We learned that some atoms tend to give electrons, while others tend to take electrons. It would seem, then, that chemical combination would be most likely to take place between atoms in which electrons could be given up by one atom and taken up by another. Metals, which tend to give up electrons, are known to show little tendency to combine with one another but they readily combine with elements (non-metals) which will take up electrons. For example, since the sodium atom readily gives up an electron and a chlorine atom readily accepts an electron, we might expect sodium and chlorine to combine chemically. They actually do combine to form the compound sodium chloride (Figure 3.22)

One electron is transferred from the sodium atom to the outer level of the chlorine atom, and in the compound formed, the two ions, $\mathrm{Na^{+}}$ and $\mathrm{Cl^{-}}$, are present. A compound in which ions exist is said to be **ionic** or **electrovalent**, and the transfer of electrons from one atom to another which results in the formation of ions is termed **electrovalence.** Similarly, the combination of one atom of magnesium with two atoms of chlorine to form the compound magnesium chloride ($MgCl_2$) may be shown as

$$\mathrm{Mg}:\!\!\!\diagdown\!\!\!\begin{array}{l}\cdot\ddot{\underset{\cdot\cdot}{Cl}}:\\[4pt] \cdot\ddot{\underset{\cdot\cdot}{Cl}}:\end{array} \rightarrow \mathrm{Mg^{2+}}, 2\ \mathrm{Cl^{-}}\ (MgCl_2)$$

In forming magnesium chloride, each of the chlorine atoms takes up one electron from the magnesium atom. The resulting compound is electrovalent or ionic, and is composed of magnesium ions and chloride ions in the ratio of one to two.

The Structure of Solid Sodium Chloride

X-ray studies have shown that a crystal of sodium chloride is composed of sodium ions and chloride ions arranged in an orderly fashion, as shown in Figure
3.23. These ions are definitely oriented in relation to one another. If a sodium
ion occupies the center of a cube, then a chloride ion appears in the center of
each of the six faces of the cube. All electrovalent compounds show an orientation of ions in the crystalline solid. When the solid is dissolved in water,
the ions simply break away from the surface of the crystal and pass into the
solution, where they are relatively widely separated and act as independent
units.

    It is incorrect to speak of a molecule of solid sodium chloride, since we
cannot say that a sodium ion and a chloride ion are paired to form a unit
particle. Each of the ions in the crystal of sodium chloride is surrounded by
six ions of opposite charge, and no one of the six ions is more closely associated with the central ion than any of the others. "NaCl" should be in-

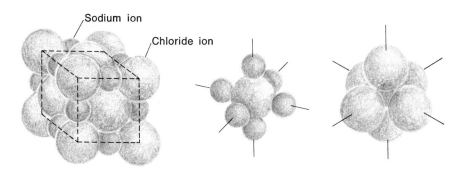

Figure 3.23

Crystal structure of NaCl is cubic.

terpreted not as a molecule of sodium chloride, but rather as a formula which
shows that this compound is composed of an equal number of sodium and
chloride ions. The formulas of all electrovalent compounds should be interpreted in this way.

Sharing of Electrons: Covalence

Chemical combination may take place without an actual transfer of electrons
from one atom to another. For example, in the compound methane (CH_4),

four atoms of hydrogen are combined with one atom of carbon. The carbon atom, with an atomic number of 6, has four electrons in the valence level, and we might assume that carbon would either take up or give up four electrons to reach a stable configuration. Actually the carbon atom does neither—as we indicated earlier, either process would be extremely difficult because of attractive or repulsive forces. Instead, the carbon atom *shares* electrons with other atoms to form its compounds. For example, in CH_4 each of the four hydrogen atoms shares its one electron with the carbon atom, and the carbon atom reciprocates by sharing its four electrons with the four hydrogen atoms. This combination is shown below by using the electronic symbols for carbon and hydrogen.

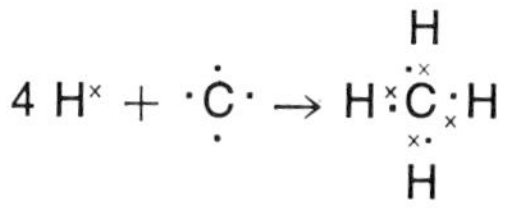

$$4\ H^{\times} + \ \cdot\overset{\displaystyle\cdot}{\underset{\displaystyle\cdot}{C}}\cdot\ \rightarrow\ H\overset{\displaystyle H}{\underset{\displaystyle H}{\overset{\times}{\underset{\times\cdot}{C}}}}H$$

Figure 3.24
The methane molecule, CH_4, is formed by a sharing of electrons between one carbon atom and four hydrogen atoms. The dots represent valence electrons of carbon, and the small crosses are valence electrons of hydrogen atoms.

Since no actual transfer of electrons takes place in this process, the compound is not ionic but molecular. Compounds formed by a sharing process are termed **covalent** compounds, and each pair of shared electrons is termed a **covalent bond.**

The Use of Electron-dot Formulas

A combination of electronic symbols to represent a compound such as methane in Figure 3.24 is termed an **electron-dot formula.** Only the valence electrons appear in the formula. Of course, in Figure 3.24 all of the electrons are alike whether they originate with carbon or hydrogen, and it is usual to show all electrons as dots.

In most cases when atoms combine, the so-called **Octet Rule** holds—that is, each atom tends to acquire a completed octet of electrons in the outer level. Although there are exceptions to the rule (these will be discussed in Chapter 5), it is a very useful rule, since it holds for most compounds.

Other examples of covalence are shown in electron-dot formulas below:

$$\overset{\cdot\cdot}{\underset{\cdot\cdot}{O}}::C::\overset{\cdot\cdot}{\underset{\cdot\cdot}{O}}$$

carbon dioxide

$$\overset{\displaystyle :\overset{\cdot\cdot}{Cl}:}{:\overset{\cdot\cdot}{Cl}:\overset{\cdot\cdot}{P}:\overset{\cdot\cdot}{Cl}:}$$

phosphorus trichloride

The different colored dots simply help show which electrons are contributed by each atom. Note that each atom in the structures has a surrounding layer of eight electrons.

Other conventions may be employed to show bonds between atoms in covalent compounds—for example, a short line called a **single bond** is frequently used to denote an electron pair. Thus methane, CH_4, may be shown as

$$
\begin{array}{c}
\text{H} \\
\text{H} \overset{..}{\text{C}} \text{H} \\
\text{H}
\end{array}
\quad \text{or} \quad
\begin{array}{c}
\text{H} \\
| \\
\text{H} - \text{C} - \text{H} \\
| \\
\text{H}
\end{array}
$$

or PCl_3 as

$$
\text{Cl} - \text{P} - \text{Cl}, \text{ etc.}
$$

$$
|
$$

$$
\text{Cl}
$$

When two pairs of electrons are shared by atoms, the bond is said to be a **double bond,** which is indicated by two parallel lines. Thus CO_2 may be shown as

$$
\text{O}=\text{C}=\text{O}.
$$

The organic compound acetylene, C_2H_2, would have an electron-dot formula

$$
\text{H} : \text{C} : : : \text{C} : \text{H} \text{ or } \text{H} - \text{C} \equiv \text{C} - \text{H}
$$

The three dashes represent three pairs of shared electrons and are called a **triple bond.** In some instances electrons may be shared between two or more atoms of the same element. For example, two chlorine atoms combine to form a molecule of chlorine by a mutual sharing of electrons:

$$
: \overset{..}{\underset{..}{\text{Cl}}} \cdot \overset{..}{\underset{..}{\text{Cl}}} :
$$

Similar cases are O_2, H_2, N_2, F_2, and Br_2. Such molecules, which consist of a pair of atoms, are known as diatomic molecules.

Electrovalent and Covalent Compounds

Although we shall find later that many compounds do not fall strictly into the electrovalent or covalent category, it will be useful now to state the general rules. (1) Compounds of metals with nonmetals tend to be electrovalent; (2) compounds of nonmetals with nonmetals tend to be covalent. Suppose we consider the electronic symbols for a number of elements:

Metals				*Nonmetals*	
Na·	Mg:	Li·	H·	:F̈·	·Ċ·
	Ȧl:			·Ö·	·S̈·
				·N̈·	:C̈l·

Any combination of a metal with a nonmetal will probably take place by a transfer of electrons, and the result will be an ionic, or electrovalent, compound:

$$Na\cdot \longrightarrow \cdot H = Na^+ H^-$$

$$Mg: \Longrightarrow \overset{..}{\underset{..}{S}}: = Mg^{2+} S^{2-}$$

$$\begin{matrix} Li\cdot \\ Li \end{matrix} \Big\rangle \overset{..}{\underset{..}{O}}: = Li_2O(2\ Li^+,\ O^{2-})$$

$$Al: \begin{matrix} \nearrow \overset{..}{\underset{..}{S}}: \\ \searrow \overset{..}{\underset{..}{S}}: \\ \end{matrix} = Al_2S_3(2\ Al^{3+},\ 3\ S^{2-})$$

In writing electron-dot covalent formulas, bring the atoms together, **each with the proper number of valence electrons, to form two-electron-pair bonds and fulfill the octet rule.** Examples:

CH_4
methane

$$4H\cdot + \cdot\overset{.}{\underset{.}{C}}\cdot \rightarrow H:\overset{\overset{H}{..}}{\underset{..}{C}}:H$$

H_2O
water

$$2H\cdot + :\overset{.}{\underset{..}{O}}: \rightarrow H:\overset{..}{\underset{..}{O}}:H$$

CS_2
carbon
disulfide

$$:C: + 2\overset{..}{\underset{..}{S}}: \rightarrow \overset{..}{\underset{..}{S}}::C::\overset{..}{\underset{..}{S}}$$

H_2SO_4
sulfuric
acid

$$2\ H\cdot + \cdot\overset{..}{\underset{..}{S}}\cdot + 4\ \overset{..}{\underset{..}{O}}: \rightarrow H:\overset{..}{\underset{..}{O}}:\overset{\overset{:\overset{..}{O}:}{}}{\underset{:\underset{..}{O}:}{S}}:\overset{..}{\underset{..}{O}}:H$$

Note that in the case of H_2SO_4 two of the covalent bonds connecting sulfur and oxygen consist of *both* of the shared electrons being furnished by *one* element, in this case sulfur. This is referred to as a "**coordinate** covalent" bond or coordinate valence. Actually, there is no difference between a coordinate covalent bond and a regular covalent bond once the bonds are formed, since each bond is formed from (or consists of) a pair of electrons and all electrons are alike regardless of where they come from. The formation of coordinate covalent bonds can be seen in the following two reactions: Ammonia may add H^+ (a proton or hydrogen ion) to form ammonium ion, NH_4^+.

$$H:\overset{\overset{H}{..}}{\underset{H}{N}}: + H^+ \rightarrow \left[H:\overset{\overset{H}{..}}{\underset{H}{N}}:H\right]^+$$

A molecule of water contains 2 pairs of unused electrons and readily adds a proton to form hydronium ion, H_3O^+.

$$H : \overset{..}{\underset{..}{O}} : H + H^+ \rightarrow \left[H : \overset{..}{\underset{\overset{\displaystyle |}{H}}{O}} : H \right]^+$$

Compounds with More than One Kind of Bonding

Numerous compounds exist in which both electrovalence and covalence are exhibited. In general, salts which contain a radical exhibit both types of bonding—consider the compound sodium sulfate (Na_2SO_4). The salt is electrovalent and exists as sodium and sulfate ions in the crystalline form. Within the SO_4^{2-} ion, however, the sulfur atom and oxygen atoms are held together by covalent linkages—electron pairs are shared by sulfur and oxygen atoms. Na_2SO_4 may be represented diagrammatically:

$$2\ Na^+ \left[\overset{:\overset{..}{O}:}{\underset{:\overset{..}{O}:}{\overset{x}{.}\overset{..}{O}: \overset{..}{S} : \overset{..}{O}\overset{x}{.}}} \right]^{2-}$$

Other examples:

$$3\ K^+ \left[\overset{:\overset{..}{O}:}{\underset{:\overset{..}{O}:}{\overset{x}{.}\overset{..}{O}: \overset{..}{P} : \overset{..}{O}\overset{x}{.}}} \right]^{3-} \qquad K^+ \left[\overset{:\overset{..}{O}:}{\underset{:\overset{..}{O}:}{:\overset{..}{Cl}: \overset{..}{O}\overset{x}{.}}} \right]^-$$

$$K_3PO_4 \qquad\qquad KClO_3$$

potassium phosphate   potassium chlorate

Note the application of the octet rule to the S, P, and Cl atoms.

Exercises

3.1. Define and illustrate each of the following: atomic number, mass number, atomic spectra, paramagnetism, diamagnetism, electron-dot formula, valence electron, orbital, nucleon, isotope, electrovalence, covalence, molecule, transition element, paired electrons, ion, electronic symbol, octet rule, double bond, coordinate valence.

3.2. Complete the following table:

Symbol	Atomic Number	Mass Number	Protons	Neutrons	Electrons
Ca	20	40			
Br		80			
	60	144			
Rn				136	86
Zr		91		51	
Ag^+				61	
	17			18	18

3.3. For each of the following elements what is (a) the number of valence electrons in the atom, (b) the valence of the most common ion formed, (c) the electronic symbol of the atom: Li, S, Ca, F, Cs, I.

3.4. Write the electronic symbols for atoms of C, Na, N, Ne, Si, and Fe.

3.5. Consider the following: $^{28}_{14}$Si atom, $^{59}_{29}$Ni atom, $^{201}_{80}$Hg atom.
   (a) How many protons are in the nucleus of the Hg atom?
   (b) How many neutrons are in the nucleus of the Hg atom?
   (c) How many electrons are in the Si atom?
   (d) What is the mass number of Ni?
   (e) What is the atomic number, Z, of Si?
   (f) How many electrons are in the L shell of Ni?
   (g) Write the electronic distribution of Si ($1s^2$, etc.).

3.6. Given the electronic configurations for the following 3 elements:
   Ca $1s^2 2s^2 2p^6 3s^2 3p^6 4s^2$
   As $1s^2 2s^2 2p^6 3s^2 3p^6 3d^{10} 4s^2 4p^3$
   Cl $1s^2 2s^2 2p^6 3s^2 3p^5$
   (a) Which are the valence electrons in Ca, As, Cl?
   (b) Give the electronic (electron dot) symbols for the 3 elements.
   (c) Calcium and chlorine react to form an ionic (electrovalent) compound. Give the formula for calcium chloride.
   (d) What is the valence of the calcium ion?
   (e) What is the valence of the chloride ion?
   (f) Arsenic and chlorine react by the formation of covalent bonds to form an arsenic chloride. Give the formula for the compound that would form.
   (g) What is the valence of the arsenic?
   (h) Write an electron dot formula for a molecule of chlorine, Cl_2.
   (i) Which of the elements would be paramagnetic?
   (j) If calcium and arsenic react to form an ionic compound, calcium arsenide, what formula would you predict for this compound?

3.7. Concerning orbitals, subshells, shells, and electron distribution:
   (a) How many electrons are allowed in an orbital?
   (b) How many orbitals in an "s" subshell?
   (c) How many orbitals in a "p" subshell?
   (d) How many subshells in an M shell (n = 3)?
   (e) How many electrons in a "d" subshell?
   (f) How many electrons in an L shell (n = 2)?

3.8. Give the electron distribution configurations ($1s^2$, etc.) for the following ions and atoms:
   (a) O^{2-}, (b) Mg^{2+}, (c) $_{31}Ga$.

3.9. The following electronic symbols or formulas are *incorrect*. In each case write a *correct* symbol or formula.
   (a) A potassium ion, K^{2-}
   (b) Water, H:O:H
   (c) A nitrogen molecule (N_2), N≡N

$$H$$

(d) Ammonia (NH_3), H·Ṅ·H

(e) Carbon dioxide (CO_2), $:\ddot{O}:C:\ddot{O}:$

(f) Sodium chloride (NaCl), Na:$\ddot{\underset{..}{C}l}$:

3.10. How many orbitals are there in each of the sublevels s, p, d, f?

3.11. Show the electronic configuration of the sublevels of each principal energy level (for example $_{16}S$ is $1s^2$; $2s^2$, $2p^6$; $3s^2$, $3p^4$) for each of the following elements: $_1H$; $_6C$; $_{10}Ne$; $_{11}Na$; $_4Be$; $_{17}Cl$; $_{18}Ar$; $_{12}Mg$; $_{21}Sc$; $_{26}Fe$; $_{23}V$; $_{38}Sr$; $_{14}Si$; $_{55}Cs$.

3.12. Draw diagrams for each of the following: Cl^-, I^-, K^+, Ca^{2+}, S^{2-}, As^{3-}. Note that all have the same outer configuration of electrons.

3.13. Identify the atoms that have the following ground-state electronic configurations in their outer shell or shells: (a) $5s^25p^2$, (b) $3s^23p^63d^54s^2$, (c) $4s^24p^64d^{10}5s^2$, (d) $4s^24p^65s^2$, (e) $5s^25p^5$.

3.14. Determine the number of unpaired electrons (if any) in each of the elements: Ne, C, S, Na, Mn, Mg, As, Si, V.

3.15. Which of the following atoms would be expected to exhibit paramagnetism? $_2He$; $_{19}K$; $_{24}Cr$; $_7N$; $_9F$; $_{16}S$; $_{38}Sr$; $_{27}Co$.

3.16. Naturally occurring carbon consists of two isotopes: ^{12}C, which is assigned a mass of exactly 12 amu; and ^{13}C, which has a mass of 13.003 amu. The atomic weight of carbon is 12.011. What is the percent abundance of each of the isotopes?

3.17. If element X consists of 60.4% of atoms with a mass of 68.9 amu each and 39.6% of atoms with a mass of 70.9 amu each, what is the atomic weight of X?

3.18. Determine the atomic weight of each element from data in the following table. Compare your answers with the more accurate values given in the table inside the front cover of this text.

	Isotope	Abundance (%)	Atomic Mass (amu)
(a)	^{79}Br	50.54	78.9183
	^{81}Br	49.46	80.9163
(b)	^{107}Ag	51.82	106.904
	^{109}Ag	48.18	108.905
(c)	^{20}Ne	90.92	19.9924
	^{21}Ne	0.26	20.994
	^{22}Ne	8.82	21.991
(d)	^{70}Ge	20.53	69.925
	^{72}Ge	27.43	71.922
	^{73}Ge	7.76	72.923
	^{74}Ge	36.54	73.922
	^{76}Ge	7.74	75.921

(e)	^{24}Mg	78.70	23.985
	^{25}Mg	10.13	24.986
	^{26}Mg	11.17	25.983

3.19. Determine the relative abundance of each isotope in naturally occurring gallium from the following data: At. wt. Ga = 69.72. Masses of isotopes: ^{69}Ga = 68.926 amu, ^{71}Ga = 70.925 amu.

3.20. Copper, with an atomic weight of 63.54, is composed of the two isotopes ^{63}Cu and ^{65}Cu with atomic masses of 62.930 amu and 64.928 amu, respectively. What is the percent of each isotope in the naturally occurring mixture?

3.21. Represent diagrammatically the formation of the ionic compounds from the elements: Na_2S, $CaCl_2$, KF, LiI, BaSe, AlF_3.

3.22. Diagram the following covalent compounds showing the arrangement of the valence electrons: CH_4, C_2H_6, C_2H_4, CCl_4, SiO_2, C_2H_2.

3.23. Using electron-dot formulas, represent the following: (a) F_2, (b) Na_2S, (c) N_2, (d) HBr, (e) CO_2, (f) KH, (g) LiOH, (h) Ga^{3+}, (i) H_3AsO_4, (j) Ca^{2+}, (k) PCl_3, (l) NH_3, (m) PH_4^+, (n) H_3O^+, (o) H_2SeO_4.

3.24. Draw electron-dot structures for the following compounds: (a) H_2NNH_2, (b) O_2NONO_2, (c) PH_3, (d) O_2NCl, (e) O_2SF_2, (f) CH_3COOH.

3.25. Draw electron-dot structures for the following ions: (a) PO_4^{3-}, (b) CO_3^{2-}, (c) NO_3^-, (d) SO_4^{2-}, (e) CN^-, (f) HCO_3^-, (g) O_2^{2-}

3.26. Draw electron-dot structures for the following molecules: (a) CO, (b) S_2Cl_2, (c) S_8 (a ring structure), (d) S_2^{2-}

Supplementary Readings

3.1. "The Background of Dalton's Atomic Theory," M. B. Hall, *Chem. in Britain*, **2**, 341 (1966).
3.2. "Atomic Orbitals," R. S. Berry, *J. Chem. Educ.*, **43**, 8 (1968).
3.3. *The Structure of Atoms*, J. J. Lagowski, Houghton Mifflin, Boston, Mass., 1964 (paper).

Chapter 4

Periodic Classification of the Elements

4.1   THE PERIODIC TABLE

According to the modern theory of atomic structure developed in the preceding chapter, electrons are added to successive energy levels of atoms in a definite and orderly manner. The electronic configurations of the elements, with the elements arranged in order of increasing atomic numbers, are found in Table 3.1 (page 50). There is a periodic assignment of electronic configurations, which is in accord with the energy level chart shown in Fig. 3.12 (page 58). In Table 3.1 the elements are listed in a vertical column with the two elements hydrogen and helium in the first grouping; eight elements, lithium through neon, in a second grouping; and eight elements, sodium through argon, in a third grouping. Further groupings include eighteen or thirty-two elements.

We may now rearrange Table 3.1 in the form shown in the back inside cover, which is termed the "long form" **periodic table.** In this arrangement elements with the same outer configuration of electrons fall in the same vertical column. The vertical columns are called **groups** and are numbered from IA to VIIIA and from IB to VIIIB. The horizontal rows are called **periods** and are numbered from 1 to 7.

Elements in the vertical groups are very much alike in chemical behavior, since they have the same outer electronic configuration. A given vertical group is often referred to as a **family** of elements. For example, elements of Group IA are termed the **alkalies;** Group IIA elements, **alkaline earths;** Group VIIA elements, **halogens;** and so on.

Periodicity of Properties of the Elements

The elements show a definite **periodicity** of properties as the atomic number increases. For example, lithium ($_3$Li) in the second horizontal row, or period, is a very active metal with one electron in the outer or valence level; the next element of higher atomic number, beryllium ($_4$Be), is a less active metal with two electrons in the outer level. As electrons are added to this second main level, the properties of the elements change. From a very active metal (Li) the period changes to less active metals, then gradually to a very active nonmetal, fluorine ($_9$F), and finally to the inert gas, neon ($_{10}$Ne). In the third horizontal row, or period, we start with sodium ($_{11}$Na), which is a very active metal and like lithium in chemical behavior. As electrons are added, reaching eight in the third main level, we again get the progression from very active through less active metals through nonmetals to argon ($_{18}$Ar), an inert gas which is like neon.

It is evident that there is a definite periodicity of properties — that is, at periods, or intervals, of eight elements there is a recurrence of like properties. For example, Li and Na are alike chemically; Ne and Ar are inert gases; F is like Cl.

Going to the next horizontal row (Period 4), starting with potassium ($_{19}$K), the situation is changed somewhat because of the energy-level sequence of electron orbitals (see the energy-level diagram, Figure 3.12). Because of the filling of $3d$ orbitals in addition to the $4s$ and $4p$ orbitals, we go through a period or interval of eighteen elements (rather than eight) before coming to rubidium ($_{37}$Rb), which is an active metal similar to potassium. Period 5 follows a similar pattern, and we go through a series of eighteen elements before coming to cesium ($_{55}$Cs), which is chemically like Rb, K, Na, and Li. After leaving Cs, we must progress through thirty-two elements before coming to francium ($_{87}$Fr), which is similar to Cs. The interval or period (Period 6) is expanded to thirty-two elements because of the filling of $4f$ orbitals.

We conclude that at certain intervals, which may be eight, eighteen, or thirty-two, we encounter elements which are very much alike in their chemical properties. This is the basis of the modern periodic law: **the properties of the elements are periodic functions of their atomic numbers.**

Main Group Elements

Elements in Groups IA—VIIIA are termed **main group,** *or* **representative group,** elements. The number of the group indicates the number of electrons in the outer energy level—for example, elements of Group VIIA have seven electrons in the outer energy level. Groups IA and IIA are called **s electron groups** because the outer configuration has only s electrons—Group IA has one s electron and Group IIA has two s electrons. Groups numbered IIIA through VIIIA are termed **p electron groups** because the p orbitals are being filled—that is, Group IIIA elements have one p electron (in addition to two s electrons); Group IVA, two p electrons; and so on to Group VIIIA elements, which have six p electrons. The block representation of the table in Figure 4.1 will clarify this.

Transition Elements

Elements in the middle portion of the periodic table constitute the B groups, and are **transition elements.** These are **d electron group** elements, since the d orbitals are being filled with electrons. It will be noted in the table that two spaces in Group IIIB are filled with fifteen elements—the first with elements numbered 57–71, and the second with elements numbered 89–103. These are termed **inner transition elements** in which the $4f$ and $5f$ orbitals are being filled. Of course, the table could be expanded laterally to thirty-two groups to include the inner transition elements, but this would be somewhat unwieldy and inconvenient to use. Instead, the inner transition elements are listed in separate rows at the bottom of the table. These two series are termed the **lanthanide** and **actinide** series, respectively.

In most cases, transition elements have one or two electrons (s electrons)

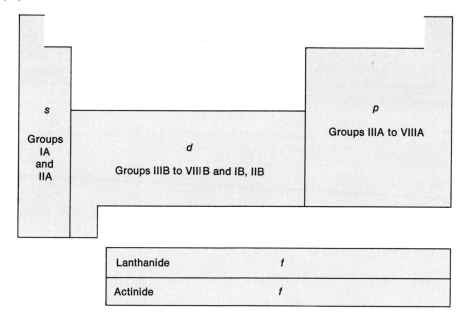

Figure 4.1
Block form of periodic table showing types of electrons most related to position of elements.

in the outermost energy level. As electrons are added to incomplete inner levels (d and f levels), the outermost configuration remains fixed at one or two electrons. Since the outermost level largely determines chemical properties, we may expect greater similarities between transition elements than between representative, or A group, elements. The transition elements will be studied later in Chapter 28 and we shall find that these elements do have many similar properties.

Historical Development of the Periodic Table

More than a hundred years ago, it had been noted that elements such as chlorine, bromine and iodine or calcium, strontium and barium, etc. exhibited very similar characteristics and these elements could be grouped together in families. In 1869, Mendeleev, a Russian chemist, was able to show a definite relationship between chemical properties and atomic weights of the elements. Mendeleev arranged the elements in order of increasing atomic weights and pointed out a recurrence of similar properties at very definite intervals or periods. These periods were not always of the same length, but in each period or interval a gradual change in properties of the elements from the beginning to the end of the interval was apparent. On the basis of this arrangement, Mendeleev stated that the **properties of the elements are periodic functions of their atomic weights.** Since atomic weights follow almost exactly the order of atomic numbers, this statement is very close to the modern periodic law which utilizes atomic numbers rather than atomic weights.

Mendeleev arranged the elements in eight main groups, usually with two sub-groups within each main group. A great disadvantage of this arrangement is the appearance of very dissimilar elements with very different electronic structures in the same group. But at the same time it was very useful in systematizing chemical properties and in predicting properties of undiscovered elements.

4.2   USES OF THE PERIODIC TABLE

Valence and the Periodic Table

Elements within any group exhibit a characteristic valence. For the A series of elements the number of the group corresponds to the number of electrons in the outermost main energy level, and hence we may predict the valence of a particular element from its position in the table. The alkali metals of Group IA exhibit a valence of $+1$, since the atoms easily lose the one electron in the outer level. The halogens of Group VIIA exhibit a characteristic valence of -1, since one electron is readily taken up, and this, added to the seven already present in the outer level, gives a stable configuration of eight. On the same basis, elements of Group IIIA would be expected to exhibit a characteristic valence of $+3$, those of group VIA a valence of -2, and so on. Typical formulas for some hydrogen and oxygen compounds of the period 2 and 3 elements are given in Table 4.1. Note that in each case

$$+ \text{ valence} = \text{group number}$$
$$- \text{ valence} = \text{group number} - 8$$

Table 4.1
Valence and the periodic table (A families)

	I	II	III	IV	V	VI	VII
				-4	-3	-2	-1
2				CH_4	NH_3	H_2O	HF
	$+1$	$+2$	$+3$	$+4$	$+5$	$+6$	$+7$
3	Na_2O	MgO	Al_2O_3	SiO_2	P_2O_5	SO_3	Cl_2O_7

The rare gases of Group VIIIA, each of which has a stable outer configuration of electrons, would not be expected to give up or take up electrons, and hence would have a zero valence. The inert or inactive characteristics of these elements are explained on this basis.

Although valences, measured by the loss or gain of electrons, are quite clear-cut for the A series of elements, they are much less well defined for the transition elements (B series). In the A series the outer energy level alone determines valence characteristics, while in the B series (in which there is at least one incomplete inner level) the incomplete level contributes also to valence characteristics. Most of the transition elements have one or two electrons in the outer level. It appears that one or two electrons from an incomplete inner level may be lost in chemical change and these, added to the one or two in the outer level, allow several possibilities of valence among the transition elements. For example, the iron atom, with the normal configuration 2, 8, 14, 2, may exhibit a valence of $+2$ by loss of the two outer electrons or a valence of $+3$ when an additional electron is lost from the incomplete third level.

Differences in Properties of Elements within Families

Although the elements in a given group show marked similarities in chemical properties, a gradual change in properties takes place as we proceed from elements of lower atomic weight to those of higher atomic weight. For example, the alkali group of metals except lithium gradually increases in activity with increasing atomic weight. Sodium is the least active of the group, followed by potassium, rubidium, and cesium, the last of which is the most active of the metals. The members of the group are very much like one another, however, in that all are quite active.

The fact that the members of a given family do not possess exactly the same chemical properties, even though the outermost levels contain the same number of electrons, is explained by the internal configuration, which will have some small effect on the properties of the atom.

The halogens in Group VIIA are very active nonmetals, but again we may observe a gradual change in activity from fluorine at the top of the group to iodine at the bottom. Fluorine is the most active and iodine is the least active; hence activity decreases as the atomic weights increase.

In Group VA we may also observe a gradual change in properties from top to bottom. Nitrogen at the top is a typical nonmetal, while bismuth at the bottom is usually classed as a metallic element. The elements in between exhibit more marked metallic than nonmetallic properties as the atomic weight increases.

Modern atomic theory explains the slight differences in properties of elements within a given family, as well as pronounced differences in properties of elements in different families. How this is done is explained in part in the next section.

Physical and Chemical Behavior of Elements

Although the student is not yet familiar with the properties of specific elements, it is appropriate to make a few generalizations relating chemical and physical behavior of atoms to certain fundamental properties.

One of the most important properties of an atom which influences its behavior is its *size*. We would expect that as energy levels are added in the structure of atoms, the size will increase, and this is true. In the periodic classification we noted that the elements in a given period (horizontal row) have the same number of main energy levels—for example, one main level for period 1 elements H and He, two main levels for period 2 elements, and so on. As a new period is started, another energy level farther from the nucleus is added to the structure. It is apparent, then, that the vertical columns, which represent families of elements, will show a gradual increase in size in going from top to bottom as electron levels are added. Thus Fr is the largest atom of Group IA, Ra the largest of Group IIA, and so on.

The change in size of atoms in a given period (horizontal row) is somewhat less obvious. In progressing from left to right in a given period, the atoms gradually decrease in size: for example, the sodium atom in Group IA is larger than the magnesium atom in Group IIA, and the latter atom is larger than the aluminum atom in Group IIIA. The explanation of this decrease in size is as follows: As the atomic number increases, the nuclear charge increases; this increased nuclear charge attracts the electrons more forcibly, pulling them closer to the nucleus and thus shrinking the atom. Figure 4.2 shows diagrammatically the relative sizes of atoms and ions in ångströms (A).°

Figure 4.3, in which atomic radii are plotted against atomic numbers, shows a definite periodicity in size as the atomic number increases. Atoms of Group IA elements, the alkali metals, are the largest, and this would account in part for their lightness or low density. Group IIA elements possess a somewhat larger density, because the atoms are somewhat smaller. In progressing from top to bottom in a given family, the density increases, since the atomic weight increases faster than atomic size. Hence the heavier elements appear toward the bottom of the table.

In ionic compounds the size of the ion is very important. The closeness of approach of an ion to a neighboring ion will determine the attractive forces between the ions, and hence the tendency to remain in close association. Obviously, an ion formed by release of electrons will be smaller than the atom, since the outer configuration is destroyed or removed. Furthermore, on removal of electrons from the outer orbitals, the positively charged nucleus now exerts a stronger attraction for the remaining electrons and tends to pull them closer to the nucleus. As we progress in a given period from left to right in the periodic table, the ions become smaller as more electrons are given up—

° The ångström unit (Å) is not an official SI unit but has been used for many years. A preferred unit is the nanometer (nm).

$$1\,\text{Å} = 0.1\,\text{nm} = 10^{-10}\text{m}$$

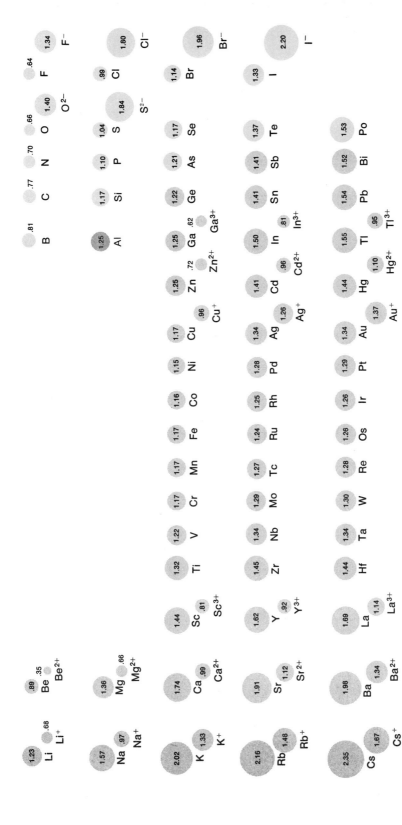

Figure 4.2

The relative sizes of metal and nonmetal atoms and ions. Nonmetals are colored. Numbers are radii in angstrom units.

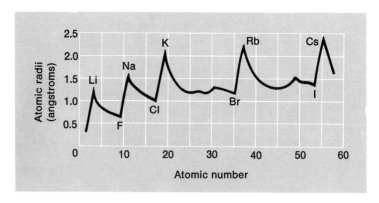

Figure 4.3

Periodicity of atomic radii.

thus Mg^{2+} is smaller than Na^+; Al^{3+} is smaller than Mg^{2+}; and so on. But when electrons are taken up to form negative ions, the size increases, since the attraction of the nucleus is distributed over a greater number of electrons than in the neutral atom, and the ion tends to swell or expand. For example, we would expect Cl^- to be smaller than S^{2-}, which in turn is smaller than P^{3-}

Ionization Energies

When a neutral atom loses or gains an electron, the product particle is an ion, and the gain or loss process is termed **ionization.** Energy is required to remove one or more electrons from a neutral atom, and this energy is referred to as **ionization energy** or **ionization potential.** It is measured in electron volts (eV)* or kilocalories per mole. One electron volt per atom equates to approximately 23 kcal per mole (6.02×10^{23} atoms).

A voltage of 10.2 volts is required to change a gaseous hydrogen atom from its ground state energy level to the next higher energy level. To remove the electron completely (that is, to an infinite distance from the nucleus) requires 13.6 volts. The latter is the ionization potential or ionization energy of hydrogen. Figure 4.4 shows the periodic change of ionization energy as the atomic number increases.

Elements in Group IA have low ionization energies, and since an electron is removed with little energy, these elements are very active chemically. As the number of electrons in the outer configuration increases, the ionization energies increase. Group VIIIA elements (inert gases) have the highest ionization energies and the least tendency to enter into chemical combination.

---

* An electron volt is the energy acquired by an electron when it is accelerated by a potential difference of one volt.

$$1 \text{ eV} = 1.602 \times 10^{-19} \text{ J} = 23.06 \text{ kcal/mole of electrons}$$

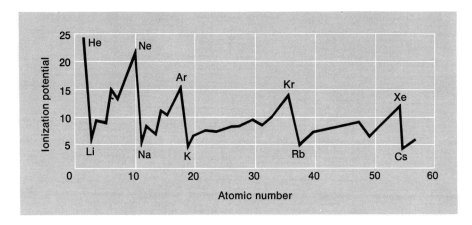

Figure 4.4

Ionization potentials (in electron volts) of the elements.

Ionization energies may be determined for the successive removal of outer electrons; for example,

$$\text{Na}\quad 1s^2;\ 2s^2,\ 2p^6;\ 3s^1 \xrightarrow{\ 5.14\ \text{eV}\ } \text{Na}^+ \ + e^-$$

$$\text{Na}^+ \xrightarrow{\ 47.3\ \text{eV}\ } \text{Na}^{2+} \ + e^-$$

$$\text{Al}\quad 1s^2;\ 2s^2,\ 2p^6;\ 3s^2,\ 3p^1 \xrightarrow{\ 5.6\ \text{eV}\ } \text{Al}^+ \ + e^-$$

$$\text{Al}^+ \xrightarrow{\ 18.8\ \text{eV}\ } \text{Al}^{2+} \ + e^-$$

$$\text{Al}^{2+} \xrightarrow{\ 28.4\ \text{eV}\ } \text{Al}^{3+} \ + e^-$$

$$\text{Al}^{3+} \xrightarrow{\ 120\ \text{eV}\ } \text{Al}^{4+} \ + e^-$$

It requires only 5.14 eV to remove the $3s^1$ electron from a sodium atom, but about nine times as much (47.3 eV) energy to remove an additional electron. The first three electrons are much more easily removed from an aluminum atom than the fourth. From the above values it is evident that sodium is a more active metal than aluminum.

Electrons may be removed from nonmetals and even from the inert gases of Group VIIIA. The ionization energies are in general higher than for the removal of the first electron from metallic elements, for example,

$$\text{Cl}\quad 1s^2;\ 2s^2,\ 2p^6;\ 3s^2,\ 3p^5 \xrightarrow{\ 13.0\ \text{eV}\ } \text{Cl}^+ \ + e^-$$

$$\text{S}\quad 1s^2;\ 2s^2,\ 2p^6;\ 3s^2,\ 3p^4 \xrightarrow{\ 10.4\ \text{eV}\ } \text{S}^+ \ + e^-$$

$$\text{S}^+ \xrightarrow{\ 23.4\ \text{eV}\ } \text{S}^{2+} \ + e^-$$

$$\text{Ar}\quad 1s^2;\ 2s^2,\ 2p^6;\ 3s^2,\ 3p^6 \xrightarrow{\ 15.8\ \text{eV}\ } \text{Ar}^+ \ + e^-$$

Examine Figure 4.4 again to check the comparative ease of the removal of one electron from the metals of Groups IA and IIA as compared with the nonmetals of Groups VIIA and VIIIA.

Electron Affinities

Nonmetals, in contrast with metals, tend to take up electrons to form negative ions. The energy released in this process is termed **electron affinity.** For example, the halogens of Group VIIA in forming negative ions all release a certain amount of energy:

$$F + e^- \rightarrow F^- + 3.45 \text{ eV}$$
$$Cl + e^- \rightarrow Cl^- + 3.61 \text{ eV}$$
$$Br + e^- \rightarrow Br^- + 3.36 \text{ eV}$$
$$I + e^- \rightarrow I^- + 3.06 \text{ eV}$$

Electron affinities are difficult to measure and have been evaluated for only a few elements. As might be expected, Group VIIA elements have the highest electron affinities, since only one electron is needed to complete a stable outer configuration of eight electrons.

Division into Metals and Nonmetals

While there is often no sharp line between metal and nonmetal, those elements which are classed as metallic always exhibit a positive valence, whereas nonmetals usually show a negative valence. Elements which exhibit properties of both metals and nonmetals are termed **metalloids.**

In the periodic table the nonmetals appear to the far right and toward the top. The metallic elements are in far greater number, including Groups IA, IIA, all of the *B* series or transition elements, and the heavier elements in Groups IIIA to VA. The transition elements (the *B* series), including the lanthanide and actinide series, would be expected to exhibit a positive valence and metallic characteristics, since none of these elements possesses more than two electrons in the outer configuration. The position of an element in the table is a good indication of its general physical properties. Metals with low melting points, such as Cd, Ga, Hg, Pb, and Sn, appear just to the left and below the nonmetallic elements; metals in the middle of the table, such as Fe, Co, Ni, Pd, Pt, and Au, are ductile and malleable; metals in Groups IIIB to VIIB tend to be hard and brittle, with high melting points; and metals in Groups IA and IIA are light and low melting.

Predicting Properties

For the chemist, the periodic table is a very useful classification of the elements. Instead of learning the properties of more than a hundred elements,

Table 4.2

Comparison of predicted properties of Eka-silicon with known properties of germanium

Property	Eka-silicon	Germanium
Atomic weight	72	72.6
Characteristic valence	4	4
Density	5.5	5.47
Color	Gray	Grayish-white
Reaction with water	Will decompose steam with difficulty	No reaction with steam
Action of acids	Slight	No reaction with HCl, oxidized by HNO_3
Action of bases	Slight	No reaction
Properties of the chloride	Boiling point below 100°C, density 1.9	Boiling point 86°C, density 1.9
Properties of the oxide	White, refractory, density 4.7	White, refractory, density 4.703

the chemist may master the general properties of each of the groups. If an element is known to be in a certain group, by association with other members of the same group the properties of which are known, its properties may be predicted. As a matter of fact, the properties of some elements were predicted accurately before they were discovered. Mendeleev predicted the properties of eka-silicon, later named germanium, many years before it was discovered. How accurate his predictions were is evident from Table 4.2.

EXAMPLE 4.1   Note the position in the periodic table of astatine, $^{210}_{85}At$, one of the last elements to be discovered, and answer the following questions about astatine.

(1) What group is At in?   7

(2) What period is At in?   6

(3) What is the name of the family At is in?   Halogen

(4) What is the electronic symbol of the At atom?   :Ät·

(5) What is the charge of the At ion?   −1

(6) What is the principle negative valence of the At ion?   −1

(7) Compared to the other members of the At family, how would you rate the activity of At?   least active

(8) In what subshell (s, p, d, or f) would the *last* e added to At go in?   p

(9) Is At a metal, metalloid, or nonmetal?   non-metal

(10) Would the ionization energy of At be high, medium or low?   medium

(11) Would the electron affinity of At be high, medium or low?   medium

(12) Would the electronegativity of At be high, medium or low? _____ medium

(13) Would the atomic radius of At (compared to other atoms) be large, medium, or small? _____ medium

(14) Would the At ionic radius be larger, the same, or smaller than the At atomic radius? _____ larger

(15) What would be the chemical formula of the calcium salt of At? _____ Ca At$_2$

EXAMPLE 4.2

	1	2	3	4	5	6	7	8
1	Ho							W
2	Ea	Ge	Rl	Y	We	/////	Co	M
3	E	To	/////	C	He	Mi	St	Ry
4	N	Ow						

Below are 18 of the first 20 elements in the periodic table, each represented by a *fictitious symbol*. Using the information and *symbols given*, fit each of these elements into the proper place in the periodic table. *Use the fictitious symbols only*, not the actual symbols.

Ea is the first member of Period 2.
Ho has the lowest atomic weight of any of the elements.
N is the most active element of those listed in Group 1.
We and He are members of group 5, with He having a larger atomic radius and atomic weight.
W has the first orbit complete but has no p electrons or other s electrons.
Ge has a structure $1s^2 2s^2$.
Y has a valence of +4 or −4.
Ge and Co unite to form GeCo$_2$, both elements being in period 2.
Element Rl is between Ge and Y in period 2.
Ge and Mi unite to form GeMi.
To is similar to Ge but more active.
M has one more proton than Co.
E forms an ion with a valence of +1.

Ge, To, and Ow, are members of the same family, Ow being the most active metal and Ge having the smallest atomic radius of the three.

C has an atomic number of 14.

Ry has an electron configuration 2, 8, 8.

St has an electronic symbol : $\ddot{\underset{\cdot\cdot}{S}}$t· .

Imperfections of the Classification

Despite its great usefulness, the periodic classification is not without its faults.

Hydrogen does not fit into the table entirely satisfactorily. It is usually placed in Group IA, since it does possess one valence electron (like the alkali metals), and combines with nonmetals, such as the halogens and sulfur. Sulfur compounds with hydrogen are covalent in character, however, and not like the electrovalent combination of alkali metals with the halogens. In some respects hydrogen resembles the members of Group VIIA, since it appears to be nonmetallic and furthermore, like the halogens, may combine with the alkali metals to form ionic compounds. For example, sodium hydride (NaH) is ionic in character, like NaCl.

In the table, a group of fifteen elements with very similar chemical and physical properties, the **rare earth** or **lanthanide** series, must be fitted into a single space. Of course, the table could be further elongated to take care of these elements, but this would make the table unwieldy without adding to its usefulness. In any event, these elements should perhaps be considered separately, since they constitute a unique group as far as structural characteristics are concerned.

Exercises

4.1. Explain how the following terms are used in connection with the periodic table. Use illustrative examples. (a) group, (b) period, (c) representative elements, (d) transition elements, (e) family of elements, (f) inert gases, (g) lanthanide series, (h) inner transition elements, (i) A and B groups, (j) alkali metals, (k) halogens.

4.2. Classify the following elements as (a) alkalies, (b) halogens, (c) noble gases, (d) main group elements, (e) transition elements, (f) lanthanides, or (g) actinides, C, I, Ni, Nd, Cs, Xe, Pt, Am, Li, Pb.

4.3. Which of the elements listed in 4.2 are (a) in Group I in the periodic table, (b) in period 4, (c) in the A family of Group IV, (d) metals?

4.4. Identify the families in the table which include the so-called s-electron elements; p-electron elements; d-electron elements; and f-electron elements.

4.5. Predict the maximum and minimum valences shown by the following A family elements in compounds: Be, I, B, Ca, S, As, Li, C, N.

4.6. How does the reactivity of elements change (a) from left to right in a given period in the periodic table, (b) from top to bottom in Group IIA, (c) from top to bottom in Group VIA?

4.7. In which of the following series would the size of atoms *decrease?* Numbers are atomic numbers. (a) 4, 5, 6; (b) 3, 19, 55; (c) 16, 34, 84; (d) 56, 20, 4; (e) 9, 35, 53.

4.8. On the basis of position in the periodic table, how would you expect the following atoms and ions to compare in size? Explain your answer in each case: (a) Be and C; (b) Li^+ and Na^+; (c) Ca^{2+} and Ba^{2+}; (d) S^{2-} and Cl^-; (e) Cu and Ca; (f) Ne and He; (g) Sr^{2+} and Ba^{2+}; (h) Al and B; (i) Si and S; (j) Zn^{2+}, Cd^{2+}, and Hg^{2+}; (k) O^{2-} and S^{2-}; (l) Fe and K; (m) Sr^{2+} and Rb.

4.9. Which member of each of the following pairs would you predict to be larger? (a) Se or Br; (b) Sb or As; (c) Te or Cd; (d) Ba or Hf; (e) Na or Ca; (f) Mg or Mg^{2+}; (g) S or S^{2-}; (h) Tl^+ or Tl^{3+}.

4.10. How do ionization potentials and electron affinities vary with decreasing atomic weights for elements in (a) the vertical groups, (b) the horizontal rows?

4.11. Which member of each of the following pairs would you predict to have the higher ionization potential? (a) K or F; (b) Al or Cl; (c) Mg or Ba; (d) Na or Al; (e) F or I; (f) Br or Cu; (g) P or O.

4.12. Describe changes in activity as atomic numbers increase in (a) metals of a given group, (b) nonmetals of a given group, (c) periods or horizontal rows.

4.13. From the data recorded for the families of elements listed below, predict the properties of the last member of each family:

Element	Density (g/ml)	Melting Point	Boiling Point	Color
F		−223	−188	pale yellow
Cl		−102	−35	green-yellow
Br	3.4	−7	59	brown
I	4.9	113	183	black
At				

Element	Density (g/ml)	Melting Point	Boiling Point	Solubility of Chloride g/100g H_2O at 0°
K	0.86	62.3	760	27.6
Rb	1.53	38.5	700	77.0
Cs	1.87	26.5	670	162.0
Fr				

Element	Density (g/ml)	Melting Point	Boiling Point	Crystalline Form
Cr	7.2	1890	2480	cubic
Mo	10.2	2600	4800	cubic
W				

				Color
S	2.07	113	445	yellow
Se	4.82	220	688	red
Te				

4.14. Where in the periodic table do we find elements with (a) metallic properties, (b) nonmetallic properties, (c) greatest activity, (d) highest ionization potentials, (e) lowest ionization potentials, (f) highest electron affinities, (g) largest atoms, (h) highest densities, (i) least chemical activity?

4.15. As regards electrons in the s, p, d, f sublevels, what is (a) a representative group element, (b) an inert group element, (c) a transition element, (d) an inner transition element?

4.16. Of the elements of the fourth period (K through Kr): (a) Which has the largest atomic radius? (b) Which has the highest ionization potential? (c) Which has the highest electron affinity? (d) Which is the most reactive metal? (e) Which is the most reactive nonmetal? (f) Which is the most unreactive chemically?

4.17. Classify the following elements as (1) representative, (2) inert, (3) transition, (4) inner transition: $_{20}Ca, _{93}Np, _{11}Na, _{87}Fr, _{33}As, _{72}Hf, _{30}Zn, _{66}Dy, _{86}Rn, _{79}Au, _{21}Sc, _{47}Ag, _{85}At.$

4.18. From each of the following lists of elements (with atomic numbers) select those which should appear in the same group in the periodic table. Work this out before consulting the table.
(a) $_{10}Ne, _{16}S, _2He, _9F, _{18}Ar;$
(b) $_{37}Rb, _{69}Tm, _{29}Cu, _{87}Fr, _{55}Cs, _3Li;$
(c) $_{29}Cu, _{35}Br, _9F, _{85}At, _{20}Ca;$
(d) $_5B, _{13}Al, _{21}Sc, _{29}Cu, _{49}In, _{81}Tl;$
(e) $_8O, _{36}Kr, _{34}Se, _{30}Zn, _{16}S, _{10}Ne, _{52}Te;$
(f) $_7N, _{15}P, _{23}V, _{31}Ga, _{33}As, _{51}Sb;$
(g) $_{34}Se, _5B, _{14}Si, _{13}Al, _{31}Ga, _{17}Cl, _3Li.$

4.19. Without referring to the periodic table, write out the electronic sublevel notation for each of the following elements: (a) the second member of Group IIA, (b) the first member of Group VIIA, (c) the third of the inert gases in Group VIIIA, (d) the first transition element.

4.20. Consider the elements $_{55}Cs, _{26}Fe, _5B, _9F, _{16}S$ and answer the following questions.
(a) What group is Fe in?
(b) What period is Cs in?
(c) What is the name of the family F is in?
(d) What is the electronic symbol (electron-dot) of the B atom?
(e) What is the charge of the S ion?
(f) What is the principle valence of the Cs ion?
(g) Which *metal* in the above list is the most active?
(h) In what subshell ($s, p, d,$ or f) would the *last* electron added to Fe go in?
(i) Is S a metal, metalloid, or nonmetal?

(j) Which of the above elements would have the lowest ionization energy?
(k) Which of the above elements would have the highest electronic affinity?
(l) Which of the above elements would have the lowest electronegativity?
(m) Which of the above elements would have the smallest atomic radius?
(n) Would the S ionic radius be larger, the same, or smaller than the S atomic radius?
(o) What would be the *chemical formula* of the fluoride salt of B (BF, B_2F, B_2F_3, BF_3, etc.)?

4.21. Consider the elements $_{11}Na$, $_{13}Al$, $_{53}I$, $_{25}Mn$, $_9F$, $_{18}Ar$ and answer the following questions. Note the periodic table in the inside back cover of the text.
(a) What group is Mn in?
(b) What period is I in?
(c) What is the name of the family to which Na belongs?
(d) Is Ar a gas, liquid, or solid?
(e) What is the electronic symbol (electron-dot) of the Al atom?
(f) Is F a metal, nonmetal, or metalloid?
(g) What is the usual charge of the F ion?
(h) Which element, if any, is a transition element?
(i) What is the principle valence of the I ion?
(j) Which element in the above list is the most active?
(k) What would be the *formula* of the iodide salt of aluminum (AlI, Al_2I, AlI_2, AlI_3, Al_2I_3, etc.)?
(l) Which of the above elements would have the *lowest* ionization energy?
(m) Which of the above would have the *smallest* atomic radius?
(n) Which would have the greatest electron affinity?
(o) Which would have the greatest electronegativity?
(p) Would the Na ionic radius be larger, smaller, or the same as the Na atomic radius?
(q) In what subshell (*s*, *p*, *d*, or *f*) would the last electron added to Ar go in?
(r) Which of the above elements is a halogen?
(s) Which one of the above elements is *not* paramagnetic?
(t) Which one of the above atoms has only $5e^-$ in the *d* subshell?

4.22.

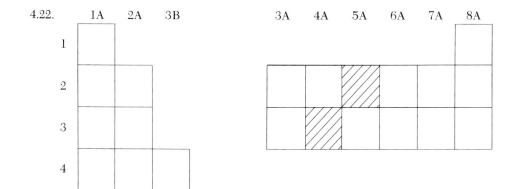

Below are 19 of the first 21 elements in the periodic table, each represented by a *fictitious symbol*. Using the information and the *symbols given*, fit each of these elements into the proper place in the periodic table. *Use the fictitious symbols* only, not the actual symbols.

Oo   is the lightest of the inert gases.

St   is the most active metal in this portion of the periodic table.

Nt   has a valence of +4 or −4.

W   has an electronic symbol of $\cdot \overset{\displaystyle \cdot \cdot}{\underset{\displaystyle \cdot}{W}} \cdot$.

G   isotopes have atomic weights 1, 2, and 3.

Ko   is in group 3 and period 2.

S   is the most active halogen.

Uc   is in the same period as S and forms a compound with the formula UcS_2.

N   is a transition element.

M   has an electron distribution $1s^2\ 2s^2\ 2p^6\ 3s^2$.

Dl   is similar to St in properties, but has a smaller atomic radius.

Io   has one less proton than N in the nucleus.

Kq   has one of the two highest electron affinities.

Ue   rarely forms compounds and has 8 valence e^- in the M shell.

H   also does not normally react with other elements and has $8e^-$ less than Ue.

Hi   is the smallest ion of valence −2.

O   is an alkali metal in period 3.

Or   forms a compound with G with the formula G_2Or.

E   has valence electrons $3s^2\ 3p^1$.

Supplementary Readings

4.1. *Electronic Structure, Properties and the Periodic Law*, H. H. Sisler, Van Nostrand, New York, 1973 (paper).

4.2. *The Periodic Table of the Elements*, R. J. Puddephatt, Clarendon, Oxford, 1972 (paper).

4.3. "A Pattern of Chemistry. Hundred Years of the Periodic Table," F. Greenaway, *Chem. in Britain*, **5**, 97 (1969).

Chapter 5

Chemical Bonds

5.1   THE NATURE OF BONDING

Chemical bonds are the forces that hold atoms together. These forces are intimately related to the arrangement of electrons in the outermost energy levels of atoms which are in combination. We have described two principal types of forces, or bonds — the **electrovalent,** or **ionic,** bond and the **covalent** bond. In an electrovalent compound the oppositely charged ions are held together simply by the electrostatic attraction of unlike charges. The attractive force between oppositely charged ions constitutes the ionic bond.

Atoms may also combine by sharing electrons, and a pair of shared electrons constitutes a covalent bond. In some cases more than one pair of electrons may be shared between atoms, as, for example, in carbon dioxide (CO_2), where the carbon atom shares two pairs of electrons with each of two oxygen atoms.

Polar and Nonpolar Bonds

Although many substances appear to fall strictly into either the electrovalent or covalent classification, others seem to exhibit a bonding intermediate between the two. This is true because elements vary in their affinity for electrons. Sodium and chlorine represent two extremes. The sodium atom has little affinity for electrons; as a matter of fact, it readily gives up an electron. Chlorine, on the other hand, has a strong affinity for electrons. The compound NaCl is definitely electrovalent or ionic, and the ions are held together simply by attraction of the opposite charges.

Consider now a situation in which two atoms with the same electron affinity combine — for example, the chlorine molecule (Cl_2). Each chlorine atom has the same affinity for electrons, and the shared pair of electrons in the chlorine molecule will lie midway between the two atoms, giving rise to an electrically symmetrical molecule (see next page). This bonding of atoms may be described as **nonpolar.**

In a combination of unlike atoms such as HCl (a covalent compound), however, attraction for electrons by the respective atoms is not equal. There is a displacement of the electron pair toward the chlorine atom, and the molecule is not electrically symmetrical. The chlorine end of the molecule tends to be negative, and the hydrogen end is positive. Thus the molecule acts like a small magnet. Such electrically unsymmetrical molecules are called **dipoles,** and the bonding between the atoms is described as **polar.** The situation is similar to ionic combinations, and a polar molecule may be considered as having some electrovalent or ionic character. The degree of polarity of the molecule depends on the relative attractions of the combining atoms for electrons. The combinations just cited (Cl_2, HCl, NaCl) represent the transition from a nonpolar to a polar and, finally, to an ionic substance.

$$:\ddot{\underset{..}{Cl}}:\ddot{\underset{..}{Cl}}: \qquad H:\ddot{\underset{..}{Cl}}: \qquad Na^+\left[:\ddot{\underset{..}{Cl}}:\right]^-$$

$$\underset{\text{(nonpolar)}}{Cl_2} \qquad \underset{\text{(polar)}}{HCl} \qquad \underset{\text{(ionic)}}{NaCl}$$

$$\text{nonpolar} \xrightarrow{\text{increasing polarity}}$$

Electronegativity

It is evident from our preceding discussion that there is a gradation between ionic and covalent compounds. Pauling and others have developed a scale of numbers, called electronegativities (Table 5.1), which represent the relative attractive forces of the elements for electrons in a covalent bond. These electronegativity values range from 4.0 for fluorine to 0.7 for cesium. As would be expected, the most electronegative elements are those in the top right-hand portion of the periodic table. It will be helpful for purposes later on to note that fluorine, oxygen, nitrogen, and chlorine have the highest electronegativities of all of the elements. Also most hydrogen compounds and organic compounds are primarily covalent in character, since H (2.1) and C (2.5) are near the middle of the scale. The least electronegative (or we might say the most electropositive) are those metals in Groups IA and IIA of the periodic table. As we would expect, the smaller atoms have a greater attraction for electrons, and electronegativity decreases as we proceed from the top of a given group in the periodic table to the bottom. Too, the attractive force should decrease with decreasing numbers of electrons in the valence level of atoms. Consequently, electronegativity decreases in going from right to left in the periodic table.

This scale of electronegativity values is particularly useful in predicting the degree of ionic character, or polarity, of compounds. The greater the difference in electronegativity values between two atoms, the more ionic, or more polar, the bond between these atoms. (See Table 5.2.) For example, in the compound NaF, the difference is $F(4.0) - Na(0.9) = 3.1$, and this compound is primarily ionic. A combination of like atoms gives a zero difference, and such bonds are entirely covalent, or nonpolar—F_2, H_2, and so on. Where the difference is small, as between boron and hydrogen $(2.1 - 2.0)$, the bonding is essentially covalent.

It is evident that when the electronegativity difference is about 1.7, the bond is about 50% ionic. When the difference is greater than this we generally call the compound ionic, and if less than 1.7 we call it covalent. This is an arbitrary rule since most compounds are both somewhat ionic and somewhat covalent in character.

Table 5.1

Electronegativities of the elements[a]

1	2	3	4	5	6	7	8	9	10	11	12	13	14	15	16	17	18
1 H 2.1																	2 He —
3 Li 1.0	4 Be 1.5											5 B 2.0	6 C 2.5	7 N 3.0	8 O 3.5	9 F 4.0	10 Ne —
11 Na 0.9	12 Mg 1.2											13 Al 1.5	14 Si 1.8	15 P 2.1	16 S 2.5	17 Cl 3.0	18 Ar —
19 K 0.8	20 Ca 1.0	21 Sc 1.3	22 Ti 1.5	23 V 1.6	24 Cr 1.6	25 Mn 1.5	26 Fe 1.8	27 Co 1.8	28 Ni 1.8	29 Cu 1.9	30 Zn 1.6	31 Ga 1.6	32 Ge 1.8	33 As 2.0	34 Se 2.4	35 Br 2.8	36 Kr —
37 Rb 0.8	38 Sr 1.0	39 Y 1.2	40 Zr 1.4	41 Nb 1.6	42 Mo 1.8	43 Tc 1.9	44 Ru 2.2	45 Rh 2.2	46 Pd 2.2	47 Ag 1.9	48 Cd 1.7	49 In 1.7	50 Sn 1.8	51 Sb 1.9	52 Te 2.1	53 I 2.5	54 Xe —
55 Cs 0.7	56 Ba 0.9	57–71 La–Lu 1.1–1.2	72 Hf 1.3	73 Ta 1.5	74 W 1.7	75 Re 1.9	76 Os 2.2	77 Ir 2.2	78 Pt 2.2	79 Au 2.4	80 Hg 1.9	81 Tl 1.8	82 Pb 1.8	83 Bi 1.9	84 Po 2.0	85 At 2.2	86 Rn —
87 Fr 0.7	88 Ra 0.9	89– Ac– 1.1–1.7															

[a] Reprinted from Linus Pauling: THE NATURE OF THE CHEMICAL BOND. © 1939, 1940, Third Edition © 1960 by Cornell University. Used by permission of Cornell University Press.

Table 5.2

Percent ionic character of bonds

Differences in Electronegativity	Percent Ionic Character	Differences in Electronegativity	Percent Ionic Character
0.1	0.5	1.7	51
0.3	2	1.9	59
0.5	6	2.1	67
0.7	12	2.3	74
0.9	19	2.5	79
1.1	26	2.7	84
1.3	34	2.9	88
1.5	43	3.1	91

5.2 ELECTRON CONFIGURATION AND BONDING

In Chapters 2 and 3, a simplified concept was given of atomic structure. The importance of valence electrons in explaining the properties of elements and compound formation was emphasized. As a result of modern research, Bohr's theory of the atom has undergone considerable revision and a modern theory treating an electron in an atom as being both a wave and a particle has emerged.

Wave Nature of an Electron

Until 1900 light had been explained by classical physics as strictly an electromagnetic wave phenomenon for which

$$\nu \text{ (frequency)} = \frac{c \text{ (speed of light)}}{\lambda \text{ (wavelength)}}$$

In 1900, Max Planck formulated a **quantum theory** of light which postulated that light can be considered to be small bundles or "quanta" of energy called **photons**. The energy, E, inherent in a given photon, is proportional to the frequency of the given light, ν.

$$E = h\nu \qquad \text{or} \qquad E = \frac{h\,c}{\lambda}$$

The proportionality constant, h, Planck's constant, has a value of 6.63×10^{-27} erg sec. Planck's quantum theory reinforces Einstein's concept, E (energy) $= mc^2$ (mass)(velocity of light)^2, that mass and energy being convertible must have both quanta and particle nature. The photon is now accepted as one of the fundamental subatomic particles (Chapter 2).

In 1924 a French physicist, Louis de Broglie, while working on his doctor's degree in quantum mechanics at the University of Paris, postulated that electrons and other moving particles had associated with them wave characteristics as given by the equation

$$\lambda \text{ (wavelength)} = \frac{h \text{ (Planck's constant)}}{m \text{ (mass of particle)} \; v \text{ (velocity of particle)}}$$

For an electron beam accelerated by 10,000 volts, the velocity of the electrons would be approximately one tenth the speed of light. The wavelength associated with this electron beam would be

$$\lambda = \frac{6.63 \times 10^{-27} \text{ erg sec}}{(9.1 \times 10^{-28} \text{ g})(0.30 \times 10^{10} \text{ cm/sec})} = 0.24\text{Å}$$

$$\text{(Since ergs} = \text{g} \left(\frac{\text{cm}}{\text{sec}}\right)^2 \text{ and 1 cm} = 10^8\text{Å.)}$$

These wavelengths of moving electrons have been experimentally verified and are, as a matter of fact, the basis of the modern **electron microscope**, the high magnifications of the instrument being made possible by the extremely short wavelengths of the accelerated electrons. It has been demonstrated that

protons, neutrons, and even molecules when accelerated have wave properties. Electron diffraction and neutron diffraction are well-known research
tools used in determining the structure of matter.

In describing the Bohr hydrogen atom model, page 40, different energy
states were designated for the electron as it orbited at different distances and
velocities. One could mentally picture this simple hydrogen atom. The new
concept of the hydrogen atom also pictures it as a proton with an electron
particle moving about it. But how does one picture a quantized wave of
energy with some particle characteristics?

The Uncertainty Principle (1927)

A German scientist, Werner Heisenberg, in a quantum mechanics mathematical treatment of the wave-particle concept of the electron, showed that
it is experimentally impossible to determine simultaneously the exact position of an electron and its velocity in an atom. Any attempt to measure the
position of so small an object as an electron, such as using light rays to photograph it, would in itself disturb the particle and cause it to occupy a new
position at a new velocity.

Much as a moving billard ball on hitting another ball on the side gives
part of its energy to the hit ball and retains part of the energy as it is deflected, a quantum (photon) of light after collision with an electron gives
some movement to the electron and the deflected quantum is of less energy
(its wavelength and frequency are changed).

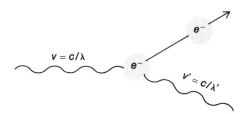

It appears, therefore, that any exact experimental determination of the path
of an electron in a Bohr orbit would be impossible.

Modern atomic theory has found a way out of this difficulty by making
use of the **Schrödinger equation** (1926).

$$\frac{d^2\psi}{dx^2} + \frac{d^2\psi}{dy^2} + \frac{d^2\psi}{dz^2} + \frac{8\pi^2 m}{h^2}\left(E + \frac{e^2}{r}\right)\psi = 0$$

The terms π and h (Planck's constant) have their usual value, $m =$ mass of
the electron, $E =$ total energy, $e^2/r =$ potential energy, x, y, and z are its coordinates and ψ **the wave function** associated with the particle. The first three

terms are expressions from calculus pertaining to the slope of the probability wave changes as a particle departs from a space point in the x, y, z directions. The Schrödinger equation is the basis for the science of **wave mechanics** and while the mathematics is too complex to consider here, the end results of this modern atomic theory need to be considered.

Probability Positioning of 1s, 2s, 2p Electrons

The solution of Schrödinger's equation relates the wave function of an electron, ψ, to the position in space where it may be found (xyz) and to its allowed energy levels (E). A useful term ψ^2 relates to the **probability** of an electron being within a given minute volume in space.

Imagine an electron in motion around the nucleus. While the Schrödinger equation cannot tell at any instant just where the electron is, it can give the **probability** (or relative chance) of the electron being in a certain location.

The probability distribution for an electron in a *1s* orbital could be likened to the holes in a rifle target. The probability would be high (many holes) near the center and decrease at greater distances (fewer holes per unit area). Likewise the electron would be found, at any instant, with a higher probability of being closer to the nucleus than at a distance farther away.

There are several ways of showing probability distribution. Three of these are shown in Figure 5.1 for a 1s electron. In curve (a) the probability of finding the electron in a unit volume of space is plotted against the distance of the electron from the nucleus. Note that even at great distances from the nucleus, the probability of an electron being there, while extremely low, is never zero.

Figure (b) is another way of representing the probability of the electron being at a given location in the atom, a spherical "electron cloud" concept. The shading is highest at the nucleus, representing the highest probability, and decreases as the distance from the nucleus becomes greater, representing lower probability. Again the probability is never zero, which means the sphere is infinite in size. For convenience, however, we often indicate a boundary contour such as Figure (c) which indicates the shape of the electron cloud and indicates a region in space where an electron is likely to be found with high probability (about 90%). Still another way commonly used to show probability distribution is shown in Figure 5.2. This is a plot of the probability of finding the electron at all possible positions (radial probability) in the sphere at each distance r. Here, while the probability per position is still highest at the nucleus, the number of possible positions is very small, and the overall probability is small. On the other hand, for a larger value of r the number of possible positions is large, but the probability per position is so low that the overall probability is low. At some intermediate value of r such as a_0 in Figure 5.2 (0.53 Å for the 1s electron) the electron will be found with highest probability. This is the same distance as the 1s electron in the Bohr atom.

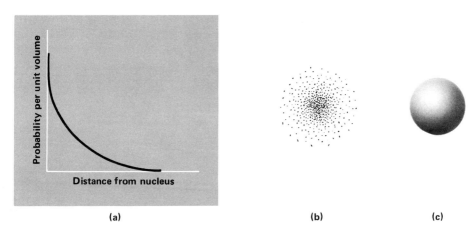

Figure 5.1

Electron probability curves.

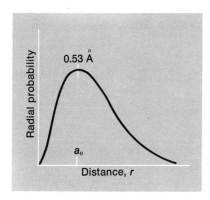

Figure 5.2
Probability of finding a 1s electron in a unit volume versus distance from nucleus, r.

These electron probability plots in the modern atomic theory now replace the electron orbit concept in the Bohr theory. The term **orbital** is used in designating the electron probability pattern. **An orbital is a region in space where an electron is likely to be found with high probability (90%).** The 1s orbital is spherical. The 2s orbital would be a concentric sphere of larger radius.

Probability plots for a p orbital show three balloon-like or dumbell-shaped orbitals oriented along each of the three mutually perpendicular axes. (Figure 5.3) Note that the meaning of a p orbital is similar to that of an s orbital; namely, that it is a region in space where an electron is likely to be found (at about 90% probability). This does not imply that an electron will never be found outside the dumbell-shaped boundary, but the probability is low.

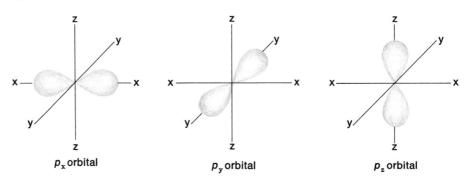

Figure 5.3

p orbitals are oriented along three axes mutually perpendicular.

Beginning with the third principle energy level, each energy level may have a set of five d orbitals. Balloon or dumbell-like pictures again designate probability contours. The shapes and orientations of d orbitals are shown in Figure 5.4.

Each of the s, p, d, orbitals pictured was represented in the energy level

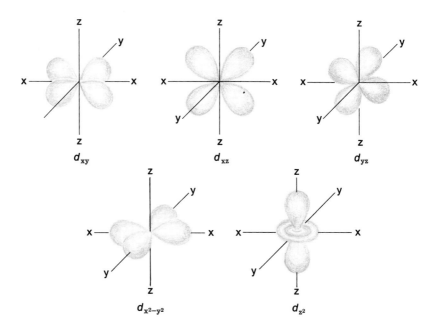

Figure 5.4

The d_{xy}, d_{yz}, and d_{xz} orbitals are alike except for the plane of their orientation. The $d_{x^2-y^2}$ orbital is like d_{xy} except that it is rotated 45° around the z axis. The d_{z^2} orbital is symmetrical around the z axis, and consists of a doughnut-shaped ring and dumbbell-shaped part.

diagram on page 58 as small squares which can hold only 2 electrons. This useful energy concept is still valid; however the squares do not give information on the directional characteristics of the orbitals, types of bonding, shapes of molecules, etc.

The quantum mechanical designation of an electron in an atom may be summarized: (a) no definite trajectory of an electron is specified, (b) an orbital probability pattern may be calculated for each electron, (c) an electron is not atomized into a "cloud." It is at some probability point; which point and momentum of the electron cannot both be precisely known at any instant. (d) various probability patterns develop for electrons of various quantum states.

5.3 ORBITALS AND BONDINGS (Shapes of molecules)

When atoms combine by sharing electrons, the bond formed results from an *overlap* of two atomic orbitals, each of which is occupied by a single electron. The bond will be oriented in the direction of the orbitals which furnished the electrons. Consequently, such bonds will influence the shape and structure of the resulting molecule. The following examples will clarify this:

EXAMPLE 5.1 In the compound H_2O, two pairs of electrons are shared between one oxygen atom and two hydrogen atoms. Oxygen ($_8O$) has the con-

figuration $1s^2$; $2s^2$, 2 | ↕ | | ↓ | | ↓ | The p_x^2 electrons are paired, the
$\qquad\qquad\qquad\quad\ p_x^2 \quad\ p_y^1 \quad\ p_z^1$

p_y^1 and p_z^1 electrons are unpaired (Figure 5.5). These p_y^1 and p_z^1 electrons each form shared pairs with electrons from two hydrogen atoms (Figure 5.5b).

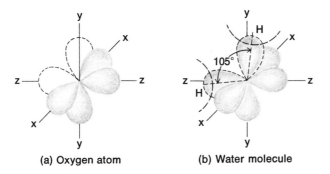

(a) Oxygen atom   (b) Water molecule

Figure 5.5
The water molecule might be expected to have an angular structure (90° between the oxygen to hydrogen bonds) produced by an overlap of orbitals from two H atoms and the p_y and p_z' orbitals of one oxygen atom. The actual bond angle of H to O to H is 105°.

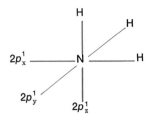

Figure 5.6
The NH$_3$ molecule. Three hydrogen atoms are oriented along x, y, and z axes.

The two bonds would be expected to be at right angles (90°) to one another in the water molecule. Actually, the angle between the bonds is about 105° because of the repulsion of the two H atoms and other factors which will be considered later in the chapter. The shape of the water molecule, then, is a bent structure, H—O, rather than linear, H—O—H. This is confirmed by

$$\overset{\displaystyle\text{H—O}}{\underset{\text{H}}{\diagdown}}$$

experiment. The H$_2$O molecule may be called a p^2 (p-two) molecule since two of the p orbitals are used in bonding.

EXAMPLE 5.2 Consider the compound ammonia (NH$_3$), in which nitrogen has the electron configuration $1s^2$; $2s^2$, $2p^3$. Nitrogen has 3 unpaired electrons in the p orbitals, which might be indicated as $1s^2$; $2s^2$, $2p_x{}^1$, $2p_y{}^1$, $2p_z{}^1$. The bonds between the nitrogen atom and the three hydrogen atoms would be expected to lie along the x, y, and z axes, which are at right angles to each other and which would give a structure like a pyramid (Figure 5.6).

The actual bond angles are about 107° instead of 90° because of other factors which widen the angles between the bonds. It should be evident, however, that the actual structure is fairly close to that predicted from the orientation of the orbitals in space. NH$_3$ may be considered a p^3 molecule. Both NH$_3$ and H$_2$O, however, are better described using the concept of hybrid orbitals (below) or electron-repulsion theory (page 110).

Hybridization of Orbitals

Although the structure of some simple compounds may be explained on the basis of electron orbitals as described above, many compounds appear to fit into a more complex pattern. For example, in considering compounds of carbon, it becomes necessary to modify our pictures somewhat. On the basis of the electronic configuration of carbon, $1s^2$; $2s^2$, $2p^2$, we expect the carbon atom to have only two unpaired electrons available for bonding—that is, $2p_x{}^1$ and $2p_y{}^1$—and therefore to form only two bonds in combination with other atoms. However, in nearly all cases, carbon exhibits a valence of four, which would call for *four* orbitals to be involved in bond formation. It appears that one of the $2s$ electrons may be "promoted" to the vacant $2p_z$ orbital of the carbon atom when the latter enters into chemical combination. The energy for this shift to a higher energy level comes from the energy of reaction resulting from the combination of atoms. In any case, four orbitals become available for bond formation — these may be represented as $2s^1$, $2p_x^1$, $2p_y^1$, and $2p_z^1$, with an unpaired electron in each orbital:

$$\boxed{\downarrow} \quad \boxed{\downarrow} \quad \boxed{\downarrow} \quad \boxed{\downarrow}$$

$$2s^1 \qquad 2p_x^1 \qquad 2p_y^1 \qquad 2p_z^1$$

Now, on the basis of our discussion of the directional qualities of covalent bonds, we may expect three of the carbon bonds to be at right angles (the p orbitals) to each other, and the fourth bond to be oriented in a random direction (the s orbital is spherical). Actually this is not the situation at all, since the four bonds are **equivalent** in most compounds. The simplest geometric pattern which has four equidirectional and equal bonds is a regular tetrahedron, where the carbon atom is at the center of the tetrahedron and each of the four orbitals is directed to a corner of the tetrahedron. Consequently, in methane (CH_4), for example, the carbon atom is at the center of a tetrahedron (Figure 5.7), with a hydrogen atom at each of the corners.

The angle between the bonds is 109° 28′, and the interatomic distance between each hydrogen and the carbon atom is 1.09 angstroms. To account for the equivalence of the four bonds, we assume that four orbitals, one s and three p, are equalized or **hybridized** in such a way that the four become exactly alike and are oriented toward the four corners of a regular tetrahedron, such as is pictured in Figure 5.8. This rearrangement of the orbitals to give four equivalent bonds is referred to as the **hybridization of orbitals,** and the bonds derived from one s and three p orbitals are referred to as sp^3 (s–p–three)

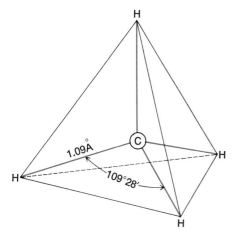

Figure 5.7
The carbon atom in methane has a tetrahedral structure — the carbon atom at the center of a tetrahedron and one hydrogen atom at each of the four corners.

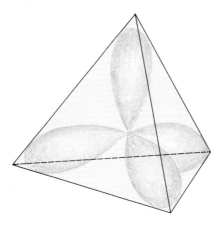

Figure 5.8
Hybridization of orbitals for carbon gives
four equivalent sp^3 orbitals oriented toward
the corners of a tetrahedron.

hybrid bonds. These bonds are found in saturated hydrocarbons, which are discussed in Chapter 31.

Several other types of hybrid bonds are possible; a few of these are listed in Table 5.3, together with the structure of the compounds resulting from the formation of these hybrid bonds. In all of these cases, the hybrid bonds result from equal contributions by s and p or s, p, and d orbitals. Some of these types are discussed in the sections to follow.

Table 5.3
Types of hybrid bonds

	Hybrid Bond Type	Number of Bonds in Compound Formation	Structure
(a)	sp	2	linear
(b)	sp^2	3	triangular planar
(c)	sp^3	4	tetrahedral
(d)	sp^2d	4	square planar
(e)	sp^3d	5	trigonal bipyramidal
(f)	sp^3d^2	6	octahedral

sp Hybrid Bonds

Unlike other fluorides of Group IIA, molten BeF_2 is a nonconductor of electric current and presumably is a covalent compound. The configuration of Be is $1s^2$; $2s^2$, with no unpaired electrons. To account for the two equivalent bonds in BeF_2 and similar compounds, it is likely that one of the $2s$ electrons is "promoted" to a $2p$ level, which would give two unpaired electrons. The resulting configuration of the Be atom would be $1s^2$; $2s^1$, $2p_x^1$. This would permit the formation of two sp hybrid bonds with a linear structure (Figure 5.9a)—that is: F—Be—F.

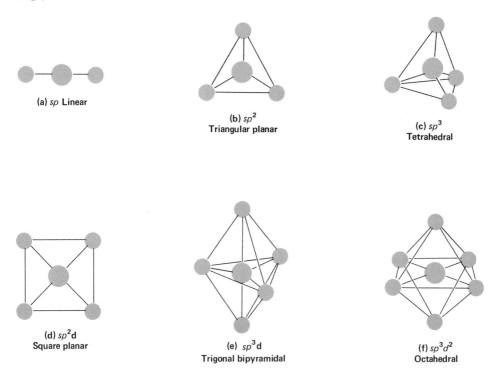

(a) *sp* **Linear**

(b) *sp*^{2}
Triangular planar

(c) *sp*^{3}
Tetrahedral

(d) *sp*^{2}d
Square planar

(e) *sp*^{3}d
Trigonal bipyramidal

(f) *sp*^{3}*d*^{2}
Octahedral

Figure 5.9
Geometric structures of hybrid bond types listed in Table 5.3.

sp^2 Hybrid Bonds

Boron forms covalent compounds in which the B atoms exhibit a valence of three—for example, BF$_3$. Apparently the boron atom with a configuration of $1s^2$; $2s^2$, $2p$ and only one unpaired electron forms three hybrid bonds by "promoting" one of the $2s$ electrons to one of the p levels. The resulting configuration on hybridization of $1s^2$; $2s^1$, $2p_x{}^1$, $2p_y{}^1$ would account for the three equivalent bonds in the trivalent compounds of boron. This is termed *sp*^{2} hybridization because one s and two p orbitals are used in the formation of bonds. The structure of such compounds is triangular, with three 120° angles between the bonds. For example,

The involvement of d and f orbitals in bond formation is complex, and a thorough discussion of the matter is beyond the scope of this text. We can

note, however, that the orientation of these orbitals in space has an important bearing on the angles between the bonds, and therefore on the structures of their compounds. How d orbitals may enter into bond formation is briefly introduced in the next section.

Deviations from the Octet Rule

Generally, when atoms combine, each atom tends to acquire in its outer level an octet of electrons consisting of two s and six p electrons. The octet rule holds well for elements in the first two periods of the periodic table, where all the elements have only s and p orbitals. Beyond the second period, where elements may have d orbitals in addition to the s and p orbitals, exceptions to the octet rule may be encountered. For example, the compounds PCl_3 and PCl_5 are well known. The latter compound requires 10 electrons to make the five bonds, five electrons from one P atom and five electrons from the five Cl atoms. The outer level of P, which has the configuration $1s^2$; $2s^2$, $2p^6$; $3s^2$, $3p^3$, has only three unpaired electrons, a fact which accounts for there being only three bonds. However, vacant $3d$ orbitals are available. If we assume that one $3s$ electron is "promoted" to a $3d$ orbital followed by hybridization, then five unpaired electrons (one s, three p, one d) are available for bonding purposes, thus:

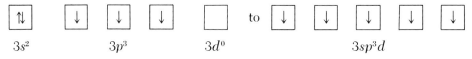

$3s^2$ $3p^3$ $3d^0$ $3sp^3d$

The bonds would be termed sp^3d bonds. The structure of PCl_5 is shown in Figures 5.9(e) and 5.10.

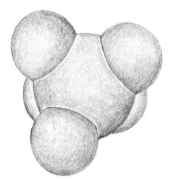

Figure 5.10
A model of PCl_5. The geometric structure for PCl_5 is trigonal pyramidal with three of the chlorine atoms in a plane — and the other two on opposite sides of the plane.

A similar situation presumably exists for such compounds as SF_6 and SeF_6, in which we may assume that two electrons are "promoted" to the d orbitals in the sulfur or Se atom. Thus, for sulfur:

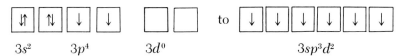

$3s^2$ $3p^4$ $3d^0$ $3sp^3d^2$

Combination with 6 F atoms would give 12 electrons around the central S or Se atom. These types of compounds with sp^3d^2 orbitals have an octahedral structure, with the single atom or ion at the center of an octahedron and one attaching group at each corner (Figure 5.11):

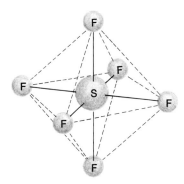

Figure 5.11
SF_6 has an octahedral structure.

Table 5.4 summarizes some of the concepts of hybrid bonds and shapes of molecules.

Table 5.4
Hybrid bonds

Type Bond	Example	Shape	Explanation
(a) sp	F—Be—F	linear	$_4$Be $1s^2 2s^2 \rightarrow 1s^2 2s^1 2p_x^1$ sp hybrid bonds
(b) sp^2	$\begin{array}{c} F \\ | \\ B \\ F \quad F \end{array}$	triangular planar	$_5$B $1s^2 2s^2 2p^1 \rightarrow 1s^2 2s^1 2p_x^1 2p_y^1$ sp^2 hybrid bonds
(c) sp^3	$\begin{array}{c} H \\ | \\ H—C—H \\ | \\ H \end{array}$	tetrahedral	$_6$C $1s^2 2s^2 2p^2 \rightarrow 1s^2 2s^1 2p_x^1 2p_y^1 2p_z^1$ sp^3 hybrid bonds
(d) sp^3d	PCl_5	trigonal bipyramidal	$_{15}$P . . . $3s^2 3p^3 \rightarrow$ . . . $3s^1 3p_x^1 3p_y^1 3p_z^1 3d$ sp^3d hybrid bonds
(e) sp^3d^2	SF_6	octahedral	$_{16}$S . . . $3s^2 3p^4 \rightarrow$ . . . $3s^1 3p_x^1 3p_y^1 3p_z^1 3d_{xy}^1 3d_{xz}^1$ sp^3d^2

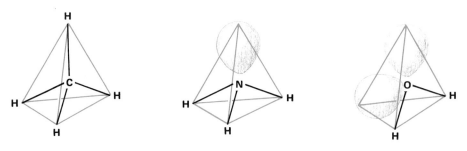

Figure 5.12

Shapes of methane (CH₄), ammonia (NH₃), and water (H₂O) molecules.

The NH_3 molecule mentioned previously might be considered to have a modified tetrahedral shape because of sp^3 hybridization. Three of the sp^3 hybrid orbitals would be used to bond with hydrogen atoms and the fourth would consist of an unshared pair of electrons, often called a **lone pair** (Figure 5.12). It could be represented as $\overset{\displaystyle \ddot{N}}{\underset{\text{H H H}}{\diagup \mid \diagdown}}$ and the experimental H—N—H bond angle of 107.3° is close to the true tetrahedral angle of 109.5° as shown by CH_4. The modified tetrahedral shape of NH_3 is trigonal pyramidal.

H_2O may similarly be described by sp^3 hybridization except that it would have two lone pairs of electrons (Figure 5.12). The experimental bond angle for the $\overset{\ddot{\text{:O:}}}{\underset{H \quad H}{}}$ atoms in H_2O is 104.5° which is closer to the tetrahedral angle of 109.5° than it is to the 90° expected for a purely p^2 molecule.

Electron-pair Repulsion Theory

A rather simple method of predicting the shape of covalent molecules has been developed based on the total number of valence electron pairs in the valence shell of the central atom. It is known as the **valence-shell electron-pair repulsion theory.** Simply stated, this theory is that **electron pairs (both bonding and lone-pairs or nonbonding electrons) tend to be distributed so as to minimize the repulsions between them.** In other words, electron pairs will take positions of greatest possible separation from each other. Qualitatively, these positions will result in the shapes outlined previously by hybridization (Table 5.4).

Consider, for example, the electron dot formula for beryllium fluoride

$$:\ddot{F}:Be:\ddot{F}: \qquad \text{or} \qquad :\ddot{F}-Be-\ddot{F}: $$

where the dash (—) represents a pair of bonding electrons (one from Be and one from F). These two bonding pairs of electrons are as far as they can get from each other when the F—Be—F angle is 180°. Consequently, the molecule is **linear.**

Similarly, the boron trifluoride molecule

$$: \ddot{F}:$$
$$|$$
$$B$$
$$: \ddot{F} \quad \ddot{F}:$$

would have maximum separation of the bonding electrons when the fluorine atoms are 120° from each other. BF$_3$ is triangular and planar in shape (trigonal planar). The four electron pairs around carbon in methane, H—C—H, have (with H above and below C) maximum separation when the molecule has the shape of a tetrahedron where the H—C—H bond angles are all 109.5°. Ammonia, H—N—H (with H below), may be related to the tetrahedral configuration by noting that the nitrogen, similar to C in CH$_4$, has 4 pairs of electrons in the valence shell. Three of these pairs are bonding pairs, however one pair is nonbonding (a lone pair). Since the electron repulsion of the lone pair of electrons would be somewhat different from the bonding pairs of electrons, the NH$_3$ molecule would be expected to have a slightly modified tetrahedral shape (trigonal pyramidal, Figure 5.12). The bent structure of water, H—Ö:, can be explained similarly. Table 5.5

Table 5.5

Electron pairs and molecular shape

Number of electron pairs in the valence shell of the central atom

Total	Bonding	Nonbonding	Shape of Molecule	Examples
2	2	0	linear	BeF_2, $HgCl_2$, CO_2, BeH_2
3	3	0	triangular planar	BF_3, BH_3, CO_3^{2-}, NO_3^-
3	2	1	bent (angular)	SO_2, NO_2^-
4	4	0	tetrahedral	CH_4, NH_4^+, SO_4^{2-}, SiF_4
4	3	1	trigonal pyramidal	NH_3, PCl_3, H_3O^+, BF_4^-
4	2	2	bent (angular)	H_2O, ClO_2^-, ICl_2^+
5	5	0	trigonal bipyramidal	PCl_5, $SnCl_5^-$
6	6	0	octahedral	SF_6, SiF_6^{2-}, PF_6^-
6	5	1	square pyramidal	IF_5, $XeOF_4$
6	4	2	square planar	XeF_4, ICl_4^-

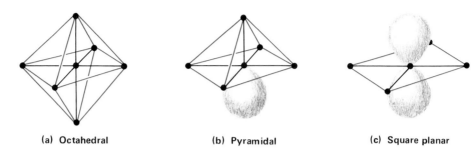

(a) **Octahedral**   (b) **Pyramidal**   (c) **Square planar**

Figure 5.13
Shapes of molecules and ions in which the central atom has six electron pairs.

summarizes a few, but by no means all, of the relationships between valence-shell electron-pair repulsions and molecular shape. Figure 5.13 shows the molecular structures for the last three entrees in Table 5.5.

Resonance

For some molecules no single electronic configuration can be drawn which will show the arrangement of the bonds between the atoms of the molecule and will be in agreement with its observed physical and chemical properties. In these cases we might assume that the actual structure lies between two or more limiting or **resonance** structures. For example, we may show the electronic configuration of ozone (O_3) to be

$$:\overset{..}{O}::\overset{..}{O}:\overset{..}{\underset{.}{O}}:\quad\text{or}\quad:\overset{..}{\underset{.}{O}}:\overset{..}{O}::\overset{..}{O}:$$

or simply

$$O{=}O{-}O\quad\text{or}\quad O{-}O{=}O$$
$$\text{I}\qquad\qquad\qquad\text{II}$$

Both structures I and II show two kinds of bonds between the oxygen atoms, a double bond (two pairs of electrons) and a single bond (one pair of electrons). Actually the properties of ozone indicate no difference in the bonds between oxygen atoms, and the bonds have equivalent strengths. We may regard the actual molecule as being something intermediate between the resonance forms I and II shown above. The intermediate structure is termed a **resonance hybrid.**

Other examples of molecules which possess a hybrid structure between resonance forms of the molecule are

O=S—O   O—S=O
sulfur dioxide

benzene

5.4   MOLECULAR ORBITAL THEORY

While the combination of atomic orbitals explains reasonably well the properties of many molecules, the theory of molecular orbitals (MO) describes the bonding characteristics more accurately. In the former theory, often called the valence-bond theory, it is assumed that each atom retains its identity in terms of electronic configuration except for the movement of a few electrons involved in forming bonds. It seems certain that in a molecule made up of two or more atoms, the nuclei of each will affect the electronic configuration of all the atoms in the molecule because of the attractive forces of the nuclei for the electrons. As a result the configuration of electrons will change. The molecular orbital theory assumes that electrons are fitted into completely new orbitals which are characteristic of the molecule as a whole. These orbitals have very definite energy levels and the electrons fit into these orbitals in an order from the lowest to higher energies in a manner similar to the filling of atomic orbitals.

The Hydrogen Molecule-ion

A good starting point in the understanding of molecular orbitals is a consideration of the attractive and repulsive forces in the hydrogen-molecule ion, H_2^+, which is the simplest possible diatomic molecule. This unit is composed of two hydrogen nuclei (protons or hydrogen ions) bonded together by one electron. It is a chemically active entity and spectral studies have ascertained its bond energy and bond length. We may consider the H_2^+ ion to be formed by a combination of a hydrogen atom and a hydrogen ion (proton), Figure 5.14.

As the hydrogen atom and hydrogen ion come together there are electrostatic attractions of protons for the electron and also repulsion between the

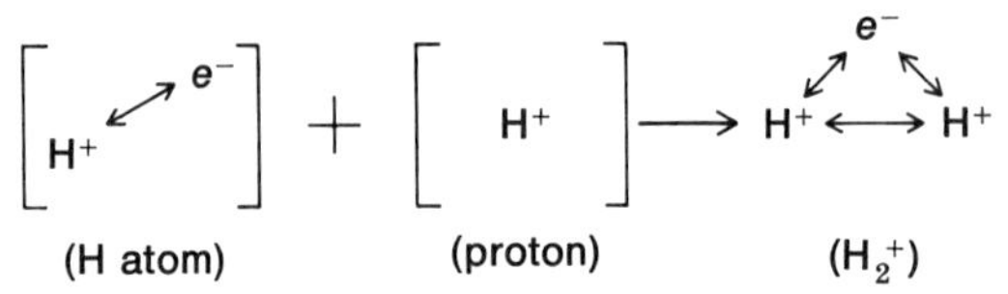

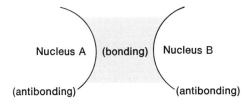

Figure 5.14
Formation of H_2^+.

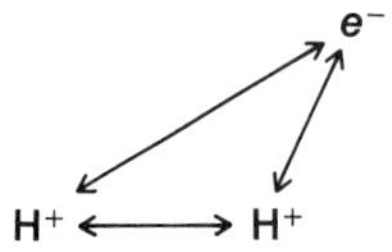

(H atom)   (proton) (H_2^+)

Figure 5.15
If the electron is not simultaneously near both nuclei, the repulsive force between the nuclei is greater than the attractive forces and no bond will be formed.

Nucleus A) (bonding) (Nucleus B

(antibonding)   (antibonding)

Figure 5.16
Bonding and antibonding zones for an electron(s). The shaded portion represents the bonding region.

protons. If the attractive forces are greater than the repulsive forces, then a bond will be formed and a net decrease in energy will occur. This will happen only if the electron is reasonably close to both nuclei simultaneously, as shown in Figure 5.14. The energy decrease for the formation of the H_2^+ ion is about 65 kcal/mole, so the ion is fairly stable.

Whether or not a bond is going to form when two atoms approach one another depends upon the location of the electron(s) or electron density relative to the two nuclei. If the electron happens to be in a position as shown in Figure 5.15, a bond will not form because the attractive forces are insufficient to offset the repulsive forces and there can be no decrease in energy.

The electron position or density is subject to probability considerations. Figure 5.16 illustrates probability density which is highest between the two positive nuclei. The shaded region is one of high probability of bond formation and may be termed the bonding region. Areas of low electron density (white) are antibonding regions.

Now consider the merging of the two $1s$ atomic orbitals of H and H^+ to form *two* molecular orbitals of H_2^+, (Figure 5.17) one of lower, the other of higher energy than either of the atomic orbitals.

The one electron fits into the orbital of lower energy since this results in a decrease in overall energy. The electron density here is high and the orbital is termed a "bonding orbital" and is designated as $\sigma 1s$. The higher energy orbital has no electron and may be considered as an antibonding orbital. Antibonding orbitals are designated with asterisks (*), in this case $\sigma^* 1s$. Sigma (σ) is used to designate orbitals in which the electron density is symmetric around a line drawn between the two nuclei.

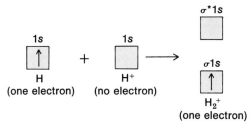

Figure 5.17
Two 1s atomic orbitals of hydrogen combine to form two molecular orbitals of different energies.

Molecular Orbitals of the H_2 Molecule

With the aid of Figure 5.18 we may examine the formation of the two molecular orbitals of the hydrogen molecule, H_2, by a coalescence of the two atomic orbitals of the single H atoms. The electron clouds formed by overlapping of the two 1s orbitals are schematically represented in Figure 5.18(b).

Two electrons are present in the H_2 molecule and both go into the lower-energy $\sigma 1s$ bonding orbital. This creates an electron-pair bond which is much stronger than the one-electron bond in the H_2^+ molecule; the bond energy in the former case is 104 kcal/mole compared to 65 kcal/mole for the H_2^+ molecule. When two atoms combine, the number of electron bonds formed or the "bond order" may be determined from the formula:

No. of bonds = $\frac{1}{2}$ (no. of bonding electrons − no. of antibonding electrons)

For H_2 with two bonding electrons in the σs bonding orbital and 0 in the $\sigma^* s$ antibonding orbital, the number of bonds is $\frac{1}{2}(2 - 0) = 1$. For H_2^+ with one

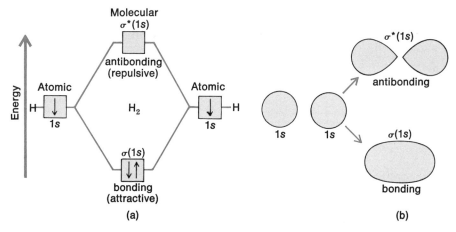

Figure 5.18

Molecular orbitals of H_2.

electron in the bonding orbital, the bond order is $\frac{1}{2}(1-0)=\frac{1}{2}$. In effect one-half a bond is formed in the latter case.

If we consider the combination of two He atoms to form He_2, a total of 4 electrons are involved. Two electrons would have to go into the σ^*1s anti-bonding orbital, which means a bond order of 0; in effect the bonding and antibonding orbitals cancel. No evidence of this molecule has been obtained experimentally. Going on to the next element, Li, with the configuration $1s^2$; $2s^1$, it would appear that Li_2 could be formed by two $2s$ electrons occupying the $\sigma2s$ bonding molecular orbital (the $\sigma1s$ and σ^*1s orbitals would cancel). Li_2 has been detected in the vapor state. The next element Be would probably not form the molecule Be_2 for the same reasons that He_2 is unstable.

Molecular Orbitals from p Atomic Orbitals

The situation becomes more complex in dealing with molecular orbitals formed from p atomic orbitals. The electronic configurations of these orbitals are very difficult to calculate and there is still some uncertainty over the arrangement of energy levels. One arrangement is shown in Figure 5.19. The energy levels are shown in part (a) and the "sausage"-like orbitals formed from overlapping of electron clouds in part (b).

The p_x atomic orbitals which are cylindrically symmetrical about the x axis form the lowest energy bonding orbital $\sigma2p_x$ and the highest energy antibonding orbital, σ^*2p_x. The p_y and p_z atomic orbitals which are at right angles to the p_x orbital and at right angles to each other behave differently — they form two bonding molecular orbitals of equal energies and two antibonding orbitals also of equal energies, all energies being intermediate to the σp_x and σ^*p_x orbitals. The latter are called pi (π) orbitals as the electron density is not symmetric about a line between the two nuclei but rather the electron clouds overlap "sidewise" instead of end to end.

As electrons are fitted into the pattern beyond the $\sigma1s$, σ^*1s and $\sigma2s$, σ^*2s levels, the order of filling is $\sigma2p_x$, $\pi2p_y$ and $\pi2p_z$, π^*2p_y and π^*2p_z, σ^*2p_x. (See Figures 5.19 and 5.20.) For example we may represent the formation of the diatomic molecules, N_2, O_2, F_2 as (also represented diagramatically in Figure 5.20):

$2 N(1s^2; 2s^2, 2p^3) \rightarrow N_2 (\sigma1s^2, \sigma^*1s^2; \sigma2s^2, \sigma^*2s^2, \sigma2p^2, \pi2p^4)$
$2 O(1s^2; 2s^2, 2p^4) \rightarrow O_2 (\sigma1s^2, \sigma^*1s^2; \sigma2s^2, \sigma^*2s^2, \sigma2p^2, \pi2p^4, \pi^*2p^2)$
$2 F(1s^2; 2s^2, 2p^5) \rightarrow F_2 (\sigma1s^2, \sigma^*1s^2; \sigma2s^2, \sigma^*2s^2, \sigma2p^2, \pi2p^4, \pi^*2p^4)$

In the inner parts of a molecule where the bonding electrons of a given level are cancelled by antibonding electrons, these electrons contribute little to the bond energy and are referred to as "non-bonding electrons." For example in the above electronic notations for N_2, O_2 and F_2, the $(\sigma1s)^2$, $(\sigma^*1s)^2$, $(\sigma2s)^2$, and $(\sigma^*2s)^2$ orbitals are non-bonding.

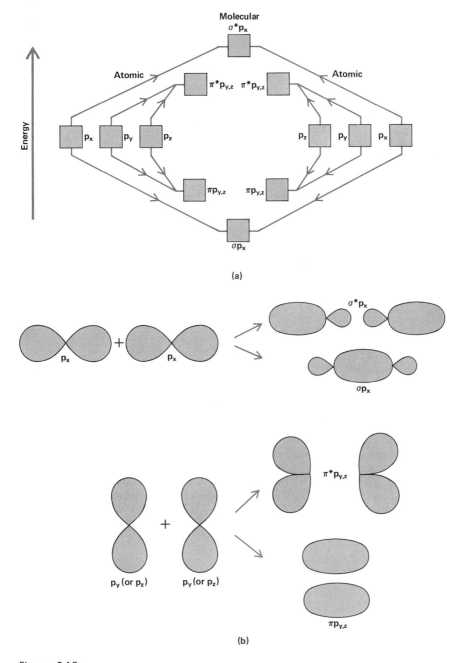

Figure 5.19

Molecular orbitals from p atomic orbitals.

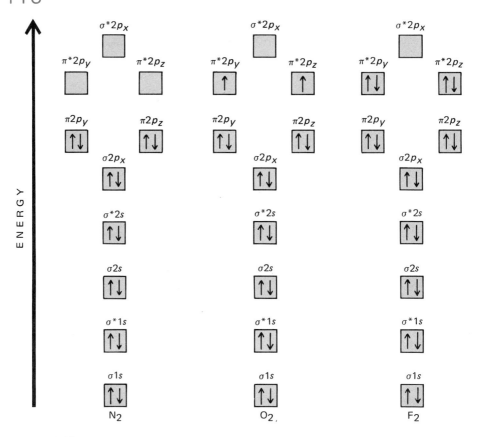

Figure 5.20

Occupancy of *p* molecular orbitals of some diatomic molecules.

We may note from Figure 5.20 that the oxygen molecule has two unpaired electrons which will explain the paramagnetic properties of oxygen gas.

Both the valence-bond and molecular orbital theories are useful in predicting bond characteristics and structure. The latter theory assumes that *all* valence electrons participate in bond formation rather than just certain ones. However since molecular orbitals are so difficult to establish and calculate, it often becomes necessary to assume that a given molecular orbital will have many of the characteristics of the atomic orbital from which it was formed.

Hydrogen Bonds

The hydrogen atom has a single electron — one *s* electron — and it seems that such an atom could form only a single bond. It could form a single ionic bond by taking up an electron (for example, in hydrides, such as NaH) or a single covalent bond by sharing the one electron with an atom of some other element or with another atom of itself. In some compounds, however, one hydrogen atom seems to be bonded to two other atoms. For example, the salt potassium hydrogen difluoride (KHF_2) is composed of K^+ and HF_2^- ions. The hydrogen difluoride ion, HF_2^-, may be represented as

$$: \ddot{\overset{..}{F}} : H : \ddot{\overset{..}{F}} : \qquad \text{or} \qquad F^- H^+ F^-$$

in which a hydrogen ion or proton acts as a bridge to connect the two fluoride ions. This situation seems to exist in all compounds between hydrogen and a highly electronegative element, such as F, O, N. The hydrogen part of the compound exerts a strong attractive force toward the negatively charged ion of the electronegative element, and long chains of molecules of the compound may result.

$$F—H—F—H—F—H—F—H \qquad \text{and so on.}$$

This attractive force is called a **hydrogen bond.** It is an electrostatic attraction and is weaker than a true ionic or covalent bond.

The formation of hydrogen bonds may be explained from a consideration of electronegativities. (See Table 5.1.) Compounds of hydrogen with highly electronegative elements (F, O, N) will be quite polar and will possess a rather high degree of ionic character. For example, in the compound hydrogen fluoride (HF) the electron pair between the atoms H : F is displaced toward the fluoride atom and away from the hydrogen atom to such an extent that the hydrogen atom has virtually lost its electron and becomes a hydrogen ion (proton). The proton, with no electrons in the $1s$ orbital, tends to attract two electrons from two fluoride ions to form the HF_2^- ion. Thus the proton forms a bridge between two fluoride ions.

The theory of hydrogen bonds is useful in explaining the abnormal behavior of certain hydrogen compounds, such as unusually high heats of vaporization and boiling points. It also helps explain the structure of certain solids containing hydrogen, for example, ice, proteins, and nucleic acids. Application of the theory to properties of water will be discussed in Chapter 10 and to biological compounds in Chapter 34.

Bond Energies

The forces of attraction between atoms in molecular and ionic combinations depend on the nature of the atoms and the kind of bond formed. The magnitude of these forces and the strength of the bond may be measured by the energy required to separate and isolate the atoms from their original grouping. Bond energies are usually expressed in kilocalories per mole — that is, the energy necessary to break Avogadro's number (6×10^{23}) of bonds. These energies are obtained from thermal data on chemical reactions. A few representative bond energies are shown in Table 5.6.

Table 5.6

Bond energies (in kcal per mole)

H—H	104	C—O	83
H—F	135	C=O	178
H—Cl	103	C—Cl	179
H—Br	88	C—F	105
H—I	71	Si—O	106
Li—H	58	Si—F	136
Cl—Cl	58	C—C	83
C—H	87	C=C	146
O—H	111	C≡C	199
O=O	118	N≡N	225
P—Cl	78		

Ionic and covalent bonds are the strongest bonds between atoms and are roughly of the same order of magnitude. We may note that bond strength increases with bond multiplicity. For example, the carbon-to-carbon triple bond (three pairs of shared electrons) has an energy of 199 kcal/mole compared with 146 for the double and 83 for a single carbon-to-carbon bond.

When bond energies are known, the approximate energy released on the formation of a gaseous molecule may be calculated. For example, the heat evolved in the reaction

$$H_2 + Cl_2 \rightarrow 2\ HCl$$

may be determined from the bond energies given in Table 5.6, as follows:

To break one H—H bond in H_2 requires   104 kcal
To break one Cl—Cl bond in Cl_2 requires   58 kcal
Total energy required   162 kcal

Energy released in forming 2 H—Cl bonds $= 2 \times 103 = 206$ kcal
Net energy released in overall reaction $= 206 - 162 = 44$ kcal
Energy released per mole of HCl formed $= \frac{44}{2} = 22$ kcal

Bond Distance and Angles

Atoms are constantly vibrating in molecules, but an *average* distance between centers of the constituent atoms and also the angle between the center of a given atom and the centers of atoms with which it is bonded may be determined by physical-chemical methods. If a substance is a crystalline solid or may be obtained in such form, these properties may be determined by X-ray diffraction; if it is not available in crystalline form, the structure may be established by spectral studies. Table 5.7 contains bond distances and angles for a few simple molecules.

As might be expected, analogous compounds of elements in the same group of the periodic table show a regular change in bond length and angles. For example, in the compounds H_2O, H_2S, H_2Se and similarly NH_3, PH_3,

Table 5.7
Bond distances and angles

Molecule	Bond Distance Between Unlike Atoms in Å	Bond Angle
H_2O	0.97	104.5°
H_2S	1.32	92°
H_2Se	1.46	91°
F_2O	1.42	103°
Cl_2O	1.70	111°
SO_2	1.43	118°
CO_2	1.16	180° (linear)
$TiCl_4$	2.18	109° (tetrahedral)
CH_4	1.09	109° (tetrahedral)
NH_3	1.01	107°
PH_3	1.42	93°
AsH_3	1.52	92°
HF	0.92	180° (linear)
HCl	1.27	180° (linear)
HBr	1.41	180° (linear)
HI	1.61	180° (linear)

AsH_3, as the atomic number increases, the bond angle decreases and the bond distance increases. With the greater bond distances of the H atoms from the central atom and the resultant change in forces between atoms in the molecule, the angle approaches 90°, the angle expected from occupancy of the p orbitals of the central element, as explained on page 104.

Bond distances are also related to bond energies: for any two atoms, the bond energy increases as the length of the bond decreases. For example (Table 5.6), in the bonding of two carbon atoms in a triple bond with a length of 1.20 Å, the energy is much higher than for a single bond with a length of 1.54 Å. In a series of compounds of related elements the bond energy decreases as the length of the bond between atoms increases; for example we may note from Tables 5.6 and 5.7 that the bond energy decreases from 135 kcal/mole in HF to 71 kcal/mole in HI, while the bond lengths increase from 0.92 Å in HF to 1.61 Å in HI.

Van der Waals Forces (Intermolecular Forces)

There are attractive forces between molecules of polar substances, since the positive end of one dipole attracts the negative end of another dipole. For nonpolar substances we would not expect an attraction of one molecule for another. Actually, however, there are weak forces of attraction between all molecules, even non-polar ones, and the attractive force is most pronounced when the molecules are close together. These weak forces are termed **van der Waals forces,** after the Dutch physicist J. D. van der Waals, who first described them.

The explanation of the weak attractive forces is based on a polar character induced by neighboring molecules. When two or more nonpolar molecules, such as He, Ar, H_2, are in close proximity, the nucleus of each atom will weakly attract electrons in another atom, resulting, at least momentarily, in an unsymmetrical arrangement of the nucleus with respect to its cloud of electrons.

In effect, some dipole character is induced in each molecule by a neighboring molecule, and the situation can be described as fluctuating dipoles, which give rise to the van der Waals forces between the molecules. These forces are extremely weak (less than 1 kcal/mole) compared with ionic and covalent bonds but are important in, for example, the liquefaction of gases, in which gas molecules are brought in close proximity on being compressed. Van der Waals forces tend to bring the molecules still closer together and aid in forming the liquid phase.

Exercises

5.1. Match the type of bond with the following descriptive statements:
(a) Bond formed when an electron is transferred from one atom to another forming positive and negative ions.
(b) A bond formed by the equal sharing of 2 electrons, one contributed from each of the similarly bonded atoms.
(c) A bond formed by the sharing of 2 electrons, both of which were contributed by one of the bonded atoms.
(d) A bond formed by the sharing of 4 electrons.
(e) A bond resulting from the sharing of 2 electrons with the bonding electrons more strongly attracted to one of the bonded atoms.
  1. Coordinate covalent   4. Ionic bond
  2. Polar covalent   5. Non-polar covalent
  3. Double bond

5.2. Draw electron-dot formulas for NaCl, HCl, Cl_2, and BrCl. Arrange in order of increasing polarity.

5.3. With the aid of the electronegativity table, indicate which of the following are primarily ionic and which primarily covalent with little polarity: KF, Br_2, H_2, H_2S, H_2O, PCl_3, CO_2, N_2, NO, OF_2, SeO_2, $MgSO_4$, ZnS.

5.4. On the basis of electronegativity state whether you think the bonds formed between the following pairs of elements would be ionic or covalent, and if covalent, estimate the degree of polarity of the bond. (a) K, Br; (b) Ca, Cl; (c) Al, Cl; (d) S, Cl; (e) Ba, O; (f) H, F; (g) Cs, I; (h) C, Cl; (i) Ca, S; (j) Ca, Br.

5.5. Referring to the periodic table, how does the electronegativity change within a given group of metals?, of non-metals?, within the horizontal groups or periods? Where in the table does one find elements with the highest electronegativities?

5.6. (a) According to the Bohr theory, the velocity of an electron in the $n = 1$ level of a hydrogen atom is 2.19×10^8 cm/s. What is the de Broglie wavelength for this electron? The mass of an electron is 9.11×10^{-28} g cm^2/s. (b) What is the de Broglie wavelength of a proton moving at the same velocity? The mass of a proton is 1.67×10^{-24} g.

5.7. Explain the tetrahedral structure of CH_4 on the basis of hybridization of orbitals.

5.8. For the following molecules describe (a) the shape of the molecule (linear, tetrahedral, etc.), (b) the bond type (p^2, sp, sp^3d, etc.), and (c) the electron distribution of the central atom after "promotion" of an electron(s) to a higher energy level just prior to hybridization ($1s^1 p_x^1$, etc.): BeF_2, CH_4, BF_3, SF_6.

5.9. For each of the following molecules, indicate the type of hybrid orbitals employed in the bonding and the geometric configuration. (a) $CdCl_2$, (b) $SnCl_4$, (c) $GaCl_3$, (d) $BeCl_2$, (e) AsF_5, (f) CCl_4.

5.10. Predict the shapes of the molecules listed in question 5.9 on the basis of the electron-pair repulsion theory.

5.11. Use the electron-pair repulsion theory to predict the shapes of the following molecules and ions: (a) SiF_6^{2-}, (b) SeO_4^{2-}, (c) $CuCl^{2-}$, (d) $TeBr_2$, (e) ICl_4^-.

5.12. From electron-pair repulsion concepts predict the configuration of the following: (a) PF_3, (b) SnF_3^-, (c) H_3O^+, (d) TeF_6, (e) BF_4^-.

5.13. What is the shape of each of the following ions? (a) ClO^-, (b) ClO_2^-, (c) ClO_3^-, (d) ClO_4^-.

5.14. Draw electron-dot formulas for the following: CO_2, SO_2, O_3, HCO_2^-. Which of these would be expected to exhibit resonance?

5.15. Draw structural formulas to show resonance forms of C_6H_6, NO_3^-, CO_3^{2-}.

5.16. The azide ion, N_3^-, is linear and symmetrical. Diagram the resonance forms of N_3^-.

5.17. In the oxalate ion, $C_2O_4^{2-}$, the two C atoms are bonded to each other, and each C atom has two O atoms bonded to it. The structure is symmetrical, and all C—O bond distances are equal. Diagram the resonance forms of $C_2O_4^{2-}$.

5.18. Write the electronic notation for the molecules, F_2, N_2, in terms of molecular orbital theory. How many antibonding electrons in each molecule?

5.19. Compare the bond strength or bond order of He_2 and He_2^+.

5.20. Fill In

(a) The symbol for a p antibonding orbital is _____.

(b) In the calculation of number of bonds σ1s and $\sigma°$1s _____ each other.

(c) Each molecular orbital may have a maximum of _____ electrons.

(d) Energy associated with molecular orbitals after like atoms combine is _____ (less than, equal to, greater than) the energy of the initial atomic orbitals.

5.21. Draw molecular-orbital energy-level diagrams for CO and NO. Use your diagrams to determine the bond order of CO, CO$^+$, CO$^-$, NO, NO$^+$, and NO$^-$ Which of these species are paramagnetic?

5.22. Draw a molecular-orbital energy-level diagram and determine the bond order of each of the following. (a) H$_2$, (b) H$_2{}^+$, (c) HHe, (d) He$_2$, (e) He$_2{}^+$.

5.23. Concerning hydrogen bonding: (a) What is it? (b) What kind of compounds show hydrogen bonding? (c) How does the bond energy compare with ionic or covalent bonding? With van der Waals forces? (d) What are some effects of hydrogen bonding?

5.24. What are the relative strengths or energies of covalent bonds, ionic bonds, hydrogen bonds, and van der Waals forces?

5.25. Using Table 5.6 of bond energies, calculate the energy change in the following:

(a) $2 \text{ H}_2(g) + \text{O}_2(g) \rightarrow 2 \text{ H}_2\text{O}(g)$

(b) $\text{CH}_4(g) + 2 \text{ O}_2(g) \rightarrow \text{CO}_2(g) + 2 \text{ H}_2\text{O}(g)$

(c) $\text{CH}_4(g) + \text{Cl}_2(g) \rightarrow \text{CH}_3\text{Cl}(g) + \text{HCl}(g)$

(d) $\text{C}_2\text{H}_6(g) + \text{Cl}_2(g) \rightarrow \text{C}_2\text{H}_5\text{Cl}(g) + \text{HCl}(g)$

5.26. For each of the following terms, formulas, or statements, mention the *scientist* with which it is most closely associated: (a) $\lambda = \dfrac{h}{mv}$, (b) wave mechanics and electron probability plots, (c) electronegativity, (d) $E = h\nu$, (e) weak bonding forces between nonpolar molecules, (f) uncertainty principle.

5.27. Fill In

(a) The German scientist _____ pointed to the fact that exact knowledge of the position of an electron and its _____ cannot be accurately determined at the same time.

(b) Electrons according to de Broglie have wave characteristics as well as _____ characteristics.

(c) The symbol inserted in Schrödingers equation for wave function is _____.

(d) The distance 0.53 Å as applied to a hydrogen atom denotes the most probable distance _____.

(e) The probability of a p electron being at or very near the nucleus is almost _____.

(f) A p electron may have three regions in space probabilities designated by the three symbols _____, _____, _____.

(g) The five d orbitals representing space probabilities are designated by the five symbols _____, _____, _____, _____, _____.

5.28. True or False
   (a) _____ A $\sigma°1s$ orbital has higher energy than a $\sigma1s$ orbital.
   (b) _____ After bonding of two hydrogen atoms the electrons favor position in $\sigma1s$.
   (c) _____ For N_2 resultant from $2N$ ($1s^2; 2s^2, 2p^3$) atoms the number of bonds calculates to two.
   (d) _____ The union of H atom $+ H^+ \rightarrow He_2^+$ is exothermic.
   (e) _____ Given two protons (H^+ and H^+); if the distance between an electron and one of the protons is greater than the distance between the protons, the e^- is an antibonding one.
   (f) _____ Bonding probability space for electrons somewhere between two protons is greater than antibonding space.
   (g) _____ The C_2 molecule is less stable than the Li_2 molecule.
   (h) _____ Molecular orbital theory predicts no He_2 molecule.
   (i) _____ For Li_2, number of bonds $= 1/2(4-2)$.
   (j) _____ pi (π) orbitals relate to $d(s, p, d)$ electron occupancy.
   (k) _____ For F_2, number of bonds $= 1/2(10-4)$.

Supplementary Readings

5.1. "Polar Bonds," L. Melander, *J. Chem. Educ.*, **49**, 687 (1972).
5.2. *Chemical Bonding Clarified Through Quantum Mechanics*, G. C. Pimentel and R. D. Spratley, Holden-Day, San Francisco, 1969 (paper).
5.3. "The Valence-Shell Electron-Pair Repulsion Theory of Directed Valence," R. J. Gillespie, *J. Chem. Educ.*, **40**, 295 (1963).
5.4. "d-Orbitals in Main Group Elements," T. B. Brill, *J. Chem. Educ.*, **50**, 392 (1973).

Chapter 6

Formulas, Equations, and Stoichiometry

The Meaning of a Formula

As pointed out on page 24, a formula is an abbreviation for a substance and at the same time shows its composition in terms of atoms of elements present. With the aid of a table of atomic weights (inside front cover), the molecular weight or formula weight may be determined. Sulfuric acid has the formula H_2SO_4 which tells us that this substance is composed of 2 parts by weight of hydrogen (2×1, where 1 is the atomic weight of hydrogen), 32 parts by weight of sulfur (1×32, where 32 is the atomic weight of sulfur), and 64 parts by weight of oxygen (4×16, where 16 is the atomic weight of oxygen). Thus the relative weight of the molecule of sulfuric acid is the sum of the relative weights of the constituent atoms, $2 + 32 + 64 = 98$. Ninety-eight is the molecular weight of the compound. Ninety-eight g of sulfuric acid would be a gram molecular weight, or mole, of the compound; one mole of sulfuric acid contains 2 g of hydrogen (2 moles of H atoms), 32 g of sulfur (1 mole of S atoms), and 64 g of oxygen (4 moles of O atoms).

A coefficient placed before a formula multiplies every constituent of the formula which it precedes. For example, $3\,C_{12}H_{22}O_{11}$ denotes three molecules of sugar — that is, 36 atoms of carbon, 66 atoms of hydrogen, and 33 atoms of oxygen.

To illustrate further the information which may be derived from a formula, consider HNO_3, which represents one molecule of nitric acid. We derive the following from this formula:

1. One molecule of nitric acid contains 1 atom of hydrogen, 1 atom of nitrogen, and 3 atoms of oxygen.
2. Nitric acid is composed of 1 part by weight of hydrogen (see atomic weights), 14 parts by weight of nitrogen, and 48 parts by weight of oxygen.
3. The molecular weight of nitric acid is 63 ($1 + 14 + 48 = 63$).
4. One gram molecular weight — that is, one mole — of nitric acid, 63 g, contains 1 g of hydrogen (1 mole of H atoms); 14 g of nitrogen (1 mole of N atoms); and 48 g of oxygen (3 moles of O atoms).
5. Since the formula of a compound is fixed, we may calculate the percentage of each element present from a consideration of the parts by weight of each element in one molecular weight of the compound. Consider the formula of water, H_2O. The atomic weights of hydrogen and oxygen are 1 and 16, respectively; hence the molecular weight of H_2O is 18. One molecular weight of water is made up of 18 parts by weight, 2 parts of which are hydrogen and 16 parts of which are oxygen. The fraction of hydrogen is $\frac{2}{18}$ and the fraction of oxygen is $\frac{16}{18}$. These fractions may, of course, be converted to percentages by multiplying by 100. Thus the percentage of hydrogen in water is $\frac{2}{18} \times 100 = 11.1\%$, and the percentage of oxygen is $\frac{16}{18} \times 100 = 88.9\%$.

The calculation of percentage composition is further illustrated in the following examples:

EXAMPLE 6.1   Calculate the percentage composition of aluminum sulfate, $Al_2(SO_4)_3$.

SOLUTION:

2 atoms of aluminum weigh	$2 \times 27 =$	54
3 atoms of sulfur weigh	$3 \times 32 =$	96
12 atoms of oxygen weigh	$12 \times 16 =$	192
	Molecular weight	342

Percent aluminum	$= \frac{54}{342} \times 100 =$	15.8%
Percent sulfur	$= \frac{96}{342} \times 100 =$	28.1%
Percent oxygen	$= \frac{192}{342} \times 100 =$	56.1%
		100.0%

The formula must always express the true composition of the substance. For example, the formula for hydrogen gas must be written H_2, not H, since in the free state two hydrogen atoms unite to form a unit particle or molecule of free hydrogen (uncombined with another element). Similarly, the formula for oxygen gas is O_2. Ozone is O_3.

Determination of Formulas from Analytical Data

The process described above may be reversed and formulas can be derived from the results of analysis for each of the elements present in a pure chemical compound. These results are usually expressed in percentage of each element present—that is, parts of each element present in 100 parts by weight of the compound. To arrive at the formula for the compound, it is necessary to know the relative number of atoms of each element present. If we divide the percentage of each element present by its atomic weight, we obtain quotients that will be in the ratio of the number of atoms of each element.

For example, by analysis a compound is found to contain 1.59% hydrogen, 22.2% nitrogen, and 76.2% oxygen. If we arbitrarily take 100 g of the compound, 1.59 g will be hydrogen, 22.2 g nitrogen, and 76.2 g oxygen. The number of moles of each of the three elements per 100 g of compound will be: $\frac{1.59}{1} = 1.59$ moles of hydrogen; $\frac{22.2}{14} = 1.59$ moles of nitrogen; and $\frac{76.2}{16} = 4.77$ moles of oxygen. The relative number of atoms will be in the same ratio as the number of moles:

H	N	O
1.59	1.59	4.77

By dividing each number by the highest common divisor (1.59) of the three values, we obtain the ratio

H	N	O
1	1	3

Hence the simplest formula for the compound is HNO_3.

A formula thus obtained from chemical analysis is an **empirical** or **simplest** formula. It tells us the ratio of the number of atoms of each element present in the compound, but it does not tell us the number of atoms in a molecule of the compound. The true molecular formula may be $H_2N_2O_6$, $H_3N_3O_9$, or any other multiple of the simplest formula. To determine which is the correct molecular formula, we must know the molecular weight of the compound. If the molecular weight of the above compound were 126, the true molecular formula would be $H_2N_2O_6$. But since the molecular weight is known to be 63 the molecular formula must be HNO_3.

To derive a formula from percentage composition data

1. Divide the percentage of each element by its atomic weight.
2. Divide the quotients obtained by the highest common divisor of all the quotients.
3. Write the simplest or empirical formula for the compound.
4. If the molecular weight or the approximate molecular weight is known, determine what multiple of the simplest formula most nearly gives the molecular weight; multiply the number of atoms of each element in the simplest formula by this multiple and write the true formula for the compound.

If the molecular weight cannot be determined or if the compound does not form molecules, the simplest empirical formula is used.

EXAMPLE 6.2   The organic compound glucose on analysis consisted of 40.0% carbon, 6.67% hydrogen, and 53.3% oxygen. The approximate experimental molecular weight was determined to be 180. What is (a) the empirical formula and (b) the molecular formula of glucose?

SOLUTION: Since % composition by weight is proportional to the relative weights of the elements, for 100 g of glucose:

$$\text{(a) moles of C} = \frac{40.0 \text{ g C}}{12.0 \text{ g/mole}} = 3.33 \text{ moles}$$

$$\text{moles of H} = \frac{6.67 \text{ g H}}{1.0 \text{ g/mole}} = 6.67 \text{ moles}$$

$$\text{moles of O} = \frac{53.3 \text{ g O}}{16.0 \text{ g/mole}} = 3.33 \text{ moles}$$

Dividing each of these numbers of moles by the smallest (3.33) gives 1 C, 2 H, 1 O. The empirical formula then is CH_2O.

(b) CH_2O has a molecular weight of 30.0 which goes into the experimental molecular weight value (180) 6 times. So the molecular formula is $C_6H_{12}O_6$.

Use of Valence in Writing Formulas

While formulas may always be calculated from analytical data as was just shown, it is generally simpler and much more convenient to make use of **valences** and **oxidation numbers** for this purpose.

Valence is a term used by the chemist to describe the nature of bonding forces between atoms and between ions. As knowledge of bonding forces has increased, the meaning of valence has undergone considerable revision.

In the beginning, valence was regarded simply as a measure of the combining capacities of atoms, and this concept has been most useful in writing formulas. But with the advent of modern atomic theory, usage of the term has changed, and we now describe more than one kind of valence or valency. Difficulties in expressing the valency of all the elements in some compounds, particularly those containing three or more elements, have led to the introduction of another term, **oxidation number**, which will be considered later.

Electrovalence and Atomic Structure

In the study of atomic structure, we learned that the outer configuration of an atom contains electrons called valence electrons. These valence electrons determine the way atoms combine with one another. We noted that all elements tend to complete the outer level of their atoms to form a stable configuration — that is, the outer level holds its capacity of electrons. The sodium atom has a single electron in its outer level and will give up this electron to other atoms. By giving up this single electron, the atom reverts to a more stable configuration. This type of valence, resulting from transfer of electrons from one atom to another, is called **electrovalence** or commonly just valence.

Since the release of an electron leaves the sodium atom with an excess of 1 positive charge, we say the electrovalence is $+1$. The charged atom which remains after the release of an electron is a sodium ion (Na^+), and the charge on the ion is the electrovalence. Calcium, which gives up the 2 electrons in its outer level and becomes a calcium ion (Ca^{2+}), has an electrovalence of $+2$; aluminum has an electrovalence of $+3$, since it gives up 3 electrons. In general, metals give up electrons as they combine with other elements. Their electrovalence is the number of electrons given up, and is positive.

Nonmetal atoms, in general, take up electrons and have a negative electrovalence. For example, an atom of chlorine takes up 1 electron to complete its outer level of 8 electrons and thus becomes a chloride ion (Cl^-). The electrovalence of chlorine is -1. Sulfur atoms accept 2 electrons from metal atoms, becoming S^{2-}, and sulfur is said to have an electrovalence of -2.

Note that an element exhibits a valence only when it is combined with some other element — in other words, when it is a part of a compound. Electrovalence means a charge on the atom; an uncombined element is without charge and must therefore have 0 valence. Thus Na by itself has 0 valence, but when combined it exhibits an electrovalence of +1.

Radicals — groups of atoms that remain together throughout a chemical change — also exhibit a valence. Since radicals are charged groups of atoms, they are, in reality, ions. In their formation one or more electrons are acquired (or lost) to give the group a net charge. Table 6.1 gives the common electro-valences of various elements and radicals.

Table 6.1
Valences and names for some common ions

	Positive Ions		
Valence $+1$	$+2$	$+3$	$+4$
Na^+	Mg^{2+}	Al^{3+}	Sn^{4+} (stannic)
K^+	Ca^{2+}	Cr^{3+} (chromic)	
NH_4^+ ammonium	Sr^{2+}	Fe^{3+} (ferric)	
Ag^+	Ba^{2+}		
Hg_2^{2+} (mercurous)	Sn^{2+} (stannous)		
°Cu^+ cuprous	Cu^{2+} (cupric)		
H^+	Fe^{2+} (ferrous)		
	Zn^{2+}		
	Mn^{2+} (manganous)		
	Pb^{2+}		
	Hg^{2+} (mercuric)		

	Negative Ions		
Valence -1	-2	-3	-4
F^- fluoride	O^{2-} oxide	PO_4^{3-} phosphate	SiO_4^{4-} silicate
Cl^- chloride	S^{2-} sulfide	AsO_4^{3-} arsenate	
Br^- bromide	SO_3^{2-} sulfite		
I^- iodide	SO_4^{2-} sulfate		
OH^- hydroxide	CO_3^{2-} carbonate		
ClO^- hypochlorite	CrO_4^{2-} chromate		
ClO_2^- chlorite	$Cr_2O_7^{2-}$ dichromate		
ClO_3^- chlorate			
ClO_4^- perchlorate			
HCO_3^- bicarbonate			
NO_2^- nitrite			
NO_3^- nitrate			
$C_2H_3O_2^-$ acetate			
CN^- cyanide			
MnO_4^- permanganate			

° The lower valence of a metal in a compound has long been designated with the suffix -ous, the higher valence with the suffix -ic: for example Cu^+ (cuprous) and Cu^{2+} (cupric) ions. Salts containing these ions — for example, Cu_2SO_4 and $CuSO_4$ — would be cuprous sulfate and cupric sulfate, respectively. A more recent method is to show the actual valence of the metal with Roman numerals; thus Cu_2SO_4 is copper(I) sulfate; $CuSO_4$ is copper (II) sulfate. Similarly, $CrCl_2$ is chromium(II) chloride and $CrCl_3$ is chromium(III) choride. In this text we will usually follow the latter convention.

The student should *learn thoroughly the contents of this table.* A knowledge of the symbols (and formulas of radicals), valences, and names of these common ions is essential to writing hundreds of correct formulas and chemical equations. The ions in the table are much like the letters of the alphabet. Knowing the alphabet allows us to write and spell words correctly. Proper use of correct words is necessary to write sentences. Likewise, proper use of the ions in the table allows us to write correct formulas and these, in turn, are necessary to write correct chemical equations.

Variable Electrovalence

A few metals exhibit two electrovalences: for example, tin $+2$ and $+4$; iron $+2$ and $+3$; copper $+1$ and $+2$. Presumably the higher valence results from the availability of one or more electrons from d orbitals. (See page 61.)

The mercury(I) ion, Hg_2^{2+}, in mercurous salts is unique in that it is a **dimer.** Experimental magnetic and X-ray studies show, however, that this formula is correct. Thus the proper formula for mercury(I) chloride is Hg_2Cl_2 rather than HgCl. Mercury(II) chloride would be $HgCl_2$.

Radicals

In many chemical changes certain groups of atoms may function as a unit, and therefore the same group or combination of atoms appears on both sides of an equation. For example, in the reaction

$$Zn + H_2SO_4 \rightarrow ZnSO_4 + H_2$$

one atom of sulfur and four atoms of oxygen make up the sulfate group (SO_4), which remains intact in the change. Such groups are termed **radicals** and may be treated as units in the chemical equation. Other radicals are hydroxide (OH), nitrate (NO_3), carbonate (CO_3), and ammonium (NH_4). If one molecular weight of a compound contains more than one unit of a radical, it is customary when writing the formula to enclose the radical in parentheses, and to indicate the number of units of the radical in the molecular weight by placing the proper subscript after the parentheses. For example, calcium hydroxide is written $Ca(OH)_2$ rather than CaO_2H_2, which would express the true composition of the compound equally well. Enclosing a radical in parentheses enables us to recognize immediately that this group functions as a unit. The subscript outside the parentheses applies to everything within the parentheses: the subscript 2 in $Ca(OH)_2$ indicates two atoms of oxygen as well as two atoms of hydrogen present in each molecular weight. A molecular weight of aluminum sulfate $-Al_2(SO_4)_3-$ contains 2 atoms of aluminum, 3 atoms of sulfur, and 12 atoms of oxygen. A coefficient placed in front of the formula necessarily multiplies everything in the formula; thus, $3\ Al_2(SO_4)_3$ would contain 6 atoms of aluminum, 9 atoms of sulfur, and 36 atoms of oxygen.

Writing formulas from valences:

1. Generally the + valence is written first. (There are a number of exceptions to this rule.)
2. The sum of the + and − valences should add up to 0.

It is reasonable that any positive ion will attract any negative ion to form a compound. In writing the formula for a given compound, it is necessary to balance the positive and negative charges, since every compound is a neutral substance. For example, in writing the formula for aluminum chloride by a combination of Al^{3+} and Cl^- ions, we must have three Cl^- ions to balance one Al^{3+} to give 0 net charge. Hence $AlCl_3$ must be the formula for aluminum chloride. Of course, we might write Al_2Cl_6, which shows the composition correctly, but it is customary to use the smaller numbers.

Consider the compound aluminum sulfate, which is composed of aluminum and sulfate ions, Al^{3+} and SO_4^{2-}. In order to balance positive and negative charges in this compound, we must use a ratio of 2 aluminum ions to 3 sulfate ions. Two aluminum ions then will possess a total of 6 positive charges, which are balanced by a total of 6 negative charges possessed by 3 sulfate ions. The formula for aluminum sulfate must then be $Al_2(SO_4)_3$.

EXAMPLE 6.3   Write correct formulas for the following compounds:

(1) sodium sulfide
(2) lead iodide
(3) copper(II) nitrate (cupric nitrate)
(4) iron(III) permanganate (ferric permanganate)
(5) ammonium arsenate
(6) mercury(I) sulfite (mercurous sulfite)

SOLUTION: (1) Na_2S, (2) PbI_2, (3) $Cu(NO_3)_2$, (4) $Fe(MnO_4)_3$, (5) $(NH_4)_3AsO_4$, (6) Hg_2SO_3.

Covalence

In the union of a chemically active metal with a nonmetal, there is a very decided transfer of electrons from the outer configuration of the metal atom to the nonmetallic atom, and the resulting compound is electrovalent. As pointed out in Chapter 3, electrons may also be shared between atoms, and the resulting compound is covalent. For example, in the compound methane (CH_4), each atom of hydrogen shares a pair of electrons with the carbon atom; in the compound carbon tetrachloride (CCl_4), each chlorine atom shares a pair of electrons with carbon:

$$\begin{array}{c} \text{H} \\ \text{H:C:H} \quad \text{(CH}_4\text{)} \\ \text{H} \end{array} \qquad \begin{array}{c} \text{:Cl:} \\ \text{:Cl:C:Cl:} \quad \text{(CCl}_4\text{)} \\ \text{:Cl:} \end{array}$$

Other examples of covalent compounds are

$$\text{H:Cl:} \qquad \begin{array}{c} \text{H} \\ \text{H:N:H} \end{array} \qquad \text{H:O:H} \qquad \text{O::C::O}$$

or

$$\text{H—Cl} \qquad \begin{array}{c} \text{H} \\ | \\ \text{H—N—H} \end{array} \qquad \text{H—O—H} \qquad \text{O=C=O}$$

The **covalence** of an element is the **number of electron pairs shared** by the atom in question. In the compounds above it is evident that hydrogen has a covalence of 1, oxygen 2, nitrogen 3, and carbon 4.

The atoms in covalent substances are tightly held together by shared electron pairs, and definite molecules are formed. This is in contrast with purely electrovalent compounds which, as crystals, are spatially placed ions, with no two ions forming an independent molecule. Electrovalent compounds, when put in a solvent such as water, indulge in a separation of the various positive and negative ions. If the covalent compound does not chemically react with a solvent—for example, with water—no separation of the molecule takes place and no electrical current-conducting solution is formed. The different bondings in electrovalent and covalent compounds give rise to distinct differences in chemical behavior.

6.1 OXIDATION NUMBER

The positive or negative nature of electrovalence is definite and easily understood. In covalent compounds where a sharing of electrons is involved, an effective charge on the atoms is not very apparent, and the covalence of an element is expressed ordinarily as a number without charge. It is often useful, however, to assign a charge to each element in a compound by an arbitrary system. This assigned charge, called **oxidation number** or **oxidation state,** is that which each element would have if it were electrovalent, and may or may not represent a real or actual charge. Oxidation numbers are determined according to the rules in Table 6.2. We may now apply these rules to determine some oxidation numbers.

Carbon in the free state has an oxidation number of 0. In CO its oxidation state will be represented by the number +2, since oxygen is assigned a number −2. In CO_2 the oxidation number of carbon is +4. In CCl_4 the oxidation number of carbon is +4; in CH_4 each hydrogen is +1 in oxidation number, and carbon is −4.

Table 6.2

Rules for determining oxidation numbers

Oxidation Number of	Magnitude of or Manner of Determining
Uncombined elements	Zero, 0
Hydrogen in most compounds	+1
Oxygen in most compounds	−2
Ions alone or in electrovalent compounds	+ or − according to their electrovalent charge.
Elements in a binary covalent compound	The charge on each atom if the shared electrons are all assigned to the more electronegative of the atoms.
The central element in a ternary compound	The apparent charge on the atom which will give the compound a neutral character when positive and negative oxidation numbers have been assigned to all other elements in the compound. The algebraic sum of the apparent charges in a compound is zero.
The first element in a radical	The sum of the oxidation numbers adds up to the charge of the radical.

In NaCl, an ionic compound, the oxidation number of sodium is $+1$, and of chlorine -1. In $AlBr_3$ the oxidation number of aluminum is $+3$; of each bromine, -1. In the ionic compound $(NH_4)_2SO_4$ the oxidation number of each NH_4^+ is $+1$; of SO_4^{2-}, -2. For the ion PO_4^{3-} the oxidation number of P must be $+5$, since the latter added to a total of -8 from the 4 oxygen atoms would leave a net charge of -3, which is the charge of the PO_4^{3-} ion. Study these examples:

Compound or Ion	*Oxidation Number*
H_2SO_4	sulfur, $+6$
$H_4P_2O_7$	phosphorus, $+5$
Al_2S_3	aluminum, $+3$; sulfur, -2
MgO	magnesium, $+2$
$NaMnO_4$	sodium, $+1$; manganese, $+7$
$Na_2Cr_2O_7$	sodium, $+1$; chromium, $+6$
$HBrO_3$	bromine, $+5$
SiO_4^{4-}	silicon, $+4$
AsO_4^{3-}	arsenic, $+5$
$C_{12}H_{22}O_{11}$	carbon, 0
C_4H_{10}	carbon, 2.5

The concept of oxidation numbers is useful not only in describing the oxidation state of an element, but also in balancing oxidation-reduction equations (Chapter 16).

Chemical Equations

Chemical equations are the chemist's way of representing chemical reactions (chemical changes) in terms of symbols and formulas. The **reactants** are on the left of the equation and the **products** are on the right. An arrow is generally used to mean "yields." In the following equation for the burning of methane (CH_4):

$$CH_4 + 2\ O_2 \rightarrow CO_2 + 2\ H_2O$$

the reactants are CH_4 and O_2, which react to yield the products CO_2 and H_2O. In order to write a correct chemical equation for a reaction, we really go through three steps in the following order:

  (1) What are the products of the reaction or *what happens?*
  (2) Write *correct formulas* for all reactants and products.
  (3) *Balance* the equation. That is, the same number of atoms of each element must appear on both sides of the equation.

  Step (1) is a difficult step and the student will be learning "what happens" as his knowledge of chemistry increases. For the time being the necessary information concerning products of a reaction will be supplied.

  Step (2) can be accomplished for most common reactants and products by proper use of valences (Table 6.1).

  Step (3) for simple equations is largely a matter of trial and error. For example, in the reaction of hydrogen with oxygen:

$$H_2 + O_2 \rightarrow H_2O \text{ (not balanced)}$$

  It is evident that this equation is not balanced, since there are two atoms of oxygen on the left side of the equation and only one atom of oxygen on the right side. Since the formula of a compound is fixed and definite, we cannot write H_2O_2, or in any other way change the formula for water. We may, however, balance the equation as follows, since a number placed in front of a formula multiplies all the constituents in the formula which it precedes:

$$2\ H_2 + O_2 \rightarrow 2\ H_2O$$

  The coefficients (the numbers in front of the formulas) represent the number of molecules or molecular weights undergoing chemical change. If a single molecule enters into a change, no number is placed in front of the formula. The above equation tells us that two molecules of hydrogen combine with one molecule of oxygen to form two molecules of water, or that two moles of hydrogen (4 g) combine with one mole of oxygen (32 g) to give two moles of water (36 g).

  Ordinary sugar, when heated, decomposes into carbon and water, and the change may be represented by the equation

$$C_{12}H_{22}O_{11} \rightarrow C + H_2O \text{ (not balanced)}$$

To balance the equation, 12 atoms of carbon are necessary; hence the number

12 is placed in front of the symbol for carbon. To balance the hydrogen and oxygen atoms, a coefficient of 11 is needed in front of the formula for water:

$$C_{12}H_{22}O_{11} \rightarrow 12\ C^* + 11\ H_2O$$

Balancing of equations is largely a matter of trial and error; small coefficients should be tried until a balance is obtained. A good general rule to follow is: balance elements other than hydrogen and oxygen first; then balance hydrogen atoms (if present); and finally balance oxygen atoms (if present). It will be found that the oxygen atoms are usually balanced when the others have been taken care of.

In equations the physical state of a substance in the reaction is frequently indicated by placing an appropriate abbreviation in parentheses immediately following the formula—(s) for solid, (l) for liquid and (g) for gas or vapor; (aq) is an abbreviation for "aqueous" and indicates that the substance is in aqueous solution. The equation for the reaction of zinc metal with dilute sulfuric acid might be shown as

$$Zn\ (s) + H_2SO_4\ (aq) \rightarrow ZnSO_4\ (aq) + H_2\ (g)$$

Some Types of Chemical Change

We are not ready at this point to discuss the many types of chemical change which may occur, but we can indicate here the more important types and study them in detail later. Most chemical changes may be classified as one of the following four types:

1. COMBINATION   Many compounds are formed by direct **combination** of two or more elements or compounds. The following equations are examples:

$$Fe + S \rightarrow FeS$$
$$CaO + CO_2 \rightarrow CaCO_3$$
$$2\ Na + Cl_2 \rightarrow 2\ NaCl$$

We note in the above that a *single* substance is formed by a combination of two substances. The combining substances are not always single elements; for example, in the second equation above, two compounds, calcium oxide and carbon dioxide, are the reactants. A **combination** reaction, then, is one in which a *single* substance is produced by the union of two or more substances.

2. DECOMPOSITION   Water may be broken down into hydrogen and oxygen by means of an electric current according to the equation†

---

* It would be incorrect to make the 12 a subscript of the carbon (C_{12}), since this would imply that twelve atoms of carbon remain together as a unit particle.

† At times the factor or agency which brings about a chemical change is indicated above the arrow of the equation—Δ for heat, e.c. for electric current.

$$\overset{\text{e.c.}}{2 \ H_2O \rightarrow 2 \ H_2 + O_2}$$

Sugar, when heated, decomposes to form a charred mass, which is essentially carbon, and water in the form of steam. The equation for the change is

$$C_{12}H_{22}O_{11} \overset{\Delta}{\rightarrow} 12 \ C + 11 \ H_2O$$

These are **decomposition** chemical changes – reactions in which two or more substances are produced from a *single* substance. This type of change is the reverse of combination. A compound may not always decompose into its constituent elements; this is illustrated in the decomposition of sugar, where water, a compound substance, is one of the products.

3. REPLACEMENT or SUBSTITUTION   In this type of chemical change one element replaces another in its compounds. For example, in

$$Zn + 2 \ HCl \rightarrow ZnCl_2 + H_2$$
hydrochloric   zinc
acid   chloride

zinc replaces, or substitutes for, the hydrogen in the compound hydrochloric acid.

If chlorine is passed into a solution of sodium bromide, the chlorine replaces the bromine:

$$2 \ NaBr + Cl_2 \rightarrow 2 \ NaCl + Br_2$$
sodium   chlo-   sodium   bro-
bromide   rine   chloride   mine

4. EXCHANGE   Compounds may be considered as composed of two parts – a positive part and a negative part. If two compounds are brought together, there is the possibility that the positive part of one compound may interact with the negative part of the second compound – in other words, the positive and negative parts of the two compounds exchange partners. A solution of silver nitrate reacts with a solution of sodium chloride according to the equation

$$AgNO_3 \ (aq) + NaCl \ (aq) \rightarrow AgCl \ (s) + NaNO_3 \ (aq)$$
silver   sodium   silver   sodium
nitrate   chloride   chloride   nitrate

There has been a double exchange, since both silver and sodium have exchanged positions. Such a chemical change is sometimes termed **double decomposition** or **double replacement.**

In the above reaction the silver chloride is insoluble and precipitates – that is, separates from the solution – as a solid. The soluble sodium nitrate remains in solution and may be separated from the insoluble AgCl by filtration (Figure 6.1).

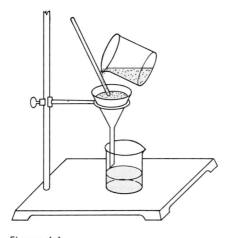

Figure 6.1

Filtration.

EXAMPLE 6.4   Write chemical equations for the following reactions:

(1) Decomposition of ammonium dichromate by heating to give nitrogen gas (N_2), chromium (III) oxide and water.
(2) Aluminum sulfate + metallic magnesium to give magnesium sulfate + metallic aluminum.
(3) Reaction of sodium with water to give sodium hydroxide and hydrogen gas.
(4) $Pb\,(C_2H_3O_2)_2 + (NH_4)_3PO_4 \rightarrow$
(5) Reaction of zinc with hydrobromic acid (HBr).

SOLUTION: Use the valences in Table 6.1 to write the necessary correct formulas.

(1) $(NH_4)_2Cr_2O_7 \xrightarrow{\text{heat}} N_2 + Cr_2O_3 + 4\ H_2O$
(2) $Al_2(SO_4)_3 + 3\ Mg \rightarrow 3\ MgSO_4 + 2\ Al$
(3) $2\ Na + 2\ H_2O \rightarrow 2\ NaOH + H_2$
(4) This is an exchange reaction. The products of the reaction may be predicted by exchanging the + ions (lead and ammonium) to give lead phosphate + ammonium acetate, the correct formulas for which are $Pb_3(PO_4)_2$ and $NH_4C_2H_3O_2$. The complete equation is $3\ Pb(C_2H_3O_2)_2 + 2\ (NH_4)_3PO_4 \rightarrow Pb_3(PO_4)_2 + 6\ NH_4C_2H_3O_2$.
(5) This is a replacement reaction. The products of the reaction may be predicted by exchanging the zinc and hydrogen. Correct formulas of the products must then be written, $ZnBr_2 + H_2$, and the equation balanced.

$$Zn + 2\ HBr \rightarrow ZnBr_2 + H_2$$

Oxidation and Reduction Reactions[*]

We have stated that reactions with oxygen are **oxidation** processes. We may note that in all of the above the oxidation number of the element combining with oxygen has been increased from 0 to some more positive value. On the basis of the oxidation number concept, **oxidation is any process in which the oxidation number of an element is increased,** and we are not restricted to the use of oxygen in bringing it about. For example, hydrogen will burn in chlorine to give hydrogen chloride:

$$H_2 + Cl_2 \rightarrow 2\ HCl$$

The oxidation number of hydrogen has been increased from 0 to +1, and we may say that hydrogen has been oxidized. Looking again at this reaction of hydrogen and chlorine, we may note that the oxidation number of chlorine has *decreased* from 0 to −1. This latter process is called **reduction.** Oxidation and reduction are mutually dependent processes — one cannot occur without the other.

Summarizing, increase of oxidation number is **oxidation;** decrease of oxidation number is **reduction.**

6.2 STOICHIOMETRY

Weight Relations from Chemical Equations

Just as a formula tells us the ratio of the weights of the different elements present in a compound, so an equation tells us the relationship of the weights of the reactants to one another and to the products. This relationship between weights or amounts of chemicals is termed the **stoichiometry** of the reaction. The **stoichiometric** amounts in terms of moles are given by the coefficients in the balanced chemical equations. If we understand that a mole represents the molecular weight of a substance in grams, pounds, and so on, the weight relations between substances in a chemical equation may be evaluated. For example, in the reaction

$$2\ C_2H_2 + 5\ O_2 \rightarrow 4\ CO_2 + 2\ H_2O$$

suppose we want to know the weight of oxygen in grams necessary to burn 104 g of C_2H_2. Molecular weights are $C_2H_2 = 26$; $O_2 = 32$.

According to the equation 2 *moles* of C_2H_2 (acetylene) react with 5 *moles* of O_2 (oxygen) to produce 4 *moles* of CO_2 (carbon dioxide) and 2 *moles* of water.

$$104\ g\ of\ C_2H_2 = \frac{104\ g}{26\ g/mole} = 4\ moles$$

---

[*] The concept of oxidation-reduction will be treated in more detail in Chapter 16.

Since 2 moles of C_2H_2 require 5 moles O_2, 4 moles C_2H_2 would require

$$\tfrac{4}{2} \times 5 = 10 \text{ moles}$$
$$10 \text{ moles of } O_2 = 10 \times 32 \text{ g} = 320 \text{ g of } O_2$$

The above method of solution of stoichiometry problems, often called the mole method, might be extended to the use of a chemical conversion factor relating the quantity sought and the quantity given as follows:

quantity sought = quantity given × conversion factor

If *correct units* are included in the quantity given and the conversion factor, the units will cancel except for the proper ones for the quantity sought. The conversion factor may be obtained by dividing the number of moles of the substance sought by the number of moles of the substance given (equal to their coefficients in the balanced chemical equation) and their molecular weights.

In the above example,

$$\text{grams } O_2 = 104 \text{ g } C_2H_2 \left[\frac{5 \text{ moles } O_2 \times 32 \text{ g/mole}}{2 \text{ moles } C_2H_2 \times 26 \text{ g/mole}} \right] = 320 \text{ g } O_2$$

The student should study carefully the following illustrative problems.

EXAMPLE 6.5   Using the equation in the preceding problem, determine the weights of CO_2 and H_2O produced on burning 104 g of C_2H_2. Molecular weights are $CO_2 = 44$ and $H_2O = 18$.

SOLUTION: Since 2 moles C_2H_2 produce 4 moles of CO_2 and 2 moles of H_2O, 4 moles C_2H_2 (104 g) will produce $\tfrac{4}{2} \times 4 = 8$ moles CO_2 and $\tfrac{4}{2} \times 2 = 4$ moles H_2O.

$$8 \text{ moles } CO_2 = 8 \times 44 \text{ g} = 352 \text{ g } CO_2$$
$$4 \text{ moles } H_2O = 4 \times 18 \text{ g} = 72 \text{ g } H_2O$$

or

$$g \ CO_2 = 104 \text{ g } C_2H_2 \left[\frac{4 \text{ moles } CO_2 \times 44 \text{ g/mole}}{2 \text{ moles } C_2H_2 \times 26 \text{ g/mole}} \right] = 352 \text{ g}$$

and

$$g \ H_2O = 104 \text{ g } C_2H_2 \left[\frac{2 \text{ moles } H_2O \times 18 \text{ g/mole}}{2 \text{ moles } C_2H_2 \times 26 \text{ g/mole}} \right] = 72 \text{ g.}$$

In all of these calculations, any consistent units of weight may be used. For example from the above, 104 tons C_2H_2 (4 ton moles) would require 320 tons O_2 and would yield 352 tons of CO_2 and 72 tons of H_2O, and so on.

EXAMPLE 6.6   In the commercial preparation of hydrogen chloride gas, what weight of HCl in grams may be obtained by heating 234 g of NaCl with excess of H_2SO_4?

$$2 \ NaCl + H_2SO_4 \xrightarrow{\Delta} 2H \ Cl + Na_2SO_4$$

The balanced equation for the reaction is

$$2 \text{ NaCl} + \text{H}_2\text{SO}_4 \rightarrow \text{Na}_2\text{SO}_4 + 2 \text{ HCl}$$
Molecular weights: NaCl = 58.5; HCl = 36.5

SOLUTION: According to the equation, 2 moles of NaCl produce 2 moles of HCl — that is, 1 mole of HCl is obtained from 1 mole of NaCl.

$$234 \text{ g NaCl is } \frac{234 \text{ g}}{58.5 \text{ g/mole}} = 4 \text{ moles NaCl}$$

4 moles of NaCl will produce 4 moles of HCl, or 4×36.5 g = 146 g HCl.

or

$$\text{g HCl} = 234 \text{ g NaCl} \left[\frac{2 \text{ moles HCl} \times 36.5 \text{ g/mole}}{2 \text{ moles NaCl} \times 58.5 \text{ g/mole}} \right] = 146 \text{ g.}$$

EXAMPLE 6.7   What weight of sulfur must combine with aluminum to prepare 4 moles of aluminum sulfide? The equation is

$$2 \text{ Al} + 3 \text{ S} \rightarrow \text{Al}_2\text{S}_3$$

SOLUTION:

1 mole of Al_2S_3 requires 3 moles of S
4 moles Al_2S_3 will require $4 \times 3 = 12$ moles of S
$$12 \text{ moles of S} = 12 \times 32 \ \frac{\text{g}}{\text{mole}} = 384 \text{ g}$$

or

$$\text{g S} = 4 \text{ moles Al}_2\text{S}_3 \left[\frac{3 \text{ moles S} \times 32 \text{ g/mole}}{1 \text{ mole Al}_2\text{S}_3} \right] = 384 \text{ g.}$$

EXAMPLE 6.8   What weight of arsenic acid (H_3AsO_4) should be used with lime (CaO) to prepare one kilogram (1000 g) of calcium arsenate ($\text{Ca}_3(\text{AsO}_4)_2$) insecticide material?

Equation: $3 \text{ CaO} + 2 \text{ H}_3\text{AsO}_4 \rightarrow \text{Ca}_3(\text{AsO}_4)_2 + 3 \text{ H}_2\text{O}$
Molecular weights: $\text{H}_3\text{AsO}_4 = 142$; $\text{Ca}_3(\text{AsO}_4)_2 = 398$

SOLUTION:

1 mole $\text{Ca}_3(\text{AsO}_4)_2$ requires 2 moles H_3AsO_4
$$1000 \text{ g Ca}_3(\text{AsO}_4)_2 \text{ is } \frac{1000 \text{ g}}{398 \text{ g/mole}}, \text{ or } 2.513 \text{ moles}$$
2.513 moles of $\text{Ca}(\text{AsO}_4)_2$ requires 5.026(2.513×2) moles H_3AsO_4
5.026 moles $\text{H}_3\text{AsO}_4 = (5.026)(142 \text{ g}) = 714 \text{ g H}_3\text{AsO}_4$

or

$$\text{g } H_3AsO_4 = 1000 \text{ g } Ca_3(AsO_4)_2 \left[\frac{2 \text{ moles } H_3AsO_4 \times 142 \text{ g/mole}}{1 \text{ mole } Ca_3(AsO_4)_2 \times 398 \text{ g/mole}}\right] = 714 \text{ g.}$$

Exercises

6.1. Define the terms symbol, formula, chemical equation, radical, percentage composition, stoichiometry, oxidation number, oxidation, reduction.

6.2. Consider the compound phenol, C_6H_6O.
(a) How many atoms of each element are present in 1 molecule of phenol?
(b) What is the weight of a mole of phenol?
(c) How many moles of each element are there in 1 mole of phenol?
(d) How many moles are in 470 g of phenol?
(e) How many moles of each element are 470 g of phenol?
(f) How many moles are in 3.76 g of phenol?

6.3. (a) How many moles are present in 100 g of each of the following: NH_4Br, H_3PO_4, Br_2, $CaCO_3$, ZnS, $Ba(ClO_4)_2$?
(b) How many grams are in 5 moles of each of the substances in part (a)?

6.4. Calculate the percentage of each element in the compounds:
(a) $NaNO_3$,   (b) C_6H_6O,   (c) $K_2Cr_2O_7$,   (d) CH_3CSNH_2.

6.5. Calculate the percentage of silicon in the minerals:
(a) orthoclase, $KAlSi_3O_8$; (b) asbestos, $Mg_3Ca(SiO_3)_4$; (c) beryl, $Be_3Al_2Si_6O_{18}$.

6.6. Calculate the percent N in the following fertilizers:
(a) ammonia, NH_3; (b) urea, $CO(NH_2)_2$; (c) ammonium sulfate, $(NH_4)_2SO_4$; (d) ammonium phosphate, $(NH_4)_3PO_4$.

6.7. Pure nicotine obtained from tobacco contains 74.03% C, 8.70% H, and 17.27% N. (a) Calculate the simplest (or empirical) formula for this compound. (b) If the molecular weight of this compound is 162, what is the molecular formula?

6.8. Given the following data, determine the empirical formula for each of the compounds listed.

Compound	Percent Composition
A	Li. 18.8; C, 16.2; O, 65.0.
B	C, 69.6; H, 10.1; N, 20.3.
C	Rb, 64.0; S, 12.0; O, 24.0.
D	K, 26.6; Cr, 35.3; O, 38.1.
E	C, 60.0; H, 13.3; O, 26.7.
F	K, 30.7; S, 25.2; O, 44.1.

BAsed on 100 grAms

Li = 18.8 g

C = 16.2 g

O = 65.0 g

6.9. The molecular weights of compounds B and E in exercise 8 are 69 and 120, respectively. Write true formulas for these compounds.

6.10. (a) An oxide of arsenic consists of 65.2% As and 34.8% O. What is the empirical formula of the compound? (b) The molecular weight of the oxide is 459.6. What is the molecular formula of the compound?

6.11. What is the empirical formula of a compound that is 35.2% S, 43.9% O, and 20.9% F?

6.12. Determine the molecular formulas of the compounds for which the following empirical formulas and molecular weights pertain. (a) S_3Cl, 263.6; (b) AsS, 428.0; (c) P_4S_7, 348.7; (d) CH_2O, 90.0; (e) BH_2, 76.8; (f) B_5H_4, 232.0; (g) SNF, 260.4; (h) TeF_5, 445.2.

6.13. Dehydroascorbic acid, a derivative of ascorbic acid (vitamin C), has the following composition:

Element	Percent by wt.
C	41.39
H	3.47
O	55.14

What is the empirical (simplest) formula?

6.14. Prepare a form like the one below and fill in the blanks with formulas of compounds composed of the indicated positive and negative ions.

	Cl^-	NO_3^-	SO_4^{2-}	PO_4^{3-}	SiO_4^{4-}	OH^-	$Cr_2O_7^{2-}$	$C_2H_3O_2^-$	AsO_4^{3-}
Na^+									
Ca^{2+}									
Al^{3+}									
NH_4^+									
Fe^{3+}									
Cu^{2+}									
Sn^{4+}									
Ag^+									
Pb^{2+}									
H^+									

6.15. Name the compounds in the previous table.

6.16. Using Table 6.1, write formulas for (a) potassium bromide, (b) ammonium carbonate, (c) iron(III) sulfate, (d) ferrous acetate, (e) chromium(III) bromide, (f) tin(II) fluoride, (g) calcium cyanide.

6.17. (a) Write formulas for potassium chromate, zinc silicate, copper(II) acetate, iron(III) sulfide, aluminum perchlorate, iron(III) chromate, copper(II) phosphate. (b) Name the compounds: $KMnO_4$, $Zn(ClO_4)_2$, Cu_2SiO_4, K_2S, $Al(C_2H_3O_2)_3$, $PbCrO_4$, $Fe_2(CrO_4)_3$, $Al(OH)_3$.

6.18. Name the following compounds: (a) K_2SO_3, (b) $Ba(ClO_3)_2$, (c) $Cu(OH)_2$, (d) $Al(C_2H_3O_2)_3$, (e) $Fe_3(PO_4)_2$, (f) NH_4ClO_4, (g) Hg_2SO_4.

6.19. Name the following compounds: (a) MnI_2, (b) $NaClO$, (c) $(NH_4)_2Cr_2O_7$, (d) Na_4SiO_4, (e) $Al(NO_2)_3$, (f) SnF_2, (g) $Hg(CN)_2$, (h) $KMnO_4$.

6.20. Determine the oxidation number of the underlined elements or radicals in each of the following compounds:

$Al\underline{Br}_3$	$NH_4\underline{Mn}O_4$	$Zn(\underline{Cl}O_2)_2$	$K_2\underline{Pt}Cl_6$
$HI\underline{O}_4$	$K_2\underline{Se}O_4$	$\underline{V}O_2CO_3$	$Na\underline{Al}Si_3O_8$
$Cu\underline{Se}O_4$	$Li_2H\underline{As}O_4$	$\underline{Mn}Cl_3$	$Ca\underline{Mn}O_4$
$Na_2\underline{Te}O_4$	$Mg_2\underline{P}_2O_7$	$Na_2\underline{Cr}_2O_7$	$H_2\underline{Mo}O_4$
$Ti\underline{Cl}_4$	$Na_3\underline{N}$	$Cu(\underline{N}O_3)_2$	$\underline{Zr}(SO_4)_2$
$\underline{V}OSO_4$	$Cd_3(\underline{PO_4})_2$	$Fe\underline{S}O_3$	$Na\underline{Bi}O_3$

6.21. What is the oxidation number of Mn in each of the following: Mn (metal), $KMnO_4$, $CaMnO_4$, MnO_2, $MnCl_2$, $Mn(NO_3)_3$, MnO_4^-, MnO_4^{2-}?

6.22. What is the oxidation number of the element combined with oxygen in each of the following: CO_2, H_2O_2, KO_2, $Cr_2O_7^{2-}$, AsO_3^{3-}, NO_2^-, P_4O_{10}?

6.23. Balance the equations:
(a) $Ag_2O \rightarrow Ag + O_2$
(b) $Zn + HCl \rightarrow ZnCl_2 + H_2$
(c) $NaOH + H_2SO_4 \rightarrow Na_2SO_4 + H_2O$
(d) $K_2O_2 + H_2O \rightarrow KOH + O_2$
(e) $SO_2 + NaOH \rightarrow Na_2SO_3 + H_2O$
(f) $PbO_2 \rightarrow PbO + O_2$
(g) $Na + H_2O \rightarrow NaOH + H_2$
(h) $Al + H_2O \rightarrow Al_2O_3 + H_2$
(i) $Bi_2O_3 + H_2 \rightarrow Bi + H_2O$
(j) $Ca(ClO_3)_2 \rightarrow CaCl_2 + O_2$
(k) $Fe_2(SO_4)_3 + Ca \rightarrow CaSO_4 + Fe$
(l) $Fe_3O_4 + H_2 \rightarrow Fe + H_2O$
(m) $PbO + C \rightarrow Pb + CO_2$
(n) $Al + H_2SO_4 \rightarrow Al_2(SO_4)_3 + H_2$
(o) $K_3PO_4 + Ba(NO_3)_2 \rightarrow Ba_3(PO_4)_2 + KNO_3$

6.24. Write balanced chemical equations for the following reactions.
(a) Hydrogen sulfide (g) and oxygen (g) yield water and sulfur dioxide (g).
(b) Sodium (s) and sulfur hexafluoride (g) yield sodium sulfide (s) and sodium fluoride (s).

(c) Titanium(IV) chloride (l) and water yield titanium(IV) oxide (s) and hydrogen chloride (g).

(d) Naphthalene, $C_{10}H_8$(s), and oxygen (g) yield carbon dioxide (g) and water.

(e) Niobium pentabromide (g) and niobium (s) yield niobium tetrabromide (g).

(f) Aluminum (s) and hydrochloric acid (aq) yield hydrogen (g) and aluminum chloride (aq).

6.25. For each of the chemical equations below, select the correct products from the list given.

(a) $Al(NO_3)_3 + NH_4OH \rightarrow$ _____ + _____

(1) Al_2O_3, (2) $AlNH_4$, (3) $Al(OH)_3$, (4) NH_4NO_3, (5) Al, (6) $NO_3(OH)$, (7) AlOH, (8) $NH_4(NO_3)_3$.

(b) $Zn + H_3PO_4 \rightarrow$ _____ + _____

(1) ZnH_3, (2) H_3Zn, (3) $ZnPO_4$, (4) H_3, (5) $Zn_3(PO_4)_2$, (6) H_2, (7) ZnP, (8) H_2O.

(c) $SO_2 + H_2O \rightarrow$ _____

(1) SOH_2, (2) H_2SO_3, (3) H_2SO_4, (4) HSO_2, (5) H_2S.

6.26. Balance the following equations (the formulas given are correct).

(a) $BaCl_2 + H_3PO_4 \rightarrow Ba_3(PO_4)_2 + HCl$

(b) $C_6H_{14} + O_2 \rightarrow CO_2 + H_2O$

6.27. Classify the following equations as (1) combination, (2) decomposition, (3) replacement, or (4) exchange.

(a) $Cu(OH)_2 + 2 HCl \rightarrow CuCl_2 + 2 H_2O$

(b) $CO_2 + BaO \rightarrow BaCO_3$

(c) $(NH_4)_2Cr_2O_7 \rightarrow Cr_2O_3 + N_2 + 4 H_2O$

(d) $Cl_2 + 2 NaBr \rightarrow Br_2 + 2 NaCl$

6.28. Classify each of the following reactions as one of the four types listed in the preceding problem.

(a) $Cu(OH)_2 + 2 HCl \rightarrow CuCl_2 + 2 H_2O$

(b) $3 Mg + 2 H_3PO_4 \rightarrow Mg_3(PO_4)_2 + 3 H_2$

(c) $2 KNO_3 \rightarrow 2 KNO_2 + O_2$

(d) $MgCO_3 \rightarrow MgO + CO_2$

(e) $2 KCl + H_2SO_4 \rightarrow K_2SO_4 + 2 HCl$

(f) $Fe_2O_3 + 6 HBr \rightarrow 2 FeBr_3 + 3 H_2O$

(g) $Pb(C_2H_3O_2)_2 + (NH_4)_2SO_4 \rightarrow PbSO_4 + 2 NH_4C_2H_3O_2$

(h) $H_2SO_3 + Ca(OH)_2 \rightarrow CaSO_3 + 2 H_2O$

(i) $CO_2 + BaO \rightarrow BaCO_3$

(j) $Na_2O + H_2O \rightarrow 2 NaOH$

(k) $N_2O_5 + H_2O \rightarrow 2 HNO_3$

(l) $Al_2(SO_4)_3 + 3 Mg \rightarrow 3 MgSO_4 + 2 Al$

6.29. Classify each of the following processes as either oxidation or reduction:

(a) $Fe^{2+} \rightarrow Fe^{3+}$

(b) $Cl^- \rightarrow Cl$

(c) $Cr_2O_7^{2-} \rightarrow Cr^{3+}$

(d) $SO_4^{2-} \rightarrow S^{2-}$

(e) $MnO_2 \rightarrow MnO_4^-$

(f) $S^{2-} \rightarrow S$

(g) $ClO_4^- \rightarrow Cl^-$

6.30. Given the balanced equation:

$$2 KNO_3 + H_2SO_4 \rightarrow K_2SO_4 + 2 HNO_3$$

(a) How many moles of K_2SO_4 are produced for each mole of KNO_3?
(b) What weight of H_2SO_4 in grams is needed for 2 moles of KNO_3?
(c) How many moles of HNO_3 will be produced from 5 moles of H_2SO_4?
(d) What weight in grams of K_2SO_4 will be produced from 5 moles of KNO_3?

6.31. Given the balanced equation:

$$4 NH_3(g) + 5 O_2(g) \rightarrow 4 NO(g) + 6 H_2O(g)$$

How many grams of NH_3 will be required to produce 100 g of H_2O?

6.32. Consider the reaction represented by the balanced equation:

$$4 NH_3 + 5 O_2 \rightarrow 4 NO + 6 H_2O$$

How many moles of NO would be formed if the necessary amount of oxygen were to react exactly and completely with 1.00 kg of ammonia?

6.33. Phosphoric acid (H_3PO_4) may be prepared by dissolving P_4O_{10} in H_2O according to the equation:

$$P_4O_{10} + 6 H_2O \rightarrow 4 H_3PO_4$$

(a) How many moles of H_3PO_4 are produced by use of 3.20 moles of H_2O?
(b) How many grams of H_2O are needed for each mole of P_4O_{10}?
(c) What weight in grams of H_3PO_4 is produced from 550 g of P_4O_{10}?

6.34. From the balanced equation:

$$2 H_2S + 3 O_2 \rightarrow 2 SO_2 + 2 H_2O$$

(a) How many moles of O_2 are necessary to react with 0.75 mole of H_2S?
(b) How many moles each of SO_2 and H_2O will be produced in part (a)?
(c) How many grams of oxygen are required in (a)?
(d) What weight of SO_2 will be produced if 68 g of H_2S are used?

6.35. From the balanced equation

$$C_3H_8 + 5 O_2 \rightarrow 3 CO_2 + 4 H_2O$$

(a) How many moles of O_2 are required for burning 4 moles of propane (C_3H_8)?
(b) What weight of oxygen in grams will be required in (a)?
(c) How many moles of carbon dioxide (CO_2) are formed for each mole of oxygen used?
(d) What weight in grams of propane would be needed to yield 25 moles of CO_2?
(e) What weight of CO_2 in grams would be obtained from 66 g of C_3H_8?

6.36.
$$2 PbS + 3 O_2 \rightarrow 2 PbO + 2 SO_2$$
$$3 SO_2 + 2 HNO_3 + 2 H_2O \rightarrow 3 H_2SO_4 + 2 NO$$

Using the above sequence of reactions, what weight of H_2SO_4 in grams would be produced from 1195 g of PbS?

6.37. Sodium azide, NaN_3, may be produced by the reaction:

$$3\ NaNH_2 + NaNO_3 \rightarrow NaN_3 + 3\ NaOH + NH_3$$

(a) What weight of NaN_3 should theoretically be obtained from the complete reaction of 15.0 g of $NaNH_2$?

(b) If 5.43 g of NaN_3 is isolated from the reaction mixture, what percentage of the theoretical yield is obtained?

Supplementary Readings

6.1. *Problems For General Chemistry and Qualitative Analysis*, C. J. Nyman and G. B. King, Wiley, New York, 1975 (paper).

6.2. *Stoichiometry*, L. K. Nash, Addison Wesley, Reading, Mass., 1966.

6.3. "History of the Chemical Sign Language," R. Winderlich, *J. Chem. Educ.*, **30**, 58 (1953).

Chapter 7

Classification and Naming of Compounds. Writing Equations

7.1   THE CLASSES OF COMPOUNDS

Thousands and tens of thousands of compounds are known to the chemist today. It would be impossible to learn the properties and behavior of even a fraction of this number if it had to be done on the basis of individual compounds. Fortunately, most chemical compounds can be grouped together in a few classes. Then, if we can properly classify a compound, we are at once aware of the general properties of the compound from knowledge of the properties of that class or group of compounds. For example, HCl is classed as an acid, and by becoming familiar with the behavior of acids as a distinct class, we are at once aware of the general properties of the compound. A great many of the compounds we are to study may be classified as acids, bases, salts, metallic oxides, or nonmetallic oxides. Of these five classes of compounds, the first three — acids, bases, and salts — are by far the most important.

When an acid, base, or salt is dissolved in water the resulting solution is a conductor of the electric current and is termed an **electrolyte.** If no conduction of current occurs, the compound is known as a **nonelectrolyte.**

The conduction of the current is attributed to the presence of ions in the solution and depends upon the relative abundance or concentration of the ions. The conductance of a solution may be qualitatively determined in an apparatus as shown in Figure 7.1. The apparatus consists of two electrodes (copper wires may be used) connected in series with a 110 volt alternating current source and an ordinary light bulb. The electrodes are immersed in the solution whose conductivity is to be tested. If a current flows across the solution between the electrodes, the bulb will light, and the degree of conduction of the electric current is roughly indicated by the intensity of light shown in the bulb. If the solution is a good conductor (that is, if the resistance

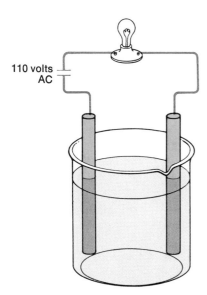

110 volts
AC

Figure 7.1
**Apparatus for measuring electrical
conductance of solutions.**

offered to the passage of the current is small), the bulb will show a bright light. If resistance is high, the current cannot pass through the solution readily, and there will be a dim light or no light at all. In Table 7.1 the results obtained with a few substances in aqueous solutions are summarized.

Table 7.1
Conductivity of aqueous solutions

Substance	Light Intensity	Classification as Conductor
HCl	bright	very good
NaOH	bright	very good
NH_3	dim	fair
$HC_2H_3O_2$	dim	fair
Sugar	none	nonconductor
HCl (in toluene)	none	nonconductor
Alcohol	none	nonconductor
Water (distilled)	none	nonconductor
Water (tap)	very dim	poor
NaCl	bright	very good
$CuSO_4$	bright	very good

It may be noted that (1) acids, bases, and salts in aqueous solution are conductors (electrolytes) while compounds such as sugar ($C_{12}H_{22}O_{11}$) and alcohol (C_2H_6O) are nonconductors (nonelectrolytes) and (2) compounds differ in degree of conductivity. For example, hydrochloric acid is a good conductor (a strong electrolyte), while acetic acid is only fair (a weak electrolyte), even though the acids are of equivalent concentrations. Although acids, bases, and salts may show different degrees of conduction, in aqueous solution these classes of compounds all conduct the current to some extent. Even tap water shows some conduction because of dissolved mineral salts.

Classification of Common Compounds

By looking at the chemical formulas we may classify many common compounds in the following way:

1. **Acids**°, in the conventional sense, may be recognized by noting that the H is written first in the formula and that the rest of the compound is generally nonmetallic. Ex., HCl, H_2SO_4, HClO.
2. Conventional **bases**° have OH radicals written last in the formula. The first part of the formula is usually a metal. Ex., NaOH, $Ca(OH)_2$, $Fe(OH)_3$.
3. A **salt** consists of a metal, written first, combined with a non-metal or radical written last in a formula. Ex., NaCl, $Fe_2(SO_4)_3$, $Ca(ClO)_2$.

° This concept of acids and bases will be extended later in the chapter.

4. Oxides are compounds containing oxygen and only *one* other element. If the element other than oxygen is a nonmetal, the oxide is classed as a **nonmetal oxide** or an **acidic anhydride.** The latter name comes about because water added to nonmetal oxides under certain conditions produces acids. For example,

$$CO_2 + H_2O \rightarrow H_2CO_3$$

Likewise, if water is removed from an oxygen containing acid, the acid anhydride (without water) results.

$$H_2SO_4 \rightarrow H_2O + SO_3$$

The other class of oxides, **metallic oxides** or **basic anhydrides,** consist of oxygen combined with a metal. When water is added under proper conditions to basic anhydrides, bases result and vice versa.

$$CaO + H_2O \rightarrow Ca(OH)_2$$

EXAMPLE 7.1   Classify the following compounds as (1) acids, (2) bases, (3) salts, (4) acidic anhydrides, and (5) basic anhydrides.

        (a) LiOH   (b) K_2O (c) HNO_3 (d) SO_2
        (e) BaI_2 (f) $SnSO_4$ (g) H_2SO_3 (h) $Fe(OH)_2$

SOLUTION:

Acids	Bases	Salts	Acidic Anhydrides	Basic Anhydrides
HNO_3	LiOH	BaI_2	SO_2	K_2O
H_2SO_3	$Fe(OH)_2$	$SnSO_4$		

EXAMPLE 7.2   Write the formulas for the anhydrides of the following acids and bases: (1) H_2SO_3 (2) $HClO$ (3) $Zn(OH)_2$ (4) $Al(OH)_3$ (5) H_3PO_4.

SOLUTION: Remove *all* of the H from each formula and *half* as many O as H (ratio H_2O). If the formulas have an odd number of H, double the number of atoms of each element in the formula. Then proceed as above.

    (1) SO_2 (2) Cl_2O (3) ZnO   (4) Al_2O_3 (5) P_2O_5

Acids

All acids in the conventional sense contain hydrogen,[*] which may be replaced by metals. The negative portion of the acid molecule is composed of a nonmetal or a radical (negative valence group). These negative valence

[*] This concept of acids will be extended later in the chapter.

groups (except oxide and hydroxide) are often referred to as acid radicals. All acids are covalent compounds in which the atoms are held together by a sharing of electrons. When an acid is dissolved in water, ions are formed as a result of the transfer of a hydrogen ion (proton) from the acid molecule to the water molecule – for example,

$$\text{H}\!:\!\ddot{\underset{..}{\text{C}}}\text{l}\!: \;+\; \text{H}\!:\!\ddot{\underset{..}{\text{O}}}\!:\!\text{H} \longrightarrow \left[\text{H}\!:\!\underset{\text{H}}{\overset{..}{\text{O}}}\!:\!\text{H}\right]^{+} + \left[:\!\ddot{\underset{..}{\text{C}}}\text{l}\!:\right]^{-}$$

which may be written

$$\text{HCl} + \text{H}_2\text{O} \rightarrow \text{H}_3\text{O}^+ + \text{Cl}^-$$

This is a case of coordinate valence, as discussed on page 71, in which an unused pair of electrons from the water molecule combines with a hydrogen ion to form a hydronium ion. The hydronium ion is a hydrated hydrogen ion or proton $(\text{H}^+ \cdot \text{H}_2\text{O})$ and, while the ionization of acids in aqueous solution depends on its formation, we shall ordinarily use the simple H^+ in writing equations. Such equations are thereby simplified and easier to balance.

The chief characteristic of an acid is its ability to furnish hydrogen ions (protons); therefore, an acid is usually defined as **a substance which may furnish protons.**

PROPERTIES OF ACIDS   In general, aqueous solutions of acids are characterized by the following properties:

1. They have a sour taste. Lemons, oranges, and other citrus fruits owe their sour taste to the presence of citric acid; the taste of sour milk is due to the presence of lactic acid.
2. They turn blue litmus paper red. Litmus is a dye which has a red color in acid solution and a blue color in basic solution; paper which has been soaked in litmus is referred to as **litmus paper.** Substances of this type, which enable us to determine whether a given solution is acid or basic, are called **indicators.** Methyl orange and phenolphthalein are other indicators frequently used by chemists.
3. They react with certain metals to produce hydrogen. Reactions of this type were studied in connection with the preparation of hydrogen.
4. They react with bases to produce salts and water.

Common **strong acids** are H_2SO_4, HNO_3, **HCl, HBr,** and **HI.** Most other acids are generally only partially ionized and consequently only moderately strong or weak.

Bases

All metallic hydroxides are classed as conventional bases. Of the common bases only NaOH, KOH, Ca(OH)_2, and Ba(OH)_2 are appreciably soluble in water. If these compounds are dissolved in water, the OH^- is common to all of their solutions:

$$NaOH \rightarrow Na^+ + OH^-$$
$$KOH \rightarrow K^+ + OH^-$$
$$Ca(OH)_2 \rightarrow Ca^{2+} + 2\ OH^-$$
$$Ba(OH)_2 \rightarrow Ba^{2+} + 2\ OH^-$$

An aqueous solution of NH_3° is also classed as a base, since OH^- ions are present in the solution.

In each of these compounds we find a combination of a metal (or NH_4) with the hydroxide group. Just as the characteristic part of an acid is hydrogen ion, so the characteristic part of a base in water solution is the hydroxide ion, OH^-. Later the concept of a base will be extended to include substances which do not furnish hydroxide ions in solution.

PROPERTIES OF BASES   In general, water solutions of metallic hydroxides (bases) exhibit the following properties:

1. Bitter taste.
2. Soapy or slippery feeling.
3. Turn red litmus paper blue.
4. React with acids to form salts and water.
5. Most metallic hydroxides are insoluble in water. Of the common ones, only $NaOH$, KOH, $Ca(OH)_2$, $Ba(OH)_2$, and NH_3 are soluble.

The common **strong bases** are **$NaOH$, KOH, $Ca(OH)_2$, and $Ba(OH)_2$.**

Salts

An acid reacts with a base to produce a salt and water. Hydrogen from the acid combines with hydroxide from the base to form water molecules:

$$NaOH + HCl \rightarrow NaCl + HOH$$
$$Mg(OH)_2 + 2\ HNO_3 \rightarrow Mg(NO_3)_2 + 2\ HOH$$
$$2\ Al(OH)_3 + 3\ H_2SO_4 \rightarrow Al_2(SO_4)_3 + 6\ HOH$$

The reaction of an acid with a base is called **neutralization.** If all the water is removed by evaporation from the solution after the reaction, the positive ions from the base and the negative ions from the acid form a crystal lattice of solid **salt.**

It was shown (Chapter 3) that the compound sodium chloride, a salt, is

---

° Although NH_4OH has never been isolated as a compound, this formula is sometimes used to represent a solution of NH_3 in H_2O in which some NH_4^+ and OH^- exist.

$$NH_3 + H_2O \rightleftharpoons (NH_4OH) \rightleftharpoons NH_4^+ + OH^-$$

an electrovalent compound and is ionized in the solid or crystalline state. The crystal is made up of positive sodium ions and negative chloride ions oriented in a definite pattern. In general, most salts in the crystalline state are electrovalent and are composed of ions oriented in a definite way.

In general, then, salts are compounds between metallic ions and non-metallic ions. They are the most numerous of inorganic compounds, and many are industrially important. NaCl, for example, is an essential food mineral; $Ca_3(PO_4)_2$ is used in making phosphate fertilizer; Na_2CO_3 is washing soda; AgBr is used in photographic film.

7.2 NOMENCLATURE

With the discovery of thousands of new inorganic compounds it has become necessary to revise the traditional rules of nomenclature. An international committee[*] has recommended a set of rules for naming compounds, and these are now being adopted throughout the world. Many of the older names are still used, however, and our ensuing discussion will include in many cases both the old and new, with emphasis on the latter. One of the principal changes is that proposed by Albert Stock and now known as the Stock system for the naming of compounds of metals (oxides, hydroxides, and salts) in which the metal may exhibit more than one oxidation state. In these cases the oxidation state of the metal is shown by a Roman numeral in parentheses immediately following the *English* name of the metal which corresponds to its oxidation number. If the metal has only *one* common oxidation number, no Roman numeral is used. Another important change is in the naming of complex ions and coordination compounds. We will defer the nomenclature of the latter until these compounds are discussed in Chapter 28.

Naming Metal Oxides, Bases, and Salts

The student should have a good start in learning nomenclature if he has learned the Valence Table 6.1 which gives both charges on ions and names for the more common ones. A compound is a combination of positive and negative ions in the proper ratio to give a balanced charge and the name of the compound follows from names of the ions, for example, NaCl is sodium chloride; $Al(OH)_3$ is aluminum hydroxide; $FeBr_2$ is iron(II) bromide or ferrous bromide; $Ca(C_2H_3O_2)_2$ is calcium acetate; $Cr_2(SO_4)_3$ is chromium(III) sulfate or chromic sulfate, and so on. Table 7.2 gives some additional examples of the naming of metal compounds. Of the two common systems used, the Stock system is preferred. Note that even in this system, however, the name of the negative ion will need to be obtained from Valence Table 6.1.

[*] Committee on Inorganic Nomenclature, "Rules for Naming Inorganic Compounds," *J. Am. Chem. Soc.*, 82:21 (1960).

Table 7.2

Names of some metal oxides, bases, and salts

Formula	Name	
FeO	iron(II) oxide	ferrous oxide
Fe_2O_3	iron(III) oxide	ferric oxide
$Sn(OH)_2$	tin(II) hydroxide	stannous hydroxide
$Sn(OH)_4$	tin(IV) hydroxide	stannic hydroxide
Hg_2SO_4	mercury(I) sulfate	mercurous sulfate
$HgSO_4$	mercury(II) sulfate	mercuric sulfate
NaClO	sodium hypochlorite	sodium hypochlorite
$K_2Cr_2O_7$	potassium dichromate	potassium dichromate
$Cu_3(AsO_4)_2$	copper(II) arsenate	cupric arsenate
$Cr(C_2H_3O_2)_3$	chromiun(III) acetate	chromic acetate

Naming Nonmetal Oxides

The older system of naming and one still widely used employs Greek pre-fixes for both the number of oxygen atoms and that of the other element in the compound. The prefixes used are (1) *mono-*, sometimes reduced to *mon-*, (2) *di-*, (3) *tri-*, (4) *tetra-*, (5) *penta-*, (6) *hexa-*, (7) *hepta-*, (8) *octa-*, (9) *nona-* and (10) *deca-*. Generally the letter *a* is omitted from the prefix (from tetra on) when naming a nonmetal oxide and often *mono-* is omitted from the name altogether.

The Stock system is also used with nonmetal oxides. Here the Roman numeral refers to the oxidation state of the element other than oxygen.

In either system, the element other than oxygen is named first, the *full name being used*, followed by oxide. Table 7.3 shows some examples.

Table 7.3

Names of some nonmetal oxides

Formula	Name	
CO	carbon(II) oxide	carbon monoxide
CO_2	carbon(IV) oxide	carbon dioxide
SO_3	sulfur(VI) oxide	sulfur trioxide
N_2O_3	nitrogen(III) oxide	dinitrogen trioxide
P_2O_5	phosphorus(V) oxide	diphosphorus pentoxide
Cl_2O_7	chlorine(VII) oxide	dichlorine heptoxide

Naming Acids

Acid names may be obtained directly from a knowledge of Valence Table 6.1 by changing the name of the acid ion (negative ion) in the table as follows:

Ion in Table 6.1	Corresponding Acid
_____ate	_____ic
_____ite	_____ous
_____ide	hydro_____ic

Table 7.4 shows examples of this relationship.

Table 7.4
Names of some acids

Formula of Acid	Acid Ion in Table 6.1	Name of Acid
$HC_2H_3O_2$	$C_2H_3O_2^-$, ace*tate*	ace*tic* acid
H_2CO_3	CO_3^{2-}, carbon*ate*	carbon*ic* acid
$HClO_2$	ClO_2^-, chlor*ite*	chlor*ous* acid
$HClO_4$	ClO_4^-, perchlor*ate*	perchlor*ic* acid
HCN	CN^-, cyan*ide*	*hydro*cyan*ic* acid
HBr	Br^-, brom*ide*	*hydro*brom*ic* acid
H_4SiO_4	SiO_4^{4-}, silic*ate*	silic*ic* acid
H_3AsO_4	AsO_4^{3-}, arsen*ate*	arsen*ic* acid
$HMnO_4$	MnO_4^-, permangan*ate*	permangan*ic* acid

There are a few cases where name of the acid is changed slightly from that of the acid radical; for example H_2SO_4 is sulfuric acid rather than sulfic. Similarly, H_3PO_4 is phosphoric acid rather than phosphic.

Even less common negative ions not included in the Valence Table 6.1 follow the relationship shown in Table 7.4. For example, BO_3^{3-} is the borate ion and H_3BO_3 is boric acid; TeO_4^{2-} is the tellurate ion and H_2TeO_4 is telluric acid, and so on.

Acid and Basic Salts

It is conceivable that in the neutralization of an acid by a base, only a part of the hydrogen might be neutralized; thus

$$Na \mid OH \mid + \begin{matrix} H \\ \diagdown \\ SO_4 \\ \diagup \\ H \end{matrix} \rightarrow \begin{matrix} Na \\ \diagdown \\ SO_4 + H_2O \\ \diagup \\ H \end{matrix}$$

The compound $NaHSO_4$ has acid properties, since it contains hydrogen, and is also a salt, since it contains both a metal and an acid radical. Such a salt containing acidic hydrogen is termed an *acid salt*. Phosphoric acid (H_3PO_4) might be progressively neutralized to form the salts, NaH_2PO_4, Na_2HPO_4, and Na_3PO_4. The first two are acid salts, since they contain replaceable hydrogen. A way of naming these salts is to call Na_2HPO_4 disodium hydrogen phosphate and NaH_2PO_4 sodium di-hydrogen phosphate. These acid phosphates are important in controlling the alkalinity of the blood. The third compound, sodium phosphate Na_3PO_4, which contains no replaceable hydrogen, is often referred to as **normal sodium phosphate**, or trisodium phosphate to differentiate it from the two acid salts.

Historically, the prefix *bi-* has been used in naming some acid salts; in industry, for example, $NaHCO_3$ is called sodium bicarbonate and $Ca(HSO_3)_2$ calcium bisulfite. Since the *bi-* is somewhat misleading, the system of naming discussed above is preferable.

If the hydroxyl radicals of a base are progressively neutralized by an acid, **basic salts** may be formed:

$$Ca\underset{OH}{\overset{OH}{<}} + H\,Cl \rightarrow CaOHCl + H_2O$$

Basic salts have properties of a base and will react with acids to form a normal salt and water; thus

$$CaOHCl + HCl \rightarrow CaCl_2 + H_2O$$

Further examples of basic salts are $BiOH(NO_3)_2$, $Bi(OH)_2NO_3$, and $Pb(OH)C_2H_3O_2$.

The OH group in a basic salt is called an **hydroxy** group. The name of $Bi(OH)_2NO_3$ would be bismuth dihydroxynitrate.

Mixed Salts

If the hydrogen atoms in an acid are replaced by two or more different metals, a **mixed salt** results. Thus the two hydrogen atoms in H_2SO_4 may be replaced with sodium and potassium to yield the mixed salt $NaKSO_4$, sodium potassium sulfate. $NaNH_4HPO_4$ is a mixed acid salt that may be crystallized from urine.

7.3 WRITING EQUATIONS

Other than organic and biochemical reactions, most chemical reactions involve acids, bases, salts, or oxides. Since the student will be faced eventually with the problem of writing several hundred equations for these reactions, it will be well to expand somewhat on the concept of equation writing introduced briefly in Chapter 6. As mentioned there, three essential steps are necessary to write an equation:

1. We must know or be able to predict the *products of the reaction*.
2. All *chemical formulas must be written correctly*.
3. The equation must be *balanced*.

The above 3 steps must be completed in the order given. For example, it is useless for the student to attempt to balance an equation if one or more of the products are not the right ones or if even a single formula is incorrect. Step 2, writing correct formulas, can generally be accomplished from a knowledge of Valence Table 6.1. Other formulas will be learned as the chemistry of specific elements is considered later. Step 3, balancing, for most equations is easily accomplished simply by inspection. There must be the same number of atoms of each element on both sides of the equation. For complicated oxidation-reduction equations special methods of balancing will be considered in Chapter 16.

The most difficult step in writing equations is step 1, predicting the products of reactions. While this step cannot always be predicted for a given set of reactants, some general helps and patterns will be considered.

In Chapter 6, four common general types of inorganic equations were mentioned: (1) combination, (2) decomposition, (3) replacement, and (4) exchange. Nearly 75% of all the inorganic equations we will write are of these types. The other 25% consist of rather complex oxidation-reduction equations which will be considered in Chapter 16 and a relatively few equations which do not fit the above types. These will need special attention when they are encountered later.

Replacement and Exchange Reactions

Of the above four general types of equations, we will consider **replacement** and **exchange** equations first and attempt to show how the products of these reactions are predicted. Over one-third of all inorganic equations are of these types.

Consider a typical replacement reaction

$$(1) \ Zn \ (s) + 2 \ HCl \rightarrow ZnCl_2 + H_2 \ (g)$$

In a replacement reaction, one element (Zn in this case) stands alone on the left and replaces one of the elements in the compound (H in this case) so that it ends up alone (as H_2 gas) on the right. Replacement reactions are of the form

$$A + BY \rightarrow B + AY$$

where A and B trade places.

Note that this process only answers the question in step 1 of writing equations — *what the products are* — and does not necessarily give correct subscripts for the formulas for B and AY. This is taken care of in step 2. Balancing, step 3, completes the equation. The rules are:

    a. Exchange A and B
    b. Write correct formulas
    c. Balance

    Consider another example:

(2) $Mg(s) + AgNO_3$ $\rightarrow Ag + Mg(NO_3)_2$
    a. $\rightarrow$ silver and magnesium nitrate
    b. $\rightarrow Ag(s) + Mg(NO_3)_2$
    c. $Mg(s) + 2\ AgNO_3$ $\rightarrow 2\ Ag(s) + Mg(NO_3)_2$

Radicals (such as NO_3^- above) do not break down in replacement and exchange reactions and remain the same on both sides of the equation.

    Exchange reactions follow much the same format as replacement reactions except that no element stands alone.

    Consider the exchange reaction

(3) $AlCl_3 + 3\ KOH$ $\rightarrow Al(OH)_3\ (s) + 3\ KCl$

This is of the form

$$AX + BY \rightarrow BX + AY$$

where again A and B trade places and the rules are the same as previously. Some other examples follow.

(4) $Ca(OH)_2 + HNO_3$ $\rightarrow$
    a. $\rightarrow$ calcium nitrate + water
    b. $\rightarrow Ca(NO_3)_2 + HOH$ (or more commonly H_2O)
    c. $Ca(OH)_2 + 2\ HNO_3$ $\rightarrow Ca(NO_3)_2 + 2\ H_2O$

(5) $Na_4SiO_4 + AgNO_3$ $\rightarrow$
    a. $\rightarrow$ silver silicate + sodium nitrate
    b. $\rightarrow Ag_4SiO_4 + NaNO_3$
    c. $Na_4SiO_4 + 4\ AgNO_3$ $\rightarrow Ag_4SiO_4(s) + 4\ NaNO_3$

(6) $H_2SO_4 + NaC_2H_3O_2$ $\rightarrow$
    a. $\rightarrow$ sodium sulfate + acetic acid
    b. $\rightarrow Na_2SO_4 + HC_2H_3O_2$
    c. $H_2SO_4 + 2\ NaC_2H_3O_2 \rightarrow Na_2SO_4 + 2\ HC_2H_3O_2$

    It might be observed that the examples of replacement and exchange reactions just given are merely representative of several common kinds of reaction. Example (1) is a typical reaction of a metal (above H in the activity series of metals) with an acid $\rightarrow$ a salt + H_2. Example (4) is the reaction of an acid + a base $\rightarrow$ a salt + H_2O. Example (5) is a solution of a salt +

a salt → a new salt + a new salt. Example (6) involves an acid + a salt → a new acid + a new salt.

Complete and Incomplete Reactions

Experiment shows that many chemical reactions continue until finally the reaction is virtually over because one or more of the reactants are consumed. We say the reaction has gone to **completion.** Example (1), for example, stops when the Zn or the HCl is all used up. Other reactions, however, only proceed part way, generally because the products of the reaction may not escape and will react with one another to yield the original reactants over again. In these cases there are some of the reactants and some of the products always present. Such reactions are called **incomplete** or **equilibrium** reactions. These important reactions will be taken up in detail in Chapter 14.

For most purposes it is desirable to have, if possible, reactions which are essentially complete. These will be reactions for which at least one of the products is (1) a **gas** (g), (2) **insoluble** (s), or (3) H_2O or a weakly ionized acid or base.

Example (1) Zn + HCl, is complete because H_2 gas is a product. Examples (3) and (5) are complete since insoluble precipitates form, and example (4) is complete because H_2O is a product.

Writing Equations in Ionic Form

Since all acids, bases, and salts in aqueous solution give ions, the chemical equations in which these classes of compounds are involved may show the ions present. An ionic equation is more instructive than a molecular equation, since the actual chemical process is more clearly indicated.

Consider, for example, the reaction of hydrochloric acid (a strong acid) with sodium hydroxide (a strong base). The **molecular equation** for the reaction is

$$HCl + NaOH \rightarrow NaCl + H_2O$$

If the equation above, however, is represented in ionic form,

$$[Na^+ + OH^-] + [H^+ + Cl^-] \rightarrow [Na^+ + Cl^-] + H_2O$$

it is evident that Na^+ and Cl^- ions may be cancelled out, so that the neutralization process really involves only the combination of H^+ and OH^- ions to form H_2O. The **ionic equation** for the reaction becomes

$$OH^- + H^+ \rightarrow H_2O$$

Na^+ and Cl^- ions are incidental to the neutralization process. Of course if the solution is evaporated, Na^+ and Cl^- form crystals of NaCl, and the latter may be recovered as a solid. So long as the NaCl remains in solution, however, there is virtually no tendency for Na^+ and Cl^- ions to combine.

In writing ionic equations, it is necessary to adopt certain rather arbitrary rules. For example, if an insoluble substance such as $BaSO_4$ is produced in a reaction, that substance is shown in molecular form followed by (s), even though the solid crystal is made up of ions. Ordinarily when ions are shown in the equation, this implies an aqueous solution.

The convention for writing ionic equations is simply *do not break up into ions* (1) *gases* (g), (2) *insoluble substances* (s), or (3) *weakly ionized substances* such as H_2O or weak acids or bases. Cancel out of the equation any ions which are *exactly the same* on both sides of the equation. When balancing an ionic equation it is important to note that not only the number of atoms of each element must be the same on both sides of the equation, but also *the algebraic sum of the charges must be the same on both sides*. For example, the ionic equation just considered

$$OH^- + H^+ \rightarrow H_2O$$

has a net charge of zero on both sides of the equation. The **charge balance** should always be checked for an ionic equation.

The subject of ionic reactions will be considered in detail later, but to familiarize the student with this method of writing equations, we will write the previous six examples of replacement and exchange reactions as ionic equations:

(1) Zn (s) $+ 2 H^+ + 2Cl^- \rightarrow Zn^{2+} + 2Cl^- + H_2(g)$
or simply Zn (s) $+ 2 H^+ \rightarrow Zn^{2+} + H_2(g)$
HCl (a strong acid) and $ZnCl_2$ (a soluble salt) are both written in their ionic form. The identical chloride ions, Cl^-, are cancelled out, but Zn (s) cannot cancel Zn^{2+} since they are not identical. Likewise, $2H^+$ cannot cancel H_2 (g). Note that the net charge of the final ionic equation is $+2$ on both sides.

(2) Mg (s) $+ 2 Ag^+ + 2NO_3^- \rightarrow 2 Ag$ (s) $+ Mg^{2+} + 2NO_3^-$
or simply Mg (s) $+ 2 Ag^+ \rightarrow 2 Ag$ (s) $+ Mg^{2+}$
Both $AgNO_3$ and $Mg(NO_3)_2$ are soluble salts and consequently are written in ionic form. Only the NO_3^- ions are identical on both sides and may be cancelled. Again the charge balance is $+2$ on both sides.

(3) $Al^{3+} + 3Cl^- + 3K^+ + 3 OH^- \rightarrow Al(OH)_3$ (s) $+ 3K^+ + 3Cl^-$
or $Al^{3+} + 3 OH^- \rightarrow Al(OH)_3$ (s)
The soluble salts $AlCl_3$ and KCl are written in ionic form as is also KOH (a strong base). $Al(OH)_3$ (s) however is insoluble. The charge balance is zero on both sides.

(4) $Ca^{2+} + 2 OH^- + 2 H^+ + 2NO_3^- \rightarrow Ca^{2+} + 2NO_3^- + 2 H_2O$
or $OH^- + H^+ \rightarrow H_2O$
This is the ionic equation for the neutralization of any strong acid with any strong base.

(5) Of the 4 salts only Ag_4SiO_4 (s) is insoluble. The ionic equation for its precipitation is

$$4\ Ag^+ + SiO_4^{4-} \rightarrow Ag_4SiO_4\ (s)$$

(6) H_2SO_4 is a strong acid and the two salts are soluble. The ionic equation for the formation of $HC_2H_3O_2$ (a weak acid), therefore, is

$$H^+ + C_2H_3O_2^- \rightarrow HC_2H_3O_2$$

A great deal more basic knowledge concerning solubilities and ionization of compounds must be acquired by the student before he can be expected to handle ionic equations with facility, however, the obvious advantage of the much simpler form of ionic equations over molecular equations is quite apparent and ionic equations are usually preferred by chemists. We will make extensive use of both kinds in future chapters.*

Combination Reactions

A surprisingly large number (about $\frac{1}{4}$) of the inorganic reactions we will be concerned with are **combination reactions.** These rather simple reactions consist of the chemical combination of two or more elements or compounds to form *one single compound* as a product. For example,

$$2\ Al\ (s) + 3\ S\ (s) \rightarrow Al_2S_3\ (s)$$
$$Cu\ (s) + Cl_2\ (g) \rightarrow CuCl_2\ (s)$$
$$CO_2\ (g) + H_2O \rightarrow H_2CO_3$$

For many of these reactions the single product that is formed is quite obvious, such as in the first reaction above where Al has only one common valence, and there remains only to write the correct formula of the product (Valence Table 6.1) and balance the equation. In other reactions, however, we may have to decide which one of two or more possible compounds is actually formed or the compound formed may not be quite so obvious. In the second reaction above we have to decide whether copper(I) chloride (CuCl) or copper(II) chloride ($CuCl_2$) is produced. This question can only be answered by more detailed study of the elements in later chapters. Similarly, in the last equation above, the formation of H_2CO_3 might not be readily predicted if we had not previously learned the relationship of acidic anhydrides and their corresponding acids (page 152). Other examples of combination reactions are

$$2\ Na\ (s) + Br_2\ (1) \rightarrow 2\ NaBr\ (s)$$
$$2\ H_2\ (g) + O_2\ (g) \rightarrow 2\ H_2O\ (g)$$
$$CaO\ (s) + H_2O \rightarrow Ca(OH)_2.$$

* The instructor may emphasize the use of ionic equations here or defer serious consideration until the writing of ordinary molecular equations has been mastered. For pedagogical reasons, some may prefer to wait until ionization as a topic is being studied.

Decomposition Reactions

These are relatively few in number (10%) and are characterized by a *single compound* as the reactant, decomposing generally upon heating, into two or more elements or compounds as products. For example,

$$2\,HgO\,(s) \xrightarrow{\text{heat}} 2\,Hg\,(l) + O_2\,(g)$$

$$2\,H_2O\,(l) \xrightarrow{\text{electrolysis}} 2\,H_2\,(g) + O_2\,(g)$$

$$H_2CO_3 \xrightarrow{\text{heat}} CO_2 + H_2O$$

$$2\,KClO_3\,(s) \xrightarrow{\text{heat}} 2\,KCl\,(s) + 3\,O_2$$

The products of many of the decomposition reactions may be readily predicted, such as the first three reactions above. Others, for example the last reaction above, will require more detailed study.

The Brönsted-Lowry Theory of Acids and Bases

In aqueous solutions, the hydroxide ion (OH^-) is responsible for basic properties. In the previous reaction (4) we may say that the OH^- ion has taken up or accepted a proton (H^+) in the neutralization process. In order to extend the ideas of acid-base phenomena to solvents other than water, Brönsted, a Danish scientist and Lowry, an Englishman independently proposed that anything that will accept a proton be regarded and termed a base, and anything which donates a proton be termed an acid. In other words, a base is a proton **acceptor,** an acid a proton **donor.** In the reaction of an acid with water, for example,

$$HCl + H_2O \rightleftharpoons H^+ \cdot H_2O + Cl^-$$
$$(H_3O^+)$$

H_2O acts as a base, since it accepts a proton to form H_3O^+. According to the Brönsted theory, the reaction of HCl with liquid ammonia would be analogous to its reaction with water:

$$HCl + NH_3 \rightleftharpoons H^+ \cdot NH_3 + Cl^-$$
$$(NH_4^+)$$

In this case NH_3 acts as the base, since it accepts a proton from HCl to form ammonium ion NH_4^+.

In general, a proton donor reacts with a base according to the general reaction, where A is an anion:

$$\underset{\text{acid}}{HA} + \underset{\text{base}}{B} \rightleftharpoons \underset{\text{acid}}{H^+ \cdot B} + \underset{\text{base}}{A^-}$$

By far the most common use of this concept of a base is in water solutions where the base is OH^- ion.

Other Acid-Base Reactions

Consider the reverse of the first equation above:

$$H^+ \cdot H_2O + Cl^- \rightleftharpoons HCl + H_2O$$

$$\text{acid} \qquad \text{base} \qquad \text{acid} \qquad \text{base}$$

The hydronium ion $H^+ \cdot H_2O$ is giving up a proton to a Cl^- ion to form HCl. Thus H_3O^+ is acting as an acid, since it is here a proton donor. Cl^- ion is a base, since it accepts a proton. The reverse of the second equation above is analogous:

$$H^+ \cdot NH_3 + Cl^- \rightleftharpoons HCl + NH_3$$

$$\text{acid} \qquad \text{base} \qquad \text{acid} \qquad \text{base}$$

According to the Brönsted-Lowry theory, once an acid gives up a proton, the species remaining is a base since it may recombine with a proton to form the acid; similarly once a base accepts a proton, it becomes an acid because it has the potential to reverse the process and give up a proton. An acid-base combination in which one is related to the other by gain or loss of a proton is often termed a **conjugate acid-base pair.** For example, HCl is the conjugate acid of Cl^-; likewise Cl^- is the conjugate base of HCl. NH_4^+ in giving up a proton to leave NH_3 is the conjugate acid of NH_3 or NH_3 is the conjugate base of NH_4^+.

It is evident from the discussion so far that acids and bases may be either molecular or ionic. From the equations above we may tabulate the following:

Acids	Bases	Conjugate acid-base pair
HCl	Cl^-	HCl, Cl^-
H_3O^+	H_2O	H_3O^+, H_2O
NH_4^+	NH_3	NH_4^+, NH_3
H_2O	OH^-	H_2O, OH^-

The Lewis Theory

A concept of acids and bases more general in its application is that introduced by G. N. Lewis. According to the Lewis theory, any substance which has an unused pair of electrons (such as NH_3) may be considered a base, and any substance which may attach itself to such an available pair of electrons is classed as an acid. For example, in the reaction of an acid anhydride with a basic anhydride to form a salt,

$$CaO + SO_3 \rightarrow CaSO_4$$

$$Ca\!:\!\ddot{\underset{\cdot\cdot}{O}}\!: \ + \ \overset{:\ddot{O}:}{\underset{:\ddot{O}:}{\overset{\cdot\cdot}{S}:\ddot{O}:}} \ \rightarrow \ Ca^{2+} \left[\overset{:\ddot{O}:}{\underset{:\ddot{O}:}{:\ddot{O}:\overset{\cdot\cdot}{S}:\ddot{O}:}} \right]^{2-}$$

CaO would be considered a base and SO_3 an acid. In the Lewis theory, a base is an **electron pair donor** and an acid an **electron pair acceptor.**

If we apply the Lewis theory to the ionization of acids — for example, to

$$HCl + H_2O \rightleftharpoons H_3O^+ + Cl^-$$

we classify H_2O as a base, since it furnishes a pair of electrons to H^+.

Exercises

7.1. Classify each of the following compounds as (1) acid, (2) base, (3) salt.
(a) $Ca(OH)_2$ (d) HI (g) $CuCrO_4$
(b) $Ba(NO_3)_2$ (e) NH_4F (h) HCN
(c) $ZnSO_4$ (f) $Al(OH)_3$ (i) H_3AsO_4

7.2. Classify the following oxides as acids or basic anhydrides. (a) SO_2, (b) CaO, (c) Na_2O, (d) NO_2, (e) Al_2O_3, (f) SnO_2, (g) Cl_2O_7.

7.3. Write formulas and names for anhydrides of the following acids and bases: H_2SO_3, $Ca(OH)_2$, $HClO_4$, $NaOH$, $HAsO_2$, HNO_3, $Zn(OH)_2$, $RbOH$, $Cr(OH)_3$, $Sn(OH)_4$, $Al(OH)_3$.

7.4. Classify each of the following as (1) acid, (2) base, (3) normal salt, (4) acid salt, (5) basic salt, (6) acid anhydride, (7) basic anhydride, (8) mixed salt.
(a) $Mg(OH)_2$ (j) $MgHPO_4$ (s) Li_2O
(b) $KClO_3$ (k) $KNaSO_4$ (t) CO_2
(c) $NaNO_3$ (l) N_2O_5 (u) $MgOHNO_3$
(d) $Fe(OH)_3$ (m) $Cu_3(AsO_4)_2$ (v) Rb_2CO_3
(e) H_2SeO_4 (n) $K_2Cr_2O_7$ (w) $KMnO_4$
(f) $Ca(ClO)_2$ (o) Na_2S (x) CdO
(g) Fe_2O_3 (p) $HAsO_3$ (y) $Ca(H_2PO_4)_2$
(h) $KHSO_4$ (q) ZnO (z) $AlBr_3$
(i) SO_3 (r) $MgNH_4PO_4$

7.5. Which of the following are (1) strong acids, (2) strong bases?
(a) H_2SO_4, (b) $HClO$, (c) $NaOH$, (d) $Cu(OH)_2$, (e) HBr, (f) $Ca(OH)_2$, (g) HNO_3, (h) NH_3, (i) HCl, (j) $HC_2H_3O_2$, (k) H_2CO_3.

7.6. Name the compounds listed in the above questions: (a) 7.1, (b) 7.2, (c) 7.4, (d) 7.5.

7.7. HIO_3, H_2TeO_4, H_3AsO_4, $HClO_3$ are iodic, telluric, arsenic, and chloric acids respectively. (a) Name the following: K_3AsO_4, $HClO_2$, CaI_2, $CrAsO_3$, Na_2HAsO_4, $Mg(IO_4)_2$, Na_2TeO_3. (b) Write formulas for hypoiodous acid; calcium arsenite; hydrotelluric acid; copper(II) tellurate; iron(III) arsenate; perchloric acid.

7.8. Name: Hg_2S, HgO, $CuBr$, $CuBr_2$, $Fe(NO_3)_2$, $Fe(NO_3)_3$, $SnCl_2$, $SnCl_4$, $Ce_2(SO_4)_3$, $Ce(SO_4)_2$, P_2O_3, P_4O_{10}.

7.9. Complete and balance the following replacement and exchange reactions. Be sure the formulas of the products are correct before you balance the equations.
(a) $Mg + HCl \rightarrow$
(b) $Zn + H_2SO_4 \rightarrow$
(c) $HCl + KOH \rightarrow$
(d) $H_2SO_4 + Ca(OH)_2 \rightarrow$
(e) $HI + Ca \rightarrow$
(f) $AgNO_3 + Na_2CO_3 \rightarrow$
(g) $FeS + HCl \rightarrow$
(h) $CaCO_3 + HNO_3 \rightarrow$
(i) $Pb(NO_3)_2 + H_3AsO_4 \rightarrow$
(j) $HC_2H_3O_2 + Ba(OH)_2 \rightarrow$

7.10. Explain why each of the above reactions in question 7.9 goes to completion.

7.11. Write ionic equations for the reactions in question 7.9.
(a) $Mg + H^+ \rightarrow$
(b) $Zn + H^+ \rightarrow$
(c) $H^+ + OH^- \rightarrow$
(d) $H^+ + OH^- \rightarrow$
(e) $H^+ + Ca \rightarrow$
(f) $Ag^+ + CO_3^{2-} \rightarrow$
(g) $FeS + H^+ \rightarrow$
(h) $CaCO_3 + H^+ \rightarrow$
(i) $Pb^{2+} + H_3AsO_4 \rightarrow$
(j) $HC_2H_3O_2 + OH^- \rightarrow$

7.12. Complete and balance the following combination or decomposition equations:
(a) $Zn + O_2 \xrightarrow{heat}$
(b) $Mg + S \xrightarrow{heat}$
(c) $CaO + CO_2 \longrightarrow$
(d) $Ag_2O \xrightarrow{heat}$
(e) $SO_3 + H_2O \longrightarrow$
(f) $Na_2O + H_2O \longrightarrow$
(g) $Al + Cl_2 \xrightarrow{heat}$
(h) $H_2SO_3 \longrightarrow$

7.13. Complete and balance the equations:
(a) $LiOH + H_2SO_4 \rightarrow$
(b) $Zn + HCl \rightarrow$
(c) $CaCO_3 + HBr \rightarrow$
(d) $K_2O + H_3AsO_4 \rightarrow$
(e) $Hg + S \rightarrow$
(f) $BiOHCl_2 + HCl \rightarrow$
(g) $KI + H_3PO_4 \rightarrow$
(h) $Al(NO_3)_3 + NaOH \rightarrow$
(i) $Zn(OH)_2 + H_3PO_4 \rightarrow$
(j) $Cu_2O + HBr \rightarrow$
(k) $As_2O_5 + H_2O \rightarrow$
(l) $KHSO_4 + KOH \rightarrow$
(m) $SO_2 + H_2O \rightarrow$
(n) $CuSO_4 + Pb(C_2H_3O_2)_2 \rightarrow$
(o) $FeCl_3 + NaOH \rightarrow$
(p) $AgNO_3 + NaCl \rightarrow$
(q) $H_2SiO_3 + LiOH \rightarrow$
(r) $C + O_2 \rightarrow$
(s) $V_2O_5 + KOH \rightarrow$
(t) $SO_3 + ZnO \rightarrow$
(u) $Ag_2SO_4 + MgCl_2 \rightarrow$
(v) $BaCl_2 + (NH_4)_3PO_4 \rightarrow$
(w) $HgCl_2 + H_2S \rightarrow$
(x) $Na_2CO_3 + Pb(NO_3)_2 \rightarrow$

7.14. Write balanced equations for reaction of the following pairs of substances in aqueous solution.
(a) Potassium hydroxide and nitric acid
(b) Acetic acid and calcium hydroxide
(c) Calcium sulfide and mercuric chloride
(d) Sodium phosphate and barium chloride

7.15. Name the compounds given as reactants for each of the equations in Exercise 7.13.

7.16. Define or illustrate (a) conjugate acid, (b) conjugate base, (c) conjugate acid-base pair, (d) Lewis acid, (e) Lewis base.

7.17. Identify the Brönsted acids and bases in the following:
(a) $NH_3 + HCl \rightleftharpoons NH_4^+ + Cl^-$
(b) $NH_4^+ + OH^- \rightleftharpoons NH_3 + H_2O$
(c) $HS^- + OH^- \rightleftharpoons S^{2-} + H_2O$
(d) $H_3O^+ + HS^- \rightleftharpoons H_2S + H_2O$
(e) $HSO_4^- + CN^- \rightleftharpoons HCN + SO_4^{2-}$

7.18. From the following pick out those species which may be regarded as conjugate acid-base pairs: $HC_2H_3O_2$, H_3O^+, HCO_3^-, HSO_4^-, H_2O, OH^-, $C_2H_3O_2^-$, NH_3, CO_3^{2-}, NH_4^+, SO_4^{2-}.

7.19. Identify the Lewis acids and bases in the following equations:
(a) $H^+ + NH_3 \rightarrow NH_4^+$
(b) $BF_3 + F^- \rightarrow BF_4^-$
(c) $Cu^{2+} + 4\ NH_3 \rightarrow Cu(NH_3)_4^{2+}$
(d) $Ag^+ + 2\ CN^- \rightarrow Ag(CN)_2^-$
(e) $BaO + SO_3 \rightarrow BaSO_4$

7.20. From the standpoint of either the Brönsted or Lewis theories of acids and bases, pick out the acids and bases from the following reactions and explain the reasons for your choices:
(a) $HNO_3 + H_2O \rightarrow H_3O^+ + NO_3^-$
(b) $NH_3 + H_2O \rightarrow NH_4^+ + OH^-$
(c) $H_3O^+ + OH^- \rightarrow H_2O + H_2O$
(d) $NH_2^- + NH_4^+ \rightarrow NH_3 + NH_3$
(e) $BaO + CO_2 \rightarrow BaCO_3$
(f) $BCl_3 + NH_3 \rightarrow NH_3BCl_3$
(g) $Ag^+ + 2\ NH_3 \rightarrow Ag(NH_3)_2^+$

Supplementary Readings

7.1. *Acids and Bases*, R. S. Drago and N. A. Matwiyoff, Heath, Boston, 1968 (paper).
7.2. "Conjugate Acid-Base and Redox Theory," R. A. Pacer, *J. Chem. Educ.*, **50**, 178 (1973).
7.3. "Notes on Nomenclature" W. C. Fernelius, K. Loening, R. M. Adams, *J. Chem. Educ.*, **48**, 730, (1971).

Chapter 8

Main Energy Levels	1	2
$^{15.9994}_{8}$O	K	s^2p^4

Main Energy Level	1
$^{1.0079}_{1}$H	s^1

Oxygen and Hydrogen

Oxygen and hydrogen are among our most important elements. Oxygen is the element which man and animals selectively use in their breathing. All plant life liberates oxygen to the atmosphere as plant tissue forms from carbon dioxide and water in the air. Oxygen is found in earth rock and also makes up a high percentage of hydrospheric water. Hydrogen is perhaps the most abundant element in stars, suns, and nebulae of the universe. With carbon it is found in an everexpanding number of known organic compounds. Our most abundant compound, water, is composed of these two elements, about 89% oxygen and 11% hydrogen by weight. Each element will now be considered in some detail.

8.1   OXYGEN

Occurrence

Oxygen is by far the most abundant of the chemical elements on earth, comprising about 50% of the earth's crust, which includes the atmosphere. It occurs in both the free (elementary) and combined states. The air we breathe contains about 20% oxygen. In the combined state, oxygen appears in many thousands of compounds and is a constituent in the protoplasm of all living things.

The Importance of Oxygen

When air is taken into the lungs, oxygen passes into the blood stream, which carries this vital substance to body tissues. The food we consume combines with oxygen, and this chemical change produces the energy which maintains our body temperature and gives strength for physical activity.

Oxygen has many other functions. Coal and wood must have oxygen to burn and produce heat. Oxygen in the air causes metals to corrode, wood to rot, paints to harden, gasoline to burn, iron to rust, and many other chemical processes to take place.

The Discovery of Oxygen

Although oxygen is plentiful, it was not discovered until 1774, after many of the less common elements were well known. Joseph Priestley, an English

NOTE: In showing the atomic number and mass number of an element in conjunction with its symbol; ex. $^{24}_{12}$Mg, the subscript (12) is the atomic number and the superscript (24) is the mass number (number of protons plus neutrons). Sometimes we will use the atomic weight (see inside front cover) to the left of the symbol in place of the mass number; ex. $^{24.305}_{12}$Mg. The numbers to the right of the symbol indicate the electronic configuration of the atom. For example, in the figure for oxygen given at the head of this chapter, 2 electrons (both s electrons) are in the first energy level and 6 electrons (two s and 4 p) are in the second and outermost level. The fact that the first level is complete is indicated by K.

clergyman, is credited with its discovery.* Priestley experimented with many kinds of gases or "airs," which he obtained by heating various substances. He did not consider these gases as essentially different substances; rather, he thought of them as different kinds of air. In heating substances, Priestley made use of heat energy from the sun by using a lens to focus rays of lights on the object to be heated. When he heated mercuric oxide, he discovered that it yielded an "air" of unusual properties in which substances burned much more vigorously than in ordinary air. Although he found that combustion proceeded much more rapidly in this new "air," he did not associate it with the component of ordinary air, in which combustion of many substances was known to take place. Following are portions of Priestley's account of his experiments:

On the 1st of August, 1774, I endeavoured to extract air from mercurious calcinatus per se (mercuric oxide); and I presently found that, by means of the lens, air was expelled from it very rapidly. Having got about three or four times as much as the bulk of my materials, I admitted water to it, and found that it was not imbibed by it. But what surprised me more than I can well express, was, that a candle burned in this air with a remarkably vigorous flame. . . .

On the 8th of this month I procured a mouse, and put it into a glass vessel, containing two ounce-measures of the air from mercurious calcinatus. Had it been common air, a full-grown mouse, as this was, would have lived in it about a quarter of an hour. In this air, however, my mouse lived a full half hour; and though it was taken out seemingly dead, it appeared to have been only exceedingly chilled; for, upon being held to the fire, it presently revived, and appeared not to have received any harm from the experiment. . . .

From the greater strength and vivacity of the flame of a candle, in this pure air, it may be conjectured, that it might be peculiarly salutary to the lungs in certain morbid cases, where the common air would not be sufficient to carry off the phlogistic putrid effluvium fast enough. But perhaps, we may also infer from these experiments, that though pure dephlogisticated air might be very useful as a medicine, it might not be so proper for us in the usual healthy state of the body; for, as a candle burns out much faster in dephlogisticated air than in common air, so we might, as may be said, live out too fast, and the animal powers be too soon exhausted in this pure kind of air. A moralist, at least, may say, that the air which nature has provided for us is as good as we deserve.

My reader will not wonder, that, after having ascertained the superior goodness of dephlogisticated air by mice living in it, and the other tests above mentioned, I should have the curiosity to taste it myself. I have gratified that couriosity, by breathing it, drawing it through a glass syphon, and by this means, I reduced a large jar of it to the standard of common air. The feeling to my lungs was not sensibly different from that of common air; but I fancied that my breast felt peculiarly light and easy for some time afterward. Who can tell but that, in time this pure air may become a fashionable

---

* It is probable that oxygen was first discovered by the Swedish chemist Scheele in about 1771. His results were not made public until 1775, and Priestley's work had received considerable attention by this time.

article in luxury. Hitherto only two mice and myself have had the privilege of breathing it. . . .°

Laboratory Preparation

BY HEATING CERTAIN OXIDES   Oxygen is conveniently prepared in the laboratory by repeating Priestley's experiment of heating certain substances. Mercuric oxide, a reddish solid, readily gives up its oxygen when heated, leaving metallic mercury (Figure 8.1):

$$2 \ HgO(s) \rightarrow 2 \ Hg(l) + O_2(g)$$

mercuric   mercury   oxygen
oxide

Oxygen is diatomic, so that its formula is O_2. Silver oxide, gold oxide, and platinum oxide react similarly to yield the metal and liberate oxygen when heated. These four metals — mercury, silver, gold, and platinum — are called the **noble** metals.

    Certain other oxides and peroxides are unstable toward heat and break down to give an oxide of the metal and free oxygen; for example,

$$2 \ PbO_2(s) \rightarrow 2 \ PbO(s) + O_2(g)$$

lead   lead
dioxide   oxide

    We must not assume from the above reactions that all metallic oxides give up oxygen when heated. As a matter of fact, most of the oxides of metals are stable and undergo no change when heated.

BY HEATING CERTAIN SALTS   Potassium chlorate, a white crystalline substance composed of the elements potassium, chlorine, and oxygen, gives

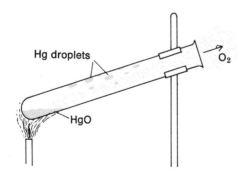

Figure 8.1
Heating mercuric oxide yields mercury and oxygen.

° Alembic Club Reprints, No. 1.

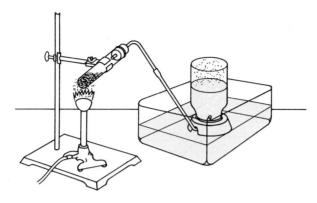

Figure 8.2

Collection of oxygen gas by displacement of water.

up all of its oxygen when heated above its melting point (368°C) according to the equation

$$2 \text{ KClO}_3(s) \rightarrow 2 \text{ KCl}(s) + 3 \text{ O}_2(g)$$
potassium   potassium
chlorate   chloride

If we wish to collect oxygen gas to study some of its properties, we may use the apparatus shown in Figure 8.2. Since oxygen is not very soluble in water, we may collect it by the displacement of water. As the gas is evolved, it displaces the water in the bottle, and we may collect a bottle full of the gas. Again it should be pointed out that not all oxygen-containing compounds give up their oxygen when heated; potassium sulfate (K_2SO_4) and silicon dioxide (SiO_2), for example are stable toward heat.

THE REACTION OF SODIUM PEROXIDE WITH WATER   Sodium peroxide, a white noncrystalline power, reacts vigorously with warm water to give oxygen and sodium hydroxide:

$$2 \text{ Na}_2\text{O}_2(s) + 2 \text{ H}_2\text{O} \rightarrow 4 \text{ NaOH}(aq) + \text{O}_2(g)$$
sodium   sodium
peroxide   hydroxide

Catalysts

The speed of many chemical reactions may be increased by the addition of certain substances. These substances do not appear to enter into the reaction —at least they are not permanently altered chemically. Such substances are called **catalysts** or **catalytic agents**. When potassium chlorate is heated, it melts to a clear liquid, from which tiny bubbles of gas may be noticed escaping. As shown on page 173, the gas is oxygen, but its evolution is slow. The addition of a small amount of manganese dioxide to the potassium chlorate greatly increases the speed at which the oxygen is evolved. That this added substance does not give up oxygen and remains unchanged in composition in the reaction may be seen by analysis of the reaction products. All of the catalyst added to the reaction mixture may be recovered and used repeatedly. Since a catalyst does not appear to enter into the reaction, its formula is usually placed over the arrow in the equation. The catalytic decomposition of potassium chlorate would be represented as

$$2 \text{ KClO}_3 \xrightarrow{\text{MnO}_2} 2 \text{ KCl} + 3 \text{ O}_2$$

A catalyst may be defined as **a substance which increases the speed of a chemical reaction but which itself remains chemically unchanged.** Catalysts are important industrially and biologically. They will be studied in more detail later.

Commercial Preparation of Oxygen

The cost of making large quantities of oxygen by the methods employed in the laboratory would be prohibitive. Instead we might naturally expect to obtain large quantities of oxygen from those sources where it occurs most extensively. Water, which is nearly 89% oxygen, is so readily available that we might expect to use it in preparing oxygen on an industrial scale. Since approximately 20% of the atmosphere is composed of oxygen, air should be a potential source, and as a matter of fact, most of the oxygen for commercial purposes is obtained from air.

FROM THE AIR   Since air is essentially a mixture of oxygen and nitrogen gases, the problem is to separate these gases. When air is cooled to a temperature of approximately $-200°C$, it becomes liquid. Actually, air is liquefied by a combination of low temperature and high pressure. If compressed air is allowed to expand rapidly through a valve, heat is absorbed and the gas is cooled. (For explanation of this see Chapter 9.) The cooled gas may be compressed and again allowed to expand to cool further. By alternate compression, cooling, and expansion of air, the temperature is continuously lowered until the air becomes a liquid. During the cooling, water vapor and carbon dioxide in the air condense to liquids and are removed.

When a mixture of liquids is heated to the boiling point, the component with the lowest boiling point boils or distills off first. Thus if an alcohol-water mixture is boiled, the first portion of the vapor will be richer in (have a higher percentage of) alcohol than the mixture from which it was distilled.

When liquid air is distilled, the first portion of vapor is richer in the lower boiling component. Liquid oxygen boils at $-183°C$, liquid nitrogen at

−195.8°C. Hence the nitrogen, with the lower boiling point, tends to distill over first, leaving behind a liquid richer in oxygen. Repetition of the distillation and liquefaction gives nearly pure oxygen. Commercially this is accomplished in fractionating columns (or towers). The oxygen gas is dried, compressed, and stored and shipped in steel bottles or cylinders.

FROM WATER   When an electric current is passed through water containing a small amount of sulfuric acid or sodium hydroxide, the water is broken down into its constituent elements, hydrogen and oxygen, which are evolved as gases:

$$2\ H_2O(l) \rightarrow 2\ H_2(g) + O_2(g)$$

This process is called **electrolysis** (Figure 8.3). Oxygen is liberated at the positive pole (anode), and hydrogen gas is liberated at the negative pole (cathode). Sulfuric acid or sodium hydroxide is used to make the solution a good conductor of electricity. Industrial oxygen is prepared by electrolysis of water containing sodium hydroxide in iron tanks. As a commercial method, electrolysis has the disadvantage of requiring large quantities of electricity, which is expensive. However, this method also produces hydrogen, a by-product which has industrial importance.

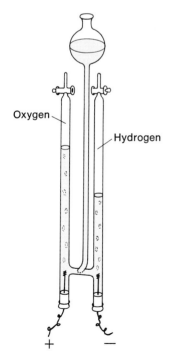

Oxygen

Hydrogen

+   −

Figure 8.3
Apparatus for the small-scale electrolysis of water.

8.2 PROPERTIES OF OXYGEN

Physical Properties

Oxygen is a colorless, odorless, and tasteless gas which may be changed to a liquid at a temperature of $-183°C$ and to a solid at $-225°C$. The gas may be liquefied at a temperature of $-118°C$ if the pressure is 50 atmospheres. Above this temperature the gas cannot be liquefied regardless of the pressure applied.

Gaseous oxygen is slightly heavier than air. It is only slightly soluble in water, but about 2% by volume dissolves at room temperature. Since oxygen gas is a little more soluble in water than nitrogen gas, dissolved air contains a slightly higher percentage of oxygen than ordinary air. Marine life and fish depend on the dissolved oxygen for respiration.

Both gaseous and liquid oxygen are paramagnetic. An electronic configuration often given for oxygen to account for this property is

$$:\overset{..}{\underset{.}{O}}:\overset{..}{\underset{.}{O}}:$$

Although this structure violates the octet rule (8 electrons in the outer configuration of each atom), it is one which follows from the molecular orbital theory which predicts three pairs of electrons in bonding orbitals and one electron in each of two antibonding orbitals, see page 114.

Chemical Properties

Many substances combine vigorously with oxygen, particularly at high temperatures. A glowing wood splint thrust into an atmosphere of oxygen bursts into flame and burns brilliantly, and this method is used as a test for oxygen gas in the laboratory. Oxygen combines slowly with most metals and nonmetals at ordinary temperatures. When iron rusts, the iron combines with oxygen present in the atmosphere to form a reddish-brown powder commonly known as iron rust. The principal chemical change is shown by the equation

$$4\ Fe(s) + 3\ O_2(g) \rightarrow 2\ Fe_2O_3(s)$$
$$\text{iron(III) oxide}$$

This reaction takes place much more rapidly in pure oxygen than in air.* Only about one-fifth of the air is made up of oxygen. Therefore, in air only

---

* The effect of the concentration of oxygen on the rate of combination of iron and oxygen may be demonstrated very convincingly by the following experiment: Steel wool heated in a flame in air burns slowly, but if the burning steel wool is placed in an atmosphere of pure oxygen, it sends off sparks of a dazzling brilliance. The reaction produces magnetic oxide of iron:

$$3\ Fe + 2\ O_2 \rightarrow Fe_3O_4$$
$$\text{magnetic}$$
$$\text{oxide of iron}$$

one-fifth of the surface of the iron is covered with oxygen molecules; the other four-fifths of the total surface is covered with nitrogen molecules. (Nitrogen, a very inactive substance, takes no part in the reaction.) Substances burn better in oxygen than in the air because (1) the concentration of oxygen is greater, and (2) an inactive substance, such as nitrogen, is not present to absorb heat and slow down the rate of reaction.

Nonmetals like sulfur, phosphorus, and carbon burn very rapidly in pure oxygen, forming **oxides** of the elements and releasing heat energy.

$$S(s) + O_2 \rightarrow SO_2(g) + 71.0 \text{ kcal/mole}$$
$$\text{sulfur dioxide}$$

$$C(s) + O_2 \rightarrow CO_2(g) + 94.1 \text{ kcal/mole}$$
$$\text{carbon dioxide}$$

$$P_4(s) + 5\ O_2 \rightarrow P_4O_{10}(s) + 720 \text{ kcal/mole}$$
$$\text{phosphorus} \qquad \text{phosphorus (V) oxide}$$

Sulfur dioxide and carbon dioxide are gases at ordinary temperatures, and phosphorus (V) oxide is a white solid.

In addition to the elementary substances, many compounds react with oxygen. Gasoline (a mixture of hydrocarbons) requires oxygen for its combustion. In the automobile carburetor, gasoline and air are mixed thoroughly and ignited in the cylinder by a spark from a spark plug. This reaction produces carbon dioxide (the same gas formed in the combustion of coal or carbon) and water. Linseed oil employed in paints and varnishes absorbs oxygen slowly to form a tough hard film which is resistant to wind and rain.

These combinations of oxygen with elements and compounds are examples of **oxidation°** reactions, and the substance which combines with the oxygen is said to be **oxidized.** Oxidation may take place at varying speeds. The formation of iron rust is normally a slow oxidation, but when iron is burned in pure oxygen, the reaction takes place much more rapidly, and heat and light are produced. An oxidation process in which heat and light are produced is called **combustion.** Usually when substances are oxidized, heat is evolved in the process. Thus when coal, gasoline, or wood are burned, the reactions produce heat. Whether a given reaction takes place rapidly or slowly, the same quantity of heat is evolved, although a much higher temperature is reached in a rapid reaction because the heat is generated in a brief period of time.

The amount of heat liberated per unit quantity of substance burned is called **heat of combustion.** For example, one gram of sulfur on burning to sulfur dioxide liberates 2200 calories; one gram of carbon burning to carbon dioxide yields 7900 calories. Heats of combustion of mixtures, such as coal, fuel oil, or foods, are called **calorific values.**

° Oxidation reactions, however are not confined to combinations with oxygen. The term "oxidation" has a much broader significance and will be discussed in more detail in Chapter 16.

Peroxides and Superoxides

With the more active metals oxygen may combine directly to form **peroxides,** which contain the peroxide ion, O_2^{2-}, for example Na_2O_2 and BaO_2. With the very active metals of group IA such as potassium and cesium, **superoxides** may be formed which contain the superoxide ion, O_2^-, for example KO_2 and CsO_2. The latter are rarely encountered, however.

Hydrogen peroxide has a tendency to decompose to water and oxygen. Various inhibitors such as acetanilide, are added to slow down the process. The decomposition is catalyzed by such substances as finely divided metals, manganese dioxide, and blood.

The mild oxidizing activity of dilute hydrogen peroxide makes it usable in bleaching such substances as feathers, wool, hair, and some silks, which would be injured by more vigorous bleaching agents. Concentrated H_2O_2 (90% or more) is produced commercially by the electrolysis of aqueous H_2SO_4 or NH_4HSO_4 solutions followed by vacuum distillation. H_2O_2 is a powerful oxidizing agent used in rocket fuels and propellants.

The structure of H_2O_2 is shown in Figure 8.4. The angle between each H atom and O—O is about 97°. The H atoms lie in planes at an angle of about 94°.

Reaction of Oxides with Water

The oxides of nonmetals, such as sulfur, carbon, and phosphorus, dissolve in water to form **acids,** a class of chemical compounds which have a sour taste and turn blue litmus paper red. (Litmus is a vegetable dye which is pink or red in acid solution.)

$$SO_2(g) + H_2O \rightarrow H_2SO_3(aq)$$
sulfur dioxide   sulfurous acid

$$CO_2(g) + H_2O \rightarrow H_2CO_3(aq)$$
carbon dioxide   carbonic acid

$$P_4O_{10}(s) + 6\ H_2O \rightarrow 4\ H_3PO_4(aq)$$
phosphorus   phosphoric acid
(V)oxide

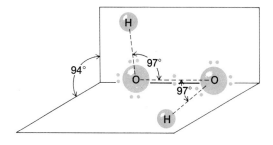

Figure 8.4
Structure of H_2O_2.

We may formulate the general rule: **If soluble, the oxide of a nonmetal reacts with water to form an acid.**

The oxides of metals, if soluble in water, form **bases,** a class of chemical compounds which are bitter in taste, have a soapy feeling, turn red litmus paper blue, and in water solution contain the hydroxide (OH^-) ion.

$$CaO(s) + H_2O \rightarrow Ca(OH)_2(aq)$$
calcium oxide   calcium hydroxide

The oxide of iron is insoluble in water, as are most of the metallic oxides. These insoluble oxides and water give no reaction to litmus paper.

Kindling Temperature

Chemical processes take place more rapidly when the temperature is raised. At ordinary temperatures iron rusts slowly, but at an elevated temperature the combination with oxygen takes place rapidly. If the temperature of a combustible substance is raised to a certain point in the presence of air or oxygen, the substance will burst into flame. The temperature at which this occurs is called the **kindling temperature** of the substance. The kindling temperature varies with the nature of the substance. This fact may be demonstrated by the following experiment. If small pieces of phosphorus, sulfur, and wood are placed in a shallow pan and the pan is heated with a burner, it will be observed that the phosphorus takes fire when the pan is barely warm, the sulfur takes fire at a much higher temperature, and at a still higher temperature the wood bursts into flame. The kindling temperature of phosphorus is lower than that of sulfur, which is lower than that of wood.

Spontaneous Combustion

We frequently hear of fires starting spontaneously. A pile of oily rags or a pile of coal may take fire spontaneously. This occurs because when the temperature of a substance has been raised to its kindling temperature, oxidation is occurring so rapidly that the substance will burst into flame.

Spontaneous combustion is the result of a cumulative process which takes place rather slowly at first. During the early stages the process is really not combustion, but a slow oxidation of the substance by the oxygen of the air. If the combustible substance is a good heat insulator—that is, if it retains most of the heat given off in the oxidation process—the combustion stage may develop. The heat which is produced as a result of the oxidation process gradually accumulates, raising the temperature of the substance to the kindling temperature, and self-combustion takes place. Two conditions, then, are essential for spontaneous combustion: (1) an existing slow oxidation with the evolution of heat, and (2) good heat insulation, or the capacity of the substance to retain a large part of the heat generated in the oxidation process.

Uses of Oxygen

Oxygen ranks 4th in the list of 50 top chemicals produced (Table 1.1), over 29 billion pounds being used annually in the U.S. Much of this use is as a cheap industrial oxidizing agent. For example, it is used in making steel to oxidize carbon, sulfur, and phosphorus impurities. These reactions are given on page 177. Oxyacetylene torches used for welding and cutting metals achieve temperatures in excess of 3000°C.

$$2 \, C_2H_2(g) + 5 \, O_2(g) \rightarrow 4 \, CO_2(g) + 2 \, H_2O(g) + 311 \text{ kcal/mole}$$

Liquid oxygen (LOX) mixed with organic compounds produce powerful explosives.

In certain diseases, such as pneumonia, tuberculosis, gas poisoning, and other respiratory impairments, pure oxygen may be administered in an "oxygen tent." Oxygen is often mixed with various anesthetic agents, such as ethylene and nitrous oxide. Flights into the stratosphere, where the concentration of oxygen is low, necessitate the use of oxygen for the respiratory needs of the passengers. Consequently, airplanes carry tanks of oxygen. Oxygen is also used in pulmotors, basal metabolism machines, submarines, and diving bells.

The reaction of oxygen with hydrogen

$$O_2(g) + 2 \, H_2(g) \rightarrow 2 \, H_2O(g) + 57.8 \text{ kcal/mole}$$

is extremely energetic and finds a number of uses. Liquid oxygen mixed with liquid hydrogen is used as a powerful rocket propellant for space rockets. The hydrogen-oxygen torch, similar to the oxyacetylene torch, is used for welding. The hydrogen-oxygen fuel cell is a modern battery used in space capsules (Chap. 17).

Ozone

If a high electrical charge is passed through oxygen gas at low temperatures, a gas with a peculiar, generator-room odor is formed.

$$3 \, O_2 \xrightarrow[\text{discharge}]{\text{electric}} 2 \, O_3$$

The odor of ozone may be detected at a concentration as low as 1 part of ozone in 100,000,000 parts of air. The composition of the substance has been proved to be that represented by the formula O_3, and it has been given the name **ozone,** from the Greek "to smell." Since its boiling point is -112°C, which is markedly higher than that for oxygen (-183°C), ozone may be condensed as a liquid from oxygen gas containing small percentages of ozone. Ozone is an **allotrope** of oxygen (see Chapter 21).

In nature O_3 is formed by lightning flashes and by the action of ultraviolet light and cosmic rays from outer space on oxygen gas in the stratosphere. It is believed that the stratospheric ozone screens the earth from too much of the sun's ultraviolet light reaching the surface of the earth where biological damage could occur, such as skin cancer. There is concern that the exhaust gases of high flying supersonic transports (SST) would react with the O_3 in the stratosphere, perhaps upsetting the delicate ozone balance in nature and removing some of the ozone shield. Similarly, there is some evidence that in the widespread use of aerosol sprays, such as hair sprays where the aerosol is a freon (CCl_2F_2, etc.), some of the freon eventually reaches the upper atmosphere where ultraviolet light breaks it down into very reactive substances which will react with stratospheric ozone, thus causing a reduction in the ozone shield.

On the other hand too much O_3 in the lower atmosphere can cause problems. Big city smog caused by automobile exhaust pollution consists in part of an excess of ozone. It seems that NO_2 from the exhaust reacts with ultraviolet light to give NO and O. The atomic oxygen is very reactive and combines with O_2 to give O_3. Eye irritation, respiratory problems, cracking of rubber products, and vegetation damage may result.

8.3 HYDROGEN

Early History

Henry Cavendish, who noted that a flammable "air" was produced from the action of metals with acids, is given credit for the discovery of hydrogen in 1776. Water was shown to be a product of its combustion in ordinary air. Lavoisier gave this new flammable gas the name *hydrogen,* which signified "water former."

Occurrence

Hydrogen stands ninth in abundance among the chemical elements, making up approximately one percent of the earth's crust. Unlike oxygen, it is seldom found in the free state (uncombined with other elements). It is found free in the atmosphere, however, in very small amounts (about 0.01 percent), and in slightly larger amounts in gases issuing from volcanoes. Spectroscopic studies indicate that large amounts of free hydrogen are present in gases surrounding the sun and many other stars.

In combination with other elements, hydrogen is widely distributed. Water is about one-ninth hydrogen by weight. Compounds of organic origin (those compounds associated with living organisms or their products), such as sugar, food products, fats, proteins, and oils, contain hydrogen. Combined with carbon, hydrogen is found in natural gas, kerosene, gasoline, and other petroleum products. It is also a constituent of all acids and hydroxides (bases), two classes of compounds which we have already discussed.

If we could count the atoms of the various elements present in the earth's crust, we would find that hydrogen atoms are exceeded in number only by atoms of oxygen, but since hydrogen is the lightest of the elements—atomic weight of 1.008—the proportion of hydrogen by weight in the earth's crust is relatively small.

Action of Hydrochloric Acid on Metals

Certain metals have the property of displacing hydrogen from acids. When hydrochloric acid (HCl) is added to granulated zinc, hydrogen gas is evolved at a rapid rate:

$$Zn(s) + 2\ HCl(aq) \rightarrow ZnCl_2(aq) + H_2(g)$$

hydrochloric   zinc
acid   chloride

In this instance the metal zinc has displaced or replaced the hydrogen in the acid. The hydrogen is set free, and the zinc combines with the chlorine to form zinc chloride. As explained in Chapter 7, a compound formed by the replacement of the hydrogen of an acid with a metal is termed a **salt**. Thus zinc chloride is a salt of hydrochloric acid, since it is derived from that acid by replacement of hydrogen. It may be noted that zinc chloride is composed of the metal zinc and the nonmetal chlorine.

Aluminum reacts with hydrochloric acid to form aluminum chloride and hydrogen:

$$2\ Al(s) + 6\ HCl(aq) \rightarrow 2\ AlCl_3(aq) + 3\ H_2(g)$$

aluminum chloride

The action, once started, is more rapid than with zinc. Aluminum chloride is a salt of hydrochloric acid, since it is derived from the acid by replacement. (All salts of hydrochloric acid are chlorides.)

The Activity Series

While many metals will displace hydrogen from acids, some—for example, copper, silver, mercury, gold—will not. The displacement of hydrogen in an acid by a metal may be regarded as taking place because the metal is more active than hydrogen—thus zinc and aluminum (see equations above) are more active than hydrogen. On the other hand, copper, silver, mercury, and gold will not displace hydrogen from acids, and are regarded as less active than hydrogen.

We may determine the relative activity of metals by a series of further

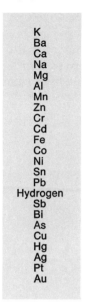

K
Ba
Ca
Na
Mg
Al
Mn
Zn
Cr
Cd
Fe
Co
Ni
Sn
Pb
Hydrogen
Sb
Bi
As
Cu
Hg
Ag
Pt
Au

Figure 8.5
Activity series.

displacement reactions. For example, a strip of zinc placed in a solution of a copper salt soon becomes coated with metallic copper:

$$Zn(s) + CuSO_4(aq) \rightarrow ZnSO_4(aq) + Cu(s)$$

On the other hand, Zn will not displace Mg from a solution of $MgSO_4$. Thus Zn is more active than Cu, but less active than magnesium.

On the basis of such series of displacement reactions, metals may be arranged in an activity series, as shown in Figure 8.5. (This list includes only the more common metals.) Hydrogen is placed in the list to divide those metals which displace hydrogen from acids from those which do not. The metals appearing above hydrogen in this series will displace hydrogen from an acid; those below hydrogen will not react with an acid to produce free hydrogen. The metals are arranged in order of decreasing activity, with the most active metal at the top. Any metal will displace a metal below it (a less active metal) from its salt in water solution.

This activity series* is sometimes referred to as the **electromotive series of metals,** since the position of the metals in the list may be determined from electromotive force measurements. (See Chap. 17.)

° The position of elements will be altered to some extent by such factors as temperature, kind of solution in contact with metal, concentrations, and secondary action of metal ions with solvent. Certain replacements which take place at one temperature and concentration may be reversed to some extent at other temperatures and concentrations; for example, hot hydrogen gas passed over iron oxide replaces iron; sodium when placed in water acts much more vigorously than calcium.

Action of Other Acids with Metals

The action of other acids on metals in most cases is similar to that of hydrochloric acid. For example, metals above hydrogen in the activity series displace the hydrogen from sulfuric acid to form a **sulfate** salt and free hydrogen:

$$Zn(s) + H_2SO_4(aq) \rightarrow ZnSO_4(aq) + H_2(g)$$

$$\text{sulfuric acid} \qquad \text{zinc sulfate}$$

$$Fe(s) + H_2SO_4(aq) \rightarrow FeSO_4(aq) + H_2(g)$$

$$\text{iron(II) sulfate}$$

Salts of sulfuric acid are called **sulfates**; $ZnSO_4$, then, is zinc sulfate, and $FeSO_4$ is iron(II) sulfate.

Phosphoric acid (H_3PO_4) acts similarly with metals above hydrogen to produce hydrogen and *phosphates:*

$$3\ Zn(s) + 2\ H_3PO_4(aq) \rightarrow Zn_3(PO_4)_2(s) + 3\ H_2(g)$$

$$\text{zinc phosphate}$$

$$2\ Al(s) + 2\ H_3PO_4(aq) \rightarrow 2\ AlPO_4(s) + 3\ H_2(g)$$

$$\text{aluminum phosphate}$$

The action of nitric acid (HNO_3) on metals is somewhat different from that of most of the other acids, because nitric acid possesses oxidizing properties in addition to acid properties. Nitric acid dissolves many of the metals appearing below hydrogen in the activity series, and its action on metals above hydrogen in the series produces substances other than free hydrogen. (These reactions will be discussed in Chapter 20.)

In general, if the acid exhibits only acid characteristics, it will act on metals appearing above hydrogen in the activity series to liberate free hydrogen. We may formulate the general rule

$$\text{Metal above hydrogen} + \text{Acid} \rightarrow \text{Salt} + \text{Hydrogen gas}$$

8.4 THE PREPARATION OF HYDROGEN

Laboratory Preparation of Hydrogen

Frequently hydrogen is prepared in laboratory quantities by the action of sulfuric acid on zinc. The apparatus shown in Figure 8.6 may be used. The acid is added through the thistle tube, and it reacts with the zinc in the flask. The hydrogen gas is collected by displacement of water, in which it is relatively insoluble.

Action of Water on Metals

The metals near the top of the activity series are sufficiently active to displace hydrogen from water. If a small piece of sodium metal is dropped into

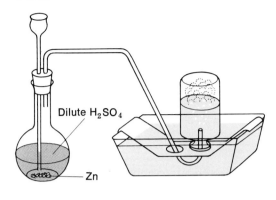

Figure 8.6
Apparatus for generating hydrogen.

water, a very vigorous reaction ensues, and hydrogen gas is liberated. (This experiment must be carried out with a great deal of caution, since the action may disperse sodium or sodium hydroxide.)

Sodium displaces only one hydrogen atom in the water molecule, forming sodium hydroxide and hydrogen. Sodium hydroxide is a base and turns red litmus blue:

$$2 \ Na(s) + 2 \ H_2O \rightarrow 2 \ NaOH(aq) + H_2(g)$$
sodium hydroxide

Similarly, potassium, barium, and calcium displace hydrogen from water. In each case the metal hydroxide (base) and hydrogen are produced. These first metals in the activity series are sufficiently active to displace hydrogen from ice water:

$$\text{Metal (K, Ba, Ca, Na)} + H_2O \rightarrow \text{Metal hydroxide} + H_2(g)$$

The next nine metals in the activity series — magnesium, aluminum, manganese, zinc, chromium, cadmium, iron, cobalt, and nickel — while not active enough to displace hydrogen from cold water, will act with water in the form of steam. These metals displace all of the hydrogen in the water molecule, yield the oxide of the metal used, and free hydrogen:

$$\text{Metal} + H_2O(g) \rightarrow \text{Metal oxide} + H_2(g)$$

Lead and tin, which follow in the series, are too inactive to act with water at any temperature to any noticeable degree.

Commercial Preparation of Hydrogen

Most of the hydrogen used commercially is produced from natural gas, CH_4 (or other hydrocarbon fuels such as propane, C_3H_8), and steam at elevated temperatures. A catalyst is used. The reaction is:

$$CH_4(g) + 2 \ H_2O(g) \xrightarrow{900°C} 4 \ H_2(g) + CO_2(g)$$

The carbon dioxide is separated from the hydrogen by absorption in an alkaline solution, leaving relatively pure hydrogen. Other methods, producing much smaller amounts are:

(a) Electrolysis of water, described on page 175. (b) When steam is passed over carbon which has been heated to a high temperature (1000°C), carbon monoxide and hydrogen gases are produced:

$$C(s) + H_2O(g) \rightarrow CO(g) + H_2(g)$$
$$\text{carbon}$$
$$\text{monoxide}$$

Since both carbon monoxide and hydrogen are combustible, the mixture has fuel value, and in many cases no attempt is made to separate the products. The mixture has been given the trade name "water gas."

Hydrogen may be separated from the mixture by cooling it to a low temperature and liquefying the carbon monoxide. The hydrogen remains in the gaseous state.

The mixture of the two gases may be treated with steam in the presence of a catalyst to oxidize the carbon monoxide to carbon dioxide and yield more hydrogen:

$$CO(g) + H_2O(g) \rightarrow CO_2(g) + H_2(g)$$

(c) If steam is passed over red hot iron filings in a tube (Figure 8.7a),

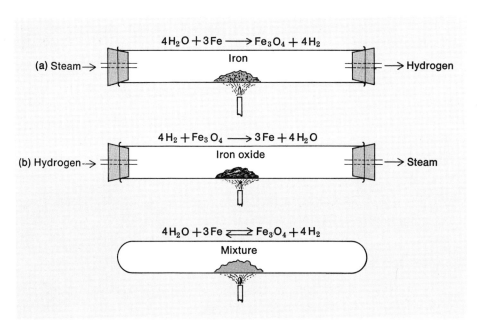

Figure 8.7

A reversible reaction.

hydrogen gas is produced and may be removed from the opposite end of this tube:

$$3 \text{ Fe(s)} + 4 \text{ H}_2\text{O(g)} \rightarrow \text{Fe}_3\text{O}_4\text{(s)} + 4 \text{ H}_2\text{(g)} \tag{1}$$
$$\text{iron oxide}$$

This reaction is **reversible** – that is, it may be made to move in the opposite direction. For example, if hydrogen gas is passed over heated iron oxide (Figure 8.7b), metallic iron and steam are produced:

$$\text{Fe}_3\text{O}_4\text{(s)} + 4 \text{ H}_2\text{(g)} \rightarrow 3 \text{ Fe(s)} + 4 \text{ H}_2\text{O(g)} \tag{2}$$

The latter is exactly the reverse of reaction (1). We shall find that many chemical reactions are reversible under certain conditions – that is, they may be made to proceed in *either* direction (see Chapter 14).

8.5   PROPERTIES OF HYDROGEN

Physical Properties

Hydrogen is a colorless, odorless gas which is relatively insoluble in water; 100 ml of water dissolves about 2 ml of the gas under ordinary conditions of temperature and pressure. It is the lightest known substance, and therefore has the smallest density. Hydrogen gas can be liquefied and solidified only with considerable difficulty; the gas liquefies at temperatures below $-253°C$ and solidifies at $-259°C$.

Certain metals, such as platinum and palladium, are able to adsorb large volumes of hydrogen. If the metal is finely divided, thus increasing its surface area, more gas will be adsorbed. These metals often adsorb many times their own volume of the gas. The hydrogen thus obtained appears to be more active than ordinary hydrogen in many reactions, and is referred to as **activated hydrogen gas.** Platinum is used as a catalyst for many reactions in which hydrogen is one of the reacting substances on the basis of this property.

Chemical Properties

A mixture of hydrogen and oxygen gases may be heated moderately without any apparent action taking place. If the mixture is ignited by a spark or flame, however, combination takes place with explosive violence to form water:

$$2 \text{ H}_2\text{(g)} + \text{O}_2\text{(g)} \rightarrow 2 \text{ H}_2\text{O(g)}$$

This action may be used as a basis for testing the purity of a sample of hydrogen. A small amount of the gas is ignited in a test tube; if a slight explosion occurs, the gas contains air or oxygen and is therefore impure. Pure hydrogen burns quietly with an almost colorless flame and does not support combustion.

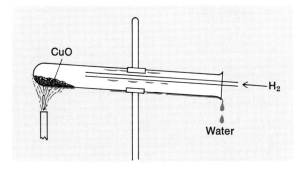

Figure 8.8

Reduction of copper oxide by hydrogen gas.

When hydrogen gas is passed over heated copper oxide in a test tube, the copper oxide gradually changes from black to a copper-colored powder (Figure 8.8). Small drops of liquid condense on the sides of the test tube. The reddish deposit remaining in the test tube is metallic copper, and the liquid formed on the sides of the tube is water:

$$CuO(s) + H_2(g) \rightarrow Cu(s) + H_2O$$

This reaction is another example of displacement, since the hydrogen has taken the place of the copper in the copper oxide. In such a chemical change, hydrogen is sometimes referred to as a **reducing agent,** since the compound has been reduced or changed to an element. Oxygen or those substances which give up oxygen in a chemical change are referred to as **oxidizing agents;** copper oxide in the above reaction is termed the oxidizing agent. The copper oxide may be said to be reduced and the hydrogen to be oxidized; and the entire process may be referred to as an oxidation-reduction reaction. Oxidation and reduction are simultaneous processes; one never takes place without the other. (The terms **oxidation** and **reduction** have a much broader significance than indicated above, and the subject will be treated in more detail in Chapter 16.)

Certain other metallic oxides may be reduced with hydrogen to form the free metal. Some of the less common metals — tungsten and molybdenum, for example — are prepared in a very pure condition in this manner. In general, the cost of production with hydrogen as a reducing agent is relatively high, and cheaper reducing agents are employed in the commercial preparation of most common metals.

Hydrides

Hydrogen will react directly with several of the more active metals to form hydrides, for example:

$$2 \text{ Na(s)} + \text{H}_2(\text{g}) \rightarrow 2 \text{ NaH(s)}$$
sodium hydride

$$\text{Ca(s)} + \text{H}_2(\text{g}) \rightarrow \text{CaH}_2(\text{s})$$
calcium hydride

These are electrovalent compounds, salt-like in character, similar to NaCl, in which the hydride ion, H^- is present in the crystal lattice. **Lithium aluminum hydride** (LiAlH_4) and **sodium borohydride** (NaBH_4) are powerful reducing agents used in organic chemistry.

Hydrides react with water to produce free hydrogen:

$$\text{CaH}_2(\text{s}) + 2 \text{ H}_2\text{O} \rightarrow \text{Ca(OH)}_2(\text{aq}) + 2 \text{ H}_2(\text{g})$$
calcium hydroxide

Uses of Hydrogen

Certain liquid fats combine with hydrogen at high temperatures in the presence of a nickel catalyst, and in the process change from the liquid state to a solid or semisolid condition. Cottonseed, corn, and soybean oils are liquid fats which may be "hydrogenated" and used as substitutes for butter and lard, and in the production of soap.

Other uses of hydrogen include the catalytic preparation of ammonia, hydrogen chloride, and methyl alcohol (CH_3OH)—three of the 50 top chemicals. As has been mentioned for oxygen, hydrogen is also used as a rocket fuel, in hydrogen-oxygen torches, and in the $\text{H}_2\text{-O}_2$ fuel cell.

Hydrogen As a Fuel

Much research is being done on the use of hydrogen as a possible replacement of gasoline and gaseous fuels.

Hydrogen itself is suitable for use in modified automobile engines and a number of experimental cars are being developed. The $\text{H}_2\text{-O}_2$ reaction is very energetic, as has been mentioned, and the product of the reaction (H_2O) is essentially pollution free. The supply of H_2 (from coal and water) is almost unlimited, and the energy necessary for manufacture might be available from nuclear reactors, but storage, handling, transportation, and safety are major problems which must be overcome.

Exercises

8.1. Oxygen occurs principally on earth in (1) the earth's crust, (2) water, and (3) air. For each of these (a) explain whether oxygen is in the form of a compound or mixture, (b) give the approximate percent of oxygen by weight, (c) explain how O_2 gas is (or might be) prepared from it.

8.2. Complete and balance the following equations for possible laboratory preparations of $O_2(g)$:

(a) $KClO_3 \xrightarrow{\text{heat}}$

(b) $Ag_2O \xrightarrow{\text{heat}}$

(c) $BaO_2 \xrightarrow{\text{heat}}$ barium oxide + oxygen gas

(d) $KNO_3 \xrightarrow{\text{heat}}$ potassium nitrite + oxygen gas

(e) $H_2O \xrightarrow{\text{electrolysis}}$

(f) $Na_2O_2 + H_2O \rightarrow$ sodium hydroxide + oxygen gas

8.3. Name: (a) HgO, (b) $KClO_3$, (c) Na_2O_2, (d) Fe_2O_3, (e) $SnCl_2$, (f) CO, (g) NaH, (h) $LiAlH_4$.

8.4. Write formulas for: (a) hydrogen peroxide, (b) manganese dioxide, (c) potassium superoxide, (d) zinc phosphate, (e) calcium hydride, (f) calcium hydroxide, (g) hydrogen gas, (h) ozone.

8.5. (a) Write equations for the combustion (in oxygen) of P, S, C, Al, and Ca. (b) Write equations for the reaction of the combustion products in (a) with H_2O.

8.6. Write chemical equations for the complete combustion, in oxygen, of (a) C_4H_{10}, (b) C_8H_{18}, (c) C_2H_6S, (d) $C_6H_{12}O_6$.

8.7. Classify the following oxygen-containing compounds as (1) peroxides, (2) superoxides, (3) oxides: (a) H_2O, (b) CsO_2, (c) ZnO, (d) H_2O_2, (e) SiO_2, (f) Na_2O_2, (g) CO_2. What is the oxidation number of oxygen in each of the above compounds?

8.8. Define the terms: superoxide, catalyst, electrolysis, heat of combustion, exothermic, kindling temperature, combustion, allotrope.

8.9. List several important uses of (a) $O_2(g)$, (b) $H_2(g)$.

8.10. Concerning ozone: (1) How is it generated in nature? (b) Of what value to us is its presence in the stratosphere? (c) Why is too much of it in the lower atmosphere objectionable?

8.11. Select the proper formula from the list on the right which best matches each question or statement. The formulas may be used more than once.

(a) What are *two* substances which separately upon heating will produce O_2?

(b) *Two* substances which separately will react with cold water to produce H_2.

(c) A substance which acts as a catalyst in the lab preparation of O_2.

1. HgO
2. O_2
3. $KClO_3$
4. KCl
5. Na_2O_2
6. NaOH

(d) A substance which, when added to water, produces an acid.
(e) Formed when an electric discharge is passed through air.
(f) A peroxide.
(g) A substance which does not react with cold water but reacts with steam at 1000°C to produce H_2.
(h) A hydride.
(i) A substance which burns in air to produce water.
(j) A substance produced commercially in large quantities from air.
(k) Used along with acetylene in high temperature welding torches.
(l) A gas which will reduce cupric oxide, when heated, to copper metal.

7. H_2
8. Fe_2O_3
9. SO_2
10. $Ca(OH)_2$
11. O_3
12. CaH_2
13. Na
14. MnO_2
15. H_2O
16. C

8.12. Complete and balance the following equations:
(a) $Zn + H_2SO_4 \rightarrow$
(b) $Fe + O_2 \rightarrow Fe_3O_4$
(c) $CuO + H_2 \rightarrow$
(d) $Zn + CuSO_4 \rightarrow$
(e) $Mg + HC_2H_3O_2 \rightarrow$
(f) $Al + HCl \rightarrow$
(g) $Na + H_2O \rightarrow$
(h) $CaO + H_2O \rightarrow$

8.13. Write the chemical equations for the preparation of hydrogen from (a) Ca and H_2O, (b) Mg and steam, (c) Fe and steam, (d) Zn and H^+, (e) C and steam.

8.14. On the basis of the activity series, which of the following reactions would not take place?
(a) $ZnSO_4 + 2\ Ag \rightarrow Zn + Ag_2SO_4$
(b) $Ag_2SO_4 + Mg \rightarrow 2\ Ag + MgSO_4$
(c) $2\ Al + 6\ HCl \rightarrow 2\ AlCl_3 + 3\ H_2$
(d) $Cu + 2\ HCl \rightarrow CuCl_2 + H_2$
(e) $Mg + ZnSO_4 \rightarrow MgSO_4 + Zn$

8.15. What action, if any, takes place when Ba (heated if necessary) is treated with (a) steam, (b) phosphoric acid, (c) oxygen, (d) hydrogen, (e) hydrochloric acid?

8.16. List four metals which react with cold water to yield hydrogen. List two metals which will not react with H_2O but will react with HCl to give hydrogen.

8.17. Define or illustrate: (a) activity series, (b) electrolysis, (c) hydrogenation of fats, (d) reducing agent, (e) water gas, (f) oxidation, (g) reversible reaction.

8.18. Write balanced equations for reactions of Zn with each of the acids HCl, H_2SO_4, $HC_2H_3O_2$, and H_3PO_4.

8.19. Consider the preparation of oxygen from $KClO_3$:

$$2\ KClO_3 \rightarrow 2\ KCl + 3\ O_2$$

(a) How many moles of oxygen are obtained per mole of $KClO_3$ used?

(b) If 6.00 moles (735 g) of $KClO_3$ were used, what weight of oxygen in grams would be obtained?

(c) If 4.50 moles of oxygen were needed, how much $KClO_3$ in grams would be required?

(d) If MnO_2 is used to catalyze the decomposition of $KClO_3$, what quantity of the latter can be decomposed theoretically with 1 g of MnO_2?

8.20. Lead dioxide gives up one-half of its oxygen on being heated, potassium chlorate all of its oxygen, and potassium nitrate one-third of its oxygen. Which of these compounds will be the cheapest source of oxygen in the laboratory if the costs per pound are PbO_2, \$1.50; $KClO_3$, \$3.50; KNO_3, \$2.00?

8.21. LiH, a white solid, is a portable source of hydrogen, since it reacts with water according to the equation

$$LiH + H_2O \rightarrow LiOH + H_2$$

Determine the weight of LiH necessary to furnish 100 kg of H_2.

Supplementary Readings

8.1. "The Oxygen Cycle," P. Cloud and A. Gibor, *Sci. American*, Sept., 1970; p. 110.

8.2. "The Hydrogen Economy," D. P. Gregory, *Sci. American*, Jan., 1973; p. 13.

8.3. "Ozone: Properties, Toxicity. and Applications," F. Leh, *J. Chem. Educ.*, **50**, 404 (1973).

Chapter 9

The Gaseous State of Matter

Familiar Properties of Gases

Matter in the gaseous state is characterized by absence of fixed volume or shape. If a gas is placed in a closed container, it expands rapidly and becomes uniformly distributed throughout the entire space in the container. When a gas is cooled sufficiently, it becomes a liquid. Although all gases may be liquefied, some are changed to the liquid state only with a great deal of difficulty. Hydrogen and helium are the most difficult gases to liquefy, since temperatures near 0°K are necessary. In contrast, such gases as chlorine and ammonia are liquefied quite easily.

If two gases are placed in a container, each gas acts independently of the other and diffuses uniformly throughout the volume of the container.

Gases may undergo expansion and compression. Air forced into an automobile tire is compressed; when the air is allowed to escape, it expands. Gases also exert pressure, and this pressure is exerted uniformly on all sides and top and bottom of the containing vessel. It is a remarkable fact that the physical behavior of gases is, to a large degree, independent of their nature or composition; for example, all gases respond in the same manner to changes in temperature and pressure. Several gas laws have been developed to describe this physical behavior, and in this chapter these laws will be discussed and utilized.

The Measurement of Air Pressure

The pressure of the atmosphere is measured with an instrument called a **barometer.** A simple barometer can be prepared by inverting a mercury-filled piece of glass tubing, about three feet long, closed at one end, in a shallow pan containing mercury (Figure 9.1). The open end of the tube lies beneath the surface of the mercury in the pan. The mercury will sink in the tube to the level at which the weight of the column of mercury is just equal to the weight of a column of air of equal cross section above the surface of the mercury in the pan. This column of air extends many miles high. The air exerts a pressure on the surface of the mercury in the container and supports the column of mercury in the tube. The exact pressure is determined by reading the difference between two mercury levels, the one in the tube and the other in the pan. The height of the mercury column is usually expressed in millimeters. The average air pressure at sea level will support a column of mercury 760 mm in height; this pressure is termed 1 atmosphere (atm) of pressure, or **standard pressure.** A pressure of 50 atm would be a pressure 50 times as great as that exerted by the atmosphere at sea level. In scientific work, **pressure** usually refers to millimeters of mercury or **torr**°; thus a pressure of 700 mm or

---

° A torr is a pressure unit of one millimeter of mercury. The unit is in honor of Evangelista Torricelli who invented the barometer.

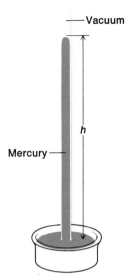

Figure 9.1
A mercury barometer.

700 torr would be a pressure equal to that exerted by a column of mercury 700 mm high. A column of water approximately 34 feet high would be supported by the atmosphere at sea level.*

The Effect of Pressure on the Volume of a Gas

When the pressure on a gas is increased, the gas is compressed — that is, an increase in pressure brings about a diminution of volume of the gas. Let us imagine a gas, such as air, enclosed in a cylinder fitted with a piston (Figure 9.2). We shall assume that the piston is weightless and frictionless. The pressure exerted on the gas will be adjusted by means of weights placed on the top side of the piston. Suppose that there are 100 ml of gas in the cylinder with a certain weight on the piston; if the weight on the piston is doubled, assuming that the temperature remains constant, the piston will move downward until the volume of gas is 50 ml. If the pressure is doubled, the volume is halved; if the pressure is increased threefold, the volume becomes one-third of the original volume, and so on.

---

* The SI derived unit of pressure is **newtons (N)** (after Isaac Newton) per square meter, for which a term **pascal (Pa)**, after the French scientist Blaise Pascal, has been proposed. A newton is defined as the force necessary to accelerate a mass of 1 kg by 1 ms^{-2}. $\left(\text{Thus one newton} = \dfrac{1 \text{ kg-m}}{s^2}.\right)$ One atmosphere = 1.01325×10^5 Pa or 1.01325 N/m^2.

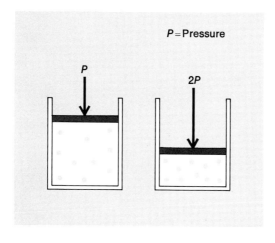

Figure 9.2
If the pressure on a gas is doubled (at a constant temperature), the volume is halved (Boyle's law).

9.1   THE GAS LAWS

The relationship between the volume of a gas and the pressure was first stated by Robert Boyle in 1662 as follows: **The volume of a given mass of gas at constant temperature varies inversely as the pressure.** This relationship, which is virtually independent of the nature of the gas, is known as **Boyle's Law.**

Boyle's law may be expressed mathematically:

$$V \propto \frac{1}{P}$$ where $\propto$ means proportional

or $PV = k$ where k is a constant

or $P_1V_1 = P_2V_2$ where V_1 is the volume at pressure P_1 and   (1)
V_2 is the volume at pressure P_2

Plots of Boyle's law are shown in Figure 9.3. Curve (a) is the P-V plot of the hyperbola defined by the equation $PV = k$, while (b) is a plot of V versus the reciprocal of the pressure — $V = k\left(\frac{1}{P}\right)$, which is a straight line.

By means of Boyle's law we may calculate the volume of a gas at any pressure, provided we know the volume at a given pressure.

EXAMPLE 9.1   100 ml of gas are enclosed in the cylinder (Figure 9.2) under a pressure of 760 torr. What would the volume be at a pressure of 1520 torr?

SOLUTION: It is immediately evident that the corrected volume will be less than 100 ml, since the pressure has been increased from 760 to 1520 torr,

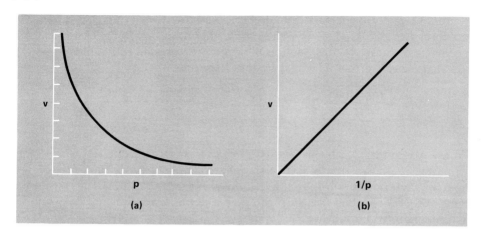

Figure 9.3

P-V and $\left(\dfrac{1}{P}\right)$ − V curves for a gas (Boyle's law).

and an increase in pressure results in a smaller volume. The volume will decrease in the same ratio that the pressure increases. The new volume then must be

$$100 \text{ ml} \times \frac{760 \text{ torr}}{1520 \text{ torr}} = 50 \text{ ml}$$

In the above calculation we have multiplied the volume by a fraction less than one—that is, $\frac{760}{1520}$. It is evident that had we multiplied by a fraction greater than one, the answer would have been greater than 100, which we know must be incorrect.

EXAMPLE 9.2   100 ml of gas at a pressure of 1000 torr will occupy what volume at standard pressure (760 torr)?

SOLUTION: We reason that since the pressure is decreased (from 1000 torr to 760 torr), the volume must be *increased*. Consequently, the new volume will be greater than 100 ml, and we must multiply the original volume by a fraction greater than one. We have two possible ratios of pressures, $\frac{760}{1000}$ or $\frac{1000}{760}$. Since the ratio to be used must be greater than one, we use the latter ratio, thus

$$\text{New volume} = 100 \text{ ml} \times \frac{1000 \text{ torr}}{760 \text{ torr}} = 132 \text{ ml}$$

In computing a new volume, determine whether or not this new volume will be greater or less than the original volume. If the pressure is increased, the volume must be less; if the pressure decreases, the volume must be greater. Then multiply the original volume by the fraction that is a ratio of the two pressures involved, so that the required condition will be satisfied. If the volume is to increase, the fraction must be greater than unity. If the volume is to decrease, the fraction must be less than one.

The Effect of Temperature on the Volume of a Gas

We are all familiar with the fact that a gas expands as the temperature rises. If the gas is placed in a closed container, expansion cannot take place, and the rise in temperature brings about an increase in pressure. If a gas is allowed to expand at a constant pressure, we note a definite relationship between the volume change and temperature change.

Let us again make use of the cylinder and movable piston shown in Figure 9.2. Assume that the volume of a given mass of gas in the cylinder is 273 ml at a given pressure and a temperature of 0°C. Now, keeping the pressure constant (by using a constant weight on the piston), let us heat the cylinder of gas. The piston will move upward, and for every centigrade degree rise in temperature, the volume will increase 1 ml. At a temperature of 1°C, then, the volume will be 274 ml; at a temperature of 10°C the volume will be 283 ml; at 273° the volume will be 546 ml, or *double* that at 0°C (Figure 9.4). If the temperature is decreased to 1° below 0°C — that is, −1°C — the volume decreases to 272 ml. A gas, then, at 0°C contracts $\frac{1}{273}$ of its volume for every degree decrease in temperature, or expands $\frac{1}{273}$ of its volume for every degree increase in temperature.

Theoretically, the volume of a gas would decrease to zero at −273°C, which is called absolute zero, and is the basis of the absolute scale of temperature explained in Chapter 1. All substances liquefy or solidify before the absolute zero temperature is reached, however. If the temperature is expressed as absolute temperature, it is evident that **at constant pressure the volume of a given mass of gas is directly proportional to the absolute temperature.** This is a statement of **Charles' Law.**

In accordance with Charles' law, we may write

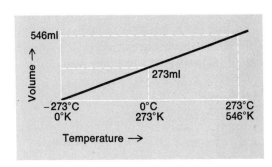

Figure 9.4

If the absolute temperature is doubled, the volume of gas is doubled (constant pressure).

$$V \propto T \quad \text{or} \quad \frac{V}{T} = k \quad \text{or} \quad \frac{V_1}{T_1} = \frac{V_2}{T_2} \tag{2}$$

where V_1 is the volume at absolute temperature T_1, and V_2 is the volume at absolute temperature T_2.

With Charles' law we can calculate the volume of a gas at any temperature if we know its volume at any given temperature, assuming that the pressure remains constant.

EXAMPLE 9.3   A certain gas occupies a volume of 100 ml at a temperature of 20°C. What will its volume be at 10°C if the pressure remains constant?

SOLUTION: First we must change these centigrade temperatures to absolute temperatures, since Charles' law states that the volume is directly proportional to the *absolute* temperature. We learned in Chapter 1 that centigrade temperature may be changed to absolute temperature by adding 273. Then

$$20°C = 20 + 273 = 293°K$$

and

$$10°C = 10 + 273 = 283°K$$

The temperature is decreased (from 293°K to 283°K), and since decrease in temperature brings about decrease in volume, the corrected volume will be less than 100 ml, and the correction factor must be less than unity. The two possible ratios of temperatures are $\frac{293}{283}$ and $\frac{283}{293}$. We must use the latter ratio, since this is the one which will give us a volume of less than 100 ml. Thus

$$\text{New Volume} = 100 \text{ ml} \times \frac{283°K}{293°K} = 96.6 \text{ ml}$$

EXAMPLE 9.4   The volume of a gas at 20°C is 100 ml. What is the volume at 100°C?

SOLUTION: Again we first convert centigrade temperatures to absolute temperatures:

$$20°C = \ \ 20 + 273 = 293°K$$
$$100°C = 100 + 273 = 373°K$$

Since the temperature is increasing, the volume must increase. If the volume is to increase, we must multiply the original volume by a fraction greater than unity — that is, $\frac{373}{293}$. Hence

$$\text{New Volume} = 100 \text{ ml} \times \frac{373°K}{293°K} = 127 \text{ ml}$$

In solving these problems we always multiply the original volume by a fraction greater or less than unity. This fraction is a ratio of the absolute temperatures involved. We reason that if the temperature is decreased, the volume must decrease; consequently, we must use the ratio less than one. If the temperature is increased, the volume must increase, and so we use the fraction greater than unity.

The Combined Gas Laws

We have learned from Boyle's law that the volume of a gas at constant temperature is inversely proportional to the pressure, and from Charles' law that the volume at constant pressure is directly proportional to the absolute temperature. The two laws may be combined and stated as follows: **For a given mass of gas, the volume is inversely proportional to the pressure and directly proportional to the absolute temperature.**

The general formula for the combined gas laws may be written

$$\frac{P_1 V_1}{T_1} = \frac{P_2 V_2}{T_2} \tag{3}$$

in which P_1 and T_1 are pressure and absolute temperature of the gas at volume V_1; and P_2 and T_2 are pressure and absolute temperature at volume V_2.

It is obvious that if five of the six variables in formulation (3) are fixed, the sixth can be calculated.

It should be noted that if $V_1 = V_2$ (volume constant), it follows from the formula that **pressure is directly proportional to absolute temperature.**

EXAMPLE 9.5   A gas occupies a volume of 1.0 l at a temperature of 27°C and 500 torr pressure. Calculate the volume of the gas if the temperature is changed to 60°C and the pressure to 700 torr.

SOLUTION: First consider the volume change with change of pressure. The pressure is increased from 500 torr to 700 torr. An increase in pressure decreases the volume, so that we must multiply the original volume by a ratio less than one, $\frac{500}{700}$. Now consider the change in volume with change in temperature. The temperature is increased from 27°C (300°K) to 60°C (333°K). Since increase in temperature causes an increase in volume, the corrected volume must be greater than one liter. The original volume must then be multiplied by a ratio greater than unity, $\frac{333}{300}$. Correcting for changes in both temperature and pressure, we have

$$\text{New Volume} = 1.0 \text{ l} \times \frac{500 \text{ torr}}{700 \text{ torr}} \times \frac{333°\text{K}}{300°\text{K}} = 0.79 \text{ l}$$

EXAMPLE 9.6   Calculate the pressure required to compress 2.00 l of a gas at 700 torr pressure and 20°C into a container of 0.100-l capacity at a temperature of −150°C.

SOLUTION: We may reason this way: The final pressure must be the original pressure corrected for the volume change and the temperature change. Since a decrease in volume means an increase in pressure, the original pressure must be multiplied by the ratio of volumes greater than one, 2.00 l/0.100 l. The temperature is decreasing from 293°K (20°C) to 123°K (−150°C), and since pressure decreases with decreasing temperature, the original pressure must be multiplied by a ratio less than one, $\frac{123}{293}$. Hence.

$$\text{New Pressure} = 700 \text{ torr} \times \frac{2.00 \text{ l}}{0.100 \text{ l}} \times \frac{123°}{293°} = 5880 \text{ torr}$$

Of course, we might substitute directly into the formula developed above to arrive at the same result.

$$\frac{P_1 V_1}{T_1} = \frac{P_2 V_2}{T_2}$$

Original Conditions

$P_1 = 700 \text{ torr}$
$V_1 = 2.00 \text{ l}$
$T_1 = 20° + 273° = 293°K$

Final Conditions

$P_2 = ?$
$V_2 = 0.100 \text{ l}$
$T_2 = -150°C + 273° = 123°K$

Substituting:

$$\frac{(700 \text{ torr})(2.00 \text{ l})}{293°K} = \frac{(P_2)(0.100 \text{ l})}{123°K}$$

$$P_2 = \frac{(700 \text{ torr})(2.00 \text{ l})(123°K)}{(0.100 \text{ l})(293°K)} = 5880 \text{ torr}$$

EXAMPLE 9.7 750 ml of gas at 300 torr pressure and 50°C is heated until the volume of gas is 2000 ml at a pressure of 700 torr. What is the final temperature of the gas?

SOLUTION: In this case, since volume is directly proportional to absolute temperature, the original temperature must be multiplied by the larger ratio, $\frac{2000}{750}$. Pressure is directly proportional to absolute temperature, so that the original temperature must be multiplied by the ratio $\frac{700}{300}$. Hence,

$$\text{New Temperature} = 323°K \times \frac{2000 \text{ ml}}{750 \text{ ml}} \times \frac{700 \text{ torr}}{300 \text{ torr}} = 2010°K$$

$$\text{Centigrade temperature} = 2010° - 273° = 1737°C$$

Or, substituting directly into the formula:

Original Conditions

$P_1 = 300 \text{ torr}$
$V_1 = 750 \text{ ml}$
$T_1 = 50°C + 273° = 323°K$

Final Conditions

$P_2 = 700 \text{ torr}$
$V_2 = 2000 \text{ ml}$
$T_2 = ?$

$$\frac{(300 \text{ torr})(750 \text{ ml})}{323°\text{K}} = \frac{(700 \text{ torr})(2000 \text{ ml})}{T_2}$$

$$T_2 = \frac{(700 \text{ torr})(2000 \text{ ml})(323°\text{K})}{(300 \text{ torr})(750 \text{ ml})} = 2010°\text{K}$$

It is convenient to use the general gas law formula for all calculations dealing with volume, temperature, and pressure changes of gases. If the temperature is constant, the T_1 and T_2 terms drop out of the equation, which then becomes the formulation for Boyle's law. If the pressure is constant, the P_1 and P_2 terms drop out to give the equation for Charles' law. Not the least of the advantages of the general formula is that it makes it unnecessary to remember three separate formulas. Let us now use the general formula for solving another problem.

EXAMPLE 9.8   A steel cylinder containing 10 l of gas at a pressure of 4.0 atmospheres and a temperature of 40°C is heated to 70°C. What is the pressure of the gas at the higher temperature?

SOLUTION:

Original Conditions	*Final Conditions*
$P_1 = 4.0$ atm	$P_2 = ?$
$V_1 = 10$ l	$V_2 = 10$ l
$T_1 = 40°\text{C} + 273° = 313°\text{K}$	$T_2 = 70°\text{C} + 273° = 343°\text{K}$

Since the volume is constant, the general formulation becomes

$$\frac{P_1}{T_1} = \frac{P_2}{T_2}$$

Substituting:

$$\frac{4.0 \text{ atm}}{313°\text{K}} = \frac{P_2}{343°\text{K}}$$

$$P_2 = \frac{(4.0 \text{ atm})(343°\text{K})}{313°\text{K}} = 4.4 \text{ atm}$$

Standard Temperature and Pressure (STP)

Since the volume of a given mass of gas changes with temperature and pressure, the volume is not defined unless the temperature and pressure of the gas are given. To get a comparison of the weights of equal volumes of gases, chemists refer all gas volumes to a standard set of conditions of temperature and pressure. We have already indicated that the standard of pressure is the average pressure of the atmosphere at sea level—760 torr, or 1 atm. The standard temperature has been selected as the melting point of ice—0°C, or 273°K. Unless otherwise specified, when referring to gas volumes we shall assume standard conditions of temperature and pressure (abbreviated STP).

The General Gas Law Equation

From Boyle's and Charles' laws we may write

$$V \propto \frac{T}{P}$$

or $PV = kT$, where k is a constant. We may evaluate the constant k by substituting known values in the equation. Experimentally, 1 mole of a gas occupies a volume of 22.4 l at 1 atm pressure and 273°K.

$$(1 \text{ atm})(22.4 \text{ l}) = k \ (273°K)$$

$$k = \frac{(1 \text{ atm})(22.4 \text{ l})}{273°K} = 0.0821 \text{ l-atm per °K mole}$$

This constant is usually designated as "R" and is termed the molar gas constant.

The general form of the equation (commonly called the equation of state) is

$$PV = nRT \quad \text{or} \quad PV = \left(\frac{g}{M}\right) RT \tag{4}$$

where P is the pressure in *atmospheres*, V is the volume in *liters*, n is the number of moles of gas, g is the weight in grams, M is the gram molecular weight, T is in degrees Kelvin, and R has the units of liter-atmospheres per degree Kelvin per mole.

Equation (4) is extremely useful in calculations involving gases, since we may easily determine the **number of moles** of a gas present under any conditions of temperature and pressure as well as the weight of gas present or its molecular weight.

EXAMPLE 9.9   How many moles of hydrogen gas are present in a 50-l steel cylinder if the pressure is 10 atm and the temperature 27°C?

SOLUTION:

$$(50 \text{ l}) \ (10 \text{ atm}) = n(0.082 \text{ l-atm/°K mole}) \ (300°K)$$

$$n = \frac{500 \text{ l-atm}}{(0.082 \text{ l-atm/°K mole}) \ (300°K)} = 20 \text{ moles}$$

EXAMPLE 9.10   Six g of He gas (molecular weight 4.00) are pumped into a 0.50-l cylinder at 60°C. What is the pressure of gas in the cylinder?

SOLUTION:

$$P = \frac{\left(\frac{6.00 \text{ g}}{4.000\text{g/mole}}\right)(0.082 \text{ l-atm/°K mole})(333°\text{K})}{0.50 \text{ l}} = 82 \text{ atm}$$

EXAMPLE 9.11   A 250-ml sample of a gas weighs 0.542 g at 100°C and 650 torr. What is the molecular weight of the gas?

SOLUTION:

$$\left(\frac{650 \text{ torr}}{760 \text{ torr}}\right)(0.250 \text{ l}) = \left(\frac{0.542 \text{ g}}{M}\right)(0.0821 \text{ l-atm/°K mol})(373°\text{K})$$

$$M = 77.6 \text{ g/mole}$$

Deviations from the Combined Gas Laws

A gas which behaves exactly according to Boyle's and Charles' laws is called an **ideal** gas. Actually, no gas is ideal. Deviations from ideal behavior are observed because (1) the molecules themselves occupy some space, thus restricting the space in which other molecules might move; and (2) slight forces of attraction, called **van der Waals forces** (mentioned on p. 121), exist between molecules, so that they do not move completely independently of one another. The first condition tends to make the volume larger, and the second tends to make the volume smaller than that calculated for ideal behavior. As the molecules of a gas get closer together, these conditions are more pronounced, and greater deviations may be expected.

The proximity of molecules will depend on both temperature and pressure — low pressures and high temperatures will keep the molecules farther apart, and closer conformity with the gas laws will follow. Under usual laboratory working pressures and temperatures (usually standard conditions), most gases conform fairly closely to ideal behavior (usually within 1%), and the relationships between volume, pressure, and temperature as expressed in Boyle's and Charles' laws are extremely useful in converting from one set of conditions to another. Corrections for deviations from ideal behavior may be made if necessary by using a modified form of the equation of state (ideal gas law equation) called the **van der Waal's equation:**

$$\left(P + \frac{n^2 a}{V^2}\right)(V - nb) = nRT$$

where a and b are constants characteristic of the gas in question.

Dalton's Law of Partial Pressures

Each of the gases in a gaseous mixture behaves independently of the other gases and exerts its own pressure, the total pressure of the mixture being the sum of the partial pressures exerted by each gas present; that is

$$P_t = p_1 + p_2 + p_3 + \ldots\ldots\ldots \tag{5}$$

where P_t is the total pressure and p_1, p_2, etc. are partial pressures of component gases. Since the atmosphere is composed of approximately 20% oxygen and 80% nitrogen by volume, the oxygen exerts a partial pressure of about $\frac{1}{5} \times 760$ torr $= 152$ torr, and the nitrogen a partial pressure of $\frac{4}{5} \times 760$ torr $= 608$ torr.

Frequently gases in the laboratory are collected by displacement of water, as shown in Figure 9.5. The pressure of the gas inside the bottle is most conveniently measured by making it equal to the pressure of the atmosphere, which can be determined with a barometer. If the level of water inside the bottle is adjusted to the same level as the water in the container outside the bottle, the pressure inside the bottle must be equal to the pressure outside the bottle — that is, the pressure inside must be equal to the barometer reading. The pressure inside the bottle is equal to the sum of the pressure of *two* gases, the gas collected and water vapor. Above the liquid in the bottle, some of the water is in the form of vapor, the amount present depending on the temperature and the space above the liquid. At any given temperature, however, the pressure exerted by the water vapor is constant.

Since the pressure of the air, as measured by the barometer, is equal to the total pressure of the two gases inside the bottle, we may write

P_{air} = partial pressure of gas + partial pressure of water vapor

Since we are interested in the actual pressure exerted by the gas, we may transpose the above:

partial pressure of gas = P_{air} − partial pressure of water vapor

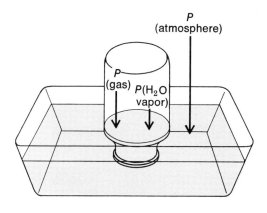

Figure 9.5
The sum of the gas and the water vapor pressures equals the atmospheric pressure.

The pressure exerted by the water vapor at the temperature of the experiment may be obtained from a standard table of vapor pressures.

EXAMPLE 9.12   Suppose 100 ml of oxygen were collected over water in the laboratory at a pressure of 700 torr and a temperature of 20°C. What would the volume of the dry oxygen gas be at STP?

SOLUTION: Vapor pressure of water at 20°C (Table 10.3, page 224) is 17.5 torr. The actual pressure of the oxygen gas is then $700 - 17.5 = 682.5$ torr. Applying the combined gas laws, we obtain

$$\text{New Volume} = 100 \text{ ml} \times \frac{682.5 \text{ torr}}{760 \text{ torr}} \times \frac{273°\text{K}}{293°\text{K}} = 83.7 \text{ ml}$$

Mole Fraction

The composition of a mixture (see Chapter 13) is often expressed as the **mole fraction** of each of the components. The mole fraction of a substance in a mixture is simply the fraction of moles of that substance present. For example, in a mixture of 2 moles of A, 5 moles of B, and 13 moles of C,

$$\text{Mole fraction of A} = \frac{2}{2 + 5 + 13} = 0.1$$

$$\text{of B} = \frac{5}{20} = 0.25$$

$$\text{of C} = \frac{13}{20} = 0.65$$

In a mixture of gases the partial pressure of each gas is proportional to its mole fraction.

EXAMPLE 9.13   What is the partial pressure of each gas in a mixture which contains 40 g He, 56 g N_2, and 16 g O_2 if the total pressure of the mixture is 5 atm?

SOLUTION:

$$\text{Moles of He} = \frac{40 \text{ g}}{4 \text{ g/mole}} = 10$$

$$\text{Moles of } N_2 = \frac{56 \text{ g}}{28 \text{ g/mole}} = 2.0$$

$$\text{Moles of } O_2 = \frac{16 \text{ g}}{32 \text{ g/mole}} = 0.50$$

$$\text{Partial pressures} - \text{He} = \frac{10}{10 + 2 + 0.5}\ (5\ \text{atm}) = 4\ \text{atm}$$

$$\text{N}_2 = \frac{2.0}{12.5}\ (5\ \text{atm}) = 0.8\ \text{atm}$$

$$\text{O}_2 = \frac{0.50}{12.5}\ (5\ \text{atm}) = 0.2\ \text{atm}$$

Effusion and diffusion of gases   If a gas is allowed to pass through a small opening, its rate of passage (effusion) depends on the mass of the molecules — that is, on the molecular weight. Heavy molecules move more slowly than light molecules. In experimenting with gases in 1832, Thomas Graham found that the **comparative rates or speeds of effusion of gases are inversely proportional to the square roots of their molecular weights.**

Expressed mathematically:

$$r \propto \frac{1}{\sqrt{M}} \quad \text{or} \quad \frac{r_1}{r_2} = \sqrt{\frac{M_2}{M_1}} \tag{6}$$

where r_1 is the rate of effusion of the gas with molecular weight of M_1 and r_2 is rate of effusion of gas with molecular weight M_2.

Examples   Oxygen gas (m.w. 32) is $\frac{32}{2} = 16$ times as heavy as hydrogen (m.w. 2). Therefore

$$\frac{\text{Speed of effusion of H}_2}{\text{Speed of effusion of O}_2} = \sqrt{\frac{32}{2}} = \sqrt{\frac{16}{1}} = \frac{4}{1}$$

An interesting lecture demonstration is to place small amounts of ammonium hydroxide and concentrated hydrochloric acid in small porcelain boats and place the boats simultaneously into opposite ends of a long large-diameter glass tube as shown in Figure 9.6.

After a short time the two gases diffusing toward each other through the air in the tube will meet and a white cloud of NH_4Cl will form at the area of contact. Approximately

$$\frac{\text{Speed of diffusion of NH}_3}{\text{Speed of diffusion of HCl}} = \sqrt{\frac{36.5}{17}} \cong \frac{6}{4} \cong \frac{3}{2}$$

The white cloud forms down the tube about $\frac{2}{3}$ of the distance toward the HCl end.

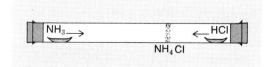

Figure 9.6
Diffusion of NH_3 and HCl through air.

9.2   THE KINETIC THEORY OF GASES

We have learned in the preceding sections that all gases behave alike under the same set of conditions. Gases may expand or be compressed; the volume of any gas is affected in the same way by changes of temperature and pressure, regardless of the nature of the gas. All gases diffuse and mix with one another in all proportions. A mixture of gases exerts a total pressure which is the sum of the partial pressures of the constituent gases, regardless of their nature. The similar behavior of all gases is explained by the kinetic theory of gases. Assumptions of the kinetic theory are

1. All gases are made up of tiny particles called molecules.
2. The molecules are small in relation to the distances between them. As molecules get closer together, attractive forces between molecules increase.
3. Gas molecules are in a constant state of motion and move in straight lines in all directions until they collide with one another or with the walls of the container.
4. Gas molecules are perfectly elastic. When they collide, they rebound with perfect elasticity and without loss of energy.
5. The velocity and kinetic energy (K.E.) of the molecules changes with the temperature, increasing with increasing temperature, and decreasing with decreasing temperatures.
6. The average energy possessed by the particles is the same, regardless of the mass. The kinetic energy (energy of motion) of the particles is proportional to the product of their mass and the square of their velocity. The particles of high mass have a lower velocity than those of low mass.

How the Theory Explains the Facts

Since gases are made up of particles which are relatively far apart, they may readily be compressed — the molecules are simply crowded closer together. Solids and liquids are more difficult to compress because of the close proximity of the molecules in the solid and liquid states.

The diffusion of gases may be explained by the motion of the particles and the relatively large distances between the particles. Since the particles of each gas are free to move about among the particles of one another, each gas is free to act relatively independently of other gases present in a mixture.

Pressure is due to the bombardment of the walls of the containing vessel with molecules of the gas. Since the molecules are moving in all directions and since, on the average, equal numbers of molecules are traveling in all directions, equal pressures are exerted on the bottom, top, and sides of the containing vessel. The combined effect of all these collisions on a unit area gives the total pressure.

If the molecules are enclosed in a smaller space (thus decreasing the volume), the collisions with the walls become more frequent, and the pressure is increased. Conversely, if the gas is expanded, fewer collisions take place, and the pressure is decreased. In other words, the volume of gas is inversely proportional to the pressure at constant temperature (Boyle's law).

If the temperature of a gas is increased, the energy of the molecules is increased and they move faster. If the volume is constant, the pressure must increase, since the number of collisions on the walls in a given time will now be increased. This explains the increase in pressure of a gas with increase in temperature at constant volume. Now, if pressure is constant when the temperature is raised, the gas must expand so that the number of collisions of the particles per unit area of the walls (pressure) will remain constant. An increase in volume with increase of temperature at constant pressure (Charles' law) is thus explained.

Since the molecules in a gaseous mixture are free to move about in an independent fashion, each exerting a pressure of its own, we would expect the total pressure of the gaseous mixture to be the sum of the pressures of all the molecular species present (Dalton's law of partial pressures).

The question might be asked, how can a light gas such as hydrogen exert as much pressure on the walls of a container as heavier oxygen gas? Since pressure is due to the sum of the magnitudes of blows of molecules on the confining walls, we are concerned with the kinetic energy of moving molecules. The energy of a moving molecule is dependent on both mass and velocity, as given by the formula

$$\text{Kinetic Energy} = \tfrac{1}{2} M v^2$$

where M is the mass of the particle and v its velocity. As indicated by Graham's law of effusion, hydrogen gas effuses 4 times as fast as oxygen, whereas their comparative masses are 2 to 32.

$$\text{K. E. of Hydrogen} = \tfrac{1}{2} \cdot 2 \cdot 4^2 = 16$$
$$\text{K. E. of Oxygen} = \tfrac{1}{2} \cdot 32 \cdot 1^2 = 16$$

The kinetic energies of hydrogen and oxygen molecules are equal. It can be shown further that at a given temperature the kinetic energy of all gas molecules is the same; the heavier gas molecules move more slowly than the lighter ones. Insofar as pressure on the walls of a container is concerned, all gas molecules behave alike and exert the same pressure.

Liquefaction of Gases

The kinetic molecular theory also explains the fact that gases may be liquefied. In slowing down molecular movement by decreased temperature and forcing molecules nearer together by increased pressure, a state is reached where intermolecular attraction (van der Waals forces) somewhat offsets the kinetic energy of particles, and the conditions of the liquid state are obtained. At still lower temperatures the kinetic energy of molecules further decreases, and the liquid changes to solid.

Both temperature and pressure, then, are factors in the condensation of a gas to liquid or, conversely, the vaporization of liquid to gas. Gaseous water at 100°C and 760 torr pressure will condense to liquid water as heat is removed. It similarly will condense to liquid if the pressure is increased. It should be emphasized that pressure must be defined to determine the condensation or boiling point. At a pressure of 3580 torr in a pressure chamber, water will boil at 150°C when heat is applied. Gaseous water will condense at 20°C under 17.4 torr pressure. (This discussion will be expanded in Chapter 10.)

Oxygen and hydrogen gases may be condensed to liquids at −183°C and −252.5°C, respectively, at 760 torr pressure. At higher pressures they will condense at higher temperatures; at 50 atmospheres pressure, for example, O_2 will condense to liquid at −118°C.

Above a definite temperature for each gas, the kinetic energy of the molecules is of such magnitude that no increase in pressure, however large, can liquefy the gas. This temperature for each gas is its **critical temperature,** and the concomitant pressure is the **critical pressure.** Oxygen cannot exist as a liquid above −118°C, even at pressures much higher than 50 atmospheres. Water cannot exist as a liquid above 374°C, no matter how high the pressure.

9.3 GASES IN CHEMICAL REACTION

Gay-Lussac's Law of Combining Volumes

When water is electrolyzed (Figure 8.3, p. 175), the compound is decomposed into its constituent elements, hydrogen and oxygen. Quantitative measurements show that the volume of hydrogen produced from a given weight of water is *twice* the volume of oxygen produced. When water is synthesized from its elements, this same relationship between the volume of oxygen and hydrogen holds—that is, twice as much hydrogen as oxygen *by volume* is required. The equation for the reaction is

$$2 H_2 + O_2 \rightarrow 2 H_2O$$

We may note in the equation that the coefficients of the formulas are in exactly the ratio (2 to 1) in which hydrogen and oxygen combine by volume. These experiments illustrate an important law of chemistry, first stated by Gay-Lussac: **The relative volumes of gases produced or consumed in a chemical reaction are in the ratio of small whole numbers.** It is assumed that all gases involved in such a reaction are at the same conditions of temperature and pressure. Ordinarily, in the change described above, the water formed as a product is allowed to condense to liquid water. If the water is allowed to remain in the form of gas (steam) and the volume measured, however, it is found that the volume of steam produced has the same volume as the hydrogen used, and twice the volume of the oxygen used, all gases being measured under the same conditions of temperature and pressure. We note again that the ratio, by volume, of the gases in the reaction is the same as that shown by the coefficients of the equation—that is, 2 to 1 to 2. The equation may be read as follows: two volumes of hydrogen react with one volume of oxygen to produce two volumes of water vapor (gases at same conditions). Of course, any unit of volume might be used; for example, we may say that two liters of hydrogen react with one liter of oxygen to produce two liters of steam.

Gay-Lussac's law applies to all substances in the gaseous state. The equation

$$N_2 + 3 H_2 \rightarrow 2 NH_3$$

tells us that one volume of nitrogen combines with three volumes of hydrogen to form two volumes of ammonia. The numbers in front of the formulas (coefficients) always give us the relationship between the gases *by volume*.

Gram Molecular Volume

Let us study further the equation for the electrolysis of water:

$$2 H_2O \rightarrow 2 H_2 + O_2$$

36 g	4 g	32 g
	44.8l	22.4l

According to the equation 2×18 or 36 g of water yields 2×2 or 4 g of hydrogen, and 2×16 or 32 g of oxygen. Experiment shows that 36 g of water decomposed by the electric current produces 44.8 l of hydrogen and 22.4 l of oxygen, the gases being measured at standard conditions. Then, 44.8 l of hydrogen at standard conditions must weight 4 g, and 22.4 l of oxygen must have a weight of 32 g. Four g of hydrogen represent 2 g molecular weights, or 2 moles, of hydrogen. If the volume of 2 moles is 44.8 l, the volume of 1 mole of hydrogen must be 22.4 l. Similarly, 32 g of oxygen represent 1 g molecular weight of oxygen or 1 mole of oxygen, and the volume of 1 mole of oxygen is also 22.4 l at standard conditions. If we should study the relationship between volume and weight of other gases, we would find that **one gram molecular weight (one mole) of any gas occupies a volume of 22.4 l at standard conditions.** This volume, 22.4 l, is called the **gram molecular volume, or molar volume,** of a gas (Figure 9.7).

This relationship between the volume and weight of a gas is used as a basis for the determination of molecular weights of substances in the gaseous state; it is simply a matter of determining the weight of 22.4 l of the gas

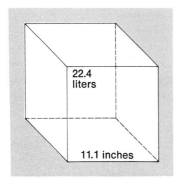

Figure 9.7
A gram molecular weight of any gas at standard conditions occupies a volume of 22.4 l. This volume is contained in a cube 11.1 inches on an edge. This volume contains 6.02×10^{23} molecules under standard conditions.

measured at standard conditions. The weight in grams is numerically equal to the molecular weight. Of course, this method is not applicable to those substances which are not readily converted into the gaseous condition.

9.4   STOICHIOMETRY INVOLVING GASES

Volume-Volume Relations

Calculations of volumes of gases based on chemical equations are very simple since the coefficients of the formulas of the gases in the balanced equation give us directly the relationship between volumes providing the gases are under the same conditions of temperature and pressure.

EXAMPLE 9.14   Calculate the volume of oxygen necessary to burn 50 l of CO.

$$2\,CO + O_2 \rightarrow 2\,CO_2$$

SOLUTION: According to the equation, 2 l of CO would require 1 l of O_2 (the numbers 2 and 1 are the coefficients of CO and O_2, respectively); therefore, 50 l of CO will require $50\,l \times \frac{1}{2} = 25\,l$ of O_2.
    What volume of CO_2 would be formed?
    The equation tells us that 2 l of CO produce 2 l of CO_2; hence, 50 l of CO produce 50 l of CO_2.

EXAMPLE 9.15   In the Ostwald process for the commercial preparation of nitric acid, ammonia gas is burned in oxygen in the presence of a Pt catalyst according to the equation

$$4\,NH_3 + 5\,O_2 \rightarrow 4\,NO + 6\,H_2O$$

(a) What volume of O_2 is required, and (b) what volume of NO (nitric oxide) is formed in the combustion of 500 l of NH_3? (All gases are under the same conditions of temperature and pressure.)

SOLUTION: According to the equation, 4 volumes of NH_3 react with 5 volumes of oxygen to form 4 volumes of NO and 6 volumes of H_2O (vapor). Therefore, 1 volume of NH_3 reacts with $\frac{5}{4}$ volumes of O_2 to form 1 volume of NO, and $\frac{6}{4} = 1\frac{1}{2}$ volumes of H_2O (vapor).

> 500 l of NH_3 will require $\frac{5}{4} \times 500 = 625$ l of O_2.
> 500 l of NH_3 will form $500 \times 1 = 500$ l of NO.

The ratio in which substances combine by volume must not be confused with the ratio in which they combine by weight. Substances do not combine in the ratio of small whole numbers by weight.

Weight-Volume Relations

Often the chemist wishes to know what *volume* of a product is the result of using a certain *weight* of a reactant. The following examples will illustrate the usefulness of **weight-volume stoichiometry** at standard conditions. When the gas is at other than standard conditions, the general gas equation (3) will need to be used to find the volume at STP before or after the stoichiometry calculation.

EXAMPLE 9.16   What volume of hydrogen at STP is produced as sulfuric acid acts on 120 g of metallic calcium. Equation for the reaction is

$$Ca + H_2SO_4 \rightarrow CaSO_4 + H_2$$

SOLUTION:

$$120 \text{ g of Ca is } \frac{120 \text{ g}}{40.0 \text{ g/mole}} = 3.00 \text{ moles}$$

According to the equation above one mole of Ca gives 1 mole of H_2. Therefore 3 moles of Ca will yield 3 moles of H_2.

> 3 moles of H_2 at STP occupy 3×22.4 l $= 67.2$ l

or

$$\text{liters } H_2 = 120 \text{ g Ca} \left[\frac{1 \text{ mole } H_2 \times 22.4 \text{ l/mole}}{1 \text{ mole Ca} \times 40.0 \text{ g/mole}} \right] = 67.2 \text{ l}$$

EXAMPLE 9.17   What volume of ammonia (NH_3) can be obtained when steam is passed over 4000 g of calcium cyanamide ($CaCN_2$)? Molecular weights $CaCN_2 = 80.0$; $NH_3 = 17$

> Equation: $CaCN_2 + 3 H_2O \rightarrow 2 NH_3 + CaCO_3$

SOLUTION:

$$1 \text{ mole CaCN}_2 \text{ produces 2 moles NH}_3$$

$$4000 \text{ g CaCN}_2 \text{ is } \frac{4000 \text{ g}}{80.0 \text{ g/mole}} = 50.0 \text{ moles CaCN}_2$$

50.0 moles $CaCN_2$ produces $(50.0 \times 2) = 100$ moles NH_3

100 moles NH_3 at STP occupy $100 \times 22.4 \text{ l} = 2240 \text{ l}$

or

$$\text{liters NH}_3 = 4000 \text{ g CaCN}_2 \left[\frac{2 \text{ moles NH}_3 \times 22.4 \text{ l/mole}}{1 \text{ mole CaCN}_2 \times 80.0 \text{ g/mole}}\right] = 2240 \text{ l}$$

Summary of Relationships in Stoichiometry

Stoichiometric calculations of the relationships between moles, weight, volume, and number of molecules are easily and quickly made if the following are kept in mind:

1 mole = 1 gram molecular weight (MW in grams) = 22.4 l of gas   (7)
at STP = 6.02×10^{23} molecules

EXAMPLE 9.18   (a) 13.2 g of CO_2 (molecular weight 44.0) occupies what volume at STP? (b) What is the weight of a million molecules (10^6) of O_2? (c) How many moles is 10.0 l of SO_2 at STP?

SOLUTION: In each of these cases note the relationships involved from (7) above in the conversion factor.

(a) liters $CO_2 = 13.2 \text{ g CO}_2 \left[\dfrac{22.4 \text{ l}}{44.0 \text{ g}}\right] = 6.72 \text{ l}$

(b) grams $O_2 = 10^6 \text{ molecules O}_2 \left[\dfrac{32.0 \text{ g}}{6.02 \times 10^{23} \text{ molecules}}\right] = 5.32 \times 10^{-17} \text{ g}$

(c) moles $SO_2 = 10.1 \text{ l SO}_2 \left[\dfrac{1.00 \text{ mole}}{22.4 \text{ l}}\right] = 0.451 \text{ mole}$

Avogadro's Principle

The similar behavior of gases under the same conditions of temperature and pressure led Avogadro, an Italian physicist, to suggest that **equal volumes of all gases under the same conditions of temperature and pressure contain the same number of molecules.** This view was first held as a speculation, but evidence for it accumulated, and now Avogadro's statement is regarded as a law.

Exercises

9.1. Define or illustrate the meaning of: (a) barometer, (b) torr, (c) STP, (d) Boyle's law, (e) partial pressure, (f) mole fraction, (g) effusion of gases, (h) kinetic energy, (i) Gay-Lussac's law, (j) molar volume, (k) Avogadro's principle.

9.2. A barometer was read as 800 torr. (a) What does 800 torr mean? (b) What is the pressure in atmospheres?

9.3. What pressure is required to compress 20 l of gas at 0.50 atm pressure to 1.0 l at constant temperature?

9.4. The temperature of 3.00 l of a gas is changed from 27.0°C to −10.0°C at constant pressure. What is the new volume in ml?

9.5. 500 l of a gas at 127°C and 600 torr would occupy what volume at STP?

9.6. A container is filled with a gas to a pressure of 5.00 atm at 30°C. (a) What pressure will develop inside the sealed container when it is warmed to 100°C? (b) At what temperature would the pressure be 100 atm?

9.7. The volume of a gas sample is 350 ml at 100°C. If the pressure is held constant, at what temperature will the sample have a volume of 150 ml?

9.8. If a sample of a gas occupies 6.72 l at STP, what volume will it occupy at 25°C and 800 torr?

9.9. If 600 ml of a gas measured at 4.50 atm and 127°C is compressed to a volume of 100 ml at 0°C, calculate the final pressure.

9.10. A sample of gas was collected over water at 27°C and 600 torr barometric pressure; volume 260 ml. (The vapor pressure of water at 27°C = 25 torr.) What volume would the dry gas occupy at STP?

9.11. A 360-ml sample of gas was collected by displacement of water at 741 torr pressure (of mercury) and 18°C. (The vapor pressure of water at 18°C is 15.5 torr.) What volume would the dry gas occupy at STP?

9.12. Eight-hundred milliliters of an unknown gas weighs 0.750 g at 27.0°C and 380 torr. What is the molecular weight of the gas?

9.13. Forty moles of hydrogen gas are pumped into a 20 l cylinder at 30°C. What is the pressure in the cylinder?

9.14. Calculate the molecular weight of a gas that has a density of 0.698 g/l at 90.0°C and a pressure of 200 torr?

9.15. A 501-ml sample of a gas collected at 20°C and a pressure of 600 torr weighs 2.13 g. What is the molecular weight of the gas?

9.16. One kilogram of nitrogen gas (N_2) is pumped into a tank of 25-l capacity. Calculate the pressure of nitrogen gas in the tank, assuming a temperature of 25°C.

9.17. A 224-ml sample of osmium oxide gas at STP weighs 2.56 g. The gas analyzes 25.2% oxygen and 74.8% osmium. What is the formula of the gas?

9.18. One-fourth mole of oxygen gas and 0.15 mole of hydrogen gas are mixed in a liter container at a temperature of 25°C. What is the total pressure of the gaseous mixture?

9.19. A cylinder contains 20 g He, 84 g N_2, and 20 g Ar. (a) What is the mole fraction of each gas in the mixture? (b) If the total pressure of the mixture is 10 atm, what is the partial pressure of He?

9.20. A cylinder contains 36.0 g of He gas, 140.0 g of N_2 gas, and 264.0 g of CO_2. (a) Compute the mole fraction of helium. (b) The total pressure of the cylinder is given as 15.0 atm. Compute the partial pressure of the He gas in atmospheres.

9.21. A mixture of 3.50 g of $O_2(g)$ and 3.50 g of $N_2(g)$ exerts a pressure of 800 torr. What is the partial pressure of each gas?

9.22. Two gases, HBr and CH_4, have molecular weights 81 and 16, respectively. The HBr effuses through a certain small opening at the rate of 0.85 ml/sec. At what rate will the CH_4 gas effuse through the same opening?

9.23. What is the molecular weight of a gas which effuses through a small opening in a glass tube one-third as fast as the gas methane (CH_4, MW = 16.0)?

9.24. Use Graham's law to calculate the molecular weight of a gas if a given volume of the gas effuses through an apparatus in 250 sec and the same volume of nitrogen, under the same conditions of temperature and pressure, effuses through the apparatus in 180 sec.

9.25. With the aid of the kinetic molecular theory, account for (a) the high compressibility of gases, (b) why gases have no definite upper surface, (c) how gases exert pressure, (d) why gases do not settle, (e) Boyle's law, (f) Charles' law, (g) Dalton's law of partial pressures, (h) Graham's law.

9.26. What volume of phosphine (measured at STP) would be formed by the reaction of 54.6 g of calcium phosphide with water?

$$Ca_3P_2(s) + 6\ H_2O \rightarrow 3\ Ca(OH)_2(s) + 2\ PH_3(g)$$

9.27. Chlorine may be prepared by the action of $KClO_3$ on HCl, and the reaction may be represented by the equation:

$$KClO_3 + 6\ HCl \rightarrow KCl + 3\ Cl_2 + 3\ H_2O$$

Calculate the weight of $KClO_3$ which would be required to produce 6.72 l of Cl_2 gas at STP.

9.28. What volume of oxygen in liters at STP would be required for the combustion of 17.0 g of PH_3?

$$PH_3 + 2\ O_2 \rightarrow HPO_3 + H_2O$$

9.29. What volume of hydrogen at STP would be produced by the action of excess dilute sulfuric acid on 40.0 g of zinc?

9.30. (a) What volume of NO (g) results from the reaction of 80.0 l of NO_2(g) with excess water if the volumes of both gases are measured at STP?

$$3\ NO_2(g) + H_2O \rightarrow 2\ HNO_3(aq) + NO(g)$$

(b) How many moles of HNO_3 are produced by the reaction?

9.31. (a) What volume of O_2 is needed to burn completely 50 l of CH_4 gas? (Consider gases at the same conditions.)
(b) What volume of products will be formed?

9.32. What is the approximate number of molecules in a drop of water which weighs 0.09 g?

9.33. What is the weight of 50.0 l of SiF_4 at STP?

9.34. Consider the compound uranium hexafluoride, UF_6 (molecular weight 352), which is a gas at STP.
(a) What would be the *actual weight in grams* of 1 molecule of UF_6?
(b) What *volume* at STP would 88 g of UF_6 occupy?

9.35. Four g of O_2:
(a) Is how many moles of O_2? (b) Occupies what volume at STP? (c) Contains how many molecules of O_2?

9.36. A volume of 56.0 ml of He at STP:
(a) Is how many moles?
(b) Weighs how much?
(c) Contains how many molecules?

9.37. For each of the following pairs of variables, which apply to measurements made on a sample of an ideal gas, draw a rough graph to show how one quantity varies with the other. (a) P vs. V, T constant; (b) T vs. V, P constant; (c) P vs. T, V constant; (d) PV vs. V, T constant.

9.38. Suppose you had 1.0-l flasks of each of the following gases at STP: H_2 (MW 2); SO_2 (MW 64); O_2 (MW 32); and UF_6 (MW 352).
(a) Which gas molecules would have the greatest average kinetic energy? (If all gases would be the *same* in these questions, write "same.")
(b) Which gas molecules would have the greatest average velocity?
(c) Which flask contains the largest number of moles?
(d) Which flask would have the largest number of molecules?
(e) Which gas would effuse through a small hole at the *lowest* rate?

9.39. Given 1.00 mole of He (g) and 1.00 mole of $H_2(g)$ measured under the same conditions. Compare quantitatively the He sample to the H_2 sample in terms of each of the following properties. (a) Number of molecules, (b) mass of a single molecule, (c) average velocity of a molecule.

Supplementary Readings

9.1. *The Gaseous State*, N. G. Parsonage, Pergamon, New York, 1966 (paper).
9.2. "Robert Boyle," M. B. Hall, *Sci. American*, Aug., 1967; p. 97.
9.3. "Graham's Laws of Diffusion and Effusion," E. A. Mason and B. Knonstadt, *J. Chem. Educ.*, **44**, 740 (1967).

Chapter 10

Water and The Liquid State

In the study of water, the most familiar of liquids, we shall discuss characteristics of the liquid state in general. In oceans, lakes, and rivers, water covers nearly three-fourths of the earth's surface, and it is a constituent of the soil and the atmosphere. The human body is more than 65% water, and many of our foods contain an even larger percentage. Table 10.1 shows the approximate percentages of water in various familiar substances.

Table 10.1
Water in familiar substances

Substance	Percent	Substance	Percent
Beef	65–70	Plant tissue	50–75
Eggs	74	Tomatoes	85
Milk	87	Bacon	19
Apples	86	Butter	17
Vegetables (green)	90–95	Peanuts	9
Potatoes	78	Bread	35
Bone	50	Wheat	11
Muscle	75	Lard	0

Physical Properties of Water

Pure water has a flat taste, is odorless and is practically colorless in thin layers. If a deep section of it is viewed, it appears blue. Water has a maximum density at 4°C, at which temperature 1 ml weighs 1 gram. Above or below this temperature, 1 ml weighs less than 1 g (Table 10.2). On the centigrade temperature scale, water freezes at 0° and boils at 100°.

Table 10.2
Density of water from 0°C to 100°C

Temp. °C	Density g/ml	Temp. °C	Density g/ml
0	0.99987	50	0.9881
2	0.99997	60	0.9832
4	1.0000	70	0.9778
10	0.9997	80	0.9718
20	0.9982	90	0.9653
30	0.9957	100	0.9584
40	0.9922		

Water has a specific heat of 1. **Specific heat** is defined as the number of calories of heat necessary to raise the temperature of one gram of a substance 1°C; thus one calorie is necessary to raise the temperature of 1 g of water 1°C. The specific heat of water is one of the highest specific heats.

When ice melts (changes from a solid to a liquid), heat is absorbed. About 80 cal of heat are required to melt 1 g of ice. When water is frozen, an equivalent amount of heat is evolved. The heat absorbed when 1 g of solid melts is called the **heat of fusion;** the heat evolved when 1 g of liquid freezes is called the **heat of solidification.**

The quantity of heat necessary to convert 1 g of a liquid into a vapor is termed the **heat of vaporization.** For water, 540 cal are necessary to change 1 g of liquid water at 100°C into vapor (steam) at 100°C. When water is condensed (changed from vapor to liquid), an equivalent amount of heat, called the **heat of condensation,** is given off. The heat of condensation is an important factor in the use of steam for heating purposes. Steam is introduced into radiators, where it is condensed to the liquid state. In this condensation process, 540 cal are given up to the radiator for every gram of water condensed.

EXAMPLE 10.1   Determine the quantity of heat required to convert 10 g of ice at 0°C to vapor at 100°C.

SOLUTION: The overall process may be considered to be the sum of the following three steps:

1. Melt 10 g ice requires $\qquad\qquad$ $10 \text{ g} \times 80 \text{ cal/g} = \quad 800 \text{ cal}$
2. Heat 10 g liquid from 0° to 100° $\quad$ $10 \text{ g} \times 100° \times 1 \text{ cal/g-deg} = 1000 \text{ cal}$
3. Vaporize 10 g liquid at 100° $\qquad$ $10 \text{ g} \times 540 \text{ cal/g} = 5400 \text{ cal}$

$$\text{Total heat energy required} \qquad\qquad\qquad 7200 \text{ cal}$$
$$\text{or} \qquad 7.2 \quad \text{kcal}$$

EXAMPLE 10.2   40 g of ice at 0°C is mixed with 100 g of water at 60°C. What is the final temperature after equilibrium has been established?

SOLUTION: In a determination of this kind, the heat lost by the water in cooling must be balanced by the heat gained by the ice in melting and in warming the resulting water (from the melted ice) to the final temperature. Let t = final temperature

Heat evolved by the 100 g of water in cooling to temp
$$t = (60° - t)(100 \text{ g})(1 \text{ cal/g-deg}) = 6000 - 100 \text{ t cal}$$
Heat absorbed by the ice in melting $= 40 \text{ g} \times 80 \text{ cal/g} = 3200 \text{ cal}$
Heat absorbed by the water (from the 40 g ice) in being warmed from
$$0° \text{ to } t° = (40 \text{ g})(t)(1 \text{ cal/g-deg}) = 40 \text{ t cal}$$
Total heat absorbed $= 3200 + 40 \text{ t cal}$
Then $\qquad\qquad (6000 - 100 \text{ t}) \text{ cal} = (3200 + 40 \text{ t}) \text{ cal}$
$$140 \text{ t} = 2800$$
$$t = 20°$$

Water is a particularly good solvent. Almost all substances are more or less soluble in water, and a study of aqueous solutions is of primary importance to the chemist.

Properties of the Liquid State

A liquid has no definite shape but, unlike a gas, does occupy a more or less definite volume. Whereas a gas completely fills any container in which it is placed, a liquid does not. Molecules in the liquid state appear to be very much closer together than in a gas — that is, the free volume surrounding each molecule is much less. Evidence for this is the relatively small compressibility of liquids. High pressures are required to compress a liquid even a small fraction of its original volume; for example, a pressure of 3000 atmospheres (about 44,100 lbs per sq in) is necessary to compress 1 l of water at 0°C to 900 ml. For additional evidence we may consider the relative volumes of 1 mole of water in the liquid and gaseous states. One mole of water (18 g) as a liquid at 100°C occupies a volume of only about 18 ml; when vaporized as steam, it occupies a volume of about 30,000 ml.

As in a gas, molecules in the liquid state are in a constant state of motion, but the movement is more restricted because of the close proximity of molecules. This motion explains the diffusion of one liquid into another. The rate of diffusion of liquids, however, is much lower than that of gases because of the restricted movement.

Molecules in a liquid presumably are held together by a large force of mutual attraction, which is a function of the distance between molecules — the smaller the distance, the greater the force. When a liquid is vaporized, the distance between molecules must be greatly increased, and energy must be expended in overcoming the attractive forces. Thus a liquid has less energy than a gas.

Evaporation of a Liquid

Evaporation of a liquid is the conversion of the liquid to vapor; condensation is the conversion of vapor to liquid. If a liquid is placed in an open container, the molecules of the liquid gradually escape into the space above the liquid — the liquid evaporates. From an open container the molecules of vapor are dispersed into the atmosphere, and more molecules change into vapor to take their place. If this process is allowed to continue, all of the liquid eventually will be converted into the vapor state; in other words, evaporation continues until the liquid has disappeared.

If a liquid is contained in a closed vessel, however, evaporation does not seem to continue indefinitely. Some of the liquid will be converted into vapor, but after the atmosphere above the liquid becomes saturated with molecules, further evaporation of the liquid appears to cease. This vapor exerts a pressure which assumes a constant value at a given temperature. As the temperature of a liquid is increased, the vapor pressure increases. Let us study carefully the causes of these phenomena.

We shall try to imagine the state of affairs existing in a liquid which is contained in a closed vessel (Figure 10.1). If we could see the individual molecules, we undoubtedly would observe that a number of molecules of the substance are present in the space above the liquid. These molecules are moving about in a random fashion and are in a constant state of motion. Collision of these molecules with the walls of the vessel and the surface of the liquid gives rise to the pressure exerted by the vapor (vapor pressure). Probably a somewhat similar state of affairs exists in the body of the liquid, but the molecules of the liquid are very close together, and movement of the molecules through the body of the liquid is more restricted. Nonetheless, these molecules possess kinetic energy and move about through the liquid constantly. The molecules are probably held together by mutual attraction. Each molecule in the body of the liquid is attracted and held by the multitude of molecules about it. Occasionally, however, a molecule near the surface of the liquid possesses sufficient energy to break through the surface and thus passes into the space above the liquid. This molecule is converted from the liquid state to the vapor state. We may note, too, that some of the molecules which are in the atmosphere above the liquid are passing back into the liquid state. A molecule strikes the surface and passes into the body of the liquid. These molecules are condensed from the vapor state to the liquid state. Two continuous processes thus are occurring simultaneously, the conversion of liquid to vapor (evaporation) and the conversion of vapor to liquid (condensation). Eventually these two processes will be taking place with exactly the same speed, and the number of molecules at any given time in the form of vapor will be constant; hence the vapor pressure of the liquid attains a constant value at a given temperature. When the rate of evaporation is exactly equal to the rate of condensation, a condition of equilibrium has been established.

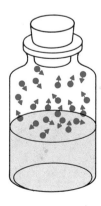

Figure 10.1
A liquid in a closed container soon comes to equilibrium with its vapor—the rate of evaporation is equal to the rate of condensation.

Equilibrium, though apparently a state of rest, actually is **a state in which two opposing processes are taking place with exactly the same speed.**

When the temperature of the liquid is increased, the kinetic energy of the molecules in the liquid increases. The increase in kinetic energy increases the relative number of molecules which have sufficient energy to escape into vapor. Since the number of molecules per unit volume determines the pressure of the vapor, the pressure must increase as the temperature is increased. Vapor pressures of liquids at various temperatures have been carefully determined; the vapor pressures of water between 0°C and 100°C will be found in Table 10.3. When the data in Table 10.3 are plotted with vapor pressure on the vertical axis and temperature on the horizontal axis, the vaporization curve for water is obtained (Figure 10.2).

Table 10.3
Pressure of water vapor, or aqueous tension (torr)

Temperature (C)	Pressure	Temperature (C)	Pressure
0°	4.6	65°	187.5
5°	6.5	70°	233.7
10°	9.2	75°	289.1
15°	12.8	80°	355.1
20°	17.5	85°	433.6
25°	23.7	90°	525.8
30°	31.7	94°	610.0
50°	92.3	97°	680.0
55°	118.0	100°	760.0
60°	149.4		

All points on this curve represent an equilibrium between liquid water and gaseous water. The vapor pressure at any temperature can be obtained from the curve by extending a horizontal line from the point of intersection of a perpendicular corresponding to the temperature. At 70°C the horizontal extension cuts the vertical axis at about 230 torr. Consequently, the vapor pressure of water at 70°C is approximately 230 torr. (See Table 10.3, which gives 233.7 torr.)

Phase Diagrams

Figure 10.2 is called the **phase diagram** for H_2O. It is a pressure-temperature diagram showing the conditions under which the three physical states gas, liquid, and solid exist as well as the equilibrium boundaries between them. The regions marked gas, liquid, or solid represent **one-phase** systems. For all temperatures and pressures falling within a boundary area of one of the phases, that is the single phase which will exist. For example, H_2O at 230 torr

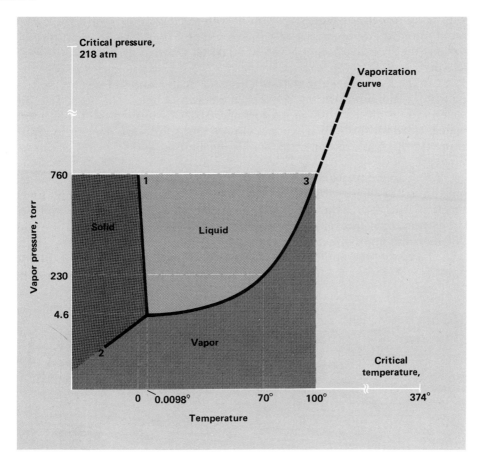

Figure 10.2
Diagram for H_2O showing equilibrium between three physical states: solid, liquid, and vapor.

and −10°C will be a solid, at +20°C it will be a liquid, and at 80°C it will be a gas.

The curves separating the regions from each other represent equilibrium conditions between two phases.

Curve 3 in the figure is the vapor pressure curve representing gas-liquid equilibria previously referred to and terminates at the **critical point** of water (374°C and 218 atm). Above 374°C water cannot exist as a liquid no matter how great the pressure. The point on curve 3, for example, corresponding to 234 torr and 70°C represents both gas and liquid phases in equilibrium with one another. Consequently 70°C is the boiling point of water at 234 torr.

Curve 1, whose slope is exaggerated in the figure, shows that as the pressure is increased the freezing point of water is lowered slightly, the freezing point being the temperature at which solid and liquid are in equilibrium. At 4.6 torr the freezing point of water is 0.01°C, whereas at 1 atm water freezes at 0.00°C.

Curve 2 is the **sublimation curve** of water and represents solid-gas equilibria, no liquid being present at pressures below 4.6 torr. The intersection of all three curves at 4.6 torr and 0.01°C is termed the **triple point** of water. Here all three phases – gas, liquid, and solid – are present in equilibrium. At the triple point water would be boiling and freezing at the same time.

Actually the phase diagram for water represented by Figure 10.2 pertains only to water in the absence of any other substances, such as air. For the practical situation where the total pressure consists of both air and water vapor pressures, the curves are only very slightly different and the phase diagrams are approximately the same.

The phase diagram for CO_2 is shown in Figure 10.3. Note that the triple point is −56.6°C at 5.11 atm. Liquid CO_2 cannot exist at pressures less than

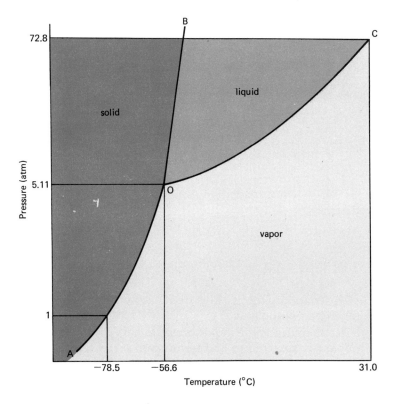

Figure 10.3
Phase diagram for carbon dioxide (not drawn to scale).

5.11 atm. so that we normally do not observe liquid CO_2. Solid CO_2 ("Dry Ice") sublimes; that is, changes directly from solid to gas at $-78°C$ and normal atmospheric pressure. It will be noted that the solid-liquid equilibrium curve for CO_2 differs from that of water in that the CO_2 curve tilts to the right. This is the usual case for most substances. H_2O is one of the very few substances which expands on freezing, and its solid-liquid curve tilts to the left.

The Boiling Point of a Liquid

When a liquid is heated in an open container, the escape of molecules from the liquid into the vapor is opposed by atmospheric pressure. As the temperature of the liquid is increased, the vapor pressure of the liquid increases and approaches the atmospheric pressure. The **temperature at which the vapor pressure of a liquid is equal to the external pressure is the boiling point of the liquid.** Thus water, which boils at a temperature of 100°C at a pressure of 760 torr, will boil at a temperature of 25°C if the external pressure is 23.7 torr. Water boils on a mountain top at a temperature below 100°C, since the pressure there is less than 760 torr. It is evident from the vaporization curve for water (Figure 10.2) that if the pressure above water is greater than 760 torr, the water will boil at a temperature above 100°C (110°C at 1075 torr, etc).

Further application of heat at the boiling point of a liquid tends only to vaporize the liquid; the liquid absorbs heat in changing from the liquid to the vapor state. The temperature of the boiling liquid remains constant during conversion to vapor, and application of more heat only increases the speed of vaporization.

Refrigeration

Since the evaporation of a liquid is an endothermic process, it may be utilized to withdraw heat from a system — for instance, in refrigeration. Heat is transferred from within the refrigerator to the outside, where it is dissipated by circulating air. This transfer of heat is accomplished by alternate condensation and evaporation of a nontoxic substance, such as dichlorodifluoromethane (CCl_2F_2). Large commercial refrigerating systems frequently use NH_3.

Surface Tension

A molecule in the center of a liquid is subject to gravitational pull in all directions by attraction of the molecules all around it. A molecule at the surface, however, is subject to attractional pull by the molecules of liquid at all

Figure 10.4
Molecules in the surface layer of a liquid have no upward attraction.

sides and below it, but not by attraction upward (Figure 10.4). This lateral attraction and downward force on surface molecules without any upward force tends to make the surface of a liquid contract, much as a sheet of rubber stretched laterally in all directions tends to contract. This helps account for the fact that a small quantity of a liquid tends to form a spherical or drop shape: there is less surface area to a sphere than to any other shape for a given weight of a liquid. With care a container can be slightly more than filled with water before it runs over because of the attractional tension between surface molecules. The surface membrane or skinlike layer of molecules is so tough that a small needle may be floated on it, and various insects walk or skate on this surface film.

Viscosity

Liquids differ in their resistance to flow, or **viscosity**. Viscosity is the ease with which the molecules of a liquid slide over and by one another. By comparing the times necessary for equal volumes of two liquids to flow through a given orifice, the relative viscosities may be determined.

Volatile Liquids

Liquids are often spoken of as being volatile or nonvolatile. Volatile means easily vaporized—that is, easily converted from a liquid to a vapor. Volatile liquids exert relatively high vapor pressures and have relatively low boiling points. Ether (b.p. 34.5°C) and alcohol (b.p. 78°C) are classed as volatile liquids, since conversion of liquid to vapor (evaporation) takes place rapidly at relatively low temperatures. Perfumes owe their use to their volatile nature. Liquids of high boiling point—for example, sulfuric acid (b.p. 330°C) and mercury (b.p. 357°C)—are classed as nonvolatile liquids.

Water Pollution

The chief categories of water pollution include the following:

1. Organic matter from food processing, paper-pulp production, domestic sewage, etc. Part of this matter becomes bottom sludge, slime, suspended solids, or dissolved material. These organic materials cause a biological oxygen demand (BOD). This results in a depletion of the normally dissolved O_2 in water which is requisite for fish life and for useful aerobic (oxygen user) microorganisms. With a reduction of oxygen, harmful anaerobic bacteria flourish.
2. Pathogenic organisms from human and animal waste.
3. Toxic materials from industry such as HF, H_2SO_4, H_2SO_3, $Ca(HSO_3)_2$, metal chlorides, and sulfates.
4. More N and P compounds as soils are more intensively fertilized.
5. Various synthetic chemicals, some of which break down slowly — includes some detergents and pesticides.
6. Radioactive wastes from nuclear energy or isotopic production plants.
7. Physical factors that involve chemical contamination such as higher water temperature, diversions of stream flow for irrigation, dams and impoundments.

Purification of Water

Natural waters usually contain dissolved minerals or gases. This dissolved material gives water its taste (pure water has a flat taste). So-called "hard" waters precipitate soap or form a curd with soap because of their mineral content.

Small amounts of mineral matter are desirable in drinking water, but suspended insoluble matter must be removed and bacteria must be killed. Filtration is the simplest process for removing suspended particles, and filtering water through prepared sand and gravel beds constitutes one of the main methods of treatment for drinking water supplies. Charcoal filter beds sometimes augment sand filters. Before passage of water through a filter bed, it may be allowed to stand in a reservoir or container to allow suspended material to settle by gravity. Chemicals, such as aluminum sulfate, may be added to form a gelatinous aluminum hydroxide, which aids in the coagulation, settling out, or filtering out of suspended matter. Since some harmful bacteria may not be retained in the filter beds, it is necessary to kill them by chemicals. Ozone is bubbled into water in some European cities; chlorine in the amount of about one part Cl_2 per million parts of water is the principal chemical used in the United States. At times bleaching powder ($CaOCl_2$) is substituted for Cl_2 because it is convenient to handle. Some $CuSO_4$ to kill algae growth in

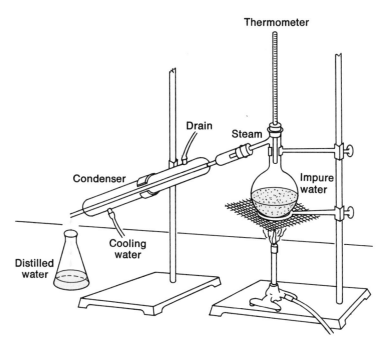

Figure 10.5

Distillation of water.

settling basin or reservoir water may also be used. In small quantities of water, bacteria may be killed by boiling or by adding one drop of household iodine per quart of water and allowing it to stand a half hour.

Water is often purified by distillation, a process in which water is boiled and the vapor subsequently condensed. The process is conveniently carried out in the apparatus shown in Figure 10.5. Cold water is circulated in the condenser, and this changes the vapor (steam) to a liquid, which is collected in a receiver. Mineral matter, which is nonvolatile, stays behind in the distilling flask. Gases contained in the water pass over with the first portion of the distillate, which may be discarded. Water purified in this way is called **distilled water.**

Distillation is used to purify or concentrate many substances in addition to water.

Deionized Water

A type of ion-exchange (see Chapter 26) utilizing certain resins may be employed to remove both cations and anions from a natural water. The procedure is somewhat like this: water is passed through a so-called cation-exchange resin containing exchangeable H^+ ions. Cations in the water are absorbed by the resin and in the process release H^+ to leave an acid solution. For example, assume that the water contains $CaSO_4$ — we may represent the change with the equation

$$Ca^{2+} + SO_4^{2-} + Resin\ (2\ H^+) \rightarrow Resin\ (Ca^{2+}) + H_2SO_4$$

The water is then passed through an anion-exchange resin containing exchangeable OH^-. The hydroxyl ions are displaced by anions (in this case SO_4^{2-}), leaving equivalent amounts of H^+ and OH^-:

$$2\ H^+ + SO_4^{2-} + Resin\ (2\ OH^-) \rightarrow Resin\ (SO_4^{2-}) + 2\ H_2O$$

The overall result is a deionized water, which contains only the relatively few OH^- and H^+ that exist in pure water.

Desalination

Enormous quantities of water exist in the oceans of the world. Unfortunately the total salt content in sea water is about 35,000 ppm (Table 10.4) which is too salty for drinking, growing crops, and most industrial uses. Considerable research has been done concerning methods of removing the salts from sea water and several desalination processes are now being used on a large scale. Several hundred desalting plants are in operation in regions where it is economically feasible, such as the Middle East. The processes in use usually involve one of the following: (1) distillation, (2) freezing, (3) osmosis or membrane separation, and (4) ion exchange.

Chemical Properties of Water

Water is relatively stable; considerable energy is required to decompose it into hydrogen and oxygen. Even at high temperatures, water is only slightly decomposed — little more than 1% at 2000°C and about 4% at 2500°C.

We learned in the study of hydrogen (Chapter 8) that some of the more active metals act with water to form hydrogen. In the study of oxygen we learned that metallic oxides, if soluble, act with water to form hydroxides (bases), while nonmetal oxides act to form acids.

The Molecular Structure of Water

We indicated in Chapter 5 that, on the basis of the orientation of p orbitals, the H_2O molecule would be expected to have an angular structure, with 90° between H atoms. Actually, the angle is about 105°, as shown in Figure 10.6.

As a result of this arrangement, the center of negative electrical charge

Table 10.4

The composition of Sea Water. The logarithmic chart shows in moles per kilogram the concentration of 40 of the 73 elements that have been identified in seawater. Only four elements are now recovered from the sea commercially: chlorine, sodium, magnesium and bromine. Recovery of the other scarce elements is not promising unless biological concentrating techniques can be developed. Manganese nodules on the sea floor are a potential source of scarce metals.

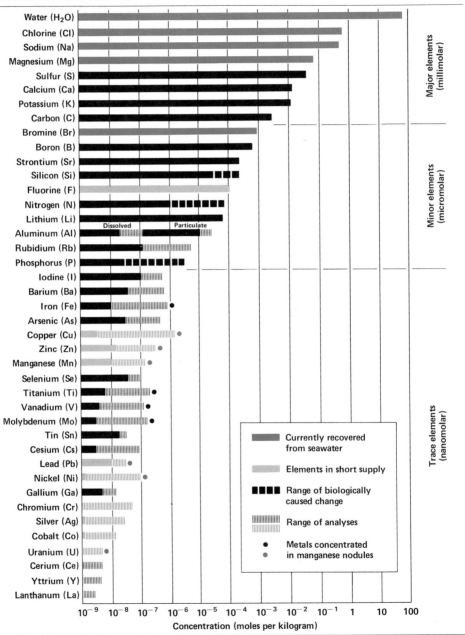

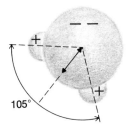

Figure 10.6
Arrangement of hydrogen atoms in the water molecule.

Figure 10.7
Diagrammatic representation of dipoles.

(at the center of the oxygen atom) is not the same point as the center of positive electrical charge, and the water molecule is a **dipole** (Figure 10.7). The polar structure of the water molecule appears to play an important role in the action of water as a solvent for various types of solutes (Chapter 13).

Because of hydrogen bonding (page 118) and attractive forces between dipoles, it is likely that water molecules in the liquid state do not exist as single dipoles; rather, two or more dipoles are associated (Figure 10.7, center and right). This association would account for such properties of water as the relatively high heat of vaporization, abnormally high boiling point, and low vapor pressure in view of its low molecular weight.

Heavy Water

In recent years it has been found that ordinary hydrogen is made up essentially of two isotopes (Figure 10.8). A third isotope of hydrogen, too unstable to be detected normally in nature, may be produced by nuclear reactions (see Chapter 36).

The names **protium, deuterium,** and **tritium** have been assigned to these isotopes, with the symbols H, D, and T. In natural hydrogen or its compounds there is only 1 D atom to 4800 H atoms. The amount of tritium is extremely small.

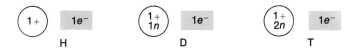

Figure 10.8
The three isotopes of hydrogen.

Consider the substitution of atoms of D and T for H in water, so that in addition to H_2O molecules of molecular weight 18, there would be D_2O, T_2O, HOD, HOT, and DOT molecules, of molecular weights 20, 22, 19, 20, 21, respectively. In the electrolysis of water, ordinary hydrogen atoms are freed much more readily than hydrogen atoms of atomic weight 2. Accordingly, on long electrolysis the water residue contains a relatively high percentage of D_2O. This water, known as "heavy water," differs somewhat from ordinary water in certain physical properties—for example, slightly higher freezing and boiling points. The principal source of heavy water has been the aqueous electrolytic baths, which have been used over a long period of time.

The Formation of Hydrates

Many compounds combine chemically with water to form **hydrates**. For example, copper sulfate combines with water to form the hydrate of the composition $CuSO_4 \cdot 5\ H_2O$:

$$CuSO_4 + 5\ H_2O \rightarrow CuSO_4 \cdot 5\ H_2O$$

The hydrate thus formed is a definite chemical compound conforming to the law of definite proportions; exactly five molecules of water, no more and no less, are combined chemically with one formula weight of $CuSO_4$. The medial dot ($\cdot$) used in the formula indicates a condition of instability; the compound is readily decomposed by heating. Water held in such a combination is called **water of hydration,** or water of crystallization. Other familiar hydrates are $Na_2CO_3 \cdot 10\ H_2O$ (washing soda), $Na_2SO_4 \cdot 10\ H_2O$ (Glauber's salt), and $KAl(SO_4)_2 \cdot 12\ H_2O$ (alum). Not all salts form hydrates; for example, NaCl, $K_2Cr_2O_7$ and KNO_3 do not form hydrates. When one buys crystalline washing soda ($Na_2CO_3 \cdot 10\ H_2O$), he buys a greater weight of water than of soda.

Hydrates may be decomposed by the application of heat. The water passes off as water vapor, leaving the **anhydrous** (without water) salt:

$$CuSO_4 \cdot 5\ H_2O \rightarrow CuSO_4 + 5\ H_2O$$
$$Na_2CO_3 \cdot 10\ H_2O \rightarrow Na_2CO_3 + 10\ H_2O$$

These changes are reversible, since the hydrates may be reformed by the addition of water to the anhydrous salt. All hydrates are crystalline, but not all crystalline salts are hydrates.

Efflorescence and Deliquescence

Some hydrates are unstable in contact with the atmosphere. Water of hydration may be released to form a lower hydrate or the anhydrous salt. A hydrated salt exerts a vapor pressure made up of water, and this vapor pressure is definite for each hydrate. If the pressure of water vapor in the atmosphere is

Material to
be dried

Drying
agent

Figure 10.9
A desiccator.

lower than the equilibrium pressure for a given hydrate, the latter will give up water to the atmosphere. This spontaneous process is termed **efflorescence.** $Na_2CO_3 \cdot 10\ H_2O$ is an example of an efflorescent salt; when exposed to dry air (low humidity), water of hydration is rapidly lost.

Certain soluble substances possess the ability to absorb moisture from the atmosphere and are said to be **hygroscopic.** In this process, water may be absorbed to form a solid hydrate, such as

$$CuSO_4 + 5\ H_2O \rightarrow CuSO_4 \cdot 5\ H_2O$$

or a solution of the substance may result. In the latter case, as moisture is absorbed by a crystal, a solution forms about the crystal as it slowly dissolves. The vapor pressure of this concentrated solution will be relatively low. If the pressure of water vapor in the atmosphere is higher than the vapor pressure of this solution, the latter will absorb water and dilute itself until its vapor pressure is equal to the pressure of water vapor in the atmosphere. A crystal which absorbs moisture to form a solution is termed **deliquescent.** Such substances may be used as drying agents, since moisture in the surrounding atmosphere is removed. $CaCl_2$ is often used for this purpose. A specific use of $CaCl_2$ is its application to roadways to keep down dust. The salt takes water from the air, and the roadbed becomes somewhat moist. In powder form it is also used for dispelling fog around airports.

Figure 10.9 shows a desiccator, an apparatus for drying chemicals or keeping them dry. Commonly used drying agents are calcium chloride, concentrated sulfuric acid, and magnesium perchlorate.

Exercises

10.1. Define and illustrate:

(a) vapor pressure	(f) distillation	(k) specific heat
(b) deliquescence	(g) volatile	(l) heat of fusion
(c) hygroscopic	(h) anhydrous	(m) heat of vaporization
(d) efflorescence	(i) viscosity	(n) dipole
(e) hydrate	(j) surface tension	(o) water of hydration

10.2. Explain the meaning of the following:

(a) evaporation
(b) boiling point
(c) freezing point
(d) calorie
(e) phase diagram

(f) triple point
(g) sublimation
(h) heavy water
(i) cation exchange resin
(j) deionized water

(k) ion exchange
(l) desalination
(m) equilibrium
(n) desiccator
(o) ppm

10.3. Contrast liquids and gases concerning the following properties: (a) upper surface, (b) compressibility, (c) density, (d) viscosity. With the aid of the kinetic molecular theory explain these differences in properties.

10.4. Explain, by means of the kinetic molecular theory: (a) why gases may be converted to liquids by lowering the temperature and increasing the pressure; (b) why drops of liquid are spherical; (c) why objects such as razor blades, needles, or paper clips will float on the surface of water; (d) evaporation; (e) the cooling effect of evaporation; (f) vapor pressure; (g) the difference between boiling and evaporation.

10.5. The following are physical properties of methyl alcohol (CH_3OH): freezing point, $-98°C$; boiling point, $65°C$; specific heat of liquid, 0.570 cal/g-degree; heat of fusion, 22.0 cal/g; and heat of vaporization, 263 cal/g. Calculate the number of kilocalories required to convert one mole of methyl alcohol solid at $-98°C$ to vapor at $65°C$.

10.6. If you had 2.00 kg of ice at $0°C$, how many calories would be required to convert this to steam at $100°C$?

10.7. A certain liquid freezes at $50°C$ and boils at $120°C$. Heat of fusion = 30 cal/g; heat of vaporization = 180 cal/g; specific heat of the liquid = 0.60 cal/g-degree. What quantity of heat energy in kilocalories is necessary to convert 52 g of solid at $50°C$ to vapor at $120°C$?

10.8. Calculate the amount of heat required to change 5 moles of H_2O from ice at $-10°C$ to steam at $110°C$. The heat of vaporization of H_2O at $100°C$ is 540 cal/g and the heat of fusion of H_2O at $0°C$ is 79.9 cal/g. The specific heat of a substance is the amount of heat required to raise the temperature of 1 g of the substance $1°C$. The specific heat of ice is 0.492 cal/g-degree, of liquid water is 1.00 cal/g-degree, and of steam is 0.484 cal/g-degree.

10.9. Any liquid left in an open container will ultimately evaporate to dryness. Which *five* of the following factors will directly influence the rate (e.g., g/day) of evaporation?

1. The volume of liquid
2. The equilibrium vapor pressure of the liquid
3. Temperature of the surroundings
4. Density of the liquid
5. The area of the liquid surface
6. Molecular weight of the compound

7. Air movement over the liquid surface
8. Insulation of the container

10.10. If one takes a bottle and fills it one-quarter full with water and then tightly closes the bottle with a stopper, which *four* of the following statements would be true concerning the system?
1. The space above the liquid would contain only air.
2. Molecules of H_2O are continuously evaporating from the surface of the liquid.
3. The percentage gas composition above the liquid would always be changing.
4. The water would dissolve small amounts of O_2 and N_2.
5. All the liquid would evaporate.
6. The space above the liquid would contain air and water vapor.
7. The space above the liquid would contain only water vapor.
8. Once equilibrium was established the gas composition would remain constant providing the temperature did not change.

10.11. Figure 10.2 is the phase diagram for water. Describe the phase changes that occur and the approximate pressures at which they occur, when the pressure on an H_2O system is gradually increased (a) at a constant temperature of $-1°C$, (b) at a constant temperature of $70°C$, and (c) at a constant temperature of $-30°C$.

10.12. Refer to Figure 10.2 and describe the phase changes that occur, and the approximate temperatures at which they occur, when water is heated from $-10°C$ to $110°C$ (a) under a pressure of 1 torr, (b) under a pressure of 600 torr, (c) under a pressure of 900 torr.

10.13. Refer to Figure 10.3 and describe the phase changes that occur and the approximate pressures at which they occur when the pressure on a CO_2 system is gradually increased (a) at a constant temperature of $-70°C$; (b) at a constant temperature of $20°C$.

10.14. Refer to Figure 10.3 and describe the phase changes that occur, and the approximate temperatures at which they occur when carbon dioxide is heated (a) at a constant pressure of 1.0 atm; (b) at a constant pressure of 10.0 atm.

10.15. Use the following data to draw a rough phase diagram for krypton. Normal boiling point, $-152°C$; normal melting point, $-157°C$; triple point, $-169°C$, 133 torr; critical point, $-63°C$, 54.2 atm; vapor pressure of solid at $-199°C$, 1.00 torr. Which has the higher density at a pressure of 1 atm: solid Kr or liquid Kr?

10.16. Suggest a means of purifying water containing the following pollution:
(a) excessive organic matter
(b) pathogenic organisms
(c) toxic soluble compounds,
(d) excessive suspended matter.

10.17. Explain how fresh water might be made from sea water (desalinization) by: (a) distillation, (b) freezing, (c) ion exchange.

10.18. A 100-lb load of washing soda ($Na_2CO_3 \cdot 10\ H_2O$) is purchased at $0.15 per lb. How much money is invested in water?

10.19. A certain hydrate analyzes as follows: 25.5% copper, 12.8% sulfur, 4.00% hydrogen, and 57.7% oxygen. Determine the formula for this hydrate.

10.20. A 5.00-g sample of hydrated zinc sulphate on being heated to expel water of hydration leaves a residue of 2.80 g of anhydrous salt. Determine the formula of the hydrate.

10.21. Sublimation is sometimes used to purify solids. The impure material is heated and the pure crystalline product condenses on a cold surface. Is it possible to purify ice by sublimation? What conditions would have to be employed?

10.22. Four grams of hydrogen and 48 g of oxygen are mixed at 0°C in a container which has a volume of 11.2 l. (a) What is the pressure in atmospheres inside the container? (b) If the mixture is ignited and the contents then returned to 0°C (no volume change), what will the pressure inside be, assuming that the water formed has a negligible vapor pressure at 0°C?

10.23. Complete and balance the following equations:
(a) $Na_2CO_3 \cdot 10\ H_2O \xrightarrow{\text{heat}}$
(b) $CuSO_4$ (anhydrous) + $H_2O \rightarrow$
(c) Precipitating Ca^{2+} from hard water:
   $CaCl_2 + Na_2CO_3 \rightarrow$
(d) Formation of the filtering aid, $Al(OH)_3(s)$:
   $Al_2(SO_4)_3 + NaOH \rightarrow$

10.24. Water is a key compound in the biological environment, yet it is a very unusual liquid. Try to explain the biological significance of the following physical properties of water:
(a) Water, unlike most liquids, expands on freezing. $H_2O(s)$ has a lower density than $H_2O(l)$.
(b) Water has the greatest specific heat of all liquids.
(c) Water has the greatest heat of vaporization.
(d) The boiling point of H_2O is unusually high when compared to similar small molecules.
(e) Water dissolves most substances to some extent.
(f) The surface tension of water is greater than that of most liquids.

Supplementary Readings

10.1. *Liquids and Solutions*, O. Dreisbach, Houghton Mifflin, Boston Mass., 1966.
10.2. "The Water Cycle," H. L. Penman, *Sci. American*, Sept., 1970; p. 98.
10.3. "Chemical Reactions and the Composition of Sea Water," K. E. Chave, *J. Chem. Educ.*, **48**, 148 (1971).

Chapter 11

The Solid State
of Matter

Properties of Solids

In general, the forces of attraction between atoms, molecules, or ions of solids are much stronger than in liquids and gases; thus a solid possesses rigidity and mechanical strength. While pressure and temperature greatly affect the volume of a gas, and to a lesser degree the volume of a liquid, they have little effect on the volume of a solid. If a piece of iron is heated, expansion or increase in volume does occur, but the change is relatively small; high pressure will diminish the volume of a piece of iron only slightly.

Solids may be classified as **crystalline** or **amorphous.** A crystalline solid is one in which the atoms are arranged in a definite geometric pattern constantly repeated. An amorphous substance does not possess this orderly arrangement. When a crystalline solid is heated, the transition from the solid to the liquid state is sharp and distinct; the solid changes state at a definite temperature, called the melting point. If an amorphous substance such as glass, is heated, it gradually softens and becomes less viscous as the temperature is raised, but no definite point of transition is recorded. In other words, a crystalline substance possesses a definite melting point; an amorphous substance does not.

Although molecules are held tightly to the surface of solids and there is little freedom of movement, a solid may exert an appreciable vapor pressure. Solid iodine, for example, evaporates slowly at ordinary temperatures.

Crystal Lattice

The study of crystal forms and their properties is termed **crystallography.** When a substance crystallizes, the crystals are bounded by plane surfaces and have a definite and characteristic form; the faces of the crystal meet at definite angles, and the edges are straight lines.

A crystal has a definite shape because of the orderly arrangement of the units that compose it. These units may be atoms, ions, or molecules, depending on the electrovalent or covalent nature of the substance. If the units are represented as points, the pattern or configuration of these points, which has all the elements of symmetry of the whole crystal, is called the **space** or **crystal lattice** for that particular substance.

Types of Crystal Lattices

Many arrangements of the units of a crystal are possible. Suppose we assume that atoms, ions, or molecules are spherical, like marbles. Imagine the ways in which a number of marbles might be grouped or packed together.

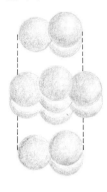

Figure 11.1
Hexagonal close packing; enlarged views showing how the
contact with the central sphere is achieved.

The tightest packing would result from the arrangement shown in Figure 11.1. In one plane, six marbles will exactly fit around a central one. Three marbles (or spheres) will seat themselves in place above and also below the initial pattern of seven spheres, so that twelve spheres are now spatially placed and touching the thirteenth or central sphere. This central sphere is said to have a **coordination number** of twelve — it has twelve spatially-placed intermass contacts. The packing arrangement* is termed **hexagonal.** The coordination number is sometimes referred to as representing the number of nearest neighbors for any given unit in the crystal.

In another arrangement, a unit of the crystal lattice may be at the center of a cube with each of the eight corners occupied by another unit, in which case the coordination number is eight. Similarly, a unit in the center of an octahedron will have a coordination number of six; a unit in the center of a tetrahedron will have a coordination number of four.

Lattice arrangements become more complex with crystals of compounds, which are composed of ions or molecules rather than of like atoms.

Six Crystal Systems

In spite of the many complex patterns which the solid state may assume, all crystal forms may be resolved into six fundamental crystal systems, as shown in Figure 11.2. In these diagrams the plane surfaces of crystals are related to imaginary lines or axes.

The structural units of the crystal system may be described by the length of three crystallographic axes (a,b,c) and the three angles between them

---

* Density differences in minerals of the same composition but having different crystal lattices indicate differences in packing arrangements. For example, the substance **quartz** (SiO_2) of density 2.66, is hexagonal, whereas the mineral **cristobalite** (also SiO_2), of density 2.32, is cubic.

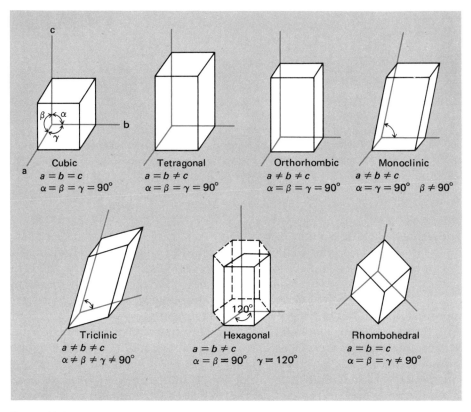

Figure 11.2

Crystal systems.

(α, β, γ). The hexagonal crystal system is sometimes divided into two systems which are closely related, hexagonal and rhombohedral.

1. **Cubic system:** Three axes of equal length, all at right angles to each other. Examples: NaCl, CaO, Ag.

2. **Tetragonal system:** Three axes, two of equal length, all intersecting at right angles. Examples: SnO_2, MgF_2.

3. **Orthorhombic system:** Three axes of unequal length, all intersecting each other at right angles. Examples: K_2SO_4, I_2.

4. **Monoclinic system:** Three axes of unequal length, two of which intersect obliquely, while the third intersects the other two at right angles. Examples: $KClO_3$, As_2S_3.

5. **Triclinic system:** Three axes of unequal length, none of which intersect at right angles. Examples: $CuSO_4 \cdot 5\ H_2O$, CuO.

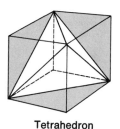

Tetrahedron

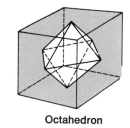

Octahedron

Figure 11.3
A tetrahedron or an octahedron may be fitted into a cube. Axes are of equal length and are at right angles.

6. **Hexagonal system:** Two axes of equal length in one plane intersecting each other at 120° angles, and a third axis at right angles to the plane of the other two. Examples: SiO_2, AgI.
   Rhombohedral system: Three axes of equal length, none of which intersect at right angles. Examples: Al_2O_3, NiS, $CaCO_3$ (calcite).

Facial Variations

An octahedron and tetrahedron belong to the cubic crystal system, since they have axes of equal length, all at right angles (Figure 11.3). NaCl, for example, may crystallize in cubes or octahedra according to certain conditions. There are some facial variations of each of the crystal systems. Although combination of two or more facial patterns of a given crystal system adds to crystallographic complexity, this is sometimes an identifying property for a given substance.

Isomorphous; Polymorphous

Two substances which crystallize in the same system and have corresponding axes of approximately the same dimensions and equal angles between corresponding faces are said to be **isomorphous** — for example, the alums (Chapter 27). A substance may sometimes crystallize in more than one form; sulfur, for example, may crystallize in the orthorhombic or the monoclinic system, depending on the temperature. Such a substance is said to be **polymorphous.** Each of the crystal forms that a substance may exhibit is termed an **allotropic form.**

Various minerals exhibit polymorphism. ZnS exists as cubic **sphalerite** or, less frequently, as hexagonal **wurtzite.** $CaCO_3$ commonly occurs as hexagonal **calcite,** and also as orthorhombic **aragonite.**

Cubic Lattices

The smallest portion of the crystal which shows the complete pattern of the particles in their relative positions is called a **unit cell.** The space lattice may be produced by repetition of the unit cell in three dimensions. While the cubic system is the simplest of the six crystal systems listed previously, three arrangements of the particles in a crystal belonging to the cubic system are possible (Figure 11.4).

1. **Simple cubic lattice,** in which atoms, ions, or molecules are located only at the corners of the cube (Figure 11.4abc).
2. **Body-centered lattice,** in which a unit particle is located at each of the corners of a cube and also in the center of the cube, equidistant from the eight corners. Iron, sodium, and potassium are examples of this type (Figure 11.4def).
3. **Face-centered lattice,** in which the unit particles are located at each corner and the center faces of a cube. Gold, silver, and copper crystallize in this pattern (Figure 11.4ghi).

In Figures 11.4 a, d, and g, the lattice points or small spheres represent only the centers of the atoms, ions, or molecules. The arrangement of the units is generally easier to follow in these open structures. In the actual crystalline solid, however, the units are so close together that they are touching each other along certain directions as shown by the other drawings (space-filling models) in Figure 11.4. In each of the three basic arrangements, the cube shown by solid lines is the unit cell. We may note that in most cases only a fraction of an atom (or other unit particle) appears within a unit cell; the rest belongs to adjacent unit cells. For example, in the simple cubic unit cell, as seen in Figure 11.4c, only $\frac{1}{8}$ of the atom at each corner is within the cube. We may say, then, that the unit cell in this case contains the equivalent of one atom — that is, $\frac{1}{8}$ atom/corner $\times$ 8 corners = 1 atom.

The following rules may be used to determine the number of unit particles (atoms, ions, etc.) associated with each type of cubic unit cell:

(a) An atom at a corner contributes $\frac{1}{8}$ of its volume to each of 8 adjacent cubes (Figure 11.4c).
(b) An atom in the face of a cube contributes $\frac{1}{2}$ of its volume to each of 2 adjacent cubes (Figure 11.4i).
(c) An atom completely within a cube contributes all of its volume to the unit cell (Figure 11.4f).

Notice that in the space-filling model of a simple-cubic unit cell (Figure 11.4 b and c) the atoms **touch along the cube edge.** In a body-centered cubic unit cell (Figure 11.4 e and f) the atoms **touch along the cube diagonal** from the lower left-hand front atom to the upper right-hand back atom. Finally, in the face-centered cubic unit cell (Figure 11.4 h and i) the atoms **touch along a face diagonal.**

Study the following sample problems.

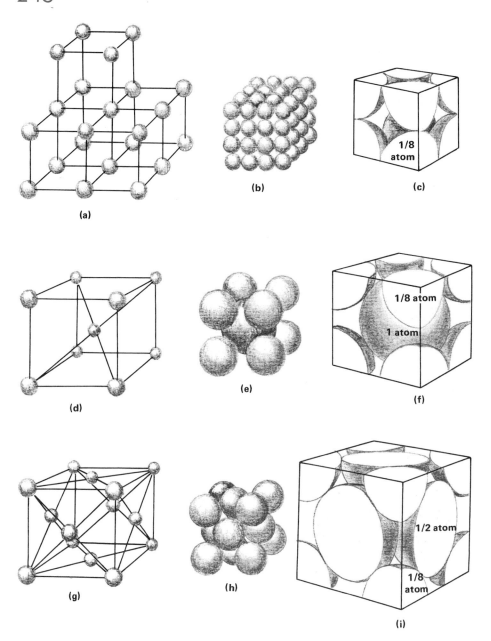

Figure 11.4
Unit cells in cubic lattices: (abc) simple cube lattice; (def) body-centered lattice; (ghi) face-centered lattice.

EXAMPLE 11.1 Refer to Figure 11.4 and determine the number of unit particles (atoms) in a unit cell of each of the three types of lattices.

SOLUTION: Apply the rules above.

Simple cubic Total atoms in a unit cell $= 8$ corners $\times \frac{1}{8}$ atom/corner $= 1$ atom.

Body-centered Total atoms in a unit cell $= 8$ corners $\times \frac{1}{8}$ atom/corner $+ 1$ atom in center $= 2$ atoms.

Face-centered Total atoms in a unit cell $= 8$ corners $\times \frac{1}{8}$ atom/corner $+ 6$ faces $\times \frac{1}{2}$ atom/face $= 4$ atoms.

EXAMPLE 11.2 Consider a unit cell of sodium chloride shown in Figure 11.10a. (a) What fraction of a Cl^- at each corner of the cube is within the unit cell? (b) What fraction of a Cl^- appears in each face? (c) What fraction of each Na^+ is within the unit cell? (Note that one complete Na^+ is in the center.) (d) Add all the fractions for each ion within the unit cell. What is the ratio of Na^+ to Cl^-?

SOLUTION: Using the rules above, (a) $\frac{1}{8}$ of a Cl^- is at each corner (b) $\frac{1}{2}$ of a Cl^- is in each face, and (c) $\frac{1}{4}$ of an Na^+ is on each edge; (d) Total ions in unit cell:

$$8 \text{ corners} - 8 \times \tfrac{1}{8} = 1 \ Cl^- \qquad 12 \text{ edges} - 12 \times \tfrac{1}{4} = 3 \ Na^+$$
$$6 \text{ faces} - 6 \times \tfrac{1}{2} = \underline{3 \ Cl^-} \qquad 1 \text{ center} - 1 \times 1 = \underline{1 \ Na^+}$$
$$\phantom{6 \text{ faces} - 6 \times \tfrac{1}{2}} 4 \ Cl^- \qquad\qquad\qquad\qquad 4 \ Na^+$$

Ratio $4/4 = 1:1$

EXAMPLE 11.3 Iron may crystallize in the face-centered cubic system. If the radius of an Fe atom is 1.26 Å, (a) Determine the length of the unit cell. (b) Calculate the density of Fe if its atomic weight is 55.85.

SOLUTION: (a) Picture the face of a cube with an iron atom at each corner and one in the center of each face

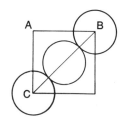

The diagonal distance BC will be equal to 4 radii of Fe atoms — that is,

$$4 \times 1.26 \text{ Å} = 5.04 \text{ Å}$$

By geometry, $$\overline{AB}^2 + \overline{AC}^2 = \overline{BC}^2$$

Since $$AB = AC, \text{ then } 2\,\overline{AB}^2 = \overline{BC}^2$$

then $$AB = \frac{\sqrt{\overline{BC}^2}}{\sqrt{2}} = \frac{BC}{\sqrt{2}} = \frac{5.04}{1.42} = 3.55 \text{ Å},$$

which is the length of the unit cell.

(b) $$\text{Density of iron} = \frac{\text{Mass of unit cell in g}}{\text{Volume of unit cell in cm}^3}$$

$$= \frac{\dfrac{4 \text{ atoms/unit cell} \times 55.85 \text{ g/mole}}{6.02 \times 10^{23} \text{ atoms/mole}}}{(3.55 \times 10^{-8} \text{ cm})^3/\text{unit cell}} = 8.30 \text{ g/cm}^3$$

Some covalent compounds crystallize in the cubic system, particularly if the molecules can approach a spherical shape. The more complex substances, however, usually crystallize in some pattern other than cubic. The majority of organic compounds crystallize in the monoclinic system. The unit cell of cellulose, asbestos, and other fibrous materials has been shown by X rays to possess a long bar or stringlike structure. A virus protein crystal is shown in Figure 11.5.

When all of the possible kinds of unit cells (simple, body-centered, face-centered, and in addition, base-centered for some crystal systems) are combined with the six crystal systems, only 14 different arrangements are possible. These are known as the 14 **Bravais space lattices** (Figure 11.6). Any other possible arrangement, for example a face-centered tetragonal, when the lines are redrawn, turns out to be identical with one of the 14 space lattices in Figure 11.6.

The Use of X rays to Study Crystal Structure

A beam of X rays striking a crystal penetrates the crystalline surface, and since the wavelength (λ) of the X rays is on the order of the spacing between the atoms, some of the X rays are diffracted by the atoms to give a symmetrical pattern of dots when recorded on a photographic film (Figure 11.7).

While it is evident that the pattern of dots on the photograph is closely related to the actual arrangement of atoms in the crystal. (note, for example,

Figure 11.5
Electron microscope view of virus protein crystal showing its molecularly well-ordered structure (50,000 x). The authors gratefully acknowledge permission to L. W. Labow and R. W. G. Wycoff for the use of the above protein microphotograph.

the two-sided appearance of the X-ray diffraction photograph of N_2H_5Cl, which is in the orthorhombic crystal system), the actual crystal structure determination is a highly sophisticated research technique.

If a beam of monochromatic X rays (X rays of a single wave length) strikes a crystal at certain angles (θ), the diffracted X rays will be in phase and will reinforce one another; at other angles, the X rays will show interference. A relationship between wave length (λ), the angle of incidence (θ), and the distance between planes (d) may be established. In Figure 11.8 consider the triangle ABR. AB $= d$; Angle RAB $= \theta$. Thus

$$RB = d \sin \theta$$
then
$$RBS = 2d \sin \theta$$

Now if the wave length $\lambda =$ RBS (waves will remain in phase)

then
$$n\lambda = 2d \sin \theta$$

where n is a small whole number, usually 1.

Simple cubic   Body-centered cubic   Face-centered cubic

Simple
orthorhombic

Base-centered
orthorhombic

Body-centered
orthorhombic

Face-centered
orthorhombic

Simple triclinic   Rhombohedral   Hexagonal

Simple
monoclinic

Base-centered
monoclinic

Simple
tetragonal

Body-centered
tetragonal

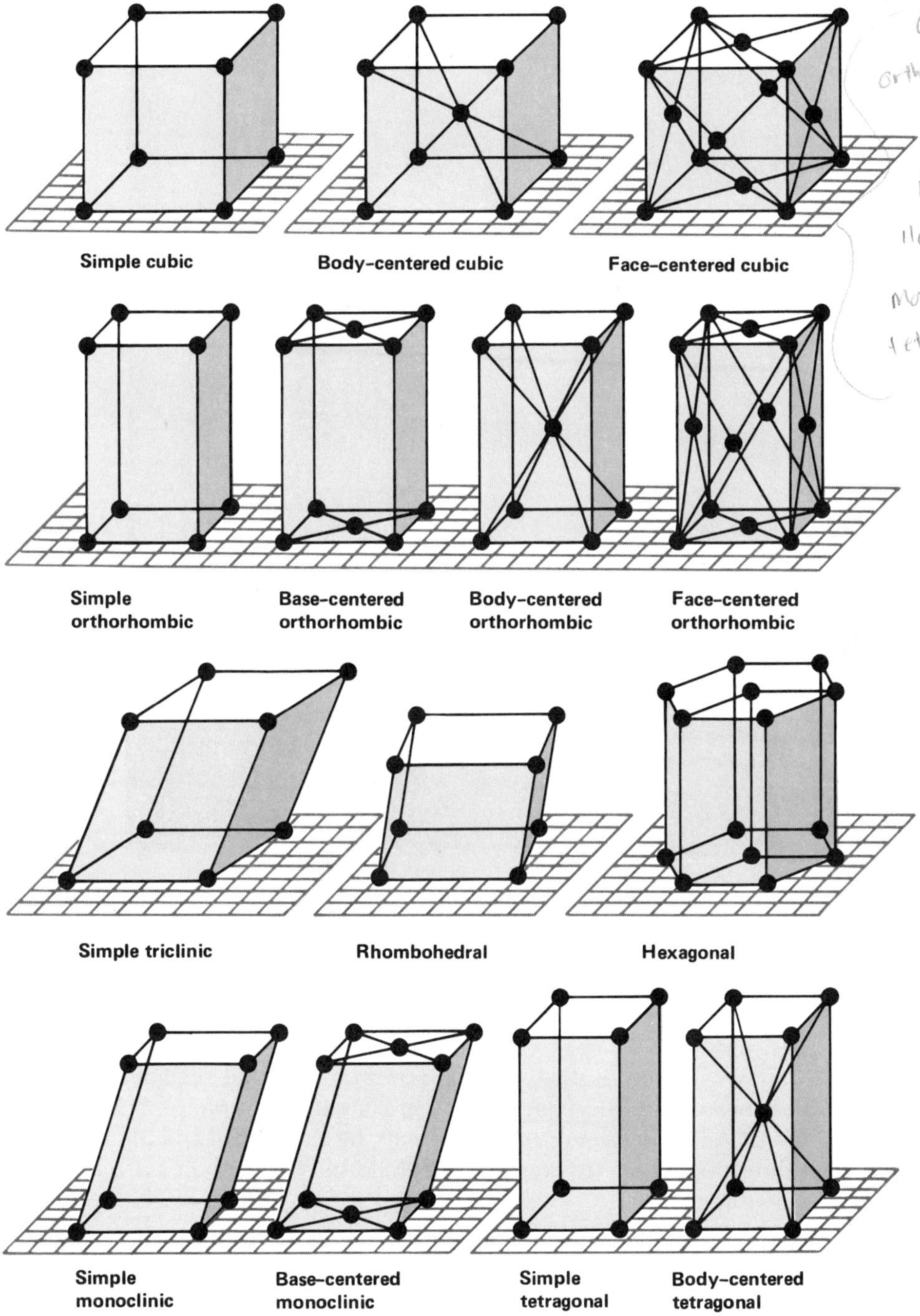

Figure 11.6
The 14 Space Lattices.

From "The Solid State" by Sir Nevill Mott, Copyright © September 1967 by Scientific American, Inc. All rights reserved.

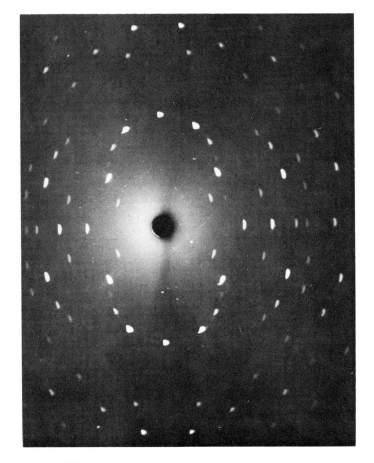

Figure 11.7
Laue photograph of hydrazine chloride, N_2H_5Cl.

This is known as **Bragg's equation.** An experiment may be set up using X rays of known wave length λ, and directed on a crystal surface at different angles. At some determined angle, θ, the rays will be in phase. If λ and θ are known, the distance between the planes may be calculated and the size of the unit cell determined, or if θ and the distance between planes are known, λ may be calculated.

The instrument used to obtain interfacial distances in crystals is called an **X-ray spectrometer.** As the angle is changed, variations of intensity of the diffracted beam are measured, and the results are plotted as a curve showing a series of peaks. Similar graphs are plotted for each of the principal planes by turning the crystal in its setting. For example, X rays may be diffracted from octahedral faces of cubic-system crystals, as well as from the normal

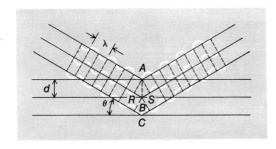

Figure 11.8
Diffraction of X rays by parallel crystal planes.

Figure 11.9

X-ray powder photograph of metallic nickel.

cubic faces. From a study of these graphs the arrangement of the crystal units (atoms, molecules, or ions) may be deduced.

A modified X-ray spectrometer may be used with a powdered sample of the crystalline material. The many crystals in a powdered sample present multiple angles for diffraction. A characteristic diffraction pattern is obtained for a given material, and so the procedure is worthwhile analytically. Figure 11.9 shows an X-ray powder photograph of nickel.

No two compounds have both the same line positions and line intensities. Consequently, X-ray powder diffraction is a powerful modern tool for the qualitative analysis of solid materials.

Examples of Crystal Structures

Figures 11.10a and b illustrate the crystal structure of NaCl as determined by X-ray diffraction. In the space-filling model (Figure 11.10a) the larger blue balls represent Cl⁻ ions (actually presumed to be touching one another along

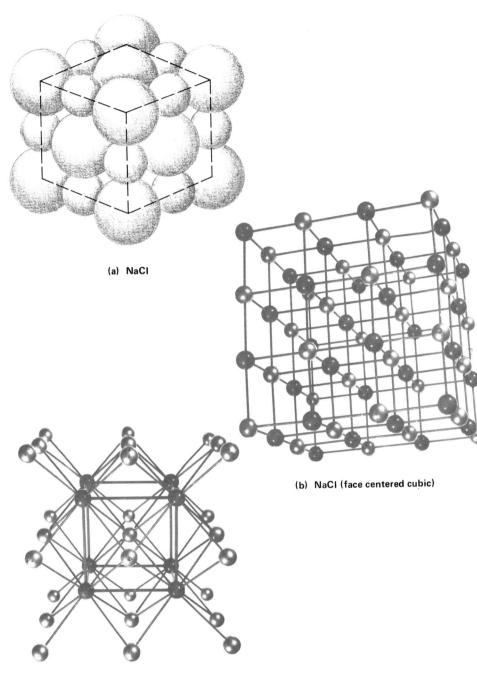

(a) NaCl

(b) NaCl (face centered cubic)

(c) CsCl (body centered cubic)

Figure 11.10

Crystal structures.

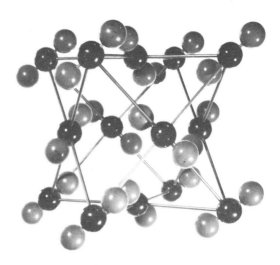

(d) Solid CO_2 (body centered cubic)

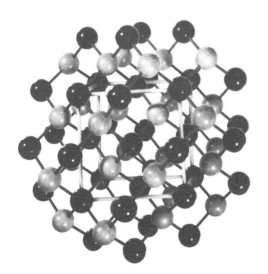

(e) C, diamond (body centered cubic)

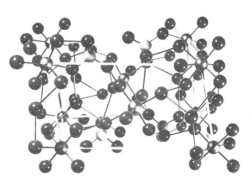

(f) $CuSO_4 \cdot 5H_2O$ (triclinic)

a diagonal of a face, in a face-centered cubic lattice). Between each pair of Cl^- ions along each edge are Na^+ ions represented by the smaller black balls, also in a face-centered cubic lattice. The crystal structure of NaCl, then, is *two* interpenetrating face-centered cubic lattices (Figure 11.10b); one, represented by lighter balls, is displaced one-half of a unit-cell length from the other (dark balls). Note that the unit cell of NaCl is the whole unit shown in Figure 11.10a rather than one of the 8 cubes shown in the figure. These small cubes have different ions at the corners of the cubes.

The crystal structure of cesium chloride, CsCl, is body-centered cubic (Figure 11.10c). Here the Cl^- ions (dark balls at the corners of the cube) form a simple-cubic lattice with a Cs^+ ion (light balls) in the center of the cube.

Solid CO_2 (Figure 11.10d) is also a face-centered cubic structure. However, the units at the lattice points (corners of the cube and centers of the faces) are CO_2 *molecules* rather than ions. The dark balls represent C atoms and the light balls oxygen atoms.

The structure of diamond (Figure 11.10e) consists of two face-centered cubic lattices of C *atoms*, one lattice (light balls) displaced from the other (dark balls) one-quarter of the way along a cube diagonal.

Metals usually have simple structures, body-centered cubic (Figure 11.4d), face-centered (Figure 11.4g), or hexagonal close-packed (Figure 11.1).

Crystal structures of most substances are much more complex than those given above. The $CuSO_4 \cdot 5H_2O$ structure, for example, is shown in Figure 14.10f. It is a triclinic crystal, the unit cell being outlined by the white wires. Cu^{2+} ions are at the corners of the unit cell and the lightest balls are S atoms to which 4 oxygen atoms are attached. The other dark balls represent H_2O molecules. A complex silicate structure (Garnet) is shown in Chapter 23, Figure 23.4. Garnet belongs to the cubic crystal system (face-centered cubic unit cell).

All crystal structures, however, even those with hundreds or thousands of atoms in a unit cell, must belong to one of the six crystal systems and possess one of the 14 space lattices.

Types of Crystalline Solids

Based on the units occupying lattice points in a unit cell, crystalline solids may be classified as ionic, molecular (polar and nonpolar), covalent (or atomic), and metallic. This is summarized in Table 11.1.

Ionic Crystals

Most simple salts of inorganic acids are composed of ions which occupy definite lattice points in a crystal; metal ions are called **cations,** and nonmetal ions or radicals are **anions.** The crystalline form assumed by an ionic sub-

Table 11.1

Types of crystalline solids

Crystal	Particles	Attractive Forces	Properties	Examples
ionic	positive and negative ions	electrostatic attractions	high m.p. hard, brittle good electrical conductor in fused state	$NaCl$, CaF_2, KNO_3
polar molecular	polar molecules	dipole–dipole	low m.p. soft poor electrical conductor	H_2O, HCl, SO_2
nonpolar molecular	nonpolar molecules	van der Waals forces	very low m.p. soft extremely poor electrical conductor	H_2, I_2, CO_2, CCl_4
covalent (atomic)	atoms	covalent bonds	very high m.p. very hard nonconductor of electricity	C (diamond) SiC, AlN, SiO_2
metallic	positive ions mobile electrons	metallic bonds	fairly high m.p. hard or soft malleable and ductile good electrical conductor	Ag, Cu, Na, Fe, Zn

(Source: Charles E. Mortimer, *Chemistry, A Conceptual Approach, Third Edition*, D. Van Nostrand Company, 1974)

stance depends on a number of factors, including (1) the ratio of the number of cation to anion units—for example, KBr, $CaBr_2$, $ScBr_3$—and (2) comparative sizes of cation and anion.

In Figure 4.2 we note that Na^+ has an ionic radius of 0.97 Å; Cs^+, 1.67 Å; and Cl^-, 1.80 Å. Crystalline NaCl is face-centered cubic, but CsCl, in which Cs^+ and Cl^- are more nearly the same size, is body-centered cubic. When the anionic size—for example, $B_4O_7^{2-}$, SO_4^{2-}, SiO_4^{4-}, $Si_2O_7^{6-}$, PO_4^{3-}—is much greater than cationic size, crystals are not likely to be cubic. Comparatively small cations tend to fit into holes throughout an anionic network—this is particularly true of most silicates.

Silicates comprise such a large percentage of the earth's rock and soil that a study of their crystal structures is important. The basic unit in silicate struc-

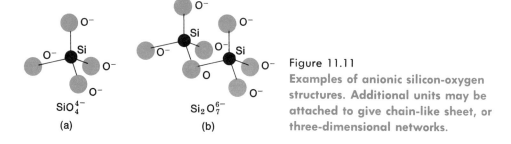

Figure 11.11
Examples of anionic silicon-oxygen structures. Additional units may be attached to give chain-like sheet, or three-dimensional networks.

tures is the SiO_4^{4-} ion, which has a tetrahedral structure (Figure 11.11a). Two such units may share an oxygen atom to form a disilicate ion, $Si_2O_7^{-6}$ (Figure 11.11b), or additional units may give larger, complex polysilicate ions. Large anionic complexes can be built from these SiO_4^{4-} tetrahedra to give chain-like structures, such as asbestos, $Mg_3Ca(SiO_3)_4 \cdot H_2O$; or planar or sheet-like structures, such as mica, $K_2O \cdot Al_2O_3 \cdot Fe_2O_3 \cdot 6\ SiO_2 \cdot 2\ H_2O$. Three-dimensional cross-linking between silicate anionic units, sometimes coupled with aluminate anions (AlO_3^{3-}), is usual in earth rocks. Various cations, such as Li^+, Na^+, K^+, Ca^{2+}, Mg^{2+}, Fe^{2+}, seem to fill holes throughout the anionic network.

Polar Molecular Solids

The structure of ice is an example of a polar molecular solid. The water molecule is unsymmetrical and therefore a polar substance (Figure 11.12a). Since oxygen is a highly electronegative element, water molecules associate

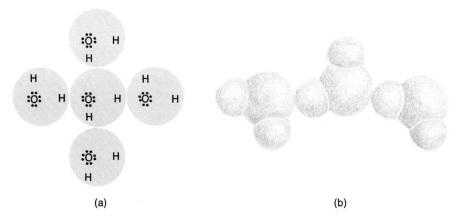

Figure 11.12
(a) Structure of ice. (b) Molecules of H_2O bound together by hydrogen bonds.

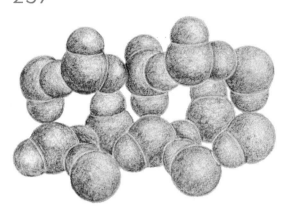

Figure 11.13
**Representation of the porous
structure of ice.**

by hydrogen bonding in both the liquid and the solid states (Figure 11.12b).
Ice has a tetrahedral structure in which each molecule of water is surrounded
by four other molecules (Figure 11.13). The result is a rather open, three-
dimensional network of tetrahedral symmetry, with channels or holes throughout the solid. This open structure accounts for the fact that ice is less dense
than liquid water. As ice melts, the molecules move closer together and the
density increases.

Nonpolar Molecular Solids

In this type of solid, simple nonpolar molecules such as those of CO_2 (Figure
11.10d) occupy the lattice points and only secondary valence forces bind these
units together in the crystal. Iodine is a good example—two iodine atoms
combine to form a simple molecule of I_2. In a crystal of iodine, the I_2 molecules are held together by weak forces. Such solids have relatively low
melting points, since little energy is required to disrupt the weak forces. In
general, the melting points of covalent substances are much lower than those
of ionic substances (Table 11.2).

Table 11.2
Comparative melting points of selected substances

Substance	Valence Type	Melting Point, °C
Diamond	close-packed, tetrahedral covalency, small atomic diameter	3500
Rhombic S; I_2	covalent	114; 113
CCl_4; $SiCl_4$	covalent	−23; −70
H_2O; CO_2	covalent	0; −56
C_8H_{18}	covalent	−57
$SnCl_4$	covalent	−33
$SnCl_2$; $CaCl_2$	ionic	246; 770
NaCl; K_2SO_4	ionic	800; 1070

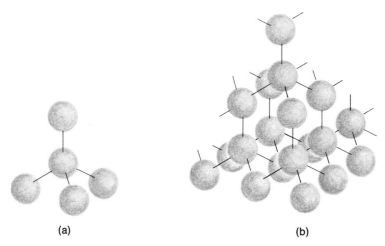

(a)   (b)

Figure 11.14
Structure of diamond. (a) Each carbon atom is bound to four other carbon atoms
in a close-packed tetrahedral structure. (b) Tetrahedrons build up to a giant
molecule.

Covalent or Atomic Solids

Nonmetallic elements are characterized by the relatively large number of
electrons in the valence level—for instance, carbon has 4, phosphorus, 5,
sulfur, 6. These electrons are available for bond formation, and as a result
many atoms of certain nonmetallic elements may become bonded together in
large aggregates or crystals, which may be considered *giant* molecules. A
crystal of diamond (discussed below) is an example. Certain compounds of
nonmetals may act in the same manner. A fragment of silicon carbide (SiC) is
composed of many covalently bonded atoms with a huge molecular weight.

Carbon crystallizes in two forms—diamond and graphite. In the diamond
(Figure 11.14 and 11.10e), each carbon atom is bound to four other carbon
atoms in a tetrahedral structure, with one atom in the center and one atom at
each of the four corners of the tetrahedron. These atoms are held together by
strong covalent forces, and thus diamond is hard, strong, and has a high
melting point.

In contrast with diamond, graphite crystals (Figure 11.15) have a two-
dimensional structure, in which carbon atoms are arranged in parallel planes
of connecting hexagons. The atoms in a given plane are held together by
strong covalent bonds, but the forces between planes are weak. As a result,
the planes can readily slide past one another, giving graphite its flaky character and lubricating properties.

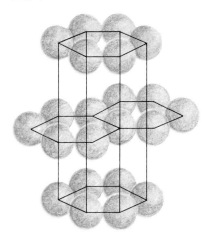

Figure 11.15
Structure of graphite — carbon atoms lie in
planes of connecting hexagons.

Metallic Solids

Metals have relatively few electrons in the valence level and do not exhibit the covalent bonding tendencies of nonmetals to form molecules (I_2 P_4, S_8, and so on). In a sense, metals are **electron deficient** — not many valence electrons are available for bond formation.

Metallic atoms are bound together in a closepacked structure, usually hexagonal or cubic. Presumably the few valence electrons permeate the entire structure as a "cloud." The nature of the bonding, called **metallic bonding** or valency, is obscure, but the ions of the metal are probably held together by attractive forces of the electron cloud.

The strength of metallic bonds varies progressively from the comparatively bulky alkali metal atoms with a single outer electron, through the alkaline earth atoms of fair size and two outer electrons, to transition atoms which are smaller and which may be multivalent. An index of this is the comparative melting points — for example, potassium, 62°C; calcium, 810°C; manganese, 1260°C; iron, 1535°C.

Lattice or Crystal Defects

A perfect crystal is one in which all lattice points are occupied by units of the substance — ions, atoms, or molecules. Most crystals are not perfect — some units are missing or misplaced. For example, consider the missing and misplaced Zn^{2+} ions in the otherwise orderly arrangement of Zn^{2+} and S^{2-} ions in a crystal of ZnS (Figure 11.16).

Because of lattice defects, substances may not quite conform to the law of definite proportions; in the above compound the ratio of Zn^{2+} to S^{2-} is

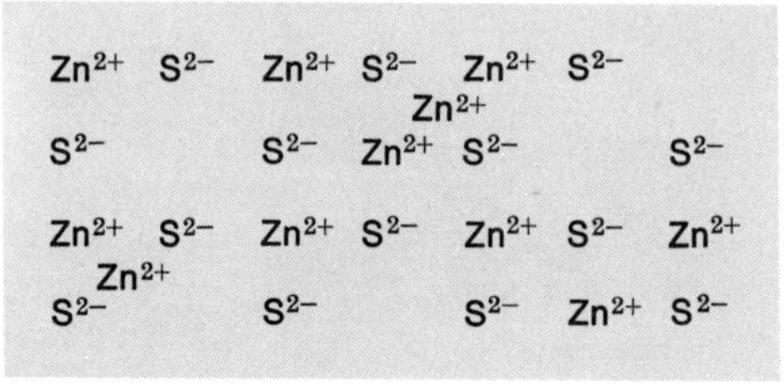

Zn²⁺ S²⁻   Zn²⁺ S²⁻   Zn²⁺ S²⁻
                  Zn²⁺
S²⁻   S²⁻ Zn²⁺ S²⁻   S²⁻

Zn²⁺ S²⁻   Zn²⁺ S²⁻   Zn²⁺ S²⁻ Zn²⁺
  Zn²⁺
S²⁻   S²⁻   S²⁻ Zn²⁺ S²⁻

Figure 11.16
Lattice defects in ZnS. Note absence of Zn^{2+} at some lattice points and also misplaced Zn^{2+}.

slightly less than 1:1. The cation is more commonly missing or misplaced, since it is the "hole filler" in anionic crystalline networks.

Lattice defects give rise to unique characteristics, such as fluorescence, phosphorescence, and semiconductivity—properties useful in the production of TV tubes, rectifiers, transistors, and other electronic devices.

Crystal Structure and Physical Properties

The physical properties of substances are related to their crystal structures. The structures of diamond and graphite, for instance, account for some of their properties.

The density of a solid is a function of the weight and radius of the atomic units in a crystal, and also of the interatomic or interionic distances between these units in the crystal lattice of the solid. Table 11.3 shows these distances in angstrom units for a few simple crystals.

Table 11.3
Distances between units in cubic crystal lattices

Substance	Type of Lattice	Length of Side of Unit Cell (angstroms)	Distance between Atoms or Ions° (angstroms)	Density in g/ml
Na	body-centered	4.30	3.72	0.97
Cu	face-centered	3.60	2.54	8.9
Diamond	face-centered	3.56	1.54	3.5
NaCl	face-centered	5.63	2.81	2.2
ZnS	face-centered	5.56	2.41	4.1

° A crystal of a metal only 1 mm on an edge would have about 4,000,000 atoms in a row.

We may note from the table that copper, for example, with its relatively high atomic weight and closer packing of atoms in a crystal, has a much higher density than sodium. Carbon (diamond), with relatively low atomic weight but very close packing and small interatomic distances, is intermediate in density.

Heat is a manifestation of atoms and molecules in motion. When a solid is melted, a considerable amount of heat is required (heat of fusion). In the solid, motion of the unit particles is highly restricted because of the quite fixed positions in the crystal lattice. There is evidence that some random motion or vibration occurs in the solid, since X-ray photos become increasingly "fuzzy" as the melting point is approached.

Specific Heats

Specific heats (number of calories required to raise the temperature of 1 g of a substance 1°C) of solids are in general much lower than those of liquids. Apparently a smaller mass motion factor is involved per 1°C change in the solid state. The specific heat of a given solid varies considerably with temperature, but the specific heat of all solids approaches zero at absolute zero temperature. The amount of heat necessary to raise the temperature of 1 g of most solid or liquid metals through 1°C is much less than that required for water.

An experimental finding of many years ago written as a law by **DuLong and Petit** states that **"the atomic weight of a metal multiplied by its specific heat approximates the constant 6.4."** (See Table 11.4.)

Table 11.4
Specific heats of a few metals

Element	At. Wt.		Specific Heat		Near Constant
Magnesium	24.30	×	0.246	=	6.0
Iron	55.85	×	0.120	=	6.70
Copper	63.5	×	0.092	=	5.84
Tin	118.7	×	0.054	=	6.41
Mercury	200.6	×	0.03325	=	6.67
Lead	207.2	×	0.0306	=	6.34

This relationship is a check on chemically-determined atomic weights; it also logically points to the like nature of the matter in all atoms in heat absorption. Note that the higher the atomic weight, the lower the specific heat.

Most solids expand when heated, and this thermal expansion is a further index of increased motion or vibration around fixed positions. Heat conduction, which varies with different solids, is also related to the transfer of energy of motion from particles in one part to particles in other parts of a solid.

Exercises

11.1, List the principal differences between the solid, liquid, and gaseous states of matter.

11.2. Distinguish between **crystalline** and **amorphous** solids in regard to (1) arrangement of the atoms or molecules, (2) melting points, (3) cleavage (breaking) direction, (4) X-ray diffraction patterns.

11.3. Define and illustrate: space lattice, unit cell, coordination number, polymorphous, lattice defect, crystal system, X rays, crystal planes, polar covalent solid.

11.4. For a base-centered monoclinic unit cell (Figure 11.6):
(a) How do a, b, c compare? α, β, γ?
(b) How many atoms are there per unit cell?

11.5. A unit cell of a body-centered space lattice has a $\neq$ b $\neq$ c and $\alpha = \beta = \gamma = 90°$.
(a) To what crystal system does it belong?
(b) What is the coordination number of the central atom?
(c) What is the number of atoms per unit cell?

11.6. Consider the lattice structures: (1) hexagonal and (2) face-centered cubic. Answer the following:
(a) Solid hydrogen (H_2) has structure (1). Which of the 5 types of solids (metallic, ionic, polar molecular, etc.) would this be?
(b) What would be present at each of the lattice points for solid H_2?
(c) Ice also has structure (1). Which of the 5 types of solids is ice?
(d) Copper is a *metallic* solid of structure (2). What units are at the lattice points?
(e) Pure Si is a covalent or atomic solid of structure (2). What would be the relative hardness of Si?
(f) Graphite has structure (1). Would it be a good conductor of electrons?
(g) NaCl has structure (2). The Na^+ ions are at the lattice points. Where are the Cl^- ions?
(h) Solid xenon has structure (2). In what direction do the atoms of Xe actually touch?

11.7. The dimensions (in angstrom units) and the axial angles of the unit cells of five minerals are listed in the following table. To what crystal system does each belong?

	a	b	c	α	β	γ
marcasite	3.35	4.40	5.35	90°	90°	90°
muscovite	5.18	9.02	20.04	90°	95°30′	90°
rhodonite	7.77	12.45	6.74	85°10′	94°4′	111°29′
rutile	4.58	4.58	2.95	90°	90°	90°
quartz	4.90	4.90	5.39	90°	90°	120°

11.8. Match each statement with the term in the list on the right with which it most closely corresponds.

(a) May be caused by dislocations in a crystal lattice
(b) Allotropic forms such as diamond and graphite
(c) The number of nearest neighboring atoms of an atom in a unit cell.
(d) (specific heat)(atomic weight) $\cong$ constant.
(e) The smallest part of a space lattice that possesses all of the symmetry and pattern of the whole lattice.
(f) The sides of a crystal.
(g) Glass.
(h) The a, b, c distances in a unit cell.
(i) NaCl and KCl, both face-centered cubic.
(j) The science of determining the structures of crystals.

1. crystalline solid
2. amorphous solid
3. crystal lattice
4. coordination number
5. crystal system
6. crystallographic axes
7. crystal faces
8. isomorphous
9. polymorphous
10. unit cell
11. X-ray crystallography
12. Bragg's equation
13. lattice defect
14. specific heat
15. law of DuLong and Petit

11.9. Make rough sketches of the following:
(a) A NaCl unit cell (label the sodium and chlorine).
(b) A lattice defect in a ZnS crystal.
(c) The structure of graphite (one of the allotropic forms of C).
(d) Solid CO_2 (face-centered cubic).

11.10. Calculate the following:
(a) Sodium crystallizes in a body-centered cubic unit cell. X-ray measurements show that the length of the unit cell side is 4.30Å. What is the *volume* of the unit cell in cm³ (cc)?
(b) If a single atom of Na weighs 3.82×10^{-23} g, what is the *mass* in grams of a unit cell of Na?
(c) What is the density of Na in g/cc?
(d) A metal, M, has an atomic radius of 2.5Å and crystallizes in a face-centered cubic lattice. What is the length of the unit cell edge in Å?
(e) A single atom of an unknown element weighs 9×10^{-23} g. What is the atomic weight of the element?

11.11. Calculate the following:
(a) Ni crystallizes in a face-centered cubic structure. The radius of the Ni atom is 1.25 Å. What is the length of the unit cell edge in Å?
(b) A unit cell of indium (face-centered tetragonal) has lattice dimensions of a = b = 4.59 Å and c = 4.94 Å. What is the volume of the unit cell in cm³?
(c) A unit cell of Ni (face-centered cubic) has a volume of 4.44×10^{-23} cm³. The density of Ni is 8.90 g/cm³. What is the weight of *one atom* of Ni?
(d) Which of the following would give Avogadro's number?

(1) $\dfrac{\text{wt. of 1 atom in g}}{\text{density}}$

(2) $\dfrac{\text{atomic wt.}}{\text{density (g/cm}^3)}$

(3) $\dfrac{\text{g atomic wt.}}{\text{wt. of 1 atom in g}}$

(4) specific heat × vol. of 1 atom

(5) $\dfrac{\text{vol. of unit cell}}{\text{mass of 1 atom}}$

11.12. Europium metal crystallizes in a body-centered lattice with the unit cell measuring 4.58 Å along each edge. (a) Determine the volume of a unit cell. (b) Calculate the volume of one mole of Eu. (c) Determine the density of Eu metal.

11.13. Silver crystallizes in the face-centered cubic system, with the closest distance between centers of silver atoms equal to 2.89 Å. (a) Determine the length of the edge of a unit cell. (b) Determine the mass of a unit cell. (c) What is the density of Ag?

11.14. NaCl crystallizes in the face-centered cubic system. If the radii of Na^+ and Cl^- ions are 0.95 Å and 1.81 Å, respectively, determine the length of each edge of the unit cell. Assume that the Cl^- and Na^+ ions touch.

11.15. Given the following data for the metal nickel: (a) crystallizes in face-centered cubic structure, (b) radius of an Ni atom is 1.246 Å, (c) density of nickel is 8.90 g/cm³, (d) atomic weight is 58.71. Assuming that nickel atoms are hard spheres and that the corner atoms touch the atom in the center of the face of the unit cell, calculate (1) the length of one edge of the unit cell, (2) Avogadro's number.

11.16. Ytterbium crystallizes in the face-centered cubic system, with the edge of the unit cell 5.48 Å. Calculate the density of solid Yb.

11.17. Tantalum crystallizes in a cubic system and the edge of the unit cell is 3.30 Å. The density of Ta is 16.6 g/cm³. How many atoms are contained in one unit cell of Ta? What type of cubic unit cell does Ta form?

11.18. The density of platinum is 21.5 g/cm³ and Pt crystallizes in a face-centered cubic lattice with the edge of the unit cell equal to 3.92 Å. Calculate the atomic weight of Pt from these data.

11.19. Potassium crystallizes in a body-centered cubic unit cell with the length of an edge equal to 5.33 Å. Calculate the dimensions of a cube that would contain exactly one gram atomic weight of K.

11.20. Aluminum crystallizes in a face-centered cubic unit cell with the length of an edge equal to 4.05 Å. Assume that the atoms are identical hard spheres and that a face-centered atom touches each of the four corner atoms of its face. (a) Calculate the length of the diagonal of a face. (b) By means of your answer to part (a), calculate the radius of the hard-sphere atoms that comprise the crystal.

11.21. Use the data given in Problem 11.20 and the answer to Problem 20 to calculate: (a) the volume of a unit cell of Al, (b) the total volume of the hard-sphere atoms contained in one unit cell, (c) the percentage of the volume of a unit cell that is empty space.

11.22. In determining the distance between planes in an NaCl crystal by diffraction of X rays, the following data were obtained: wavelength of X-ray beam, 0.52 Å; angle of incident beam, 5°18′. Using the Bragg equation, calculate the interplanar distance of Na^+ and Cl^-.

11.23. In the diffraction of a gold crystal using X rays with a wavelength equal to 1.54 Å, a first-order reflection is shown at an angle of 22°10′. What is the distance between the diffraction planes?

11.24. In the diffraction of a crystal using X rays with a wavelength equal to 1.54 Å, a first-order reflection is found at an angle of 16°. What is the wavelength of X rays that show this same reflection at an angle of 20°20′?

11.25. At what angle would a first-order reflection be observed in the X-ray diffraction of a set of crystal planes for which d is 3.03 Å if the X rays used have a wavelength of 0.710 Å? At what angle would a second-order reflection from this same set of planes be observed?

11.26. Describe the crystal structures of each of the following and outline the unit cells:
(a) copper (face-centered cubic)
(b) NaCl
(c) $CO_2(s)$
(d) diamond
(e) graphite

11.27. Explain the appearance of an X-ray diffraction photograph such as Figure 11.9.

11.28. The specific heat of antimony is 0.0504 cal/g°C. Calculate the approximate atomic weight of Sb.

11.29. Calculate the approximate specific heat of bismuth.

Supplementary Readings

11.1. "The Solid State," N. Mott, Sci. American, Sept., 1967; p. 80.
11.2. "X-Ray Crystallography," L. Bragg, Sci. American, July, 1968; p. 58.
11.3. "Positrons as a Probe of the Solid State," W. Brandt, Sci. American, July, 1975; p. 34.
11.4. "Other Views of Unit Cells," L. Suchow, J. Chem. Educ., 53, 226 (1976).

Chapter 12

Energy
Relationships
in Chemical
Changes

Energy changes accompany all chemical and most physical changes. Usually the energy is manifest as heat energy — those changes in which heat is evolved are termed **exothermic;** those in which heat is absorbed are **endothermic.**

Energy change of a given chemical change is termed the heat of reaction or **enthalpy** of that change, and is usually designated by the symbol ΔH. Its calorific value may be determined in a calorimeter, such as shown in Figure 12.1.

The usual type of calorimeter is an insulated container of water, equipped with a thermometer and a stirrer. Reactants in stoichiometric amounts are placed in a reaction vessel inserted into the water bath. When the reaction proceeds, the heat energy evolved or absorbed either warms or cools the water. The temperature before and after the chemical change is recorded. Once we know the weight of water present, the temperature change, and the specific heat of the reaction vessel and its contents, we can calculate the heat energy of the reaction.

Other types of calorimeters may be employed, depending upon the reaction to be carried out. For example, in a reaction of gases where the reaction may occur with explosive violence, the reactants are enclosed in a heavy steel container or "bomb."

Study the following sample calculations.

EXAMPLE 12.1 The following reaction using hydrogen and oxygen is carried out in a bomb calorimeter:

$$2\ H_2(g) + O_2(g) \rightarrow 2\ H_2O(l)$$

The following data are recorded: Weight of water in calorimeter = 2.650 kg. Initial temperature of water = 24.442°C. Final temperature of water after reaction = 25.635°C. Weight of reaction vessel = 1.060 kg. If the specific heat of the reaction vessel is 0.200 kcal/°-kg and the specific heat of water is 1.00 kcal/°-kg, calculate the heat of reaction. Assuming that 0.050 mole of water was formed in this experiment, calculate the heat of reaction per *mole* of liquid water formed. Neglect the specific heat of the thermometer and stirrer.

SOLUTION: The heat of reaction must be equal to the heat absorbed by the system, including the water and the reaction vessel.

No. of kcal absorbed by water =

$$2.650\ kg \times 1.0\ kcal/kg \times (25.635° - 24.442°) = 3.161\ kcal$$

No. of kcal absorbed by reaction vessel =

$$1.060\ kg \times 0.200\ kcal/kg \times (25.635° - 24.442°) = 0.253\ kcal$$

Total heat absorbed = 3.161 + 0.253 = 3.414 kcal

This is for the formation of 0.050 mole of water; therefore for each mole of $H_2O(l)$ formed, heat energy = 3.415/0.05 = 68.3 kcal

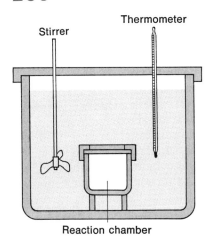

Figure 12.1

A calorimeter. The heat of reaction may be determined from the change in temperature of the system.

Various conventions are employed for showing heat changes in reactions, and we need to become familiar with these. For the exothermic union of H_2 and O_2 above, we may write

$$2 \, H_2(g) + O_2(g) \rightarrow 2 \, H_2O(l) + 136.6 \text{ kcal}$$

or we may use fractional coefficients so that we show the heat change per *mole* of the product, H_2O:

$$H_2(g) + \tfrac{1}{2} \, O_2(g) \rightarrow H_2O(l) + 68.3 \text{ kcal}$$

It is evident that the energy value shown in the equation is directly related to the *moles* of product formed — that is, 68.3 kcal evolved per mole of $H_2O(l)$ formed. A more frequent convention is to record the value of ΔH for the reaction apart from the equation:

$$2 \, H_2(g) + O_2(g) \rightarrow 2 \, H_2O(l) \qquad \Delta H = -136.6 \text{ kcal}$$

or

$$H_2(g) + \tfrac{1}{2} \, O_2(g) \rightarrow H_2O(l) \qquad \Delta H = -68.3 \text{ kcal}$$

with ΔH having a negative value for an exothermic reaction and a positive value for an endothermic process. In an exothermic process the products possess less energy than the reactants, and hence a negative value for ΔH is indicated.

Heats of Formation

The energy change involved in the formation of one mole of a compound from its elements in their normal state, is called the ~~heat~~ of formation, designated ΔH_f° (or enthalpy of formation). In the reaction discussed above between H_2

and O_2, the heat evolved per mole of $H_2O(l)$ is the heat of formation of the
latter. <u>Heats of formation of elementary substances are assigned a zero value.</u>
Some heats of formation are shown in Table 12.1.

[handwritten: ? Single elements Such as O₂ H₂]

Table 12.1

~~Heats~~ of formation* (substances are gases unless otherwise indicated)

Compound	$\Delta H°$ (kcal/mole)	Compound	$\Delta H°$ (kcal/mole)
$H_2O(g)$	− 57.8	C_3H_8	− 24.8
$H_2O(l)$	− 68.3	C_2H_2	54.3
$H_2O_2(l)$	− 44.5	C_2H_4	9.6
CO	− 26.4	CH_3OH (l)	− 60.0
CO_2	− 94.1	$C_2H_5OH(l)$	− 65.9
HCl	− 22.1	$C_6H_6(l)$	11.7
HF	− 64.0	$COCl_2$	− 53.3
SO_2	− 71.0	NO	21.6
SO_3	− 94.5	NO_2	8.1
H_2S	− 5.3	$P_4O_{10}(s)$	−720
$BaSO_4(s)$	−350	NH_3	− 11.0
$HNO_3(aq)$	− 49.8	ZnO(s)	− 84.4
$H_2SO_4(aq)$	−217	$Fe_2O_3(s)$	−197
CH_4	− 17.9	$Fe_3O_4(s)$	−267
C_2H_6	− 23.4	FeO(s)	− 63.7

° These are for the substances in their so-called *standard states*, 25°C and 1 atm.

Heats of Reaction

These may be calculated from heats of formation by simply subtracting the
sum of the heats of formation for all reactants from the sum of the heats of
formation of all products.

EXAMPLE 12.2   Determine $\Delta H°$ for the following reaction of burning ethyl
alcohol in oxygen: *[handwritten: ↳ enthalpy]*

SOLUTION:
$$C_2H_5OH(l) + 3\ O_2(g) \rightarrow 2\ CO_2(g) + 3\ H_2O(l) \qquad \Delta H° = ?$$

$\Delta H° =$	$\Delta H_f°$(products)	minus	$\Delta H_f°$(reactants)
	2 $CO_2(g)$ −188.2 kcal		$C_2H_5OH(l)$ −65.9 kcal
	3 $H_2O(l)$ −204.9 kcal		3 $O_2(g)$ 0.0 kcal
	−393.1 kcal		−65.9 kcal

$\Delta H° = (-393.1) - (-65.9) = -327.2$ kcal

$\Delta H°$ in this case may be termed the **heat of combustion** of C_2H_5OH. A few
other heats of combustion are tabulated in Table 12.2.

Table 12.2

Heats of combustion (in oxygen)

Compound	$\Delta H\,^{\circ}$ (kcal/mole)	Compound	$\Delta H\,^{\circ}$ (kcal/mole)
C_2H_6	−368	$C_6H_6(l)$	−782
C_3H_8	−526	$CH_3OH(l)$	−171
C_2H_2	−311	$C_2H_5OH(l)$	−327.2
C_2H_4	−332		

The Hess Law of Constant Heat Summation

According to this law, the heat of a given chemical change is the same whether the reaction proceeds in one or several steps; in other words, the energy change is independent of the path taken by the reaction. The heat for a given reaction is the algebraic sum of the heats of any sequence of reactions which will yield the reaction in question. For example, we may calculate ΔH for the reaction

$$C(s) + \tfrac{1}{2} O_2(g) \rightarrow CO(g)$$

from the algebraic sum of the two equations shown below. (Note that since the equation for the combustion of CO is reversed, the sign of ΔH is also reversed. Note also the cancellation of CO_2 and $\tfrac{1}{2} O_2$.)

$-94.1-(-26.4)$

$C(s) + O_2(g) \rightarrow CO_2(g)$	$\Delta H^{\circ} = -94.1$ kcal
$CO_2(g) \rightarrow CO(g) + \tfrac{1}{2} O_2(g)$	$\Delta H^{\circ} = \ \ \ 67.7$ kcal
$C(s) + \tfrac{1}{2} O_2(g) \rightarrow CO(g)$	$\Delta H^{\circ} = -26.4$ kcal

This method of arriving at a heat of reaction is very useful, since the reaction in question may be difficult to measure experimentally.

As a second example we may consider the combustion of methane as taking place in three steps

(1)	$CH_4(g) + O_2(g) \rightarrow C(s) + 2\ H_2O(l)$	$\Delta H^{\circ} = -118.7$ kcal
	$-17.9 + 0.0 \quad \rightarrow 0.0 + 2(-68.3)$	
(2)	$C(s) + \tfrac{1}{2} O_2(g) \rightarrow CO(g)$	$\Delta H^{\circ} = -26.4$ kcal
(3)	$CO(g) + \tfrac{1}{2} O_2(g) \rightarrow CO_2(g)$	$\Delta H^{\circ} = -67.7$ kcal
	$CH_4(g) + 2\ O_2(g) \rightarrow CO_2(g) + 2\ H_2O(l)$	$\Delta H^{\circ} = -212.8$ kcal

12.1 CHEMICAL THERMODYNAMICS

Chemical thermodynamics is concerned with the relationships between all forms of energy and includes a number of concepts in addition to enthalpy of reaction. These relationships are embodied in three fundamental laws:

First Law of Thermodynamics

In effect this law is simply a statement of the law of conservation of energy, i.e. energy may be converted from one form to another but may neither be created nor destroyed.

Every system possesses energy called **internal energy**, designated by E, which includes all of the energy of the system such as chemical or potential energy, bond energy, lattice energy, etc. When a system changes from one set of conditions to another (from State 1 to State 2), a change in the internal energy, ΔE, occurs which is equal to the heat absorbed (q) in the change minus any work (w) done by the system on the surroundings, i.e.

$$\Delta E = q - w$$

If the work done is so-called pressure-volume work, such as expansion of a gas against the atmosphere at constant pressure, i.e. $w = P\Delta V$ where P = pressure and ΔV = change in volume, then

$$\Delta E = q - P\Delta V$$

or

$$q = \Delta E + P\Delta V$$

The heat absorbed at constant pressure, q (heat of reaction), has previously been termed the change in **enthalpy** (ΔH), page 267; hence we may write

$$\Delta H = \Delta E + P\Delta V$$

In considering $P\Delta V$ work, the volume change in liquids and solids is relatively small and ordinarily may be neglected, hence $P\Delta V = 0$. However when gases are present in the reaction, there may be considerable volume change and the work may be calculated from the expression

$$\text{work} = P\Delta V = \Delta n\, RT$$

where Δn = number of moles of gaseous products minus the number of moles of gaseous reactants as shown by the balanced equation; R is the molar gas constant of 1.99 cal/°K mole, T is the absolute or kelvin temperature.

EXAMPLE 12.3   Calculate ΔH for the reaction at 25°C

$$CO(g) + \tfrac{1}{2} O_2(g) \rightarrow CO_2(g) \qquad \Delta E = -67.4 \text{ kcal}$$

SOLUTION:
$$\Delta H = \Delta E + P\Delta V = \Delta E + \Delta n\, RT$$
$$\Delta n = 1 - (1 + 0.5) = -0.5$$
$$\Delta n\, RT = (-0.5 \text{ mole})(1.99 \text{ cal/°K mole})(298°K) = -298 \text{ cal}$$
$$= \text{approximately } -0.3 \text{ kcal}$$
$$\Delta H = -67.4 \text{ kcal} + (-0.3 \text{ kcal}) = -67.7 \text{ kcal}$$

Obviously where the volume change is small as in this example, ΔE and ΔH have nearly the same values.

The Second Law

The first law is concerned only with the conservation of energy during changes in a system and not with the conditions under which changes might or might not occur. The second law does specify that only certain processes will take place spontaneously, e.g. heat flows from a hot object to a cooler one. A simple statement of the second law might be: A spontaneous process occurs only if there is an increase in **entropy** of a system and its surroundings.*
Entropy, denoted by S, is a measure of the disorder or randomness of a system. The entropy of a gas is greater than that of a liquid which in turn is greater than that of a solid; that is, there is more disorder in a gas than a liquid, in a liquid than in a solid, etc.

When an equilibrium process occurs at constant temperature, the change in entropy, ΔS, is equal to the heat absorbed divided by the absolute temperature at which the change occurs, i.e.

$$\Delta S = \frac{\Delta H}{T} \tag{1}$$

EXAMPLE 12.4   Calculate ΔS for the conversion of one mole of liquid water to vapor at 100°C. Heat of vaporization = 540 cal/g.

SOLUTION:
Molar heat of vaporization = 540 cal/g × 18 g/mole = 9720 cal/mole

$$\Delta S = \frac{\Delta H}{T} = \frac{9720 \text{ cal/mole}}{100 + 273°} = 26.1 \text{ cal/°K mole}$$

Since ΔS is positive, the entropy is increasing as the water passes from the liquid to the vapor state.

Free Energy

Another important property in considering the second law is the change in free energy of a system. Free energy, denoted by G, is the energy of a system

---

* This can be compared to the situation which exists when pool balls are "racked" in a triangular compact arrangement (a highly-ordered state) and the situation which exists immediately after the "break" (a highly-disordered state). If one were simply to throw fifteen pool balls onto a table, one would not expect them to achieve the ordered state. They are much more likely to achieve one of an almost infinite number of disordered states. The change from an ordered to a disordered state on the "break" corresponds to an increase in entropy.

which is *available* for useful work. The change in free energy, ΔG, of a system is related to ΔH, ΔS, T by the equation

$$\Delta G = \Delta H - T\Delta S \qquad (2)$$

ΔG is a measure of the tendency of a reaction to proceed spontaneously – it turns out that a negative value for ΔG means the process will occur in the direction shown by the chemical equation; if ΔG is positive, it will not proceed spontaneously. Table 12.3 includes values for free energy of formation, and entropy for a number of substances in their standard states at 1 atm pressure and 25°C, designated ΔG_f°, and S° respectively.

Table 12.3
Thermodynamic properties
    Values are for substances in their **standard states** at 25°C and 1 atm pressure. Free energies of formation (ΔG_f°) are expressed in kilocalories per mole; entropy (S°) values in calories per mole per degree absolute temperature.

Substance	ΔG_f°	S°	Substance	ΔG_f°	S°
$NH_3(g)$	−4.0	46.0	$CH_4(g)$	−12.1	44.5
$HCl(g)$	−22.8	44.6	$COCl_2(g)$	−50.3	69.1
$CO_2(g)$	−94.3	51.1	$SO_2(g)$	−71.8	59.4
$CO(g)$	−32.8	47.3	$O_2(g)$	0	49.0
$H_2O(g)$	−54.6	45.1	$H_2(g)$	0	31.2
$H_2O(l)$	−56.7	16.7	$Cl_2(g)$	0	53.3
$NO(g)$	20.7	50.3			

In a manner similar to determining ΔH° for a reaction, ΔG° may be obtained by subtracting the sum of the free energies of reactants from the sum of the free energies of the products, assuming that elements in the free state have ΔG° values of 0. Similarly ΔS° for a reaction may be determined (note though that entropy values for elements are not zero). See Table 12.3.

EXAMPLE 12.5   Determine ΔG° for the reaction

$$4\ NH_3(g) + 5\ O_2(g) \rightarrow 4\ NO(g) + 6\ H_2O(l)$$

SOLUTION:
$\Delta G^\circ = \Delta G_f^\circ$ (products) $- \Delta G_f^\circ$ (reactants). Obtain ΔG_f° values from Table 12.3.

$$\Delta G^\circ = [(4 \times 20.7) + (6 \times -56.7)] - [(4 \times -4.0) + (0)]$$
$$= -257.4 + 16.0 = -241.4\ \text{kcal}$$

We could also determine ΔG° for the reaction from the equation

$$\Delta G^\circ = \Delta H^\circ - T\Delta S^\circ$$

Obtain ΔH° from $\Delta H^\circ = \Delta H_f^\circ$ (products) $- \Delta H_f^\circ$ (reactants) See Table 12.1.

$$\Delta H° = [(4 \times 21.6) + (6 \times -68.3)] - [(4 \times -11.0)]$$
$$= -323.4 - (-44.0) = -279.4 \text{ kcal}$$

Obtain $\Delta S°$ from $\Delta S° = S°$ (products) $- S°$ (reactants). See Table 12.3.

$$\Delta S° = [(4 \times 50.3) + (6 \times 16.7)] - [(4 \times 46.0) + (5 \times 49.0)]$$
$$= 301.4 \text{ cal} - 429.0 \text{ cal} = -127.6 \text{ cal} = -0.1276 \text{ kcal}$$

Then substituting in $\Delta G° = \Delta H° - T\Delta S°$

$$\Delta G° = -279.4 - (298)(-0.1276) = -279.4 + 38.0 = -241.4 \text{ kcal}$$

The Third Law

This law states that the entropy of a perfectly ordered system at 0°K is zero. Such a system is a perfect crystalline substance where the units making up the crystal are arranged in a very definite pattern. There is no disorder or randomness and hence such a substance may be regarded as having a minimum value of entropy, i.e., zero. A few standard entropy values are recorded in Table 12.3.

EXAMPLE 12.6   Determine $\Delta S°$, $\Delta H°$, and $\Delta G°$ for the following reaction at 25°C.

$$CO(g) + Cl_2(g) \rightarrow COCl_2(g)$$

SOLUTION: Obtain $S°$ and $\Delta H_f°$ values from Tables 12.1 and 12.3

$$S° = S°(COCl_2) - [S°(CO) + S°(Cl_2)]$$
$$= 69.1 - (47.3 + 53.3) = -31.5 \text{ cal/°K-mole}$$
$$\Delta H° = \Delta H_f°(COCl_2) - [\Delta H_f°(CO) + \Delta H_f°(Cl_2)]$$
$$= (-53.3) - (-26.4 + 0)$$
$$= -26.9 \text{ kcal}$$
$$\Delta G° = \Delta H° - T\Delta S° = -26.9 - \frac{(298)(-31.5 \text{ cal})}{1000 \text{ cal/kcal}} = -26.9 + 9.4 = -17.5 \text{ kcal}$$

Since $\Delta G°$ is negative the reaction will proceed spontaneously.

Exercises

12.1. The heating value of a sample of coal was determined from the following data: weight of coal = 1.5000 g; weight of water in calorimeter = 2000 g; weight of calorimeter and contents except water = 2.250 kg; initial temperature of water = 20.892°C; final temperature of water = 24.326°C; specific heat of calorimeter = 0.2000 cal/g°C. Calculate the heating value (heat of reaction) of the coal in calories per gram.

12.2. A 2.50-g sample of sucrose ($C_{12}H_{22}O_{11}$) was burned in excess oxygen in a calorimeter which contained 1.95 kg of water. The temperature of the calorimeter and contents increased from 18.22°C to 22.73°C. The calorimeter had a water equivalent of 240 g. Determine the heat of combustion of sucrose in kcal/mole.

$$C_{12}H_{22}O_{11}(s) + 12\ O_2(g) \rightarrow 12\ CO_2(g) + 11\ H_2O(1)$$

12.3. The heat of combustion of glutaric acid ($H_2C_5H_6O_4[s]$) is 514.9 kcal/mole. The heat from the combustion of 1.50 g of glutaric acid is evolved in a calorimeter with a water equivalent of 530 g and containing 2250 g of water. What is the final temperature of the calorimeter and contents if the original temperature was 25.20°C?

12.4. From the heats of formation in Table 12.1, calculate the heat of reaction for the combustion of the following compounds in oxygen, assuming the products in each case to be carbon dioxide and water: (a) C_2H_6, (b) C_2H_4, (c) C_6H_6. Compare the values of $\Delta H°$ which you calculated with those recorded in Table 12.2 for heats of combustion of these compounds.

12.5. Using information from Table 12.1, calculate $\Delta H°$ for the following reactions:
   (a) $ZnO(s) + CO(g) \rightarrow Zn(s) + CO_2(g)$
   (b) $2\ CH_3OH(1) + 3\ O_2(g) \rightarrow 2\ CO_2(g) + 4\ H_2O(1)$
   (c) $3\ H_2S(g) + 8\ HNO_3(aq) \rightarrow 3\ H_2SO_4(aq) + 8\ NO(g) + 4\ H_2O(1)$

12.6. Determine $\Delta H°$ for the following reactions:
   (a) $SO_2(g) + 2\ H_2S(g) \rightarrow 3\ S(s) + 2\ H_2O(g)$
   (b) $SO_2(g) + H_2O_2(1) \rightarrow SO_3(g) + H_2O(1)$
   (c) $C_2H_5OH(1) \rightarrow C_2H_4(g) + H_2O(g)$

12.7. Using information from Table 12.1, determine $\Delta H°$ for the reactions:
   (a) $3\ Fe_2O_3(s) + CO(g) \rightarrow 2\ Fe_3O_4(s) + CO_2(g)$
   (b) $Fe_3O_4(s) + CO(g) \rightarrow 3\ FeO(s) + CO_2(g)$
   (c) $FeO(s) + CO(g) \rightarrow Fe(s) + CO_2(g)$

12.8. From the heats of reaction in the previous exercise, calculate $\Delta H°$ for the reaction:

$$Fe_2O_3(s) + 3\ CO(g) \rightarrow 2Fe(s) + 3\ CO_2(g)$$

12.9. Given the following reactions:

$$S(s) + O_2(g) \rightarrow SO_2(g) \qquad \Delta H = -71.0\ \text{kcal}$$
$$SO_2(g) + \tfrac{1}{2}O_2(g) \rightarrow SO_3(g) \qquad \Delta H = -23.5\ \text{kcal}$$

calculate ΔH for the reaction:

$$S(s) + 1\tfrac{1}{2}O_2(g) \rightarrow SO_3(g)$$

12.10. Given the following heats of reaction:

$$4\ NH_3(g) + 5\ O_2(g) \rightarrow 4\ NO(g) + 6\ H_2O(1) \qquad \Delta H = -279.4\ \text{kcal}$$
$$4\ NH_3(g) + 3\ O_2(g) \rightarrow 2\ N_2(g) + 6\ H_2O(1) \qquad \Delta H = -365.8\ \text{kcal}$$

Calculate the formation of NO(g).

12.11. (a) Calculate the heat of formation of hydrazine, $N_2H_4(l)$, from the following data:

$$
\begin{aligned}
2\,NH_3(g) + 3\,N_2O(g) &\rightarrow 4\,N_2(g) + 3\,H_2O(l) & \Delta H &= -241.4 \text{ kcal} \\
N_2O(g) + 3\,H_2(g) &\rightarrow N_2H_4(l) + H_2O(l) & \Delta H &= -\ 75.7 \text{ kcal} \\
2\,NH_3(g) + \tfrac{1}{2}\,O_2(g) &\rightarrow N_2H_4(l) + H_2O(l) & \Delta H &= -\ 34.2 \text{ kcal} \\
H_2(g) + \tfrac{1}{2}\,O_2(g) &\rightarrow H_2O(l) & \Delta H &= -\ 68.3 \text{ kcal}
\end{aligned}
$$

12.12. Calculate ΔE° and ΔH° for the reactions at 25°C:
(a) $4\,NH_3(g) + 5\,O_2(g) \rightarrow 4\,NO(g) + 6\,H_2O(l)$
(b) $H_2S(g) + 1\tfrac{1}{2}\,O_2(g) \rightarrow SO_2(g) + H_2O(l)$

12.13. Calculate ΔE° for the combustion of ethene, $C_2H_4(g)$, to $CO_2(g)$ and $H_2O(l)$ at 25°C. The enthalpy of combustion, ΔH°, of ethene is -332 kcal/mole.

12.14. (a) Use the values given in Table 12.1 to calculate ΔH° for the reaction

$$2\,NO(g) + O_2(g) \rightarrow 2\,NO_2(g)$$

(b) What is ΔE° for the reaction?

12.15. Solid mercury melts at $-39°C$. If the heat of fusion is 2.82 cal per g, calculate the entropy change per mole, assuming no change in free energy.

12.16. At the melting point of a solid (or freezing point of a liquid), the free energy of the solid state and the liquid state is the same and $\Delta G = 0$. Likewise at the boiling point of a liquid where there is an equilibrium between the liquid and vapor phases, the free energy is equal in the two states. Calculate the change in entropy for the following process at 0°C if the heat of fusion of $H_2O =$ 80 cal/g:

$$H_2O(s) \rightarrow H_2O(l)$$

12.17. Match the appropriate thermodynamic change with the statement given below:

(a) The overall energy considerations indicate that the reaction will proceed spontaneously – reactants converted to products.
(b) The reaction proceeds with the release of heat energy.
(c) The whole system would be perfectly ordered – only attained at 0°K.
(d) As the reactants are converted to products there is no net increase in the number of moles of gas.
(e) The reactants are less ordered than the products.
(f) The reaction is at equilibrium – equal tendency to form reactants and products.

1. ΔS positive
2. ΔH positive
3. ΔG positive
4. $\Delta G = 0$
5. $S = 0$
6. $\Delta H = \Delta E$
7. $\Delta E = 0$
8. ΔS negative
9. ΔH negative
10. ΔG negative

12.18. If $AlCl_3$ is produced by the direct combination of the elements, the values of ΔG° and ΔH° at 25°C are respectively -152 and -166 kcal per mole. Calculate the entropy of formation per mole of $AlCl_3$.

12.19. From entropy values in Table 12.3 (a) Calculate the entropy change in the reaction:

$$2 H_2(g) + O_2(g) \rightarrow 2 H_2O(l) \qquad \Delta H° = -136.6 \text{ kcal}$$

(b) Determine the change in free energy. (c) What do you conclude regarding the spontaneity of the reaction?

12.20. The heat of vaporization of chloroform ($CHCl_3$) is 62.8 cal/g at 61.7°C, which is the normal boiling point of chloroform. For the vaporization of one mole of liquid $CHCl_3$, calculate (a) ΔH, (b) ΔE, (c) ΔS, (d) ΔG.

12.21. For the reaction

$$2 CO(g) + O_2(g) \rightarrow 2 CO_2(g)$$

calculate: (a) $\Delta H°$, (b) $\Delta G°$, (c) $\Delta S°$. (d) Is the reaction spontaneous at 25°C?

12.22. The $\Delta G_f°$ of methanol (CH_3OH) is −39.73 kcal/mole. Is it possible to make methanol by a reaction of C(s) with $H_2(g)$ and $O_2(g)$ at 25°C?

12.23. A chemist claims that the following reaction occurs at 25°C:

$$2 CO_2(g) + 2 Cl_2(g) \rightarrow 2 COCl_2(g) + O_2(g)$$

Is it thermodynamically possible? Use the data in Table 12.3 to calculate $\Delta G°$ for the reaction.

12.24. Consider the transformation:

$$2 Cl_2(g) + 2 H_2O(l) \rightarrow 4 HCl(g) + O_2(g)$$

(a) Use the $\Delta H_f°$ values of Table 12.1 to calculate $\Delta H°$. (b) Use the $\Delta G_f°$ values of Table 12.3 to calculate $\Delta G°$. (c) Calculate $\Delta S°$ from your answers to parts (a) and (b). (d) Is the reaction spontaneous at 25°C?

12.25. The dissaccharide sucrose is hydrolyzed to glucose and fructose as follows:

$$\text{Sucrose} + H_2O \rightarrow \text{glucose} + \text{fructose}$$

At 25°C the enthalpy change is −4,800 cal/mole and the entropy change is 2.35 cal/°K mole.
(a) Calculate the free energy change for the reaction.
(b) Would the reduction be considered spontaneous? Why?

12.26. Calculate $\Delta G°$ for the reaction

$$PCl_3(g) + Cl_2(g) \rightarrow PbCl_5(g)$$

The absolute entropies of reactants and products (in cal/°K mole) are: $PCl_3(g)$, +74.49; $Cl_2(g)$, +53.286; $PCl_5(g)$, +84.3. The enthalpy of formation of $PCl_3(g)$ is −73.22 kcal/mole, and $\Delta H_f°$ of $PCl_5(g)$ is −95.35 kcal/mole.

12.27. What is the absolute entropy, $S°$, at 25°C of carbon disulfide ($CS_2(l)$)? For

$CS_2(1)$, ΔH_f° is $+21.0$ kcal/mole and ΔG_f° is $+15.2$ kcal/mole. The absolute entropy of C (graphite) is $+1.36$ cal/$^\circ$K mole and of S (rhombic) is $+7.62$ cal/$^\circ$K mole.

Supplementary Readings

12.1. *Chemical Thermodynamics*, J. R. Goates and J. B. Ott, Harcourt Brace Jovanovich, New York (1971).
12.2. "Perpetual Motion Machines," S. W. Augrist, *Sci. American*, Jan., 1968; p. 114.
12.3. "The First Law," H. Bent, *J. Chem. Educ.*, **50**, 323 (1973).

Chapter 13

Solutions

What is a Solution?

Solutions are familiar to all of us. The water we drink is a solution of various minerals dissolved in water. Plants derive their food, in part, from the water-soluble constituents of the soil. In the digestion of food, a chemical change takes place whereby constituents of food are converted into soluble substances which can be absorbed by the blood and carried to various tissues in the body. The atmosphere we breathe is a solution of nitrogen and oxygen gases. Fish and marine life derive the oxygen necessary for their life processes from oxygen dissolved in water.

Many chemical changes are brought about in solution, since the rate of reaction in a solution may be faster than between the substances in the pure state. This is probably due to the better contact of the reacting substances.

The Definition of Solution

A solution is formed when sugar is dissolved in water. It is perfectly uniform or homogeneous; if we taste samples taken from various parts of the solution, we find that each sample has the same degree of sweetness. The amount of sugar which we can dissolve in 100 ml of water is variable; we may dissolve 1 g, 10 g, or 50 g of sugar in the water. There is a limit, though, to the amount of sugar we can dissolve in a given amount of water at a certain temperature. This limiting amount is termed the **solubility** of sugar in water at that temperature.

If we could observe the molecules in the sugar solution, we would probably see that individual molecules of sugar are uniformly distributed throughout the solution—each unit volume of solution would contain the same number of sugar molecules. The particles of sugar are dispersed in the solution as individual molecules, and the solid is said to be **molecularly dispersed.** Molecular dispersion is a characteristic of a true solution. We shall learn later (Chapter 24) that aggregates or groups of molecules may be dispersed in a colloidal state.

The properties of a solution may be summarized in the definition: **A solution is a homogeneous mixture of two or more substances, the relative proportions of which may vary continuously within certain limits.**

Solute and Solvent

The dissolved material in a solution is termed the **solute** and the dissolving medium, the **solvent.** Thus in a solution of sugar in water, the sugar is the solute and the water the solvent. These terms, however, are arbitrary and lose their significance when applied to certain solutions. For example, in a solution of two liquids, such as alcohol and water, either the water or the alcohol may properly be classified as the solute. In situations of this kind, the constituent of the solution present in the smaller amount is usually designated as the solute.

Types of Solutions

When we speak of a solution, we usually think of a solid dissolved in water. While water is the most common of solvents, other liquids are frequently employed as solvents for certain substances—for example, wax may be dissolved in gasoline, oil in turpentine, and sulfur in carbon disulfide. Solutions, however, are not confined to the solution of solids in liquids. Since there are three physical states of matter—gas, liquid, and solid—nine possibilities suggest themselves:

1. GASES IN GASES   Gases mix in all proportions, and since the particles of the gases exist as individual molecules, any mixture of gases may be regarded as a solution. Air is essentially a solution of nitrogen and oxygen gases. Actually, other gases are present in the atmosphere, and the entire system constitutes a complex solution.

2. GASES IN LIQUIDS   The solution of oxygen in water and the solution of carbon dioxide in water (carbonated water) are examples of this type.

3. GASES IN SOLIDS   Certain metals (platinum, palladium) absorb large volumes of hydrogen gas. In these cases, the gas may be considered as being dissolved in a solid.

4. LIQUIDS IN GASES   Doubtful; if liquid particles were reduced to molecular size, this would be a gas-in-gas type.

5. LIQUIDS IN LIQUIDS   Examples of this type are well known. Alcohol and water are completely miscible (mix with one another in all proportions); gasoline and kerosene dissolve in one another in all proportions. Certain liquids show incomplete solubility in one another. For example, ether and water are only partially miscible.

6. LIQUIDS IN SOLIDS   The solution of mercury in certain metals (amalgams) may be considered an example of this type.

7. SOLIDS IN GASES   Doubtful; same argument as (4).

8. SOLIDS IN LIQUIDS   This is the most common type of solution.

9. SOLIDS IN SOLIDS   If two solids are held tightly together, diffusion of each solid into the other occurs very slowly. To demonstrate this, a piece

of lead and a piece of gold were tightly clamped together for several years. It was found on analysis that the lead contained gold in the part that had been adjacent to the piece of gold, and the gold contained a small amount of lead. This experiment indicates that diffusion in the solid state, while extremely slow, does occur. Certain alloys are sometimes classed as solid solutions.

Polar and Nonpolar Solvents

We know that liquids differ greatly in solvent action. Water is a good solvent for acids, bases, and salts—those substances that are ionic or that produce ions in solution—but it is almost completely immiscible with such liquids as benzene, carbon disulfide, and carbon tetrachloride. These three liquids, however, readily dissolve such substances as sulfur, phosphorus, and wax, which are insoluble in water.

On the basis of molecular structure, solvents are classed as **polar** and **nonpolar.** A polar solvent consists of dipoles—molecules in which constituent atoms are unsymmetrically arranged so that the center of positive and negative electrical charge is not located at the same point in the molecule. (See Chapter 10.) (Of course, the molecule as a whole is electrically neutral.) Water and liquid ammonia are polar solvents. Nonpolar solvents consist of symmetrical molecules, with the center of positive and negative electrical charge at the same point within the molecule; benzene and carbon tetrachloride are nonpolar liquids.

Cohesive forces between molecules of polar liquids are large compared with forces tending to hold molecules of nonpolar liquids together, since positive ends of the dipoles will tend to attract negative ends of other dipoles and hold the molecules together.

Water acts as a good solvent for electrovalent compounds, since there is a definite attraction between the ions present and the dipoles of the water molecules. In crystalline sodium chloride, sodium and chloride ions are oriented in a definite pattern by electrostatic forces acting between the oppositely charged ions; thus the crystals of the compound exhibit a definite geometric pattern (NaCl crystallizes in the cubic system; see Chapter 11). If NaCl is to dissolve, the forces holding the ions together must be overcome so that the ions will be free to move in solution. Water weakens interionic forces to some extent, probably by attraction between the dipoles and ions in some such fashion as is represented by Figure 13.1.

These electrovalent compounds are insoluble in nonpolar liquids like benzene, since there is no electrical attraction between the ions in the compounds and the solvent molecules. Summarizing, polar solvents are best for dissolving ionic or electrovalent substances; nonpolar solvents are best for nonpolar and covalent substances. In other words, like dissolves like.

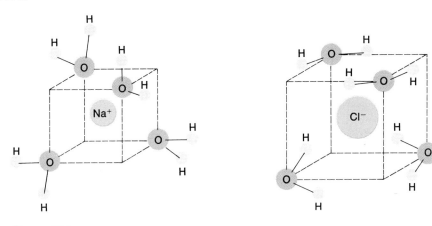

Figure 13.1

Attraction between ions and water dipoles.

Equilibrium in Solutions

When a solid is dissolved in a liquid, molecules (or ions) break away from the surface of the solid and pass into the solvent, where they diffuse until a uniform distribution has been effected. Stirring or agitation will speed up the solution process. As the molecules find their way into the solvent, some are deflected back toward the solid by collision with other molecules. Some of the particles which strike the solid will deposit out on the solid surface and be converted back to the solid form. There are, then, two processes occurring: (1) the solution of molecules from the surface of the solid, and (2) the deposition of molecules on the surface of the solid. If the amount of solid present is sufficient to saturate the solution, processes (1) and (2) will eventually equalize each other, and the number of molecules passing into solution will be the same as the number coming out of solution. When the speed of these two processes is the same, a condition of equilibrium is reached and the solution is said to be saturated (Figure 13.2). **A saturated solution is one in which the dissolved solute is in equilibrium with the undissolved solute.**

The passage of a solid into solution is analogous to the passage of liquid into vapor. Just as a certain amount of vapor (vapor pressure) can exist in equilibrium with the liquid, in a solid dissolved in a liquid a certain amount of solid can be present in the solution in equilibrium with the undissolved solid.

We must not think of the processes of solution and deposition (precipitation) as stopping when equilibrium has been reached. Actually, the equilibrium is a dynamic one; the two opposing processes are occurring continuously. Thus the solubility of a substance is a constant value at a given temperature. As the temperature is raised, relatively more molecules of the solid may

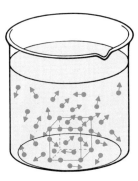

Figure 13.2
Equilibrium between solid and solution.

pass into solution than precipitate out, and a new condition of equilibrium is established at the higher temperature.[*]

Supersaturation

Ordinarily when a saturated solution is cooled, excess solute will crystallize out at the lower temperature. With certain substances, however, it is possible to cool a saturated solution *in the absence of solid solute* without the separation of the excess solute. The cooled solution then contains more of the solute than is present in a saturated solution at that particular temperature. Such a solution is said to be **supersaturated.** These solutions are very unstable, and the excess solute immediately separates out if a tiny crystal of the solid solute is added to the solution.

To determine whether a given solution is saturated, unsaturated, or supersaturated, we can add a crystal of the dissolved substance. If the crystal dissolves, the solution is unsaturated; if it remains the same, the solution is saturated; and if it causes the formation of more crystals, the solution is supersaturated.

13.1   SOLUBILITY

The weight of a substance dissolved by a given weight or volume of solvent at a given temperature is termed the **solubility** of the substance. Solubility varies greatly with the nature of the dissolved substance: 100 g of water will dissolve 179 g of sugar, 35.7 g of sodium chloride, 3.0 g of potassium chlorate, or 0.00009 g of silver chloride at 0°C. Sugar is classed as very soluble, sodium chloride as moderately soluble, potassium chlorate as slightly soluble, and silver chloride as practically insoluble.

The general rules for solubility of compounds are summarized in Table 13.1.

[*] The solubility of a few substances decreases with increasing temperature.

Table 13.1

Solubility of common compounds

Soluble

1. All Na⁺, K⁺, NH₄⁺ compounds.
2. All NO₃⁻, C₂H₃O₂⁻, ClO₃⁻ compounds (nitrates, acetates, chlorates).
3. All Cl⁻, Br⁻, I⁻ (except Ag⁺, Hg₂²⁺, Pb²⁺).
4. All SO₄²⁻ (except Ba²⁺, Pb²⁺, Sr²⁺) (Ag⁺ and Ca²⁺ sparingly soluble).

Insoluble

1. Most O²⁻, OH⁻, PO₄³⁻, CO₃²⁻, S²⁻ (except Na⁺, K⁺, NH₄⁺), (oxides, hydroxides, phosphates, carbonates, sulfides).

The Solubility of Gases in Liquids

The solubility of a gas is usually expressed as the volume of gas dissolved by a certain volume of liquid. The solubility of gases in liquids varies over a wide range and depends on the nature of the solute and solvent. At 0°C, 1 vol of water may dissolve 0.021 vol of hydrogen, 1.71 vol of carbon dioxide, 506 vol of hydrogen chloride, or 1175 vol of ammonia. Gases which are very soluble in water may be nearly insoluble in some other solvent.

The weight of a gas dissolved by a given weight of liquid at constant temperature is directly proportional to the pressure of the gas over the liquid. This is a statement of **Henry's Law.** If a certain weight of gas is dissolved by a liquid at a pressure of 1 atmosphere, then ten times that weight will dissolve if the pressure is increased to 10 atmospheres. A practical application of this law is the preparation of carbonated beverages. Carbon dioxide gas is dissolved in the liquid under pressure. When the cap of the bottle is removed, the solution effervesces (gives off gas from solution) rapidly, since the pressure has been reduced.

The solubility of a gas in a liquid is also dependent on the temperature. The quantity of gas dissolved by a given weight of liquid usually decreases as the temperature is raised. If water containing dissolved air is warmed, bubbles of the air escape from the solution as the temperature increases. At the boiling point of the solvent, the gas is practically insoluble.

Effect of Temperature on Solubility of Solids in Liquids

The solubility of a solid in a liquid is usually (but not always) increased with an increase in temperature. The solubilities of several salts in water are shown in Table 13.2 and represented graphically in Figure 13.3.

It may be noted that in some cases an increase in temperature greatly changes the solubility, while in other cases there is very little change. The solubility of potassium nitrate increases from about 13 g per 100 g of water at 0°C to about 139 g at 70°C. Sodium chloride increases only from 35.7 g to

Table 13.2

Solubility of salts at various temperatures (g/100 g H₂O)

Tempera-ture °C	Potassium Nitrate	Sodium Chloride	Potassium Alum	Calcium Chromate	Potassium Chloride	Sodium Sulfate	Sodium Nitrate
0	13	35.7	4	13.0	28	4.8	73
10	21	35.8	10	12.0		9.0	80
20	31	36.0	15	10.4	34	19.5	85
30	45	36.3	23	9.4		40.9	92
40	64	36.6	31	8.5	40	48.8	98
50	86	37.0	49	7.3		46.7	104
60	111	37.3	67	6.0	46	45.3	
70	139	37.9	101	5.3		44.4	
80		38.4	135	4.4	51	43.7	133
90		39.1		3.8		43.1	
100	249	39.8		3.0	57	42.5	163

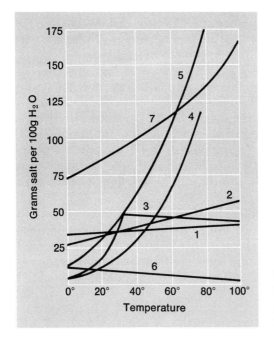

Figure 13.3
Solubility curves: (1) NaCl; (2) KCl;
(3) Na₂SO₄; (4) alum; (5) KNO₃;
(6) CaCrO₄; (7) NaNO₃.

37.9 g for the same temperature range. Calcium chromate actually decreases in solubility as the temperature is raised. A break in the solubility curve for sodium sulfate appears at a temperature of 32.7°C. The solubility curve below this temperature is for the hydrated salt and above this temperature, for the anhydrous salt. At this temperature, known as a **transition point,** a transition from the hydrate to the anhydrous form takes place. The transition point of sodium sulfate is so definite that it is often used for the standardization of thermometers.

Fractional Crystallization

Differences of solubility at various temperatures form the basis for the selective separation of salts from a solution containing two or more dissolved salts, a process known as fractional crystallization. Consider the evaporation of a solution containing 50 g KNO_3, 50 g KCl, and 200 g H_2O. If the solution is partially evaporated at 100°C to 100 g H_2O, no crystallization is effected, since, according to Table 13.2, the solubilities are 249 g KNO_3 and 57 g KCl, respectively. Suppose, however, that the solution is further evaporated at 100°C to a weight of 50 g H_2O. All of the KNO_3 remains in solution, but only $\frac{57}{2}$ or 28.5 g KCl remains in solution. The excess KCl above a saturated solution $(50 - 28.5 = 21.5$ g KCl) crystallizes out. If the hot liquid is poured off from the crystallized KCl, 21.5 g of fairly pure KCl is left. (It is not entirely pure, since the adhering liquid still contains a small amount of KNO_3.) If the crystals are redissolved in water and the process repeated, a purer fraction of KCl crystals can be obtained. Such a process is an example of fractional crystallization.

Concentration

Sometimes we speak of a solution as being **dilute** or **concentrated.** These are qualitative terms to designate whether a solution contains a relatively little or a relatively large amount of solute. For example, a solution containing 1 g of sugar in 100 ml of water is dilute, while a solution containing 50 g of sugar in 100 ml of water is concentrated. The concentration of a solution represents the weight of solute dissolved in a given weight or volume of the solution. Several methods are available for expressing the concentration of a solution. The weight of a solute in a given weight of the solution may be expressed as a percentage. Thus a solution containing 5 g of solute and 95 g of solvent is a 5% solution, since 5 parts of 100 are solute.

Molarity

Another means of expressing concentration, termed **molarity**, involves expression of the concentration as moles of solute in 1 l of the solution. A solution containing 1 mole of solute in 1 l of solution is termed a **molar** (1 *M*) solution; a solution containing 2 moles of solute in 1 l of solution is 2 molar (2 *M*), one containing 0.5 mole in 1 l is 0.5 molar (0.5 *M*), and so on. In other words,

$$\text{Molarity} = \frac{\text{Number of moles of solute}}{\text{Number of liters of solution}} \qquad (1)$$

Since moles $= \dfrac{\text{grams}}{\text{gram molecular weight (GMW)}}$, we may substitute in the numerator and obtain equation (2)

$$\text{Molarity} = \frac{\text{grams of solute}}{(\text{GMW of solute})(\text{liters of solution})} \qquad (2)$$

Consider the following illustrative problems:

EXAMPLE 13.1   Calculate the molarity of a solution containing 10.0 g of sulfuric acid in 500 ml of solution. MW of $H_2SO_4 = 98.1$;

SOLUTION:
$$10.0 \text{ g is } 10.0/98.1 = 0.102 \text{ mole}$$

0.102 mole of sulfuric acid in 500 ml is the same as 0.204 (2×0.102) mole in 1000 ml or 1 l. The molarity is thus 0.204. OR by substituting in formula (2) above

$$\text{Molarity} = \frac{10.1 \text{ g}}{(98.1 \text{ g/mole})(0.500 \text{ l})} = 0.204$$

EXAMPLE 13.2   Calculate the weight in grams of sulfuric acid in 2.00 l of 0.100 molar (0.100 *M*) solution.

SOLUTION:
    One liter of 1 molar solution contains 1 mole or 98.1 g H_2SO_4.
    One liter of 0.100 molar solution contains $0.1 \times 98.1 = 9.81$ g.
    Two liters of 0.100 molar solution then will contain $2 \times 9.81 = 19.6$ g.

OR substituting in equation (2)

$$\text{Molarity} = 0.100 \text{ mole/l} = \frac{\text{grams of solute}}{(98.1 \text{ g/mole})(2.00 \text{ l})}$$
$$\text{grams of solute} = 0.100 \text{ mole/l})(98.1 \text{ g/mole})(2.00 \text{ l}) = 19.6 \text{ g}$$

Normal Solutions

The equation

$$H_2SO_4 + 2\,NaOH \rightarrow Na_2SO_4 + 2\,H_2O$$

shows that one mole of sulfuric acid reacts with two moles of sodium hydroxide. It is evident that if the solutions of H_2SO_4 and NaOH were of the same molarity, it would require twice the volume of NaOH as of H_2SO_4 for the neutralization. One liter of 1 M H_2SO_4 will require 2 l of 1 M NaOH, or 1 l of 2 M NaOH.

In the reaction of NaOH and HCl, equal volumes of molar solutions will exactly neutralize one another.

$$HCl + NaOH \rightarrow NaCl + H_2O$$

The difficulty of working with equal volumes in some reactions and different volumes in other reactions can be avoided by making up the solutions in terms of equivalent weights rather than molecular weights.

One equivalent (gram equivalent weight (GEW)) of a substance is the weight (1), which (as an acid) contains 1 mole of replaceable hydrogen; or (2) which (as a base) reacts with 1 mole of hydrogen; or (3) which (as a salt) is produced in a reaction involving 1 mole of hydrogen; or (4) as an oxidizing or reducing agent depends upon the number of electrons transferred per mole (see Chapter 16). Summarizing:

Equivalent weight for

Acids	MW/no. of replaceable H
Bases	MW/no. of replaceable OH
Salts	MW/total + valence of metal
Oxidant or Reductant	MW/total oxidation number change

Thus 36.5 g HCl contains 1 mole of replaceable hydrogen and is 1 equivalent (gram equivalent); 40 g NaOH will react with 36.5 g of HCl which contains 1 mole of hydrogen, so 40 g of NaOH is 1 equivalent; 58.5 g of NaCl is formed in the reaction involving 36.5 g HCl and is 1 equivalent. 98 g H_2SO_4 (MW 98) contains two moles of hydrogen, hence $98/2 = 49$ g $= 1$ equivalent.

A solution containing one equivalent of a solute in 1 l of solution is a 1 normal solution; one containing 2 equivalents in 1 l is 2 normal (2 N) and so on. Normality of any solution is given by the expression:

$$\text{Normality}(N) = \frac{\text{equivalents of solute}}{\text{liters of solution}} \qquad (3)$$

Since equivalents $= \dfrac{\text{grams}}{\text{GEW}}$, we may substitute for the numerator in equation (3) above and obtain

$$\text{Normality} = \frac{\text{grams of solute}}{\text{GEW} \times \text{liters of soln}} \qquad (4)$$

EXAMPLE 13.3   Calculate the normality of a solution containing 2.45 g of sulfuric acid in 2.00 l of solution.

SOLUTION: One equivalent of H_2SO_4 (MW 98.1) is 98.1/2 = 49.0 g, since one molecule of H_2SO_4 contains two atoms of replaceable hydrogen. The solution in this example contains 2.45 g in 2.00 l or 1.225 g/l. The normality is 1.225/49.0 = 0.025 N

OR by substituting in equation (4) above

$$\text{Normality} = \frac{2.45 \text{ g}}{(49.0 \text{ g/equiv})(2.00 \text{ l})} = 0.025 \text{ equiv/l}$$

EXAMPLE 13.4   How many grams of sulfuric acid are contained in 3.00 l of 0.500 N solution?

SOLUTION: There are 49.0 g of H_2SO_4 (one equivalent) in one l of 1 normal soln. In 1.00 l of 0.500 N solution there are 24.5 g($\frac{1}{2} \times$ 49.0); in 3.00 l, 3.00 $\times$ 24.5 g = 73.5 g

OR substituting in equation (4)

$$0.500 \text{ equiv/l} = \frac{\text{grams of solute}}{(49.0 \text{ g/equiv})(3.00 \text{ l})}$$
$$\text{grams of solute} = (0.500 \text{ equiv/l})(49.0 \text{ g/equiv})(3.00 \text{ l}) = 73.5 \text{ g}.$$

Equal volumes of solutions of the same normality are equivalent; for example, 1 l of 1 N solution of any acid will react with exactly 1 l of 1 N solution of any base.

Standard Solutions and Titrations

Very often in analytical work it is necessary to determine the acidity or basicity of an unknown solution. This may be accomplished with the aid of a **standard solution** of an acid or a base—a solution whose concentration is accurately known. The operation of analyzing an unknown solution by means of a standard solution is known as **titration**. The determinations are usually carried out with the aid of **burets** (Figure 13.4), devices for accurately measuring the volumes of solutions used. Acids neutralize bases, and the endpoint of a titration is reached when equivalent quantities of acid and base have been brought together. A certain volume of solution from one buret is collected in a beaker, and solution from the other buret is added carefully until equivalent quantities of the two solutions have been brought together. The point at which equivalent quantities have been brought together is determined by an indicator, which changes in color at the endpoint. Knowing the volumes of the two solutions used and the normality of either the acid or base, we may calculate the normality of the other solution and thereby determine its acidity or basicity.

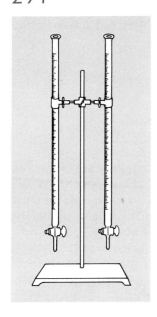

Figure 13.4
Burets.

The concentration of a standard solution is usually expressed in normality:

$$\text{Normality } (N) = \frac{\text{Equivalents } (E)}{\text{Liters } (L)}$$

or

$$E = N \times L$$

Since equivalent quantities of acid and base must be present at the endpoint of the titration, $E_{\text{acid}} = E_{\text{base}}$; therefore,

Normality of acid $\times$ Volume of acid = Normality of base $\times$ Volume of base

or
$$N_1 V_1 = N_2 V_2 \tag{5}$$

If three of the above four factors are known, the fourth may be calculated. Volumes may be expressed in milliliters or liters. Study the following sample calculations.

EXAMPLE 13.5   Calculate the volume of 0.3 N base necessary to neutralize 3 l of 0.01 N acid.

SOLUTION: Substituting in the formula
$$N_1 V_1 = N_2 V_2$$
$$0.01 \text{ equiv/l} \times 3 \text{ l} = 0.3 \text{ equiv/l} \times V_2$$
$$V_2 = \frac{0.01 \text{ equiv/l} \times 3 \text{ l}}{0.3 \text{ equiv/l}} = 0.1 \text{ l or } 100 \text{ ml}$$

EXAMPLE 13.6   If 20 ml of 0.50 N salt solution is diluted to 1 l, what is the new concentration?

SOLUTION:

$$N_1V_1 = N_2V_2$$
$$0.50 \text{ equiv/l} \times 20 \text{ ml} = N_2 \times 1000 \text{ ml}$$
$$N_2 = \frac{0.50 \text{ equiv/l} \times 20 \text{ ml}}{1000 \text{ ml}} = 0.010$$

Various solid substances which can be weighed accurately are employed for standardization of acid and basic solutions. Anhydrous Na_2CO_3 is frequently used for standardization of acid solutions, oxalic acid crystals for basic solutions. A good reagent for the latter is potassium hydrogen phthalate, $KHC_8H_4O_4$, with a relatively high molecular weight of about 204. With only one replaceable hydrogen used in the neutralization process, a full molecular weight in grams per liter is required for a 1 normal solution. Slight errors in weighing become less important for such substances of high equivalent weights, and the accuracy of a determination is improved.

Since at the equivalence point in a titration the number of equivalents of one substance used (E_1) must be equal to the number of equivalents of the other substance (E_2), an expression similar to equation (4) may be used for calculations involving titrations of a weighed amount of one substance with a solution of another substance,

$$\frac{\text{grams of solute}}{\text{GEW of solute}} = LN \tag{6}$$

where either half of the equation equals the number of equivalents $(E_1$ or $E_2)$.

EXAMPLE 13.7   A 0.3060-g sample of $KHC_8H_4O_4$ (MW 204.1) was dissolved in water and titrated with 0.1050 N NaOH. What volume of the base solution would be required to reach the equivalence point?

SOLUTION: From equation (6)

$$\frac{0.3060 \text{ g}}{204.1 \text{ g/equiv}} = (L)(0.1050 \text{ equiv/l})$$
$$L = 0.01428 \text{ l or } 14.28 \text{ ml}$$

EXAMPLE 13.8   A 0.2850-g sample of lye (impure NaOH, MW 40.00) was weighed out and dissolved in water. 44.54 ml of 0.1500 N HCl was required to titrate the sample to the end point. What was the percent purity of the lye?

SOLUTION: From equation (6)

$$\frac{x \text{ g of pure NaOH}}{40.00 \text{ g/equiv}} = (0.04454 \text{ l})(0.1500 \text{ equiv/l})$$

$$x = 0.2672 \text{ g of pure NaOH in the } 0.2850\text{-g}$$

sample of lye. The percent NaOH in the lye (percent purity) would be

$$\text{percent NaOH} = \frac{(0.2672 \text{ g})(100)}{0.2850 \text{ g}} = 93.77\%$$

A simple relationship exists between molarity and normality of a solution

$$N = (M)(n) \tag{7}$$

where n = a whole number 1, 2, 3, ... For an acid, n is the number of replaceable H's, for a base the number of replaceable OH's, etc. For example, a 0.050 M H_2SO_4 solution would be 0.100 N and a 0.0024 N $Ca(OH)_2$ solution would be 0.0012 M.

Molal Solutions

In the study of **colligative** properties – properties of solutions which depend on the ratio of number of solute particles to number of solvent particles – concentrations are expressed as **molality**, usually given the symbol m.

$$\text{Molality} = \frac{\text{moles of solute}}{\text{number of kg of solvent}} \tag{8}$$

Since moles of solute = grams/GMW and substituting in (6)

$$\text{Molality} = \frac{\text{grams of solute}}{\text{GMW} \times \text{kg of solvent}} \tag{9}$$

Note the similarity of equations (2) and (7). Equation (2) contains liters of solution whereas equation (7) contains kilograms of solvent. Weight is independent of temperature whereas volume is not, hence the molarity of a solution changes with temperature, the molality does not. In dilute solutions, molar and molal solutions are very nearly identical, but in more concentrated solutions, wide differences may be expected.

EXAMPLE 13.9   What is the molality of a solution in which 49 g of H_2SO_4 (MW 98) is dissolved in 250 g of water?

SOLUTION: Since the solution contains 0.50 mole($\frac{1}{2} \times 98$) in 0.25 kg(250/1000)

$$\text{Molality} = 0.50/0.25 = 2.0$$

OR by substituting in equation (9) on page 293,

$$\text{Molality} = \frac{49 \text{ g}}{(98 \text{ g/mole})(0.25 \text{ kg})} = 2.0 \text{ moles/kg}$$

Mole Fraction

(See also p. 206). In considering colligative properties, the concentration of a solution is often expressed as a **mole fraction,** where

$$\text{Mole fraction of solute} = \frac{\text{Moles of solute}}{\text{Moles of solute} + \text{Moles of solvent}} \quad (10)$$

and

$$\text{Mole fraction of solvent} = \frac{\text{Moles of solvent}}{\text{Moles of solute} + \text{Moles of solvent}} \quad (11)$$

The sum of the two fractions must be 1.

EXAMPLE 13.10   What are the mole fractions of solute and solvent in a solution prepared by dissolving 98 g H_2SO_4 in 162 g H_2O?

SOLUTION:

$$\text{Moles } H_2SO_4 = 98 \text{ g}/98 \text{ g/mole} = 1 \text{ mole}$$
$$\text{Moles } H_2O = 162 \text{ g}/18 \text{ g/mole} = 9 \text{ moles}$$
$$\text{Mole fraction of } H_2SO_4 = \frac{1}{1+9} = 0.1$$
$$\text{Mole fraction of } H_2O = \tfrac{9}{10} = 0.9$$

EXAMPLE 13.11   What is the mole fraction of H_2SO_4 in a 7.0 molar solution of H_2SO_4 which has a density of 1.39 g/ml?

SOLUTION:

one liter of this solution weighs   1390 g
7 moles of H_2SO_4 weigh 7 × 98 g   686 g
Weight of H_2O in 1 l of soln =   704 g
Moles of $H_2O = 704 \text{ g}/18 \text{ g/mole} = 39 \text{ moles}$
Mole fraction of $H_2SO_4 = \tfrac{7}{46} = 0.15$

13.2 PHYSICAL PROPERTIES OF SOLUTIONS

Vapor Pressure of Solutions of Non-Volatile Solutes

The addition of a non-volatile solute (like sugar) lowers the vapor pressure of the liquid because the solute reduces the fraction of solvent present, Figure

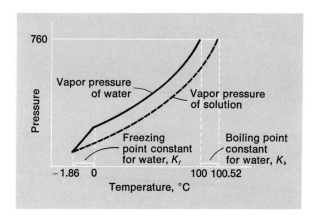

Figure 13.5
(Not drawn to scale.) A solution of non-volatile solute has a lower vapor
pressure and a higher boiling point than the pure solvent (in this case H_2O). Also
the freezing point is lowered. A one molal solution (aqueous) boils at 100.52°C
(standard pressure of 760 torr or 1 atm) and freezes at −1.86°C.

13.5. With relatively fewer solvent molecules present, the rate of their escape
from solution is diminished, resulting in a decreased vapor pressure. At constant temperature, the **lowering of the vapor pressure by a non-volatile solute
is proportional to the concentration of the solute in the solution.** This is a
statement of Raoult's Law, which may be expressed mathematically as

$$\Delta P = P° - P = P°X_2 \tag{12}$$

where ΔP is the lowering of the vapor pressure, $P°$ is the vapor pressure of the
pure solvent, P the vapor pressure of the solution, and X_2 the mole fraction of
solute. Alternatively, the vapor pressure of the solution is proportional to the
mole fraction of solvent,

$$P = P°X_1 \tag{13}$$

where X_1 is the mole fraction of solvent.

EXAMPLE 13.12   A solution of 20.0 g of a non-volatile solute in 100 g of
benzene at 30°C has a vapor pressure 13.4 torr lower than the vapor pressure
of pure benzene. What is the mole fraction of solute? Vapor pressure of
benzene at 30°C = 121.8 torr.

SOLUTION: Substituting in equation 10,

$$\Delta P = P°X_2$$
$$13.4 \text{ torr} = (121.8 \text{ torr})(X_2)$$
$$X_2 = 13.4 \text{ torr}/121.2 \text{ torr} = 0.111$$

The effect of a non-volatile solute on the vapor pressure of water is shown in Figure 13.5 which is similar to the vapor pressure curve for water in Figure 10.2.

The Boiling Point of Solutions

A decreased vapor pressure means a rise in the boiling point of the liquid (Figure 13.5), since a higher temperature will be necessary to bring the vapor pressure up to the external pressure. From a kinetic standpoint, it would seem that the rise of the boiling point would depend directly on the number of solute particles, since the escape of solvent molecules (the vapor pressure) is influenced by the number of solute particles. The following facts indicate the validity of this speculation. Three-hundred and forty-two grams of sugar (1 gram molecular weight) dissolved in 1000 g of water will raise the boiling point 0.52° (in other words, the solution will boil at 100.52°C, Figure 13.5); 60 g of urea (1 gram molecular weight) dissolved in 1000 g of water will give exactly the *same* elevation of the boiling point. It has been found that 1 gram molecular weight of other substances gives the same elevation of the boiling point. There is no relation between the weight in grams of the substance and the boiling point elevation, since 60 g of urea produce the same effect as 342 g of sugar. However, a gram molecular weight of each of these substances contains exactly the same number of molecules. This seems to indicate beyond doubt that the boiling point of a solution is dependent on the number of solute particles in solution. The normal elevation of the boiling point of a liquid containing 1 gram molecular weight of dissolved substance in 1000 g is called the **boiling point constant,** often designated as K_b. For water, this constant is 0.52°C. The elevation of the boiling point of water containing 3 moles of solute in 1000 g of water would be $3 \times 0.52°$, or 1.56°C, and so on.

We shall find that solutions of acids, bases, and salts have abnormal boiling points. This abnormal behavior is attributed to the greater number of particles(ions) in solution.

Solutions of Volatile Solutes

Solutions containing volatile solutes, in which both the solvent and solute exert their own vapor pressure, do not follow the pattern described above. For example, a solution of alcohol and water boils below the boiling point of water. The vapor pressures of alcohol and water probably are separately reduced by the presence of each other, but the additive reduced vapor pressures exceed the vapor pressure of water alone.

In Figure 13.6 the liquid and vapor curves for solutions of two volatile liquids A and B are shown. The lower curve shows the boiling points of solu-

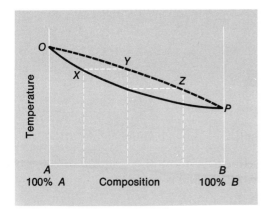

Figure 13.6
Boiling points of solutions of two volatile liquids.

tions of A and B of varying composition from 100 percent A to 100 percent B. O is the boiling point of pure A, and P the boiling point of pure B. The broken curve shows the compositions of vapor in equilibrium with various solutions of A and B. The composition can be determined by a perpendicular drawn to the composition axis.

Suppose that a solution of composition X is distilled. The composition of the vapor is represented by Y—the vapor is richer in the lower boiling component B. As the solution is boiled, the temperature rises and the solution becomes richer in the higher boiling component A. If the vapor at Y is condensed to liquid and subsequently redistilled, the composition of the vapor is that represented by point Z. Through a series of successive vaporizations and condensations in a fractionating column, two volatile liquids in a mixture can be effectively separated. This process, termed **fractional distillation**, is invaluable in separation of the components of such complex mixtures as petroleum.

The Freezing Point of Solutions

Solutes lower the freezing point of water. Alcohol and other antifreeze mixtures are added to water in car radiators to prevent freezing. Just as the boiling point of a solution is dependent on the number of solute particles, so the freezing point of a solution is a function of the number of particles in solution. One mole of a substance (except those substances which ionize) dissolved in 1000 grams of water lowers the freezing point 1.86°. This number, 1.86, is called the **molal freezing point constant**, designated as K_f, for water, Figure 13.5. Thus a solution containing 1 mole of alcohol (46 g) in 1000 g of water will freeze 1.86°C below the normal freezing point of water (0°C)—that is, at −1.86°C; a solution containing 2 moles of alcohol in 1000 g of water will freeze at $-(2 \times 1.86°) = -3.72°C$, and so on. In other words the freezing point depression, Δt_f, of a solution can be calculated by

$$\Delta t^*{}_f = mK_f \tag{14}$$

or, combining with equation (9)

$$\frac{\text{grams of solute}}{\text{GMW of solute}} = (\text{kg of solvent})\left(\frac{\Delta t_f}{K_f}\right). \tag{15}$$

A similar expression holds for boiling point elevation of solution, Δt_b.

EXAMPLE 13.13　What will be the freezing point of a solution of 200 g of ethylene glycol (CH_2OHCH_2OH, MW 62) in 500 g of H_2O?

SOLUTION:

$$\frac{200 \text{ g}}{62 \text{ g/mole}} = (0.50 \text{ kg})\left(\frac{\Delta t_f}{1.86° \text{ kg/mole}}\right)$$

$\Delta t_f = 12°C =$ the lowering of the freezing point. Therefore the freezing point of the solution will be

$$0°C - 12°C = -12°C.$$

EXAMPLE 13.14　One-fourth (0.250) g of an unknown nonvolatile solute was dissolved in 40.0 g of CCl_4. The solution was found to boil at 77.2°C. What is the molecular weight of the unknown solute? The boiling point of CCl_4 is 76.8°C and the molal boiling point constant (K_b) for CCl_4 is 5.02°C.

SOLUTION:　$\Delta t_b = 77.2°C - 76.8°C = 0.40°C$. Using the boiling point equation similar to (15)

$$\frac{0.250 \text{ g}}{\text{GMW}} = (0.0400 \text{ kg})\left(\frac{0.40°C}{5.02°C \text{ kg/mole}}\right)$$

$$\text{GMW} = 78.4 \text{ g/mole and MW} = 78.4$$

Osmotic Pressure

If the U-shaped tube shown in Figure 13.7 were filled with water on one side and a sugar solution on the other, in time the sugar molecules would diffuse throughout, so that a uniform concentration of sugar would be present on both sides of the tube. If, however, a semipermeable membrane — such as an animal or vegetable membrane that will allow water to pass through it but not the solute molecules — is placed across the bottom of the U-tube, and

---

° It should be pointed out that this relationship between freezing point and concentration does not hold very well for concentrated solutions.

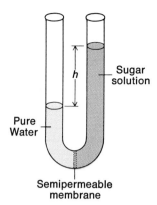

Figure 13.7
Apparatus for demonstrating osmotic pressure.

the U-tube is again filled with water on one side and sugar solution on the other to the same level, it will be noted that water will flow through the semipermeable membrane from the water side of the tube to dilute the sugar solution on the other side. As a result, the level of solution on the sugar side will rise, while the level on the pure water side will be depressed. If evaporation is prevented, the system will remain indefinitely at these new levels. The difference in levels between the two arms of the U-tube (h) is a measure of the pressure tendency of the water to go through the membrane and dilute the solution. This pressure—the pressure sufficient to prevent the flow of pure solvent through a semipermeable membrane into a solution—is called **osmotic pressure.** At constant temperatures this pressure is proportional to the concentration of the solution. Like the boiling point elevation and freezing point depression of a solvent and the vapor pressure of a solution, the osmotic pressure is a property dependent on the *number* of dissolved particles and is independent of the nature of the dissolved particles. Such properties, which are dependent on *number* but not *kind* of dissolved particles, are termed **colligative** *properties.*

Osmosis is the passage of a solvent through a semipermeable membrane from a solution of lesser concentration to one of higher concentration. Osmosis through plant and animal membranes is of prime importance in many physiological processes.

Colligative Properties of Electrolytes (Acids, Bases, Salts) in Aqueous Solution

When one mole of an acid, base, or salt is dissolved, more than one mole of ions is present in the solution; for example 1 mole of NaCl yields two moles of ions in solution—one mole of Na^+ and one mole of Cl^-. Since more dissolved particles (ions) are present the colligative properties—vapor pressure, boiling and freezing points and osmotic pressure, which depend upon *num-*

ber of dissolved particles, are affected more than for a non-electrolyte. One mole of NaCl which gives 2 moles of ions would be expected to have double the effect of 1 mole of something like sugar, a nonelectrolyte, which yields just 1 mole of dissolved particles (molecules). Actually the effects are not quite what would be predicted on theoretical grounds since the ions in solution do not act quite independently and their influence on one another diminishes the total expected effect. This gives rise to an apparent degree of ionization which may differ from the true degree of ionization. An apparent degree of ionization may be calculated from colligative properties observed with electrolytes. Study the following examples:

EXAMPLE 13.15   A solution of 10.0 g of HF in 500 g H_2O freezes at $-1.98°C$. Calculate the degree of ionization of HF. MW HF $= 20.0$

SOLUTION:

$$\text{Molality of HF} = \frac{10.0 \text{ g}}{(20.0 \text{ g/mole})(0.500 \text{ kg})} = 1.00$$

But the HF is partially ionized according to the equation

$$HF \rightleftharpoons H^+ + F^-$$

and we need to know the concentration of each species, H^+, F^- and HF molecules. Let $x =$ degree(fraction) of ionization.

Then $x =$ molality of H^+
and $x =$ molality of F^- since in the ionization one F^-
                 is formed for each H^+ produced.
and $1.00 - x =$ molality of HF remaining after the ionization.

(At equilibrium)
Effective molality of all species $= x + x + (1.00 - x) = 1.00 + x$

and from equation (14)

$$1.00 + x = \frac{1.98°}{1.86°} = 1.06$$
$$x = 0.06 \text{ or } 6\%$$

EXAMPLE 13.16   Determine the freezing point of a 0.50 molal aqueous solution of a weak acid HA which is 10% ionized.

SOLUTION:
$$HA \rightleftharpoons H^+ + A^-$$

Moles of H^+ at equilibrium $= (0.10)(0.50) = 0.050$
Moles of A^- at equilibrium $= (0.10)(0.50) = 0.050$
Moles of HA at equilibrium $= (0.90)(0.50) = \underline{0.45}$
Total molality of all species 0.55

Freezing point depression $= (0.55 \text{ moles/kg})(1.86° \text{ kg/mole}) = 1.0°C$
Freezing point of solution $= -1.0°C$

Exercises

13.1. Define and illustrate the terms: solution, solute, solvent, solubility, polar and non-polar solvents, saturated, supersaturated, Henry's law, non-volatile solute, colligative property, osmosis, titration, buret, standard solution, Raoult's law.

13.2. Match the statement with the appropriate term.

(a) A solution.
(b) The dissolved material in a solution.
(c) An example of a polar solvent.
(d) A solution in which the dissolved solute is in equilibrium with the undissolved solute.
(e) The solubility of a gas dissolved in a liquid is proportional to the pressure of the gas over the liquid.
(f) Concentration of a solution in equivalents of solute/liters of solution.
(g) Concentration in moles of solute per kilogram of solvent.
(h) A solution whose concentration is accurately known.
(i) P solv $= (P°_1)$(mole fraction of solvent)
(j) The operation of determining the concentration of an unknown solution by means of a measured volume of a standard solution.
(k) Passage of solvent through a semipermeable membrane.

1. H_2O C
2. CCl_4
3. benzene
4. standard solution h
5. titration J
6. a homogeneous A mixture
7. osmosis K
8. solute B
9. solvent
10. saturated solution D
11. unsaturated solution
12. supersaturated solution
13. Hess's law
14. Raoult's law i
15. Henry's law e
16. M
17. N F
18. m G
19. mole fraction
20. percent

13.3. Give examples of the various types of true solutions; e.g., gas in gas, gas in liquid, etc. Which combinations of physical states are unlikely or impossible?

13.4. Consider a solution of $NaC_2H_3O_2$ in which no solid phase is present. How could you determine whether the solution is unsaturated, saturated, or supersaturated? Add a crystal

13.5. From the following list of water-soluble compounds, select (a) 5 pairs which, when brought together in aqueous solution, will not form a precipitate, (b) 5 pairs which will give a precipitate. Write equations for part (b). NaOH, K_2CO_3, $Pb(NO_3)_2$, K_3PO_4, $(NH_4)_2CO_3$, LiCl, $CaCl_2$, $CuSO_4$, $AgC_2H_3O_2$, $AlCl_3$, $NaNO_3$, $(NH_4)_2SO_4$, $BaBr_2$, K_3AsO_4, $Cr(C_2H_3O_2)_3$, $ZnCl_2$.

13.6. Underscore in the following list those compounds which are relatively insoluble in water. $BaCl_2$, $Cu(NO_3)_2$, $Zn_3(PO_4)_2$, Na_2CO_3, CuS, $Fe(OH)_3$, $(NH_4)_2$ CrO_4,

Ag_2SO_4, $CdCO_3$, $Pb_3(AsO_4)_2$, $BaSO_4$, $Al(C_2H_3O_2)_3$, $AgBr$, KI, Na_3BO_3, $MgCO_3$, NH_4I, CaS.

13.7. Refer to the solubility data on page 286. A solution containing 50 g of NaCl, 50 g of KNO_3, and 150 g H_2O is evaporated to 100 g H_2O at 100°C. (a) What solid will crystallize from solution and how much? (b) If the solution is allowed to cool to 50°C, what solid will crystallize and how much? (c) If the solution from (b), after filtering off the solid, is concentrated to 50 g of H_2O and cooled to 20°C, what will be the composition of the solid crystallizing out?

13.8. With the aid of the solubility curves on page 286, answer the following questions:
(a) At what approximate temperature will the solubilities of NaCl and KCl be the same?
(b) At what approximate temperature will the solubilities of KCl and KNO_3 be the same?
(c) Under what conditions will KCl exhibit a greater solubility than KNO_3?
(d) If separate saturated solutions of $NaNO_3$ and NaCl are made up at 40°C and cooled to 0°C, will the weight of $NaNO_3$ separating out be less or greater than the weight of NaCl crystallizing out?
(e) Which of the salts in the diagram is most soluble at 40°C?
(f) If a saturated solution of $CaCrO_4$ at 40°C is cooled to 10°C, will a solid crystallize? If this solution is heated to 100°C, what happens?
(g) If separate saturated solutions of the salts NaCl, KCl, and KNO_3 are made up at 100°C in 100 g H_2O, what weight of each salt will be crystallized from solution on cooling to 10°C?

13.9. A 30% solution of ethyl alcohol in H_2O at 20°C has a density of 0.95 g/ml. What weight of alcohol is contained in 750 ml of solution?

13.10. Using NaCl and H_2O explain what is meant by the following concentrations: (a) 10% NaCl solution, (b) 3 M NaCl, (c) 0.1 N NaCl, (d) 2.5 m NaCl, (e) mole fraction of NaCl = 0.3.

13.11. Determine the molarity and normality of the following aqueous solutions which contain the indicated amounts of solute per liter of solution:
(a) 120 g NaOH   (d) 5.0 g H_3PO_4
(b) 29.4 g of H_2SO_4 (e) 18.5 g $Ca(OH)_2$
(c) 378 g HNO_3 (f) 100 g $Al_2(SO_4)_3$

13.12. What weight of solute is needed to prepare each of the following solutions?
(a) 3 l of 0.5 M NaOH   (d) 350 ml 0.6 M Na_2CO_3
(b) 200 ml of 0.3 N H_2SO_4 (e) 25 ml 0.20 N $AgNO_3$
(c) 400 ml of 0.05 N $Ba(OH)_2$ (f) 36.9 ml 0.153 M $CuSO_4$.

13.13. (a) What is the molarity of a 0.050 N solution of $Ba(OH)_2$ (calculated on the basis of complete neutralization of the alkali)? (b) What is the normality of 0.050 M H_3PO_4 (based on neutralization of the acid to the HPO_4^{2-} ion)?

13.14. What volume of concentrated HNO_3 should be used to prepare 750 ml of

0.500 M HNO_3? The concentrated acid is 70.0% HNO_3 and has a density of 1.42 g/ml.

13.15. If 50 ml of 6.0 N sulfuric acid is diluted to one liter, what is the normality of the resulting solution?

13.16. A 35.42-ml sample of a solution of 0.1010 N HCl is required to neutralize 25.00 ml of a solution of NaOH. What is the concentration of NaOH?

13.17. A solution of NaOH is standardized by titrating a weighed sample of oxalic acid dihydrate, $H_2C_2O_4 \cdot 2H_2O$. If 45.22 ml of NaOH is required to neutralize 0.3500 g of acid, what is the normality of the NaOH?

13.18. A 0.250-g sample of a pure solid acid requires 25.1 ml of 0.120 N NaOH for neutralization. What is the equivalent weight of the acid?

13.19. A 30.0-ml sample of vinegar requires 40.0 ml of 0.500 N NaOH for neutralization. The density of the vinegar is 1.00 g/ml. What is the concentration of acetic acid ($HC_2H_3O_2$) in vinegar in terms of (a) normality, and (b) weight percent?

13.20. Potassium acid phthalate ($KHC_8H_4O_4$) functions as a monoprotic acid. If a 1.50-g sample of impure $KHC_8H_4O_4$ requires 35.5 ml of 0.150 N NaOH for neutralization, what is the percentage of $KHC_8H_4O_4$ in the material?

13.21. Sodium carbonate reacts with acids:

$$Na_2CO_3 + 2\ HCl \rightarrow 2\ NaCl + H_2O + CO_2(g)$$

(a) What is the equivalent weight of Na_2CO_3 for this reaction? (b) If a 0.420-g sample of impure Na_2CO_3 reacts with exactly 3.45 ml of 0.1066 N HCl, what is the weight percentage of Na_2CO_3 in the material?

13.22. A 40.0-ml sample of rhubarb juice is titrated against 0.270 N NaOH, and 18.4 ml of NaOH solution is required for neutralization. Assuming the acidity of the juice as due to oxalic acid ($H_2C_2O_4$), determine (a) the normality of the juice, (b) the weight of oxalic acid per liter of juice.

13.23. Calculate the mole fraction of solute and solvent in a solution prepared by dissolving 500 g of C_2H_5OH (ethyl alcohol) in 500 g of H_2O.

13.24. A solution is prepared by dissolving 400 g of NaOH in water and then diluting to one liter. The density of the resulting solution is 1.31 g/ml. Express the concentration of NaOH as (a) percentage by weight, (b) molarity, (c) normality, (d) molality, (e) mole fraction.

13.25. Consider a solution of oxalic acid, $H_2C_2O_4$, consisting of 45.0 g of $H_2C_2O_4$ per 100 g H_2O. The density of the solution is 1.20 g/ml. The volume of the solution is 109 ml.
(a) What is the molality of the solution?
(b) What is the mole fraction of $H_2C_2O_4$?

(c) What is the molarity of the solution?

(d) What is the normality of the solution?

13.26. Consider a solution of sodium carbonate made up as follows:

Wt. of Na_2CO_3 (MW 106) = 212 g

Wt. of H_2O (MW 18) = 180 g

Total volume of solution = 327 ml

(a) Calculate the mole fraction of H_2O in the solution.

(b) Calculate the molarity of the solution.

(c) Calculate the molality of the solution.

(d) Calculate the normality of the solution.

(e) What is the percent solute by weight in the solution?

13.27. The vapor pressure of water at 25°C is 23.76 torr. Determine the vapor pressure of a solution of 50.00 g of glucose ($C_6H_{12}O_6$) in 100.0 g of H_2O.

13.28. Consider solutions A and B. Solution A consists of 90.0 g of glucose ($C_6H_{12}O_6$) dissolved in 180 g of H_2O. Solution B consists of 100 g of an unknown compound dissolved in 150 g of H_2O (boiling point 100.85°C).

(a) What is the molecular weight of the unknown compound in solution B?

(b) What is the vapor pressure of solution A at 85°C? (The vapor pressure of H_2O at 85°C = 500 torr.)

13.29. Consider a solution of galactose ($C_6H_{12}O_6$, MW 180) in water (MW 18) made up as follows:

weight of galactose	36 g
weight of water	36 g
temperature of solution	40°C
vapor pressure of pure H_2O at 40° = 55 torr	

(a) At what temperature (°C) would the solution boil?

(b) What would be the vapor pressure in torrs of the *solution* at 40°C?

13.30. What is the freezing point of a solution of 100 g of alcohol (C_2H_5OH) and 200 g of H_2O?

13.31. Determine (a) the boiling point (at 1 atm) and (b) the freezing point of a solution which contains 35.0 g of sucrose ($C_{12}H_{22}O_{11}$) in 370 g of H_2O.

13.32. The boiling point of a solution prepared from 8.530 g of nonvolatile, non-dissociating solute in 500.0 g of chloroform ($CHCl_3$) is 0.558°C higher than the boiling point of pure chloroform. What is the molecular weight of the solute? K_b for $CHCl_3$ = 3.63°/mole.

13.33. A solution containing 4.96 g of a nonvolatile, nonelectrolyte in 220 g of water freezes at −0.480°C. What is the molecular weight of the solute?

13.34. The freezing-point constant of toluene is 3.33° per mole per 1000 g. Calculate the freezing point of a solution prepared by dissolving 0.500 mole of solute in 480 g of toluene. The freezing point of toluene is −95.0°C.

13.35. The boiling-point constant for carbon tetrachloride is 5.03°. The boiling point of carbon tetrachloride is 77°. Determine the boiling point (at 1 atm) of a solution made by dissolving 0.35 mole of a substance in 250 g of carbon tetrachloride.

13.36. (a) What is the boiling point elevation of a solution prepared from 10.5 g of glycerol $(C_3H_5(OH)_3)$ and 250 g of CCl_4? (b) What is the freezing point lowering of this solution? $K_b = 5.02°$/mole, $K_f = 29.8°$/mole.

13.37. Explain by means of a diagram how you might separate a mixture of two miscible liquids, A and B, if A boils at 90°C and B boils at 130°C, assuming that the liquid and vapor curves are similar to those in Figure 13.6.

13.38. A solution consists of 3.00 g of a non-volatile solute and 90.10 g of water. At 60°C the vapor pressure of this solution is 147.2 torr. What is the molecular weight of the solute? The vapor pressure of water at 60°C is 148.9 torr.

13.39. Explain (a) osmosis, (b) osmotic pressure, (c) how the osmotic pressure of a solution depends on its concentration.

13.40. Explain why colligative properties, such as vapor pressure and boiling and freezing points, indicate that there must be a greater number of particles per mole in solutions of acids, bases, and salts than in solutions of non-electrolytes.

13.41. A solution prepared by adding a given weight of a solute to 20.0 g of benzene freezes at a temperature 0.384°C below the freezing point of pure benzene. The freezing point of a solution prepared from the same weight of the solute and 20.00 g of water freezes at −0.4185°C. Assume that the solute is undissociated in benzene solution but is completely dissociated into ions in water solution. How many ions result from the dissociation of one molecule in water solution? K_f for benzene = 5.12°/mole.

13.42. What is the freezing point of a 0.180 m aqueous solution of a weak acid HX if the acid is 7.30% ionized?

13.43. Ammonia ionizes in water as follows:

$$NH_3 + H_2O \rightleftharpoons NH_4^+ + OH^-$$

A 0.0100 m solution of NH_3 in water freezes at −0.0193°C. Calculate the percentage ionization of the ammonia.

Supplementary Readings

13.1. "Molecular Motions," B. J. Alder and T. E. Wainwright, Sci. American, Oct., 1959; p. 133.
13.2. "Colligative Properties," F. Rioux, J. Chem. Educ., 50, 490 (1973).
13.3. "The Osmotic Pump," O. Levenspiel and N. de Nevers, Science, 183, 157 (1974).

Chapter 14

Reaction
Rates and
Chemical
Equilibrium

We have learned that much information may be derived from a chemical equation—for example, the products we can expect from a given set of reactants, and the weight relations between the substances involved in the reaction. The equation, however, does *not* tell us anything about the conditions under which the reaction will take place, nor does it give us any information about the rate at which the reaction proceeds. Chemical thermodynamics likewise indicates whether or not a reaction will be spontaneous under a given set of conditions, but it says nothing about the rate of the reaction.

We have observed that some chemical changes—for example, the explosion of dynamite—take place very rapidly, while others—like the rusting of iron—take place very slowly over a period of days or even years. All chemical reactions require a certain amount of time, although in some cases it may be very difficult to measure the time interval. When we speak of reaction rate, we mean the amount of chemical change which takes place in a given interval of time. For example, we may measure the volume of oxygen gas obtained in one minute by heating a given amount of potassium chlorate at a given temperature. This quantity of oxygen produced in a definite period of time will be a measure of the rate or speed of the reaction at that temperature. Alternately, the reaction rate may be expressed as the rate of disappearance of potassium chlorate.

These reaction rates are of great importance to the chemist; often the feasibility of a chemical reaction, particularly a commercial chemical process, depends on the reaction rate—if the process is too slow, it may not be suitable for commercial exploitation.

The study of reaction rates and the mechanism or step sequence under which reactions occur is called **chemical kinetics.**

Factors which Influence the Speed of Reactions

Many factors are involved in reaction rates; the more important ones may be summarized as follows:

1. NATURE OF THE REACTING SUBSTANCES   Substances differ in activity and hence in the speed with which they react with other substances. The active metals displace hydrogen vigorously and rapidly from acids, while the less active metals act slowly, if at all. Metals differ in their rates of corrosion because of differences in speed of combination with oxygen and other elements. Elements like nitrogen combine very slowly with other elements; in contrast, the halogens combine with most of the other elements readily.

2. TEMPERATURE   The speed of all chemical changes increases as the temperature rises. Hydrogen and oxygen combine very slowly at ordinary temperatures, but rapidly at high temperatures. A piece of coal will burn readily in air when the temperature is raised sufficiently. In general, **the**

speed of a chemical change is approximately doubled to tripled for each ten degrees rise in temperature.

3. CATALYSTS   Catalysts are substances which increase the rate of certain reactions without being used up, and they may be recovered unchanged upon the completion of the reaction. For example, the decomposition of potassium chlorate when heated is speeded up by the addition of the catalyst manganese dioxide

$$2 \text{ KClO}_3 \xrightarrow{\text{MnO}_2} 2 \text{ KCl} + 3 \text{ O}_2$$

The catalyst is written over the arrow since it does not affect the overall balanced equation, only the rate.

Enzymes are organic catalysts for specific biological reactions. The protein casein, for example, is readily hydrolyzed to its constituent amino acids by the enzymes in the gastric juices of the stomach.

Some substances may reduce the speed of chemical change, for example, acetanilide decreases the speed of decomposition of hydrogen peroxide. These substances are termed **inhibitors**. Catalysis is extremely important in the preparation of many industrial products. The addition of a catalyst may speed up a slow chemical change enough to make it commercially feasible.

4. CONCENTRATION   Generally, an increase in concentration of the reactants will increase the reaction rate. There are many exceptions, however, and the exact relationship must be determined by actual experiment. For a generalized reaction

$$aA + bB + \ldots \rightarrow$$

an expression showing the relationship of reaction rate and concentration of the reactants is

$$\text{Rate} = k[A]^x[B]^y \ldots \tag{1}$$

where the brackets stand for **molar concentrations** (moles/l or M) of reactants A and B. The term k is a constant for a particular reaction and is known as the **specific rate constant.** It includes the effects of temperature, pressure, catalyst, and nature of the reactants—all the factors other than concentration which influence the rate of reaction. The exponents x and y, while usually integers, may be fractions, may be zero, or may even occasionally be negative in value. **The actual values of x and y must be determined by experiment and may or may not be the coefficients a and b of the balanced chemical equation.**

Equation (1) is known as the **rate law.** It was discovered experimentally by two Norwegian chemists, Cato Guldberg and Peter Waage, in 1864 and is often referred to as the **Law of Mass Action.**

Consider, as an example, the reaction between hydrogen gas with nitrogen (II) oxide at a high temperature:

$$2\ H_2(g) + 2\ NO(g) \rightarrow 2\ H_2O(g) + N_2(g)$$

Experimentally, the rate law for this reaction is

$$\text{Rate} = k[H_2][NO]^2$$

where $[H_2]$ and $[NO]$ represent the molar concentrations of H_2 and NO, respectively. Note that the coefficient (a and b) for H_2 and NO in the balanced chemical equation is 2 in both cases, while the exponents (x and y) in the rate equation experimentally are 1 for $[H_2]$ and 2 for $[NO]$.

The **order** of a reaction is the sum of the exponents of the concentrations in the rate equation. The above reaction is a **third order** reaction, **first order** with respect to H_2 and **second order** with respect to NO, but third order over-all. For this reaction, note that if the concentration of NO is held fixed but the concentration of H_2 is varied, the rate of reaction will vary in direct propor-tion to $[H_2]$. This amounts to saying that the exponent of $[H_2]$ must be 1. If the H_2 concentration is doubled, the rate will double. When the $[H_2]$ is tripled, the rate is tripled, and so on. On the other hand, if $[H_2]$ is held constant and $[NO]$ is varied, the reaction rate goes up as the *square* of the $[NO]$; that is, if $[NO]$ is doubled, the rate *quadruples*, and if $[NO]$ is tripled, the rate increases by a factor of *nine*. When this is observed experimentally, we know that the exponent of $[NO]$ must be 2.

The rate constant, k, for the reaction can be evaluated experimentally by measuring the amount of N_2 or H_2O produced in unit time—let us say one minute—from known concentrations of H_2 and NO. For example

$$k = \frac{\text{Moles of } H_2O \text{ produced per minute}}{[H_2][NO_2]^2}$$

Once the rate constant is evaluated, it can be used to determine the amount of H_2O produced in one minute from any given concentrations of H_2 and NO.

EXAMPLE 14.1   Experimental data concerning a reaction between gases A and B to form C is as follows:

Experiment	Initial $[A]$	Initial $[B]$	Initial rate of C formation, (moles/liter)/min
1	1.0×10^{-2}	0.50×10^{-3}	0.25×10^{-6}
2	1.0×10^{-2}	$1.0\ \times 10^{-3}$	0.50×10^{-6}
3	1.0×10^{-2}	$1.5\ \times 10^{-3}$	0.75×10^{-6}
4	2.0×10^{-2}	0.50×10^{-3}	$1.0\ \times 10^{-6}$
5	3.0×10^{-2}	0.50×10^{-3}	2.25×10^{-6}

What is: (a) the rate law expression?
(b) the order of the reaction?
(c) the specific rate constant, k?
(d) the initial rate of the reaction when the initial concentration of A is 4.0×10^{-2} mole/l and B is 2.0×10^{-3} mole/l?

SOLUTION:

(a) From the data, when [A] is held constant, the rate is directly proportional to [B]. So [B] *is to the first power* in the rate equation. For example, compare data in experiments 1 and 2. Here [A] is the same (1.0×10^{-2}), but in 2 [B] is twice that in 1. The rate in 2 is also twice that in 1. Compare 1 and 3. Again [A] remains constant (1.0×10^{-2}) but in 3 where [B] is tripled, the rate is also tripled.

On the other hand, when [B] is held constant, the rate is seen to go up as the *square* of [A]. Consequently, [A] *is to the second power* in the rate law. For example, compare 1 and 4 where [B] is the same (0.50×10^{-3}). In 4, when [A] is double what it is in 1, the rate is quadrupled. Likewise, compare experiments 1 and 5. [B] is constant (0.50×10^{-3} in both cases), but when [A] is tripled, the rate is increased 9 times. From the foregoing, therefore, the rate law expression is

$$\text{Rate} = k[A]^2[B]$$

(b) The reaction is *third order* since the sum of the exponents in the rate law $(2 + 1)$ is 3.

(c) The specific rate constant, k, may be calculated from the data of any of the five experiments. Take, for example, experiment 2. For this

$$k = \frac{\text{Rate}}{[A]^2[B]} = \frac{0.50 \times 10^{-6} \text{ (mole/l)/min}}{(1.0 \times 10^{-2} \text{ mole/l})^2 (1.0 \times 10^{-3} \text{ mole/l})} = (5.0 \text{ mole/l})^{-2}/\text{min}$$

Likewise for experiment 5:

$$k = \frac{2.25 \times 10^{-6}}{(3.0 \times 10^{-2})^2 (0.50 \times 10^{-3})} = 5.0$$

(d) $\text{Rate} = (5.0)(4.0 \times 10^{-2})^2(2.0 \times 10^{-3})$
$= 1.6 \times 10^{-5} \text{ (moles/l)/min}$

First-order reactions   For radioactivity (Chap. 35) and quite a number of chemical reactions, the rate of reaction is proportional to the concentration of a single reactant. Such reactions are called "first-order" reactions. For example, if substance A decomposes into the products B and C

$$A \rightarrow B + C$$

the rate is proportional to the concentration of A which is present at any time. Expressed mathematically,

$$\text{Rate} \propto [A] \quad \text{or} \quad \text{Rate} = k[A] \tag{1}$$

where the constant, k, is the rate constant which was discussed above. The rate at which the reaction proceeds is equal to the decrease in concentration of A with change of time and may be denoted as $-\dfrac{d[A]}{dt}$ where $d[A]$ is a small change in concentration of A and dt is a small change in time. Equation (1) then becomes

$$\text{Rate} = -\frac{d[A]}{dt} = k[A] \tag{2}$$

By methods of the calculus, the differential equation (2) may be integrated to give

$$2.303 \log \frac{[A]}{[A_0]} = -k\,t \tag{3}$$

where $[A_0]$ is the concentration of A at the beginning of the reaction when the time is zero, and $[A]$ represents the concentration of A at time t. Once the rate constant has been determined, equation (3) is very useful, since one may calculate the quantity of material decomposing in a given time – or the time required for any given amount of substance to decompose.

The rate for a first-order reaction is often described in terms of a "half-life" – that is, the time required for one-half of any given amount of the reactant to decompose. For a half-life period of time $t_{1/2}$, $\frac{[A]}{[A_0]} = \frac{1}{2}$. Substituting in equation (3),

$$2.303 \log \tfrac{1}{2} = -k\,t_{1/2}$$
$$(2.303)(-0.301) = -k\,t_{1/2}$$

or

$$t_{1/2} = \frac{0.693}{k}$$

EXAMPLE 14.2 The half-life for the first-order reaction: $SO_2Cl_2 \rightarrow SO_2 + Cl_2$ is 8.0 min. In what period of time would the concentration of SO_2Cl_2 be reduced to 1.0% of the original?

SOLUTION: First calculate the rate constant – that is, k – from $t_{1/2} = \dfrac{0.693}{k}$

$$k = \frac{0.693}{8.0 \text{ min}} = 0.087/\text{min}$$

Now use the expression $2.303 \log \dfrac{[A]}{[A_0]} = -k\,t$, where $\dfrac{[A]}{[A_0]} = 1.0\%$ or 0.010

$$2.303 \log 0.010 = -(0.087/\text{min})(t)$$

$$t = -\frac{(2.303)(-2) \text{ min}}{0.087} = 53 \text{ min}$$

The chemical reactions which have been used so far in this section are known as **homogeneous** reactions, where the reaction occurs in a single phase – all substances are gases, or all are liquids, etc. **Heterogeneous** reactions involve more than one phase and the reactions take place on a boundary between two phases. Here the rate of reaction may be markedly altered by **stirring** or the **state of subdivision**; for example, the rusting of iron. Rusting is a slow reaction involving in part an iron object, the solid phase, reacting with oxygen of the air, a gas phase. If the iron is finely divided, however, the reaction is rapid because of the greater surface area of iron. Combustible dusts are often explosive for this same reason. Solid catalysts are usually porous and finely divided for greatest effectiveness.

Molecular Collision Theory

From a kinetic standpoint, chemical change takes place as a result of collisions of molecules. The greater the number of collisions per unit time, the greater the conversion of initial substances into products per unit time — that is, the greater the speed of reaction. If the concentrations of reactants are high (a large number of molecules in a given volume), the chance for collision is much greater, just as the chance for collision between couples on a dance floor is much greater when the floor is crowded.

The student should not infer from the preceding discussion that all collisions of molecules necessarily result in chemical change. According to collision theory, two conditions must be met in order to have an effective collision. (1) The orientation of the colliding molecules must be favorable for the making and breaking of bonds and (2) before molecules (or other reacting units) can react, they must possess a certain minimum energy, termed **activation energy.** We may look at it this way: of the billions of molecules in a system, some will have a higher kinetic energy and some a lower kinetic energy than the average of all molecules in the system. Only those with the necessary activation energy will undergo reaction. Apparently the reactants do not proceed directly to products; instead, an intermediate species, frequently referred to as an **activated complex,** is first formed. This activated complex then decomposes into the products of the chemical change. Such an activated complex is represented in Figure 14.1, which is a plot of potential energy versus progress of reaction or **reaction coordinate.** The intermediate complex has the minimum energy required for reaction to occur. Collisions of reactants which do not possess sufficient energy to reach this minimum will not be effective, and the reactants will remain unchanged. However, if the energy of the reactants is sufficient to form the activated complex and to surmount the minimum energy barrier which it represents, the reaction proceeds to the formation of products. When the energy of activation is not high, a relatively high percentage of collisions may result in atomic rearrangements (chemical change), while in cases where the energy requirements are high, only a small percentage of collisions will be effective.

Suppose for example, molecules of reactants A and B approach each other on a collision course, (Figures 14.1 and 14.2(a)). While still relatively far apart, A and B will have some total average potential energy, E_1. This will be all of the molecules' energy except their energy of motion. As the molecules get closer together (Figure 14.2 (b)) work must be done to overcome the mutual repulsion of their electron clouds; that is, some of the kinetic energy of the molecules will be transformed into potential energy of repulsion. If the molecules have sufficient energy, E_2 (and if the orientation is favorable), an intermediate high energy molecule AB^* will be formed momentarily (Figure 14.2 (c)). AB^* is the **activated complex** and the energy required to form it is the **activation energy.** Once the unstable, high energy AB^* is formed, it decomposes or splits apart to form new substances, C and D, which then repel

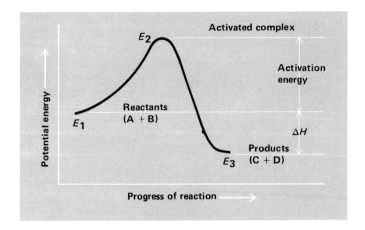

Figure 14.1
Potential energy change during a reaction. Before reaction takes place, reactants must attain a certain minimal energy level represented by E_2. The average energy of the reactants is shown at E_1 and of the products at E_3. The reaction is exothermic ($\Delta H-$).

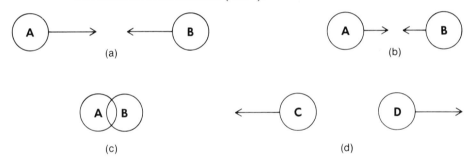

Figure 14.2
Four moments in bimolecular collision between A and B.

one another; their potential energy finally drops to E_3 when C and D are relatively far apart (Figure 14.2 (d)). It will be noted in Figure 14.1 that the difference in energy between the products (E_3) and the reactants (E_1) is the heat of reaction, ΔH. In the figure, the energy of the products is lower than that of the reactants and the sign of ΔH is negative. This corresponds to an **exothermic** reaction. A similar diagram for an endothermic reaction is shown in Figure 14.3.

The equation for the reaction is

$$A + B \rightarrow AB^\circ \rightarrow C + D$$

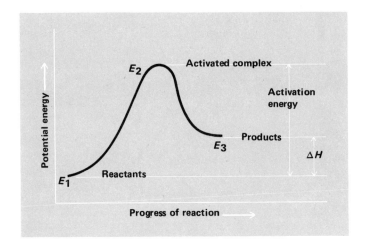

Figure 14.3
Potential energy diagram for an endothermic reaction.

However, AB^* may be so short-lived that all we really observe is

$$A + B \rightarrow C + D$$

While it is not entirely clear what happens when a catalyst speeds up a reaction, it must in some way reduce the activation energy for the reaction, so that more reactive units undergo change in a given period of time. One theory is that the catalyst forms an unstable intermediate with one or more of the reacting species with a lower activation energy than the normal activated complex without a catalyst present. Since the catalyst is not permanently changed, somewhere in the process the catalyst intermediate must decompose to regenerate the catalyst. The catalyst presumably functions by providing a different and easier path (lower activation energy) along which the reactants can proceed to products of the reaction, Figure 14.4.

Reaction Mechanism

Few chemical reactions occur in a single step. There are usually one or more intermediate products which form and the step-wise sequence of reactions is known as the **reaction mechanism.** Consider the nitrogen (II) oxide-hydrogen reaction mentioned previously. The overall net reaction is

$$2 H_2(g) + 2 NO(g) \rightarrow 2 H_2O(g) + N_2(g)$$

and the experimental rate law is

$$Rate = k[H_2] [NO]^2$$

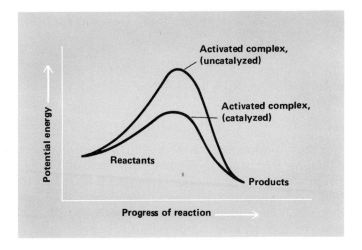

Figure 14.4
Potential energy diagram comparing catalyzed and
uncatalyzed reactions.

What is a possible mechanism? One possibility is

$$H_2(g) + 2\ NO(g) \rightarrow \quad N_2O(g) + H_2O(g) \quad (1)\ \text{slow step}$$
$$\underline{H_2(g) +\ \ N_2O(g) \rightarrow \quad\quad N_2(g) + H_2O(g)} \quad (2)\ \text{fast step}$$
$$\text{Net: } 2\ H_2(g) + 2\ NO(g) \rightarrow 2\ H_2O(g) + \quad N_2(g)$$

In the above mechanism, reaction (1) produces an intermediate compound
(N_2O) which then reacts with more H_2 in reaction (2). Generally one of the
steps in the mechanism has a rate considerably slower than the others and is
called the **slow step** or **rate-determining** step. It is this step which will
determine the overall reaction rate. In the above example, reaction (1) is the
slow step and for this step the coefficients of H_2 and NO will be the exponents
of $[H_2]$ and $[NO]$ in the rate law. Note also that the sum of reactions (1) and
(2) must equal the overall net reaction and that the intermediate, N_2O, cancels
out. The intermediate compounds are often unstable and very short-lived. If
they can be detected experimentally, however, they give good evidence in
support of the theoretical mechanism.

  Another possible mechanism of the above reaction of H_2 with NO might
be

$$2\ NO(g) \rightarrow N_2O_2(g) \quad\quad\quad\quad\quad\quad (1)\ \text{fast step}$$
$$N_2O_2(g) + H_2(g) \rightarrow N_2O(g) + H_2O(g) \quad\quad (2)\ \text{slow step}$$
$$N_2O(g) + H_2(g) \rightarrow N_2(g) + H_2O(g) \quad\quad (3)\ \text{fast step}$$

Net: $2\ H_2(g) + 2\ NO(g) \rightarrow 2\ H_2O(g) + N_2(g)$

Step (2) would be the rate-determining step and for this the rate law would be

$$\text{Rate} = k[N_2O_2][H_2]$$

or, since the formation of N_2O_2 from 2 NO molecules is fast and $[N_2O_2]$ would be proportional to $[NO]^2$,

$$\text{Rate} = k[NO]^2[H_2]$$

which matches the experimental rate law. Actually this second mechanism is more logical than the first, since the first mechanism would involve the simultaneous collision of *three* molecules and this is of low probability.

The actual mechanism of a reaction is often very complicated and some simple-looking reactions still do not have satisfactory mechanisms. To be acceptable, the mechanism must agree with the experimental rate law, and the various steps must add up to the overall net equation. The experimental confirmation of at least some of the intermediate compounds should be shown, and the mechanism should agree with all of the known facts concerning the reaction.

Photochemical and Chain Reactions

Reactions which are initiated by light are called photochemical reactions. The most important of these is **photosynthesis,** wherein plants convert CO_2 and H_2O into carbohydrates, chlorophyll being a catalyst. The reaction of $H_2(g)$ and $Cl_2(g)$ to form HCl in the presence of sunlight proceeds as follows:

First the light energy dissociates Cl_2 molecules into Cl atoms. This is a **chain initiating** step

$$Cl_2 \xrightarrow{\text{light}} 2\,Cl$$

Cl atoms are very reactive and start a **chain reaction** involving H_2 and Cl_2 molecules

$$Cl + H_2 \rightarrow HCl + H$$
$$H + Cl_2 \rightarrow HCl + Cl$$
$$Cl + H_2 \rightarrow HCl + H$$
$$H + Cl_2 \rightarrow HCl + Cl$$

which is repeated over and over again. Note that the sum of the steps of the chain reaction gives the overall net equation

$$H_2 + Cl_2 \rightarrow 2\,HCl$$

and all of the Cl and H atom intermediates cancel out. Also note that only a relatively few Cl atoms are necessary to keep the reaction going, since it is regenerated and used again many times. Chain reactions are generally very rapid and often are explosive. The above reaction of H_2 with Cl_2 is stable in the dark, but explodes in the presence of ultraviolet light.

Other examples of chain reactions are H_2—O_2 explosions, flame combustion, nuclear fusion and fission, and the formation of smog initiated by the photochemical reaction

$$NO_2 \xrightarrow{\text{sunlight}} NO + O_2$$

The very reactive atomic oxygen reacts readily with O_2 to form O_3 and with hydrocarbons from automobile exhausts to form eye irritating aldehydes, ketones, and peroxides (Chapter 19).

14.1 CHEMICAL EQUILIBRIUM

We have encountered several examples of physical equilibrium in our study thus far—for example, saturated solutions. In such solutions, two opposing processes are occurring with the same speed: molecules or ions of solid are continually passing into solution, and at the same time molecules or ions from the solution are precipitating or depositing out as solid. Since these processes are taking place at the same speed, no apparent change in the solution is observed, and at a given temperature the solubility is a constant.

A liquid placed in a closed container soon comes to an equilibrium with its vapor. Molecules from the liquid are continually escaping from the surface of the liquid and passing into the vapor state, and molecules from the vapor state are continually passing into the liquid state. When equilibrium has been attained, these two processes are taking place with the same speed, and the rate of evaporation is equal to the rate of condensation.

A chemical equilibrium is like a physical equilibrium in that two opposing processes are occurring simultaneously with the same speed. There is this difference between physical equilibrium and chemical equilibrium, however: in physical equilibrium, only a physical change occurs and there is no change in composition of the substances involved; in chemical equilibrium, a chemical change does occur—the reactants undergo a change in composition to form the products, which in turn are reconverted to the original reacting substances.

Equilibrium in Reversible Changes

In our discussion of the action of steam on iron filings (page 186), it was shown that the following reaction takes place:

$$3 \text{ Fe} + 4 \text{ H}_2\text{O (steam)} \rightarrow \text{Fe}_3\text{O}_4 + 4 \text{ H}_2 \tag{1}$$

This reaction is reversible, since the passage of hydrogen gas over Fe_3O_4 results in the formation of iron and steam:

$$Fe_3O_4 + 4 H_2 \rightarrow 3 Fe + 4 H_2O \tag{2}$$

In both of these cases, a gas has been passed over a solid; in the first, steam over iron; in the second, hydrogen over iron oxide. Let us now modify the conditions by enclosing iron filings and steam in a container so that no gaseous substance can escape. Reaction (1) above will proceed fairly rapidly at first, but reaction (2) cannot take place, since no Fe_3O_4 and H_2 are initially present. But as these two products are formed as a result of reaction (1), reaction (2) will start. At first, reaction (2) is slow because the concentration of reactants is small, but gradually the speed increases, until reactions (1) and (2) are taking place with the same speed. Under these conditions, the amounts of Fe, H_2O, Fe_3O_4, and H_2 remain unchanged, and the system is said to be in **equilibrium.**

To discuss the general case of equilibrium, consider the reaction

$$A + B \rightarrow C + D$$

If the reaction is reversible, C must react with D to reform the reactants A and B:

$$C + D \rightarrow A + B$$

If the two reactions above occur at the same time and at the same speed, a chemical equilibrium results and may be represented thus:

$$A + B \rightleftharpoons C + D$$

When this condition results, no apparent change is taking place; analysis of the system will show that at all times constant amounts of reactants and products will be present. **Chemical equilibrium is an apparent state of rest in which two opposing chemical reactions are proceeding in opposite directions at the same speed.**

It should be pointed out that an equilibrium may be approached from either direction. In the reaction above, it makes no difference whether we start with A and B or C and D as reactants; the same final condition of equilibrium will result.

Often the student has much difficulty in grasping the concepts of chemical equilibrium. First of all, he must realize that all of the reactants and products are mixed together in the same reaction vessel. The chemical equation for the equilibrium shows the mole ratio in which the substances react, but it does not show the relative amounts of substances actually present in the equilibrium mixture. For example, suppose we consider the equilibrium between NO, O_2, and NO_2 gases as shown by the equation

$$2\ NO + O_2 \rightleftharpoons 2\ NO_2$$

At equilibrium all three of these gases are present in the same reaction vessel, and the equation shows that when reaction occurs toward the right, 2 moles of NO will combine with 1 mole of O_2, or for the reverse reaction, 2 moles of

NO_2 will produce 2 moles of NO and 1 mole of O_2. We might, however, put NO and O_2 together in any proportion—for example, 10 moles of NO to 1 mole of O_2, or 50 moles of NO to 1 mole of O_2, or any other ratio. Of course, analysis of the equilibrium mixtures would yield different concentrations of the component gases in each case, but equilibrium conditions would be established no matter what the ratio in which the gases were originally mixed together. The equation tells us only in what ratio the substances actually *react*, but gives us no information about the relative amounts of substances actually present in an equilibrium mixture.

The Equilibrium Constant

Again consider the reaction

$$A + B \rightleftharpoons C + D$$

in which an equilibrium exists between the four substances A, B, C, and D. Two reactions are proceeding, one to the right (called the **forward reaction**), in which A and B react to form C and D, and the other to the left (called the **reverse reaction**), in which C and D react to form A and B. According to the law of mass action,

$$\text{Speed of forward reaction} = k_1 \, [A] \, [B]$$

and

$$\text{Speed of reverse reaction} = k_2 \, [C] \, [D]$$

At the beginning of the reaction, if we start with A and B, the concentrations of A and B will be high, and consequently the forward reaction will proceed rapidly. At the start, no reverse reaction is possible, since C and D are not yet present. But as C and D are formed, the reverse reaction will begin slowly and gradually increase in speed as C and D accumulate. Meanwhile the forward reaction is slowing down as A and B are being used up. Equilibrium results when the speed in the two directions is equal—that is,

$$\overrightarrow{\text{Speed of forward reaction}} = \overleftarrow{\text{Speed of reverse reaction}}$$

Consequently, at equilibrium,

$$k_1 \, [A] \, [B] = k_2 \, [C] \, [D]$$

or

$$\frac{k_1}{k_2} = \frac{[C][D]}{[A][B]}$$

or since the ratio of the two constants $\dfrac{k_1}{k_2}$ is also a constant, we may write

$$K = \frac{[C][D]}{[A][B]}$$

in which K is termed the **equilibrium constant.** We must keep in mind that equilibrium conditions may be approached from either direction. If we had started with C and D as reactants instead of A and B, the same final situation of equilibrium would have resulted.

If more than one molecule of a substance is used as a reactant or is formed as a product, such as

$$N_2 + 3\,H_2 \rightleftharpoons 2\,NH_3$$

the expressions for the rates of reaction in the two directions are more complicated,* but it can be shown that the equilibrium constant is given by the expression:

$$K = \frac{[NH_3]^2}{[N_2][H_2]^3}$$

in which **each concentration is raised to that power which corresponds to the coefficient of that substance in the balanced equation irrespective of the rate laws or mechanisms of the reactions.**

This principle can be demonstrated in the laboratory by analyzing equilibrium systems of nitrogen, hydrogen, and ammonia gases. For a general reaction:

$$aA + bB + \cdots \rightarrow cC + dD + \cdots$$

in which a, b, c, d represent the coefficients of A, B, C, D, respectively, in the balanced equation, the formulation of K becomes

$$K = \frac{[C]^c[D]^d \cdots}{[A]^a[B]^b \cdots}$$

Summarizing, we may say: **In a chemical equilibrium, the product of the concentrations of products divided by the product of concentrations of reactants — each concentration raised to that power whose exponent is the coefficient of the substance in the balanced chemical equation — is a constant.**

K has a constant numerical value for any given reaction at a given temperature. Actual analysis of an equilibrium mixture will give the concentration of each of the substances, and thus will determine the value of K. If the concentration of one of the substances in the mixture is changed, the concentrations of other substances must change in such a way that the ratio

$$\frac{[C][D]}{[A][B]}$$

retains the same numerical value.

Suppose, for example, in the equilibrium

$$A + B \rightleftharpoons C + D$$

---

* In general, the exact form of the rate expression for a reaction must be found by experiment because a reaction may proceed in several steps rather than in the single step shown by the equation.

the concentration of A is increased. We should expect the speed of reaction between A and B to increase, since now there are more possible collisions between A and B molecules. As a result, the equilibrium is shifted to the right; the concentrations of C and D become larger. From an inspection of the formula for the equilibrium constant, it is evident that an increase of [A] must result in an increase of [C] and [D] and a decrease in [B] if K is to retain its value. In similar manner, changes in concentration of other reactants or products would result in concentration adjustments so that K would retain its constant value.

It should be evident from the expression for an equilibrium constant that if K has a large value, the numerator must be large compared with the denominator; consequently, the concentrations of products are relatively high. This indicates that the reaction is more nearly complete to the right. On the other hand, a small value of K shows that the chemical change to the right has not proceeded very far.

The equilibrium constant° is very important in chemistry, since it shows the extent to which a reaction may take place. In the commercial preparation of substances, it is desirable to choose a reaction in which K has a high value, since this indicates a high proportion of products.

The extent to which a reaction takes place is often roughly indicated by varying the length of the arrows appearing in the equilibrium equation; $\rightleftharpoons$ indicates an equilibrium which is displaced far to the right and so has a high proportion of products, $\rightleftharpoons$ one far to the left, with a high proportion of reactants; intermediate stages may be represented by arrows of approximately the same length, $\rightleftharpoons$. A single arrow $\rightarrow$ indicates a complete reaction to the right; $\leftarrow$ indicates no reaction.

Equilibrium Calculations

Students often have considerable trouble in handling equilibrium calculations. Usually the following steps should be undertaken and in the order given.

Write: (1) the **balanced** chemical equation,
   (2) the K expression,
   (3) the **initial** or starting concentrations, mole/l (if given),
   (4) the **equilibrium** concentrations, mole/l,
Finally substitute (4) in (2) and solve.

EXAMPLE 14.3   Nitrogen gas reacts with hydrogen gas at elevated temperatures to produce ammonia as follows:

° The following simple relationship exists between the change in free energy (ΔG) of a reaction and the equilibrium constant:

$$\Delta G = -2.303 \, RT \log K$$

where R is the molar gas constant (1.99 cal/mole-deg) and T is the absolute temperature.

$$N_2 + 3\,H_2 \rightleftharpoons 2\,NH_3$$

The equilibrium concentrations at 500°C are as follows: $N_2 = 0.90$ mole/l, $H_2 = 2.0$ mole/l, and $NH_3 = 0.66$ mole/l. Calculate the value of the equilibrium constant at 500°C.

SOLUTION:

(1) The balanced chemical equation is given in the statement of the problem

$$N_2 + 3\,H_2 \rightleftharpoons 2\,NH_3$$

(2) From the above equation, the K expression is

$$K = \frac{[NH_3]^2}{[N_2][H_2]^3}$$

(3) Since there are no initial concentrations given in this simple case, only equilibrium concentrations, step 3 is omitted and we proceed to step 4.

(4) The equilibrium concentrations are shown beneath the formulas in the chemical equation

$$N_2 + 3\,H_2 \rightleftharpoons 2\,NH_3$$

at equilibrium:	0.90	2.0	0.66
	mole/l	mole/l	mole/l

(5) Substitute the **equilibrium** concentrations in the K expression and solve for K.

$$K = \frac{(0.66 \text{ mole/l})^2}{(0.90 \text{ mole/l})(2.0 \text{ mole/l})^3} = \frac{0.44}{(0.90)(8.0)} = 0.060 \text{ (mole/l)}^2$$

It is common practice, however, to omit the units for K and express it simply as $K = 0.060$.

EXAMPLE 14.4   Hydrogen gas reacts with carbon dioxide at high temperatures as follows

$$H_2 + CO_2 \rightleftharpoons CO + H_2O$$

7.0 moles of H_2 and 3.0 moles of CO_2 are placed in a one-liter reaction vessel and the gas mixture heated to 986°C. At equilibrium, analysis showed the presence of 2.3 moles of H_2O in the vessel. What is the value of K at 986°C?

SOLUTION:

(1) The chemical equation is given above.

(2) The K expression is

$$K = \frac{[CO][H_2O]}{[H_2][CO_2]}$$

(3) The starting concentrations, **before any reaction takes place** are shown **above** the formulas in the chemical equation.

$$\text{Initial: 7.0 mole/l} \quad \text{3.0 mole/l} \qquad 0 \qquad 0$$
$$H_2 \quad + \quad CO_2 \quad \rightleftharpoons \quad CO + H_2O$$

(4) The equilibrium concentrations will be shown beneath the equation; however only the $[H_2O]$ is given in the problem and the others must be reasoned out.

$$\text{Initial: 7.0 mole/l} \quad \text{3.0 mole/l} \qquad 0 \qquad 0$$
$$H_2 \quad + \quad CO_2 \quad \rightleftharpoons \quad CO \quad + \quad H_2O$$
$$\text{At equilibrium: 4.7 mole/l} \quad \text{0.70 mole/l} \quad \text{2.3 mole/l} \quad \text{2.3 mole/l}$$

The balanced chemical equation infers that for every mole of H_2O formed, 1 mole of CO is also formed, and 1 mole each of H_2 and CO_2 will be used up. Actually, however, 2.3 moles of H_2O has been produced at equilibrium. Consequently, 2.3 moles of CO will also be formed, and 2.3 mole of H_2 and 2.3 moles of CO_2 will have reacted. The amount of H_2 left at equilibrium will be $7.0 - 2.3$ or 4.7 moles. Likewise, the amount of CO_2 remaining at equilibrium will be $3.0 - 2.3$ or 0.7 mole.

(5) Substituting the **equilibrium** concentrations (never the initial concentrations) in the K expression and solving, we have

$$K = \frac{(2.3)(2.3)}{(4.7)(0.70)} = 1.6.$$

EXAMPLE 14.5   Five moles of HBr were placed in a 1.00-liter flask and the flask was heated to 1025° C where the equilibrium 2 HBr (g) $\rightleftharpoons$ $H_2(g)$ + $Br_2(g)$ was established. Assuming no volume change of the flask at the high temperature, what would be the concentration of H_2 and Br_2 at equilibrium? The equilibrium constant for HBr at 1025° C is 7.32×10^{-6}.

SOLUTION:
   (1) and (2) For the equation 2 HBr $\rightleftharpoons$ H_2 + Br_2

$$K = \frac{[H_2][Br_2]}{[HBr]^2}$$

(3) and (4)

$$\text{Initial:} \quad \text{5.00 mole/l} \qquad 0 \qquad 0$$
$$2 \text{ HBr} \quad \rightleftharpoons \quad H_2 \quad + \quad Br_2$$
$$\text{Equil.: } (5.00 - 2x)\text{mole/l} \quad x \text{ mole/l} \quad x \text{ mole/l}$$

(4) may be reasoned as follows:
   Let x be the concentration of Br_2. Then x will also be the concentra-

tion of H_2, since for every mole of Br_2 formed one mole of H_2 will also be formed (or if x moles of Br_2 are produced there must also be x moles of H_2 formed). According to the balanced chemical equation, however, it will take 2 moles of HBr to produce 1 mole of Br_2 (or in order to produce x moles of Br_2, $2x$ moles of HBr will be used up). The amount of HBr left at equilibrium will be the initial amount (5.00 moles) minus the amount of HBr used up ($2x$ moles) or the equilibrium concentration of HBr will be $(5.00 - 2x)$mole/l.

(5) The equilibrium concentrations are substituted in the K expression, which then becomes

$$K = 7.32 \times 10^{-6} = \frac{(x)(x)}{(5.00 - 2x)^2} = \frac{x^2}{(5.00 - 2x)^2}$$

This is a quadratic equation and could be solved by the quadratic formula. It is simpler, however, for this particular expression, to take the square root of both sides

$$\sqrt{7.32 \times 10^{-6}} = 2.71 \times 10^{-3} = \frac{x}{(5.00 - 2x)} \text{ and } x = 0.0135 \text{ mole/l}$$

An approximation which greatly simplifies calculations like this is to note that when K is small, x will also generally be small, and when x is added to or subtracted from a relatively larger number ($5.00 - 2x$ in the above equation), the larger number is not changed appreciably. Consequently, little error will result if x is neglected in the $(5 - 2x)$ term. In the above situation, where the actual value of $x = 0.0135$, the value of $5.00 - 2(0.0135)$ is 4.97, which for practical purposes is so close to 5.00 that little difference would result in the final calculated value of x. This is the usual situation for the important ionic equilibria calculations (Chapter 15).

When the $2x$ is omitted in the $(5.00 - 2x)$ term, the K expression becomes

$$K = 7.32 \times 10^{-6} = \frac{x^2}{(5.00)^2}$$

Then $\qquad x^2 = 183 \times 10^{-6}$

and $\qquad x = 13.5 \times 10^{-3} = 0.0135 \text{ mole/l}$

Consequently, at equilibrium

$$x = [Br_2] = [H_2] = 0.0135 \text{ mole/l}$$

Le Chatelier's Principle

A principle of universal application to systems in equilibrium was stated by Le Chatelier as follows: **If a stress is placed on a system in equilibrium whereby the equilibrium is altered, that change will take place which tends**

to relieve or neutralize the effect of the added stress. Consider the equilibrium reaction

$$A + B \rightleftharpoons C + D$$

We learned that an increase in concentration of any of the reactants or products will result in a shift in the equilibrium. An increase in concentration may be considered an added stress or force. For example, if more A is added to the system, thereby increasing the concentration or stress of A, the equilibrium is shifted to the right. Since A is used up in the reaction, the stress (increase in concentration of A) is relieved or diminished. On the other hand, if A is removed (the concentration of A is decreased), the equilibrium will shift to the left, resulting in the formation of more A. The stress in this case is the removal of A, and the system tends to relieve or diminish this stress by forming more A. Other stresses which may be imposed on a system are the addition of heat (increase of temperature) and the application of pressure.

Effect of Pressure on a System in Equilibrium

Consider a reaction in which a relatively large change in volume occurs — for example,

$$N_2 + 3\ H_2 \rightleftharpoons 2\ NH_3$$

The reactants N_2 and H_2 and the product NH_3 are all gaseous at ordinary temperatures. According to the equation, one volume of nitrogen reacts with three volumes of hydrogen to produce two volumes of ammonia. It is evident, then, that four volumes of reactants produce only two volumes of products, a decrease in volume as the reaction proceeds from left to right. We may readily predict the effect of increased pressure on this system from Le Chatelier's principle. If the pressure on the system is increased, the reaction will proceed to the right in the direction of smaller volume, because a decrease in volume will tend to relieve the added stress (pressure). If the added stress is a decrease of pressure, the reaction in which there is a volume increase will be favored, since an increase in volume will tend to offset or neutralize the decrease in pressure. This will result in a displacement of equilibrium to the left. A high pressure will be favorable in the production of ammonia by the above process, since the higher the pressure, the greater the displacement of the equilibrium to the right. A decrease in pressure will have the opposite effect, and the equilibrium will be displaced to the left.

Certain gaseous reactions involve no change in volume — for example,

$$H_2 + I_2 \rightleftharpoons 2\ HI$$

A total of two volumes (one of hydrogen and one of iodine) of reactants produces two volumes of products (two volumes of hydrogen iodide). Pressure has no effect on such a system, since there is no volume change in the reaction.

The Effect of Temperature on an Equilibrium

All chemical changes are accompanied by changes in energy. Certain reactions are exothermic—produce heat during the reaction; others are endothermic—absorb heat. In the reaction

$$N_2 + 3\ H_2 \rightleftharpoons 2\ NH_3 \qquad \Delta H = -21.8\ \text{kcal}$$

nitrogen and hydrogen, in combining to form ammonia, produce heat. When ammonia undergoes decomposition into nitrogen and hydrogen (the reverse of the reaction), an equivalent amount of heat must be taken up or absorbed. When the system is at equilibrium, no heat change is taking place.

Let us apply Le Chatelier's principle to an equilibrium in which heat is added to the system (increase of temperature). To relieve the strain (increase of temperature), the system will absorb heat. For example, a solid, in melting, absorbs heat. If an equilibrium mixture of ice and water is heated, the ice melts and absorbs heat, and the temperature does not rise above 0°C until all of the ice is melted. The equilibrium proportions of ice and water shift, so that the newly applied heat is used.

In the reaction above in which ammonia is produced, an increase of temperature would favor the reaction to the left—that is, the formation of nitrogen and hydrogen from ammonia, since this is the reaction which absorbs heat. Part of the heat is used to raise the temperature of the system and part to shift the equilibrium.

It should be realized that Le Chatelier's principle has considerable significance from a practical standpoint. If ammonia were being produced commercially by the method above, it would be desirable to have the equilibrium displaced as far to the right as possible, since this would mean a greater yield of ammonia. In order to accomplish this, a high pressure and a low temperature are desirable. Ammonia is actually produced commercially (Haber process) by this reaction at a pressure of several hundred atmospheres and a temperature of 400°–500°C. This temperature is rather high, but it is the minimum temperature feasible for the process, since the reaction between H_2 and N_2 to attain equilibrium takes place too slowly at lower temperatures.

If N_2 and H_2 gases are mixed at ordinary temperatures, reaction between them is so slow that thousands of years might elapse before equilibrium proportions of N_2, H_2, and NH_3 would be reached. However, at 400°–500°C, equilibrium proportions are reached in a matter of minutes. Although the equilibrium proportions at the higher temperature are not quite so favorable as at ordinary temperature, time to attain equilibrium is very much reduced. It is economically feasible to obtain a 17.6% yield of NH_3 in minutes by using 400°–500°C temperature and 200 atm pressure, and then to condense out the NH_3 and recycle the unused N_2 and H_2. It is not economically feasible to use room temperature and 200 atm pressure to attain a much greater percentage of NH_3, for it would require ages of time.

Catalysts and Equilibrium

A catalyst cannot change the numerical value of the equilibrium constant and hence the relative amounts of reactants and products present at equilibrium. However, it may greatly reduce the time necessary for the establishment of equilibrium. This is extremely important from an industrial standpoint, since the speed at which a product can be produced is a primary consideration. Catalysts may be effectively used in many reactions which allow conversion of only a small percentage of reactants into products (equilibrium far to the left). From a production standpoint, it is more important to obtain a small yield in a few minutes than to obtain a large yield in several days. A catalyst adds no energy to a system, and thus cannot alter equilibrium proportions.

Exercises

14.1. List four factors which influence reaction rates and explain on a molecular level how each exerts its influence.

14.2. Define or illustrate the terms: (a) catalyst, (b) rate law, (c) reaction order, (d) rate constant, (e) heterogeneous reactions, (f) activated complex, (g) energy of activation, (h) reaction mechanism, (i) rate determining step, (j) chain reaction, (k) chemical equilibrium, (l) equilibrium constant, (m) Le Chatelier's principle.

14.3. At 25°C, the rate of a given chemical reaction is "a." What will be the approximate rate of this reaction at 55°C?

14.4. What is the half-life of an unstable substance if 75% of any given amount of the substance decomposes in one hour?

14.5. At a certain temperature, the half-life for the decomposition of SO_2Cl_2 is 4.00 hr. (a) In what period of time would the concentration of SO_2Cl_2 be reduced to 10% of the original concentration? (b) Starting with 100 mg of SO_2Cl_2, what weight would be left at the end of 6.50 hr? Use the first-order rate equation on page 311 to solve this problem.

14.6. For the general reaction $A + B \rightarrow C + D$, what is the effect on the number of collisions between A and B of (a) tripling the concentration of each, (b) quadrupling the concentration of each?

14.7. The effect of concentration of reactants on the following reaction

$$2\ H_2(g) + 2\ NO(g) \rightarrow 2\ H_2O(g) + N_2(g)$$

is summarized by the following experimental data obtained at 800°C.

Expt.	Initial molar concentration		Rate of formation of N_2 (moles/l)/min
	NO	H_2	
I	6.00×10^{-3}	1.00×10^{-3}	1.5×10^{-3}
II	6.00×10^{-3}	2.00×10^{-3}	3.0×10^{-3}
III	6.00×10^{-3}	3.00×10^{-3}	4.5×10^{-3}
IV	1.00×10^{-3}	6.00×10^{-3}	2.2×10^{-3}
V	2.00×10^{-3}	6.00×10^{-3}	8.8×10^{-3}

Answer the following:
(a) What happens to the rate of the reaction if the concentration of NO is doubled?
(b) Tripling the concentration of H_2 increases the rate by what factor?
(c) What would be the rate law for the reaction?

14.8. For the following hypothetical reaction

$$A(g) + 2\ B(g) + 3\ C(g) \rightarrow 4\ D(g)$$

the rate law was found by experiment to be:

$$\text{Rate of formation of } D = k[A]^2[B]$$

Answer the following.
(a) Doubling the concentration of A would increase reaction rate by what factor, if any?
(b) Doubling the concentration of B would increase reaction rate by what factor?
(c) Tripling the concentration of C would increase reaction rate by what factor? The reaction would be considered to be what order?
(d) Calculate k from the following data:
   Initial molar concentrations: $A = 2.0 \times 10^{-2}$; $B = 3.0 \times 10^{-2}$;
   $$C = 5 \times 10^{-2};$$
   Rate of formation of $D = 4.0 \times 10^{-3}$ (moles/l)/min

14.9. The data show the effect of concentration of reactants on the rate of the formation of NOCl gas by the equation:

$$Cl_2(g) + 2\ NO(g) \rightarrow 2\ NOCl(g) \text{ at } 450°C$$

Expt.	Cl(g), moles/l	NO(g), moles/l	Rate of NOCl(g) formed, (moles/l)/sec
1	1.0×10^{-3}	2.5×10^{-4}	2.6×10^{-10}
2	1.0×10^{-3}	5.0×10^{-4}	5.2×10^{-10}
3	1.0×10^{-3}	10.0×10^{-4}	10.4×10^{-10}
4	2.0×10^{-3}	10.0×10^{-4}	41.6×10^{-10}
5	3.0×10^{-3}	10.0×10^{-4}	93.6×10^{-10}
6	4.0×10^{-3}	10.0×10^{-4}	166.4×10^{-10}

How is the reaction rate affected by:
(a) doubling the concentration of NO?
(b) increasing the concentration of NO four times?
(c) doubling the concentration of Cl_2?
(d) increasing the concentration of Cl_2 three times?

(e) What would be the rate law expression for the above reaction?

(f) What would be the order of the reaction?

14.10. The rate equation for the reaction

$$2 \text{ NO(g)} + 2 \text{ H}_2\text{(g)} \rightarrow \text{N}_2\text{(g)} + 2 \text{ H}_2\text{O(g)}$$

is second order in NO(g) and first order in H_2(g). (a) Write an equation for the rate of appearance of N_2(g). (b) If concentrations are expressed in moles/liter, what units would the rate constant, k, have? (c) Give an equation for the rate of disappearance of NO(g). Would k in this equation have the same numerical value as k in the equation of part (a)?

14.11. For a reaction of A and B to form C, the following data were obtained from three experiments:

Expt.	$[A]$	$[B]$	Rate (Formation of C)
1	0.60 M	0.15 M	6.3×10^{-3} M/min
2	0.20 M	0.60 M	2.8×10^{-3} M/min
3	0.20 M	0.15 M	7.0×10^{-4} M/min

(a) What is the rate equation for the reaction? (b) What is the numerical value of the rate constant, k?

14.12. Assume that the rate-determining step of a reaction is

$$2 A\text{(g)} + B\text{(g)} \rightarrow C\text{(g)}$$

and that 2 moles of A(g) and 1 mole of B(g) are mixed in a 1-l container. Compare the following to the initial reaction rate of this mixture: (a) rate when half of both A(g) and B(g) has been consumed, (b) rate when two thirds of both A(g) and B(g) has been consumed, (c) initial reaction rate of a mixture of 2 moles of A(g) and 2 moles of B(g) in a 1-l container, (d) initial reaction rate of a mixture of 4 moles of A(g) and 2 moles of B(g) in a 1-l container.

14.13. Draw a potential energy diagram for the hypothetical reaction $A\text{(g)} + B\text{(g)} \rightleftharpoons 2 C\text{(g)}$ for which the energy of activation for the forward reaction is 42 kcal and for the reverse reaction is 20 kcal. Indicate the magnitude of ΔH on the diagram. Is the reaction exothermic or endothermic? Draw a curve for a catalyzed reaction on the diagram.

14.14. The mechanism for the sulfuric acid catalyzed degradation of formic acid is as follows:

$\text{HCOOH} + \text{H}_2\text{SO}_4 \rightarrow \text{HSO}_4^- + \text{H}_2\text{COOH}^+$	Step 1
$\text{H}_2\text{COOH}^+ \rightarrow \text{H}_2\text{O} + \text{HCO}^+$	Step 2
$\text{HCO}^+ + \text{HSO}_4^- \rightarrow \text{H}_2\text{SO}_4 + \text{CO}$	Step 3

(a) What would be the net reaction for the overall process?

(b) The reaction profile below summarizes the energy factors for the catalyzed reaction. Answer the following:

(1) What would be the activation energy (kilocalories) for step 1?

(2) What would be the activation energy (kilocalories) for step 3?

(3) What would be the slowest or rate-limiting step?

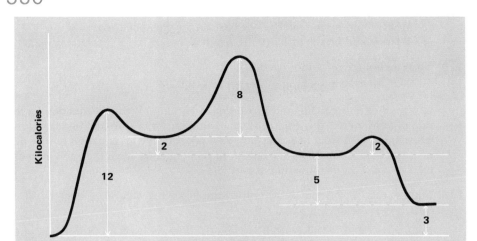

(4) What would be the overall ΔH change (kilocalories) for the reaction?
(5) What would be the ΔH change (kilocalories) for step 3?
(c) Which *two* of the following statements would correctly describe the catalytic effect of H_2SO_4?
1. H_2SO_4 decreases the enthalpy change for the overall reaction.
2. H_2SO_4 increases the rate by increasing the activation energy.
3. H_2SO_4 participates in the reaction and is regenerated.
4. H_2SO_4 changes the mechanism of the degradation reaction.
5. None of these.

14.15. Below is a potential energy curve for a chemical reaction, $A + B \rightarrow AB^* \rightarrow C + D$

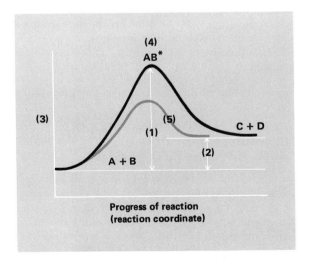

Identify briefly, with a word(s) or symbol, the parts of the diagram marked (1), (2), (3), etc.

14.16. Suppose that the step mechanism for a reaction is as follows:
(1) $N_2 + H_2 \rightarrow N_2H_2$ slow step
(2) $N_2H_2 + H_2 \rightarrow NH_3 + NH$ fast
(3) $NH + H_2 \rightarrow NH_3$ fast
(a) Write the over-all net chemical equation for the reaction.
(b) What is the rate-law equation for the reaction?

14.17. The reaction

$$CH_4(g) + Cl_2(g) \rightarrow CH_3Cl(g) + HCl(g)$$

proceeds by a chain mechanism. The chain propagators are Cl atoms and CH_3 radicals, and it is believed that free H atoms are not involved. Write a series of equations showing the mechanism and identify the chain-initiating, chain-sustaining, and chain-terminating steps.

14.18. At a certain temperature, K for the reaction $3\,C_2H_2 \rightleftarrows C_6H_6$ is 4. If the equilibrium concentration of C_2H_2 is 0.60 mole/l, what is the concentration of C_6H_6?

14.19. Consider the gaseous equilibrium $2\,A \rightleftarrows 2\,B + C$. Four moles of A are put into a one-liter flask. At equilibrium, the flask is found to contain 1.0 mole of C. Calculate the value for the equilibrium constant.

14.20. The equilibrium constant for the gaseous equilibrium $X \rightleftarrows 3\,Y$ is 2.0. A 2-l flask contains 3.0 moles of Y at equilibrium. How many moles of X does the flask contain at equilibrium?

14.21. Consider the following gaseous equilibrium: $3\,A \rightleftarrows 2\,B + C$. Three moles of A were put into a one-liter flask. At equilibrium it was found that 0.5 mole of C was present. What is the value of K?

14.22. Four moles of PCl are placed in a 2-l flask. When equilibrium is established

$$PCl_5 \rightleftarrows PCl_3 + Cl_2$$

the flask is found to contain 0.4 mole of Cl_2. What is the equilibrium constant?

14.23. If HI is 25% dissociated according to the equation

$$2\,HI \rightleftarrows H_2 + I_2$$

at a given temperature, (a) calculate the equilibrium concentrations of HI, H_2, and I_2 if we start with 10 moles of HI in a 1.0-l container; (b) calculate the value for the equilibrium constant.

14.24. A mixture consisting of 1.000 mole of $H_2(g)$ and 1.000 mole of $CO_2(g)$ is placed in a 1.000-l container at 800°C. At equilibrium 0.491 mole of CO(g) and 0.491 mole of $H_2O(g)$ are present.

$$H_2(g) + CO_2(g) \rightleftarrows H_2O(g) + CO(g)$$

(a) What are the equilibrium concentrations of $H_2(g)$ and $CO_2(g)$?

(b) What is the value of K?

14.25. A mixture containing 0.075 mole/l of $HCl(g)$ and 0.033 mole/l of $O_2(g)$ is allowed to come to equilibrium at 480°C.

$$4\,HCl(g) + O_2(g) \rightleftarrows 2\,Cl_2(g) + 2\,H_2O(g)$$

At equilibrium, the concentration of $Cl_2(g)$ is 0.030 mole/l. (a) What are the equilibrium concentrations of $HCl(g)$, $O_2(g)$, and $H_2O(g)$? (b) What is the value of K at 480°C?

14.26. At 100°C, the equilibrium constant, K, for the reaction

$$CO(g) + Cl_2(g) \rightleftarrows COCl_2(g)$$

is 4.57×10^9 l/mole. If 1.00 mole of $COCl_2$ is confined in a 1.00-l container, what is the concentration of CO after equilibrium has been established?

14.27. Analysis of an equilibrium mixture of $N_2 + 3\,H_2 \rightleftarrows 2\,NH_3$ showed moles per liter yields of 0.60 H_2 and 0.20 NH_3. The value of $K = 0.37$. Compute moles/liter of N_2.

14.28. Consider the reaction $CO + H_2O \rightleftarrows CO_2$. Into a 1-l vessel are placed 5.0 moles CO and 3.0 moles H_2O and the mixture is heated to a reacting temperature. Analysis shows the presence, at equilibrium, of 1.5 moles of CO_2. Calculate K for the reaction.

14.29. Consider the reaction $2\,A + B \rightleftarrows 2\,C + D$. Initially 3 moles/l A and 5 moles/l B are placed in a vessel and reacted. At equilibrium 0.5 mole of D is found. What is the value of K?

14.30. Consider the following equilibrium at a high temperature: $2\,CH_4(g) + 3\,O_2(g) \rightleftarrows 2\,CO(g) + 4\,H_2O(g)$, $K = 0.045$. The concentrations of O_2, CO, and H_2O at equilibrium are as follows: $[O_2] = 2.00$ moles/l, $[CO] = 3.00$ moles/l, $[H_2O] = 1.00$ mole/l. What would be the concentration of CH_4 at equilibrium in moles/l?

14.31. Carbon monoxide and water react at 1000°C to give the following equilibrium:

$$CO + H_2O \rightleftarrows CO_2 + H_2$$

Into a one-liter vessel is placed *initially* 0.750 mole of CO and 0.500 mole of H_2O. Analysis at equilibrium shows the presence of 0.200 mole of H_2. Calculate K for the reaction.

14.32. Consider the following reaction at 600°C:

$$N_2(g) + 3\,H_2(g) \rightleftarrows 2\,NH_3(g)$$

Initially 4.0 moles of NH_3 is placed in a 1.0-l container. *At equilibrium* 3.0 moles of H_2 are found in the container. Calculate K.

14.33. What is the effect of increased temperature and pressure on the following equilibrium systems? All reactants and products are in the vapor state.

(a) $3 X_2 + 2 Y_3 \rightleftarrows 6 XY$ $\Delta H = 5000$ cal
(b) $H_2 + Br_2 \rightleftarrows 2 HBr$ $\Delta H = -17.3$ kcal
(c) $CO + 2 H_2 \rightleftarrows CH_3OH$ $\Delta H = -21.7$ kcal
(d) $N_2O_4 \rightleftarrows 2 NO_2$ $\Delta H = 13.9$ kcal

14.34. Given the following equilibrium mixture of gases at 500°C where ΔH is negative, what will be the effect of

$$2 CO + O_2 \rightleftarrows 2 CO_2$$

(a) adding more O_2 at constant pressure?
(b) increasing the temperature?
(c) cooling below 100°C?
(d) adding a catalyst?
(e) doubling the pressure at 200°C?
(f) adding a base to absorb CO_2?

14.35. State the direction in which each of the following equilibrium systems would be shifted upon the application of the stress listed beside the equation:

(a) $CO(g) + H_2O(g) \rightleftarrows CO_2(g) + H_2(g)$ increase temperature
     (exothermic)
(b) $N_2(g) + O_2(g) \rightleftarrows 2 NO(g)$ (endothermic)   decrease temperature
(c) $2 NOBr(g) \rightleftarrows 2 NO(g) + Br_2(g)$ decrease total pressure
(d) $3 Fe(s) + 4 H_2O(g) \rightleftarrows Fe_3O_4(s) + 4 H_2(g)$ increase total pressure
(e) $C(s) + CO_2(g) \rightleftarrows 2 CO(g)$ decrease total pressure
(f) $C(s) + CO_2(g) \rightleftarrows 2 CO(g)$ add $C(s)$
(g) $N_2O_4(g) \rightleftarrows 2 NO_2(g)$ increase concentration
      of NO_2
(h) $N_2(g) + 3 H_2(g) \rightleftarrows 2 NH_3(g)$ add a catalyst
(i) $BaCl_2 \cdot H_2O(s) \rightleftarrows BaCl_2(s) + H_2O(g)$ remove $H_2O(g)$

14.36. Given the following equilibrium mixture of *gases* at 500°C where ΔH is *negative:*

$$2 CO(g) + O_2(g) \rightleftarrows 2 CO_2(g)$$

What will be the effect on the equilibrium (1. shift to the right →, 2. shift to the left ←, 3. no shift) of:
(a) adding more O_2 at constant pressure?
(b) increasing the temperature?
(c) cooling below 100°C?
(d) adding a catalyst?
(e) doubling the total pressure?
(f) adding a base to absorb the CO_2?
(g) stirring the mixture rapidly?
(h) increasing the volume of the container but leaving the total weight of gases the same?

14.37. Given the following equilibrium system of 300°C and 1 atm:

$$2 N_2(g) + 5 O_2(g) \rightleftarrows 2 N_2O_5 \qquad \Delta H = +89.5 \text{ kcal}$$

Indicate what effect the following changes would have on the composition of the equilibrium system (1. Shift to the right (products), 2. shift to the left (reactants), 3. no effect.):
(a) Adding additional N_2?
(b) Cooling the system below 100°C?
(c) Increasing the pressure to 2 atmospheres?
(d) Reducing the pressure to 0.5 atmospheres?
(e) Doubling the concentration of O_2?
(f) Adding a catalyst?
(g) Increasing the temperature to 600°C?
(h) Removing N_2O_5 as fast as it forms?

14.38. The following reaction is used commercially for the production of ammonia:

$$N_2 + 3 H_2 \rightleftarrows 2 NH_3 \qquad \Delta H = -22{,}000 \text{ calories}$$

Indicate by placing a check in the appropriate column what effect if any the following changes would have on the equilibrium.

	No Change	Shifts Equilibrium to: the left	the right	
(a)	_____	_____	_____	increase in temperature.
(b)	_____	_____	_____	decrease in temperature.
(c)	_____	_____	_____	decrease in volume.
(d)	_____	_____	_____	decrease in temperature.
(e)	_____	_____	_____	addition of more H_2.
(f)	_____	_____	_____	addition of a catalyst.

Supplementary Readings

14.1. *How Chemical Reactions Occur*, E. L. King, Benjamin, New York, 1963 (paper).
14.2. *Why Do Chemical Reactions Occur?*, J. A. Campbell, Prentice-Hall, New Jersey, 1965 (paper).
14.3. "Catalysis," V. Haensel and R. L. Burwell, Jr., *Sci. American*, Dec., 1971; p. 46.

Chapter 15

Ionic Equilibria

Most reactions of acids, bases, and salts are brought about in water solutions, and these necessarily involve the interaction of ions. Such processes are usually reversible, at least to some extent and, as a result, equilibrium conditions are quickly established. To get a clear understanding of these reactions, it is necessary to understand the equilibria involved. In this chapter we shall consider the ionic equilibria in aqueous solutions, the factors influencing these, and particularly the relationships of ionic concentrations in solutions of weak electrolytes and relatively insoluble substances. We shall find that ionic equilibria conform to the general principles of equilibrium processes.

The concepts of ionic equilibria are particularly valuable in analytical chemistry, but they also find application in many other areas—for example, ion exchange in resins and gels (see Chap. 26), biological processes involving body fluids, acidity or basicity of soils, and so on.

Ionization Constants

The Law of Mass Action is applicable to ionic as well as molecular reactions, and we may write an expression for the equilibrium constant for ionic reactions. In these cases the equilibrium constant, usually written as K_a for acids and K_b for bases, is termed the **ionization constant.** Thus in the ionization of acetic acid, which is a weak electrolyte,

$$HC_2H_3O_2 + H_2O \rightleftharpoons H_3O^+ + C_2H_3O_2^-$$

or, ignoring the hydration of H^+,

$$HC_2H_3O_2 \rightleftharpoons H^+ + C_2H_3O_2^-$$

we may write
$$K_a = \frac{[H^+][C_2H_3O_2^-]}{[HC_2H_3O_2]}$$

in which the bracket [ ] represents the molar concentration of each ion* or molecule. In other words, at a given temperature the product of the concentrations of H^+ and $C_2H_3O_2^-$ ions divided by the concentration of undissociated acetic acid must have a constant numerical value.† A small value for K_a means that the numerator is small compared with the denominator, which, of course, means that the concentrations of H^+ ions and $C_2H_3O_2^-$ ions are relatively small compared with the concentration of undissociated acetic acid. Thus the ionization constant is a measure of the extent of the dissociation; in the case

---

* In referring to concentration of ions, the term **gram ion** may be used instead of mole. A gram ion is the weight of the ion in grams and is synonymous with a gram mole or mole of the ion.
† In the equilibrium

$$HC_2H_3O_2 + H_2O \rightleftharpoons H_3O^+ + C_2H_3O_2^-$$

the concentration of H_2O in dilute solutions is essentially a constant, and this concentration term is embodied in the ionization constant, K_a.

of acids it is a measure of their strength, since the strength of an acid is de-termined by its ability to produce H^+ ions.

Obviously, if the concentrations of ions and undissociated molecules are known, K_a may be calculated. For example, in 0.100 molar acetic acid solution at 25°C, the acid ionizes to the extent of about 1.34%. Since each molecule of acetic acid which ionizes produces one H^+ ion and one $C_2H_3O_2^-$ ion, the concentrations in the solution are

$$[H^+] = 0.100 \times 0.0134 = 0.00134 \text{ (mole/l)}$$

and

$$[C_2H_3O_2^-] = 0.100 \times 0.0134 = 0.00134 \text{ mole/l}$$
$$[HC_2H_3O_2] = 0.100 - 0.00134 = 0.09866 \text{ mole/l}$$

Substituting in the above formulation for K_a,

$$K_a = \frac{[H^+][C_2H_3O_2^-]}{[HC_2H_3O_2]} = \frac{0.00134 \times 0.00134}{0.09866} = 0.0000182$$

usually written as 1.82×10^{-5}.

Table 15.1 and the appendix give the ionization constants of a number of weak acids.

Table 15.1

Ionization constants of some weak acids

Acid	K_a
Acetic, $HC_2H_3O_2$	1.8×10^{-5}
Benzoic, HC_6H_5O	6.4×10^{-5}
Carbonic, H_2CO_3, K_1	4.3×10^{-7}
Carbonic, HCO_3^-, K_2	5.6×10^{-11}
Formic, $HCOOH$	1.6×10^{-4}
Iodic, HIO_3	1.9×10^{-1}
Phosphoric, H_3PO_4, K_1	7.5×10^{-3}
Phosphoric, $H_2PO_4^-, K_2$	6.2×10^{-8}
Phosphoric, HPO_4^{2-}, K_3	1×10^{-12}
Trichloracetic, $HC_2Cl_3O_2$	2×10^{-1}

The larger the value of K_a, the stronger the acid; thus, of the acids listed, trichloracetic acid is the strongest and phosphoric acid (HPO_4^{2-}) the weakest.

It may be noted that two constants are given for carbonic acid and three for phosphoric acid. Acids with more than one replaceable hydrogen are termed **polyprotic** acids and ionize in steps; therefore, each successive step of the ionization is an equilibrium and has a value for K. K_1 is for the first ionization step, K_2 for the second, and so on. In the case of H_3PO_4

$$H_3PO_4 \rightleftharpoons H^+ + H_2PO_4^- \qquad K_1 = 7.5 \times 10^{-3}$$
$$H_2PO_4^- \rightleftharpoons H^+ + HPO_4^{2-} \qquad K_2 = 6.2 \times 10^{-8}$$
$$HPO_4^{2-} \rightleftharpoons H^+ + PO_4^{3-} \qquad K_3 = 1 \times 10^{-12}$$

Thus H_3PO_4 is a stronger acid than $H_2PO_4^-$, which is stronger than HPO_4^{2-}.

The ionization of bases and their equilibrium constants follows a similar pattern: the larger the value for the ionization constant, the stronger the base. The ionization constants of several weak bases are included in Table 15.2 and the appendix.

Table 15.2

Ionization constants of some weak bases

Base	K_b
Ammonia, $NH_3 + H_2O$	1.8×10^{-5}
Aniline, $C_6H_5NH_2 + H_2O$	4.6×10^{-10}
Hydrazine, $N_2H_4 + H_2O$	9.8×10^{-7}
Methylamine, $CH_3NH_2 + H_2O$	5.0×10^{-4}

The ionization of a solution of ammonia in water may be represented as follows:

$$NH_3 + H_2O \rightleftharpoons NH_4^+ + OH^-$$

The $[H_2O]$ is omitted from the K expression since its value does not change appreciably and can be incorporated in the overall value of K_b. Other nitrogen bases may be handled similarly. The K_b expression for the above is

$$K_b = \frac{[NH_4^+]\,[OH^-]}{[NH_3]}$$

EXAMPLE 15.1   What is the H^+ concentration in a 0.10 molar solution of acetic acid?

SOLUTION: Recall the 5 steps used in handling equilibrium calculations (page 321). Applying these to the present calculation we have the following:
(1) the equilibrium equation is

$$HC_2H_3O_2 \rightleftharpoons H^+ + C_2H_3O_2^-$$

(2) the K expression is

$$K_a = \frac{[H^+]\,[C_2H_3O_2^-]}{[HC_2H_3O_2]} = 1.8 \times 10^{-5}$$

(3) By the phrase, "0.10 molar solution of acetic acid," it is understood to mean that the $HC_2H_3O_2$ concentration would be 0.10 M *before any ionization takes place* — in other words, the initial $[HC_2H_3O_2]$ is 0.10 M and the initial $[H^+]$ and $[C_2H_3O_2^-]$ are (near) zero.

$$\text{Initial:} \quad \begin{array}{ccc} 0.10\ M & 0 & 0 \\ HC_2H_3O_2^- \rightleftharpoons H^+ & + & C_2H_3O_2^- \end{array}$$

(4) After ionization takes place and equilibrium is established, some H^+ and $C_2H_3O_2^-$ are formed and some $HC_2H_3O_2$ used up. If we let $x = [H^+]$ at equilibrium, then, as shown beneath the equation, the other concentrations at equilibrium are:

$$\text{Initial:} \quad 0.10\ M \qquad 0 \qquad 0$$
$$HC_2H_3O_2 \rightleftharpoons H^+ + C_2H_3O_2^-$$
$$\text{At equilibrium:} \quad 0.10 - x \qquad x \qquad x$$

(5) Substituting the equilibrium concentrations in the K_a expression, we have

$$\frac{(x)\,(x)}{(0.10 - x)} = 1.8 \times 10^{-5}$$

Since K_a is small (as is usually the case) x will be small and can be neglected in the $(0.10 - x)$ term. The K_a expression then becomes

$$\frac{x^2}{0.10} = 1.8 \times 10^{-5}$$

from which $x = 1.3 \times 10^{-3}$ mole/l $= [H^+]$.

15.1 HYDROGEN ION CONCENTRATION

In scientific work the pH method* for expressing hydrogen ion concentration is employed. pH is defined as the logarithm of the reciprocal of the H^+ concentration—in other words, as the negative logarithm of the H^+ concentration expressed in moles per liter.

$$pH = \log \frac{1}{[H^+]} = -\log\,[H^+]$$

or

$$[H^+] = 10^{-pH} = \text{antilog}(-pH)$$

We may best illustrate this method with examples. Let us consider the hydrogen ion concentration in solutions of hydrochloric acid. Hydrochloric acid, being a strong acid, is almost completely ionized. Assuming complete ionization, a solution of 0.01 M HCl will produce a 0.01 molar concentration of hydrogen ions, a solution of 0.001 M HCl will give a hydrogen ion concentration of 0.001 molar, and so on. These concentrations may be expressed as 10^{-2} and 10^{-3} ($0.01 = 10^{-2}$ and $0.001 = 10^{-3}$), respectively. Since the logarithm of $10^{-2} = -2$, then $-\log 10^{-2} = 2$, and pH is 2. For the second solution,

---

* pH was originally proposed as a measure of hydrogen ion concentration. Although the term "hydrogen ion concentration" is still in common use, the ion which is actually responsible for acidity is the hydronium ion—that is, the hydrated H^+, H_3O^+.

where the hydrogen ion concentration is 10^{-3}, the pH will be 3, and so on. Note that a change of one unit in pH corresponds to ten times the change in hydrogen ion concentration. The concentration of hydrogen ions in solution of pH 2 is ten times the concentration in solution pH 3. Table 15.3 indicates further the relationship between hydrogen ion concentration and pH. It will be noted that the pH of the solution increases as the concentration of hydrogen ion decreases.

Table 15.3
Hydrogen ion concentration of solutions of HCl

Concentration of HCl (in moles/liter)	Concentration of Hydrogen Ion		pH
0.1	0.1	$= 10^{-1}$	1
0.01	0.01	$= 10^{-2}$	2
0.001	0.001	$= 10^{-3}$	3
0.0001	0.0001	$= 10^{-4}$	4
0.00001	0.00001	$= 10^{-5}$	5
0.000001	0.000001	$= 10^{-6}$	6

In similar manner, the hydroxide ion concentration of a solution may be expressed as pOH, where

$$pOH = \log \frac{1}{[OH^-]} = -\log [OH^-]$$

As pOH increases, the basicity of the system decreases. While pH is much more widely used than pOH, the latter appears to be gaining favor in describing the basicity of a solution. It will be shown later (page 342) that for any aqueous solution,

$$pH + pOH = 14$$

Study the sample calculations below.

Calculations of pH and pOH

EXAMPLE 15.2   Assuming complete ionization, calculate (a) the pH of 0.0001 N HCl, (b) the pOH of 0.0001 N KOH.

SOLUTION:

(a) $pH = \log \dfrac{1}{[H^+]} = -\log [H^+] = -\log (1 \times 10^{-4}) = 4$

(b) The calculation of pOH is similar, and for 0.0001 N KOH is 4.

EXAMPLE 15.3   Calculate the pH of 0.0036 N HCl (this would be approximately the acidity of swimming pool water which has been chlorinated).

SOLUTION:

$$p\text{H} = -\log [\text{H}^+] = -\log (3.6 \times 10^{-3}) = -(\log 3.6 + \log 10^{-3})$$
$$= -(0.56 - 3) = 2.44$$

EXAMPLE 15.4   What is the pOH of a dilute lye (NaOH) solution of 0.000072 N?

SOLUTION:

$$p\text{OH} = -\log [\text{OH}^-]$$
$$p\text{OH} = -(\log 7.2 + \log 10^{-5}) = -(0.86 - 5) = 4.14$$
$$p\text{H} = 14 - p\text{OH} = 14 - 4.14 = 9.86$$

Note that a pH of 9.86 represents only *two* significant figures, since any number to the left of the decimal place in a logarithm of a number merely establishes the decimal place in the original number.

EXAMPLE 15.5   What is the $[\text{H}^+]$ of a solution which has a pH of 12.60?

SOLUTION:

$$[\text{H}^+] = 10^{-p\text{H}} = 10^{-12.60} = 10^{0.40-13}$$
$$= 10^{0.40} \times 10^{-13}$$

and since the antilog of 0.40 is 2.5

$$[\text{H}^+] = 2.5 \times 10^{-13} \ M.$$

The student may obtain further practice in calculating pH from $[\text{H}^+]$ and vice versa by using the data in Table 15.5.

The Ionization of Water

Pure water is a very poor conductor of electricity; liquid mercury, for example, is about 25,000,000 times as good a conductor as water. The low conductivity of water may be attributed to its very slight ionization in accordance with the reaction

$$\text{H}_2\text{O} \rightleftharpoons \text{H}^+ + \text{OH}^-$$

Evidence indicates that but one molecule of water in 555,000,000 is ionized.

This corresponds to 1 mole of H^+ and also 1 mole of OH^- per 10,000,000 l of water. Consequently in pure water,

$$[H^+] = [OH^-] = \frac{1 \text{ mole}}{10,000,000 \text{ l}} = \frac{1}{10^7} = 1 \times 10^{-7} \text{ mole per l.}$$

Hence the pH and the pOH of pure water is 7, and since equal numbers of H^+ and OH^- ions are present, this pH or pOH represents a neutral solution.

Applying the Law of Mass Action to the above equilibrium equation,

$$K_i = \frac{[H^+][OH^-]}{[H_2O]}$$

The concentration of H_2O can be considered a constant in dilute solutions; hence $K_i \times [H_2O]$ is a constant and may be represented by another constant, K_w, called the **ion product constant** for water. The value of K_w at ordinary temperatures is 10^{-14}. This value is the product of the concentrations of hydrogen ion and hydroxyl ion in any water solution.

$$K_w = [H^+][OH^-] = 1 \times 10^{-14}$$

Let us suppose that an acid is added to water until the hydrogen ion concentration is 10^{-3} ($pH = 3$). The increase in hydrogen ion concentration displaces the equilibrium above to the left and results in a decrease in the concentration of OH^-. The value of the concentration of OH^- in the solution is 10^{-11}, since $10^{-3} \times 10^{-11} = 10^{-14}$. Students frequently think that an acid contains hydrogen ions but no hydroxyl ions, but it is evident from the above that any aqueous solution of an acid also contains hydroxyl ions. Similarly, any basic or alkaline solution contains not only hydroxyl ions but hydrogen ions as well. A solution in which the concentration of OH^- is 10^{-4} must have a hydrogen ion concentration of 10^{-10}. The pH of this latter solution is 10, and since the $[OH^-]$ is greater than $[H^+]$, the solution is basic. Summarizing: a solution of pH 7 is neutral; a solution of pH less than 7 is acid; and one of pH greater than 7 is basic.

$$\xleftarrow{\text{Increasing acidity}} \quad 7 \quad \xrightarrow{\text{Increasing basicity}}$$
$$\text{Neutral}$$

Since, for any aqueous solution,

$$[H^+][OH^-] = 1 \times 10^{-14}$$

then $$\log [H^+] + \log [OH^-] = \log (1 \times 10^{-14})$$

but $$\log [H^+] = -pH; \text{ and } \log [OH^-] = -pOH$$

therefore $$pH + pOH = 14$$

Another term frequently used in connection with equilibrium systems is pK, which is defined:

$$pK = -\log K$$

where K is the equilibrium constant. For ionic systems, K is the ionization constant.

For example, K_a for $HC_2H_3O_2 = 1.8 \times 10^{-5}$ or $pK_a = 4.74$.

For H_2O, where

$$K_w = [H^+][OH^-] = 1 \times 10^{-14}$$
$$pK_w = -\log (1 \times 10^{-14}) = 14$$

From the above it follows that

$$pH + pOH = pK_w$$

Applications of pH

A careful control of pH is important in many biological processes. Most of the body fluids in a healthy person have a pH very near 7. The gastric juices are an exception, with a pH of about 2. pH is important in the growth of bacteria, and certain pH values represent optimum conditions for growth.

The utility of a soil for growing certain crops depends on its acidity or basicity (its pH or pOH). In very humous soil with decay and carbonic acid formation, a slight acidity is built up, and this is suitable for certain plants, such as rhododendrons, azaleas, and blueberries. Most grasses require a pH above 5.5 for optimum growth. "Liming" of soil by ground limestone ($CaCO_3$) increases the pH by reaction with H^+ to form CO_2 and H_2O. Very few plants survive in soils of pH greater than 8.5.

The pH of aqueous solutions is also important in cooking foods. Meats may be tenderized with acidic juices, such as tomato and vinegar. In baking, a dough of pH above 7 yields a crumbly and less moist cake or bread.

To measure pH, the so-called pH meter may be used with an accuracy approaching 0.01 in the range from 0 to 14. This device is essentially an electrolytic cell in which the voltage varies with H^+ concentration (Figure 15.1).

Indicators

Certain organic substances undergo a change in color at definite pH values. For example, methyl orange has a red or orange color in solutions with a pH of 3 or less and a yellow color in solutions of pH more than 4. Phenolphthalein changes from colorless to red violet at a pH of 9 to 10. Such substances, called **indicators**, may be used to determine the approximate pH of a solution. Some of the more common indicators and color changes at various pH values are shown in Table 15.4.

To illustrate the use of indicators in determining approximate pH, suppose a solution gives no color with p-nitrophenol and a red color with congo

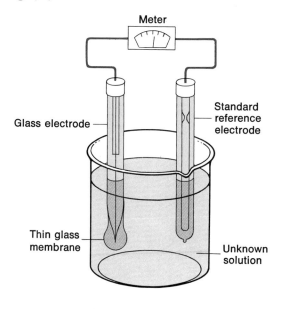

Figure 15.1
pH meter with a glass electrode and standard calomel reference electrode.

Table 15.4
Color change in indicators at various pH values°

Indicator	0	1	2	3	4	5	6	7	8	9	10	11	12	13	14
Methyl violet	Y	‖‖‖	BV	BV	BV	BV	V	V	V	V	V	V	V	V	V
Meta-cresol purple	R	R	‖‖‖	Y	Y	Y	Y	Y	‖‖‖	Pu	Pu	Pu	Pu	Pu	Pu
Thymol blue	R	R	‖‖‖	Y	Y	Y	Y	Y	Y	‖‖‖	B	B	B	B	B
Methyl orange	R	R	R	R	‖‖‖	Y	Y	Y	Y	Y	Y	Y	Y	Y	Y
Congo red	B	B	B	B	‖‖‖	R	R	R	R	R	R	R	R	R	R
p-Nitrophenol	C	C	C	C	C	C	‖‖‖	Y	Y	Y	Y	Y	Y	Y	Y
Rosolic acid	A	A	A	A	A	A	A	P	P	P	P	P	P	P	P
Phenolphthalein	C	C	C	C	C	C	C	C	C	‖‖‖	V	V	V	V	V
Malachite green	BG	BG	BG	BG	BG	BG	BG	BG	BG	BG	BG	BG	‖‖‖	C	C

A—amber	C—colorless	R—red
B—blue	P—pink	V—violet
BG—blue green	Pu—purple	Y—yellow
BV—blue violet		

° The vertical lines signify that the indicator is changing in color and shows the pH range over which the color change takes place.

red indicator. From Table 15.4 we find that colorless p-nitrophenol shows the pH to be 6 or less, and congo red shows the pH to be 5 or greater. Hence the pH range is between 5 and 6.

To choose the proper indicator for any given titration of acids and bases, we must know the pH of the solution when equivalent amounts of the acid and base are present. For example, the pH of the solution resulting from neutralization of acetic acid with sodium hydroxide is 8.5, the same as would be obtained if pure sodium acetate were dissolved in water. Phenolphthalein changes color at a pH of about 9 and may be used satisfactorily in this titration. Actually, in the titration of strong acids and strong bases, any indicator which changes in a pH range of 5 to 9 may be used, but in the titration of weak acids and weak bases, the indicator must be carefully chosen to insure accuracy of determination.

Table 15.5 contains data showing what happens to the $[H^+]$, $[OH^-]$, and pH during a titration of 50.0 ml of 0.100 N HCl as successive amounts of 0.100 N NaOH are added.

Table 15.5
Titration of 50.0 ml of 0.100 N HCl with 0.100 N NaOH

Ml of NaOH	$[H^+]$	$[OH^-]$	pH
0.0	1.00×10^{-1}	1.00×10^{-13}	1.00
10.0	6.67×10^{-2}	1.50×10^{-13}	1.18
20.0	4.29×10^{-2}	2.33×10^{-13}	1.37
30.0	2.50×10^{-2}	4.00×10^{-13}	1.60
40.0	1.11×10^{-2}	9.00×10^{-13}	1.95
49.0	1.01×10^{-3}	9.90×10^{-12}	3.00
49.9	1.00×10^{-4}	1.00×10^{-10}	4.00
50.0	1.00×10^{-7}	1.00×10^{-7}	7.00
50.1	1.00×10^{-10}	9.99×10^{-5}	10.00
51.0	1.01×10^{-11}	9.90×10^{-4}	11.00
60.0	1.10×10^{-12}	9.09×10^{-3}	11.96
70.0	6.00×10^{-13}	1.67×10^{-2}	12.22
80.0	4.33×10^{-13}	2.31×10^{-2}	12.36
90.0	3.50×10^{-13}	2.86×10^{-2}	12.46
100.0	3.00×10^{-13}	3.33×10^{-2}	12.52

Figure 15.2 is the titration curve for the data in Table 15.5.

Common Ion Effect

At a given temperature the relationship given for acetic acid

$$K_a = \frac{[H^+][C_2H_3O_2^-]}{[HC_2H_3O_2]}$$

must hold for all solutions of acetic acid irrespective of concentrations. If the equilibrium is disturbed by changing the concentration of either H^+, $C_2H_3O_2^-$ or $HC_2H_3O_2$, the concentrations of the other two will adjust themselves in such a way that K retains its constant value. For example, if more acetate ion is added, the value for the concentration of H^+ necessarily decreases which causes an increase in the value of concentration of $HC_2H_3O_2$ in order that K remain constant.

The addition of an ion which is the same as one already present in the equilibrium is termed the **common ion effect.** Thus in the example above, acetate is the common ion and its effect is to shift the equilibrium, resulting

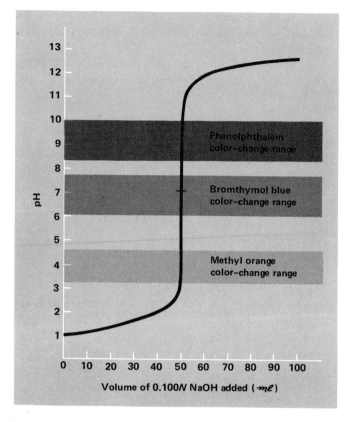

Figure 15.2
Titration curve for 50.0 ml of 0.100 N HCl with 0.100 N NaOH.

in a decrease of concentration of H^+ ions or an increase in pH. Assuming that the salt sodium acetate is the source of added acetate ion, the equations may be represented:

$$HC_2H_3O_2 \rightleftharpoons H^+ + \boxed{C_2H_3O_2^-}$$
$$NaC_2H_3O_2 \rightarrow Na^+ + \boxed{C_2H_3O_2^-}$$

The ionization of acetic acid is repressed and, since the hydrogen ion concentration is decreased, the pH of the solution is increased. Regulating the amount of sodium acetate added to a solution of acetic acid controls the pH. Careful control of pH is extremely important in many chemical processes, and the **common ion effect** is often the basis for such control.

Similarly, the addition of a common ion to a weak base decreases the pH of the solution. For example, if the salt NH_4Cl is added to a solution of NH_3, the concentration of OH^- is decreased and the pH becomes lower.

$$NH_3 + H_2O \rightleftharpoons \boxed{NH_4^+} + OH^-$$
$$NH_4Cl \rightarrow \boxed{NH_4^+} + Cl^-$$

Study the following sample calculations.

EXAMPLE 15.6   Calculate the concentration of H^+ ions and pH in a 0.10 M solution of acetic acid in which the concentration of acetate ions has been increased to 1.0 M by the addition of sodium acetate.

SOLUTION:

(1) We must be sure to write the *correct* equilibrium chemical equation; in this case the $HC_2H_3O_2$ ionization, not the $NaC_2H_3O_2$. The latter is a salt, which like most salts is completely ionized. When in doubt concerning which equation to use, keep in mind it will always be the one for which K is given or is involved.

(1) $$HC_2H_3O_2 \rightleftharpoons H^+ + C_2H_3O_2^-$$
for which

(2) $$K_a = \frac{[H^+][C_2H_3O_2^-]}{[HC_2H_3O_2]} = 1.8 \times 10^{-5}$$

(3)   Initial: 0.10 mole/l   0   1.0 mole/l
$$HC_2H_3O_2 \rightleftharpoons H^+ + C_2H_3O_2^-$$

The initial concentration of acetate ion, before any $HC_2H_3O_2$ ionizes, is 1.0 mole/l, coming entirely from the added $NaC_2H_3O_2$. The 1.0 mole/l of Na^+ is not part of the equilibrium equation and consequently may be ignored.

(4) At equilibrium, some of the acetic acid has ionized to produce more H^+ and $C_2H_3O_2^-$. Let $x = [H^+]$ at equilibrium. Then

Initial: 0.10 mole/l   0   1.0 mole/l
$$HC_2H_3O_2 \rightleftharpoons H^+ + C_2H_3O_2^-$$
Equilibrium: $(0.10 - x)$mole/l x mole/l $(1.0 + x)$mole/l

(5) Substituting the equilibrium concentration in the K expression

$$\frac{(x)(1.0 + x)}{(0.10 - x)} = 1.8 \times 10^{-5}$$

which becomes, when x is neglected in both the $(0.1 + x)$ and $(0.10 - x)$ terms,

$$\frac{(x)(1.0)}{0.10} = 1.8 \times 10^{-5}$$

and
$$x = 1.8 \times 10^{-6} \text{ mole/l} = [H^+]$$
$$pH = -\log (1.8 \times 10^{-6}) = 5.75$$

The concentration of H^+ has been decreased from 1.3×10^{-3} to 1.8×10^{-6} mole per l by the common ion effect.

EXAMPLE 15.7   The ionization constant for NH_3 is 1.8×10^{-5}.
   (a) Calculate the concentration of OH^- ions and pH in a 1.0 M solution of NH_3.

SOLUTION:
(1) and (2)
$$NH_3 + H_2O \rightleftharpoons NH_4^+ + OH^-$$

$$\frac{[NH_4^+][OH^-]}{[NH_3]} = K_b = 1.8 \times 10^{-5}$$

(3) and (4)

Initial: 1.0   0   0
$$NH_3 + H_2O \rightleftharpoons NH_4^+ + OH^-$$
Equilibrium: $(1.0 - x)$ x x

(5)
$$\frac{x^2}{1.0 - x} \cong^* \frac{x^2}{1.0} = 1.8 \times 10^{-5}$$

$$x^2 = 1.8 \times 10^{-5} = 18 \times 10^{-6}$$
$$x = 4.2 \times 10^{-3} \text{ mole/l} = [OH^-]$$
$$pOH = -\log (4.2 \times 10^{-3}) = 2.38$$
$$pH = 14.00 - 2.38 = 11.62$$

   (b) Calculate the concentration of OH^- ions and pH in a 1.0 M solution of NH_3 to which 0.10 mole of NH_4Cl is added to one liter of solution.

(1) and (2) are the same as in part (a).

   Since NH_4Cl is a salt and is 100% ionized, 0.10 mole of NH_4Cl would give 0.10 mole of NH_4^+.
   Then

(3) and (4)

Initial: 1.0   0.10   0
$$NH_3 + H_2O \rightleftharpoons NH_4^+ + OH^-$$
At equilibrium: $(1.0 - x)$ $(0.10 + x)$ x

(5)
$$\frac{(x)(0.10 + x)}{(1.0 - x)} \cong \frac{(x)(0.10)}{1.0} = 1.8 \times 10^{-5}$$

---

* $\cong$ means "approximately equal to."

$$\text{and } x = 1.8 \times 10^{-4} \text{ mole/l} = [OH^-]$$
$$pOH = -\log 1.8 \times 10^{-4} \text{ mole/l} = 3.74$$
$$pH = 14.00 - 3.74 = 10.26$$

The concentration of OH^- has been decreased from 4.2×10^{-3} to 1.8×10^{-4} moles per l by the common ion effect.

Buffer Solutions

An important application of the common ion principle is the preparation of buffer solutions. These may be prepared from a combination of a weak acid and a salt of the weak acid—acetic acid and sodium acetate, for instance— or a weak base and a·salt of the base—for example, NH_4OH and NH_4Cl. Such solutions are resistant to changes in acidity or basicity. Consider the addition of a strong acid like HCl to a solution of $HC_2H_3O_2$ and $NaC_2H_3O_2$.

$$NaC_2H_3O_2 \rightarrow Na^+ + \boxed{C_2H_3O_2^-}$$
$$HCl \rightleftharpoons Cl^- + \boxed{H^+}$$
$$\downarrow$$
$$HC_2H_3O_2$$

Even though a relatively high concentration of H^+ ions is added, the pH of the solution is not altered appreciably, since the excess $C_2H_3O_2^-$ ions in the solution react with the added H^+ ions to form slightly ionized acetic acid. Thus the acidity remains essentially unchanged. In a somewhat similar way the addition of OH^- ions to this same solution does not appreciably alter the pH, since the added OH^- ions combine with H^+ ions to form undissociated H_2O molecules.

$$HC_2H_3O_2 \rightleftharpoons \boxed{H^+} + C_2H_3O_2^-$$
$$NaOH \rightarrow \boxed{OH^-} + Na^+$$
$$\downarrow$$
$$H_2O$$

Although the number of H^+ ions in the solution is relatively small, they are produced as needed by the dissociation of the acetic acid present.

(The student should figure out for himself how a solution of a weak base and a salt of that base acts in resisting a change in pH on the addition of either H^+ or OH^- ions.)

Buffers are important in many biological processes. Phosphates in the blood act as buffer salts, since excess acid is neutralized by the formation of relatively undissociated phosphoric acid. Thus the pH of the blood is maintained at a nearly constant value.

15.2 SOLUBILITY PRODUCT PRINCIPLE

When a relatively insoluble salt, such as AgCl, is shaken with water, ions from the crystal lattice of the solid pass into solution until the solution becomes saturated. In the saturated solution thus formed, an equilibrium exists between the ions in solution and the ions present in the solid. For AgCl,

$$AgCl(s) \rightleftharpoons Ag^+ + Cl^-$$

The amount of AgCl dissolved depends on the temperature; at a given temperature the solubility of a substance is a constant.

Applying the Law of Mass Action to the above equilibrium, we may write

$$K = \frac{[Ag^+][Cl^-]}{[AgCl]}$$

Since the concentration of solid AgCl is independent of the amount of solid AgCl present, this term may be incorporated in the constant K to give a new constant K_{sp}

$$K_{sp} = [Ag^+][Cl^-]$$

in which K_{sp} is the **solubility product constant.** In other words, **the product of the concentrations of ions in a saturated solution of a relatively insoluble salt,** such as AgCl, **at a given temperature is a constant.** This is known as the **solubility product principle.**

Solubility Product Constant, K_{sp}

The numerical value for K_{sp} can readily be calculated if the concentrations of ions in the saturated solution are known. In a saturated solution of AgCl prepared at 25°C, the solubility by analysis is 1.06×10^{-5} mole per l. AgCl is completely ionized; hence each AgCl will give one Ag^+ and one Cl^- ion in solution. Consequently, the concentration of Ag^+ ions at equilibrium will be 1.06×10^{-5} mole per l, and the concentration of Cl^- ions will be 1.06×10^{-5} mole per l. Substituting these concentrations in the solubility product formulation,

$$K_{sp} = [Ag^+][Cl^-]$$
$$K_{sp} = (1.06 \times 10^{-5})(1.06 \times 10^{-5}) = 1.12 \times 10^{-10}$$

The value for K_{sp} is a measure of the solubility of a substance; the larger the value of K_{sp}, the greater must be the concentrations of ions and hence the greater the solubility of the substance. The smaller the value of K_{sp}, the more insoluble the substance.

If two or more ions of one kind are produced from an ionization, the concentration of that particular ion must be raised to that power which corresponds to the coefficient of the ion in the balanced equation called for by the Law of Mass Action. To illustrate,

$$PbCl_2(s) \rightleftharpoons Pb^{2+} + 2\ Cl^-$$
$$K_{sp} = [Pb^{2+}][Cl^-]^2$$

The solubility of $PbCl_2$ in water at 25°C is 4.41 g per l. Since the molecular weight of $PbCl_2$ is 278,

$$4.41\ g = 0.0159\ \text{mole}$$
$$[Pb^{2+}] = 1.59 \times 10^{-2}\ \text{mole/l}$$

Since each $PbCl_2$ yields two Cl^-,

$$[Cl^-] = 2 \times 0.0159 = 3.18 \times 10^{-2}\ \text{mole/l}$$
$$K_{sp} = (1.59 \times 10^{-2})(3.18 \times 10^{-2})^2 = 1.61 \times 10^{-5}$$

A partial list of solubility product constants is given in Table 15.6 and in the appendix.

Table 15.6

Solubility product constants

Compound	K_{sp}	Compound	K_{sp}
AgCl	1.1×10^{-10}	$BaSO_4$	1.0×10^{-10}
$PbCrO_4$	2×10^{-16}	HgS	1.6×10^{-54}
$Fe(OH)_3$	6×10^{-38}	CuS	8×10^{-37}
$BaCrO_4$	8.5×10^{-11}	CdS	1.0×10^{-28}
CaC_2O_4	2.6×10^{-9}	PbS	7×10^{-29}
$SrSO_4$	3.6×10^{-7}	ZnS	1.2×10^{-23}
$PbCl_2$	1.6×10^{-5}	FeS	4×10^{-19}
AgBr	4.0×10^{-13}	MnS	1.4×10^{-15}

EXAMPLE 15.8   The solubility of Ag_2CrO_4 is 7.8×10^{-5} mole/l. Calculate the K_{sp} for Ag_2CrO_4.

SOLUTION:

(1) The chemical equation is

$$Ag_2CrO_4\ (s) \rightleftharpoons 2\ Ag^+ + CrO_4^{2-}$$

from which the K_{sp} expression is

(2) $$K_{sp} = [Ag^+]^2[CrO_4^{2-}]$$

(3) and (4) Since 7.8×10^{-5} mole/l of Ag_2CrO_4 go into solution

Initial: --   0   0

$$Ag_2CrO_4 \text{ (s)} \rightleftarrows \quad 2 Ag^+ \quad + \quad CrO_4{}^{2-}$$

At equilibrium: -- $2(7.8 \times 10^{-5})$ 7.8×10^{-5}

(5) Substitute the equilibrium ion concentrations in the K_{sp} expression:

$$K_{sp} = (15.6 \times 10^{-5})^2 (7.8 \times 10^{-5})$$
$$= 1.9 \times 10^{-12}$$

EXAMPLE 15.9   What is the solubility (mole/l) of $PbCl_2$ in water?

SOLUTION:

(1) $PbCl_2 \text{ (s)} \rightleftarrows Pb^{2+} + 2 Cl^-$

(2) $K_{sp} = [Pb^{2+}][Cl^-]^2 = 1.6 \times 10^{-5}$

(3) and (4) Let $x =$ the moles of $PbCl_2$ dissolved per liter of solution (the solubility).

Then

Initial: --   0   0

$$PbCl_2 \text{ (s)} \rightleftarrows Pb^{2+} + 2 Cl^-$$

At equilibrium: -- x $2x$

(5) $(x)(2x)^2 = 1.6 \times 10^{-5}$
$4x^3 = 1.6 \times 10^{-5}$
$x^3 = 4.0 \times 10^{-6}$
$x = 1.6 \times 10^{-2}$ mole/l $= Pb^{2+} =$ solubility of $PbCl_2$

EXAMPLE 15.10   What is the solubility (mole/l) of AgCl in 0.10 M NaCl solution? K_{sp} of AgCl $= 1.1 \times 10^{-10}$.

SOLUTION: Note that Cl^- is the **common ion** of the two salts. The chemical equation involves the AgCl equilibrium, however (since the K_{sp} is for AgCl).

(1) $AgCl \text{ (s)} \rightleftarrows Ag^+ + Cl^-$

(2) $K_{sp} = [Ag^+][Cl^-] = 1.1 \times 10^{-10}$

(3) and (4) Let x be the solubility of AgCl in the NaCl solution.

Initial: --   0   0.10

$$AgCl \text{ (s)} \rightleftarrows Ag^+ + \quad Cl^-$$

At equilibrium: -- x $(0.10 + x)$

(5) Neglecting x in the $(0.10 + x)$ term

$$(x)(0.10) = 1.1 \times 10^{-10}$$
$$x = 1.1 \times 10^{-9} \text{ mole/l} = [Ag^+]$$
$$= \text{AgCl solubility}$$

Application of Solubility Product Principle
to Formation of Precipitates

The formation of a precipitate depends first on the formation of a saturated solution; only when the solubility of the substance is exceeded can a precipitate form. K_{sp} is directly related to the solubility of a substance, since it is equal to the product of ion concentrations in a saturated solution. Any solution in which the product of the ion concentrations exceeds K_{sp} is supersaturated, and precipitation will occur until the solution is saturated, in which case the product of ion concentrations attains the value of K_{sp}. Any solution in which the ion product is less than K_{sp} is unsaturated and is capable of dissolving more of the solute. (See Figures 15.3, 15.4, and 15.5.)

In the case of AgCl, when the product of the concentration of Ag^+ ions and the concentration of Cl^- ions exceeds 1.1×10^{-10}, the salt will precipitate. In a water solution of AgCl the concentrations of Ag^+ and Cl^- are the same — 1.06×10^{-5} mole per l. It is not necessary, however, that the concentrations of the ions have equivalent values to bring about precipitation so long as the product of these ion concentrations exceeds K_{sp}.

Suppose that we know the concentration of Ag^+ ions in a certain solution to be 1.0×10^{-7} mole per l. We can calculate the concentration of Cl^- ions necessary to bring about precipitation. Substituting in the equation

$$K_{sp} = [Ag^+][Cl^-] = 1.1 \times 10^{-10}$$

and solving for $[Cl^-]$,

$$[Cl^-] = \frac{1.1 \times 10^{-10}}{1.0 \times 10^{-7}} = 1.1 \times 10^{-3}$$

This concentration of Cl^- ions will just form a saturated solution, but any value of the concentration of Cl^- ions larger than 1.1×10^{-3} will start precipitation of AgCl.

It was shown above that the solubility of AgCl in water is 1.06×10^{-5} mole per l, while the solubility in a solution in which the concentration of Cl^- ions is 1.1×10^{-3} is only 1.0×10^{-7} mole per l; in other words, using an excess of Cl^- decreases the solubility of the salt nearly a hundredfold. In the precipitation of ions from solution in qualitative and quantitative analysis, a large excess of the precipitating ion is used to reduce the solubility of the substance markedly and cause it to precipitate. Study the following illustrations.

EXAMPLE 15.11   With the aid of Table 15.6 calculate the minimum concentration of Br^- ion necessary to bring about precipitation of AgBr from a solution in which the concentration of Ag^+ ion is 1×10^{-5} mole per l.

SOLUTION:

$$K_{sp} \text{ for AgBr} = [Ag^+][Br^-] = 4 \times 10^{-13}$$

Substituting for $[Ag^+]$,

$$(1 \times 10^{-5})[Br^-] = 4 \times 10^{-13}$$

$$[Br^-] = \frac{4 \times 10^{-13}}{1 \times 10^{-5}} = 4 \times 10^{-8}$$

This value for the concentration of Br^- just forms a saturated solution; any value larger than this would cause precipitation of AgBr.

EXAMPLE 15.12   Will precipitation occur if 0.01 mole of Ba^{2+} is added to a liter of solution containing 0.05 mole of SO_4^{2-}?

SOLUTION:
K_{sp} for $BaSO_4 = [Ba^{2+}][SO_4^{2-}] = 1 \times 10^{-10}$
$\quad\quad [Ba^{2+}] = 0.01$ mole/l $= 1 \times 10^{-2}$ mole/l
$\quad\quad [SO_4^{2-}] = 0.05$ mole/l $= 5 \times 10^{-2}$ mole/l
Trial ion product $(1 \times 10^{-2})(5 \times 10^{-2}) = 5 \times 10^{-4}$.

5×10^{-4} is larger than the K_{sp}, 1×10^{-10}; hence precipitation of $BaSO_4$ will take place.

EXAMPLE 15.13   After precipitation of $BaSO_4$ from a liter of solution which originally contained 0.01 mole of Ba^{2+}, the concentration of SO_4^{-2} is found to be 0.1 mole per l. How much $BaSO_4$ is precipitated?

SOLUTION: First calculate the concentration of Ba^{2+} in the solution after precipitation.

$$K_{sp} = [Ba^{2+}][SO_4^{2-}] = 1 \times 10^{-10}$$

$$[Ba^{2+}] \text{ left in solution} = \frac{1 \times 10^{-10}}{0.1} = 1 \times 10^{-9} \text{ mole/l.}$$

Number of moles precipitated $= 0.01 - 0.000000001 = 0.009999999 \cong 0.01$

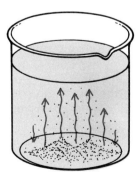

Figure 15.3
When the product of cation and anion concentrations is less than the solubility product, the solid will continue dissolving.

Figure 15.4
When the ion product of cation and anion concentrations in solution exceeds the solubility product, precipitation takes place.

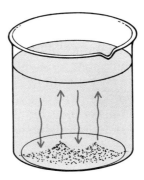

Figure 15.5
When the product of cation and anion concentrations equals the solubility product, equilibrium between the solid and solution exists; dissolving and crystallization (precipitation) proceed at the same rate.

The solubility product principle is applicable only to solutions of slightly soluble salts. Experimental studies show that it does not hold for soluble or moderately soluble salts.

The Dissolving of Precipitates

Just as a precipitate is formed when the product of ion concentrations exceeds K_{sp}, so a precipitate must dissolve when the ion concentrations are decreased sufficiently to give an ion product less than K_{sp}. Anything that lowers the concentrations of ions in the solution will disturb the equilibrium with the undissolved solid. If the supply of ions is replenished to reestablish equilibrium, more solid will pass into solution (Figure 15.6). This process will continue

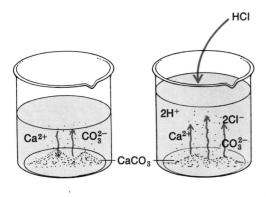

Figure 15.6
If HCl is added to a solution containing undissolved $CaCO_3$, more of the $CaCO_3$ will dissolve.

until all the solid has dissolved, provided the concentrations of ions are continuously lowered. Decreasing the concentration of ions in solution may be accomplished in the following ways:

(a) Dilution. As a saturated solution in contact with undissolved solute is diluted, the concentrations become less and more solid passes into solution.

(b) Addition of a reagent which forms an undissociated substance with one of the ions of the solid. For example, addition of HCl to a precipitate of calcium oxalate forms slightly dissociated oxalic acid:

$$CaC_2O_4(s) \rightleftharpoons Ca^{2+} + \boxed{C_2O_4^{2-}}$$
$$2\ HCl \rightarrow 2\ Cl^- + \boxed{2\ H^+}$$
$$\longrightarrow H_2C_2O_4$$

As $C_2O_4^{2-}$ ions are removed by combination with H^+ ions, more CaC_2O_4 passes into solution to replenish the supply of $C_2O_4^{2-}$ ions. If sufficient HCl is present, all of the precipitate will dissolve.

(c) Addition of a reagent which reacts with the solid to form a gas that is insoluble and escapes from solution. For example, iron(II) sulfide is soluble in HCl:

$$\rightarrow H_2S$$
$$FeS(s) \rightleftharpoons Fe^{2+} + \boxed{S^{2-}}$$
$$2\ HCl \rightarrow 2\ Cl^- + \boxed{2\ H^+}$$

15.3 HYDROLYSIS

Although water is only slightly ionized according to the equation

$$H_2O \rightleftharpoons H^+ + OH^- \tag{1}$$

the few H^+ ions and OH^- ions produced may become of primary importance in aqueous solutions of certain salts. As long as the number of H^+ ions is equal to the number of OH^- ions, the solution is neutral. If something is added which furnishes either of the ions or which uses up either of the ions, however, then one of the two ions is present in excess. If excess H^+ ions are present, the solution is acidic; if excess OH^- ions are present, the solution is basic.

Ions of salts may disturb the equilibrium in aqueous solution. As a result, the solution is actually acidic or basic, depending on whether H^+ or OH^- ions are present in excess. This reaction of the ions of a salt with water to form an acid + a base is termed **hydrolysis.**

SODIUM CARBONATE   An aqueous solution of this salt tests basic toward litmus, indicating an excess of OH^- ions present in the solution. This may be explained as follows: Sodium carbonate, a salt derived from the strong base sodium hydroxide and the weak acid carbonic acid, is completely ionized into Na^+ and CO_3^{2-}. CO_3^{2-} ions have a great affinity for protons to form the weak (slightly ionized) acid HCO_3^-; consequently, protons are removed from solution by the reaction

$$H^+ + CO_3^{2-} \rightleftharpoons HCO_3^-$$

As a result, the equilibrium in (1) above is shifted in the direction of a higher concentration of OH^- ions to give the solution a basic reaction. Meanwhile, there is very little tendency for Na^+ to react with H_2O or its ions, since NaOH is a strong base. The equation for the reaction is

$$2\,Na^+ + CO_3^{2-} + H_2O \rightleftharpoons 2\,Na^+ + OH^- + HCO_3^- \tag{2}$$

or

$$CO_3^{2-} + H_2O \rightleftharpoons OH^- + HCO_3^-$$

since Na^+ ions may be canceled out. A solution of Na_2CO_3 is only weakly alkaline, because the equilibrium in (2) above is far to the left. The extent of hydrolysis is said to be slight.

AMMONIUM CHLORIDE   The ions present in the solution are H^+, OH^-, NH_4^+, Cl^-. There is very little tendency for H^+ to combine with Cl^-, since HCl is a strong acid (or, according to the Bronsted theory, Cl^- is a weak base). However, NH_4^+ may furnish protons to the solution by forming weakly basic NH_3:

$$NH_4^+ + H_2O \rightleftharpoons NH_3 + H_3O^+ \tag{3}$$

Accordingly, an excess of H^+ (or H_3O^+) ions are left in solution, which reacts acidic to litmus.

SODIUM CHLORIDE   Ions present in the solution are Na^+, Cl^-, H^+, OH^-. There is very little tendency for Na^+ or Cl^- to hydrolyze, because NaCl is the salt of the strong base NaOH and the strong acid HCl. Consequently, the concentrations of H^+ and OH^- remain about equal, and the resulting solution is approximately neutral. The hydrolysis of a salt derived from a strong acid and a strong base is negligible.

AMMONIUM ACETATE   The aqueous solution contains NH_4^+, $C_2H_3O_2^-$, H^+, OH^-. This salt is derived from the weak acid acetic acid and the weak base $NH_3 + H_2O$, so that the following reactions occur:

$$NH_4^+ + H_2O \rightleftharpoons NH_3 + H_3O^+ \ (H^+ + H_2O) \tag{4}$$

and

$$C_2H_3O_2^- + H_2O \rightleftharpoons HC_2H_3O_2 + OH^- \tag{5}$$

Reactions (4) and (5) proceed to about the same extent, since K_a for $HC_2H_3O_2$ and K_b for $NH_3 + H_2O$ are about equal. Hence the solution is approximately neutral.

COPPER SULFATE Salts of most of the heavy metals contain hydrated ions, and such ions may furnish protons to the solution and give it an acidic reaction. For example, copper sulfate in solution gives $Cu(H_2O)_4^{2+}$ and SO_4^{2-} ions. There is little tendency for SO_4^{2-} to combine with H^+, since H_2SO_4 is a strong acid. However, the hydrated copper(II) ions may react with water to form H^+ ions:

$$Cu(H_2O)_4^{2+} + H_2O \rightleftharpoons Cu(H_2O)_3OH^+ + H_3O^+(H^+ + H_2O) \qquad (6)$$

Thus a solution of this salt is slightly acidic because of the excess H^+ ions present.

It is evident from the above that whether or not the aqueous solution of a salt is acidic, basic, or neutral depends on the strength of the acid and base from which the salt may be considered to be derived. A knowledge of the strength of acids and bases is essential, therefore, in predicting the effects of hydrolysis. **Of the common acids only HNO_3, H_2SO_4, and the hydrohalogen acids (except HF) are strong. The strong bases in aqueous solution are the hydroxides of the metals in Groups IA and IIA of the periodic table. Students should memorize these strong acids and bases.**

We may summarize the possible hydrolysis effects as follows:

1. Salt of strong acid and strong base. Solution is approximately neutral, pH about 7. Examples: $NaCl$, K_2SO_4, $CaCl_2$
2. Salt of weak acid and weak base. Solution is approximately neutral, pH about 7. The pH will depend on the relative strengths of the weak acid and the base. Examples: NH_4F, $Cu(C_2H_3O_2)_2$
3. Salt of weak acid and strong base. Basic solution, pH greater than 7. Examples: Na_2CO_3, K_2S, $Ca(C_2H_3O_2)_2$
4. Salt of strong acid and weak base. Acid solution, pH less than 7. Examples: NH_4Cl, $FeSO_4$, $AlBr_3$

HYDROLYSIS CONSTANTS Since hydrolysis is an equilibrium process, an equilibrium constant may be determined. The value of this constant will be a measure of the degree or extent of the hydrolysis process. Consider the hydrolysis of $NaC_2H_3O_2$:

$$C_2H_3O_2^- + H_2O \rightleftharpoons HC_2H_3O_2 + OH^- \qquad (1)$$
$$K = \frac{[HC_2H_3O_2][OH^-]}{[C_2H_3O_2^-][H_2O]}$$

Since $[H_2O]$ is nearly constant in dilute solutions, we may incorporate its constant value into K and obtain an expression for K_h called the hydrolysis constant:

$$K[H_2O] = K_h = \frac{[HC_2H_3O_2][OH^-]}{[C_2H_3O_2^-]} \qquad (2)$$

The larger the value of K_h, the greater the degree or extent of hydrolysis — that is, the farther the equilibrium is displaced to the right side of the equation as written. K_h has a value of 5.5×10^{-10} for the hydrolysis of $C_2H_3O_2^-$. The small number indicates that the reaction as written does not proceed very far to the right — that is, the hydrolysis is not high.

K_h may be expressed as a ratio of K_a for the weak acid, acetic acid, and K_w as follows:

Multiply both the numerator and denominator of equation (2) by $[H^+]$; then

$$K_h = \frac{[HC_2H_3O_2][OH^-][H^+]}{[C_2H_2O_2^-][H^+]} \tag{3}$$

but

$$[OH^-][H^+] = K_w$$

and

$$\frac{[HC_2H_3O_2]}{[C_2H_3O_2^-][H^+]} = \frac{1}{K_a}$$

hence

$$K_h = \frac{K_w}{K_a} \tag{4}$$

Similarly, we can arrive at the expression

$$K_h = \frac{K_w}{K_b} \tag{5}$$

for the hydrolysis of the cation of a weak base. If the hydrolysis is for a salt of both a weak acid and a weak base,

$$K_h = \frac{K_w}{K_a K_b} \tag{6}$$

The student should verify equations (5) and (6).

Study the problem examples below.

EXAMPLE 15.14   Determine K_h for sodium nitrite ($NaNO_2$). K_a for $HNO_2 = 4.5 \times 10^{-4}$.

SOLUTION:
(1) The ionic chemical equation for the hydrolysis of the salt $NaNO_2$ is

$$NO_2^- + H_2O \rightarrow HNO_2 + OH^-$$

from which the K_h expression is

(2) $\dfrac{[HNO_2][OH^-]}{[NO_2^-]} = K_h = \dfrac{K_w}{K_a} = \dfrac{1 \times 10^{-14}}{4.5 \times 10^{-4}} = 2.2 \times 10^{-11}$

EXAMPLE 15.15   Determine the pH of a 1.0 M solution of sodium formate (NaHCOO) if K_h for the formate ion $= 5.0 \times 10^{-11}$.

SOLUTION:

(1)
$$HCOO^- + H_2O \rightleftharpoons HCOOH + OH^-$$

(2)
$$K_h = \frac{[HCOOH][OH^-]}{[HCOO^-]} = 5.0 \times 10^{-11}$$

(3) and (4)

Initial: 1.0 M $-$ 0   0

$$HCOO^- + H_2O \rightleftharpoons HCOOH + OH^-$$

At equilibrium: $(1.0 - x)\,M$ $-$ x x Where $x = [OH^-]$

(5) Neglecting x in the $(1.0 - x)$ term

$$\frac{x^2}{1.0} = 5.0 \times 10^{-11} = 50 \times 10^{-12}$$

and
$$x = 7.1 \times 10^{-6} \, M = [OH^-]$$
$$pOH = -\log{(7.1 \times 10^{-6})} = 5.15$$

and
$$pH = 14.0 - 5.15 = 8.85.$$

Exercises

15.1. Given the ionization constants for a number of acids as follows:

Acid	K_a	Acid	K_a
(a) Trichloroacetic	2.0×10^{-1}	(e) Nitrous	4.5×10^{-4}
(b) Hypobromous	2.1×10^{-9}	(f) Formic	1.6×10^{-4}
(c) Butyric	1.5×10^{-5}	(g) Bromoacetic	2.0×10^{-3}
(d) Benzoic	6.4×10^{-5}	(h) Chlorous	1.1×10^{-2}

List these acids in order of *increasing* strength.

15.2. K_a for ascorbic acid, $HC_6H_7O_6$, $= 8.1 \times 10^{-5}$. Calculate the H^+ concentration of 0.10 M ascorbic acid.

15.3. A certain acid HX is 4.0% ionized in 0.30 M solution. Calculate K_a.

15.4. What are the concentrations of H^+, $C_7H_5O_2^-$, and $HC_7H_5O_2$ in a 0.50 M solution of benzoic acid? $K_a = 6.4 \times 10^{-5}$.

15.5. The ionization constant for lactic acid, a weak monobasic acid, is 8.4×10^{-4}. What is the concentration of H^+ in a 0.30 M solution of lactic acid?

15.6. Propanoic acid, $HC_3H_5O_2$, is 0.72% ionized in 0.25 M solution. What is the ionization constant for this acid?

15.7. Oxalic acid ($H_2C_2O_4$) is a polyprotic acid and ionization constants listed are $K_1 = 5.9 \times 10^{-2}$ and $K_2 = 6.4 \times 10^{-5}$.
(a) Write the equations for the two equilibrium processes which would correspond to the two constants.
(b) Write the equilibrium expression for K_2.

(c) Give the formulae of the four different molecules and/or ions other than H_2O that would be present in a solution of oxalic acid.

(d) Which molecule or ion would be present in the lowest proportion?

15.8. K_b for the base C_5H_5N, pyridine, is 1.5×10^{-9}. Calculate the concentration of OH^- of a 0.035 molar aqueous solution.

15.9. Calculate the H^+ concentration in a 0.050 M HX solution after 0.50 mole of the soluble salt NaX is also added. $K_{HX} = 7.4 \times 10^{-6}$

15.10. A solution is prepared by adding 0.020 mole of sodium formate, $Na(CHO_2)$, to 1 l of 0.050 M formic acid ($HCHO_2$). Assume that no volume change occurs. Calculate the H^+ concentration of the solution. $K_a = 1.6 \times 10^{-4}$.

15.11. K_a for $HC_2H_3O_2 = 1.8 \times 10^{-5}$. If 10 g of $NaC_2H_3O_2$ is added to 500 ml of 0.20 N acid, what is the H^+ ion concentration of the solution?

15.12. A certain hypothetical base MOH is 10% ionized in 0.005 M aqueous solution. (a) Calculate K_b. (b) If 0.04 mole of the salt M_2SO_4 is added to 1 l of 0.001 M MOH, what is the concentration of OH^-?

15.13. A buffer solution is made up by dissolving 0.25 mole of $NaC_2H_3O_2$ in 1 l of 0.20 M $HC_2H_3O_2$; $K_a = 1.8 \times 10^{-5}$. Calculate the H^+ concentration of the solution.

15.14. What is the OH^- concentration of a buffer solution composed of 0.20 M NH_3 and 0.75 M NH_4Cl? $K_b = 1.8 \times 10^{-5}$.

15.15. What is the pH of a buffer prepared from 0.30 mole of trimethylamine $((CH_3)_3N)$ and 0.50 mole of trimethylammonium chloride $((CH_3)_3NH^+Cl^-)$ in 250 ml of solution?

15.16. Assuming 100% ionization, what are the pH and pOH of each of the following solutions?
(a) 0.03 M HNO_3
(b) 0.001 N H_2SO_4
(c) 0.1 N NaOH
(d) 4×10^{-4} N $Ca(OH)_2$

15.17. What pH corresponds to each of the following? (a) $[H^+] = 0.65$ M; (b) $[OH^-] = 6.0 \times 10^{-4}$ M; (c) $[OH^-] = 3.0 \times 10^{-10}$ M; (d) $[H^+] = 4.0 \times 10^{-4}$ M.

15.18. What concentration of H^+ corresponds to each of the following? (a) $pH = 4.40$; (b) $pOH = 4.40$; (c) $pOH = 12.22$; (d) $pH = 5.55$.

15.19. Complete the following table (fill in the blanks):

Acid or Base	$[H^+]$ mole/l	$[OH^-]$ mole/l	pH	pOH
X	10^{-5}			
0.0010 N HCl				
X		5.7×10^{-9}		
X				2.8

15.20 The $[OH^-]$ of a 0.20 M BOH (weak base) solution is 0.0010 mole/l. What is
(a) the pH of the solution?
(b) the K_b of the base?

15.21. Complete the following table (fill in the blanks):

Acid or Base	$[H^+]$ mole/l	$[OH^-]$ mole/l	pH	pOH
X	_____	1.0×10^{-7}	_____	_____
0.010 N NaOH	_____	_____	_____	_____
X	7.0×10^{-4}	_____	_____	_____
X	_____	_____	_____	12.4

15.22. Calculate the pH of (a) 0.0030 M $HC_2H_3O_2$ solution; (b) a 0.20 M $HC_2H_3O_2$
solution after 0.50 mole per l of $NaC_2H_3O_2$ is also added.

15.23. Below is listed some information on a series of indicators and the color a solution of unknown pH gave when samples of the solution were tested with each
indicator.

Indicator	Color at lowest pH	pH range of color change	Color at highest pH	Sample of unknown solution
Bromphenol blue	yellow	3.0–4.6	blue	blue
Bromcresol green	yellow	3.8–5.4	blue	blue
Bromcresol purple	yellow	5.2–6.8	purple	purple
m-cresol purple	red	7.5–8.5	purple	red
Phenolphthalein	colorless	8.3–10.0	red	colorless

What would be the most likely pH of the unknown solution?

15.24. An indicator is a weak acid and the pH range of its color change is 4.3–5.7.
Assuming that the neutral point of the indicator is the center of this range,
calculate the ionization constant of the indicator.

15.25. An indicator, HIn, has an ionization constant of 9.0×10^{-9}. The acid color of
the indicator is yellow, and the alkaline color is red. The yellow color is visible
when the ratio of yellow form to red form is 30 to 1, and the red color is predominant when the ratio of red form to yellow form is 2 to 1. What is the pH
range of color change for the indicator?

15.26. Determine pH values at the following stages of a titration of 30.0 ml of 0.100 N
NaOH with 0.100 N HCl: (a) after 10.00 ml of HCl have been added, (b) 20.00
ml, (c) 29.90 ml, (d) 30.10 ml, (e) 40.0 ml.

15.27. Determine pH values for the titration curve pertaining to the titration of 30.00
ml of 0.100 N benzoic acid, $HC_7H_5O_2$, with 0.100 N NaOH (a) after 10.00 ml of
NaOH solution has been added, (b) after 30.00 ml of NaOH solution has been
added, (c) after 40.00 ml of NaOH solution has been added.

15.28. From the following K_{sp} values, list the following compounds in order of *increasing* solubility:

Compound	K_{sp}	Compound	K_{sp}
(a) AgBr	4×10^{-13}	(e) $BaCO_3$	1.6×10^{-9}
(b) CaC_2O_4	1.3×10^{-9}	(f) $SrSO_4$	7.6×10^{-7}
(c) CdS	1.0×10^{-28}	(g) $PbCrO_4$	2×10^{-16}
(d) FeS	4×10^{-19}	(h) AgI	8.5×10^{-17}

15.29. The solubility in moles per liter for several compounds is given below. Calculate the K_{sp} for each compound.

Compound	Solubility (mole)l)
(a) TlBr	1.84×10^{-3}
(b) MnS	3.74×10^{-8}
(c) CaF_2	2.1×10^{-4}
(d) Ag_3PO_4	1.6×10^{-5}

15.30. Solubility data for several compounds are given below in grams per liter of solution at 20°C. For each compound calculate K_{sp}.

Compound	Solubility	Compound	Solubility
(a) Ag_3AsO_4	8.5×10^{-3}	(g) MnS	1.7×10^{-2}
(b) AgCN	2.2×10^{-4}	(h) $PbBr_2$	8.5
(c) Ag_2S	1.6×10^{-4}	(i) PbF_2	0.64
(d) As_2S_3	5.2×10^{-4}	(j) Sb_2S_3	1.7×10^{-3}
(e) Hg_2Br_2	4.0×10^{-5}	(k) SrC_2O_4	0.044
(f) HgI_2	5.9×10^{-2}	(l) TlI	0.036

15.31. From the following K_{sp} values, calculate the molar solubility of each compound.

Compound	K_{sp}	Compound	K_{sp}
(a) $CaCO_3$	4.7×10^{-9}	(e) $PbCl_2$	1.6×10^{-5}
(b) CaC_2O_4	1.3×10^{-9}	(f) $Zn(OH)_2$	1.8×10^{-14}
(c) Li_2CO_3	1.7×10^{-3}	(g) AgI	8.5×10^{-17}
(d) $Mg(OH)_2$	3×10^{-11}	(h) $MgNH_4PO_4$	2.5×10^{-13}

15.32. K_{sp} for $SrSO_4$ is 7.6×10^{-7}. (a) Calculate the solubility of $SrSO_4$ in H_2O. (b) What would be the solubility of $SrSO_4$ in a solution which is 0.01 M with respect to sulfate ion? (c) By what factor is the solubility decreased from part (a) to part (b)?

15.33. At 25°C, 3.4×10^{-6} mole of $Ni(OH)_2$ dissolves in 1.0 l of water. Calculate the K_{sp} of $Ni(OH)_2$.

15.34. In a saturated solution of calcium phosphate, the concentration of PO_4^{3-} ion is 3.3×10^{-7} M. Calculate the K_{sp} of $Ca_3(PO_4)_2$.

15.35. The K_{sp} of Ag_3PO_4 is 1.8×10^{-18}. What is the molar solubility of Ag_3PO_4?

15.36. Calculate the minimum concentration of CrO_4^{2-} necessary to precipitate $BaCrO_4$ from a solution in which the concentration of Ba^{2+} is 6.0×10^{-4}. K_{sp} for $BaCrO_4 = 8.5 \times 10^{-11}$.

15.37. (a) A solution is $0.15\ M$ in Pb^{2+} and $0.20\ M$ in Ag^+. If solid Na_2SO_4 is very slowly added to this solution, which will precipitate first, $PbSO_4$ or Ag_2SO_4? (b) The addition of Na_2SO_4 is continued until the second cation just starts to precipitate as the sulfate. What is the concentration of the first cation at this point? K_{sp} for $PbSO_4 = 1.3 \times 10^{-8}$, $Ag_2SO_4 = 1.2 \times 10^{-5}$.

15.38. Will a precipitate form in each of the following cases? Prove each answer by calculating the ion product and comparing to the K_{sp}. (a) 0.0050 mole of solid NaCl is added to a liter of $4.0 \times 10^{-5}\ M$ $AgNO_3$, (b) 0.0020 mole of KCl is added to $0.0060\ M$ $Pb(NO_3)_2$, (c) $6.5 \times 10^{-4}\ M$ $FeCl_3$ solution is adjusted to pH 11.

15.39. How would water solutions of the following salts test with litmus?
(a) Na_2CO_3 (f) KNO_3
(b) K_3PO_4 (g) KCN
(c) $CuCl_2$ (h) Na_2S
(d) $NH_4C_2H_3O_2$ (i) $Al_2(SO_4)_3$
(e) ZnI_2 (j) $AgNO_3$

15.40. Calculate (a) the hydrolysis constant for the following reaction, and

$$C_2H_3O_2^- + H_2O \rightarrow HC_2H_3O_2 + OH^- \qquad K_a \text{ for } HC_2H_3O_2 = 1.8 \times 10^{-5}$$

(b) determine the pH of a $0.05\ M$ solution of $NaC_2H_3O_2$.

15.41. K_h for the reaction

$$Be^{2+} + H_2O \rightarrow BeOH^+ + H^+$$

is 2×10^{-7}. Calculate the pH of $0.02\ M$ $BeCl_2$.

15.42. What is the pH at the endpoint of a titration of $0.50\ M$ HCOOH and $0.50\ M$ NaOH? K_a for HCOOH $= 1.6 \times 10^{-4}$

15.43. What is the pH of a $0.30\ M$ solution of sodium benzoate ($NaC_7H_5O_2$)?

15.44. What is the pH of a $0.50\ M$ solution of hydrazine hydrochloride (N_2H_5Cl)?

15.45. Consider the hydrolysis of KCN.
(a) Would an aqueous solution be acidic, basic, or neutral?
(b) Write the hydrolysis equation for KCN.

$$CN^- + H_2O \rightarrow$$

(c) What is the numerical value of K_h for KCN? K_b for HCN $= 4.0 \times 10^{-10}$.
(d) What is the $[H^+]$ of a $0.20\ M$ KCN solution?

15.46. Consider the hydrolysis of NaF:
  (a) Would an aqueous solution be acidic, basic, or neutral?
  (b) Write the hydrolysis equation for NaF.

$$F^- + H_2O \rightarrow$$

  (c) What is the numerical value of K_h for NaF? K_a for HF $= 6.7 \times 10^{-4}$.
  (d) What is the pH of a 0.30 M NaF solution?

Supplementary Readings

15.1. *Ions in Aqueous Systems*, T. Moeller and R. O'Connor, McGraw-Hill, New York, 1972 (paper).
15.2. "Development of the pH Concept," F. Szabadvary and R. E. Oesper, *J. Chem. Educ.*, **41**, 105 (1964).
15.3. "Hydrolysis of Sodium Carbonate," F. S. Nakayama, *J. Chem. Educ.*, **47**, 67 (1970).

Chapter 16

Oxidation-Reduction

Chemical changes may be classified in two large groups: (1) those involving no change in oxidation number of the elements, and (2) those in which a change of oxidation numbers occurs. The latter are termed **oxidation-reduction** or simply **redox** reactions.

Most of the reactions which we encounter fall into the first class, such as exchange reactions between acids, bases, and salts in which simple combinations of ions take place. Redox reactions in general are more complex and may involve more than two reactants or products. The student should review the oxidation number concept as discussed in Chapter 6 before proceeding further.

An Example of Oxidation-Reduction

Oxidation is defined simply as a change in which the oxidation number of an element increases. Such a change can take place only at the expense of a decrease in the oxidation number of some other element. The latter is termed **reduction.** For example, consider the following reaction:

$$\overset{0}{2\ Al} + \overset{0}{3\ S} \rightarrow \overset{+3\ -2}{Al_2S_3}$$

in which the oxidation number of each element is written above the symbol. Aluminum and sulfur in the free state have, as indicated by the rules (page 135), an oxidation number of 0, whereas in the compound aluminum sulfide the oxidation number of aluminum is +3 and that of sulfur is −2. In this reaction, aluminum gains in oxidation number from 0 to +3 and hence is **oxidized;** sulfur loses in oxidation number from 0 to −2 and is **reduced.**

Let us examine this reaction in another way. As it proceeds, each atom of aluminum gives up 3 electrons and each atom of sulfur takes up 2 electrons. In the reaction as a whole, 6 electrons given up by the 2 aluminum atoms are taken up by the 3 sulfur atoms. A transfer of electrons has taken place, and the resultant compound, Al_2S_3, is electrovalent; it is composed of Al^{3+} and S^{2-} ions. In this particular case, aluminum gains in oxidation number or loses electrons, and sulfur loses in oxidation number or gains electrons. It is often convenient to discuss these changes as a gain or loss in electrons as well as changes in oxidation number. To summarize: for reactions involving a transfer of electrons,

> Oxidation is **gain** in oxidation number or **loss** of electrons.
> Reduction is **loss** in oxidation number or **gain** of electrons.

Oxidation and reduction are mutually dependent processes; if electrons are taken up by one substance, they must be given up by another. There can be no oxidation without reduction, no reduction without oxidation.

The union of aluminum and sulfur may be written as taking place in two steps, with e^- representing an electron:

$$2 \text{ Al} \rightarrow 2 \text{ Al}^{3+} + 6 \text{ } e^- \text{ (oxidation)} \tag{1}$$

$$6 \text{ } e^- + 3 \text{ S} \rightarrow 3 \text{ S}^{2-} \text{ (reduction)} \tag{2}$$

By adding the two steps and cancelling the electrons, we obtain the net equation which was written above.

Oxidizing and Reducing Agents

Oxygen, sulfur, or other nonmetals, as they combine with metals, are recognized as **oxidizing agents;** therefore, the metals must be the **reducing agents.** In all oxidation-reduction chemical changes, the following relationship exists:

Oxidizing agent = Electron receiver = Loser in oxidation number
= Substance reduced

Reducing agent = Electron giver = Gainer in oxidation number
= Substance oxidized

In the above example, Al is the element oxidized, while S is the oxidizing agent. Likewise S is reduced, and Al is the reducing agent.

EXAMPLE 16.1   For the redox equation

$$\overset{+5}{2 \text{ HNO}_3} + \overset{-1}{6 \text{ HCl}} \rightarrow \overset{+2}{2 \text{ NO}} + \overset{0}{3 \text{ Cl}_2} + 4 \text{ H}_2\text{O}$$

what is (1) the element oxidized, (2) the element reduced, (3) the oxidizing agent, and (4) the reducing agent?

SOLUTION: Note the oxidation numbers above the elements in the equation which change oxidation numbers (N and Cl).

(1) The *element oxidized* is Cl (-1 to 0).
(2) The *element reduced* is N ($+5$ to $+2$).
(3) The *oxidizing agent* is HNO_3 (since N is reduced). The convention usually used in writing the oxidizing agent or reducing agent is to write the formula of the *whole compound* in the equation rather than that of the element involved, although it is really an individual element which is oxidized or reduced. Hence, in the above equation HNO_3 is usually written as the oxidizing agent, rather than N.
(4) The *reducing agent* is HCl (since Cl^- is oxidized).

16.1   BALANCING

Oxidation-Reduction Equations

To balance these equations, we make use of the rule that the total gain in electrons must equal the total loss in electrons of the elements. Consider the reaction

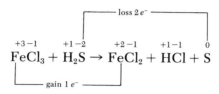

in which the oxidation number of each element is placed above the symbol. Two of the elements undergo a change in oxidation number; each iron atom gains $1 \ e^-$, and each sulfur atom loses $2 \ e^-$. To balance the gain and loss, 2 iron atoms will be required for each sulfur atom, and this will call for a ratio of $2 \ FeCl_3$ to $1 \ H_2S$. This must be the ratio in which these two substances react:

$$2 \ FeCl_3 + 1 \ H_2S \rightarrow 2 \ FeCl_2 + 2 \ HCl + S$$

Having obtained the ratio of the oxidizing agent, $FeCl_3$, to the reducing agent, H_2S, it is easy to complete the balancing of the equation on the right side. In this equation the loss of electrons by sulfur (2) is made the coefficient of $FeCl_3$, while the gain in electrons by iron (1) becomes the coefficient of H_2S.

    In general, we may employ the following steps in balancing oxidation-reduction equations:

1. Find the 2 elements changing oxidation number and indicate the *total change* for each.
2. Make the gain of electrons equal to the loss of electrons by *cross multiplying* and balance these elements on the *other side* of the equation.
3. Complete the balancing by inspection. Often this will include:
     *a. Balance any *remaining metals*.
     *b. Balance any *common anions*.
     c. Balance H and O.
     d. *Check charges* (ionic equations only).

Let us illustrate the outlined steps by balancing an equation:

$$\text{(1)} \quad \overset{+2}{FeCl_2} + \overset{+7}{KMnO_4} + HCl \rightarrow \overset{+3}{FeCl_3} + KCl + \overset{+2}{MnCl_2} + H_2O$$

with loss $1e^-$ from Fe and gain $5e^-$ by Mn.

° Usually absent in ionic equations.

(2) The 2 elements changing oxidation number, Fe and Mn, are balanced.

$$5 \, FeCl_2 + 1 \, KMnO_4 + HCl \rightarrow 5 \, FeCl_3 + KCl + 1 \, MnCl_2 + H_2O$$

(3a) K is balanced.

$$5 \, FeCl_2 + 1 \, KMnO_4 + HCl \rightarrow 5 \, FeCl_3 + 1 \, KCl + 1 \, MnCl_2 + H_2O$$

(3b) Cl is balanced.

$$5 \, FeCl_2 + 1 \, KMnO_4 + 8 \, HCl \rightarrow 5 \, FeCl_3 + 1 \, KCl + 1 \, MnCl_2 + H_2O$$

(3c) H and O are balanced.

$$5 \, FeCl_2 + 1 \, KMnO_4 + 8 \, HCl \rightarrow 5 \, FeCl_3 + 1 \, KCl + 1 \, MnCl_2 + 4 \, H_2O \ \text{(balanced)}$$

Oxidation-reduction equations are more difficult to balance by inspection methods than those which involve no change in oxidation number. The "change in oxidation number" method for balancing oxidation-reduction equations, however, is rapid and accurate. Note these examples:

EXAMPLE 16.2   Balance the redox equation

$$H_2S + KMnO_4 + HCl \rightarrow S + KCl + MnCl_2 + H_2O$$

SOLUTION:

step (1)

$$\overset{-2}{H_2}S + K\overset{+7}{Mn}O_4 + HCl \rightarrow \overset{0}{S} + KCl + \overset{+2}{Mn}Cl_2 + H_2O$$

loss $2 \, e^-$

gain $5 \, e^-$

(2) Mn in $KMnO_4$ appears to gain $5 \, e^-$, while the sulfur atom in H_2S loses $2 \, e^-$. Thus the number 5 is made the coefficient of H_2S, and the number 2 is made the coefficient of $KMnO_4$.

The ratio of H_2S to $KMnO_4$ must be 5 to 2, as determined by the rule of gain and loss of electrons. S and Mn should be balanced immediately on the right side of the equation.

$$5 \, H_2S + 2 \, KMnO_4 + HCl \rightarrow 5 \, S + KCl + 2 \, MnCl_2 + H_2O$$

(3) K, Cl, O, and H are balanced, in this order.

$$5 \, H_2S + 2 \, KMnO_4 + 6 \, HCl \rightarrow 5 \, S + 2 \, KCl + 2 \, MnCl_2 + 8 \, H_2O \ \text{(balanced)}$$

EXAMPLE 16.3   Balance the redox equation

$$K_2Cr_2O_7 + S_8 \rightarrow K_2O + Cr_2O_3 + SO_2$$

SOLUTION:

(gain 6 e⁻ per molecule)

$$\text{K}_2\text{Cr}_2\text{O}_7 + \text{S}_8 \rightarrow \text{K}_2\text{O} + \text{Cr}_2\text{O}_3 + \text{SO}_2$$

gain 3 e⁻ per atom

loss 4 e⁻ per atom

(loss 32 e⁻ per molecule)

Since 2 atoms of Cr are present per molecule of $\text{K}_2\text{Cr}_2\text{O}_7$, the total change per molecule is 6 (3 for each atom). Likewise, the total change for the S_8 molecule is 32 (4 for each S atom) and the ratio of $\text{K}_2\text{Cr}_2\text{O}_7$ to S_8 becomes 6 to 32, or 3 to 16. Generally, if the total change for each of the two elements can be divided by 2, or some other small whole number, it should be done at this point. Otherwise the final completed equation will need to be reduced to smaller numbers.

(2) $16\ \text{K}_2\text{Cr}_2\text{O}_7 + 3\ \text{S}_8 \rightarrow 16\ \text{K}_2\text{O} + 16\ \text{Cr}_2\text{O}_3 + 24\ \text{SO}_2$

EXAMPLE 16.4   Balance the redox equation

$$\text{Cu} + \text{HNO}_3 \rightarrow \text{Cu(NO}_3)_2 + \text{NO} + \text{H}_2\text{O}.$$

SOLUTION:

loss 2 e⁻

$$\overset{0}{\text{Cu}} + \overset{+5}{\text{HNO}_3} \rightarrow \overset{+2}{\text{Cu}}\overset{+5}{(\text{NO}_3)_2} + \overset{+2}{\text{NO}} + \text{H}_2\text{O}$$

no change

gain 3 e⁻

(2) From change in oxidation numbers, the ratio of the reducing agent (Cu) to the oxidizing agent (HNO_3) must be 3 to 2:

$$3\ \text{Cu} + 2\ \text{HNO}_3 \rightarrow 3\ \text{Cu(NO}_3)_2 + 2\ \text{NO} + \text{H}_2\text{O}$$

(3) We note that part of the HNO_3 undergoes no change in oxidation number in forming the salt, copper(II) nitrate; however, $3\ \text{Cu(NO}_3)_2$ calls for 6 more atoms of nitrogen, which must be furnished by the nitric acid (this is the only source of nitrogen atoms). This requires 6 additional molecules of HNO_3 to be added to the 2 molecules which have been reduced to nitric oxide. The complete equation becomes

$$3\ \text{Cu} + 2\ \text{HNO}_3 + 6\ \text{HNO}_3 \rightarrow 3\ \text{Cu(NO}_3)_2 + 2\ \text{NO} + 4\ \text{H}_2\text{O}$$

or

$$3\ \text{Cu} + 8\ \text{HNO}_3 \rightarrow 3\ \text{Cu(NO}_3)_2 + 2\ \text{NO} + 4\ \text{H}_2\text{O}$$

The change in oxidation numbers gives us information only about the parts which undergo oxidation and reduction. No information about the amount of a reagent used in some capacity other than oxidation-reduction is indicated. Thus in the above the ratio of reducing agent to oxidizing agent is 3 to 2, but because part of the nitric acid is used in another capacity, the number of molecules of HNO_3 used in this other capacity must be added to those used for oxidation purposes. It is evident from the equation that only $\frac{1}{4}$ of the HNO_3 (2 molecules of 8) act in an oxidizing capacity; the other $\frac{3}{4}$ act as a salt former (to form copper(II) nitrate).

After a little practice the student may determine the oxidation number changes by inspection and place the correct coefficients in the equation. Occasionally it may be necessary to double the numbers corresponding to the changes; however, the ratio must be maintained, and if one number is doubled, the second must also be doubled. For example, in the reaction

$$\overset{\text{gain } 5e^-}{KMnO_4 + FeSO_4 + H_2SO_4 \rightarrow K_2SO_4 + MnSO_4 + Fe_2(SO_4)_3 + H_2O}$$

loss $1e^-$

an even number of atoms of Fe is necessary on the left side, so that the numbers used in the final balancing must be 10 to 2 (which is the same ratio as 5:1). Thus

$$2\,KMnO_4 + 10\,FeSO_4 + 8\,H_2SO_4 \rightarrow K_2SO_4 + 5\,Fe_2(SO_4)_3 + 2\,MnSO_4 + 8\,H_2O$$

Ionic Oxidation-Reduction Equations

Having balanced the above molecular equation by changes in oxidation numbers, we may then write in ionic form:

$$2\,MnO_4^- + 10\,Fe^{2+} + 16\,H^+ \rightarrow 2\,Mn^{2+} + 10\,Fe^{3+} + 8\,H_2O \qquad (1)$$

Note that the ions not taking part in the chemical change have been omitted; they would merely cancel each other on either side of the equation if written. Other equations in ionic form are

$$3\,Cu + 2\,NO_3^- + 8\,H^+ \rightarrow 3\,Cu^{2+} + 2\,NO + 4\,H_2O$$
$$2\,Fe^{3+} + H_2S \rightarrow 2\,Fe^{2+} + 2\,H^+ + S$$
$$5\,H_2S + 2\,MnO_4^- + 6\,H^+ \rightarrow 5\,S + 2\,Mn^{2+} + 8\,H_2O$$

It is often preferable to write and balance a chemical change involving ions without first writing the molecular equation. The equation is balanced by a consideration of change in oxidation numbers as explained above for molecular equations. For example,

$$3 \text{ Ag} + 4 \text{ H}^+ + \text{NO}_3^- \rightarrow 3 \text{ Ag}^+ + \text{NO} + 2 \text{ H}_2\text{O}$$

with: gain 3 e^- (top), loss 1 e^- (bottom)

or

$$2 \text{ MnO}_4^- + 5 \text{ AsO}_3^{3-} + 6 \text{ H}^+ \rightarrow 2 \text{ Mn}^{2+} + 5 \text{ AsO}_4^{3-} + 3 \text{ H}_2\text{O}$$

with: loss 2 e^- (top), gain 5 e^- (bottom)

Note that in all of these balanced ionic equations the net positive or negative charge on one side of an equation balances that on the other side. For example, in equation (1) above, the net charge on the left is $(2-) + (20+) + (16+) = 34+$, and on the right side is $(4+) + (30+) = 34+$. The equation is said to be balanced *electrically*, as well as atomically. The charges of ionic equations should always be checked as part of the balancing procedure (Step 3d).

More Difficult Oxidation-Reduction Equations

Occasionally an equation is encountered in which more than two elements change in oxidation numbers.

EXAMPLE 1

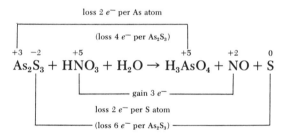

loss 2 e^- per As atom
(loss 4 e^- per As$_2$S$_3$)

$$\overset{+3\ \ -2}{\text{As}_2\text{S}_3} + \overset{+5}{\text{HNO}_3} + \text{H}_2\text{O} \rightarrow \overset{+5}{\text{H}_3\text{AsO}_4} + \overset{+2}{\text{NO}} + \overset{0}{\text{S}}$$

gain 3 e^-
loss 2 e^- per S atom
(loss 6 e^- per As$_2$S$_3$)

Note that both As and S atoms in As_2S_3 are oxidized—each As atom from +3 to +5 and each S atom from −2 to 0. Since there are 2 As atoms per As_2S_3 molecule, the total loss in As is $2 \times 2\ e^- = 4\ e^-$. The total loss in S in As_2S_3 is $3 \times 2\ e^- = 6\ e^-$. The net total loss in As_2S_3 is $(4e^-) + (6\ e^-) = 10\ e^-$. The change in N in HNO_3 is simply +5 to +2, a gain of $3\ e^-$. Hence the ratio of As_2S_3 to HNO_3 is 3 to 10. On balancing the hydrogen and oxygen atoms, the equation becomes

$$3 \text{ As}_2\text{S}_3 + 10 \text{ HNO}_3 + 4 \text{ H}_2\text{O} \rightarrow 6 \text{ H}_3\text{AsO}_4 + 10 \text{ NO} + 9 \text{ S}$$

To arrive at the proper coefficients for equations of this type, it is necessary only to determine the total net change of electrons in each molecule of reacting substances.

EXAMPLE 2

$$\overset{+2\quad +5}{Mn(NO_3)_2} + KOH + \overset{0}{S} \rightarrow \overset{+6}{K_2MnO_4} + \overset{+6}{K_2SO_4} + \overset{+2}{NO} + H_2O$$

loss 6 e^-

loss 4 e^-

gain 3e^-

(gain 6 e^- per formula)

Net change in $Mn(NO_3)_2$ = loss 4 e^- + gain 6 e^- = gain 2 e^-
Net change in S   = loss 6 e^-

This gives a ratio of 6 to 2, or 3 to 1; therefore,

$$3\ Mn(NO_3)_2 + 8\ KOH + S \rightarrow 3\ K_2MnO_4 + K_2SO_4 + 6\ NO + 4\ H_2O$$

The Step Ion-Electron Method of Balancing Equations

This method is applicable only to ionic reactions, but since most reactions in solution are between ions, the method proves convenient and useful. The method takes into account only those ions in solution which actually enter into reaction; in other words, only the net reaction is shown. Those ions which remain unaltered in the solution do not appear in the equation. The oxidation-reduction equation is broken down into two partial equations or steps, called **half reactions;** one equation represents the oxidation step, the other the reduction step. The number of electrons transferred in each step is indicated, and each equation is balanced both atomically and electrically. The gain and loss of electrons is then balanced by multiplying each step by the appropriate number:

EXAMPLE 1

$$FeCl_3 + SnCl_2 \rightarrow FeCl_2 + SnCl_4 \text{ (not balanced)}$$

From an ionic standpoint, we should write down only the ions which enter into the chemical change: Fe^{3+}, which is reduced to Fe^{2+}; and Sn^{2+}, which is oxidized to Sn^{4+}. The two half-reactions are

$$Sn^{2+} \rightarrow Sn^{4+} + 2\ e^- \text{ (oxidation)}$$
$$Fe^{3+} + e^- \rightarrow Fe^{2+} \text{ (reduction)}$$

Each step is now balanced electrically as well as atomically. Since 2 electrons appear in the first step, we may balance electrons by multiplying the second equation by 2:

$$2\ Fe^{3+} + 2\ e^- \rightarrow 2\ Fe^{2+}$$

Now the two steps may be added (electrons are cancelled), and we obtain the net ionic reaction

$$Sn^{2+} \rightarrow Sn^{4+} + 2e$$
$$\underline{2\ Fe^{3+} + 2e \rightarrow 2\ Fe^{2+}}$$
$$Sn^{2+} + 2\ Fe^{3+} \rightarrow Sn^{4+} + 2\ Fe^{2+}$$

Note that the final step is also balanced electrically; the net charge on each side of the equation is +8.

EXAMPLE 2   Copper reacts with dilute nitric acid to form copper(II) ions, water, and nitric oxide. If we apply the ion-electron method to this reaction, copper is oxidized from the free state to copper(II) ion:

$$Cu \rightarrow Cu^{2+} + 2\ e^- \text{ (oxidation)} \qquad (1)$$

Meanwhile, nitric acid is reduced to NO and H_2O, the latter two substances being un-ionized:

$$H^+ + NO_3^- \rightarrow NO + H_2O$$

The latter equation is balanced atomically as follows:

$$4\ H^+ + NO_3^- \rightarrow NO + 2\ H_2O$$

This equation is now balanced atomically but not electrically, since the net charge on the left is +3, and the net charge on the right side is 0. The addition of 3 e^- to the left side will balance the step electrically:

$$3\ e^- + 4\ H^+ + NO_3^- \rightarrow NO + 2\ H_2O \qquad (2)$$

To balance the gain and loss of electrons, equation (1) may be multiplied by 3 and equation (2) by 2. Then

$$3\ Cu \rightarrow 3\ Cu^{2+} + 6e$$
$$\underline{6e + 8\ H^+ + 2\ NO_3^- \rightarrow 2\ NO + 4\ H_2O}$$
$$3\ Cu + 8\ H^+ + 2\ NO_3^- \rightarrow 3\ Cu^{2+} + 2\ NO + 4\ H_2O$$

EXAMPLE 3   It is further assumed in the ion-electron method that H^+ ions and H_2O molecules may be used to balance each half-reaction atomically if the reaction is being carried out in acid solution, or OH^- ions and H_2O molecules may be used if the reaction is in basic solution. To illustrate, permanganate ion will oxidize sulfide ion to free sulfur in acid solution:

$$S^{2-} \rightarrow S + 2\ e^- \qquad (1)$$

and the permanganate ion is reduced to manganese(II) ion:

$$MnO_4^- \rightarrow Mn^{2+}$$

To balance the latter step atomically, we may add 8 hydrogen ions to the left side, in which case 4 molecules of water will appear on the right side:

$$8 \text{ H}^+ + \text{MnO}_4^- \rightarrow \text{Mn}^{2+} + 4 \text{ H}_2\text{O}$$

5 electrons are then necessary on the left side to balance electrically:

$$5 \text{ } e^- + 8 \text{ H}^+ + \text{MnO}_4^- \rightarrow \text{Mn}^{2+} + 4 \text{ H}_2\text{O} \tag{2}$$

Combining steps (1) and (2) and balancing gain and loss of electrons.

$$5 \text{ S}^{2-} \rightarrow 5 \text{ S} + \cancel{10 \text{ } e^-}$$
$$\cancel{10 \text{ } e^-} + 16 \text{ H}^+ + 2 \text{ MnO}_4^- \rightarrow 2 \text{ Mn}^{2+} + 8 \text{ H}_2\text{O}$$
$$\overline{5 \text{ S}^{2-} + 16 \text{ H}^+ + 2 \text{ MnO}_4^- \rightarrow 5 \text{ S} + 2 \text{ Mn}^{2+} + 8 \text{ H}_2\text{O}}$$

16.2 PREDICTING PRODUCTS OF OXIDATION-REDUCTION REACTIONS

An important objective of the chemist is to learn how to predict the probable results of chemical changes. We cannot summarize here all the information needed to predict chemical behavior over the broad field of inorganic chemistry; however, certain basic patterns of behavior may be outlined from which some reasonable predictions may be made.

Oxidizing and reducing agents usually behave in fairly definite ways, and if we learn these patterns of behavior for the common oxidizing and reducing agents, we should be able to predict many reactions with a better than fair degree of success. Thus concentrated nitric acid, in acting as an oxidizing agent, is usually reduced to NO_2 and H_2O, whereas dilute nitric acid usually yields NO and H_2O. Similarly, permanganate ion (MnO_4^-), which is a common laboratory oxidizing agent ($KMnO_4$), yields Mn^{2+} and H_2O when the oxidation-reduction reaction is carried out in acid solution, or MnO_2 and H_2O in a neutral or basic solution. Some of the common oxidizing agents and their usual reduction products are tabulated below.

Oxidizing Agent	Products
HNO_3 (conc)	$NO_2 + H_2O$
HNO_3 (dilute)	$NO + H_2O$
MnO_4^- (acid soln)	$Mn^{2+} + H_2O$
MnO_4^- (basic soln)	MnO_2
$Cr_2O_7^{2-}$ or CrO_4^{2-}	$Cr^{3+} + H_2O$
F_2, Cl_2, Br_2, I_2	F^-, Cl^-, Br^-, I^-
Fe^{3+}	Fe^{2+}
MnO_2	Mn^{2+}
$KClO_3$	KCl

Reducing agents also follow fairly definite patterns. The elements, particularly the metals, constitute the largest single group of reducing agents. Nearly all may be converted into oxides, and these changes all represent reducing properties of the elements. If an element like Na, K, Ca, Mg, or Al exhibits a single valence, the element can form only a compound in which that particular valence is shown – that is, these elements may exhibit only one oxidation state. If more than one oxidation state is possible, the element will generally acquire the higher oxidation state whenever it undergoes oxidation – thus iron is likely to be oxidized to Fe^{3+} rather than to Fe^{2+}. Some of the more common reducing agents and their oxidation products are tabulated below.

Reducing Agents	*Products*
Metals	Metallic ions (cations)
H_2S	S, or possibly SO_2 or SO_4^{2-}
S	SO_2, SO_3^{2-} or SO_4^{2-}
HCl, HBr, HI	Free halogen
Fe^{2+}	Fe^{3+}
Sn^{2+}	Sn^{4+}
$C_2O_4^{2-}$ (oxalate)	$CO_2 + H_2O$

To illustrate the use of this summarized data on oxidizing and reducing agents, consider a reaction in acid solution between oxalate ion ($C_2O_4^{2-}$) and permanganate ion (MnO_4^-). From the tables above we note that the products are CO_2, Mn^{2+}, and H_2O; hence

$$5\ C_2O_4^{2-} + 2\ MnO_4^- + 16\ H^+ \rightarrow 10\ CO_2 + 2\ Mn^{2+} + 8\ H_2O$$

Or consider the reaction of H_2S with concentrated HNO_3. We may represent the reaction as producing S, NO_2 and H_2O.

$$H_2S + 2\ HNO_3 \rightarrow S + 2\ NO_2 + 2\ H_2O$$

With excess HNO_3 and heat, sulfur perhaps would be oxidized further, and we might reasonably expect something like this:

$$H_2S + 8\ HNO_3 \rightarrow H_2SO_4 + 8\ NO_2 + 4\ H_2O$$

Whether or not a reaction will proceed between an oxidizing agent and a reducing agent depends primarily on what we call **oxidation potentials,** a subject which we shall develop in some detail in Chapter 17. Meanwhile, the general information recorded above will allow us to make some reasonable predictions regarding oxidation-reduction processes.

16.3 EQUIVALENT WEIGHTS OF OXIDIZING AND REDUCING AGENTS

As stated on page 289, the equivalent weight of oxidizing and reducing agents depends upon the total oxidation number change or the number of electrons transferred per mole in the reaction. For example, in the following:

$$10\ FeSO_4 + 8\ H_2SO_4 + 2\ KMnO_4 \rightarrow 5\ Fe_2(SO_4)_3 + K_2SO_4 + 2\ MnSO_4 + 8\ H_2O$$

(or ionically [and divide by 2])

$$5\ Fe^{2+} + 8\ H^+ + MnO_4^- \rightarrow 5\ Fe^{3+} + Mn^{2+} + 4\ H_2O$$

we obtain the equivalent weight of the oxidizing or reducing agent by simply dividing the molecular weight by the number of electrons *transferred* per mole. In the above, the equivalent weight of $KMnO_4$ is MW/5 or $\frac{158}{5} = 31.6$ and the equivalent weight of $FeSO_4$ is MW/1 or $\frac{152}{1} = 152$.

One must know the reaction taking place to obtain equivalent weights since the number of electrons transferred per mole in redox reactions may vary. For example, in basic solution, $KMnO_4$ is converted to MnO_2, a change of only 3 electrons per mole. Hence if $KMnO_4$ reacts in basic solution, the equivalent weight is MW/3 or $\frac{158}{3} = 52.7$, instead of 31.6 as was calculated in the equation above for the reaction of $KMnO_4$ in acid solution. But once the equation for the reaction is known, the electron change is easily ascertained for both the oxidizing and reducing agent. Then it is simply a matter of dividing molecular weights by the number of electrons transferred per mole to obtain the equivalent weights.

$$\text{Equivalent weight} = \frac{\text{molecular weight}}{\text{total oxidation number change}}$$

Exercises

16.1. From the standpoint of oxidation numbers, define the following terms: (a) oxidation, (b) reduction, (c) oxidizing agent, (d) reducing agent.

16.2. State the oxidation number of:
(a) Sb in Sb_4O_6
(b) Ti in $K_2Ti_2O_5$
(c) I in H_5IO_6
(d) S in $S_2O_5Cl_2$
(e) Fe in $BaFeO_4$
(f) N in N_2H_4
(g) N in O_2NF
(h) Ge in Mg_2GeO_4

16.3. State the oxidation number of:
(a) Pb in $PbCl_6^{2-}$
(b) Sn in $Sn_2F_5^{-}$
(c) Re in ReO_4^{-}
(d) Xe in $HXeO_4^{-}$
(e) Bi in BiO^{+}
(f) N in $(NH_3OH)^{+}$
(g) Mo in $(Mo_6Cl_8)^{4+}$
(h) W in $(H_2W_{12}O_{40})^{6-}$

16.4. Consider the following redox equation:

$$5\ Mo_2O_3 + 6\ KMnO_4 + 9\ H_2SO_4 \rightarrow 10\ MoO_3 + 3\ K_2SO_4 + 6\ MnSO_4 + 9\ H_2O$$

(a) What is the element oxidized?
(b) Which element is reduced?
(c) Which compound is the oxidizing agent?
(d) Which compound is the reducing agent?

16.5. Answer the questions below concerning the following redox reaction:

$$4\ AsCl_3 + 6\ H_3PO_3 + 6\ H_2O \rightarrow As_4 + 6\ H_3PO_4 + 12\ HCl$$

(a) Give the symbol of the element oxidized.
(b) Give the symbol of the element reduced.

(c) Give the formula of the compound which is the reducing agent.

(d) Give the formula of the compound which is the oxidizing agent.

16.6. Balance the following redox equations:

(a) $Cr(OH)_3 + Cl_2 + KOH \rightarrow K_2CrO_4 + KCl + H_2O$

(b) $I^- + Br_2 + H_2O \rightarrow IO_3^- + H^+ + Br^-$

16.7. Balance the following redox equations:

(a) $Mn(OH)_2 + KNO_3 + K_2CO_3 \rightarrow K_2MnO_4 + NO + CO_2 + H_2O$

(b) $MnO_4^- + H^+ + C_2O_4^{2-} \rightarrow Mn^{2+} + H_2O + CO_2$

16.8. From a consideration of change of oxidation number or electron transfer, balance the following equations and indicate the element oxidized and the element reduced.

(a) $Ag_2S + HNO_3 \rightarrow Ag_2SO_4 + NO_2 + H_2O$

(b) $S + HNO_3 \rightarrow SO_2 + NO_2 + H_2O$

(c) $FeBr_2 + Br_2 \rightarrow FeBr_3$

(d) $As_2O_3 + HIO_3 + H_2O \rightarrow H_3AsO_4 + I_2$

(e) $Mn(OH)_2 + KNO_3 + K_2CO_3 \rightarrow K_2MnO_4 + NO + CO_2 + H_2O$

(f) $AuCl_3 + H_2C_2O_4 \rightarrow Au + CO_2 + HCl$

(g) $SO_2 + AuCl_3 + H_2O \rightarrow H_2SO_4 + HCl + Au$

(h) $Pt + HCl + HNO_3 \rightarrow H_2PtCl_6 + NO + H_2O$

(i) $SnCl_2 + SO_2 + HCl \rightarrow SnCl_4 + SnS_2 + H_2O$

(j) $AuCl_3 + Sb_2O_3 + H_2O \rightarrow Au + Sb_2O_5 + HCl$

(k) $SbH_3 + S \rightarrow Sb_2S_3 + H_2S$

(l) $Sb_2O_3 + Zn + H_2SO_4 \rightarrow ZnSO_4 + SbH_3 + H_2O$

(m) $Ag_3Sb + AgNO_3 + H_2O \rightarrow Ag + H_3SbO_3 + HNO_3$

(n) $AsCl_3 + H_3PO_2 + H_2O \rightarrow As_4 + H_3PO_4 + HCl$

(o) $Mo_2O_3 + KMnO_4 + H_2SO_4 \rightarrow MoO_3 + K_2SO_4 + MnSO_4 + H_2O$

16.9. Write balanced equations for:

(a) Arsenic pentasulfide plus nitric acid yielding arsenic acid (H_3AsO_4), sulphur, nitrogen dioxide, and water.

(b) Hydrogen sulfide plus potassium dichromate plus hydrochloric acid to yield sulphur, chromium(III) chloride, potassium chloride, and water.

(c) Manganese dioxide plus lead dioxide plus nitric acid to yield permanganic acid ($HMnO_4$), lead nitrate, and water.

16.10. Balance the following equations for oxidation-reduction reactions by the oxidation-number method:

(a) $H_2O + MnO_2^- + ClO_2^- \rightarrow MnO_2 + ClO_4^- + OH^-$

(b) $H^+ + Cr_2O_7^{2-} + H_2S \rightarrow Cr^{3+} + S + H_2O$

(c) $H^+ + IO_3^- + SO_3^{2-} \rightarrow I_2 + SO_4^{2-} + H_2O$

(d) $H_2O + P_4 + HOCl \rightarrow H_3PO_4 + Cl^- + H^+$

(e) $OH^- + Cl_2 \rightarrow ClO_3^- + Cl^- + H_2O$

16.11. Balance the following ionic equations considering change in oxidation number or electron transfer.

(a) $CrO_4^{2-} + H^+ + Cl^- \rightarrow Cr^{3+} + Cl_2 + H_2O$

(b) $Cd + H^+ + NO_3^- \rightarrow Cd^{2+} + NO + H_2O$

(c) $AsO_3^{2-} + H^+ + MnO_4^- \rightarrow AsO_4^{3-} + Mn^{2+} + H_2O$
(d) $Cr_2O_7^{2-} + CH_4 + H^+ \rightarrow Cr^{3+} + CO_2 + H_2O$
(e) $Mn^{2+} + BiO_3^- + H^+ \rightarrow HMnO_4 + Bi^{3+} + H_2O$
(f) $SO_3^{2-} + MnO_4^- + H^+ \rightarrow HSO_4^- + Mn^{2+} + H_2O$
(g) $Ag + H^+ + NO_3^- \rightarrow Ag^+ + NO + H_2O$
(h) $MnO_4^- + H^+ + C_2O_4^{2-} \rightarrow Mn^{2+} + H_2O + CO_2$
(i) $Sn^{2+} + MnO_4^- + H_2O \rightarrow SnO_2 + Mn^{2+} + H^+$
(j) $Fe(OH)_2 + MnO_4^- + H_2O \rightarrow Fe(OH)_3 + MnO_2 + OH^-$
(k) $Au(CN)_2^- + Zn \rightarrow Au + Zn(CN)_4^{2-}$
(l) $As_2O_3 + Zn + H^+ \rightarrow AsH_3 + Zn^{2+} + H_2O$

16.12. Complete and balance the following equations in the steps indicated, using the ion-electron method; then add the two half reactions to obtain the complete single-step equation. Make cancellations wherever possible.

(a) $Al \rightarrow Al^{3+}$
$Cl_2 \rightarrow Cl^-$
(b) $F_2 \rightarrow F^-$
$Br^- \rightarrow Br_2$
(c) $CuS \rightarrow Cu^{2+} + S$
$H^+ + NO_3^- \rightarrow H_2O + NO$
(d) $MnO_2 + H_2O \rightarrow MnO_4^- + H^+$
$H^+ + BiO_3^- \rightarrow H_2O + Bi^{3+}$
(e) $AsO_4^{3-} + H^+ \rightarrow AsH_3 + H_2O$
$Zn \rightarrow Zn^{2+}$

(f) $Bi \rightarrow Bi^{3+}$
$H^+ + NO_3^- \rightarrow H_2O + NO$
(g) $C_2O_4^{2-} \rightarrow CO_2$
$H^+ + Cr_2O_7^{2-} \rightarrow H_2O + Cr^{3+}$
(h) $Bi(OH)_3 \rightarrow Bi + OH^-$
$OH^- + HSnO_2^- \rightarrow H_2O + HSnO_3^-$
(i) $Fe(OH)_2 + OH^- \rightarrow Fe(OH)_3$
$MnO_2 + H_2O \rightarrow Mn(OH)_2 + OH^-$
(j) $AsO_2^- + OH^- \rightarrow AsO_3^- + H_2O$
$MnO_4^{2-} + H_2O \rightarrow MnO_2 + OH^-$

16.13. The following ionic equations are to be balanced. Break the reaction down into its two half reactions, the oxidation part and the reduction part, and balance each part by the ion-electron method. Where necessary, add H^+, OH^-, or H_2O. (Remember that H^+ and OH^- are not compatible and must not be used in the presence of one another.) Then add the two parts to get the single-step equation. Make cancellations wherever possible.

(a) $Sn + H^+ + NO_3^- \rightarrow SnO_2 + NO_2$
(b) $Br_2 + Cl_2 + H_2O \rightarrow BrO_3^- + H^+ + Cl^-$
(c) $CoS + H^+ + NO_3^- + Cl^- \rightarrow Co^{2+} + S + NOCl$
(d) $Cr_2O_7^{2-} + H_2O_2 + H^+ \rightarrow Cr^{3+} + O_2 + H_2O$
(e) $Au + CN^- + O_2 + H_2O \rightarrow Au(CN)_2^- + OH^-$
(f) $Ag^+ + AsH_3 + H_2O \rightarrow Ag + H_3AsO_3 + H^+$
(g) $AsO_3^{3-} + Cu^{2+} + OH^- \rightarrow AsO_4^{3-} + Cu_2O + H_2O$
(h) $As_2O_3 + BrO_3^- + H_2O \rightarrow H_3AsO_4 + Br^-$
(i) $Cr_2O_7^- + I^- + H^+ \rightarrow Cr^{3+} + I_2 + H_2O$
(j) $Mn^{2+} + ClO_3^- + H_2O \rightarrow MnO_2 + Cl_2 + H^+$
(k) $MnO_4^- + S_2O_3^{2-} + H_2O \rightarrow SO_4^{2-} + MnO_2 + OH^-$
(l) $S_2O_8^{2-} + Mn^{2+} + H_2O \rightarrow HMnO_4 + HSO_4^-$
(m) $MnO_4^- + NH_4OH \rightarrow MnO_2 + N_2 + H_2O$

16.14. Complete and balance the following equations for oxidation-reduction reactions by the ion-electron method.

(a) $ReO_2 + Cl_2 + H_2O \rightarrow HReO_4 + Cl^- + H^+$
(b) $HgI_4^{2-} + N_2H_4 \rightarrow Hg + I^- + N_2 + H^+$

(c) $H^+ + Te + NO_3^- \rightarrow TeO_2 + NO + H_2O$

(d) $H^+ + UO^{2+} + Cr_2O_7^{2+} \rightarrow UO_2^{2+} + Cr^{2+} + H_2O$

(e) $H^+ + Zn + H_2MoO_4 \rightarrow Zn^{2+} + Mo^{3+} + H_2O$

(f) $AsH_3 + Ag^+ + H_2O \rightarrow As_4O_6 + Ag + H^+$

(g) $H^+ + MnO_4^- + HCN + I^- \rightarrow Mn^{2+} + ICN + H_2O$

16.15. Balance the equations given below. Note that three elements change in oxidation number.

(a) $MnI_2 + HNO_3 + NaBiO_3 \rightarrow HMnO_4 + I_2 + Bi(NO_3)_3 + NaNO_3 + H_2O$

(b) $CrBr_3 + NaOH + Cl_2 \rightarrow Na_2CrO_4 + NaBrO + NaCl + H_2O$

(c) $As_2S_3 + MnO_2 + K_2CO_3 \rightarrow K_3AsO_4 + K_2SO_4 + MnO + CO_2$

(d) $Ag_3AsO_4 + Zn + H_2SO_4 \rightarrow AsH_3 + Ag + ZnSO_4 + H_2O$

(e) $MnBr_2 + PbO_2 + HNO_3 \rightarrow HMnO_4 + Pb(BrO_3)_2 + Pb(NO_3)_2 + H_2O$

(f) $Cr_2O_3 + Mn(NO_3)_2 + Na_2CO_3 \rightarrow Na_2CrO_4 + Na_2MnO_4 + NO + CO_2$

16.16. Balance the following equations by any method (Add H_2O where needed):

(a) $Sn + HNO_3 \rightarrow SnO_2 + NO + H_2O$

(b) $Mn(OH)_2 + BiO_3^- + H^+ \rightarrow HMnO_4 + Bi^{3+} + H_2O$

(c) $HgS + HNO_3 + HCl \rightarrow HgCl_2 + NOCl + H_2O$

(d) $F_2 + H_2O \rightarrow HF + O_3$

(e) $Ni(CN)_4^{2-} + Br_2 + OH^- \rightarrow NiO_2 + Br^- + CN^- + H_2O$

(f) $MnO_4^- + U^{4+} + H_2O \rightarrow UO_2^{2+} + H^+ + Mn^{2+}$

(g) $Cr_2O_7^{2-} + H^+ + C_2H_5OH \rightarrow Cr^{3+} + HC_2H_3O_2 + H_2O$

(h) $As + H^+ + NO_3^- \rightarrow AsO_4^{3-} + NO_2 + H_2O$

(i) $As_2S_3 + H^+ + NO_3^- \rightarrow AsO_4^{3-} + SO_4^{2-} + NO_2 + H_2O$

(j) $As + BrO^- + OH^- \rightarrow AsO_4^{3-} + Br^-$

(k) $S + NO_3^- + H^+ \rightarrow HSO_4^- + NO_2$

(l) $Zn + NO_3^- + H^+ \rightarrow Zn^{2+} + NH_4^+$

(m) $Mn^{2+} + ClO_3^- \rightarrow MnO_2 + ClO_2$

(n) $Co(OH)_2 + O_2 + H_2O \rightarrow Co(OH)_3$

(o) $Cr^{3+} + ClO_3^- + H_2O \rightarrow Cr_2O_7^{2-} + ClO_2 + H^+$

(p) $CNS^- + NO_3^- + H^+ \rightarrow CO_2 + SO_4^{2-} + NO$

(q) $ClO_3^- + NO_2^- \rightarrow NO_3^- + Cl^-$

(r) $Cl_2 + OH^- \rightarrow ClO_3^- + Cl^-$

16.17. Predict the products of the following reactions; then complete and balance each one.

(a) $Cl^- + H^+ + MnO_2 \rightarrow$

(b) $C + HNO_3(conc) \rightarrow$

(c) $AsO_2^- + Cr_2O_7^{2-} + H^+ \rightarrow$

(d) $Ag + HNO_3(dil) \rightarrow$

(e) $C_2O_4^{2-} + H^+ + Cr_2O_7^{2-} \rightarrow$

(f) $C_2H_5OH + H^+ + Cr_2O_7^{2-} \rightarrow C_2H_4O_2$

(g) $PbS + HNO_3(dil) \rightarrow$

(h) $H_2S + MnO_4^- \rightarrow OH^- + S +$

(i) $Cu + HNO_3(conc) \rightarrow$

16.18. What fraction of a mole is the equivalent weight of each of the following substances?

(a) H_5IO_6 in a reaction in which IO_3^- is produced.
(b) H_5IO_6 in a reaction in which $Na_3H_2IO_6$ is produced.
(c) $K_2Cr_2O_7$ in a reaction in which Cr^{3+} is produced.
(d) $Ca(OCl)_2$ in a reaction in which Cl^- is produced.
(e) $K_4[Fe(CN)_6]$ in a reaction in which $K_3[Fe(CN)_6]$ is produced.

16.19. What fraction of a mole is the equivalent weight of each of the following substances?
(a) H_3PO_3 in a reaction in which $CaHPO_3$ is produced.
(b) KIO_3 in a reaction in which I^- is produced.
(c) $Na_2S_2O_3$ in a reaction in which $S_4O_6^{2-}$ is produced.
(d) Mo_2O_3 in a reaction in which H_2MoO_4 is produced.
(e) CaC_2O_4 in a reaction in which CO_2 is produced.

16.20. Determine equivalent weights of the oxidizing agent and reducing agent in each of the reactions in exercise 8.

16.21. A mixture of calcium and magnesium carbonates is analyzed for calcium by precipitation of CaC_2O_4 followed by titration with $KMnO_4$ solution. Reactions are

$$Ca^{2+} + C_2O_4^{2-} \rightarrow CaC_2O_4(s)$$

$$5\ CaC_2O_4 + 8\ H_2SO_4 + 2\ KMnO_4 \rightarrow 5\ CaSO_4 + 10\ CO_2$$
$$+ 2\ MnSO_4 + K_2SO_4 + 8\ H_2O$$

If a 1.420-g sample of the mixture requires 37.22 ml of 0.1120 N $KMnO_4$ for the titration, calculate the percentage of $CaCO_3$ in the sample.

16.22. A 4.00-g sample of an ore of Bi_2S_3 was reacted with acid and a quantity of H_2S released equivalent to the amount of Bi_2S_3 present in the ore. The H_2S was reacted with I_2 yielding S and I^-; the complete reaction required 2.85 g of I_2. (a) What are the equivalent weights of H_2S and I_2 for the second reaction? (b) What weight of H_2S is equivalent to 2.85 g of I_2? (c) What percent of the ore is Bi_2S_3?

Supplementary Readings

16.1. "Interpretation of Oxidation-Reduction," M. P. Goodstein, *J. Chem. Educ.,* **47**, 452 (1970).
16.2. "The Stoichiometry of an Oxidation-Reduction Reaction," W. C. Child, Jr., and R. W. Ramette, *J. Chem. Educ.,* **44**, 109 (1967).

Chapter 17

Electro-Chemistry

17.1 THE NATURE OF ELECTROCHEMISTRY

Electrochemistry is the division of chemistry which involves (1) use of an electric current to cause chemical changes, and (2) generation of an electric current by chemical change. Applied electrochemistry is important in the manufacture of many industrial products. For example, hydrogen and oxygen may be produced by electrolysis of water; electrolysis of an aqueous solution of sodium chloride yields hydrogen, sodium hydroxide, and chlorine; many metals are purified by electrodeposition; aluminum and the alkaline earth metals are produced by electrolysis of nonaqueous solutions of their respective compounds. Electrochemical cells or batteries are widely used by individuals and in industrial operations.

Conduction of Electric Current

Electric current may be carried by either of two types of conductors: (1) a metallic conductor, such as a piece of copper wire, or (2) a solution of an electrolyte — for example, a solution of sodium chloride in water. In a metallic conductor, electricity is carried simply by movement of electrons through the metal, while in an electrolyte the current is carried by the ions in solution moving toward the electrodes of opposite charge. In the latter case, since an ion is a charged particle of matter, the passage of current involves the transport of matter, which is essentially different from conduction of metals.

Electrolysis

This process consists of passing a direct current through a solution of an electrolyte between two electrodes° causing chemical changes to take place.

In the electrolysis of aqueous hydrogen chloride (hydrochloric acid), for example, the principal ions in the solution are H_3O^+ (H^+) and Cl^- (Figure 17.1). As direct current is passed chloride ions (anions) move toward the anode, where they give up their electrons. These electrons move through the metallic conductor and through the battery or generator of the electric current to the cathode, where hydrogen ions (cations) are discharged by taking up these electrons. The battery or generator is simply a device for moving electrons from one electrode to the other. Since the chemical change occurring involves the transfer of electrons, oxidation-reduction takes place. At the anode, electrons are given up by anions — *oxidation takes place at the anode;* at the cathode, electrons are taken up by cations — *reduction takes place at the cathode.*

° An electrode is a solid conductor immersed in a conducting solution.

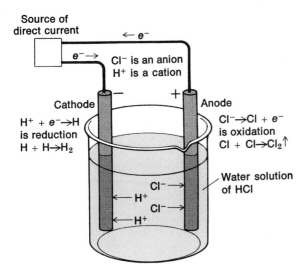

Figure 17.1
Electrolysis of water solution
of HCl.

The **anode half reaction** (**oxidation** occurs, electrons **out**) is

$$2 \text{ Cl}^- \rightarrow \text{Cl}_2 + 2 \ e^-$$

The **cathode half reaction** (**reduction** occurs, electrons **in**) is

$$2 \text{ H}^+ + 2 \ e^- \rightarrow \text{H}_2$$

The overall reaction is obtained by properly adding the two half-reactions and cancelling the electrons.

$$2 \text{ H}^+ + 2 \text{ Cl}^- \rightarrow \text{H}_2 + \text{Cl}_2$$

Salts may also undergo electrolysis, for example common table salt, NaCl, may be melted in a suitable container and nonreactive electrodes immersed. Mentally substitute in Figure 17.1 Na^+ for H^+; there will be no water. The fluid mixture of mobile Na^+ and Cl^-, prior to application of electrical current is neutral as to totality of charge (equal number of Na^+ and Cl^-). After a source of direct current adds electrons at the cathode and withdraws them from the anode electrolysis takes place. Cl_2 gas is the anode product ($2 \text{ Cl}^- \rightarrow \text{Cl}_2 + 2 \ e^-$); liquid sodium metal is the cathode product ($2 \text{ Na}^+ + 2 \ e^- \rightarrow 2 \text{ Na}$).

The electrolysis of *aqueous NaCl solution* is a more complex example, since there are present not only Na^+ and Cl^-, but also H_2O molecules and a relatively few H^+ and OH^- from the ionization of water. The possible **anode** half reactions are

(1) $2 \text{ Cl}^- \rightarrow \text{Cl}_2 + 2 \ e^-$
(2) $2 \text{ H}_2\text{O} \rightarrow \text{O}_2 + 4 \text{ H}^+ + 4 \ e^-$
(3) $2 \text{ OH}^- \rightarrow \text{O}_2 + 2 \text{ H}^+ + 4 \ e^-$

Since Cl_2 is observed experimentally at the anode, reaction (1) is the half reaction which actually occurs.

The possible *cathode* half reactions are

(1) $Na^+ + e^- \rightarrow Na$
(2) $2 H_2O + 2 e^- \rightarrow H_2 + 2 OH^-$
(3) $2 H^+ + 2 e^- \rightarrow H_2$

Experimental studies show that (2) is the half reaction which really occurs in preference to (1) or (3).

The over-all reaction for the electrolysis of aqueous NaCl is

$$2 Cl^- \rightarrow Cl_2 + 2 e^-$$
$$\underline{2 H_2O + \quad 2 e^- \rightarrow H_2 + 2 OH^-}$$
$$2 Cl^- + 2 H_2O \rightarrow Cl_2 + H_2 + 2 OH^-$$

anode   cathode
oxidation reduction

The anode product is chlorine gas, the cathode product is hydrogen gas; the Na^+ ion concentration remains unchanged, but an OH^- ion concentration builds up in the electrolyte; some water is used up. The electrolysis of $NaCl-H_2O$ brine is a big industrial process with products as specified above; chlorine and hydrogen gases and sodium hydroxide (Na^+, OH^-, caustic soda).

Electrical Units

When a current passes through a conductor a certain **resistance** to its passage is offered. The resistance depends on the nature of the conductor. The greater the resistance to a conductor, the poorer the conductance; thus water offers a high resistance to passage and is a poor conductor, and hydrochloric acid has low resistance and is a good conductor. Copper is a good conductor, sulfur practically a nonconductor. The SI derived unit of electrical resistance R is the **ohm** (Ω).

The *SI* base unit of electric current (I) is the **ampere** (A). The SI derived unit quantity of electricity (Q) is the **coulomb** (C) defined as **a current of one ampere flowing for one second (an ampere-second).**

For a current to pass from one point to another through a conductor, a **difference of potential** or **electromotive force** (*emf* or E) must exist. Just as water will flow only from a higher to a lower level, so electricity will pass only from a region of high potential to one of lower potential. The SI derived unit of electrical potential, or electromotive force, is the **volt** (V) — the potential which will drive a current of one ampere through a resistance of one ohm. The relationships between these units are:

$$Q \text{ (coulombs)} = I \text{ (amperes) } t \text{ (seconds)} \quad \text{or} \quad I = Q/t \qquad (1)$$

Ohm's law is described by the formula relation between I (amperes), E (volts), and R (ohms).

$$I = E/R \quad \text{or} \quad E = IR \qquad (2)$$

Resistance to electrical current flow in a wire increases with temperature, probably due to increased vibrations of the metal atoms-ions about their lattice positions. At progressively lower temperatures metal conductance increases. The conductance of an electrolytic solution depends on the mobility of the ions. Factors relating to their mobility are: interionic attractions (which are greater in concentrated solutions than dilute), solvation of ions, viscosity of the solvent, ionic charge. With increased temperature the kinetic energy of the ions is increased and less resistance is offered to their movement. In an electrolytic solution there is an equal charge of anions and cations to give electrical neutrality. During electrolysis there is some concentration built up near the respective electrodes—of anions at the anode and cations about the cathode.

Faraday's Laws

In 1833 Michael Faraday, an English scientist, discovered that during electrolysis (1) **the quantities of substances produced at the electrodes are directly proportional to the quantity of electricity passing through solution.**

Grams produced α *Q*

(2) **When a given quantity of electricity is passed through solutions of several electrolytes, the weights of substances formed at the electrodes are directly proportional to their equivalent weights.** The quantity of electricity required to liberate one gram equivalent weight of an element was found to be 96,487 coulombs, called a **faraday** (F). If one faraday is passed through a series of solutions as shown in Figure 17.2, one equivalent of substance is liberated at each electrode—that is, 1.008 g of hydrogen, 107.87 g of silver, 31.77 g of copper (63.54 ÷ 2), and 65.66 g of gold from gold(III) chloride (196.97 ÷ 3). One faraday (96,487 coulombs) represents 6.02×10^{23} electrons in movement according to our theoretical explanation of the nature of an electrical current. These electrons are responsible for chemical changes at the electrodes in electrolytic cells. For example,

$$6.02 \times 10^{23} \text{ electrons} + 6.02 \times 10^{23} \text{ H}^+ \rightarrow 1.008 \text{ g hydrogen}$$
$$6.02 \times 10^{23} \text{ electrons} + 3.01 \times 10^{23} \text{ Cu}^{2+} \rightarrow 31.77 \text{ g copper}$$
$$6.02 \times 10^{23} \text{ electrons} + 2.01 \times 10^{23} \text{ Au}^{3+} \rightarrow 65.66 \text{ g gold}$$

Note that electrode reactions are always oxidation-reduction reactions, consequently for the substance deposited or liberated at the electrode

$$EW = \frac{MW}{\text{total change in oxidation number}}$$

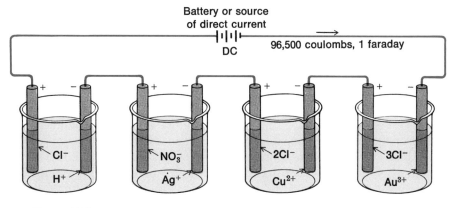

Figure 17.2
Faraday's Law. One gram equivalent weight of an element is deposited for each faraday of electricity passed through the solution.

For example, in the electrolysis of HCl the equivalent weight of the Cl_2 liberated ($2 Cl^- \rightarrow Cl_2 + 2 e^-$) is $\frac{71}{2} = 35.5$. For the liberation of O_2 from the half reaction $2 H_2O \rightarrow O_2 + 4 H^+ + 4 e^-$, the equivalent weight of O_2 is $\frac{32}{4} = 8$, and so on.

Faraday's two laws of electrolysis may be combined into a single expression

$$\text{grams deposited} = \frac{(GEW)(I)(t)}{96,487}$$

EXAMPLE 17.1　In the electroplating of nickel from a solution containing Ni^{2+} ions, what weight of nickel metal will be produced at the cathode by a current of 4.02 amp flowing for 100 min?

SOLUTION:

$$96,487\ C \text{ will deposit } 58.71 \div 2 = 29.36 \text{ g Ni}$$
$$4.02 \text{ amps} \times 100 \times 60 \text{ sec} = 24,120\ C$$
$$24,120/96,487 \times 29.3 = 7.34 \text{ g Ni deposits}$$

or using the expression above

$$\text{grams of Ni deposited} = \left(\frac{58.71 \text{ g/mole}}{2 \text{ equiv/mole}}\right) \frac{(4.02 \text{ amp})(100 \text{ min})(60 \text{ sec/min})}{(96,487\ C/\text{equ})(1 \text{ amp sec}/C)}$$
$$= 7.34 \text{ g}$$

EXAMPLE 17.2　An Na_2SO_4 solution is electrolyzed with a current of 1.00 amp flowing for 5.00 min. The anode half reaction is

$$2 H_2O \rightarrow O_2 + 4 H^+ + 4 e^-$$

Calculate (1) the grams of O_2 produced, (2) the volume of O_2 produced, (3) the equivalents of O_2 liberated, (4) the number of coulombs used, and (5) the number of faradays used.

SOLUTION:

(1) grams of O_2 produced

$$= \frac{\left(\dfrac{32.0 \text{ g/mole}}{4 \text{ equiv/mole}}\right)(1.00 \text{ amp})(5.00 \text{ min} \times 60 \text{ sec/min})}{(96{,}487 \ C/\text{equiv})(1 \text{ amp sec}/C)}$$
$$= 0.0249 \text{ g}$$

(2) volume of O_2 produced

$$= \frac{\left(\dfrac{22.4 \text{ l/mole}}{4 \text{ equiv/mole}}\right)(1.00 \text{ amp})(5.00 \text{ min} \times 60 \text{ sec/min})}{(96{,}487 \ C/\text{equiv})(1 \text{ amp sec}/C)}$$
$$= 0.0174 \text{ l or } 17.4 \text{ ml}$$

(3) equivalents of O_2 liberated $= \dfrac{(1.00 \text{ amp})(5.00 \text{ min} \times 60 \text{ sec/min})}{(96{,}487 \ C/\text{equiv})(1 \text{ amp sec}/C)}$
$$= 0.00311 \text{ equiv}$$

(4) coulombs used $= (1.00 \text{ amp})(5.00 \text{ min} \times 60 \text{ sec/min})\dfrac{(1 \ C)}{(1 \text{ amp})(\text{sec})} = 300 \ C$

(5) faradays used $= \dfrac{300 \ C}{96{,}487 \ C} = 0.00311 \text{ F}$

Voltaic Cells

Since an electric current may cause a chemical change, can the process be reversed—can a current be generated from a chemical reaction? The answer is yes, if certain conditions are established.

Chemical changes are always attended by energy changes, usually in the form of heat. To produce an electrical current, chemical energy must be transformed to electrical energy instead of to heat energy. As in electrolysis, electrons must be transferred—the reaction must be one of oxidation-reduction.

A cell that is used to generate electric current from chemical reactions is known as a **voltaic cell**, after Alessandro Volta (1800), or a **galvanic cell,** after Luigi Galvani (1780) (both were early experimenters in this field). Commercial votaic cells are also commonly known as **batteries**. An example is the **Daniell cell.**

If a strip of metallic zinc is dipped into a solution of copper(II) sulfate, zinc goes into solution as zinc ion (zinc is more active than copper), and metallic copper is formed. This may be represented by

$$Cu^{2+} + Zn \rightarrow Cu + Zn^{2+}$$

or by two half reactions

$$Zn \rightarrow Zn^{2+} + 2\ e^- \text{ (oxidation)}$$
$$Cu^{2+} + 2\ e^- \rightarrow Cu \text{ (reduction)}$$

Electrons are transferred directly from zinc to copper ions. The action is exothermic; the heat energy released is absorbed by the solution and container. No electric current has been generated in this case, but if the oxidation and reduction reactions could be made to take place at electrode surfaces in contact with the solution, and the transfer of electrons were made through a metallic conductor, then at least a part of the chemical energy would be transformed into electrical energy. Thus if we set up an arrangement as shown in Figure 17.3, in which metallic zinc is in contact with its ions, and at another point a strip of copper is in contact with copper ions, an electric current is generated by the flow of electrons through the wire which connects the two metals outside the solution. This will be indicated by a deflection of the needle of a voltmeter connected in series between the two electrodes. A porous diaphragm serves to prevent mixing of the solutions of zinc and copper ions, but allows the passage of the SO_4^{2-} under the influence of the current. To generate an electric current, the reaction must be one of **oxidation-reduction,** and **these two processes must be made to take place at electrode surfaces in a solution.**

Note from the previous equations that it is at the zinc electrode that electrons are released to be available to flow through the wire; consequently the Zn electrode (anode) is designated as the negative pole. These electrons fed into the copper electrode are taken up by Cu^{2+}, and Cu deposits. The

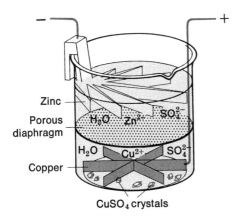

Figure 17.3

The Daniell cell. Electrons flow in wire (external) circuit from the Zn electrode to the Cu electrode, thus an electric current.

copper electrode (cathode) is the positive pole of the cell. During the operation of the cell SO_4^{2-} ions move toward the anode and neutralize the effect of the positive zinc ions being produced — electrical neutrality in the solution must always be maintained.

If $1\,M$ $ZnSO_4$ and $1\,M$ $CuSO_4$ were used in the Daniell cell, the cell notation would be represented as:

$$\text{Zn (s)} \mid \text{Zn}^{2+}\,(1M) \parallel \text{Cu}^{2+}\,(1M) \mid \text{Cu (s)}$$

By convention, the anode is given first followed by the active ion concentration in the anode compartment, the ion concentration in the cathode compartment, and finally the cathode. The single vertical lines represent phase boundaries and the double vertical lines denote the porous diaphragm or salt bridge separating the two solutions.

Electrode Potentials

When a metal is placed in contact with water, the metal shows a tendency to pass into solution producing ions of the metal and consequent liberation of electrons to the metal surface. This tendency of a metal to lose electrons sets up an electrical potential difference between the metal and the solution, and the resulting voltage is termed the **electrode potential** (E) for that particular metal. After ions of the metal are formed in solution and are in contact with the metal, the reverse of ion formation — that is, deposition of metal from ions — occurs. Eventually an equilibrium exists between the metal and its ions. For example:

$$\text{Zn metal immersed in } 1M\ \text{Zn}^{2+}\ \text{solution}$$
$$\text{Zn} \rightleftarrows \text{Zn}^{2+} + 2\ e^{-}$$

develops an electrode potential of 0.76 volt. When the temperature is 25°C, any gases are at atmospheric pressure, and the ion concentration is $1M$, the electrode potential is known as the **standard electrode potential**, and the symbol $E°$ (called E zero) is used. The above conditions are standard conditions for electrochemistry. Actually, for exact work, standard concentration is unit **activity**, which is a corrected concentration for deviations from ideal behavior caused by attraction between the ions. For practical purposes, the molarity of the solution may be used for the approximate activity.

Electrode potentials may be written two ways:

or
$$(1)\ \text{Zn} \rightleftarrows \text{Zn}^{2+} + 2\ e^{-} \qquad E = +0.7628\ \text{V}$$
$$(2)\ \text{Zn}^{2+} + 2\ e^{-} \rightleftarrows \text{Zn} \qquad E = -0.7628\ \text{V}$$

The first equation is an oxidation reaction and the electrode potential is called the **oxidation potential**. The second equation is written as the reverse of the first and is a reduction. The E value is called the **reduction potential**.

The International Union of Pure and Applied Chemistry (IUPAC) has recommended the use of *reduction potentials* as *standard electrode* potentials. Table 17.1 lists some common standard electrode potentials, and a more extensive table is found in the appendix. The table contains the reduction half reactions and the corresponding electrode potentials ($E°$) at standard conditions. The position of a substance in the table gives an indication of the tendency of the half reaction to occur, those higher in the table

Table 17.1

Standard electrode potentials at 25°C (reduction potentials)

Half-Cell Reaction	$E°$
$Li^+ + e^- \rightleftarrows Li$	−3.05
$K^+ + e^- \rightleftarrows K$	−2.95
$Ba^{2+} + 2\ e^- \rightleftarrows Ba$	−2.90
$Ca^{2+} + 2\ e^- \rightleftarrows Ca$	−2.87
$Na^+ + e^- \rightleftarrows Na$	−2.71
$Mg^{2+} + 2\ e^- \rightleftarrows Mg$	−2.37
$Al^{3+} + 3\ e^- \rightleftarrows Al$	−1.66
$Zn^{2+} + 2\ e^- \rightleftarrows Zn$	−0.763
$Cr^{3+} + 3\ e^- \rightleftarrows Cr$	−0.744
$Fe^{2+} + 2\ e^- \rightleftarrows Fe$	−0.440
$Ni^{2+} + 2\ e^- \rightleftarrows Ni$	−0.250
$Sn^{2+} + 2\ e^- \rightleftarrows Sn$	−0.136
$Pb^{2+} + 2\ e^- \rightleftarrows Pb$	−0.126
$2H^+ + 2\ e^- \rightleftarrows H_2$	0.000
$Cu^{2+} + 2\ e^- \rightleftarrows Cu$	+0.337
$Ag^+ + e^- \rightleftarrows Ag$	+0.7991
$Hg^{2+} + 2\ e^- \rightleftarrows Hg$	+0.854
$Pt^{2+} + 2\ e^- \rightleftarrows Pt$	+1.2
$Au^{3+} + 3\ e^- \rightleftarrows Au$	+1.50
$2\ H_2O + 2\ e^- \rightleftarrows H_2 + 2\ OH^-$	−0.828
$I_2 + 2\ e^- \rightleftarrows 2I^-$	+0.536
$Br_2 + 2\ e^- \rightleftarrows 2\ Br^-$	+1.065
$O_2 + 4\ H^+ + 4\ e^- \rightleftarrows 2\ H_2O$	+1.229
$Cl_2 + 2\ e^- \rightleftarrows 2\ Cl^-$	+1.360
$F_2 + 2\ e^- \rightleftarrows 2\ F^-$	+2.87

being more readily oxidized, those lower being more readily reduced. In Table 17.1, for example, Li metal at the top of the table has the greatest tendency to be oxidized to Li^+ (or is the strongest reducing agent) and has the highest negative $E°$ value. On the other hand F_2 at the bottom of the table has the greatest tendency to be reduced (and is the strongest oxidizing agent). It has the highest positive $E°$ value.

Evaluation of Electrode Potentials

Absolute electrode potentials can not be experimentally determined, since the potential difference between a metal and a solution of its ions cannot be recorded on a voltmeter directly. But if the metal and a salt solution of it are made one half of a cell in which the other half is also made up of a metal and a solution of its ions, the potential difference of the two half-cells can be measured and a relative value or relative potential may be assigned. Thus in the zinc and copper cell, while we do not know the absolute potential difference of either of the half-cells, or **couples** – that is, $Zn^{2+} \rightleftarrows Zn$ and $Cu^{2+} \rightleftarrows Cu$ – the relative potential of one to the other can be said to be 1.100 volts.

In order to arrive at a working system of electrode potentials, all values are referred to a standard electrode – the hydrogen electrode, which has been assigned the arbitrary value of zero potential. The electrode (Figure 17.4) consists of a piece of platinum foil which has been coated with a spongy deposit of platinum black in contact with a solution containing hydrogen ions at a concentration of 1 M. Hydrogen gas at one atmosphere pressure is bubbled over the platinum electrode, setting up the equilibrium:

$$2\ H^+ + 2\ e^- \rightleftarrows H_2 \qquad E° = 0.00$$

By using the hydrogen half-cell in conjunction with a half-cell composed of a metal and its ions at a concentration of 1 M, the potential of the metal compared with the standard can be measured. Thus if the cell indicated in Figure 17.4 is set up, the voltage is 0.763, and since zinc gives up electrons more easily than hydrogen, the standard potential of the Zn^{2+}/Zn electrode is recorded as -0.763.

The electrode potential of Cu^{2+}/Cu may be determined in a similar manner. Since, in the cell reaction, hydrogen gives up electrons to form H^+,

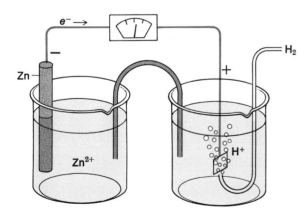

Figure 17.4
The voltage of the cell is
0.763 V. Since E^0 for the
hydrogen electrode is
arbitrarily taken as 0, the
E^0 for the zinc electrode is
taken as -0.763 V.

and Cu^{2+} takes up these electrons, copper gives up electrons less readily than hydrogen, and the voltage of the Cu^{2+}/Cu couple is recorded as +0.337.

By properly combining the two half-cells, Zn^{2+}/Zn and Cu^{2+}/Cu, it is evident that the potential difference of such a cell should be $0.337 - (-0.763) = 1.100$ V.

Information can be derived by combining couples, writing the cell reaction, and considering the voltage of the cell. If the cell potential is positive the cell reaction will proceed spontaneously as written. Consider the example

$$Pb^{2+} + Zn \rightarrow Pb + Zn^{2+}$$

involving the two half-cells $Zn^{2+} + 2e^- \rightleftarrows Zn$, and $Pb^{2+} + 2e^- \rightarrow Pb$. The cell may be represented as $Zn|Zn^{2+}$ (1 M)$\|Pb^{2+}$ (1 M)$|Pb$. The potential of the first couple is -0.763; that of the second is -0.126. Since in the above equation Zn gives up electrons (oxidation), the sign of the potential and direction of the half-cell equation will be reversed and given the value +0.763.

$$Zn \rightarrow Zn^{2+} + 2\ e^- \qquad E° = +0.763 \text{ (oxidation)}$$
$$2\ e^- + Pb^{2+} \rightarrow Pb \qquad E° = -0.126 \text{ (reduction)}$$

Adding the two steps we obtain the cell reaction,

$$Zn + Pb^{2+} \rightarrow Zn^{2+} + Pb \qquad E°_{cell} = +0.637$$

Since the potential for the cell is positive, the reaction will proceed to a great extent as written — that is, Pb^{2+} will oxidize Zn or Zn will reduce (replace) Pb^{2+}.

Consider a second example: Will Ag^+ oxidize Au? In other words, will the following reaction occur?

$$Au + 3\ Ag^+ \rightarrow Au^{3+} + 3\ Ag$$

Setting up the 2 half reactions (one oxidation, one reduction),

$$3\ e^- + 3\ Ag^+ \rightarrow 3\ Ag \qquad E° = 0.7991 \text{ (reduction)}$$
$$Au \rightarrow Au^{3+} + 3\ e^- \qquad E° = -1.50 \text{ (oxidation)}$$

The latter potential is negative because the step is oxidation (the table gives reduction potentials). The fact that 3 electrons are involved in balancing does not change the potential. Adding the two steps, we obtain

$$Au + 3\ Ag^+ \rightarrow Au^{3+} + 3\ Ag \qquad E_{cell} = -0.70$$

The negative value for the voltage of the cell indicates that the reaction will *not* proceed to an appreciable extent as written. The reverse change will take place spontaneously.

The replacement of bromine by chlorine in chemical change may be predicted by study of their comparative electrode potentials.

$$2\ Br^- \rightarrow Br_2 + 2\ e^- \qquad E° = -1.065 \text{ (oxidation)}$$

$$\frac{2\ e^- + Cl_2 \rightarrow 2\ Cl^- \qquad\qquad E^\circ = 1.360\ \text{(reduction)}}{Cl_2 + 2\ Br^- \rightarrow Br_2 + 2\ Cl^- \qquad E_{\text{cell}} = 0.295}$$

EXAMPLE 17.3   If F_2 is added to Ag, (1) Will the reaction be spontaneous? (2) What will be the E° value? (3) What are the anode and cathode?

SOLUTION:

$$\frac{\begin{array}{ll} 2\ Ag \rightleftarrows 2\ Ag^+ + 2\ e^- & -0.80\ V \\ F_2 + 2\ e^- \rightleftarrows 2\ F^- & +2.87\ V \end{array}}{2\ Ag + F_2 \rightleftarrows 2\ Ag^+ + 2\ F^- \quad +2.07\ V}$$

Note that if the half reaction with the lower E° value is written first (higher in the table) and then reversed, followed by the half reaction with the higher E° value, when the electrons in the two half reactions are made the same and the two equations added, the E cell will always be positive—indicating a spontaneous reaction as written. Ag is the anode (oxidation occurs) and F_2 is the cathode (reduction).

Electrode Potentials and Concentration Changes

The standard electrode potentials recorded in Table 17.1 are for one-molar solutions of the ions. For concentrations other than one molar, the voltage of a half reaction or a voltaic cell may be calculated from the **Nernst Equation,** derived by the German thermodynamicist Walther Nernst (1889):

$$E = E^\circ - \frac{0.0592}{n} \log Q \tag{1}$$

where

$E =$ the voltage for the reaction for the new concentrations
$E^\circ =$ the voltage at standard concentrations
$n =$ the number of electrons transferred in the reaction
$Q =$ a concentration expression which takes the same form
  as the equilibrium constant expression, K(page 320).

In this equation, if all other conditions are standard ($t = 25°C$, gas pressures = 1 atm), Q will involve only the concentrations of the ions present (each raised to the power equal to its coefficient in the balanced equation). The concentrations of any solids may be considered constant and are omitted from the expression.

Consider, for example, the potential of the Zn/Zn²⁺ electrode where the concentration of Zn^{2+} is 0.010 M. From Table 17.1

$$Zn^{2+} + 2\ e^- \rightleftarrows Zn \qquad E^\circ = -0.763\ V$$

Substituting in the Nernst equation:

$$E = E° - \frac{0.0592}{2} \log \frac{1}{[Zn^{2+}]}$$

$$= -0.763 - \frac{0.0592}{2} \log \frac{1}{1.0 \times 10^{-2}}$$

$$= -0.763 - (0.0296)(2) = -0.822 \text{ V}$$

EXAMPLE 17.4   What is the potential of the following cell?

$$Zn|Zn^{2+}(0.100 \ M)||Ag^+(0.0100 \ M)|Ag$$

SOLUTION:

The cell reaction is: $Zn + 2 \ Ag^+ \rightarrow Zn^{2+} + 2 \ Ag$
From Table 17.1, $E°_{cell} = +1.562$ V

Substituting in the Nernst equation

$$E_{cell} = E°_{cell} - \frac{0.0592}{2} \log \frac{[Zn^{2+}]}{[Ag^+]^2}$$

$$= 0.637 - 0.0296 \log \frac{1.00 \times 10^{-1}}{(1.00 \times 10^{-2})^2}$$

$$= 0.637 - (0.0296)(3) = 0.548 \text{ V}$$

The Nernst equation may also be used to calculate the potential developed when two electrodes, both of the same material, are dipped into two different concentrations of the same salt separated by a semipermeable partition. This is called a concentration cell.

For example,

$$Zn|Zn^{2+}(0.0100 \ M)||Zn^{2+}(0.100 \ M)|Zn$$

$$E = 0.000(\text{same electrode}) - \frac{0.0592}{2} \log \frac{1.00 \times 10^{-2}}{1.00 \times 10^{-1}}$$

$$= -0.0296(-1) = +0.0296 \text{ V}$$

Equilibrium Constants from Reduction Potentials

For an oxidation-reduction reaction at equilibrium, the voltage of the cell will be zero $(E = 0)$, $Q = K$, and the Nernst equation (1) becomes

$$E°_{cell} = \frac{0.0592}{n} \log K \tag{2}$$

where $E°_{cell}$ is the difference in standard electrode potentials of two half-reactions making up the cell and K is the equilibrium constant. For example, consider the cell reaction:

$$Zn + Pb^{2+} \rightarrow Zn^{2+} + Pb$$
in which $E°_{cell} = +0.637$ V, and $n = 2$
Substituting in equation (2)

$$0.637 = \frac{0.0592}{2} \log K$$

$$\log K = \frac{0.637}{0.0296} = 21.52$$

$$K = 3.31 \times 10^{21}$$

In arriving at the antilog where $\log K = 21.52$, the 21 is the characteristic denoting an exponent of 10, and the .52 is the mantissa which from a log table gives the antilog 3.31. Thus, the large value for K shows that the reaction as written proceeds to the right to a very great extent.

Relationship of $E°$ and the Change in Free Energy

Standard free energy changes, $\Delta G°$, may be determined for electrochemical processes by the relationship

$$\Delta G° = -nFE° \tag{3}$$

where F is the value of the faraday expressed in calories. It has a value of 23,061 cal/V.

EXAMPLE 17.5. What is the standard free energy change for the following reaction?

$$Zn + Cu^{2+} \rightleftarrows Cu + Zn^{2+} \qquad E° = 1.100 \text{ V}$$

SOLUTION: Substituting in (3)

$$\Delta G° = -(2)(23,061 \text{ cal/V})(1.100 \text{ V})$$
$$= -50,730 \text{ cal or } -50.73 \text{ kcal}$$

It will be recalled that a negative value of ΔG indicates that the reaction is spontaneous. Consequently the above reaction will proceed to the right as written. This will always be the case when the value of $E°$ is positive.

17.2   PRACTICAL ELECTROCHEMISTRY

The Dry Cell

For small intermittent currents, such as for flashlights and door bells, the dry cell is convenient and effective (see Figure 17.5). The cell consists of a zinc

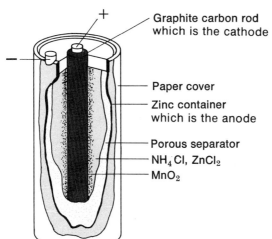

Graphite carbon rod
which is the cathode

Paper cover

Zinc container
which is the anode

Porous separator

NH_4Cl, $ZnCl_2$

MnO_2

Figure 17.5
Cross section of the dry cell.

container which acts as the negative electrode; a graphite rod as the positive electrode; and a paste of ammonium chloride, zinc chloride, manganese dioxide, and water. Although the substances are present in a semi-solid (paste) state, the cell is not really dry, and a wax seal on the top of the cell is used to prevent evaporation of water. At the negative electrode, Zn passes into solution as Zn^{2+}, giving up electrons to the electrode.

$$Zn \rightarrow Zn^{2+} + 2\ e^-$$

At the carbon electrode (positive pole), ammonium ion is converted into ammonia and hydrogen by taking up electrons:

$$2\ e^- + 2\ NH_4^+ \rightarrow 2\ NH_3 + H_2$$

The products of this reaction are gases which must be removed, since their accumulation would result in a swelling and eventual bursting of the cell. Ammonia gas is absorbed by the zinc ions present, forming the complex ion, $Zn(NH_3)_4^{2+}$.

$$Zn^{2+} + 4\ NH_3 \rightarrow Zn(NH_3)_4^{2+}$$

Hydrogen gas formed at the carbon electrode tends to remain in contact with the electrode. An accumulation of the gas forms an insulating layer, which increases the resistance of the cell, thereby reducing the voltage. Such a cell is said to be **polarized.** The manganese dioxide functions as a **depolarizer,** since it oxidizes the hydrogen to water by the reaction.[*]

$$MnO_2 + H_2 \rightarrow MnO + H_2O$$

[*] The cathode reactions are complex and vary with cell current. A variety of Mn compounds are actually formed.

If current is drawn from a dry cell continuously, the voltage soon drops because of accumulation of hydrogen on the carbon electrode. After a brief period of disuse, the hydrogen is slowly oxidized by the manganese dioxide, and the voltage of the cell is restored to 1.5 V, the normal value.

One of the new types of dry cells now used for small devices uses, as does the above described dry cell, a zinc container and carbon rod cathode, but the filler material is mercury(II) oxide (HgO) with moist KOH. A potential of about 1.35 V develops.

$$2\ e^- + \mathrm{HgO} + \mathrm{H_2O} \rightarrow \mathrm{Hg} + 2\ \mathrm{OH^-}$$
$$\mathrm{Zn} + 2\ \mathrm{OH^-} \rightarrow \mathrm{Zn(OH)_2} + 2\ e^-$$

The Storage Battery

The storage battery is a cell which, after discharge, can be restored to its original state by passage of a direct electric current through the cell from an outside source. Theoretically, the reaction which takes place in *any* cell producing an electric current may be reversed by passing an electric current through the cell in the opposite direction. Practically, this cannot be done with most cells; a storage cell is one in which a cell reaction may be carried out reversibly.

The lead storage battery (Figure 17.6) is the most common of this type. The electrodes consist of plates or grids arranged in pairs; the negative electrode is a sheet of pure lead, and the positive pole a framework of lead in which lead dioxide has been deposited. The electrolyte is sulfuric acid. When the cell is discharging — producing an electric current — lead from the negative electrode produces lead ions which immediately combine with sulfate ions from the sulfuric acid to precipitate insoluble lead sulfate.

$$\mathrm{Pb} \rightarrow \mathrm{Pb^{2+}} + 2\ e^-$$
$$\underline{\mathrm{Pb^{2+}} + \mathrm{SO_4^{2-}} \rightarrow \mathrm{PbSO_4(s)}}$$
$$\mathrm{Pb} + \mathrm{SO_4^{2-}} \rightarrow \mathrm{PbSO_4(s)} + 2\ e^-$$

Positive post $+$ $-$ Negative post

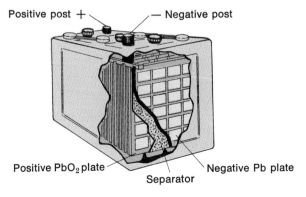

Positive PbO$_2$ plate   Negative Pb plate

Separator

**Figure 17.6
A lead storage battery.**

At the positive pole of the battery, lead dioxide is reduced to Pb^{2+} by taking up electrons according to the equations

$$2\ e^- + PbO_2 + 4\ H^+ \rightarrow Pb^{2+} + 2\ H_2O$$
$$\underline{ Pb^{2+} + SO_4^{2-} \rightarrow PbSO_4(s)}$$
$$2\ e^- + PbO_2 + 4\ H^+ + SO_4^{2-} \rightarrow PbSO_4(s) + 2\ H_2O$$

The sum of the reactions taking place at the two electrodes gives the overall reaction occurring during **discharge** of the battery.

$$Pb + PbO_2 + 2\ H_2SO_4 \rightarrow 2\ PbSO_4(s) + 2\ H_2O$$

During **charge** of the battery this reaction is reversed

$$2\ PbSO_4 + 2\ H_2O \rightarrow Pb + PbO_2 + 2\ H_2SO_4$$

It is evident that the concentration of sulfuric acid in the cell decreases during discharge; during the charge phase the sulfuric acid is regenerated and its concentration in the cell increased. The state of charge of a battery may be determined approximately by a hydrometer, which gives the specific gravity of the acid. In a fully charged battery the specific gravity of the acid is about 1.30. The voltage of the lead storage cell is about two volts. For automobile batteries, several cells are used in series, to give a total potential of six or twelve volts.

The Edison storage battery is a lighter, more rugged and more expensive battery than the lead storage battery. The anode is iron, and the cathode has a surface coating of nickel(III) oxide. The electrolyte is KOH or NaOH solution. Reactions occurring in the cell are

Anode $Fe + 2\ OH^- \rightarrow Fe(OH)_2 + 2\ e^-$
Cathode $\underline{2\ e^- + Ni_2O_3 + 3\ H_2O \rightarrow 2\ Ni(OH)_2 + 2\ OH^-}$

Overall cell $Fe + Ni_2O_3 + 3\ H_2O \underset{\text{charge}}{\overset{\text{discharge}}{\rightleftharpoons}} Fe(OH)_2 + 2\ Ni(OH)_2$

Another storage battery that uses more expensive metals, but has longer life, makes use of the potential developed by the following half-cells. The reaction may be reversed on a recharge. It is called the **Nicad** battery and is commonly used for portable tools. The electrode reactions are

Cathode $2\ e^- + NiO_2 + 2\ H_2O \rightarrow Ni(OH)_2 + 2\ OH^-$
Anode $\underline{Cd + 2\ OH^- \rightarrow Cd(OH)_2 + 2\ e^-}$
Overall $Cd + NiO_2 + 2\ H_2O \rightarrow Cd(OH)_2 + Ni(OH)_2\ E = 1.4\ V.$

Fuel Cells

A type of voltaic cell currently receiving considerable attention is the fuel cell, a cell which converts the energy of conventional fuels, such as natural gas and diesel oil, directly into electricity. The usual method of burning the fuel in a heat engine—converting the heat energy to mechanical energy and finally to electrical energy—is a very inefficient process, with a conversion of less than 50%. A direct conversion is a great deal more efficient.

The simplest type of cell so far developed is the hydrogen-oxygen, or air, cell (Figure 17.7), which utilizes the energy of the reaction

$$2 \ H_2 + O_2 \rightarrow 2 \ H_2O + \text{energy}$$

Hydrogen and oxygen gases are adsorbed onto the porous carbon electrodes, which contain finely divided platinum or palladium metal as a catalyst. At the hydrogen electrode, hydrogen reacts with OH^- from the 30% NaOH or KOH electrolyte.

$$H_2 + 2 \ OH^- \rightarrow 2 \ H_2O + 2 \ e^-$$

Electrons flowing through the circuit toward the oxygen electrode constitute the current. At the oxygen electrode, the following reaction occurs:

$$\tfrac{1}{2} \ O_2 + H_2O + 2 \ e^- \rightarrow 2 \ OH^-$$

The cell operates between 25–60°C and gives an EMF of 0.9 V with a current density of about 35 milliamperes per cm² of electrode surface. Note the half-cell standard electrode potentials related to H_2 and to O_2 in Table 17.1. The potential is favorable for a cell, but practically hard to attain.

In order to use natural gas or diesel oil, the cells must be operated at fairly high temperatures (500°C or more), a fact which calls for a molten salt electrolyte. Electrodes of metals such as nickel and silver have been tried with high temperature and high pressure equipment with some success. Melted alkali carbonates may replace the hydroxide-water electrolyte. Much development work needs to be done, but fuel cells hold considerable promise for the future.

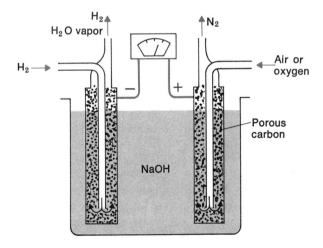

Figure 17.7
Fuel cell.

Electrode reactions for a natural gas (CH_4, methane) and oxygen fuel cell can be represented with these equations.

$$CH_4 + 10\ OH^- \rightarrow CO_3^{2-} + 7\ H_2O + 8\ e^-$$
$$8\ e^- + 4\ H_2O + 2\ O_2 \rightarrow 8\ OH^-$$
$$\overline{CH_4 + 2\ O_2 + 2\ OH^- \rightarrow CO_3^{2-} + 3\ H_2O}$$

The Solar Battery

In the case of photoelectric cells, light energy is converted to a weak electric current, which activates a larger store of energy to cause mechanical action. A solar battery is a photoelectric cell producing electric current from solar radiation.

One experimental solar battery makes use of silicon which, like carbon, has a diamond crystalline structure. A small amount of arsenic is incorporated in the silicon. Arsenic has five outer electrons, whereas silicon has four. Thus an arsenic atom is a slight misfit in the tetrahedral space patterning of crystal-line silicon atoms – the extra electron of the arsenic is somewhat free to move. The surface of the silicon crystal is treated with boron. Boron atoms have but three valence electrons. The boron atoms occupying positions in the space lattice of silicon are one-electron deficient as compared with silicon. A small electrostatic force exists in a prepared silicon crystal between the electron-deficient boron-containing surface region and the arsenic-containing interior with its excess electrons. An equilibrium of electron flow between the interior and surface of the treated silicon crystal establishes itself, and can be disturbed by outside influences.

In a solar battery, one lead wire is connected with the interiors of a series of the prepared silicon crystals or wafers and the other lead wire with the surfaces. Energy from sunlight disturbs the electron equilibrium and causes an electron movement into the arsenic-containing central or body positions. This electron movement generates an electric current in the lead wires. The positive terminal of this silicon battery is at the boron-treated surface and the negative terminal contacts the arsenic-containing central section. No large source of electrical current has yet been developed, but the power is adequate to operate some communication equipment in space satellites or rockets.

Some photoelectric cells differ in the principle of their operation. As light hits a section of limited conducting substance it increases the ease of electron flow. Thus the photoelectric cell may function in control of current.

Other Modern Batteries

Zinc-silver oxide batteries have been used in artificial satellites and space probes. The electrode reactions are:

Anode	$Zn + 2 OH^- \rightarrow Zn (OH)_2 + 2 e^-$	
Cathode	$2 AgO + H_2O + 2 e^- \rightarrow Ag_2O + 2 OH^-$	
Overall	$Zn + 2 AgO + H_2O \rightarrow Zn (OH)_2 + Ag_2O$	$E = 1.86$ V

The **zinc-mercuric oxide battery** is the one most commonly used for heart pacemakers. For this purpose the mercury battery must be small but dependable. The electrode reactions are:

Anode	$Zn + 2 OH^- \rightarrow Zn(OH)_2 + 2 e^-$	
Cathode	$HgO + H_2O + 2 e^- \rightarrow Hg + 2OH^-$	
Overall	$Zn + HgO + H_2O \rightarrow Zn(OH)_2 + Hg$	$E = 1.34$ V

Much research has been done on batteries for possible use in electric automobiles. For example, Ford Motor Company has investigated the **sodium-sulfur battery** which operates at a temperature of 300°C and uses $NaAl_{11}O_{17}$ as an electrolyte.

Anode	$4 Na \rightarrow 4 Na^+ + 4 e^-$	
Cathode	$3 Na_2S_5 + 4 Na^+ + 4 e^- \rightarrow 5 Na_2S_3$	
Overall	$4 Na + 3 Na_2S_5 \rightarrow 5 Na_2S_3$	$E = 2.0$ V

General Motors Corporation has carried out research on a lithium-chlorine battery which operates at 650°C. Molten LiCl is the electrolyte.

Anode	$2 Li \rightarrow 2 Li^+ + 2 e^-$	
Cathode	$Cl_2 + 2 e^- \rightarrow 2 Cl^-$	
Overall	$2 Li + Cl_2 \rightarrow 2 Li^+ + 2 Cl^-$	$E = 3.5$ V

Whether these or any batteries prove to be successful for large scale use would also depend, of course, on the availability of an enormous increase in the number of electric power plants which would be needed to keep the batteries charged.

Electroplating

To prevent a metal from corroding, or to give it a more pleasing appearance, very often a second metal is coated over it. This is most effectively accomplished electrolytically. The metal object which is to be coated or plated is made the cathode in an electrolytic cell (Figure 17.8). The anode usually consists of a strip of the plating metal dipping into the electrolyte. As the current is passed through the cell, the plating metal dissolves at the anode, and the ions thus produced migrate to the cathode, where they are discharged and plated out on the object. Any desire thickness of deposit may be obtained. If one metal does not adhere well to another, a series of platings may be necessary; for example, an auto bumper may be first plated with copper, then with nickel, and finally with chromium.

In much electroplating the plating metal is made the anode in an elec-

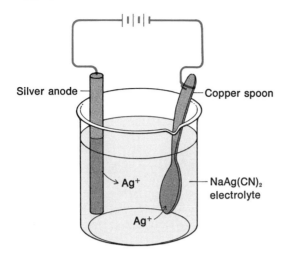

Figure 17.8
A spoon may be silver plated by making it the cathode in which a bar of pure silver acts as the anode. A $NaAg(CN)_2$ solution makes a good electrolyte.

trolytic cell. For example when plating an iron object with nickel, the iron is made the cathode, an electrolyte of $NiSO_4$ may be used, and nickel rod is made the anode. It is not merely the Ni^{2+} of the electrolyte which deposits. At the anode three half-cell actions are possible.

$$2\ SO_4^{2-} \rightarrow 2(SO_4) + 4\ e^- \qquad 2\ (SO_4) + 2\ H_2O \rightarrow 2\ H_2SO_4 + O_2$$
$$2\ H_2O \rightarrow O_2 + 4\ H^+ + 4\ e^-$$
$$Ni \rightarrow Ni^{2+} + 2\ e^-$$

The nickel is observed to be used up and keeps furnishing new Ni^{2+} in solution to migrate and plate at the cathode. The Ni-Ni^{2+} half-cell reaction is the favored one of the three to take place.

17.3   ELECTROMETALLURGY

Consult Table 17.1 and note that the metals Au, Pt, Ag, and Cu are readily electroplated from water solution because of their positive reduction potential (negative oxidation potential). It is a fortunate fact that $H^+ \rightarrow H + e^-$ (deposition plating) slows down at most metal cathode surfaces; however on platinum and certain other metals it discharges readily with follow up of $2\ H \rightarrow H_2(g)$. At other metal cathodes a sort of reverse voltage builds up, called **overvoltage**, and differs with factors of each cathode metal, concentration, and current density. With a moderate current density hydrogen overvoltage builds up at a zinc cathode above the $+0.762$ V of the half-reaction $Zn \rightleftarrows Zn^{2+} + 2\ e^-$. With an applied potential of slightly over this voltage the above half-cell action can be reversed and Zn plated from Zn^{2+} solution. Metals above zinc in the electrode potential table cannot be plated in competition with hydrogen from a water solution. Thus Al, Mg, Na, Ca, Ba, K require **non-aqueous** electrometallurgy; that is electroplating from melts of their salts, salt mixes, or hydroxides. Consult Chapter 27 for electrometallurgy of Al, Chapter 26 for Na, as representative of the metals below zinc in electrode potential.

The ability to plate metal ions, above Zn^{2+} in electrode potentials, from water solution is the basis for electroplating of metals for surface covering, or for purification recovery of various metals. Thus copper or zinc can be electro-deposited from, respectively, $CuSO_4$ or $ZnSO_4$ aqueous solutions. The metal sulfates can come from impure anodes in electrolytic metal purification tanks or as products of ore leaching with dilute H_2SO_4. Impure anode lead is immersed in a $PbSiF_6$–water electrolyte and pure lead is deposited on a pure lead cathode starter sheet. Other less common metal ions of electrode potential more than zinc may be cathodically deposited from water solutions of one of their salts. Examples are Cr^{3+}, Ni^{2+}, Sn^{2+}, Cd^{2+}, U^{4+}.

Exercises

17.1. Define or illustrate the terms: electrolysis, ohm, ampere, coulomb, volt, electromotive force, faraday, electrode potential, voltaic cell, depolarizer.

17.2. In electrochemistry the terms: (1) anode, (2) cathode, (3) anion, and (4) cation are used extensively. Answer the following questions.
(a) Which electrode would have a negative charge in an electrolytic cell?
(b) Which ion moves toward the cathode?
(c) Which ion has a positive charge?
(d) Reduction takes place at which electrode?
(e) Oxidation takes place at which electrode?
(f) Which electrode would show a negative charge in a galvanic cell?

17.3. Volts, coulombs, ohms, joules, and amps are units used for the definition of electrical phenomena. Define the units and the relation between them by matching them with the statements below.
(a) Unit of electrical charge.
(b) Unit of energy.
(c) Unit of electrical potential.
(d) Unit of electrical current.
(e) Volts/amperes.
(f) Amperes × seconds.
(g) Coulombs/second.

17.4. Match the following statements with the appropriate term or symbol.

(a) Process where electrical energy produces a chemical reaction.	1. H_2
	2. Faraday
(b) Standard electrode potential.	3. $\Delta G°$
(c) Amount of chemical change at an electrode depends on equivalent weight and current.	4. O_2
	5. Fuel cell
(d) 96,487 coulombs.	6. PbO_2

(e) A system where a chemical reaction produces electrical energy.

(f) A cell which produces electrical energy — reacting chemical stored *outside* electrode compartment

(g) Oxidant and reductant of fuel cell used in satellites.

7. Faraday's Law
8. $E°$
9. N_2
10. Electrolysis
11. Galvanic Cell
12. Nernst Equation

17.5. Consider the electrolysis of molten KCl:
(a) What chemical is produced at the anode?
(b) What is the cation?
(c) What is the cathode half reaction?

17.6. Consider the electrolysis of an aqueous solution of NaCl:
(a) Write the cathode half reaction.
(b) Write the anode half reaction.
(c) Write the overall reaction.
(d) What gas is formed at the cathode?

17.7. Sketch a cell for the electrolysis of $CdCl_2$ between inert electrodes. On the sketch indicate: (a) the signs of the electrodes, (b) the cathode and anode, (c) the directions in which the ions move, (d) the direction in which the electrons move, (e) the electrode reactions.

17.8. Sketch a voltaic cell in which the reaction is

$$Cd + Cl_2 \rightarrow Cd^{2+} + 2\ Cl^-$$

On the sketch indicate (a) the signs of the electrodes, (b) the cathode and anode, (c) the directions in which the ions move, (d) the direction in which the electrons move, (e) the electrode reactions, (f) the cell voltage.

17.9. A sodium sulfate solution was electrolyzed using inert Pt electrodes. The cathode reaction was:

$$2\ H_2O + 2\ e^- \rightarrow H_2 + 2\ OH^-$$

If a current of 3.0 amp was used for 30 min, what weight of H_2 gas would be produced?

17.10. A solution of $CrCl_3$ is electrolyzed and Cr is plated out on the cathode. If a steady current of 1.85 amp is used for 4.50 hr, what weight of Cr in grams will be deposited?

17.11. Calculate the quantity of electricity in faradays necessary to deposit 20.0 g of Ca from melted $CaCl_2$.

17.12. What volume of H_2 gas at STP would be produced during the passage of 30,000 C in the electrolysis of H_2O?

17.13. Calculate the quantity of electricity (in coulombs) necessary to deposit 100.0 g of copper from a $CuSO_4$ solution.

17.14. How many minutes will it take to plate out 40.0 g of Ni from a solution of $NiSO_4$ using a current of 3.45 amp?

17.15. What is the equivalent weight of a metal if a current of 0.250 amp causes 0.524 g of the metal to plate out of a solution undergoing electrolysis in 1.00 hr?

17.16. How many hours will it take to plate out all of the copper in 200 ml of a 0.150 M Cu^{2+} solution using a current of 0.200 amp?

17.17. Give the notation for a cell that utilizes the reaction
(a) $\qquad$ $Cl_2(g) + 2\ I^-(aq) \rightarrow 2\ Cl^-(aq) + I_2(s)$
(b) What is $E°$ for the cell? (c) Which electrode is the cathode?

17.18. Consider the table of electrode potentials in the appendix and answer the following:
(a) For the voltaic cell at 25°C:

$$Cd|Cd^{2+}\ (1M)||Cu^{2+}\ (1M)|Cu$$

(1) Which electrode is the anode?
(2) What is the $E°$ value of the cell as written?
(3) What would be the overall cell reaction when the battery is operating?
(4) What would be the electrode potential (E) for the cell if the concentration of Cd^{2+} were changed to $10^{-5}\ M$?
(5) What would be the value of ΔG in the previous question?
(6) Calculate the equilibrium constant for the cell reaction.
(b) (1) Would MnO_4^- react spontaneously with Cu? Explain your answer briefly.
(2) Write the overall balanced equation for the combination of the following half reactions in such a way as to show a spontaneous reaction to the right:
$I_2 + 2\ e^- \rightleftarrows 2\ I^-$
$MnO_4^- + 8\ H^+ + 5\ e^- \rightleftarrows Mn^{2+} + 4\ H_2O$

17.19. Use this abbreviated table of oxidation potentials to answer the questions below.

$Mg^{2+} + 2\ e^-$	$\rightleftarrows Mg(s)$	-2.37 V
$2\ H^+ + 2\ e^-$	$\rightleftarrows H_2(g)$	0.00
$S + 2\ H^+ + 2\ e^-$	$\rightleftarrows H_2S(g)$	$+0.14$
$Sn^{4+} + 2\ e^-$	$\rightleftarrows Sn^{2+}$	$+0.15$
$Cu^{2+} + 2\ e^-$	$\rightleftarrows Cu(s)$	$+0.34$
$Fe^{3+} + e^-$	$\rightleftarrows Fe^{2+}$	$+0.77$
$NO_3^- + 2\ H^+ + e^-$	$\rightleftarrows 2\ H_2O + NO(g)$	$+0.79$
$Br_2 + 2\ e^-$	$\rightleftarrows 2\ Br^-$	$+1.07$
$MnO_4^- + 8\ H^+ + 5\ e^- \rightleftarrows 4\ H_2O + Mn^{2+}$		$+1.51$

(a) (1) Give the formula for the strongest oxidizing agent.
(2) Give the formula for the strongest reducing agent.
(3) Which two of the following would reduce Fe^{3+} to Fe^{2+}? Br^-, H_2S, Mn^{2+}, NO, Sn^{2+}

(4) Which of the following would oxidize Fe^{2+} to Fe^{3+}?
Sn^{4+}, MnO_4^-, NO_3^-, Cu^{2+}, S

(b) Write a balanced ionic equation for (1) reduction of Cu^{2+} to Cu by Sn^{2+}, (2) Oxidation of Br^- to Br_2 with MnO_4^-.

(c) (1) What would be the voltage of the following galvanic cell? (1) $Mg|Mg^{2+}$ (1.0 M)||Cu^{2+} (1.0 M)|Cu

(2) Write the cell reaction.

(3) Using the Nernst equation calculate the voltage of the same cell for $Mg^{2+} = 1 \times 10^{-3}$ M.

17.20. From the table of standard electrode potentials that appears in the appendix, select a suitable substance for each of the following transformations (assume that all soluble substances are present in 1 M concentrations): (a) an oxidizing agent capable of oxidizing Ni to Ni^{2+} but not H_2 to H^+; (b) an oxidizing agent capable of oxidizing H_2S to S but not Ag to AgCl; (c) a reducing agent capable of reducing Fe^{3+} to Fe^{2+} but not I_2 to I^-; (d) a reducing agent capable of reducing $Cr_2O_7^{2-}$ to Cr^{3+} but not HNO_3 to NO_2.

17.21. Use $E°$ values to predict whether or not the following skeleton equations represent reactions that will occur in acid solution with all soluble substances present in 1 M concentrations. All reactants and products other than H^+ and H_2O are shown below. Complete and balance the equation for each reaction that is predicted to occur.

(a) $I^- + NO_3^- \rightarrow I_2 + NO$

(b) $Ni + Cr^{3+} \rightarrow Cr^{2+} + Ni^{2+}$

(c) $Br_2 + MnO_4^- \rightarrow Br^- + Mn^{2+}$

(d) $Cr_2O_7^{2-} + Fe^{2+} \rightarrow Fe^{3+} + Cr^{3+}$

17.22. Predict whether or not each of the following reactions will occur spontaneously in acid solution, and write a balanced chemical equation for each reaction that is predicted to occur. Assume that soluble reactants and products are present in 1 M concentrations. (a) Oxidation of Hg to Hg^{2+} by Br_2 (reduced to (Br^-); (b) reduction of Br_2 to Br^- by H_2O_2 (oxidized to O_2); (c) oxidation of Mn^{2+} to MnO_4^- by HNO_3 (reduced to NO); (d) oxidation of Cl^- to Cl_2 by NO_3^- (reduced to NO); (e) reduction of S to H_2S by H_2 (oxidized to H^+).

17.23. From Table 17.1 of standard electrode potentials, (1) determine the voltage of the following cells (all cells are at "standard" conditions), (2) write the cell reaction, (3) calculate $\Delta G°$.

(a) $Mg|Mg^{2+}||Ag^+|Ag$

(b) $Sn|Sn^{2+}||Cu^{2+}$, Cu

(c) $Ca|Ca^{2+}||Zn^{2+}|Zn$

(d) $Fe|Fe^{2+}||H^+|H_2$

17.24. (a) What is $E°$ for the cell

$$Ni|Ni^{2+}||Ag^+|Ag$$

(b) Write an equation for the cell reaction. (c) Which electrode is positive?

17.25. For the cell

$$Zn|Zn^{2+}||In^{3+}|In$$

$E°$ is $+0.4198$ V. Use the *emf* of the cell and the $E°$ for the Zn^{2+}/Zn couple to calculate $E°$ for the In^{3+}/In half reaction.

17.26. (a)According to the $E°$ value, should the following reaction proceed spontaneously at 25°C with all soluble substances present at 1.00 M concentration?
$$MnO_2(s) + 4\ H^+ + 2\ e^- \rightleftharpoons Mn^{2+} + 2\ H_2O,\ E° = 1.23\ V$$

$$MnO_2(s) + 4\ H^+(aq) + 2\ Cl^-(aq) \rightarrow Mn^{2+}(aq) + 2\ H_2O + Cl_2(g)$$

(b) Would the reaction be spontaneous if the concentrations of H^+ and Cl^- were increased to 10.0 M?

17.27. Using the Nernst equation and Table 17.1, determine the voltage of the following cells:
(a) $Ag|Ag^+$ $(0.1M)||Au^{3+}$ $(0.01M)|Au$
(b) $Zn|Zn^{2+}$ $(0.1M)||H^+$ $(10^{-4}M)|H_2$
(c) $Al|Al^{3+}$ $(0.01M)||Zn^{2+}$ $(0.01M)|Zn$
(d) $Pb|Pb^{2+}$ $(2M)||Ag^+$ $(0.1\ M)|Ag$

17.28. For a half reaction of the form

$$M^{2+} + 2\ e^- \rightarrow M$$

what would be the effect on the electrode potential if (a) the concentration of M^{2+} were doubled, (b) the concentration of M^{2+} were cut in half?

17.29. For the half reaction

$$Cr_2O_7^{2-} + 14\ H^+ + 6\ e^- \rightarrow 2\ Cr^{3+} + 7\ H_2O$$

$E°$ is $+1.33$ V. What would the potential be if the concentration of $H^+(aq)$ were reduced to $1.00 \times 10^{-3}M$?

17.30. Using Table 17.1, determine the equilibrium constants for the following reactions:
(a) $Ni + Cu^{2+} \rightleftharpoons Ni^{2+} + Cu$
(b) $Pb + Cl_2 \rightleftharpoons Pb^{2+} + 2\ Cl^-$
(c) $3\ Zn^{2+} + 2\ Al \rightleftharpoons 3\ Zn + 2\ Al^{3+}$
(d) $Fe + Sn^{2+} \rightleftharpoons Fe^{2+} + Sn$

17.31. The standard electrode potential for the $Fe^{2+} \rightleftharpoons Fe^{3+} + e^-$ half reaction is -0.770 V. (a) Using the Nernst equation $\left(E = E° - \dfrac{0.059}{n} \log \dfrac{[Oxid]}{[Red]}\right)$ calculate the voltage of this half-cell when the $[Fe^{3+}]$ is 1.00×10^{-3} and $[Fe^{2+}]$ is 1.00×10^{-1}. (b) Calculate the approximate equilibrium constant for the half reaction.

17.32. Describe the construction of a flashlight dry cell. List the materials used and the function of each. Write equations for the half reactions occurring at the electrodes and the equation for the overall cell reaction.

17.33. List materials for the construction of a lead storage battery as used in automobiles. Write electrode reactions and the overall cell reaction. What changes occur during "charging" of the battery?

17.34. An iron spoon may be electroplated with silver in an apparatus such as that shown in Figure 17.8. A battery moves electrons from the silver anode to the iron spoon, where Ag^+ ions from the solution accept these electrons to form free silver. The latter plates out on the spoon. Write equations for the half reactions which occur at the electrodes.

Supplementary Readings

17.1. "Electrochemical Machining," J. P. Hoare and M. A. LaBoda, *Sci. American,* Jan., (1974); p. 30.
17.2. "Fuel Cells," L. G. Austin, *Sci. American,* Oct., 1959; p. 72.
17.3. *Elementary Electrochemistry,* A. R. Denaro, Butterworth, Washington, D.C., 1965 (paper).

Chapter 18

Main Energy Levels	1	2	3	4	5	6
$^{19.00}_{9}\text{F}$	K	s^2p^5				
$^{35.453}_{17}\text{Cl}$	K	L	s^2p^5			
$^{79.90}_{35}\text{Br}$	K	L	M	s^2p^5		
$^{126.91}_{53}\text{I}$	K	L	M	$s^2p^6d^{10}$	s^2p^5	
$^{210}_{85}\text{At}$	K	L	M	N	$s^2p^6d^{10}$	s^2p^5

The Halogens

Halogen	Symbol	Color and Physical State at STP	Melting Point °C	Boiling Point °C	Formula	Common Oxidation State	Reduction Potential, E° $x_2 + 2e^- \rightleftarrows 2x^-$
Fluorine	F	pale-yellow gas	−223.0	−187.0	F_2	−1	+2.87
Chlorine	Cl	greenish-yellow gas	−101.6	−34.5	Cl_2	−1	+1.36
Bromine	Br	deep red-brown liquid	−7.3	58.7	Br_2	−1	+1.07
Iodine	I	gray solid	113.0	184.0	I_2	−1	+0.54

These elements constitute a family called the **halogens,** a term which means "salt formers." As indicated above, all have an outer configuration of s^2p^5, with seven valence electrons. A number of properties of these elements may be predicted solely from the electronic configuration. Since only one electron is needed to complete the quota of eight, these elements are all very active nonmetals, with high electronegativities. The smallest of these atoms, fluorine, is the most active of all nonmetals, with the highest electronegativity. With only one unpaired electron, the common valence or oxidation state is −1. However, with seven electrons available for bond formation, these elements exhibit a number of oxidation states, ranging from −1 to +7.

Although these elements are markedly alike chemically, a gradual change in properties is shown as the atomic weight increases. Fluorine, with the lowest atomic weight, is the most active of the family, followed by chlorine, bromine, and iodine.

The order of reactivity of the free halogens is illustrated by their oxidation potentials. Thus F_2 ($E° = +2.85$ V) is the strongest oxidizing agent while fluoride ion, F^-, is the most difficult of the halides to oxidize. On the other hand I_2 ($E° = +0.54$ V) is the weakest oxidizing agent and least active of the free halogens, whereas iodide ion, I^-, is the most easily oxidized and most active of the halides.

In physical properties, too, a gradual change may be noted: fluorine and chlorine are gases at ordinary temperatures, bromine is a liquid, and iodine is a solid.

Because of their extreme activity, the halogens are not found in the free state, but are widely distributed in nature as halide salts. All are found in sea water in the form of salts.

Chlorine, the most important of the halogens, will be studied in more detail than the others. Much of what we say about chlorine will apply also to the other halogens.

18.1 CHLORINE

History

Scheele, a Swedish chemist, is credited with the discovery (1774) of chlorine. Scheele observed that a heavy, greenish-yellow gas is obtained when manganese dioxide is allowed to act on hydrochloric acid. He believed the gas to be a compound, and it remained for Davy in 1810 to establish the substance as an element and to name it "chlorine" because of its color (Greek *chloros*, greenish yellow).

Occurrence

Chlorine is found in combination with many of the metals as chlorides, the most abundant being NaCl. Sea water contains nearly 3% NaCl, and evaporation of inland bodies of water has produced large and extensive deposits of it. The Great Salt Lake in Utah contains nearly 20% of the compound. Large deposits of rock salt, which is impure NaCl, are found in New York, Michigan, Louisiana, Kansas, Oklahoma, and Texas.

In addition to sodium chloride, sea water contains smaller percentages of magnesium chloride, potassium chloride, sodium bromide, and calcium chloride. Compounds of chlorine are found in the body — sodium chloride in the blood and hydrochloric acid in the gastric juices.

Preparation

ELECTROLYSIS OF BRINE   Chlorine is prepared commercially by the electrolysis of an aqueous sodium chloride solution (page 385).

$$2 \text{ NaCl} + 2 \text{ H}_2\text{O} \xrightarrow[\text{current}]{\text{electric}} 2 \text{ NaOH} + \text{H}_2 + \text{Cl}_2$$

In the actual commercial process it is necessary to keep the chlorine separated from the aqueous sodium hydroxide solution, since chlorine reacts with the latter to form sodium hypochlorite. This may be accomplished by separating the anode and cathode compartments by a porous diaphragm.

Cl^- ions are discharged at the anode to form Cl_2 gas; H_2O is reduced at the cathode to form H_2 gas and OH^-. When the solution is evaporated, solid NaOH (lye) is formed. Cl_2, H_2, and NaOH are all important industrial products.

OXIDATION OF CHLORIDES   Free chlorine may be obtained by the removal of an electron from the chloride ion:

$$2 \text{ Cl}^- \rightarrow \text{Cl}_2 + 2 \text{ } e^- \text{ (oxidation)}$$

In the laboratory, manganese dioxide is usually employed as the oxidizing agent. Chlorine is collected by displacement of air (Figure 18.1).

$$\text{MnO}_2(\text{s}) + 4 \text{ HCl}(\text{aq}) \rightarrow \text{MnCl}_2(\text{aq}) + 2 \text{ H}_2\text{O} + \text{Cl}_2(\text{g})$$

or ionically,

$$\text{MnO}_2 + 4 \text{ H}^+ + 2 \text{ Cl}^- \rightarrow \text{Mn}^{+2} + 2 \text{ H}_2\text{O} + \text{Cl}_2$$

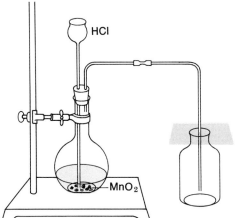

Figure 18.1
Laboratory preparation of chlorine.

Instead of hydrochloric acid, sodium chloride and sulfuric acid may be used with the manganese dioxide. This change is essentially the same as the one above, since we may assume that the sulfuric acid first acts with the salt to form hydrogen chloride, which is then oxidized by the manganese dioxide.

$$2 \text{ NaCl(s)} + 2 \text{ H}_2\text{SO}_4(\text{l}) + \text{MnO}_2(\text{s}) \rightarrow$$
$$\text{MnSO}_4(\text{aq}) + \text{Na}_2\text{SO}_4(\text{aq}) + \text{Cl}_2(\text{g}) + 2 \text{ H}_2\text{O}$$

Other oxidizing agents which may be used with HCl are KMnO_4, $\text{K}_2\text{Cr}_2\text{O}_7$, KClO_3, PbO_2, and HNO_3.

Physical Properties

At ordinary temperatures chlorine is a greenish-yellow gas with a sharp, disagreeable odor. It is very irritating to the mucous lining of the lungs, and its inhalation in large quantities may be fatal. The gas is moderately soluble in water: 2.26 vol dissolve in 1 vol of water at 20°C. The gas condenses to a liquid at a temperature of −34.5°C at a pressure of one atmosphere, and freezes to a solid at −101.6°C. Chlorine is easily liquified under pressure and is stored as a liquid in steel cylinders. Liquid chlorine is transported in railway tank cars.

Chemical Properties

Because of its great activity, chlorine reacts with a great number of elements and compounds. Examples are shown below.

METALS   Chlorine combines directly with practically all of the metallic elements to form chlorides. The more active metals unite vigorously. Powdered antimony, when sprinkled into a jar containing chlorine, bursts into flame, with the formation of a white cloud of antimony chloride. A few typical unions with metals are

$$2\ Na(s) + Cl_2(g) \rightarrow 2\ NaCl(s)$$
$$Cu(s) + Cl_2(g) \rightarrow CuCl_2(s)$$
$$2\ Sb(s) + 3\ Cl_2(g) \rightarrow 2\ SbCl_3(s)$$

NONMETALS   Most of the nonmetals combine directly with chlorine. Phosphorus burns in chlorine, forming phosphorus trichloride (PCl_3), a colorless liquid. With an excess of chlorine it forms PCl_5 a yellow solid:

$$P_4(s) + \ \ 6\ Cl_2(g) \rightarrow 4\ PCl_3(l)$$
$$P_4(s) + 10\ Cl_2(g) \rightarrow 4\ PCl_5(s)$$

COMPOUNDS OF HYDROGEN   Chlorine readily removes hydrogen from hydrocarbons (compounds of carbon and hydrogen) to form hydrogen chloride and carbon in each case. If a lighted jet of methane gas (the main constituent of natural gas) is introduced into an atmosphere of chlorine, combustion continues, with the formation of a black sooty deposit of carbon:

$$CH_4(g) + 2\ Cl_2(g) \rightarrow C(s) + 4\ HCl(g)$$

Filter paper which has been saturated with turpentine ($C_{10}H_{16}$) and inserted into a jar of chlorine rapidly takes fire, with the formation of a deposit of carbon on the sides of the vessel:

$$C_{10}H_{16}(l) + 8\ Cl_2(g) \rightarrow 10\ C(s) + 16\ HCl(g)$$

Other hydrocarbons act similarly with free chlorine.

WATER   Chlorine with water forms a solution of hydrochloric and hypochlorous ($HClO$) acids.

$$Cl_2 + H_2O \rightleftarrows HCl + HClO$$

    Hypochlorous acid is quite unstable and breaks down on exposure to sunlight:

$$2\ HClO \rightarrow 2\ HCl + O_2$$

As a result, the equilibrium above is displaced to the right, and the reaction eventually is completed to the right when the hypochlorous acid has completely decomposed.

BASES   Since a solution of two acids is produced when chlorine is dissolved in water, we can expect that the addition of a strong base to such a solution will result in the neutralization of the acids to form salts. This is true in the formation of a chloride salt and a hypochlorite salt:

$$Cl_2 + 2\ NaOH \rightarrow NaCl + NaClO + H_2O$$

or ionically

$$Cl_2 + 2\ OH^- \rightarrow Cl^- + ClO^- + H_2O$$

The resulting solution of two salts is widely used in bleaching.

Uses

Among the 50 top chemicals produced in the United States (Table 1.1), five contain chlorine. Cl_2 (seventh in the list, 18,000,000,000 pounds produced annually) is used extensively as a bleaching agent. If a piece of colored calico is dipped in a water solution of chlorine, bleaching of the color takes place rapidly. The hypochlorous acid present in the solution seems to be the active bleaching agent, since dry chlorine is not very effective. Chlorine is too corrosive to be used with fibers of animal origin (wool, silk, and so on). Much of the chlorine produced commercially is used for bleaching wood pulp, utilized in the production of paper and rayon. Pathogenic organisms are destroyed in water containing as little as one or two parts of chlorine per million; hence chlorine finds extensive use in the sterilization of drinking water.

Some recent concern is being expressed, however, that chlorinating water supplies also chlorinates some of the organic compounds present in trace quantities in many water supplies. A few of these chlorinated compounds may be carcinogenic.

Cl_2 is used extensively in making a number of everyday products such as paper, dyestuffs, textiles, medicines, automotive antifreeze and antiknock compounds, insecticides, solvents, propellants, and plastics.

Ethylene dichloride, $C_2H_4Cl_2$ (number 16 in the list), is used principally (70%) to make vinyl chloride. It is used also as a scavenging agent for leaded gasoline, as a solvent, fumigant, and an ingredient in paint removers.

Vinyl chloride, $CH=CHCl$ (number 24) is used principally to make polymers, the most important being polyvinyl chloride (PVC). Hydrochloric acid, HCl (number 25), sometimes called **muriatic acid,** commercially is used in the petroleum, chemical, food, and metal industries. Its use in increasing the activation of oil wells results from the acid dissolving parts of the limestone and other carbonate formation in the oil-bearing rock, thus allowing greater oil flow. Other uses include metal cleaning, starch hydrolysis, and making chemicals. Calcium chloride, $CaCl_2$ (number 34) is used to settle dust on highways (since it is deliquescent and remains moist) and in low-temperature refrigeration.

18.2 BROMINE

History and Occurrence

In 1826 Balard isolated a heavy, dark-brown liquid from sea salt. The liquid proved to be an element and was given the name "bromine," from the Greek word *bromos*, meaning stench. Bromides are found in natural salt brines and in deposits of such brines. The Stassfurt salt deposits in Germany and brines from wells near Midland, Michigan, furnish small quantities of the element. Bromides are also found in sea water, from which the element is extracted on a commercial scale.

Preparation

Since chlorine is more active than bromine, the latter may be liberated from its salts by treatment with chlorine water.

$$2 \, Br^- + Cl_2 \rightarrow 2 \, Cl^- + Br_2 \tag{1}$$

Like chlorine, it may be prepared by electrolysis of bromides, or by the oxidation of hydrobromic acid.

It is frequently prepared in the laboratory by treating a bromide salt with an oxidizing agent and sulfuric acid. For example,

$$2 \, NaBr(s) + MnO_2(s) + 2 \, H_2SO_4(l) \rightarrow$$
$$Na_2SO_4(aq) + MnSO_4(aq) + Br_2(g) + 2 \, H_2O$$

Commercially, bromine is obtained from sea water by displacing the bromide present with free chlorine according to equation (1) above. Sea water is pumped into large towers, where it is circulated in contact with chlorine. The liberated Br is swept upward in the towers by an air stream and concentrated by reacting with sulfur dioxide and water

$$Br_2 + SO_2 + 2 \, H_2O \rightarrow 2 \, HBr + H_2SO_4$$

the HBr being extremely soluble in water. Cl_2 is added to the HBr solution to liberate Br_2

$$2 \, HBr + Cl_2 \rightarrow Br_2 + 2 \, HCl$$

which is removed by steam to yield pure Br_2 (l). About 1 g of Br_2 is produced from 12 l of sea water.

Properties and Uses

At ordinary temperatures bromine is a heavy, red-brown liquid with a high vapor pressure, as evidenced by the brown vapor always present above a sample of the liquid. The vapor is particularly irritating to the eyes and respiratory system. Liquid bromine, if spilled on the skin, produces severe burns that are very slow to heal.

Bromine boils at 58.7°C and freezes at −7.3°C. Bromine water is produced by dissolving bromine in water, in which it is moderately soluble. Bromine dissolves readily in carbon disulfide, carbon tetrachloride, ether, and alcohol. Since bromine is much more soluble in these liquids than in water, its presence is detected by adding one of them to a solution. The element is concentrated in the added liquid and the color is intensified.

Bromine is used in the manufacture of dyes, drugs, and medicines. Potassium bromide added to silver nitrate precipitates silver bromide, which, since it is very sensitive to light, is used extensively in the preparation of photographic film.

Most of the bromine, however, is used in the manufacture of ethylene dibromide, $C_2H_4Br_2$, for antiknock gasolines. This compound prevents PbO deposits in the engines which would otherwise result from burning the tetraethyl lead in the gasoline.

In chemical properties the element is very similar to chlorine. Although bromine is somewhat less active than chlorine, it combines directly with most of the metals and nonmetals to form bromides.

18.3   IODINE

History and Occurrence

Iodine was discovered in 1812 by Courtois, who extracted crystals of the violet solid from the ashes of seaweed. Later Gay-Lussac proved this substance to be an element and named it "iodine," from the Greek word meaning "violet."

The principal source of iodine is sodium iodate ($NaIO_3$), which occurs in small amounts in Chile saltpeter ($NaNO_3$). Free iodine is liberated from the former compound by reduction with $NaHSO_3$.

Iodine in the form of sodium iodide is found in very small amounts in sea water. Certain seaweeds called **kelp** also contain small quantities and have been employed as a commercial source of the element.

Preparation

Iodine is prepared in the laboratory by the methods usually employed for the preparation of chlorine and bromine.

$$2\ NaI + MnO_2 + 2\ H_2SO_4 \rightarrow Na_2SO_4 + MnSO_4 + I_2 + 2\ H_2O$$

Iodides are more easily oxidized than either chlorides or bromides.

Since iodine is less active than chlorine or bromine, both of these elements displace iodine from iodides.

$$2\ NaI + Br_2 \rightarrow 2\ NaBr + I_2$$
$$2\ NaI + Cl_2 \rightarrow 2\ NaCl + I_2$$

Properties and Uses

Iodine is a dark gray or black solid at ordinary temperatures. When heated below its melting point of 113°C, it vaporizes rapidly. The vapor may be condensed as crystals on a cold surface, without apparent formation of liquid. The property of changing directly from solid to gas and from gas to solid without apparently liquefying is termed **sublimation.**

Iodine is only slightly soluble in water, but readily dissolves in a solution which contains iodine ion, I^-, because of the formation of I_3^-:

$$I_2(s) + I^- \rightarrow I_3^-$$

It is readily soluble in alcohol, carbon tetrachloride, and ether and, as is the case with bromine, these solvents are often used in extracting the element from water solutions. Starch forms an unstable blue compound with iodine and is used as a test for the presence of free iodine.

Although less active than the other halogens, iodine combines with most of the metals and many of the nonmetals to form iodides.

Everyone is familiar with the use of iodine as an antiseptic. A solution of iodine in alcohol is known as "tincture" of iodine. Silver iodide, like silver bromide, is light-sensitive and is used in the photographic industry.

The presence of iodine in the body seems to be essential for good health. Iodine appears to perform a vital function in the thyroid gland, in which it is found in the compound iodothyrin. If the body is not furnished with iodine in some form, this vital gland does not function properly. In a normal diet we derive sufficient iodine from such foods as butter, spinach, beans, and, in particular, sea foods.

Most of the elemental I_2 is converted into inorganic compounds, primarily KI used in medicines, photography, and pharmaceuticals, and a variety of organic compounds.

18.4   FLUORINE

History and Occurrence

Although fluorine compounds are fairly common and have been known for several centuries, it was not until 1886 that Moissan, a French chemist, isolated the element. Among the more common minerals of fluorine are **fluorspar** or **fluorite** (CaF$_2$), **cryolite** (Na$_3$AlF$_6$), and **apatite** [CaF$_2 \cdot$ 3 Ca$_3$(PO$_4$)$_2$]. It was shown in the seventeenth century that fluorspar treated with acids yields a substance capable of etching glass. This substance has since been proved to be hydrogen fluoride.

Preparation and Properties

Fluorine was the last of the halogens to be prepared in the free state because of its extreme activity. This activity gives very stable compounds, and the methods used in the liberation of the other halogens from their compounds are not applicable. Attempts to liberate fluorine from aqueous solutions by means of oxidizing agents or electrolysis fail, since fluorine reacts vigorously with water according to the equation*

$$2 \text{ F}_2(g) + 2 \text{ H}_2\text{O} \rightarrow 4 \text{ HF(aq)} + \text{O}_2(g)$$

Moissan was finally able to isolate the element by electrolyzing an an-hydrous solution of potassium fluoride in liquid hydrofluoric acid. The mix-ture is a good conductor of the current, and fluorine is liberated at the anode, while hydrogen is produced at the cathode. The composition of the container and the electrode to be used in the electrolysis process present a problem, since fluorine is extremely active. At present the containers used are made of Monel metal, an alloy of copper and nickel, which is resistant to the action of fluorine. The anode is composed of graphite.

In chemical properties the element is similar to the other halogens. It is the most active of the family and consequently will displace any of the others from their salts.

At ordinary conditions fluorine is a pale yellow gas which is very irritat-ing and extremely poisonous. Its principal use is in the preparation of fluorine compounds, such as hydrofluoric acid (HF) and a number of organic fluorine compounds.

Fluorocarbons

Increasing amounts of fluorine and hydrofluoric acid are being used in the preparation of **fluorocarbons**, compounds in which fluorine atoms (sometimes with chlorine atoms) have been substituted for all of the H atoms in hydro-carbons (Chapter 31). These compounds are inert and are extremely resistant to the action of chemicals.

The most widely used fluorocarbon is CCl$_2$F$_2$ (fluorocarbon-12) in the form of an aerosol propellent in spray cans and as a refrigerant. CCl$_3$F (fluoro-carbon-11) is usually used along with CCl$_2$F$_2$ in spray cans and in air condi-

---

* Ozone (O$_3$) is a product along with O$_2$.

tioners. These compounds are commonly called **freons.** There are over 100 different uses for these and other fluorocarbons such as plastics, films, rubbers, lubricants and fire extinguishers ($CBrF_3$).

There is widespread concern that some of the fluorocarbons which reach the upper atmosphere may eventually be broken down by the high-energy ultraviolet light present there. The chlorine atoms thus produced may enter into a chain reaction (page 316) that destroys the ozone in the stratosphere which acts to shield the earth from harmful UV radiation.

By polymerization of C_2F_4, an analogue of ethylene, a plastic called "Teflon" is produced. It is resistant to even the most active chemicals, such as ozone, chlorine, and aqua regia.

Interhalogen Compounds

Since the halogens form diatomic molecules in the free state, we might expect an atom of one halogen to combine with an atom of another halogen to form a diatomic molecule. This is so, and a number of such interhalogen compounds have been prepared. ICl, iodine monochloride, which is a reddish solid, is formed by direct combination. $BrCl$ and IBr may be prepared similarly. A few more complex types have also been made, such as ClF_3, ICl_3, BrF_5, and IF_7.

18.5   THE HYDROHALOGENS

The compounds HF, HCl, HBr, and HI are termed the **hydrohalogens,** and their water solutions are known as **hydrofluoric, hydrochloric, hydrobromic,** and **hydriodic** acids, respectively. All are colorless gases at ordinary temperatures, and all are very soluble in water.

These compounds may be prepared by direct combination of the elements. The rate of reaction varies from an extremely rapid reaction with fluorine to a slow and incomplete reaction with iodine. In general this method of preparation is not practical in the laboratory.

Preparation of Hydrogen Chloride and Hydrogen Fluoride

The usual method for the preparation of any volatile acid is to treat a salt of the desired acid with a less volatile acid and distill over the desired acid. If $NaCl$ is treated with concentrated sulfuric acid (a nonvolatile acid), hydrogen chloride, a gas at ordinary temperatures and therefore volatile, passes into the delivery tube and may be collected by displacement of air (Figure 18.2).

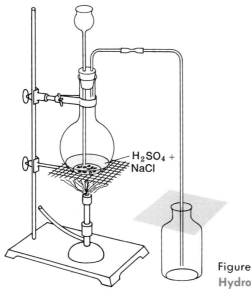

Figure 18.2
Hydrogen chloride generator.

The reaction may proceed in two steps:

$$NaCl(s) + H_2SO_4(l) \rightarrow NaHSO_4(s) + HCl(g)$$

If excess salt is present, and the mixture is heated, a second reaction takes place:

$$NaCl(s) + NaHSO_4 \rightarrow Na_2SO_4(s) + HCl(g)$$

Hydrogen fluoride is prepared from fluorspar (CaF_2) and concentrated sulfuric acid:

$$CaF_2(s) + H_2SO_4(l) \rightarrow CaSO_4(s) + 2\ HF(g)$$

The reaction is usually carried out in a lead container which is resistant to the action of HF. Glass vessels cannot be used because the acid attacks the constituents of the glass. Hydrofluoric acid is stored in wax or plastic containers.

Action of Concentrated Sulfuric Acid on Bromides and Iodides

We might expect the reaction of concentrated H_2SO_4 with bromides and iodides to be similar to its action with chlorides and fluorides:

$$NaBr + H_2SO_4 \rightarrow NaHSO_4 + HBr \qquad (1)$$

$$NaI + H_2SO_4 \rightarrow NaHSO_4 + HI \qquad (2)$$

The above reactions do take place, but the products are impure because of secondary reactions. Both HBr and HI are better reducing agents than HCl, and after their formation in reactions (1) and (2), they react with the excess sulfuric acid present. Some of the HBr is oxidized to free bromine, and as a result the product of reaction (1) is somewhat brown in color:

$$2 \text{ HBr} + \text{H}_2\text{SO}_4 \rightarrow \text{Br}_2 + \text{SO}_2 + 2 \text{ H}_2\text{O} \tag{3}$$

Hydrogen iodide is a still better reducing agent than hydrogen bromide and is almost completely oxidized to free iodine; the sulfuric acid is reduced to hydrogen sulfide:

$$8 \text{ HI} + \text{H}_2\text{SO}_4 \rightarrow 4 \text{ I}_2 + \text{H}_2\text{S} + 4 \text{ H}_2\text{O} \tag{4}$$

That HI is a better reducing agent than HBr is evident from reactions (3) and (4). While HBr reduces the oxidation number of sulfur from +6 in sulfuric acid to +4 in sulfur dioxide, HI reduces the oxidation number of sulfur from +6 to −2 in hydrogen sulfide.

Physical Properties

The hydrohalogens are colorless gases at ordinary temperatures, and are extremely soluble in water.* All possess a sharp, irritating odor. For other physical properties, see Table 18.1.

Table 18.1
Physical properties of hydrohalogens

Formula	Molecular Weight	Boiling Point °C	Freezing Point °C	Solubility in Water: Volumes of Gas in 1 Volume of Water
HF	20.0 (above 90°C)	19.4	−92	507 at 0°C
HCl	36.5	−84	−112	610 at 0°C
HBr	80.9	−67	−89	425 at 10°C
HI	127.9	−35	−51	

At temperatures above 90°C, determinations of densities of hydrogen fluoride indicate the formula to be HF; at temperatures below 90°C, the molecules are associated to correspond with the formula H_2F_2. It is likely that an equilibrium between the two forms exists:

* The extreme solubility of the hydrohalogens in water may be demonstrated by the following experiment (Figure 18.3): The upper flask is filled with hydrogen chloride gas and the lower flask with water. When the stopcock is opened, hydrogen chloride begins dissolving in water. The solution process is slow at first because a small surface of water is exposed to the gas. As the gas dissolves, a partial vacuum is produced in the upper flask and water flows up the vertical tube into the upper flask. The rate of solution increases until finally water sprays rapidly into the upper flask, and the gas seems to dissolve almost at once.

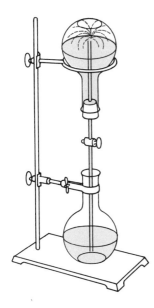

Figure 18.3
Hydrogen chloride fountain.

$$2 \text{ HF} \rightleftharpoons \text{H}_2\text{F}_2$$

Above 90°C the equilibrium is displaced almost completely to the left, but at ordinary temperatures the reverse is true. A still higher degree of association is indicated in liquid hydrogen fluoride.

Uses of Hydrochloric Acid

As an acid, hydrochloric acid is second only to sulfuric acid in importance. It finds extensive use in "pickling baths," which clean the layer of oxide from iron and steel objects before the latter are plated. Large quantities of HCl are used in the hydrolysis of starch to glucose in the manufacture of corn syrup. Other major uses are in the manufacture of textiles, dyes, and chemicals. Hydrochloric acid is important in the stomach, where it aids in digestion.

Hydrofluoric Acid and Fluorides

Hydrofluoric acid possesses the unique property of reacting with silica (SiO_2) to form silicon tetrafluoride (SiF_4), a gas at ordinary temperatures:

$$SiO_2(s) + 4 \text{ HF(aq)} \rightarrow SiF_4(g) + 2 \text{ H}_2\text{O}$$

Glass, which is a mixture of sodium and calcium silicates, is readily attacked by the acid, and the following reactions take place:

$$Na_2SiO_3 + 6\ HF \rightarrow SiF_4 + 2\ NaF + 3\ H_2O$$
$$CaSiO_3 + 6\ HF \rightarrow SiF_4 + CaF_2 + 3\ H_2O$$

Because of these reactions, hydrofluoric acid finds extensive use in the etching of glass. A glass object to be etched is coated with paraffin, and a design is cut into the paraffin. The object is then exposed to the action of hydrofluoric acid. Burets, pipets, graduated cylinders, and other pieces of chemical apparatus are etched in this way.

Hydrofluoric acid is a weak acid—the other hydrohalogen acids are very strong—and, as a result, its reactions with metals and bases are very slow. Salts of the acid are extremely poisonous and find limited use. Sodium fluoride and sodium aluminum fluoride are used as insecticides; sodium fluosilicate (Na_2SiF_6) is used as a germicide and deodorant. Fluorides in very small quantities in drinking water (about one part per million) have been shown to inhibit tooth decay.

18.6   OXYGEN COMPOUNDS OF THE HALOGENS

Oxides

Although the halogens do not combine directly with oxygen, several oxides have been prepared indirectly. Generally speaking, these compounds are unstable and may decompose with explosive violence. Cl_2O (chlorine(I) oxide), a yellowish gas which may be condensed to a liquid at 4°C, is the anhydride of HClO; ClO_2 (chlorine(IV) oxide) is a yellowish gas, which liquefies at 10°C and is the anhydride of chlorous acid, $HClO_2$. ClO_2 is an example of a so-called "odd" molecule—that is, one which contains an odd number of valence electrons (chlorine has 7 and the two oxygen atoms each have 6, to total 19). Any electronic formula drawn for this substance includes one unpaired electron. Presumably this compound is a hybrid of two or more resonance forms (see page 112).

Cl_2O_7 (chlorine heptoxide), the anhydride of perchloric acid, is a colorless, oily liquid with a boiling point of 82°C. It readily explodes when struck or heated above its boiling point. It may be prepared by dehydration of $HClO_4$. A few other oxygen compounds of the halogens—for example F_2O, F_2O_2, Br_2O, BrO_2, I_2O_3, I_2O_5 and Br_3O_8—have been prepared, but they are unstable.

Oxyacids of the Halogens

The halogens, with 7 electrons in the outer level of their atoms, are capable of exhibiting several oxidation states by sharing electrons with oxygen; consequently, they form a series of oxygen-containing acids with several acids in each series (Table 18.2).

Table 18.2

Halogen oxyacids

Name	Oxidation Number of Halogen	Fluorine	Chlorine	Bromine	Iodine
Per. . . .ic	+7	–	$HClO_4$	$HBrO_4$	HIO_4
. . . .ic	+5	–	$HClO_3$	$HBrO_3$	HIO_3
. . . .ous	+3	–	$HClO_2$	–	–
Hypo. . . .ous	+1	HFO	HClO	HBrO	HIO

The electronic configurations of the oxyacids of chlorine are given below:

$$H:\overset{\cdot\cdot}{\underset{\cdot\cdot}{O}}:\overset{\cdot\cdot}{\underset{\cdot\cdot}{Cl}}: \qquad H:\overset{\cdot\cdot}{\underset{\cdot\cdot}{O}}:\overset{\cdot\cdot}{\underset{\cdot\cdot}{Cl}}:\overset{\cdot\cdot}{\underset{\cdot\cdot}{O}}: \qquad H:\overset{\cdot\cdot}{\underset{\cdot\cdot}{O}}:\overset{:\overset{\cdot\cdot}{O}:}{\underset{\cdot\cdot}{Cl}}:\overset{\cdot\cdot}{\underset{\cdot\cdot}{O}}: \qquad H:\overset{\cdot\cdot}{\underset{\cdot\cdot}{O}}:\overset{:\overset{\cdot\cdot}{O}:}{\underset{:\overset{\cdot\cdot}{O}:}{Cl}}:\overset{\cdot\cdot}{\underset{\cdot\cdot}{O}}:$$

Hypochlorous Acid

HClO is formed when chlorine is dissolved in water:

$$Cl_2 + H_2O \rightleftharpoons HCl + HClO$$

The bleaching action of chlorine, which is effective only in the presence of water, presumably is due to the oxidizing properties of the hypochlorous acid produced.

Hypochlorous acid is a weak acid, in contrast with hydrochloric acid, which is also produced when chlorine is dissolved in water. Advantage is taken of this difference in acid strengths in separation of the two acids; the strong acid may be neutralized with $NaHCO_3$ or $CaCO_3$, leaving free hypochlorous acid in solution.

$$Cl_2 + NaHCO_3 \rightarrow NaCl + HClO + CO_2$$

A dilute solution of HClO may then be distilled from the mixture.

Bleaching Powder

Although liquid chlorine is now used extensively in bleaching, available chlorine as a solid powder of composition $CaOCl_2$ (called **bleaching powder,** or **chloride of lime**) has long been used. It is easily made and transported, and is convenient to use.

Bleaching powder is produced when chlorine gas is passed over layers of slaked lime, $Ca(OH)_2$:

$$Ca\begin{matrix}OH\\\\OH\end{matrix} + \begin{matrix}Cl\\|\\Cl\end{matrix} \rightarrow Ca\begin{matrix}Cl\\\\OCl\end{matrix} + HOH$$

Note that the compound is a mixed salt, containing both the chloride and hypochlorite acid radicals. It may be considered a mixture of the two salts $CaCl_2$ and $Ca(ClO)_2$ present in equal molecular proportions. Chlorine is readily liberated by treatment with an acid:

$$Ca\begin{matrix}Cl\\\\OCl\end{matrix} + H_2SO_4 \rightarrow CaSO_4 + \underbrace{HOCl + HCl}_{Cl_2 + H_2O}$$

If lime is treated with HClO, calcium hypochlorite is produced:

$$CaO + 2\ HClO \rightarrow Ca(ClO)_2 + H_2O$$

This compound has twice the oxidizing capacity of $CaOCl_2$. It is called "high test" hypochlorite, HTH.

Chlorates

Sodium and potassium chlorates find extensive use as laboratory oxidizing agents and in the production of matches, fireworks, and explosives. An oxidizable substance, such as wood or charcoal, burns brilliantly and rapidly when dropped into the molten salt. Potassium chlorate may be prepared by passing chlorine gas into a hot concentrated solution of KOH or by electrolysis of KCl solution. The reactions occurring in both cases may be summarized in the equation

$$6\ KOH(aq) + 3\ Cl_2(g) \rightarrow KClO_3(aq) + 5\ KCl(aq) + 3\ H_2O$$

Since potassium chlorate is much less soluble than the potassium chloride produced, it may be recovered by crystallization from the cooled solution.

The structure of the chlorate ion, ClO_3^-, is pyramidal (Figure 18.4a). Oxygen atoms form the base of the pyramid with a chlorine atom at the apex.

Perchloric Acid and Perchlorates

Perchloric acid is a colorless liquid which may explode spontaneously. It is a very hazardous chemical, and great care must be exercised in handling and using it. It is prepared by treatment of a perchlorate (preferably barium

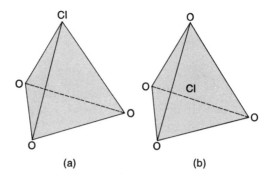

Figure 18.4
Structures of the ClO_3^- (a) and ClO_4^- (b) ions. In the formation of each ion, one electron has been gained to give each group or ion a charge of -1.

perchlorate) with dilute sulfuric acid, followed by distillation under reduced pressure. It is one of the strongest acids, comparable in strength to HCl and HNO_3. As a 60–70% solution, it is used in the analytical separation of potassium, since potassium perchlorate is one of the few slightly soluble salts of potassium.

Magnesium perchlorate is an effective desiccant or drying agent, capable of absorbing up to 30% of its own weight of water.

The perchlorate ion, ClO_4^-, has a tetrahedral structure (Figure 18.4b) with a chlorine atom at the center of the tetrahedron and one oxygen atom at each of the four corners.

Exercises

18.1. Answer as many of the following questions as you can concerning the halogens. (a) What is the meaning or origin of each halogen name? (b) Who discovered each halogen? (c) What is the physical state, color, and formula of each of the free halogens and the hydrogen halides? (d) What is the order of reactivity of the free halogens and the hydrogen halides? (e) How does each halogen generally occur in nature? (f) What is the order of stability of the various sodium halides?

18.2. Write chemical equations to show how $Cl_2(g)$ may be prepared from $Cl^-(aq)$ by use of (a) $MnO_2(s)$, (b) $PbO_2(s)$, (c) $MnO_4^-(aq)$, (d) $Cr_2O_7^{2-}(aq)$. (e) Analogous equations may be written for reactions in which bromine and iodine are prepared from bromide ions and iodide ions. However, fluorine cannot be prepared from fluoride ions by use of these oxidizing agents. Why not?

18.3. Write chemical equations to show how to prepare the following: (a) F_2 from CaF_2, (b) Cl_2 from NaCl, (c) Br_2 from sea water, (d) I_2 from KI.

18.4. How could you distinguish between: (a) KI and AgI, (b) I_2 and Cl_2, (c) HI and HBr, (d) HF and HCl, (e) $CHCl_3$ and CCl_2F_2, (f) NaCl and NaBr, (g) HCl and HClO, (h) $KClO_3$ and KCl?

18.5. Write correct formulas for:
(a) Solid iodine
(b) The most active free halogen
(c) Hydrofluoric acid
(d) Brine
(e) The most abundant natural compound of chlorine
(f) Calcium chloride
(g) Freon
(h) Hypochlorous acid

18.6. Name the following compounds:
(a) $K_2Cr_2O_7$
(b) SnF_2
(c) $(C_2F_4)_n$
(d) NaClO
(e) $C_2H_4Cl_2$
(f) HF(aq)
(g) $HClO_4$

18.7. Write formulas for compounds in which iodine exhibits oxidation numbers of $-1, +1, +3, +5, +7$.

18.8. Complete and balance the equations:
(a) $Al(s) + HCl(aq)$
(b) $NaOH(aq) + HBr(aq)$
(c) $H_2(g) + F_2(g)$
(d) $Br_2(g) + H_2O$
(e) $Cl_2(g) + NaOH(aq)$
(f) $Br_2(g) + KOH(aq)$
(g) $AlCl_3(s) + H_3PO_4(aq)$
(h) $CuO(s) + HCl(aq)$
(i) $SiO_2(s) + HF(aq)$
(j) $Mg(s) + Cl_2(g)$
(k) $NaF(s) + H_2SO_4(l)$
(l) $C_{10}H_{16}(l) + Cl_2(g)$

18.9. Show by equations how you would prepare each of the following compounds starting with elemental bromine: (a) $MgBr_2$, (b) HBr, (c) KBrO, (d) $KBrO_3$.

18.10. Write chemical equations for the reactions of Cl_2 with (a) H_2, (b) Zn, (c) P, (d) S, (e) H_2S, (f) I^-(aq), (g) cold H_2O.

18.11. Since HCl(g) can be prepared from NaCl and H_2SO_4, why is it that the reaction of NaBr and H_2SO_4 cannot be used to prepare HBr(g)?

18.12. Write the formulas and names of the products for each of the following reactions:
(a) $K_2Cr_2O_7 + HI \rightarrow$
(b) $MnO_2 + HCl \rightarrow$
(c) $HBr + H_2SO_4 \rightarrow$

18.13. Complete and balance the following:
(a) $Cl_2 + Al \rightarrow$
(b) $Br^- + Cl_2 \rightarrow$
(c) $Br_2 + SO_2 + H_2O \rightarrow$
(d) $KI + I_2 \rightarrow$
(e) $HI + H_2SO_4 \rightarrow$
(f) $HBr + H_2SO_4 \rightarrow$
(g) $NaCl + H_2SO_4 \rightarrow$
(h) $Cl_2O_5 + H_2O \rightarrow$

18.14. What is one important industrial use for (a) $Cl_2(g)$, (b) $C_2H_4Cl_2$, (c) $CH{=}CHCl$, (d) HCl, (e) $CaCl_2$, (f) $C_2H_4Br_2$, (g) AgBr, (h) HF, (i) CCl_2F_2, (j) $(C_2F_4)_n$, (k) NaClO?

18.15. Describe an experiment which would enable one to differentiate between the four colorless gases: HF, HCl, HBr, and HI.

18.16. Describe how you would differentiate between the four white salts: NaF, NaCl, NaBr, and NaI.

18.17. A 213-g sample of Cl_2 was mixed with 10.0 g of H_2 and the mixture exploded. What gases remain after the explosion, and how many grams of each?

18.18. Calculate the theoretical quantity of chlorine obtainable by the electrolysis of 5.4 kg of an 18% sodium chloride solution. What other products would be obtained and what weight of each?

18.19. The average bromine content of sea water is 0.0064%. (a) How much sea water in kilograms would be required to obtain 15 kg of bromine? (b) What volume of chlorine gas measured at STP would be required to liberate the bromine from one ton of sea water?

Supplementary Readings

18.1. *The Representative Elements,* M. J. Bigelow, Bogden and Quigley, Croton-on-Hudson, New York, 1970 (paper).
18.2. "Why the Sea is Salt," F. Macintyre, *Sci. American,* Nov., 1970; p. 104.
18.3. "Fluorine Chemistry," A. K. Barbour, *Chem. in Britain,* **5,** 250 (1969).

Chapter 19

Nitrogen and the Atmosphere

Nitrogen gas as encountered in the free state is composed of diatomic molecules, N_2, in which two atoms of nitrogen are bonded together by a sharing of the three unpaired $2p$ electrons in each atom. This combination is quite stable; consequently, molecular nitrogen is relatively inert, and considerable energy is required to break the bond as compared to most other diatomic molecules.

$$N_2 \rightarrow N + N \qquad \Delta H = 225,000 \text{ cal}$$

The bond between the two nitrogen atoms involves the $2p^3$ electrons from each atom, that is, six electrons or three electron pairs.

Importance of Nitrogen

The air we inhale is approximately 80% nitrogen, 20% oxygen, and 0.04% carbon dioxide. Exhaled air contains about 80% nitrogen, 16% oxygen, and 4% carbon dioxide. We use, then, only about 4% of the air we breathe in; our lungs pump a large amount of inert nitrogen in and out.

Combustion of fuel would be extremely rapid were it not for nitrogen in the air. A glowing cigarette placed in pure oxygen bursts into flame and quickly burns up. Smoking would be out of the question if the atmosphere were pure oxygen. Corrosion of metals would proceed at such a rapid rate that many metals would be impractical for use.

The Occurrence of Nitrogen

Tremendous quantities of free nitrogen are found in the atmosphere; it has been estimated that there are more than 20 million tons over each square mile of the earth's surface. Certain natural gases contain a small percentage of the free element, but little nitrogen is found in compounds. The main concentrated source of combined nitrogen is sodium nitrate (Chile saltpeter), deposits of which are found in Chile.

Commercial nitrogen is prepared by liquefaction of air and subsequent fractional distillation (see page 174). The nitrogen obtained by this process contains about 1% of other substances, mainly argon. Argon is also inactive chemically, so that its presence does not ordinarily interfere in the use of the nitrogen.

Properties of Nitrogen

Nitrogen is an odorless, colorless, and tasteless gas at ordinary temperatures. The boiling point at atmospheric pressure is $-195.8°C$, and the freezing point $-209.9°C$. Because of its low boiling point, nitrogen is difficult to liquefy. At standard conditions 22.4 l of the gas weigh 28 g; hence its formula is N_2. The gas is relatively insoluble in water: a little more than 2 ml dissolves in 100 ml of water at standard conditions.

Chemically, nitrogen is an inert substance; it combines with few elements and then only with difficulty. We shall refer to these reactions in the next chapter, in discussing the compounds of nitrogen.

Uses of Nitrogen Gas

Nitrogen gas ranks 8th in production in the list of 50 top chemicals (page 4), some 17 billion pounds being produced annually in the United States alone. The greatest use of N_2 is in the manufacture of ammonia, NH_3, by the Haber process (page 444). Large amounts of N_2 are also used, because of its unreactive nature, in the manufacture of electronic components such as transistors, in missile work as a purge, in the oil industry to form high pressures underground which result in an increased flow of crude oil, as a shield to prevent oxidation in the annealing of metals, as a preservative to prevent rancidity in packaged foods and so on.

19.1   OTHER GASES OF THE ATMOSPHERE

Although nitrogen and oxygen are the most abundant constituents of the atmosphere, small quantities of water vapor, carbon dioxide, dust particles, argon, and other inert gases are present. The average composition of dry air at sea level by volume is shown in Table 19.1 and Figure 19.1.

Table 19.1
Composition of dry air

Component	Percentage by Volume	Component	Percentage by Volume
Nitrogen	78.1	Hydrogen	0.00005
Oxygen	20.9	Neon	0.0018
Argon	0.93	Helium	0.00052
Carbon		Krypton	0.0001
dioxide	0.03–.04	Xenon	0.000008
Nitrogen oxides		Carbon monoxide	
(variable)	0.00003	(variable)	0.00001
Ozone (variable)	0.000002		

Carbon Dioxide in the Air

Analysis of air shows an average carbon dioxide content of about 0.035%. This carbon dioxide is produced by the respiration processes of man and animals, decay and rotting, and the combustion of coal, wood, gasoline, and other

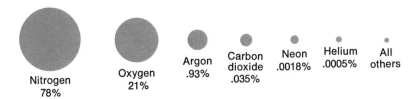

Figure 19.1
Proportions of the gaseous constituents of the atmosphere.

fuels. The content of carbon dioxide is usually appreciably higher in urban centers than in the country because of the concentration of population and industrial activity.

On the basis of the continued processes mentioned above, we might ask why the content of carbon dioxide in the air does not increase. Actually, a nearly constant balance is maintained between carbon dioxide and the other constituents of the atmosphere because plants utilize carbon dioxide of the atmosphere in building up sugars, starches, cellulose, and plant tissue. In plant tissue, water and carbon dioxide somehow combine to form complex compounds — a process called **photosynthesis.** The reactions are catalyzed by chlorophyll, the green coloring matter of plants. The reaction producing starch may be represented as

$$6 \; CO_2 + 5 \; H_2O \rightarrow C_6H_{10}O_5 + 6 \; O_2$$

Note that oxygen is also produced in the reaction and is released to the atmosphere. Energy for the above reaction is derived from the sun.

Water Vapor in the Atmosphere

Water is present in the atmosphere as a result of evaporation from oceans, lakes, and streams over the earth's crust. For every temperature there is a saturation pressure of water vapor in the air, and this saturation pressure is the vapor pressure of water at the temperature in question. Thus at $20°C$ the air is saturated with water when the pressure of water vapor in the air is 17.5 torr (Table 10.3, page 224). Usually the pressure of water vapor in the air is less than this saturation pressure. The ratio of the pressure of the water vapor in the air to the saturation pressure at any given temperature is called the **relative humidity.** If the pressure of water vapor in the air is 10 torr at $20°C$, the humidity is $\frac{10}{17.5} = 57.1\%$. Of course, if the pressure of water vapor were 17.5 torr, the humidity would be 100%.

When the atmosphere is cooled during the night, the cooling may progress to the point where the pressure of water vapor in the air becomes greater than the saturation pressure. When this occurs, dew or fog is formed as a result of condensation of water vapor from the air. Clouds result from the rising and subsequent cooling of moist warm air, and the formation of large droplets may result in precipitation of rain.

19.2 THE INERT GASES

Lord Rayleigh and William Ramsey in 1894 discovered an inert gas in the atmosphere which they called **argon,** meaning "inert." The discovery came as a result of studies of the atomic weight of nitrogen. Nitrogen derived from the atmosphere invariably gave a higher atomic weight than nitrogen derived by chemical means. These studies led to the conclusion that air contains, besides nitrogen, some inert substance more dense than nitrogen. Further investigation resulted in the isolation of argon from the atmosphere. Argon is present in the air to the extent of nearly 1%. Since the discovery of argon, other rare gases have been isolated from the atmosphere, including helium (Gr. *helios*, the sun), neon (Gr. *neosm*, new), krypton (Gr. *kryptos*, hidden), xenon (Gr. *xenon*, stranger), and radon (L. *nitens*, shining).

Gas	Boiling Point, °C	Freezing Point, °C	Critical Temp.	Atomic Radius, Å	Quantity in Air, Parts per Million
Helium	−269	−269.7°	−268	0.93	5.2
Neon	−246	−249	−229	1.12	18
Argon	−186	−189	−122	1.54	9300
Krypton	−152	−169	−63	1.69	1.0
Xenon	−109	−140	−17	1.90	0.08
Radon	−62	−71	−	2.2	trace

° At 10 atm pressure.

The inactivity of these elements (Group VIIIA of the periodic table) may be explained on the basis of their atomic structure. Helium has two electrons in the first energy level (its outer level), which represents a stable configuration. The other elements have eight electrons in the outermost level, which also represents a stable configuration. The atom of each of these elements is completely satisfied in its electronic configuration and so has little tendency to give up, take up, or share electrons.

Helium is unique among the chemical elements in that it was discovered in the atmosphere around the sun before being found on the earth. As much as 2% helium is found in certain natural gas wells in Kansas and Texas. Its source is the radioactive decay of certain elements found in minerals. Helium has the lowest boiling point of all substances known; consequently, it may be separated from the other constituents of natural gas by lowering the temperature sufficiently to liquefy or freeze the other components. The boiling point of helium is −269°C; its freezing point (−272.2°C at 26 atm pressure) is just a degree above the absolute zero temperature.

Uses of Inert Gases

Prior to 1946 helium was used mainly in balloons and other lighter-than-air craft. While helium is somewhat heavier[*] than hydrogen, it has the important advantages of being noncombustible and having a lower rate of diffusibility. It is used in diving bells to lessen the danger of caisson disease (**the bends**), which results from a too-rapid escape of nitrogen from the blood stream of divers coming to the surface. Since the solubility of helium in blood is much less than that of nitrogen, it may advantageously be mixed with oxygen for use under high pressures. It is also used for shielded-arc welding and low-temperature research. Liquid helium boils at 4.25°K, so that this extremely cold liquid can condense any other gases. A large use of He has been for pressurizing liquid fuel rockets. The Saturn booster, for example, used on the Apollo lunar missions required about 13 million cubic feet of He for this purpose for each firing.

One of the important uses of the inert gases, particularly neon, is to produce colored light in discharge tubes. Although these gases do not form ions in ordinary chemical changes, ions may be produced by removal of electrons in a discharge tube under a high alternating current potential. The gas is enclosed in a tube at a low pressure (5–10 torr), where the application of several thousand volts produces ions and renders the gas a conductor of the current. Light of a characteristic color is emitted because of the "activation" of the atoms through a displacement of electrons.

Argon is used in filling electric light bulbs and electronic tubes; it has certain advantages over nitrogen in prolonging the life of the filaments, mainly chemical inertness.

Radon will be considered in Chapter 35. Its only uses depend on its radioactivity; it is collected in small tubes as emanation from radium samples; these tubes may be used in radiotherapy for malignant growths.

---

[*] Although helium is twice as heavy as hydrogen, its lifting power is almost as good. The lifting power of a gas in a balloon depends on the difference between the density of the gas on the inside and the air on the outside.

$$22.4 \text{ ft}^3 \text{ H}_2 = 2 \text{ oz}$$
$$22.4 \text{ ft}^3 \text{ He} = 4 \text{ oz}$$
$$22.4 \text{ ft}^3 \text{ air} = 29 \text{ oz}$$

Lifting power:
$$22.4 \text{ ft}^3 \text{ H}_2 = 29 - 2 = 27 \text{ oz}$$
$$22.4 \text{ ft}^3 \text{ He} = 29 - 4 = 25 \text{ oz}$$

Thus helium is $\frac{25}{27}$, or about 93%, as effective as hydrogen in lifting power.

Compounds of the Inert Gases

Although the inert gases are characterized by their chemical inactivity, a few compounds of these elements have been prepared in recent years. For example, at an elevated temperature, xenon and fluorine form xenon tetrafluoride, XeF_4, with a planar square structure. To explain the bonding in such compounds, one may assume that one or more electrons are promoted from the valence configuration to outer d orbitals, giving rise to unpaired electrons which would be available for bonding. For example, if two electrons are promoted (one s and one p),

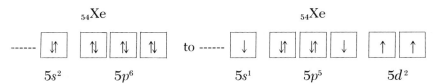

followed by hybridization to spd^2 orbitals, then four electrons are available for pairing and bond formation to account for the formation of XeF_4.

Preparation of colorless crystals of XeF_4 (melting point about 90°) in 1962 stimulated further research for additional compounds of the previously named inert or noble gases. The name in general still applies, and it is mainly the rare heavier elements, krypton and xenon, that have their outer orbitals electrons excited by a strong oxidizing element such as F_2. Examples of some of the new compounds are XeF_2, KrF_2, $XePtCl_6$, XeF_6, XeO_3, $RbXeF_7$, $RbXeF_8$, $XeF_2 \cdot 2SbF_5$, and XeO_4.

19.3   POLLUTION

Air Pollution

Pollutants are basically either particulate or gaseous. The particulates are termed dusts, smokes, aerosols, and fogs. Smog is comprised of smoke and fog. The particulates, although bothersome, are not as difficult to deal with as gaseous air contaminants. Dusts and smokes are mechanically removable from air, and in themselves are not usually chemically active. Table 19.2 includes the major contaminants and their sources.

Automotive exhaust is a major pollutant in urban regions. The CO content from incomplete gasoline or fuel oil combustion is not the major problem, except in confined spaces. In the air it dilutes and slowly oxidizes to CO_2.

Except for the air in large cities, most of the CO in the atmosphere (80%) results from the oxidation of methane produced by decaying organic materials. Some CO is also released from the oceans in which a certain amount

Table 19.2

Magnitude of air pollution (U.S.)

Categories
Transportation ....144 million tons/year
Industry... 36
Elec. generating plants and
Building heating................................... 45
Refuse disposal.................................... 11
Field burning, forest fires, etc. 28
Specific contaminants
CO..147
SO_2, SO_3 34
Hydrocarbons..................................... 35
Particulates 25
NO, NO_2... 23

of the gas is always dissolved. CO is removed from the atmosphere, however, by natural oxidation processes such as OH^- ions and soil bacteria which convert the CO to CO_2.

Oxides of nitrogen, the chemistry of which is emphasized in the following chapter, are produced and released into the atmosphere, mainly by the automobile and electric power plants (except for the fixation processes in nature, which are part of the nitrogen cycle). The nitrogen compounds are not present appreciably in the fuels used but are formed from the nitrogen and oxygen gases of the air as a by-product of the fuel combustion. High temperatures (as in a car cylinder explosion) cause a small percentage of fixation followed by other chemical action.

$$N_2 + O_2 \xrightarrow{\text{high temp.}} 2\ NO \qquad 2\ NO + O_2 \longrightarrow 2\ NO_2 \text{ (brownish gas)}$$

$$3\ NO_2 + H_2O \longrightarrow 2\ HNO_3 + NO \qquad NO_2 \xrightarrow{\text{sunlight}} NO + (O)$$

$$(O) + O_2 \longrightarrow O_3 \text{ (ozone)}$$

The brown color of NO_2 indicates its capability to absorb ultraviolet rays of sunlight for energy to aid in ozone formation. O_3 and NO_2 are strong oxidants and aided by light energy partially oxidize hydrocarbons. Hydrogen-saturated hydrocarbons and double-bonded hydrocarbon vapors are also present in car exhaust, as cylinder combustion is never 100% complete. Partial oxidation of these hydrocarbon vapors, some in the hot air of exhaust, some in the air by NO_2 and O_3, yields identifiable smoke constituents of an aldehyde or organic acid nature. For example a peroxyacetyl nitrate (PAN to the smog scientist, $CH_3\overset{\|}{\underset{O}{C}}-O-O-NO_2$) is a known eye irritant and tear producer. Aldehydes

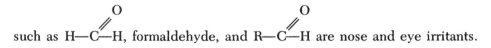

such as $H-\overset{O}{\overset{\|}{C}}-H$, formaldehyde, and $R-\overset{O}{\overset{\|}{C}}-H$ are nose and eye irritants.

Automotive fuels have been purified so as to contain little sulfur; thus oxides of sulfur are not important air contaminants from car exhaust.*

Exercises

19.1. Where and in what form are the following usually found in nature?
(a) N_2, (b) N compounds, (c) He, (d) CO_2, (e) H, (f) Ar, (g) O.

19.2. How could you distinguish between:
(a) N_2 and O_2, (b) N_2 and He, (c) He and Ar, (d) N_2 and CO_2, (e) N_2 and H_2?

19.3. Consider a cubic meter of dry air at STP (Table 19.1). Calculate:
(a) the partial pressure of N_2, (b) the volume of O_2 at STP, (c) the number of molecules of Ne, (d) the weight of Xe, (e) the number of moles of Kr.

19.4. What arguments could you use that air is a solution of gases rather than a compound?

19.5. Explain why the percentage of CO_2 in air stays at nearly a constant amount, about 0.035%. What factors in nature tend to increase or decrease this?

19.6. The pressure of water vapor in a sample of air at 10°C was 7.6 torr.
(a) Calculate the relative humidity of the sample by referring to Table 10.3.
(b) Calculate the relative humidity of the sample if heated to 30°C.

19.7. A balloon contains 15,000 l of He at STP. What weight in kg could be lifted in air, assuming the balloon itself to be weightless. Refer to the footnote on page 436.

19.8. Hydrogen and helium are used to fill balloons. (a) At STP, what is the weight of 22.4 l of each of the following: H_2, He, and air? (Assume that air is 80.0% N_2 and 20.0% O_2 by vol.) (b) Compare the lifting power of a balloon filled with H_2 with that of a balloon filled with He. What must the volume of an He-filled balloon be in order to compare in lifting power with a 100-l balloon filled with H_2? Why should He be used in place of H_2?

19.9. Argon is used to fill electric light bulbs and to provide an atmosphere under which high-temperature welding may be carried out. Justify these uses.

19.10. Write separate equations for the formation of a dilute acid fog as NO and SO_2 pollutants react with air moisture.

19.11. Write equations showing:
(a) the formation of both CO and CO_2 simultaneously when heptane (C_7H_{16}) in gasoline is burned;

* Some doctors specify asbestos fibre dust in air, from brake lining wear, as being a lung irritant and a possible precursor of lung cancer and emphysema.

(b) the formation of sulfur dioxide when organic sulfur compounds, such as C_2H_5SH, are burned;

(c) photosynthesis.

19.12. Assume that an octane (C_8H_{18}) commercial gasoline is subject to air oxidation in an automotive cylinder. List a number of gaseous molecules that may exit from exhaust pipe in marked or trace amounts.

19.13. The weight of 6.34 l of nitrogen gas at 21°C and 720 torr pressure is 7.00 g. Calculate the molecular weight of nitrogen from this data.

19.14. Calculate the approximate number of argon atoms you inhale and exhale with each 1.12 l of air you breathe. Assume STP conditions.

19.15. (a) Write a balanced equation for the oxidation, in acid solution, of Mn^{2+} to MnO_4^- by XeO_3, which is reduced to $Xe(g)$ [$H_2O + XeO_3 + Mn^{2+} \rightarrow Xe + MnO_4^- + H^+$]. (b) In a reaction between XeO_3 and Mn^{2+} that occurs in 200 ml of solution, the $Xe(g)$ produced occupied a vol of 448 ml at STP. What are the normality and molarity of the MnO_4^- solution resulting from the reaction? (c) What weight of XeO_3 was consumed by the reaction?

Supplementary Readings

19.1. "The Chemistry of Planetary Atmospheres," W. T. Huntress, Jr., *J. Chem. Educ.*, **53**, 204 (1976).

19.2. "A Decade of Xenon Chemistry," G. J. Moody, *J. Chem. Educ.*, **51**, 628 (1974).

19.3. *The Noble Gases*, H. H. Claassen, Health, Boston, 1966 (paper).

19.4. "The Automobile and Air Pollution," T. R. Wideman, *J. Chem. Educ.*, **51**, 290 (1974).

Chapter 20

Main Energy Levels	1	2	3	4	5	6
$^{14.007}_{7}\text{N}$	K	s^2p^3				
$^{30.974}_{15}\text{P}$	K	L	s^2p^3			
$^{74.92}_{33}\text{As}$	K	L	M	s^2p^3		
$^{121.75}_{51}\text{Sb}$	K	L	M	$s^2p^6d^{10}$	s^2p^3	
$^{208.98}_{83}\text{Bi}$	K	L	M	N	$s^2p^6d^{10}$	s^2p^3

Nitrogen and Other Group VA Elements

Group VA elements are characterized by the outer s^2p^3 electron configuration. Nitrogen furnishes a pattern for the similar elements phosphorus, arsenic, antimony, and bismuth. As the atomic weight increases, a gradual change from distinct nonmetallic properties to a combination of metallic and nonmetallic properties in arsenic and antimony is observed. Bismuth is decidedly metallic. The latter may either give up three electrons to form Bi^{3+} or share three or five electrons to exhibit oxidation states of +3 or +5. There is virtually no tendency for bismuth to accept electrons. On the other hand nitrogen, antimony, arsenic, and especially phosphorus accept three electrons to pair up electrons in the outer p orbitals. Hence antimonides, arsenides, phosphides, and nitrides may be prepared. Sharing of three or five outer electrons accounts for the +3 and +5 oxidation states.

Compounds of the elements are usually covalent — that is, the atoms are held together by a sharing of electrons. The elements form many similar compounds.

Hydrides	Chlorides	Oxides	Acids
NH_3	NCl_3	N_2O_3; N_2O_5	HNO_2; HNO_3
PH_3	PCl_3	P_4O_6; P_4O_{10}	H_3PO_3; H_3PO_4
AsH_3	$AsCl_3$	As_4O_6; As_2O_5	H_3AsO_3; H_3AsO_4
SbH_3	$SbCl_3$	Sb_4O_6; Sb_2O_5	$HSbO_2$; H_3SbO_4
BiH_3	$BiCl_3$	Bi_2O_3	$HBiO_3$

20.1 NITROGEN FIXATION

Nitrogen in the combined state — that is, in compounds — is far more useful and important than in the free state. Animals and plants (with few exceptions) cannot use free nitrogen, but must depend on nitrogen compounds for this essential element. It is the task of the chemist to convert free nitrogen into compounds which can be utilized. Any process by which elementary nitrogen is converted into nitrogen compounds is termed **nitrogen fixation.**

Nitrogen compounds are essential to plant and animal life in building proteins, which are complex organic substances containing the elements carbon, hydrogen, oxygen, nitrogen, and sometimes phosphorus and sulfur. Animals must eat protein directly, because the body is incapable of building up proteins from other nitrogen compounds. Beans, peas, and meat contain considerable protein. The plant, on the other hand, can synthesize plant protein using ammonium salts or nitrate salts, which it derives from the soil. The soil may get some nitrogen compounds from the decay of plants, leaves, and other organic matter. Farm soil must be replenished frequently with fertilizer in the form of animal refuse matter, manure, or compounds of nitrogen, such as ammonium sulfate or calcium nitrate.

Methods for the fixation of nitrogen can be divided into two groups: (1) natural fixation and (2) artificial fixation.

Natural Fixation

Certain plants, such as peas, beans, alfalfa, and clover, called **legumes,** are equipped with nodules which contain nitrogen-fixing organisms. These organisms are able to assimilate nitrogen directly from the atmosphere and convert free nitrogen into nitrogen compounds. To replenish the soil with nitrogen, legumes may be grown on the depleted soil and plowed under; decomposition occurs, furnishing fixed nitrogen to the soil. A system of crop rotation using leguminous plants for maintaining soil fertility is now common practice.

One of nature's principal methods of replenishing the supply of nitrogen compounds in the soil occurs as a result of electrical discharges in the air and subsequent rainfall. At the high temperature associated with lightning bolts or other arc discharges, nitrogen and oxygen gases combine to form nitrogen oxides. These oxides, combined with air moisture, form nitrous and nitric acids, which come to the earth in rainfall. A person out in a thunder shower is being subjected to a shower of very dilute nitric and nitrous acids. As these acids act on soil minerals, soluble nitrogen-containing salts are formed in soil waters. For example, as very dilute nitric acid acts on limestone ($CaCO_3$), calcium nitrate is formed; this salt is a common soil mineral. Considering the bulk of the atmosphere as a factory for HNO_3 production and the geologic time of operation, we can account for millions of tons of nitrate compound formation by the equations

$$N_2 + O_2 \xrightarrow[\text{electrical discharges}]{\text{lightning arc}} 2\ NO$$

$$2\ NO + O_2 \rightarrow 2\ NO_2$$
$$3\ NO_2 + H_2O \rightarrow 2\ HNO_3 + NO$$
$$CaCO_3 + 2\ HNO_3 \rightarrow Ca^{2+} + 2\ NO_3^- \text{ (in soil)} + H_2O + CO_2$$

Artificial Fixation of N_2

ARC PROCESS   Many years ago man learned how to hasten artificially one of nature's methods of nitrogen fixation. Cavendish discovered that nitrogen and oxygen will combine if air is passed through an electric discharge:

$$N_2 + O_2 \xrightarrow{\text{arc}} 2\ NO \qquad \Delta H = 43{,}200\ \text{cal}$$

Since this reaction is endothermic, it is favored by high temperatures (Le Chatelier's principle). Thus air passed through an electric arc or through a highly heated checkerwork of bricks is converted in small percentage to oxides of nitrogen. (See page 444 for subsequent equations in which man patterns after nature in large scale industrial production of HNO_3.) Birkeland and

Eyde of Norway first perfected the arc process, which depends on the production of cheap electricity if it is to be commercially successful.

HABER PROCESS   When nitrogen and hydrogen gases are heated together in the presence of certain catalysts, some ammonia is formed, according to the reaction

$$N_2 + 3\ H_2 \rightleftharpoons 2\ NH_3 \qquad \Delta H = -22.1\ \text{kcal}$$

The reaction is very incomplete at ordinary conditions, with equilibrium established when the ammonia content is less than 1%. Since the reaction is exothermic, it is evident from a consideration of Le Chatelier's principle that a low temperature is favorable to the formation of ammonia. If the temperature is below 500°C, the combination of nitrogen and hydrogen occurs too slowly to be practical as a commercial means of preparation. Applying Le Chatelier's principle, we note that an increase of pressure will shift the equilibrium to the right — that is, toward the formation of ammonia. An increase of pressure from 1 atmosphere at 500°C to 200 atmospheres at 500°C increases the equilibrium yield of ammonia in the above reaction from 0.13% to 17.6%. The process is ordinarily carried out at a temperature of approximately 500°C and a pressure of 200 atmospheres in the presence of a catalyst mixture, such as Fe_3O_4 and $K_2O \cdot Al_2O_3$. Ammonia is liquefied from the equilibrium mixture by rapid refrigeration of the gas mix, and the residual N_2 and H_2 is counterpassed back to the high-pressure catalyst chamber. Haber, a German chemist, perfected the process just before World War I began. In World War II Germany again had to rely on synthetic nitrogen compounds.

A fellow German chemist, Ostwald, worked out the chemistry and industrial details to convert NH_3 into HNO_3, a compound essential to explosive manufacture. Ammonia and air, in the proportions of one volume of ammonia to about ten volumes of air, are passed over a platinum catalyst at a temperature of about 700°C. The ammonia is oxidized to nitric oxide (a colorless gas):

$$4\ NH_3 + 5\ O_2 \rightarrow 4\ NO + 6\ H_2O \qquad \Delta H = -216.0\ \text{kcal}$$

NO readily combines with excess oxygen to form nitrogen dioxide (NO_2), a brownish-red gas:

$$2\ NO + O_2 \rightarrow 2\ NO_2 \qquad \Delta H = -27.0\ \text{kcal}$$

Nitrogen dioxide is then dissolved in water to produce nitric acid and nitric oxide:

$$3\ NO_2 + H_2O \rightarrow 2\ HNO_3 + NO$$

The NO produced is again mixed with air and converted to NO_2. The acid produced by this method is about 50% HNO_3 and may be concentrated by distillation to a 68% HNO_3 solution as currently used in industry. For certain specific uses, the latter may be concentrated to 95–98% by adding concentrated H_2SO_4 and distilling from a retort.

20.2 EARTH SOURCES OF NITROGEN COMPOUNDS

Coal, a product of the decomposition of vegetation, contains a small percentage of combined nitrogen. In the production of coke, when coal is heated in the absence of air, an ammoniacal liquor is produced as a by-product. Ammonia is recovered from this liquor by distillation and subsequent absorption in water. The ammonium hydroxide solution thus produced may be converted into ammonium salts by the addition of acids. Most of the ammonium sulfate used as a fertilizer is prepared in this manner.

The export of Chile saltpeter, $NaNO_3$ (some KNO_3, $NaIO_3$), from bird guano deposits has long been important to the economy of Chile. Very few other arid region deposits of nitrate are available for mining. A nitrate may be readily converted into nitric acid, which may be used to prepare other nitrate compounds.

The Nitrogen Cycle

The cycle of changes which nitrogen undergoes — the **nitrogen cycle** of nature — can best be shown by a diagram (Figure 20.1). Although the process involves more steps than are indicated, the diagram represents the essential changes taking place.

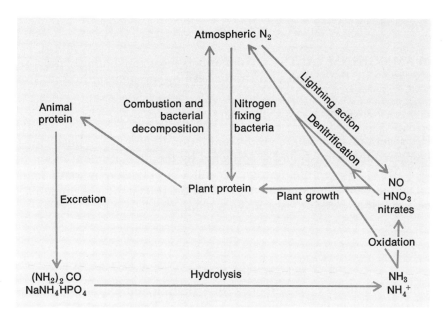

Figure 20.1

The nitrogen cycle.

We may start with atmospheric nitrogen and trace some of the possible changes. A nitrogen atom in the free state may be fixed by (1) legumes into plant protein, (2) lightning discharges into nitrate salts, or (3) artificial processes into ammonium or nitrate salts. A plant may then utilize the salts for production of plant protein. The plant may serve as a food for animals, and the original nitrogen atom may become a part of animal protein, or degradation of the plant may again produce ammonium salts. Animal protein, in undergoing decay or degradation, may be converted either into ammonium salts or into free nitrogen, which finds its way back into the atmosphere where it is ready to start another cycle of changes.

20.3   AMMONIA AND AMMONIUM SALTS

Decay of nitrogenous matter in the absence of air produces ammonia. The decomposition may be brought about with the aid of heat or in the presence of putrefying bacteria. As a result of the action of bacteria on nitrogen compounds, small quantities of ammonia gas may be present in air. Because of the great solubility of ammonia in water, however, it rapidly finds its way, by the action of rain and snow, to the soil, where it may be converted into other compounds.

Laboratory Preparation of Ammonia

FROM AMMONIUM SALTS   Ammonia is most conveniently prepared in the laboratory by the action of a strong base on an ammonium salt. For example,

$$NH_4Cl + NaOH \rightarrow NH_3(g) + H_2O + NaCl$$

The reaction is caused to go to completion by the use of heat, since the ammonia escapes as a gas and is collected by the displacement of air. Since the gas is extremely soluble in water, it cannot be collected by water displacement. The reaction given above is typical for the preparation of ammonia. Any ammonium salt and strong base may be employed.

Properties of Ammonia

Ammonia is a colorless gas with a sharp, irritating odor at ordinary temperatures. It is easily liquefied, forming a colorless liquid the boiling point of which is $-33°C$. The gas is extremely soluble in water; one liter of water dissolves more than one thousand liters of the gas at $0°C$.

When ammonia is dissolved in water, it reacts with the water to form a basic solution:

$$NH_3 + H_2O \rightleftharpoons NH_4^+ + OH^-$$

This is an equilibrium reaction; the change to the right takes place to only a slight extent. Consequently, ammonia is referred to as a weak base. Concentrated aqueous ammonia contains about 28% ammonia at ordinary conditions of temperature and pressure. Since it has basic properties, aqueous ammonia or ammonia gas reacts with acids to form ammonium salts.

$$NH_3 + HCl \rightarrow NH_4Cl$$
$$2\ NH_3 + H_2SO_4 \rightarrow (NH_4)_2SO_4$$

The reaction of ammonia with acid gases in the laboratory causes the familiar white salt deposits on laboratory glassware and windows.

Ammonium salts have one characteristic generally not possessed by most salts. If an ammonium salt such as NH_4Cl is heated, it is dissociated according to the reaction

$$NH_4Cl(s) \rightarrow NH_3(g) + HCl(g)$$

The ammonia and hydrogen chloride gases recombine on cooling, forming solid ammonium chloride.

Uses of Ammonia and Ammonium Salts

Ammonia (NH_3, number 3 in the list of the 50 top chemicals; 31,600,000,-000 pounds produced annually) has its greatest use in the manufacture of fertilizers such as NH_4NO_3 (number 12 in the list), $(NH_4)_2SO_4$ (number 27), $(NH_4)_3PO_4$, and urea (number 15), and in the direct application of NH_3 to the soil as a fertilizer. The equipment consists essentially of tractor-mounted tanks of liquid ammonia and a battery of several steel tubes welded to the backs of cultivator shovels for injecting the gas into the soil. These tubes are connected by hoses to the ammonia tanks, from which the flow of ammonia gas can be controlled with a meter. Very little loss occurs if the gas is placed about six inches below the surface. The feeding of ammonia gas directly into irrigation streams is also successful. A solution of NH_3 in water is used in the household as a cleaning agent and water softener.

Ammonia is used extensively as a refrigerant. If liquid ammonia is allowed to evaporate, it absorbs much heat from its surroundings. The gas may then be compressed and liquefied again, after which it once more goes through the cycle of evaporation and liquefaction.

Other uses for ammonia include the manufacture of nitric acid (number 11 of the 50 top chemicals), explosives, plastics, rubber, textiles, food additives, and drugs.

Other Compounds of Nitrogen and Hydrogen

Two additional nitrogen-hydrogen compounds merit mention. A compound $H_2N—NH_2$ or N_2H_4 is a liquid called hydrazine; it finds specific use as a fuel in some rockets. Its water solution acts as a weak base and on neutralization forms salts such as N_2H_5Cl and $N_2H_6Cl_2$. A compound HN_3 (note formula difference from NH_3) is called **hydrazoic acid.** The water-insoluble salts of this acid, silver azide (AgN_3) and lead azide (PbN_6), are very sensitive explosives; consequently, they are used as detonators to initiate the explosion of less sensitive explosives.

20.4   NITRIC ACID AND NITRATES

Nitric acid (HNO_3) was prepared by the alchemists from saltpeter (KNO_3) and concentrated sulfuric acid (H_2SO_4). It was used for separating gold from silver, the silver being soluble and the gold insoluble. The acid was called **aqua fortis,** meaning "strong water."

Since pure HNO_3 (hydrogen nitrate) is volatile (b.p. 86°C) and can be distilled easily, one may use cheap nonvolatile H_2SO_4 for its preparation. Sodium nitrate (Chile saltpeter) is the salt commonly employed.

$$2 \text{ NaNO}_3 + \text{H}_2\text{SO}_4 \rightarrow \text{Na}_2\text{SO}_4 + 2 \text{ HNO}_3(g) \text{ (gas at 86°C)}$$

Since the pure gaseous or liquid hydrogen nitrate is active and attacks rubber, an all-glass apparatus is essential for the distillation in a laboratory preparation; iron retorts are used in industry, as concentrated HNO_3 acts very slowly on iron.

The Haber process to make NH_3 followed by the Ostwald process to catalytically burn NH_3 to NO, then to NO_2 plus H_2O to give HNO_3, has been adequately covered under nitrogen fixation. Likewise the Birkeland–Eyde arc process was discussed. A bulk quantity of commercial HNO_3 is prepared by the chemistry of these processes.

Properties of Nitric Acid

Nitric acid is a colorless liquid when pure, with a specific gravity of 1.5 and a boiling point of 86°C. Commercial nitric acid is about 68% HNO_3 by weight and has a specific gravity of about 1.4.

The chemistry of HNO_3 is slightly complicated because it can act in several ways according to its concentration and the material with which it is reacting. As an acid it is highly ionized in dilute aqueous solution. It exhibits typical acid reactions, reacting with metallic oxides and bases to form salts. In addition to its acidic properties, nitric acid acts as a powerful oxidizing agent and also exhibits a nitrating property.

OXIDATION ACTION WITH METALS   Nitric acid reacts with practically all of the metals, even those below hydrogen in the activity series except Au and Pt. This would seem to be contradictory to the rule that only metals above hydrogen react with acids; however, nitric acid reacts with metals below hydrogen not because of its acid properties but because of its oxidizing action:

$$3\ Cu + 8\ HNO_3 \rightarrow 3\ Cu(NO_3)_2 + 2\ NO + 4\ H_2O$$

or

$$3\ Cu + 8\ H^+ + 2\ NO_3^- \rightarrow 3\ Cu^{2+} + 2\ NO + 4\ H_2O$$

Other metals react similarly. The reduction products of nitric acid depend on concentration and temperature, and on the chemical nature of the reducing agent and its degree of subdivision if it is a metal. Actually, it is only in rare cases that a single product such as NO is obtained; usually both NO and NO_2 are products in varying proportions.

Thus dilute acid produces more nitrogen(II) oxide, while concentrated acid produces more nitrogen(IV) oxide:

$$Cu + 4\ HNO_3 \rightarrow Cu(NO_3)_2 + 2\ NO_2 + 2\ H_2O$$

Very active metals may react with nitric acid to produce hydrogen only if the acid is very dilute and cold.

OXIDATION ACTION WITH NONMETALS   Hot concentrated nitric acid oxidizes several of the nonmetals, as shown by the equations

$$S + 4\ HNO_3 \rightarrow SO_2 + 4\ NO_2 + 2\ H_2O$$
$$P + 5\ HNO_3 \rightarrow H_3PO_4 + 5\ NO_2 + H_2O$$
$$C + 4\ HNO_3 \rightarrow CO_2 + 4\ NO_2 + 2\ H_2O$$

ACTION WITH HYDROCHLORIC ACID   A mixture of three moles of hydrochloric acid to one mole of nitric acid is called **aqua regia**, meaning "royal water."

$$3\ HCl + HNO_3 \rightarrow 2\ H_2O + NOCl + Cl_2$$

Aqua regia dissolves gold and platinum, metals inactive in either of the acids separately. It is probably the elemental Cl_2 that acts with the metals.

$$2\ Au + 3\ Cl_2 \rightarrow 2\ AuCl_3 \qquad AuCl_3 + HCl \rightarrow HAuCl_4$$
$$Pt + 2\ Cl_2 \rightarrow PtCl_4 \qquad PtCl_4 + 2\ HCl \rightarrow H_2PtCl_6$$

EXPLOSIVES MANUFACTURE (NITRATION REACTIONS)   Nitroglycerin is prepared by the action of a concentrated nitric acid-sulfuric acid

Table 20.1
Oxides of nitrogen

Name	Formula	State	Preparation	Use or Relationship
Nitrogen(I) oxide (nitrous)	N_2O	colorless gas	$$NH_4NO_3 \xrightarrow{\text{low } \Delta} N_2O + 2\,H_2O$$	Pleasant odor, caused exhilaration, was named laughing gas, early use as anesthetic in dentistry, minor surgery. Gas supports burning.
Nitrogen(II) oxide (nitric)	NO	colorless gas	$$\text{(dilute)}$$ $$3\,Cu + 8\,HNO_3 \rightarrow 2\,NO + 3\,Cu(NO_3)_2 + 4\,H_2O$$ $$\xrightarrow{\text{arc}}$$ $$N_2 + O_2 \rightarrow 2\,NO$$	Colorless gas, water insoluble, a nitrogen fixation product.
Nitrogen(III) oxide (trioxide)	N_2O_3	blue liquid boils 3°C	$$2\,HNO_2 \rightarrow N_2O_3 + H_2O$$	Anhydride of HNO_2.°°
Nitrogen(IV) oxide (dioxide)	NO_2.°	red-brown gas	$$\text{Nitric acid prep.}$$ $$Cu + 4\,HNO_3\ (\text{conc.}) \rightarrow 2\,NO_2 + Cu(NO_3)_2 + 2\,H_2O$$	
Nitrogen(V) oxide (pentoxide)	N_2O_5	white solid melts 30°C	$$4\,HNO_3 + P_4O_{10} \rightarrow 4\,HPO_3 + 2\,N_2O_5\ \text{(anhydride of } HNO_3)$$	

° NO_2 dimerizes to N_2O_4; $2\,NO_2 \underset{\text{red-brown}}{\overset{\text{colorless}}{\rightleftarrows}} N_2O_4 + \Delta$. When heated the red-brown color becomes more intense, when cooled the color fades to pale yellow; makes good lecture demonstration on equilibrium. N_2O_4 is a good oxidant for use in rocket propellant.

°° HNO_2, nitrous acid, can exist in water solution; otherwise is unstable. It may be prepared as follows: $2\,KNO_3 \rightarrow 2\,KNO_2$ (potassium nitrite) $+ O_2$, dilute $H_2SO_4 + 2\,KNO_2 \rightarrow 2\,HNO_2 + K_2SO_4$.
HNO_2 has little use compared to HNO_3; in organic chemistry the nitrite radical $(-NO_2)$ nears the importance of the nitrate $(-NO_3)$ radical.

mix on glycerin. The concentrated sulfuric acid acts as a dehydrating agent (page 474).

$$C_3H_5(OH)_3 + 3\ HNO_3 \rightarrow C_3H_5(NO_3)_3 + 3\ H_2O$$

glycerin $\qquad\qquad\qquad$ nitroglycerin

TRINITROTOLUENE   (TNT) is produced by a similar reaction:

$$C_6H_5CH_3 + 3\ HNO_3 \rightarrow C_6H_2CH_3(NO_2)_3 + 3\ H_2O$$

Most of the compounds formed by the action of nitric acid on organic substances are explosive. Dynamite consists of nitroglycerin absorbed in a mixture such as sawdust and nitre.

An explosion occurs as a result of a rapid chemical reaction attended by the formation of a large volume of gas. When nitroglycerin explodes, the volume of gas produced is many times the volume of the liquid nitroglycerin.

The crystalline substance NH_4NO_3, an explosive, is mixed in high percentage with other explosive materials. As the other materials explode, the shock brings about the instant decomposition of solid NH_4NO_3 into gaseous nitrogen, water, and oxygen.

Examples of the use of HNO_3 in nitration as given above pertain mainly to explosive manufacture. Numerous other uses of nitration are found in the general field of organic chemistry (Chapter 32).

Nitric acid ranks 11 in the list of top chemicals, its primary use being in the manufacture of ammonium nitrate, NH_4NO_3 (12 in the list), which in turn is used as a fertilizer, an industrial explosive, and for making nitrous oxide, N_2O. Nitric acid is also reacted with phosphate rock, $Ca_3(PO_4)_2$, to produce mixed fertilizers. Further uses include the cleaning of stainless steel, recovery of uranium by solvent extraction and ion exchange, and the manufacture of dyes, plastics, and synthetic fibers.

20.5   OXIDES OF NITROGEN

The chemistry of five oxides of nitrogen is outlined in Table 20.1.

Nitrogen has merited much more attention than the other four Group VA elements, which will now be individually considered; but note first Table 20.2 showing changes in physical properties with increasing atomic weight.

20.6   PHOSPHORUS

History and Occurrence

Brand, a German alchemist, accidentally prepared the element phosphorus in 1669 while experimenting with urine. Crystals obtained from the evapora-

Table 20.2

Element	Symbol	Color and Physical State	Sp. Gr.	Melting Point, °C	Boiling Point, °C	Atomic Radius, Å	Oxidation States
Phosphorus	P	yellow solid	1.8	44	280	1.10	−3, +3, +5
Arsenic	As	gray solid	5.7	sublimes	–	1.21	−3, +3, +5
Antimony	Sb	silvery solid	6.7	631	1440	1.41	−3, +3, +5
Bismuth	Bi	silvery solid	9.8	271	1450	1.52	+3, +5

tion of urine were mixed with sand and charcoal. This mixture was heated in a retort and thus out of contact with air. One product obtained was a substance which glowed in the dark. The substance was subsequently shown to be an element and was given the name **phosphorus**, which means "light bearer."

It is now known that the human body eliminates $NaNH_4HPO_4$ in urine as a product of metabolism of proteins and phosphoproteins. This compound was historically called microcosmic salt because it was recognized that small perfect crystals of a substance could be obtained from within a larger cosmos — man. If the crystalline compound is heated with carbon and sand, it can yield the element phosphorus. (See equations for commercial preparation of phosphorus below.)

The element is never found in the free state, but its compounds are widely distributed. The most abundant mineral is **phosphate rock,** or **phosphorite,** which is largely $Ca_3(PO_4)_2$. Deposits are found in Montana, Idaho, and several of the southern states. The bones of man and animals have a high percentage of calcium phosphate, and many proteins contain phosphorus,

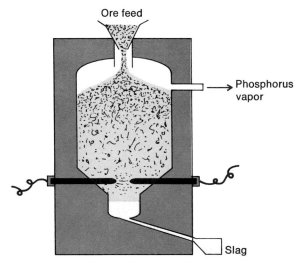

Figure 20.2
Electric furnace for the manufacture of elementary phosphorus.

particularly nerve tissues and brain cells. The human body contains a total of about 1% of the element.

Preparation

Phosphorus is prepared by heating a mixture of phosphate rock, sand, and coke to a high temperature in an electric furnace (Figure 20.2). Heat is produced by the resistance to the passage of current between the electrodes near the bottom of the furnace. The reaction probably takes place in steps.

$$2\ Ca_3(PO_4)_2 + 6\ SiO_2 \rightarrow 6\ CaSiO_3 + P_4O_{10}$$
$$P_4O_{10} + 10\ C \rightarrow P_4(g) + 10\ CO$$

Molten calcium silicate (slag) collects at the bottom and may be drawn off. Phosphorus leaves the furnace as a vapor; the vapor is condensed to a liquid, which is further cooled and cast into sticks. The white variety of phosphorus that results is known to be tetraatomic, P_4.

Properties

When it is prepared by this process, phosphorus is a soft, white, wax-like solid which turns yellow gradually. On exposure to air, oxidation takes place slowly with the evolution of heat, and the element spontaneously ignites at a temperature of 40–45°C. The glowing-in-the-dark action of phosphorus is due to light emitted during the slow oxidation.

The element is insoluble in water but readily dissolves in carbon disulfide and certain other organic solvents. For storage and handling it is placed under water. Great care should be exercised in handling phosphorus; it should always be picked up with tongs or forceps, never with the fingers, since warmth from the fingers may be sufficient to raise the temperature to the kindling point and cause spontaneous ignition. Phosphorus burns are painful and slow to heal.

In contrast with nitrogen, phosphorus is fairly active and combines readily with oxygen, the halogens, and certain metals. It burns brightly in the air, producing dense white fumes of the (III)oxide and (V)oxide. In moist air these fumes form droplets of phosphorous and phosphoric acids, which appear as a fog or mist. On the basis of this behavior, the element has been used in production of smoke screens in warfare.

When white phosphorus is heated to a temperature of 250–300°C in the absence of air, it changes to a modification called **red phosphorus**. In contrast to the extremely poisonous yellow form, red phosphorus is nonpoisonous; it ignites in air only at a comparatively high temperature and is insoluble in carbon disulfide.

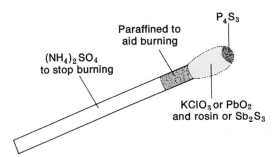

Figure 20.3
Composition of a wooden match.

Uses

Because of its low kindling temperature, phosphorus has been used in the manufacture of matches. Yellow phosphorus, once used in match manufacture, is now prohibited because factory workers inhaling air containing poisonous phosphorus vapor developed a disease causing decay of the bones of the jaw. A nonpoisonous compound—tetraphosphorus trisulfide (P_4S_3)—is now used. (See Figure 20.3.) Because it ignites so readily, further use is in incendiaries and tracer bullets.

The essential ingredients of a match are (1) tetraphosphorus trisulfide, which ignites at a low temperature; (2) a combustible substance, such as paraffin, rosin, or antimony sulfide; (3) an oxidizing agent like potassium chlorate or lead dioxide; (4) glue to bind the ingredients; and (5) a wood stick. For the "strike anywhere" match, these ingredients constitute the head, and friction with a rough surface produces enough heat to ignite the phosphorus compound and initiate the reaction of the combustible substance and oxidizing agent. The flame produced is transmitted to a paraffined wood stick. The "safety" match contains no phosphorus in the head. Instead, the phosphorus compound is placed on the striking surface of the box along with ground glass and glue, and friction causes a bit of the phosphorus to ignite and thence the ingredients of the head.

20.7 COMPOUNDS OF PHOSPHORUS

Phosphine

The hydride of phosphorus (PH_3), a colorless gas with an offensive odor resembling that of decaying fish, is the analog of ammonia (NH_3). Unlike ammonia, it cannot be prepared by direct combination of the elements but may be obtained by the action of water on calcium phosphide:

$$Ca_3P_2 + 6\ H_2O \rightarrow 3\ Ca(OH)_2 + 2\ PH_3(g)$$

The gas is only slightly soluble in water and ignites spontaneously* on contact with the air to form smoke-ring clouds of phosphoric acid.

$$PH_3 + 2\ O_2 \rightarrow H_3PO_4$$

Oxides of Phosphorus

The most important oxides are phosphorus(III) oxide (P_4O_6) and phosphorus(V) oxide (P_4O_{10}), both white solids. They are obtained by the combustion of phosphorus in air; the (III)oxide is formed in a limited supply of air and the (V)oxide in an excess of air. The oxides are anhydrides of phosphorous and phosphoric acids, respectively.

$$P_4O_6 + 6\ H_2O \rightarrow 4\ H_3PO_3$$
$$P_4O_{10} + 6\ H_2O \rightarrow 4\ H_3PO_4$$

The reaction of phosphorus(V) oxide and water is particularly vigorous, and this behavior makes phosphorus(V) oxide an effective dehydrating agent.

Phosphoric Acids

Phosphorus(V) oxide is the anhydride of three phosphoric acids which differ from one another only in degree of hydration of the oxide. The relationship among the three acids is shown below.

$$P_4O_{10} + 6\ H_2O \rightarrow 4\ H_3PO_4 \qquad \text{(orthophosphoric acid)}$$
$$P_4O_{10} + 4\ H_2O \rightarrow 2\ H_4P_2O_7 \qquad \text{(pyrophosphoric acid)}$$
$$P_4O_{10} + 2\ H_2O \rightarrow 4\ HPO_3 \qquad \text{(metaphosphoric acid)}$$

The most hydrated of the acids is the **ortho** acid and the least hydrated the **meta** acid; the **pyro** acid represents an intermediate degree of hydration. Orthophosphoric acid is obtained when phosphorus(V) oxide is dissolved in water under ordinary conditions. When the ortho acid is heated to relatively high temperatures, the pyro and meta varieties of the acid are formed. The term "phosphoric acid" denotes the ortho variety unless otherwise specified.

Phosphoric acid is obtained commercially either by ignition of phosphorus in the air and subsequent solution of the resulting oxide in water or by the treatment of phosphate rock with 60% sulfuric acid.

$$Ca_3(PO_4)_2 + 3\ H_2SO_4 \rightarrow 3\ CaSO_4 + 2\ H_3PO_4$$

After the insoluble calcium sulfate is filtered off, the solution is concentrated to approximately 85% phosphoric acid by evaporation.

---

* It is believed that small amounts of P_2H_4 produced by a secondary reaction are responsible for initiating the combustion.

Phosphoric acid is 9th in the list of top chemicals. It or one of its derivatives is used in a vast array of industries and products. These might be grouped as follows: (1) fertilizers, (2) water treatment and detergents (see below), (3) food industry; for example, self-rising flours and baking powders, yeast culture, soft drink industry, and tooth powders, (4) miscellaneous – making enamels, glazes for pottery, matches, petroleum refining, catalyst, etc.

Ionization of Phosphoric Acid

The acid is triprotic – that is, it has three replaceable hydrogen atoms – and thus ionizes in three steps.

<table>
<tr><td></td><td>Ionization Constant</td></tr>
<tr><td>$H_3PO_4 \rightleftharpoons H^+ + H_2PO_4^-$</td><td>$K_1 = 7.5 \times 10^{-3}$</td></tr>
<tr><td>$H_2PO_4^- \rightleftharpoons H^+ + HPO_4^{2-}$</td><td>$K_2 = 6.2 \times 10^{-8}$</td></tr>
<tr><td>$HPO_4^{2-} \rightleftharpoons H^+ + PO_4^{3-}$</td><td>$K_3 = 1 \times 10^{-12}$</td></tr>
</table>

As with all acids which ionize in stages, each stage of the ionization is less complete than the preceding one (evidenced by the ionization constants). H_3PO_4 is the strongest of the three, yet it is a relatively weak acid; $H_2PO_4^-$ is classed as a very weak acid, and HPO_4^{2-} as a very, very weak acid. Salts of all three acids are known – for example, the three sodium salts:

NaH_2PO_4 sodium dihydrogen phosphate
Na_2HPO_4 disodium hydrogen phosphate
Na_3PO_4 normal or trisodium phosphate

Trisodium phosphate is used extensively as a water softener and in detergents. A mixture of the sodium acid phosphates plays an important part in maintaining a constant pH of the blood.

Phosphorus Fertilizers

Phosphorus is vital to the growth of plants and therefore must be present in soils for successful plant propagation. One of the most important of the phosphorus fertilizers is **superphosphate,** a mixture of acid calcium phosphate and calcium sulfate, which is prepared by the action of sulfuric acid on rock phosphate according to the equation

$$Ca_3(PO_4)_2 + 2\ H_2SO_4 \rightarrow Ca(H_2PO_4)_2 + 2\ CaSO_4$$
superphosphate

Rock phosphate is too insoluble to be utilized by plants, but the acid salt is soluble, though not soluble enough to be quickly washed from soil; thus it is available for plant metabolism.

Note that three moles H_2SO_4 to one mole $Ca_3(PO_4)$ would convert all the phosphorus into H_3PO_4, which would be too water soluble for a good fertilizer.

Triple phosphate, a more effective fertilizer per pound, is made as follows:

$$Ca_3(PO_4)_2 + 4\ H_3PO_4 \rightarrow 3\ Ca(H_2PO_4)_2$$

Ammonium phosphate, $((NH_4)_3PO_4)$, another fertilizer, contains two important soil elements.

20.8   ARSENIC

Occurrence and Preparation

Although arsenic is found in small quantities in the free state, it occurs principally as sulfides — **arsenopyrite** (FeAsS), **orpiment** (As_2S_3), and **realgar** (As_2S_2). Small quantities of arsenic are associated with most sulfide ores. As a matter of fact, nearly all the arsenic produced is recovered as a by-product from the flues of smelters utilizing sulfide ores of zinc, copper, and lead. During the roasting process, the arsenic is converted to volatile arsenic(III) oxide (As_4O_6), which is condensed in the flues as a finely divided solid. The element is easily obtained from the arsenic(III) oxide by reduction with carbon:

$$2\ As_2S_3 + 9\ O_2 \rightarrow As_4O_6 + 6\ SO_2$$
$$As_4O_6 + 6\ C \rightarrow 4\ As + 6\ CO$$

Properties and Uses

Arsenic is a gray crystalline solid. The solid sublimes to a yellowish, poisonous vapor which has the odor of garlic. The principal use of elemental arsenic is in hardening lead shot, made by pouring molten lead through a screen which breaks up the liquid into small drops. The presence of arsenic lengthens the time of solidification, thereby allowing the drops to attain a more nearly spherical shape. At the same time the arsenic forms an alloy with the lead to produce a harder product.

Compounds of Arsenic

Arsenic is very similar to phosphorus in many of its compounds.

Arsenic(III) oxide, a white solid commonly called "white arsenic," or simply "arsenic" is the starting material in the preparation of most of the compounds of arsenic. It has a sweet taste and is highly poisonous.* As it

---

° One of the most notorious women of history, Lucrezia Borgia, used arsenic oxide to poison many men. Roman ladies ate minute quantities of "white arsenic" to enhance the whiteness of their skin and pink of their cheeks. They gradually built up an immunity to the poisonous effects of the arsenic oxide and ultimately could safely take a dose that would "kill a horse." A dose of 0.1 g As_4O_6 will usually kill a person.

reacts with water, it forms H_3AsO_3, orthoarsenous acid, or $HAsO_2$, meta-arsenous acid. The acids formed from arsenic(V) oxide and water are H_3AsO_4, ortho-, and $HAsO_3$, meta- arsenic acids. Salts of these various acids have formerly been used in the preparation of insecticides, but organic insecticides have largely replaced these arsenic compounds.

20.9 ANTIMONY

Occurrence and Preparation

The chief source of antimony is the mineral **stibnite** (Sb_2S_3), though small quantities of the element are found in the free state.

Stibnite, similar to As_2S_3, may also be roasted to the oxide, followed by reduction with carbon.

Properties, Uses, and Compounds

Antimony is more metallic in its appearance and properties than arsenic. The element is brittle and a poor conductor of heat and electricity.

The principal use of antimony is in the preparation of alloys. The addition of antimony to lead produces a much harder product, which is utilized in making bullets and shrapnel. **Type metal** is an alloy of lead, tin, and antimony which expands on solidification, thereby filling all parts of the mold to produce a sharp and distinct casting of the type characters.

ANTIMONY CHLORIDE $SbCl_3$, called "butter of antimony" because of the soft, creamy appearance of the crystals, is made by direct combination of the elements. This salt, as well as most of the other salts of antimony, is extensively hydrolyzed when one attempts to prepare an aqueous solution of it. An insoluble basic salt results.

$$SbCl_3 + H_2O \rightleftharpoons SbOCl(s) + 2\ HCl$$

When hydrogen sulfide is passed into a solution containing antimony ions, a reddish-orange precipitate of the sulfide is formed. This reaction is often used in testing for the presence of antimony.

$$2\ Sb^{3+} + 2\ S^{2-} \rightarrow Sb_2S_3(s)$$

20.10 BISMUTH

Occurrence and Preparation

Bismuth occurs in both the free and the combined states. Naturally occurring compounds are **bismuth glance** (Bi_2S_3) and **bismite** ($Bi_2O_3 \cdot 3\ H_2O$). Like arsenic and antimony, bismuth may be obtained from the sulfide ore by roasting, followed by reduction with carbon. A good deal of bismuth is recovered as a by-product of lead smelting and refining.

Properties and Uses

Bismuth is distinctly a metallic element. It is hard and brittle and has a silvery luster with a reddish tinge. It burns when highly heated in air to form the (III)oxide (Bi_2O_3) and combines directly with the halogens to form halide salts, though these reactions are not vigorous. Unlike the other members of the Group VA elements, bismuth dissolves in nitric acid to form a nitrate salt.

Bismuth is used in the preparation of several low-melting alloys, such as **Rose's metal** (Sn, Pb, Bi), melting point 94°C, and **Wood's metal** (Pb, Cd, Sn, Bi), melting point 71°C. These alloys are used in electrical fuses, automatic fire sprinklers, safety plugs for boilers, automatic fire alarms, and so on.

Exercises

20.1. Discuss how the properties of the elements of Group VA and their compounds change with increasing atomic number.

20.2. Keeping in mind the formulas of (1) ammonia, (2) nitrogen(III) oxide, (3) nitrous acid, and (4) nitric acid, write formulas and name corresponding compounds of (a) P, (b) As, (c) Sb, (d) Bi.

20.3. Uses and properties of compounds of N, P and S — Select from the list in the right-hand column the formula of a compound corresponding to the descriptions on the left.

(a) A strong acid
(b) A colorless gas which in air is converted to a brown gas
(c) An explosive
(d) A very unreactive colorless gas
(e) A weak acid
(f) A solid fertilizer
(g) Major component of bone
(h) Anhydride of HNO_3
(i) A gas which dissolves very readily in water to form a base
(j) Phosphine
(k) Laughing gas

1. NO
2. N_2
3. N_2O_5
4. NH_3
5. HNO_2
6. PH_3
7. NO_2
8. N_2O

9.

$$NO_2 - \overset{\displaystyle CH_3}{\underset{\displaystyle NO_2}{\bigcirc}} - NO_2$$

10. $Ca_3(PO_4)_2$
11. HNO_3
12. $(NH_4)_3PO_4$

20.4. Write correct formulas for:
    (a) nitrogen gas
    (b) nitric oxide
    (c) ammonium sulfate
    (d) ammonia
    (e) orthophosphoric acid

    (f) lead arsenate
    (g) phosphate rock
    (h) metaarsenic acid
    (i) potassium dihydrogen phosphate

20.5. Give one important commercial *use* for:
    (a) nitrogen gas
    (b) ammonium sulfate
    (c) ammonia

    (d) NH_4NO_3
    (e) HNO_3
    (f) H_3PO_4

20.6. Name:
    (a) H_3AsO_4
    (b) $Ca(H_2PO_4)_2$
    (c) NH_4NO_2

    (d) NO_2
    (e) $C_3H_5(NO_3)_3$

20.7. Write equations to summarize the industrial preparation of HNO_3 from air and H_2O.

20.8. (a) List and name the oxides of nitrogen. (b) Write a chemical equation for the preparation of each of them.

20.9. Experimentally, using simple tests or observations, how could you distinguish between:
    (a) O_2 and NO_2
    (b) HNO_3 and HCl
    (c) N_2 and NH_3
    (d) solid $NaNO_3$ and $NaBr$
    (e) white and red phosphorus

20.10. Write formulas for orthoarsenic acid, metaarsenic acid, and pyrostibnic acid. Write formulas for copper(II) metaarsenate and sodium pyroarsenate.

20.11. Write all the equations that you can for the reactions of elementary nitrogen. Explain the low order of chemical reactivity of N_2.

20.12. Write balanced equations for the following reactions:
    (a) Ammonium sulfate heated with sodium hydroxide solution.
    (b) Calcium nitride plus steam.
    (c) Combustion of warm ammonia gas in oxygen in presence of Pt catalyst with NO as a product.
    (d) Formation of white smoke as vapors intermingle from open bottles of concentrated HNO_3 and concentrated ammonia solution.

20.13. Complete and balance the following equations:
    (a) The Haber process for nitrogen fixation
    (b) Qualitative analysis test for NH_4^+
       $NH_4Cl +$ _____ $\rightarrow NH_3(g) +$ _____ $+$ _____

(c) Ammonia + acids
$$NH_3 + HCl \rightarrow$$
(d) A step in the Ostwald process for making HNO_3
$$NO_2 + H_2O \rightarrow HNO_3 + \underline{\hspace{2cm}}$$
(e) Action of dilute HNO_3 on metals (ionic equation)
$$Cu + H^+ + NO_3^- \rightarrow$$
(f) Making phosphine
$$Ca_3P_2 + H_2O \rightarrow PH_3 + \underline{\hspace{2cm}}$$

20.14. Complete and balance:
(a) NH_4NO_2 (heat) $\rightarrow$ (d) $S + HNO_3$ (conc.) $\rightarrow$
(b) $(NH_4)_3PO_4 + NaOH \rightarrow$ (e) $Ca_3(PO_4)_2 + H_2SO_4 \rightarrow$
(c) $Ag + HNO_3$ (dilute) $\rightarrow$

20.15. Write ionic equations for the reaction of dilute HNO_3 with
(a) Al (to give N_2O)   (d) $CaCO_3$
(b) $Ca(OH)_2$ (e) Hg
(c) $Ca_3(PO_4)_2$ (f) $NaHCO_3$

20.16. Write balanced equations for
(a) calcium chloride + orthoarsenic acid
(b) arsenic(V) oxide + H_2O
(c) combustion of arsenic in oxygen
(d) hydrolysis of bismuth(III) chloride

20.17. Write equations for the following reactions of ammonia:

(a) $NH_3 + Ag^+ \rightarrow$ (d) $NH_3(g) + O_2(g) \xrightarrow{\text{heat}} N_2 +$

(b) $NH_3 + H^+ \rightarrow$ (e) $NH_3(g) + O_2(g) \xrightarrow[\text{heat}]{\text{Pt}}$

(c) $NH_3 + H_2O + CO_2 \rightarrow$ (f) $NH_3(g) + HCl(g) \longrightarrow$

20.18. Write an equation for the reaction of concentrated HNO_3 with (a) Cu, (b) Zn, (c) P_4O_{10}, (d) NH_3, (e) $Ca(OH)_2$.

20.19. What is the acid anhydride of each of the following? (a) $H_2N_2O_2$, (b) HNO_2, (c) HNO_3, (d) $H_4P_2O_7$, (e) $HAsO_3$, (f) H_3PO_3, (g) H_3AsO_4.

20.20. Describe how you would test a sample of fertilizer for (a) ammonium salts, (b) a nitrate.

20.21. Describe simple tests which would enable you to distinguish among the following gases: (a) N_2, (b) NH_3, (c) NO, (d) N_2O, (e) NO_2.

20.22. Give the formula of the nitrogen-containing (a) gas evolved when dried blood from a slaughter house is heated with $Ca(OH)_2$, (b) ions formed when a farmer jets ammonia into moist soil, (c) vapor above 86°C when a nitrate is heated with concentrated H_2SO_4. Write equations to represent the reaction in (c).

20.23. Complete the following table summarizing information on some compounds commonly used as fertilizers.

Name	Chemical Formula	Percent N
(a) Urea	$(NH_2)_2CO$	_____
(b) Ammonium sulfate	_____	21.2
(c) _____	$(NH_4)_3PO_4$	_____

20.24. From your knowledge of nitrogen chemistry, make an "educated guess" concerning the following questions about arsenic:
  (a) What do you predict would be the highest oxidation number As would show?
  (b) Write the formula for calcium arsenate.
  (c) Of the following formulas, which would correspond to an oxide of arsenic: AsO_2, As_2O, As_4O_6, AsO, AsO_3?
  (d) Write the formula for arsine.
  (e) Reaction of metaarsenic acid with aluminum.

20.25. A sack of fertilizer usually has three numbers on it representing (1) percent N, (2) percent P expressed as percent P_2O_5, and (3) percent K represented as percent K_2O, in this order. Calculate what these numbers will be for the following fertilizers: (a) $(NH_4)_2SO_4$; (b) urea, $(NH_2)_2CO$; (c) ammonium phosphate, (d) triple phosphate, $Ca(H_2PO_4)_2$; (e) NH_4NO_3; (f) anhydrous ammonia, NH_3.

20.26. When lead azide (PbN_6) is detonated, metallic lead and nitrogen gas are products. Write the equation. What volume of N_2 gas at STP would be obtained from 100 g of PbN_6?

20.27. Solid ammonium nitrate, **amitol**, is widely used as an explosive; it requires a detonator explosive to cause its decomposition to hot gases.

$$NH_4NO_3 \rightarrow N_2O + 2\ H_2O \qquad 2\ NH_4NO_3 \rightarrow 2\ N_2 + 2\ H_2O + O_2$$

Assume 900 g (about 2 lbs) NH_4NO_3 almost instantaneously decomposed (use first equation). What total volume of gas in liters forms at 1092°C?

20.28. (a) What weight of 100% HNO_3 may be obtained from 1.00 kg of Chile saltpeter ($NaNO_3$) by treatment with concentrated H_2SO_4? (b) How many grams of 70% HNO_3 could be obtained from this quantity of concentrated HNO_3?

20.29. Consider the following nitrogen compounds as fertilizers and the current approximate price per pound in bulk quantities: NH_3, $0.05; NH_4NO_3, $0.035; $(NH_4)_2SO_4$, $0.025; $(NH_4)_3PO_4$, $0.085; $CaCN_2$, $0.027; $NaNO_3$, $0.02. Arrange in order of increasing cost per pound of nitrogen.

20.30. A match mixture probably functions in part according to the equation

$$2\ Sb_2S_3 + 6\ KClO_3 \rightarrow Sb_4O_6 + 6\ SO_2 + 6\ KCl$$

Calculate the quantities of Sb_2S_3 and $KClO_3$ needed per 25.0 kg of mix if these two substances make up 90% of the mixture.

Supplementary Readings

20.1. "The Nitrogen Cycle," C. C. Delwiche, *Sci. American*, Sept. 1970; p. 136.
20.2. "Chemical Fertilizers," C. J. Pratt, *Sci. American*, June 1965; p. 62.
20.3. "The Chemistry of Orthophosphoric Acid and its Sodium Salts," H. A. Neidig, T. G. Teates, and R. T. Yingling, *J. Chem. Educ.*, **45,** 57 (1968).

Chapter 21

Main Energy Levels	1	2	3	4	5
$^{15.9994}_{8}$O	K	s^2p^4			
$^{32.06}_{16}$S	K	L	s^2p^4		
$^{78.96}_{34}$Se	K	L	M	s^2p^4	
$^{127.6}_{52}$Te	K	L	M	$s^2p^6d^{10}$	s^2p^4

The Sulfur Family, Group VIA

Element	Color	Sp. Gr.	Melting Point °C	Boiling Point °C	Atomic Radius (Å)	Ionic Radius (Å)
Oxygen	colorless	–	−225	−118	0.66	1.4
Sulfur	pale yellow	2.07	113	445	1.04	1.84
Selenium	steel gray	4.8	220	668	1.17	1.98
Tellurium	metallic	6.24	452	1390	1.37	2.21

Although there appears to be little in common between yellow solid sulfur and the colorless gas oxygen, these elements form many similar compounds. Consider the s^2p^4 outer electron configuration of Group VIA elements. A sharing of either 4 or 6 electrons accounts for the +4 and +6 oxidation states. An acceptance of 2 electrons by the two half-filled p orbitals gives rise to S^{2-}, Se^{2-}, and Te^{2-} of −2 oxidation state. Electronic symbols are

$$\cdot \ddot{\underset{\cdot\cdot}{O}} \cdot \qquad \cdot \ddot{\underset{\cdot\cdot}{S}} \cdot \qquad \cdot \ddot{\underset{\cdot\cdot}{Se}} \cdot \qquad \cdot \ddot{\underset{\cdot\cdot}{Te}} \cdot$$

Note the progressive changes in listed physical properties from the lightest to heaviest element. The acceptance of two electrons to negative ions makes for the increase of ionic over atomic radii.

21.1   SULFUR

Occurrence of Sulfur

Sulfur was known to the ancients as brimstone – that is, "burning stone." It is often referred to in the Bible; ancient priests taught of hell as a place not only of fire but of burning brimstone (choking odor of SO_2). The Greek philosophers considered sulfur an elementary substance, though their concept of an element was different from the view held today. The Greeks used sulfur extensively in the treatment of disease and as a fumigant.[*] Deposits of sulfur are found in Sicily and in Louisiana and Texas, but reserves in these areas are being rapidly depleted. Recent discoveries in the Vera Cruz area of Mexico indicate large deposits that will be an important source of supply.

In the combined state, sulfur occurs in the form of metallic sulfides, such as **sphalerite** (ZnS), **galena** (PbS), **chalcocite** (Cu_2S), **chalcopyrite** ($CuFeS_2$), **iron pyrites** (FeS_2), **cinnabar** (HgS), and several other minerals. These compounds are economically important, since they are the sources of the metals Zn, Cu, Pb, Hg, and others. Some combined sulfur exists as sulfates of calcium (**gypsum**, $CaSO_4 \cdot 2\,H_2O$) and barium (**barite**, $BaSO_4$). Many organic compounds contain sulfur: proteins in the body contain small amounts, as do

[*] In the *Odyssey* it is related that Penelope ordered brimstone burned in the marble halls of her home after her long-absent husband Ulysses came home and killed the men who had wasted his flocks and sought to marry his wife. The brimstone was scattered over the bodies. It is interesting to know that the rudiments of fumigation were known in this early time.

certain compounds in onion, garlic, horseradish, and mustard. Sulfur appears to be one of the elements essential to vital processes.

The Extraction of Sulfur

In volcanic regions, sulfur is mixed with soil and rocks. It can be separated by heat, which melts the sulfur and leaves the impurities in a solid condition. In Louisiana and Texas, sulfur deposits are found several hundred feet below the surface of the ground. Because these deposits are covered with a layer of quicksand, the sulfur cannot be mined by ordinary means.

Herman Frasch, a German-American chemist and engineer, devised an ingenious method for obtaining sulfur from underground deposits (Figure 21.1). His method is based on the relatively low melting point of sulfur (113°C). A hole is bored to the bed of sulfur, and the well is then lined with four concentric pipes, one fitted inside another. Superheated water and air

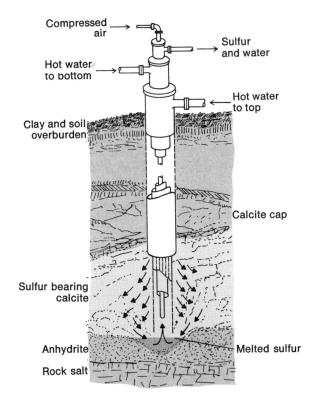

Figure 21.1

Frasch process.

under pressure are forced down the well through two of the pipes. The heat melts the sulfur, which is then forced to the surface in a liquid state by compressed air. The sulfur is allowed to solidify in large blocks at the surface. Each well can produce several hundred tons of sulfur daily, and the product is about 99.5% pure (Figure 21.4).

Different Forms of Sulfur

Free sulfur may exist in several allotropic forms. The form of sulfur found in nature is called the **rhombic** variety because of the shape of the crystals (Figure 21.2). If these crystals are heated to a temperature of 96°C, they slowly change into another, more needle-like crystalline variety, which is called **monoclinic** sulfur. The temperature at which a given solid will change its crystalline form is known as the **transition point** or **transition temperature.** Thus 96°C is the transition point of sulfur. This property of a substance of existing in more than one physical form in the same physical state is termed **allotropy,** and the different forms are referred to as **allotropes.** Other elements, such as phosphorus and carbon, may exist in allotropic forms.

When rhombic sulfur is heated, it melts at a temperature of 113°C to a straw-colored liquid, called **mobile** sulfur because it flows freely. When mobile sulfur is allowed to cool slowly and solidify, needlelike crystals of monoclinic sulfur are obtained. If the heating of mobile sulfur is continued, the liquid darkens and becomes thick and sticky, like molasses. This form of sulfur is called **viscous** sulfur. Continued heating changes the liquid sulfur to vapor at a temperature of about 445°C. If viscous sulfur is cooled rapidly by immersion in cold water, a rubbery, plastic mass of solid sulfur is obtained, called **plastic** or **amorphous** (noncrystalline) sulfur. Under ordinary conditions, this variety of sulfur will gradually change into rhombic sulfur, the stable form at ordinary temperatures. The relationship which exists between the various forms of sulfur may be represented by a diagram.

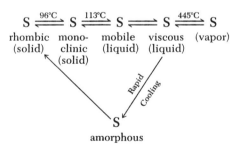

The transformations shown in this diagram are reversible, except that of amorphous sulfur to rhombic sulfur. All forms finally revert to the stable rhombic form under ordinary conditions.

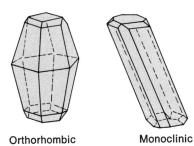

Orthorhombic   Monoclinic   Figure 21.2
                                    Crystals of sulfur.

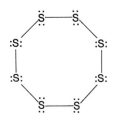

Figure 21.3
The ring structure of sulfur.

The Structure of Sulfur Molecules

Molecular structure studies reveal some interesting configurations for elemental sulfur. Rhombic, monoclinic, and mobile sulfur have the formula S_8, where eight atoms of sulfur are connected in a ring structure (Figure 21.3).

When sulfur is melted, the ring structures do not offer much resistance to the passage of one ring past another, and the resulting liquid is quite fluid (mobile sulfur). At higher temperatures the rings are broken and chains of sulfur atoms are formed:

When the ring is broken, one end of each chain will have an unsatisfied electronic structure, to which an additional chain may attach. As a result, many of the original eight sulfur-atom units polymerize into very long chains. These long-chain polymers become interwined and entangled with one another, so that passage of one unit past another is difficult. Consequently, the liquid becomes very viscous and resistant to flow. On cooling, this liquid becomes plastic or amorphous sulfur. The chain-like structure gradually changes back to the ring structure of rhombic sulfur below 96°C.

Sulfur vapor will have formulas varying from S_8 to S_2 as the temperature is increased.

Figure 21.4
Loading sulfur, which after
extraction, has been solidified in
the walled enclosure.

Chemical Properties

Sulfur combines directly with most metals to form sulfides. When iron is heated with sulfur, a great deal of heat is evolved. A vigorous reaction ensues, and iron(II) sulfide is formed. Other metals combine similarly. In some cases the combination of a metal with sulfur may take place at ordinary temperatures. For example, when mercury and sulfur are ground together in a mortar, black mercury(II) sulfide is formed.

$$Hg + S \rightarrow HgS$$

Sulfur will combine directly with certain of the nonmetals. Sulfur burns with a blue flame in air or oxygen to form sulfur dioxide.

$$S + O_2 \rightarrow SO_2$$

Examples of other compounds formed by sulfur with nonmetals are carbon disulfide (CS_2), tetraphosphorus trisulfide (P_4S_3), and sulfur chloride (S_2Cl_2).

As a member of Group VIA, it tends to acquire two electrons to complete a stable octet configuration in the sulfide ion, S^{2-}. Sulfur also forms covalent bonds with various other nonmetals to assume oxidation numbers of +4, as in sulfur dioxide and sulfites, and +6, as in sulfuric acid and sulfates.

Uses of Sulfur

Eighty-seven percent of the 72 million tons of sulfur produced annually in the free world goes to produce sulfuric acid, the most important industrial chemical (page 475), but sulfur is also extremely important in the **vulcanization** of rubber.

Pure rubber tends to become sticky when slightly warmed and brittle when cold. In the vulcanization process, sulfur is incorporated in the rubber, and a product is obtained which has the desirable resiliency and toughness. Heated with lime, sulfur forms "lime sulfur spray," which is used as an insecticide. Black gunpowder is a mixture of carbon, sulfur, and saltpeter (KNO_3). Sulfides of phosphorus and antimony have taken the place of free phosphorus in the production of matches. Mixed with rock phosphate fertilizer, sulfur is oxidized slowly in soil to sulfuric acid, which converts the fertilizer into a more soluble form.

Other uses of sulfur include the making of paper by the sulfate and sulfite processes and the manufacture of carbon disulfide, sulfur dioxide, and a large number of sulfur-containing organic compounds.

21.2 HYDROGEN SULFIDE

Hydrogen sulfide is an odorous gas produced during the destructive distillation of proteins and other sulfur-containing compounds. It is a constituent of coal and sewer gases. The offensive, rotten-egg odor of these gases is due largely to the presence of hydrogen sulfide.

Preparation of Hydrogen Sulfide

Sulfur does not combine appreciably with hydrogen at ordinary temperatures, but at an elevated temperature combination takes place readily to form hydrogen sulfide:

$$H_2 + S \rightleftharpoons H_2S$$

The reaction is reversible, however, and since hydrogen sulfide itself is quite unstable toward heat, the yield is small. In the laboratory, H_2S is most conveniently prepared by the action of a nonoxidizing acid on certain metallic sulfides.

$$FeS + (dilute) \; H_2SO_4 \rightarrow FeSO_4 + H_2S(g)$$
$$ZnS + 2 \; HCl \rightarrow ZnCl_2 + H_2S(g)$$

These equations are further examples of the general method of preparing an acid by treating a salt of the desired acid with a less volatile acid. Though volatile, hydrochloric acid may be used, since it is less volatile than hydrogen sulfide in water solution.

A particularly convenient source of H_2S in qualitative analysis is the reaction of thioacetamide with water

$$CH_3CSNH_2 + H_2O \rightarrow CH_3CONH_2 + H_2S$$
$$\text{thioacetamide} \qquad\qquad \text{acetamide}$$

When the thioacetamide solution is heated, the release of H_2S is slow enough that the H_2S level remains low in the solution and very little of the gas escapes into the air.

Properties

At ordinary temperatures hydrogen sulfide is a colorless gas with a foul odor like that of rotten eggs. It may be condensed to a liquid at $-59.6°C$ and converted to a solid at $-83°C$. It is not very soluble in water – at ordinary temperatures, one liter of water dissolves about three liters of the gas. It is very poisonous because it paralyzes the nerve centers of the heart and lungs – more than one-half of one percent in the air may be fatal to persons and animals. Fortunately, it is easily detected even in small quantities because of its characteristic odor.

Hydrogen sulfide reacts with water to form a solution with acidic properties. It ionizes in two steps:

$$H_2S \rightleftharpoons H^+ + HS^- \qquad K_1 = 1.1 \times 10^{-7}$$
$$HS^- \rightleftharpoons H^+ + S^{2-} \qquad K_2 = 1 \times 10^{-14}$$

Since ionization is very slight, hydrosulfuric acid is classed as a weak acid. It exhibits typical weak-acid reactions, reacting with bases to form salts. For example,

$$2\ NaOH + H_2S \rightleftharpoons Na_2S + 2\ H_2O$$

Hydrogen sulfide burns in the air with a blue flame, producing sulfur dioxide and water.

$$2\ H_2S + 3\ O_2 \rightarrow 2\ H_2O + 2\ SO_2$$

If combustion takes place in a limited supply of air – for example, in an open bottle – free sulfur is deposited.

$$2\ H_2S + O_2 \rightarrow 2\ H_2O + 2\ S$$

The water of certain natural hot springs smells of H_2S, and the partial oxidation of the H_2S in the water accounts for many elemental sulfur deposits. Volcanic gases contain H_2S, and the equation above indicates why elemental sulfur is deposited in the cracks and crevasses of a volcanic throat.

Hydrogen sulfide is a good reducing agent. For example,

$$3 \text{ H}_2\text{S} + \text{(dilute) } 2 \text{ HNO}_3 \rightarrow 3 \text{ S} + 2 \text{ NO} + 4 \text{ H}_2\text{O}$$
$$\text{H}_2\text{S} + \text{I}_2 \rightarrow \text{S} + 2 \text{ HI}$$

$$\text{Na}_2\text{SO}_3 + 2 \text{ HCl} \rightarrow 2 \text{ NaCl} + \text{H}_2\text{SO}_3 \; \underset{}{\overset{\longrightarrow}{\rule{0pt}{1.5ex}}} \; \text{SO}_2 + \text{H}_2\text{O}$$

or $\qquad\qquad$ $$\text{NaHSO}_3 + \text{HCl} \rightarrow \text{NaCl} + \text{H}_2\text{SO}_3 \; \underset{}{\overset{\longrightarrow}{\rule{0pt}{1.5ex}}} \; \text{SO}_2 + \text{H}_2\text{O}$$

The sulfurous acid is unstable and decomposes into sulfur dioxide and water. As a result of the unstable character of sulfurous acid, sulfite salts effervesce on addition of acids.

21.3   SULFUR DIOXIDE

Sulfur dioxide at ordinary temperatures is a gas with a disagreeable, penetrating odor. The gas is about twice as heavy as air and may be liquefied easily by cooling in a salt-ice mixture. The boiling point of the liquid is $-10°\text{C}$. The gas is quite soluble in water: 1 l of water dissolves about 80 l at $0°\text{C}$. Sulfur dioxide reacts with water, forming a solution with weakly acidic properties, to which the formula of H_2SO_3 is given.

$$\text{SO}_2 + \text{H}_2\text{O} \rightleftharpoons \text{H}_2\text{SO}_3 \rightleftharpoons \text{H}^+ + \text{HSO}_3^- \qquad K_1 = 1.7 \times 10^{-2}$$
$$\text{HSO}_3^- \rightleftharpoons \text{H}^+ + \text{SO}_3^{2-} \qquad K_2 = 6.2 \times 10^{-8}$$

Pure H_2SO_3 has not been isolated, but its salts (sulfites) are well known. A solution of sulfurous acid gives typical weak-acid reactions, reacting with metals above hydrogen in the activity series and with bases.

$$\text{Zn} + \text{H}_2\text{SO}_3 \rightarrow \text{ZnSO}_3 + \text{H}_2$$
$$2 \text{ NaOH} + \text{H}_2\text{SO}_3 \rightarrow \text{Na}_2\text{SO}_3 + 2 \text{ H}_2\text{O}$$

Sulfurous acid slowly absorbs oxygen from the air to form sulfuric acid.

$$2 \text{ H}_2\text{SO}_3 + \text{O}_2 \rightarrow 2 \text{ H}_2\text{SO}_4$$

Sulfur dioxide is a good reducing agent, as illustrated by the following reactions:

$$5 \text{ SO}_2 + 2 \text{ H}_2\text{O} + 2 \text{ KMnO}_4 \rightarrow \text{K}_2\text{SO}_4 + 2 \text{ MnSO}_4 + 2 \text{ H}_2\text{SO}_4$$
$$2 \text{ H}_2\text{O} + 3 \text{ SO}_2 + 2 \text{ HNO}_3 \rightarrow 2 \text{ NO} + 3 \text{ H}_2\text{SO}_4$$

Use of Sulfur Dioxide and Sulfites

The most important use of SO_2 is in the preparation of sulfuric acid. A solution of sulfurous acid is a mild bleaching agent which may be used for bleaching straw, silk, or wool. The bleaching action is probably due to the reducing property of the acid. Sulfur dioxide is also used as a refrigerant. Water solutions of **calcium bisulfite** [$Ca(HSO_3)_2$], **ammonium bisulfite** (NH_4HSO_3), and **magnesium bisulfite** [$Mg(HSO_3)_2$] are used extensively in the paper and pulp industry for dissolving lignin, a resinous substance which holds the wood fibers together.

A new drug, dimethyl sulfoxide (($CH_3)_2SO$), referred to as DMSO, apparently has a number of uses. It penetrates the skin and lessens the healing time of sprains and bruises, such as a black eye. It appears to be effective also in relieving bursitis. Because of some undesirable side effects on experimental animals, more research is needed before it becomes available for general use.

21.4 SULFUR TRIOXIDE AND SULFURIC ACID

The preparation of sulfuric acid involves the preparation of its anhydride, sulfur trioxide, and the solution of the anhydride in water to form H_2SO_4 by a process called the **contact process.**

Sulfur dioxide does not combine with oxygen appreciably at room temperature, but at a temperature of about 400°C in the presence of a catalyst the reaction is about 98% complete:

$$2\ SO_2 + O_2 \rightleftharpoons 2\ SO_3 \rightleftharpoons S_2O_6$$

Platinum is used as a catalyst, but because of the ease with which the platinum becomes poisoned (rendered inactive), this metal may be replaced by a mixture of various metallic oxides with vanadium pentoxide (V_2O_5). The sulfur trioxide is absorbed in concentrated sulfuric acid* in which it is readily soluble, forming pyrosulfuric acid.

$$SO_3 + H_2SO_4 \rightarrow H_2S_2O_7$$
pyrosulfuric acid
(fuming sulfuric acid)

Pyrosulfuric acid is then converted into sulfuric acid by adding water.

$$H_2S_2O_7 + H_2O \rightarrow 2\ H_2SO_4$$

The acid produced by this process may be 100% or any other desired percentage.

Properties of Sulfuric Acid

Pure sulfuric acid is a heavy, viscous liquid (sometimes called "oil of vitriol"), with a density of 1.84. It exhibits three distinct sets of chemical properties.

---

* SO_3 tends to **dimerize** (unite two molecules) to the solid S_2O_6, which as a smoke in air is not easily dissolved in water; thus it is initially dissolved in concentrated H_2SO_4, followed by the addition of water.

ACID PROPERTIES   Dilute sulfuric acid exhibits typical acid reactions. It reacts with metals above hydrogen

$$Fe + H_2SO_4 \rightarrow FeSO_4 + H_2$$

and neutralizes bases to form salts.

$$2\ NaOH + H_2SO_4 \rightarrow Na_2SO_4 + 2\ H_2O$$
$$Ca(OH)_2 + H_2SO_4 \rightarrow CaSO_4 + 2\ H_2O$$

Concentrated sulfuric acid reacts with salts by exchange to produce salts and acids. (This is the general method of preparing volatile acids.)

$$2\ NaCl + H_2SO_4 \rightarrow Na_2SO_4 + 2\ HCl(g)$$
$$2\ NaNO_3 + H_2SO_4 \rightarrow Na_2SO_4 + 2\ HNO_3(g)$$

These equations are practical in industry.

OXIDIZING AGENT   Hot concentrated sulfuric acid acts as an oxidizing agent, reacting with some metals below hydrogen in the activity series.

$$Cu + 2\ H_2SO_4 \rightarrow CuSO_4 + SO_2 + 2\ H_2O$$
$$2\ Ag + 2\ H_2SO_4 \rightarrow Ag_2So_4 + SO_2 + 2\ H_2O$$

Some metals above hydrogen act in the same manner with hot concentrated acid.

$$Zn + 2\ H_2SO_4 \rightarrow ZnSO_4 + SO_2 + 2\ H_2O$$

Sulfuric acid reacts with metals below hydrogen in the activity series, not because of its acid property, but because of its oxidizing action.

Hot concentrated H_2SO_4 will also oxidize nonmetals.

$$C + 2\ H_2SO_4 \rightarrow CO_2 + 2\ SO_2 + 2\ H_2O$$
$$2\ S + 4\ H_2SO_4 \rightarrow 6\ SO_2 + 4\ H_2O$$

The action of $H_2SO_4 + S$ is an example of autooxidation; the same element is involved in both gain and loss of oxidation number.

DEHYDRATING AGENT   Concentrated sulfuric acid has a great affinity for water. If an open container nearly full of the acid is allowed to stand in contact with the air for some time, the liquid in the container increases in volume and may even flow over the top. Concentrated sulfuric acid is often used in a closed space to produce a virtually dry atmosphere.

When concentrated sulfuric acid is added to sugar, a charred mass of carbon is formed. Sugar is a compound in which the ratio of hydrogen to oxygen is the same as in water—that is, two atoms of hydrogen to one atom of oxygen.

$$C_{12}H_{22}O_{11} \xrightarrow{H_2SO_4} 12\ C + 11\ H_2O$$

The hydrogen and oxygen are removed from the sugar molecule in the form of water because of the affinity of the sulfuric acid for water. The acid probably is not changed chemically in the reaction, but is diluted by the water formed. Wood is blackened in contact with sulfuric acid because of a similar reaction.

$$(C_6H_{10}O_5)_x \xrightarrow{H_2SO_4} (6\ C)_x + (5\ H_2O)_x$$

The manufacture of explosives (discussed in Chapter 20) involves the use of concentrated sulfuric acid. The action of concentrated nitric acid on such compounds as glycerin or nitrocelluloses seems to be markedly hastened by concentrated sulfuric acid because water is removed from the reaction by the acid.

$$\underset{\text{glycerin}}{C_3H_5(OH)_3} + 3\ HNO_3 \xrightarrow{H_2SO_4} \underset{\text{nitroglycerin}}{C_3H_5(NO_3)_3} + 3\ H_2O$$

Uses of Sulfuric Acid

Sulfuric acid stands at the head of the list of the 50 top chemicals. The annual production of 66 billion pounds in the U.S. easily outstrips the next most abundantly produced chemical, lime, by 25 billion pounds. There is hardly an industry that is not directly or indirectly dependent on the use of sulfuric acid. Annual consumption in the various industries of the United States is shown in Table 21.1.

Table 21.1
Use of sulfuric acid in the United States per year

Industry	Tons (100% acid)	Use
Fertilizer	19,180,000	Production of ammonium sulfate and superphosphate
Petroleum refining	1,880,000	Purifier of petroleum by removing dark colored products, especially sulfur compounds
Chemicals	3,830,000	Preparation of hydrochloric and nitric acids, metal sulfates, ether, and so on
Iron and steel	530,000	Removal of rust on iron and steel before galvanizing or enameling
Metallurgy	1,300,000	Refining of metals by electrolysis
Paints and pigments	1,160,000	
Explosives and miscellaneous	4,260,000	Dehydrating agent in the nitration of organic compounds
Rayon and cellulose film	860,000	
Total	33,000,000	

21.5 SELENIUM AND TELLURIUM

Selenium and tellurium closely resemble sulfur, though tellurium exhibits more metallic properties than other members of the group. This is in accord with the general tendency of nonmetallic elements to show increasingly metallic properties as the atomic weight increases.

Selenium is obtained as a by-product of the roasting of pyrites. Selenium and tellurium are obtained as by-products of the electrolytic refining of copper, lead, and other metals. Selenium is used in the production of red glass, and selenium cells are used in the activation of automatic doors. The element becomes an increasingly better conductor of an electric current according to the strength of light impinging on it.

Tellurium has no major use at present. Some use is made of relatively insoluble selenates and tellurates by incorporating them into paints for ship bottoms to inhibit the growth of barnacles and other organisms.

Examples of formulas for some compounds of selenium and tellurium follow; note the similarity to the family member sulfur.

$$H_2Se, \ H_2SeO_3, \ H_2SeO_4, \ CuSeO_4, \ FeSe, \ SeO_2, \ SeO_3$$
$$H_2Te \ (more \ miserable \ odor \ than \ H_2S), \ CuTeO_4, \ Ca_3Te_2$$

Exercises

21.1. Since the text has largely covered sulfur chemistry, it is good mental practice to consider selenium as a similar element.

   (a) Complete the electron distribution for $_{34}Se - 1s^2; \ 2s^2, \ 2p^6;$ _____
   (b) The N quantum state of Se has in total how many electrons? Properly fill in orbital designations, using a single arrow for each electron.
   (c) What is the maximum oxidation number for Se?
   (d) What is the minimum oxidation number for Se?
   (e) What are the oxidation numbers for elemental selenium and for Se in SeO_2?
   (f) What element lighter than sulfur does selenium resemble in outer structure, particularly in its lower oxidation state behavior?

21.2. Recalling the formulas of typical oxygen compounds, write the *formulas* or *names* for the following sulfur family compounds: (a) hydrogen selenide, (b) tellurium dioxide, (c) sodium thiosulfate, (d) CH_3CSNH_2, (e) CuTe.

21.3. From the first column pick a number that designates a name or formula that matches a name or formula in the second column.

1. FeS_2	$CaSO_4 \cdot 2 \ H_2O$	_____
2. hydrosulfuric acid	gunpowder	_____
3. cinnabar	conc. H_2SO_4	_____
4. $CuFeS_2$	galena	_____
5. gypsum	pyrite	_____

6. PbS	HgS	_____
7. C, S, KNO$_3$	chalcopyrite	_____
8. a sulfite	match head	_____
9. P$_4$S$_3$ or Sb$_2$S$_3$	H$_2$S in water	_____
10. dehydrating agent	CaSO$_3$	_____

21.4. Match the numbers with a related phrase.

1. 113	density of conc. H$_2$SO$_4$	_____
2. 1.84	boiling point of H$_2$S	_____
3. 33,000,000	oxidation number of S in CuSO$_4$	_____
4. 6	1st ionization constant of H$_2$S	_____
5. 1.2×10^{-7}	g H$_2$SO$_4$ in 1 liter 0.1 N H$_2$SO$_4$	_____
6. 4.9	melting temp. of S	_____
7. 445	tons H$_2$SO$_4$ used per year	_____
8. -59.6	boiling point of S	_____

21.5. Describe the Frasch process for mining elementary sulfur.

21.6. Describe the changes in sulfur that occur as the temperature is increased.

21.7. What is meant by (a) rhombic sulfur, (b) monoclinic sulfur, (c) plastic sulfur, (d) thioacetamide, (e) rotten egg gas?

21.8. Select the formula from the list on the right which best applies.

(a) Product obtained by burning sulfur compounds
(b) Used in paper industry to dissolve lignin in wood
(c) "Oil of vitriol"
(d) A product from the contact process
(e) Fool's gold
(f) Elements other than S with a maximum oxidation number of +6
(g) The Frasch process
(h) "Brimstone"
(i) An unstable weak acid
(j) Fuming sulfuric acid (pyrosulfuric acid)
(k) A powerful dehydrating agent

1. S
2. FeS$_2$
3. PbS
4. H$_2$S
5. H$_2$SO$_3$
6. conc. H$_2$SO$_4$
7. dil. H$_2$SO$_4$
8. Ca(HSO$_3$)$_2$
9. SO$_3$
10. H$_2$S$_2$O$_7$
11. NO$_2$
12. Se and Te
13. As and Sb
14. N and P
15. SO$_2$

21.9. (a) Write formulas for:
1. telluric acid
2. zinc telluride
3. sodium selenate

(b) Name:
1. H$_2$SeO$_3$
2. CuSeO$_4$
3. TeO$_3$

21.10. Write balanced molecular or ionic equations for:
(a) precipitation of a divalent metal ion by H$_2$S.
(b) what takes place as H$_2$S in emitted volcanic gas or in mineral water is partially oxidized by O$_2$ of the air or by dissolved O$_2$ in water.

(c) action of Cl_2 bubbled into H_2S containing water.

(d) reaction of an alkaline metal sulfide plus dilute H_2SO_4.

(e) $H_2S + Fe^{3+} \rightarrow FeS + S + H^+$.

21.11. Concerning SO_2 chemistry, write balanced chemical equations for:

(a) roasting 4 $CuFeS_2$ (chalcopyrite) in air to copper(II) and iron(III) oxides and sulfur dioxide.

(b) reaction of calcium hydroxide with sulfurous acid.

(c) oxidation of SO_2 by $K_2Cr_2O_7$ in acid solution. $K_2Cr_2O_7 + SO_2 + H_2SO_4 \rightarrow K_2SO_4 + Cr_2(SO_4)_3 + H_2O$

21.12. Write equations for the reactions of sulfur with (a) O_2, (b) S^{2-}(aq), (c) SO_3^{2-}(aq), (d) Fe, (e) HNO_3.

21.13. Write equations for the reactions of H_2SO_4 with (a) $C_{12}H_{22}O_{11}$, (b) $NaNO_3$, (c) Cu, (d) Zn, (e) ZnS, (f) Fe_2O_3.

21.14. Write a sequence of equations representing reactions that could be used to synthesize each of the following and that start with elementary S. (a) H_2S (not by direct union of the elements), (b) H_2SO_3, (c) $Na_2S_2O_3$, (d) $NaHSO_4$, (e) $H_2S_2O_7$.

21.15. Write balanced equations for:

(a) the action of H_2SO_4 on rust (consider as $Fe_2O_3 \cdot H_2O$) removal as in "picking" steel in the iron and steel industry.

(b) the action of H_2SO_4 on calcium phosphate in the preparation of H_3PO_4.

(c) the action of ammonia gas bubbled into a water solution of H_2SO_4.

21.16. Paper pulp (almost 100% cellulose) is prepared commercially by digesting wood chips in a hot aqueous solution of calcium bisulfite, $Ca(HSO_3)_2$. The latter dissolves lignin and resins in the wood, leaving nearly pure cellulose. The sulfite solution is prepared by the following reactions:

$$S + O_2 \rightarrow SO_2$$
$$SO_2 + H_2O \rightarrow H_2SO_3$$
$$CaCO_3 + 2\ H_2SO_3 \rightarrow Ca(HSO_3)_2 + CO_2 + H_2O$$

(a) For each kg of limestone ($CaCO_3$) used in the process, what weight of sulfur would be required?

(b) What weight of $Ca(HSO_3)_2$ is produced from the limestone and sulfur in part (a)?

21.17. A mined rock is 70.0% $Ca_3(PO_4)_2$ and 30.0% insoluble silicates. In making "superphosphate"

$$Ca_3(PO_4)_2 + 2\ H_2SO_4 \rightarrow 2\ CaSO_4 + Ca(H_2PO_4)_2$$

how much H_2SO_4 would be used per kg of this rock?

21.18. A tomato weighing 50.0 g is analyzed for sulfur content by digestion in concentrated HNO_3 followed by precipitation of 0.425 g of $BaSO_4$. If all of the sulfur is converted to $BaSO_4$, what is the percentage of sulfur in the tomato?

21.19. What volume of H_2S at STP is required for reaction with $CdSO_4$ solution to produce 1250 g of CdS to be used as a yellow paint pigment?

21.20. A 15.0-ml sample of auto storage battery acid (H_2SO_4) is diluted to 100 ml and titrated with 0.450 N NaOH. From the latter solution 45.50 ml is required for neutralization. (a) What is the normality of the diluted acid? (b) If the battery acid is H_2SO_4, what is its concentration in grams of H_2SO_4 per liter of solution?

21.21. Refer to Table 21.1 concerning the use of H_2SO_4 and explain the role of this acid in each of the uses listed. Illustrate, when possible, with balanced equations.

Supplementary Readings

21.1. "Mineral Cycles," E. S. Deevey, Jr., *Sci. American*, Sept. 1970; p. 148.
21.2. "The Sulfur Cycle," W. W. Kellogg et al., *Science*, **175**, 587 (1972).
21.3. "Sulfur," C. J. Pratt, *Sci. American*, May 1970; p. 62.

Chapter 22

Main Energy Levels   1   2

$^{12.011}_{6}$C   *K* s^2p^2

Carbon and Its Oxides

22.1   CARBON

Carbon, in Group IVA of the periodic table, is centrally located between those elements that act distinctly as metals and those that act distinctly as non-metals. Carbon in some forms possesses certain properties characteristic of metals and in others exhibits properties usually associated with nonmetals. It is the key element present in all of the millions of organic compounds.

Occurrence

Carbon is nineteenth in abundance among the chemical elements, yet its percentage in the earth's crust is only 0.027. It is widely distributed in both free and combined states. Elementary carbon is found as diamond and graphite, and in various forms of coal. Its compounds are almost innumerable. Every living cell, plant or animal, contains carbon compounds; petroleum is a mixture of compounds of carbon and hydrogen (hydrocarbons); most of our food and clothing consists of mixtures of carbon compounds; and many carbonate minerals occur on the earth.

Carbon forms compounds only by sharing electrons — all of its compounds are covalent. Not only does it share electrons with other elements but it shares electrons with itself: many carbon atoms may cluster together to form what may be called giant molecules. The chemistry of these compounds is so extensive that it comprises the special field of organic chemistry (discussed in Chapters 31–33).

An environmental pollution panel in 1965 estimated that there is 40,000 times as much carbon in rocks of the earth's crust and in carboniferous deposits as there is in the air as carbon dioxide. We can infer from this that the percentage of carbon dioxide in the air in earlier geologic times was much higher than the 0.035% now present.

Nearly three-fourths of the CO_2 once present in the air has combined with oxides of calcium, magnesium, and other elements during the weathering of silicate rocks to form carbonates. The other one-fourth has been largely converted to organic matter of past life material buried under sediment and thus protected from oxidation.

Under normal conditions, the ratio of CO_2 to O_2 in the atmosphere remains approximately constant in an equilibrium on a year basis. In 1963, Helmut Leuth reported that the annual use of carbon in photosynthesis is about 150 billion tons per year, roughly divided between land plants and marine plants. This is about one-fifth of the carbon present in the atmosphere as CO_2. This is matched by the annual release from oxidation of organic matter in rotting, respiration, and the burning process.

Man's increasing activity in land clearing and burning of fossil fuels is gradually increasing the CO_2 content of air. World weather changes and rate of photosynthesis may be effected in the decades to come as air CO_2 content increases by even some hundredths or tenths of 1%.

Russian scientific literature reports that the temperatures near the surface of the planet Venus are very high—over 720°F—and its atmosphere contains more than 75% CO_2. The polar ice caps of Mars probably consist of some solid CO_2 (dry ice) as well as solid H_2O.

Allotropic Forms

Carbon exists in three allotropic forms. Two of these are definitely crystalline—**graphite** and **diamond** (which will be considered in more detail later). The third allotropic form, **amorphous** carbon, has a number of common names—coke, charcoal, soot, lampblack, carbon black, sugar charcoal, and boneblack—as we encounter it with various impurities. These forms of carbon have been shown by X-ray analysis to be minutely crystalline, but the crystals are too small to give the substance apparent crystalline properties.

Carbon black is of such industrial importance that it ranks 29th among all of the chemicals produced. Over 3,350,000,000 pounds are produced annually in the U.S. Most of this is obtained from natural gas, CH_4, or various petroleum fractions by partial burning or thermal decomposition. For example,

$$CH_4 + O_2 \xrightarrow{\text{heat}} C(\text{carbon black}) + 2\ H_2O$$

About 93% of the carbon black used is in rubber products such as tires, heels, and industrial rubber goods. The carbon imparts toughness to the rubber. Carbon black combined with synthetic rubber gives 30 to 50% better wear than natural rubber and carbon. A modern tire contains about 40% carbon black. Carbon black is also used for printing inks, black plastics and paints, typewriter ribbons, carbon paper, shoe and stove polishes, drawing inks, etc.

Destructive Distillation of Wood and Coal

Heating a substance in the absence of air to form volatile and nonvolatile products is called **destructive distillation.** If wood (covered with sand to exclude air) is heated, volatile material is given off and charcoal remains. The properties of the charcoal depend largely on the kind of wood used. The more modern process is to heat wood in airtight ovens in which it decomposes into charcoal, water vapor, and a poor-burning gas. Wood alcohol, acetic acid, and acetone may be condensed from this gas.

Industrially, a destructive distillation consists of heating suitable coal in ovens to yield coke and volatile products. The coke retains the nonvolatile mineral matter of the coal and, since it also contains a high percentage of carbon, is often used in metallurgical operations and in making gas. Volatile products from the destructive distillation of coal are coal gas, benzene, coal tar, and ammonia. Benzene and coal tar are raw materials for chemical industries which produce medicines, dyes, explosives, certain kinds of plastics, and many other organic compounds.

Formation of Coal

Coal is the result of change of fossil plant material which has been protected from complete decay by overlying water-washed earth deposits. During the Carboniferous Era, the air had a much higher carbon dioxide content than at the present time, the average earth surface temperature was higher, and vegetation grew in profusion. Much carboniferous plant material ended beneath a cover of earth, and in time changes were effected by pressure, heat, and other factors. The original plant material probably had a composition somewhat like that of wood. Preliminary decomposition of such material gives a product called **peat,** which may be dug from its deposits and burned, though it makes a rather poor fuel. In peat formation, carbon dioxide and methane are the principal gases evolved. The methane, CH_4, is known as marsh gas, since it occurs in the bubbles coming up through the water in boggy regions where organic decomposition takes place. Further loss of carbon dioxide, water, and minor amounts of other gases from the underground organic mass brings about an increase of the percentage of carbon, with resultant formation of coal. Various grades of coal differ, according to the amount of metamorphosis the carboniferous deposit has undergone during geological time.

Fuels

Wood, coal, and oil, man's most useful fuels, are mixtures largely of carbon or carbon-containing compounds. Their combustion gives usable heat energy, with carbon dioxide and water as the main products. The results of chemical analysis of a number of varieties of fuels are given in Table 22.1.

Ultimate analysis aims at ascertaining the percentages of the elements in a mixture such as wood or coal. **Proximate analysis,** which is much more rapid, is a practical means of evaluating coal for its different uses. Bituminous (soft) coal, with a high percentage of volatile matter, gives a higher yield of coal gas and coal tar than an anthracite coal. The lignite coals, with high percentages of moisture and low percentages of carbon, make mediocre fuels.

Petroleum is a naturally occurring mixture of compounds of carbon and hydrogen (Chapter 31). Wood is composed largely of lignin and cellulose, compounds containing carbon, hydrogen, and oxygen, and a very small percentage of mineral matter that was required in plant growth.

In general, the higher the hydrogen content of fuel, the greater the heat

Table 22.1
Fuels

Fuel	Ultimate Analysis (%)						Proximate Analysis (%)				cal/g
	C	**H**	**N**	**O**	**S**	Si, Fe, Mg Oxides	Mois- ture	Vola- tile Matter	Fixed **C**	Ash	
Anthracite	82.0	0.5	0.1	1.8	0.9	14.7	4.5	3.0	78.7	13.8	7100
Pa. Bituminous	71.5	5.3	1.3	9.1	3.1	9.7	1.8	32.8	47.3	18.2	7200
Ore. Subbituminous	51.1	5.5	1.2	28.2	0.8		16.1	31.1	39.6	13.2	5500
N. D. Lignite	37.4	6.4	0.6	45.0	0.2		36.9	24.9	27.7	10.4	5000
Coke									89.0	10.2	7200
Charcoal									97.0	3.0	7700
Wood	40.0	7.2	0.8	50.7		1.3					4500
Petroleum	84.0	13.0									10400

liberated per unit burned; and the higher the oxygen content of fuel, the lower the heating value. **Calorific value** is the measured amount of heat liberated per unit of a substance burned; it may be measured in calories or British thermal units.[*]

Gaseous fuels are in high demand because they can be conveniently delivered by pipe. Chemically, **natural gas,** which is obtained from bore holes in many parts of the country, is composed of the lighter hydrocarbon molecules. Methane (CH_4) is a high-percentage constituent of natural gas. Propane (C_3H_8) and butane (C_4H_{10}) gases find extensive use as delivered in pressurized tanks. **Coal gas,** from the destructive distillation of bituminous coal, is used for heating and cooking in urban areas and as an industrial fuel. It is carefully treated to remove hydrogen sulfide, ammonia, and small particles of solid or liquid matter. As distributed for use, it is about 50% hydrogen, 30 to 35% methane, 8% carbon monoxide, 2% other hydrocarbons, and 5 to 10% nitrogen.

Properties and Uses of Carbon

Since a small weight of charcoal has a large surface area because of its high porosity, it is widely used as an adsorbent. Adsorption is the tendency of all solids to condense a layer of gas or liquid on their surfaces. It has long been a practice of farmers to suspend a gunnysack of charcoal in a cistern to adsorb gas odors. Gas mask cannisters contain specially prepared charcoal of high adsorptive capacity. A brownish, impure sugar solution is rendered colorless by filtering it through charcoal. It should be stressed that adsorp-

---

[*] A British thermal unit (BTU) is the amount of heat required to raise the temperature of one pound of water one degree Fahrenheit.

tion by a given solid is selective. For example, charcoal adsorbs coloring matter but not sugar from solution; charcoal adsorbs a high percentage of hydrogen sulfide gas from an atmosphere but only a small percentage of nitrogen.

Other uses for various kinds of amorphous carbon are printer's ink, filler for rubber, shoe polish, paints, and enamels.

Chemical Properties

At ordinary temperatures, carbon does not readily unite with other elements. At higher temperatures, however, it combines with many metals and non-metals. Carbon heated in air burns to form CO_2, with the evolution of about 7900 cal per gram of carbon.

$$C + O_2 \rightarrow CO_2$$

If it is heated in a limited amount of air or at a very high temperature, carbon monoxide rather than CO_2 is produced, according to the equation

$$2\,C + O_2 \rightarrow 2\,CO$$

Carbon monoxide, a dangerous gas, results from incomplete combustion of carbon or its compounds. If a large volume of oxygen is available, CO, if first formed, readily burns to CO_2.

Carbon disulfide, CS_2, a liquid compound formed by sulfur vapor in contact with hot carbon, is used to make "viscose," an intermediary substance in rayon production. It is also used to kill rodents. Since it is a readily volatile and flammable liquid, it must be kept away from open flames.

Carbon tetrachloride, CCl_4, is a nonflammable liquid used as a solvent for oils and greases. Vapors of both CS_2 and CCl_4 are toxic, and their inhalation is to be avoided.

Reducing Action

Carbon and carbon monoxide, the cheapest reducing agents available, are used extensively to obtain metals from metal ores in the metallurgical industry. For example,

$$ZnO + C \rightarrow Zn + CO$$
$$Fe_2O_3 + 3\,CO \rightarrow 2\,Fe + 3\,CO_2$$

Carbides

The more active metals cannot be prepared from their oxides in this manner because of their tendency to unite with hot carbon to form carbides. Aluminum would be a much cheaper metal if it could be obtained by heating aluminum oxide with carbon. The chemical change that takes place at high temperatures is

$$2 \, Al_2O_3 + 9 \, C \rightarrow 6 \, CO + Al_4C_3$$

Similarly, heating sand and coke in an electric furnace will produce silicon carbide, called **carborundum.**

$$SiO_2 + 3 \, C \rightarrow SiC + 2 \, CO$$

The impure blue-black, iridescent, crystalline carborundum is almost as hard as diamond and is widely used as an abrasive in the form of powder, whetstones, or grinding wheels.

In certain other compounds of carbon with metals, the oxidation state of carbon seems to be a number other than four. One group of compounds, called carbides, would better be called acetylides. For example, carbon and lime heated together in an electric furnace form calcium carbide, which will react with water to produce acetylene gas, the basis of an important industry.

$$CaO + 3 \, C \rightarrow CO + CaC_2$$
$$CaC_2 + 2 \, H_2O \rightarrow Ca(OH)_2 + C_2H_2$$

In the iron and steel industry, other carbides are encountered. Steels may contain some Mn_3C or some Fe_3C (**cementite**), a hard crystalline substance largely accountable for the hardness and brittleness of a high-carbon steel.

Graphite

This shining, black, soft, slippery-feeling, flaky, crystalline form of carbon occurs in widely separated deposits. It is mined in Ceylon, Siberia, and to some extent on the North American continent. Naturally occurring graphite has the mineral name **plumbago.** Plumbago deposits have probably resulted from heating of carbonaceous material to high temperature underneath a covering of rock. Simulating nature, man produces graphite by heating hard coal in an electric furnace, under cover of sand, to a temperature of 3500°C. At this temperature carbon seems to vaporize quite readily, and the condensing vapor molecules form hexagonal crystals of graphite.

Figure 11.15, page 259, shows the hexagonal structure of carbon atoms of graphite in one plane, in contrast with the less closely interbonded third dimension. This structure helps account for the flaky nature of graphite and the fact that it can be oxidized to various six-carbon-atom organic compounds. The interplanar bonding electrons allow graphite to exhibit a relative softness and some degree of electrical conductivity.

Since graphite is a better conductor of electric current than other forms of carbon and is resistant to heat and chemical change, it has numerous uses. Electrodes and crucibles are made of it. The scalelike crystals readily slip over each other, and thus graphite is a good lubricant. A suspension of graphite powder in water for lubricating purposes is known as "aquadag." Stove polish contains graphite to inhibit oxidation of the iron. Lead pencils contain a mixture of graphite and clay as a core; the hardness is determined by the percentage of clay and the texture of the mix.

Diamond

Diamond is a colorless solid capable of being cut into brilliant stones, though it occurs naturally as a lusterless crystal. Burning diamond in oxygen at 700°C yields only carbon dioxide as a product, proving that it is pure carbon. See Figure 22.1.

Diamond is the hardest substance known to man. Its conduction of heat and electricity is very poor. These properties may be attributed to the tetrahedral structure of the crystals (see Figure 11.14, page 258). The closely packed atoms account for the hardness and relatively high density; the four covalent bonds per atom effectively tie down the valence electrons, a fact which explains the insulating properties.

Diamonds are found in several regions, but the principal deposits are in South Africa, the East Indies, and Brazil. About one-quarter of those mined

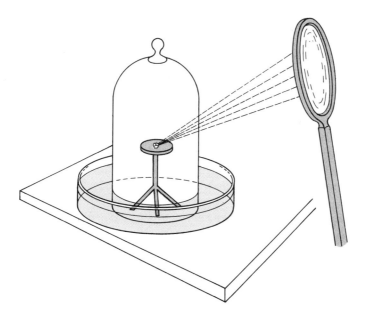

Figure 22.1
Newton discovered that diamond was simply carbon by burning a small diamond to give off CO_2.

are suitable for gemstones; the rest are used industrially for rock drill bits, for polishing and cutting, and as pivot supports in precision instruments.

Man-made diamonds of a quality suitable for certain industrial applications became commercially available in 1957. Pressure greater than 1,500,000 pounds per square inch and temperatures of about 2700°C must be utilized to produce diamond artificially.

Dr. Guy Suits, of the General Electric Company, stated in 1961:

Our first diamonds were dark in color, and only thousandths of a carat in size—about the size of fine grains of sand. But this size is needed for many cutting, grinding, and polishing applications in industry. We succeeded in making the transition from laboratory to full-scale production of these small diamonds in two years.

Millions of carats of these industrial diamonds have been manufactured. Now $\frac{1}{10}$ carat stones are made, and some even up to 1 carat (0.2 g); but the larger, man-made diamonds are still dark in color and subject to mechanical flaws.

The weight of a diamond is expressed in carats. A one-carat diamond has an average value of about $600.

The compounds SiC, silicon carbide (carborundum), and BN, boron nitride, are nearly as hard as diamond and thus find much use. Note that multiple covalent bonding, as in C_x, is inherent in these compounds.

22.2 CARBON DIOXIDE

Carbon dioxide is present in the air to the extent of 3.5 parts in 10,000. This amount is not constant in geological time because of additions from burning carbon compounds, decay, fermentation, respiration, and volcanic gases. The amount is reduced because it is absorbed by plants, dissolved by water, and fixed into such minerals as calcium or magnesium carbonate. There are some gas wells in the central United States and Mexico from which carbon dioxide issues at considerable pressure. Escaping carbon dioxide from ground waters in a cave near Naples, Italy, builds up such a concentration that a small dog walking into the quiet air of the cave will suffocate. A man walking upright, however, will have little discomfort, since carbon dioxide is about 1.5 times as heavy as air, and the concentration at the cave floor is much higher than at the five-foot level. Carbon dioxide is not a poison, but a moderately high concentration of it causes lack of oxygen and inability to rid the body of carbon dioxide.

Preparation

(*a*) Carbon dioxide is the product of the complete burning of carbon or its compounds. Most vaporizable or gaseous carbon compounds form an explosive mixture when mixed with air. Occasionally, serious explosions occur when a suspension of finely divided solid carbon compounds in air (coal dust, flour dust, and so on) is subjected to a spark or flame. Precautions must be taken in mills for grinding organic material so that sparks or flames do not contact the organic dust-laden air. The slow oxidation processes of rotting and respiration form carbon dioxide as a product.

(b) Fermentation of carbohydrate compounds results in the formation of alcohol and carbon dioxide. **Zymase,** an organic secretion of yeast cells, hastens the fermentation process. (An organic substance produced by living cells which catalyzes specific chemical changes is called an **enzyme.**)

$$C_6H_{12}O_6 \xrightarrow{\text{zymase}} 2\ C_2H_5OH + 2\ CO_2$$
$$\text{glucose} \qquad\qquad \text{alcohol}$$

(c) Carbon dioxide is produced when metal carbonates (except those of the alkali metals, such as Na_2CO_3 or K_2CO_3) are heated. Heating limestone to produce lime is the basis of a large industry.

$$CaCO_3 \rightarrow CaO + CO_2$$

This chemical change requires a temperature of about 900°C. Since the reaction is reversible, a draft is used to keep removing the CO_2.

(d) The most convenient laboratory method of preparing carbon dioxide is by action of an acid on a carbonate or bicarbonate.

$$CaCO_3 + 2\ H^+ \rightarrow Ca^{2+} + H_2O + CO_2$$
$$NaHCO_3 + H^+ \rightarrow Na^+ + H_2O + CO_2$$

Physical Properties

Carbon dioxide gas may be liquefied under moderate pressures below its critical temperature of 31.4°C. It is usually transported as a liquid in steel cylinders for commercial purposes. If the nozzle of such a cylinder is opened, with the nozzle end lower than the rest of the cylinder, liquid carbon dioxide is ejected into the air. Under the reduced pressures, the liquid immediately boils. Since a boiling liquid absorbs its heat of vaporization from the surroundings, a marked cooling takes place. If the liquid is allowed to squirt into a cloth bag so that the cooling is confined, some of the carbon dioxide liquid will cool to the solid state. Solid carbon dioxide, commercially called "dry ice," is made in this fashion. Since the vapor pressure of the solid is one atmosphere at −79°C, the solid sublimes at this temperature without melting (see Figure 10.3).

Carbon dioxide gas is about 1.5 times as heavy as air and, since it is non-flammable, may be used to blanket and extinguish a flame. It is not necessary to attain a high percentage of CO_2 about a fire. If the oxygen content of the air, which is only 20% to start with, is reduced by dilution with CO_2 to 17% or less, the flame will go out. The CO_2 functions in putting out a fire primarily because it lessens the oxygen concentration.

CO_2 is moderately soluble in water under ordinary conditions and markedly more soluble under pressure. Its solubility is enhanced by the reaction

$$CO_2 + H_2O \rightleftharpoons H_2CO_3$$

Carbonic acid, H_2CO_3, which has a slightly sour or biting taste, is used in carbonating beverages.

22.3   CHEMICAL PROPERTIES AND USES

Carbon dioxide is a very stable substance, but at exceedingly high temperatures it decomposes.

$$2\ CO_2 \xrightarrow{\text{high temperature}} 2\ CO + O_2$$

22.3   USES

Carbon dioxide is 29 in the list of top chemicals. It is used in large quantities as a solid refrigerant "dry ice" for food transportation. For example, 1000 pounds of solid CO_2 will adequately refrigerate an average railroad car in a trip across the country, whereas several times this amount of ordinary ice, with frequent replenishing, would be necessary. Compressed carbon dioxide (liquid CO_2) in steel cylinders has its greatest use in the carbonated beverage industry and in some fire extinguishers. Gaseous carbon dioxide is used to manufacture soda ash, Na_2CO_3, and organic compounds such as salicylic acid for aspirin production.

Carbon dioxide from the atmosphere is constantly dissolving in water on the earth's surface. Thus natural water is actually a dilute solution of carbonic acid, which effects disintegration of rocks. Minerals are acted on by the dilute acid. Enormous deposits of limestone ($CaCO_3$) and dolomitic limestone ($CaCO_3 \cdot MgCO_3$) are the result of such action.

Carbon dioxide is tested for on the basis of the chemical change

$$Ca(OH)_2 + CO_2 \rightarrow CaCO_3(s) + H_2O$$

When a clear calcium hydroxide solution (limewater) is subjected to carbon dioxide, the solution becomes milky as calcium carbonate precipitates. If carbon dioxide is added to a fine suspension of calcium carbonate, the milkiness will clear up, since calcium bicarbonate, which is water-soluble, is formed.

$$CaCO_3(s) + H_2CO_3 \rightleftharpoons Ca(HCO_3)_2$$

This chemical change is very important as the main cause of "hardness" in water. Naturally occurring limestone is constantly being slowly dissolved by carbonic acid in ground water to form soluble calcium bicarbonate. (See Chapter 26.)

Photosynthesis, the process by which plants synthesize carbohydrates from carbon dioxide and water, would be much faster if a higher concentration of carbon dioxide than the 0.035% in air were available. Increased temperature and increased sunlight would also hasten the growing process. Growing corn in a hothouse atmosphere containing more carbon dioxide than is present in ordinary air has produced phenomenally rapid rates of plant growth.

Our exhaled breath is about 4% CO_2. Oxidation of body tissue or food to largely CO_2 and H_2O gives rise to a higher CO_2 content of venous blood, from which CO_2 is released in the lungs.

22.4 CARBON MONOXIDE

We most commonly hear of carbon monoxide as a product of the incomplete burning of gasoline:

$$2 \ C_8H_{18} + 25 \ O_2 \rightarrow 16 \ CO_2 + 18 \ H_2O \text{ (complete combustion)}$$
$$2 \ C_8H_{18} + 17 \ O_2 \rightarrow 16 \ CO \ + 18 \ H_2O \text{ (incomplete combustion)}$$

With insufficient oxygen, a gasoline compound such as octane (C_8H_{18}) will give off a large quantity of carbon monoxide. Air intake into an automotive cylinder is adjusted so that almost complete combustion is effected.

Pure carbon monoxide may be prepared by heating formic acid in the presence of concentrated sulfuric acid which acts as a catalyst.

$$HCOOH \rightarrow H_2O + CO$$

Carbon monoxide is a light gas, which is very slightly soluble in water. It is an insidious poison, since it has no odor to warn of its presence. While oxygen combines with hemoglobin to form a compound that readily decomposes again to yield oxygen to body tissue, carbon monoxide unites with hemoglobin to form a stable compound. When carbon monoxide is inhaled, hemoglobin combines with it, and insufficient hemoglobin remains to function in its oxygen-carrying capacity. A person suffocates for lack of enough body oxygen when about a third of his hemoglobin has been combined with carbon monoxide.

As carbon burns, carbon monoxide is almost always an intermediate oxidation product.

$$2 \ C + O_2 \rightarrow 2 \ CO$$

The blue flame surrounding burning carbon and observable in the inner cone of a Bunsen burner flame is due to the oxidation of carbon monoxide.

$$2 \ CO + O_2 \rightarrow 2 \ CO_2$$

Good draft in a furnace allows for complete burning of carbon; a damped furnace may yield considerable carbon monoxide, which escapes up the chimney.

Since carbon monoxide is formed by incomplete combustion of carbon or its compounds, and since it is a cumulative poison, some means should be employed to warn of its presence. Canaries are sometimes kept in garages, since they are more sensitive to suffocation than man. Rather than relying on a canary, it is better to ensure sufficient ventilation of the building and to minimize the running of automobile motors indoors.

22.5   CYANIDES

Hydrogen cyanide gas, called **prussic acid**, is very poisonous and has been used in the lethal gas chambers for capital punishment in some states. It may be prepared by adding an acid to a cyanide salt.

$$2\ NaCN + H_2SO_4 \rightarrow Na_2SO_4 + 2\ HCN$$

The gas is used in fumigation. Since HCN has the rather pleasant smell of bitter almonds or crushed peach leaves, it is customary in fumigation to add some cyanogen chloride, CNCl, which by its irritating odor warns of the presence of poison. The ancient Egyptians boiled peach leaves and seeds to obtain a poisonous solution which was dilute HCN. Certain tropical trees emit small concentrations of HCN into the air.

Cyanides are used in dissolving gold and silver from ores, in cleaning silverware, as insecticides and rodenticides, in electroplating, and in case-hardening steel. Alkali cyanides have a marked tendency to unite with higher-valence metals.

$$6\ KCN + FeCl_2 \rightarrow 2\ KCl + K_4Fe(CN)_6$$

potassium ferrocyanide   or   potassium hexacyano-ferrate(II)

An antidote for a cyanide, if taken soon enough, is a solution of an iron(II) compound. The $Fe(CN)_6{}^{4-}$ ion does not have the poisonous qualities of CN^-.

Exercises

22.1. For each of the following forms of carbon explain (1) how it occurs or is prepared, (2) its importance or uses, (3) its crystal structure or lack of structure. (a) graphite, (b) diamond, (c) carbon black, (d) charcoal, (e) coke.

22.2. On destructive distillation, which of the following fuels gives the highest yield of volatile matter: anthracite coal, bituminous coal, graphite? Which is easiest to start burning? Give formulas of three compounds you might expect to find in ashes from burned anthracite or bituminous coal. As indicated by Table 22.1, which type of mined fuel is most like wood in ultimate analysis? Which form of coal would seem to result from most dehydration and degradation in geological time?

22.3. Complete and balance the equations:

(a) $C_8H_{18} + O_2 \xrightarrow{\Delta} CO +$

(b) $Fe_2O_3 + CO \xrightarrow{\Delta}$

(c) $Ca(HCO_3)_2 \xrightarrow{\Delta}$

(d) $Ca(OH)_2 + CO_2$ (excess) $\rightarrow$

(e) $Ca(CN)_2 + HCl \rightarrow$

(f) $CuO + C \xrightarrow{\Delta}$

(g) $CaCO_3 + H^+ \rightarrow$

22.4. Name the following carbon compounds: (a) CH_4, (b) CO, (c) H_2CO_3, (d) $NaHCO_3$, (e) KCN, (f) $CaCO_3$, (g) CaC_2, (h) SiC, (i) CCl_4, (j) CS_2, (k) HCN.

22.5. Name the compounds: NaCN, $K_4Fe(CN)_6$, $Ca(CN)_2$, C_2H_2, HCOOH.

22.6. Match each of the following with the formula which best applies.

(a) Coke.
(b) Limestone.
(c) A poisonous liquid with a very low kindling temperature.
(d) The gas in the gas mains in lab.
(e) Acetylene gas.
(f) Carborundum.
(g) A solid which in solution is a good solvent for Ag and Au.
(h) A poisonous gas.
(i) A liquid which will not burn and can be used as a fire extinguisher.
(j) Dry ice.
(k) An unstable acid present in soda pop.
(l) Baking soda.

1. C
2. CH_4
3. CO
4. CO_2
5. CCl_4
6. CS_2
7. Al_4C_3
8. SiC
9. CaC_2
10. C_2H_2
11. $C_6H_{12}O_6$
12. $CaCO_3$
13. $NaHCO_3$
14. Na_2CO_3
15. H_2CO_3
16. $Ca(HCO_3)_2$
17. C_8H_{18}
18. NaCN
19. $K_4Fe(CN)_6$
20. $H_2C_2O_4$

22.7. What weight of 0.10% aqueous $Ca(HCO_3)_2$ solution must have dripped and decomposed in a cave to form a 10-kg stalagmite?

$$Ca(HCO_3)_2 \rightarrow \underset{\text{(stalagmite)}}{CaCO_3} + CO_2 + H_2O$$

22.8. What weight of CaC_2 will be required to generate 6.72×10^4 liters of acetylene, C_2H_2, at STP?

22.9. The density of liquid CO_2 is 0.81 g/ml as it exists under pressure in a cylinder. Thirty liters of this liquid on gasifying and exiting from a cylinder at STP will occupy what volume?

22.10. Gold will dissolve in a cyanide solution as follows

$$4\ Au + 8\ NaCN + 2\ H_2O + O_2 \rightarrow 4\ NaAu(CN)_2 + 4\ NaOH$$

How much NaCN would be required to extract 1 kg of Au?

22.11. A piece of dry ice weighs 110 g. This (a) is how many moles? (b) will occupy what volume at 25°C and 740 torr after all of the solid changes to gas? (c) consists of how many molecules of CO_2?

22.12. What is the oxidation number of carbon in: (a) CO, (b) CO_2, (c) $CHCl_3$, (d) $C_{12}H_{22}O_{11}$, (e) C_4H_{10}, (f) $Ca(CN)_2$, (g) HCO_3^-?

22.13. Show electronic structures of CO_2, CO, $COCl_2$, CO_3^{2-}. Are any of these resonating structures?

22.14. Evaluate a girl's 0.1-carat (0.02-g) diamond ring of 12-g metal weight. The gold ring is marked 12 carat. (Carat for gold is different from that for diamond; 24 carat is pure gold. Gold is about $4.50 per gram and diamond $600 per carat.)

Supplementary Readings

22.1. "The Carbon Cycle," B. Bolin, *Sci. American*, Sept. 1970; p. 124.
22.2. "Synthesis of Diamond at Low Pressures," B. V. Derjaguin and D. B. Fedoseev, *Sci. American*, Nov. 1975; p. 102.
22.3. "The Chemistry and Manufacture of the Lead Pencil," F. L. Encke, *J. Chem. Educ.*, **47**, 575 (1970).

Chapter 23

Cloud is water as is also snow. What
is the composition of biosphere
and earth below?

Geological
Chemistry,
Silicates

23.1 NEBULAR ORIGIN OF ATOMS

The origin of the elements which comprise the rocks, liquids, and gases of this earth, and of these same elements comprising suns, planets, moons, and asteroids, is a subject of theorizing and conjecture. Spectral evidence tells us that the composition of the sun's outer region consists of hydrogen, helium, and metallic elements low in atomic weight. Nonmetallic elements, although present, are more difficult to detect in solar spectra as they do not readily give up electrons to become excited atoms in the cooler, outer-solar, low-density region. Almost all of the known elements are known to exist in our sun. The evaluation of an accurate percentage of the various elements constituting the total sun is not possible, as the total mass of the sun's atmosphere down to the level at which it becomes too dense to see further (to obtain spectral lines) is only one ten-billionth part of the sun's total mass.

It seems certain that somewhere in the vastness of space and time, atom forming has taken place from subatomic particles that constitute a major part of gaseous nebulae. In our own galaxy there are many gaseous nebulae and obscuring clouds of elemental particulates. The spiral nebulae that astronomers report are separate galaxies or island universes in space, each containing sufficient matter to make billions of stars and possible concomitant planets (Figure 23.1).

What subatomic particles are involved in atom formation in the nebular volumes of space? Our most prominent scientists admit that they do not know the part that most of the about 200 named new particles, that physicists have experimentally detected, play in atom building. Endeavor to classify 35 light-weight mesons, 56 heavyweight baryons, and over 70 other identified particles into a periodic-like table of nuclear particles is in progress. We must retreat into our admittedly inadequate theory of atomic structure as related to protons, neutrons, and electrons as atom components. Until this theory is expanded, we can but suggest that these are the essential sub-atom particles that are binding together, with emission of binding energy (Figure 23.2), to form the atoms of nebular gases and dust and, somehow, suns, planets, and moons.

The cosmic abundance of the elements from solar and stellar spectral data and calculated densities are tabulated in geochemical textbooks, but a few pertinent generalities are listed below:

(1) The abundances show a rapid decrease for elements of the lower atomic numbers (to about atomic number 30), followed by an almost constant, very small value for the heavier elements.

(2) Elements of even atomic number are more abundant than those of odd.

(3) Only ten elements, H, He, C, N, O, Na, Mg, Si, S, and Fe, all with atomic numbers less than 27, show any appreciable abundance. Hydrogen and helium far exceed the others.

The **solar system** is essentially a closed system in the universe and its

Figure 23.1

The great nebula in Orion.

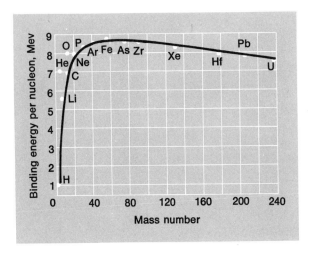

Figure 23.2
Binding energy per nucleon as a function of the mass number.

elemental composition is the same as when formed, except as it has been modified by the conversion of hydrogen to helium and other nuclear fusions in the sun, and by decay of radioactive elements. Acknowledging that the same elements constitute the mass of all suns, planets, and moons, it is appropriate to limit this study to the composition of the earth's material. The sun contains 99.8% of the mass of the solar system, but to man, earth crust composition is of most importance.

23.2   ABUNDANCE OF VARIOUS ELEMENTS COMPOSING EARTH

Geophysical evidence has our earth composed of a metallic iron alloy inner core of density ±11. Geochemically, our total earth contains far more iron than any other element. See Figure 23.3.

A great bulk of the earth is an intermediate layer, density ±6. Densities of the inner and intermediate cores are somewhat higher because of excessive pressures than for similar solid or liquid matter in the crust. However, it is believed there is little of the pure compound SiO_2, of density 2.65, in the intermediate zone. The chemical composition of this zone is probably high in metallic oxides, mainly of iron, manganese, chromium, titanium, and also higher density metal silicates. Scientists would like to drill into this zone; thus, the Mohole project was formulated but has never been carried out.

No mine or oil well is over a few miles deep. Thus, man has sampled only the outer part of the approximately 20 miles of earth's crust. Since the crust is judged to be thinner beneath certain parts of ocean bottoms, the Mohole project contemplated drilling beneath a platform or big ship stabilized on a sea surface.

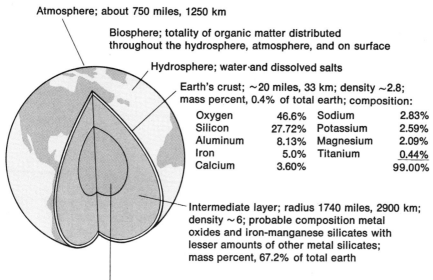

Atmosphere; about 750 miles, 1250 km

Biosphere; totality of organic matter distributed throughout the hydrosphere, atmosphere, and on surface

Hydrosphere; water·and dissolved salts

Earth's crust; ~20 miles, 33 km; density ~2.8; mass percent, 0.4% of total earth; composition:

Oxygen	46.6%	Sodium	2.83%
Silicon	27.72%	Potassium	2.59%
Aluminum	8.13%	Magnesium	2.09%
Iron	5.0%	Titanium	0.44%
Calcium	3.60%		99.00%

Intermediate layer; radius 1740 miles, 2900 km; density ~6; probable composition metal oxides and iron-manganese silicates with lesser amounts of other metal silicates; mass percent, 67.2% of total earth

Inner core; radius 2082 miles, 3470 km; density ~11; enormous pressure lessens distinction between liquid and solid; probable composition mainly iron with some nickel; mass percent, 32.4% of total earth

Figure 23.3

Structure and composition of the Earth.

The earth's crust, with a density of about 2.8, is composed principally of low-atomic-weight elements, especially oxygen and silicon.

Note from inspection of Figure 23.3 that nine elements comprise about 99.0% of all the earth's materials.

Many metallic elements and oxygen are united with the relatively small percentages of sulfur, phosphorus, carbon, nitrogen, and halogen elements in compounds of marked importance but totalling less than 1.0% of the earth's rocks.

Hydrogen and helium atoms are known to be abundant in our sun's atmosphere. Hydrogen atoms are also abundant on earth (water alone), but it takes fifty-six $^{1}_{1}H$ atoms to equal one $^{56}_{26}Fe$ atom in mass. Thus, hydrogen in percentage composition by weight of the earth's crust and hydrosphere is only 0.9%. About 0.14% hydrogen is affixed in rock composition as covalently combined water; the rest is largely in the hydrosphere.

Whether one accepts the origin of the earth as a condensation from incandescent gaseous material, or as a gradual accretion of solid particles, the internal structure seems to point to a time when the total earth was sufficiently hot for metallic iron to liquefy and sink because of its higher density over silicate material to form the theorized metallic iron (small nickel percent) core. The chemical content of early earth also included metal oxides, which are of higher density than earth gabbro and basalt silicates, which in turn are heavier than silicate granites and rhyolites (compositions of which are discussed later). Thus, there is some layering of the earth's mantle and crust. Much faulting, deformation, and volcanic movement has rendered the earth's crust heterogeneous in its form and make-up. The outer rocks, up to 3–4 miles in depth, are mainly igneous rock or the result of igneous silicate rock weathering to altered igneous or sedimentary consolidated rock, or soil.

23.3   LUNAR MATERIAL

The impossible task for man to accurately sample the total earth's crust in depth points to the inadequacy of a few moon samples to definitely ascertain elemental and compound composition. It has long been the belief of geochemists that the surface of the moon would be of unweathered igneous silicate rock, augmented by some meteoric dust or particle material of iron alloy or stony silicate. This belief is confirmed now after analysis of lunar material. From a percentage basis for the major elements present there is a very close parallelism between the composition of moon rock and that of gabbro, basalt, granitic earth, and igneous rock. Compare percentages in Table 23.1 with those in Figure 23.3.

Table 23.1

Average percent abundance of elements in lunar rock samples

O = 39.0%	Mg = 3.6%	Mn = 0.18%
Si = 20.1%	Ca = 10.4%	Cr = 0.16%
Al = 5.6%	Na = 0.33%	Zr = 0.04%
Fe = 14.5%	K = 0.15%	Ni = 0.01%
Ti = 5.5%	(lesser % of many other elements)	

The percentages of elements in Table 23.1 total 99.6%, and the remaining composition is mainly of small percentages of other metals and of S, C, N, and P nonmetals. These nonmetals are, of course, in compounds with metals, and the trace metals are mainly oxide or silicate minerals.

23.4   SILICON DIOXIDE, SiO$_2$, AND SILICATE IONS

The second most abundant element, silicon, bonds with oxygen to form SiO$_2$, which greatly exceeds H$_2$O as the most abundant earth compound (this includes the SiO$_2$ in silicates; see below). Depending on the crystallinity, purity, and other physical aspects, naturally occurring SiO$_2$ has various names.

Sand and sandstone are very high in **silica**, a common name related to small SiO_2 granules (note Si, silicon; SiO_2 silica). A clear crystalline form of SiO_2 is called **quartz**. When quartz has a purple color from trace impurity of a manganese oxide it is **amethyst. Rose quartz, rock crystal, smoky quartz, milky quartz,** and **quartzite** are other terms applied to SiO_2.

Theoretically a molecule of SiO_2 will bond with one or two molecules of water to form two compounds: H_2SiO_3, **metasilicic** acid, and H_4SiO_4, **ortho-silicic** acid. These acids have not been prepared in a pure state, and further hydration of them appears possible. These silicic acids are weak and very slightly water soluble. Heat and pressure enhance their solubility so they are very important in geochemical mineral formation.

Evaporation of water from the silicic acids accounts for **agate** and **onyx**, which may be banded by color impurities. **Chalcedony, flint, jasper,** and **opal** are other names for mainly SiO_2 minerals made opaque by impurities, with small amounts of covalently bonded water.

23.5   SILICATE MINERAL STRUCTURES

X-ray examination of silicate minerals reveals that the basis of their structure is the silicate ion. Metal ions are interspaced with the larger negative silicate ions. The SiO_4^{4-} ion has a tetrahedral structure, with an atom of silicon at the center of the tetrahedron and an oxygen atom at each of the four corners (see Figure 23.4a). The silicon atom is much smaller than the oxygen atoms and fits into a so-called "tetrahedral hole." Thus silicate tetrahedra may arrange themselves in a number of ways, as shown in Figure 23.4. Some of these are as follows:

(1) Separate SiO_4^{4-} and metallic ions arranged in a lattice much like the Na^+Cl^- structure. For example, **garnet**, $Ca_3Al_2(SiO_4)_3$, is in the cubic crystal system, Figure 23.5. Note the SiO_4^{4-} tetrahedra (large black balls = Si, small balls = oxygen). The lighter balls are the metals. Garnets are very hard and are used as industrial abrasives.

(2) Two SiO_4^{4-} tetrahedra joined at a corner to give the **disilicate** ion, $Si_2O_7^{6-}$, Figure 23.4(b). Scandium silicate, $Sc_2Si_2O_7$, is an example. Likewise **trisilicate**, Figure 23.4(c), and **hexasilicate** cyclic anions are formed by three and six SiO_4^{4-} tetrahedra joining at the corners, respectively. The jewel emerald, $Be_3Al_2Si_6O_{18}$, is an example of a hexasilicate.

(3) Silicon-oxygen chains, Figure 23.4(d), characteristic of fibrous silicates such as asbestos, $Mg_6(Si_4O_{11})(OH)_6$.

(4) Silicon-oxygen sheets, Figure 23.4(e), in which each tetrahedron shares three corners with other tetrahedra. Micas, such as muscovite $KAl_3Si_3O_{10}(OH)_2$, cleave readily into thin sheets.

(a) Tetrahedral
SiO$_4^{4-}$.

(b) SiO$_7^{6-}$.

(c) Si$_3$O$_4^{6-}$.

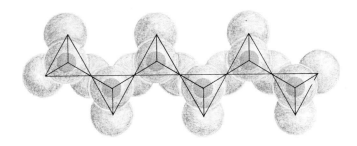

(d) Silicon-oxygen chains.

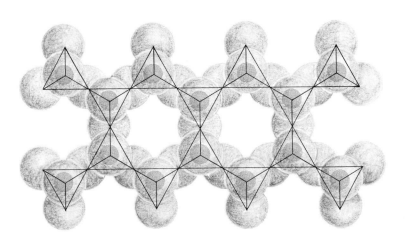

(e) Silicon-oxygen sheets.

Figure 23.4

Silicate structures.

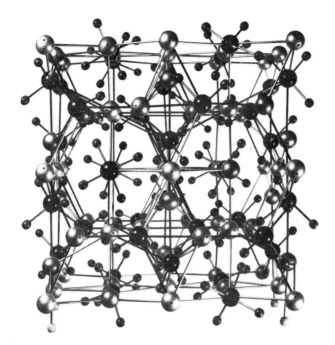

Figure 23.5

Crystal structure of garnet.

Table 23.2

The eight major rock-forming minerals

Mineral Name	Formulas	
Quartz	SiO_2	SiO_2
Olivine	$(Mg,Fe)_2SiO_4$	$2(Mg,Fe)O \cdot SiO_2$
Hornblende (an amphibole)	$Ca_2Na(Mg,Fe)_4(Al,Fe,Ti)_3 \cdot Si_6O_{22}(F,OH)_2$	Ca, Na, Mg, Fe, Al, Ti metasilicate
Augite (a pyroxene)	$Ca(Mg,Fe,Al)(SiAl)_2O_6$	Ca, Mg, Fe, Al silicate
Muscovite (a mica)	$KAl_3Si_3O_{10}(OH)_2$	$K_2O \cdot 3Al_2O_3 \cdot 6SiO_2 \cdot 2H_2O$
Biotite (a mica)	$K(Mg,Fe)_3AlSi_3O_{10}(OH)_2$	$K_2O \cdot 6(Mg,Fe)O \cdot Al_2O_3 \cdot 6SiO_2 \cdot 2H_2O$
Orthoclase (a feldspar)	$KAlSi_3O_8$	$K_2O \cdot Al_2O_3 \cdot 6SiO_2$
Plagioclase (a feldspar)	$NaAlSi_3O_8 \cdot CaAl_2Si_2O_8$	$Na_2O \cdot Al_2O_3 \cdot 6SiO_2$ and $CaO \cdot Al_2O_3 \cdot 2SiO_2$ mixtures

(5) Silicon-oxygen three-dimensional networks such as ultramarine, $Na_8(Si_6Al_6O_{24})(S_2)$, a very hard mineral.

Metal ions in spatial arrangement fit into holes in the silicate ion structures and in such numbers that there is an equalization of electrostatic charges.

Relatively few minerals make up the bulk of rocks. Table 23.2 shows the 8 major rock-forming minerals of the earth. Listed are the mineral names and chemical formulas. The formulas of most silicates at first glance appear to be very complex. However, when written as a combination of oxides, they do not look so formidable. Both kinds of formulas are in common use and are included in the table. In addition to the above formulas, sometimes ionic formulas are also used. For example, the formula for orthoclase may be written $KAlSi_3O_8$ or $K_2O \cdot Al_2O_3 \cdot 6\ SiO_2$ or $2\ K^+\ 2\ Al^{3+}\ Si_6O_{16}^{8-}$. It should be noted that

Figure 23.6
Some common minerals: Cassiterite, Quartz, and Arsenopyrite. From Portugal.
Courtesy of the American Museum of Natural History.

minerals rarely have definite chemical compositions and often consist of isomorphous mixtures, solid solutions, or colloidal dispersions. For example, the formula of olivine, $(Mg,Fe)_2SiO_4$, means that it may consist of Mg_2SiO_4, Fe_2SiO_4, a double salt $Mg_2SiO_4 \cdot Fe_2SiO_4$, or various isomorphous mixtures of magnesium and ferrous silicates. X-ray crystallography has done much to simplify our understanding of mineral structures. Figure 23.6 shows the appearance of some of the common minerals.

Gems

Gems or precious stones are silicates or other minerals highly prized for their beauty. They are rare, chemically quite inert, have high hardness, and generally possess a high refractive index which promotes brilliance. Table 23.3 gives the composition and appearance of several gems.

Table 23.3
Some common gems

Name	Chemical Formula	Appearance
Emerald, aquamarine	$Be_3Al_2(SiO_3)_6$	Green-blue, green
Ruby, sapphire	Al_2O_3	Red, blue
Diamond	C	Colorless to black
Jade	$NaAl(SiO_3)_2$	Green
Opal, amythest	SiO_2	White, purple
Pearl	$CaCO_3$	White to black
Topaz	$(AlF)_2SiO_4$	Pink to blue
Turquoise	$H[Al(OH)_2]_2PO_4$	Blue, green
Garnet, peridot	$Ca_3Al_2(SiO_4)_3$	Red, yellow-green

23.6 MAGMA AND LAVA

In present and past geological time liquid rock masses, in what is termed volcanic action, came near the earth's surface and sometimes flowed onto the earth's surface. **Lava** is the name given to surface flow liquid and thence its solidified rock. The term **magma** applies to below-surface liquid rock or its solidification product. **Igneous** rock is a general name for any rock that is an unaltered product of magma or lava solidification and as such constitutes by far the greatest percentage of earth crust rock.

The minerals $Na_2O \cdot Al_2O_3 \cdot 6 SiO_2$ and $K_2O \cdot Al_2O_3 \cdot 6 SiO_2$, called the **feldspars,** are found in abundance in igneous rocks. They, along with mica, $K_2O \cdot Al_2O_3 \cdot Fe_2O_3 \cdot 6 SiO_2 \cdot 2 H_2O$, and free SiO_2 compose most of the rock **granite.** The gabbro and basalt rocks do not contain SiO_2 that is not chemically combined and are mainly metal meta- and orthosilicate mixtures.

Geologists have estimated that the upper 10 miles of the earth's crust consist of 95% igneous rocks and 5% sedimentary. Of the sedimentary rocks, 4% is shale, 0.75% sandstone, and 0.25% limestone. Sedimentary rocks in general form relatively thin layers on an igneous base, except for some locally thick layers.

23.7   CHEMICAL PRODUCTS FROM SiO$_2$

Aside from its use as sand and in siliceous ground stone for building purposes, a considerable bulk of SiO$_2$ is chemically converted to useful materials.

In an electric furnace with controlled proportions the following simple equation is the basis of an industrial operation:

$$SiO_2 + 2\ C \rightarrow Si + 2\ CO$$

The liquid silicon from the furnace solidifies to a shiny grey material that resembles graphite, but is harder. One use of the element is in alloying aluminum, as a small percentage of it gives added strength. An iron-silicon alloy, **duriron,** is 15% Si and resists chemical action. It is widely used in laboratory drain pipes.

SiC, **carborundum,** used on abrasive paper or in grinding wheels, is produced from carbon and ground quartz (SiO$_2$) by electric induction heating.

$$SiO_2 + 3\ C \rightarrow SiC + 2\ CO$$

Silica appears to be important in certain biological processes. The ashes of certain grains, such as oats and barley, and the feathers of certain birds contain appreciable quantities. The supporting structure of the **diatom,** a tiny organism which thrives in sea water, is composed largely of silica. In certain areas skeletons of diatoms have accumulated in finely divided silica deposits several hundred feet deep. This **diatomaceous earth,** or **infusorial earth,** is used as a filtering medium for certain dye products and as a mild abrasive in scouring powders.

Silica possesses very definite acidic properties, reacting with bases and metallic oxides to form silicate salts. If silica is fused with sodium carbonate,

$$Na_2CO_3 + SiO_2 \rightarrow Na_2SiO_3 + CO_2$$

carbon dioxide escapes as a gas, and the Na$_2$SiO$_3$ formed is easily soluble in water.

Actually there are some 40 varieties of sodium silicates dependent on various ratios of Na$_2$O to SiO$_2$. Compounds with ratios of 1 Na$_2$O/1.6 SiO$_2$ up to 1 Na$_2$O/4 SiO$_2$ are called **colloidal** silicates, and their solutions are called **water glass.** They are used extensively as adhesives for materials such as corrugated boxes, plywood, wallboard, and flooring. Na$_2$SiO$_3$ is sodium metasilicate (1 : 1 ratio), and Na$_4$SiO$_4$ (2 : 1 ratio) is sodium orthosilicate. Intermediate compounds such as the sesquisilicate, Na$_4$SiO$_4$ · Na$_2$SiO$_3$ · 11 H$_2$O

($1\frac{1}{2}$: 1 ratio), are also known. The silicates are also used as detergents, for metal cleaning, fireproofing, and sizing. The higher-ratio silicates are strongly alkaline due to hydrolysis. Sodium silicates are 46th in the list of top chemicals.

When an acid is added to a solution of sodium silicate, it produces a jelly-like colloidal suspension of hydrated SiO_2, which gradually loses water when warmed or allowed to stand in the open air. The acid H_2SiO_3 does not appear to exist, and the formula for the jelly-like mass should be written $xSiO_2 \cdot yH_2O$. When this product is dehydrated to 5% water or less, a granular, porous solid remains. This solid, called **silica gel,** is widely used as an adsorbent for gases and as a carrier for platinum and other catalysts in certain contact catalytic reactions.

Silica, while resistant to the action of the strong mineral acids HCl, H_2SO_4, and HNO_3, is rapidly attacked by hydrofluoric acid, with the formation of gaseous SiF_4.

$$SiO_2 + 4 \text{ HF} \rightarrow SiF_4(g) + 2 \text{ H}_2O$$

SiF_4 is readily hydrolyzed with water to form fluosilicic acid, H_2SiF_6.

$$3 \text{ SiF}_4 + 4 \text{ H}_2O \rightarrow H_4SiO_4 + 2 \text{ H}_2SiF_6$$

Fluosilicic acid is used as an analytical reagent in the determination of sodium and potassium, since Na_2SiF_6 and K_2SiF_6 are among the few water-insoluble salts of these elements. **Sodium fluosilicate** finds use as a germicide, **lead fluosilicate** as an electrolyte in the electrolytic refining of lead.

Silicon tetrachloride, which may be formed by the direct action of chlorine on silicon, is a volatile liquid which fumes strongly in moist air, with the formation of a smoke or cloud of finely divided orthosilicic acid and fog of HCl. Liquid $SiCl_4$ is sometimes jet ejected from tanks by airplanes for smoky sky writing.

$$SiCl_4 + 4 \text{ H}_2O \rightarrow 2 \text{ H}_2O \cdot SiO_2 + 4 \text{ HCl}$$

If ammonia gas is added, the density of the smoke is increased by the formation of ammonium chloride.

23.8 GLASS

When a mixture of limestone, sodium carbonate, and sand is melted together, a clear, homogeneous mixture of sodium and calcium silicates is produced. When the fused liquid is cooled, it becomes more and more viscous and finally hardens to a transparent rigid mass called **glass.**

$$CaCO_3 + Na_2CO_3 + 2 \text{ SiO}_2 \xrightarrow{\Delta} CaSiO_3 + Na_2SiO_3 + 2 \text{ CO}_2$$

Proportions may be varied, but a cheap soda-lime glass is the product. During the cooling process the transition from liquid to solid does not take place at any definite temperature. Similarly, in the reverse process, when the glass is warmed, it gradually softens and shows no definite transition temperature — a characteristic of the change from the solid state. X-ray examination reveals no definite pattern or orientation of atoms characteristic of crystalline solids. It may be considered to be an amorphous solid and as such is a supercooled liquid with very high viscosity.

The proportions of raw materials in glass may be varied within wide limits and, by the substitution of other metal oxides, a great variety of glasses may be obtained. Common glass is a soda-lime glass consisting of a mixture of sodium and calcium silicates with an excess of dissolved silica. It is often called soft glass, since it melts at a relatively low temperature. The substitution of potassium for sodium results in a harder glass, which has a higher melting point and is more insoluble in water and alkalies. The substitution of lead oxide for calcium oxide yields **flint** glass, which has a high refractive index and is used in the production of lenses and other optical glass. If a part of the silica is replaced by boric oxide (B_2O_3) and the metal oxide content reduced, a product with a low coefficient of heat expansion is obtained. **Pyrex** is one variety of the borosilicate glasses. Colored glasses may be obtained by incorporating small amounts of certain substances either in true solution or as a colloidal suspension. A colloidal dispersion of gold or selenium produces a red glass; cobalt oxide gives a blue glass; chromium(III) or iron(II) oxide produces a greenish-colored glass; colloidal carbon in glass gives a brown color as to beverage bottles.

Table 23.4 shows the composition and use of some of the modern commercial glasses.

Photochromic glass (Table 23.4) used in "automatic" sunglasses darkens in sunlight and returns to a clear glass in the dark. The AgCl (or AgI) molecules dissociate to free I and metallic Ag (dark when finely divided) under the influence of sunlight

$$AgI \xrightarrow{\text{light}} Ag + I$$

In the dark, Ag atoms recombine with the halogen atoms and the reaction is reversed.

The general procedure for the manufacture of glass is as follows: the ingredients are heated in a furnace until a homogeneous solution is obtained. The glass is removed from the furnace as a semifluid mass and fashioned into various shapes by molding, rolling into sheets, blowing, or pressing. Many of these operations were formerly done by hand, but nearly all glass shapes are now turned out by machinery. To relieve strains set up by rapid cooling, the glass must be annealed by a very slow and gradual cooling over a period of hours or days. Annealing the glass of the huge 200-inch reflecting telescope lens on Mount Palomar, California, took almost a year.

Safety glass is produced by cementing together two thin sheets of ordi-

Table 23.4

Composition of several commercial glasses

Glass	Approximate Composition	Some Uses
Soda-lime	70% SiO_2, 15% Na_2O, 10% CaO	Common soft glass, windows, jars, light bulbs
Borosilicate (Pyrex)	60–80% SiO_2, 10–25% B_2O_3, 1–4% Al_2O_3	Cookware, laboratory glassware, pipes
Lead-alkali (Flint)	30–70% SiO_2, 18–65% PbO, 5–20% Na_2O or K_2O	Optical glass, "crystal" for tableware, neon-sign tubes
Pure silica	100% SiO_2	Transparent to UV light, low thermal expansion laboratory ware, laser beam reflectors
Lead glass	20% SiO_2, 80% PbO	Shield for X-rays or γ-rays
Photochromic	60% SiO_2, 10% Na_2O, 10% Al_2O_3, 19% B_2O_3, 0.6% Ag, 0.3% Cl, 0 9% I	"Automatic" sunglasses

nary glass with a transparent resinous plastic. When the glass breaks it does not shatter, since the broken pieces stick to the plastic binding.

Soils

Plant and animal life are dependent on soil which, for its bulk part, is composed of silica particles, silicate particles, or product compounds from the weathering of silicate minerals. It is from the weathered minerals that most of the metal ions such as K^+, Ca^{2+}, Mg^{2+}, Fe^{3+}, and trace quantities of other ions get into soil water. Organic matter from past plant life in decomposition gives slight acidity which aids in dissolving the surface of mineral particles to ions in soil water. Some PO_4^{3-}, NO_3^-, S^{2-}, and I^- ions that are needed come from decomposed past plant material, or trace soil minerals, or added fertilizer. The soil of a region relates to the geochemistry of the silicate rocks underlying or near it.

Exercises

23.1. Use the list of elements below to answer the questions concerning the abundance and distribution of elements in the earth.

Al, Ti, Ni, O, Cu, P, Si, N, Fe, Cl, H

(a) What 4 elements would you expect to be present in the highest proportions in a sample of soil?

(b) If you were able to obtain a sample of the inner core of the earth, what two elements would you expect to be the most abundant?

(c) The analysis of lunar rocks show which three major elements to be present in much higher proportions than in the earth's crust?

(d) What is the most abundant metal in the whole earth?

(e) What is the most abundant element in the universe?

23.2. (a) About what percent of the solar system mass is the total mass of all the planets and their moons? (b) Covalent union of what two atoms to form what compound is predominant for the earth's crust and intermediate layer (and probably also for other planets and moons)?

23.3. List the symbols and atomic numbers for the elements whose mass numbers are, 4, 12, 16, 20, 24, 28, 32, and 56. How many of these elements whose mass number is divisible by four show appreciable abundance as indicated by solar and stellar spectral data? Give symbols for four that are abundant in earth crustal rock.

23.4. From the list on the right select the item which most closely matches.

( ) 1. A feldspar	1. A silicate containing Be
( ) 2. The most abundant element in rocks	2. SiO_2
( ) 3. Emerald	3. Ti
( ) 4. Silica	4. Silica gel
( ) 5. Quartz	5. A copper ore
( ) 6. Asbestos	6. Fe and Ni
( ) 7. Silicosis	7. Biosphere
( ) 8. Granite	8. Na
( ) 9. A fluosilicate	9. K_2SiF_6
( )10. Glass	10. Magma
( )11. The gem Jade	11. Si
( )12. $xH_2O \cdot ySiO_2$ or $SiO_2 \cdot xH_2O$	12. Na_4SiO_4
( )13. Water glass	13. Limestone
( )14. A **meta**silicate	14. O
( )15. Soil	15. H_2SiO_3
	16. C
	17. Gypsum
	18. K
	19. A disease of the lungs
	20. A sodium aluminum silicate
	21. Complex silicate(s)
	22. Mixture of mica, feldspar, and quartz
	23. Mixture of sodium and calcium silicates

23.5. Referring to Figure 23.3 depicting the earth's structure, what division of the total earth has the highest percent of earth mass? Write five formulas for probable compounds that in part comprise it.

23.6. Suggest the method of analysis used for the valuable moon rock samples. Why would this method be used?

23.7. Classify the following common substances as (a) free silicon, (b) a form of SiO_2, (c) a metal silicate mixture or (d) a nonsilicon substance.
   1. Rose quartz
   2. Garnet
   3. Solid product obtained when sand and limited amount of coke are reacted in an electric furnace
   4. Silica gel
   5. Orthoclase
   6. Soil
   7. Granite
   8. Diatomaceous earth (used in scouring powders)
   9. Diamond
   10. Asbestos

23.8. For each of the following select the formula from the list below which matches.
   (a) Common glass (soda-lime glass)
   (b) Diatomaceous earth
   (c) Pyrex
   (d) Colloidal silicates (water glass)
   (e) Feldspar-orthoclase
   (f) Photochromic glass
   (g) Formula of substance with lowest coefficient of thermal expansion

   1. SiO_2
   2. SiO_2, 60%; Na_2O, 10%; Al_2O_3, 10%; B_2O_3, 19%; AgBr, 1.0%
   3. $SiO_2 \cdot xH_2O$
   4. SiO_2, 70%; Na_2O, 15%; CaO, 10%
   5. SiO_2, 80%; B_2O_3, 18%; Al_2O_3, 2%
   6. $KAlSi_3O_8$, $K_2O \cdot Al_2O_3 \cdot 6 SiO_2$
   7. Various ratios of Na_2O to SiO_2

23.9. Write balanced equations for:
   (a) Making glass by fusing a mixture of limestone, soda ash, and sand
   (b) Precipitation of orthosilicic acid
$$HCl + Na_4SiO_4 \rightarrow$$
   (c) $CaSiO_3 + HF \rightarrow$
   (d) Fusion of PbO, K_2CO_3, and excess SiO_2

23.10. Which of the following structures correspond to the statement given?
   (a) Mica sheets
   (b) Structural unit of all silicates
   (c) Glass
   (d) Asbestos fiber
   (e) Feldspar

   1. SiO_4^{4-} tetrahedron
   2. No definitive crystalline structure
   3. SiO_4^- planar
   4. Three-dimensional network

5.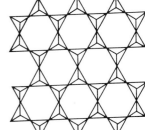

6.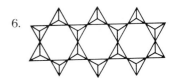

23.11. Concerning rock mineral formulas:
(a) What is the meaning of the formula for augite, $Ca(Mg, Fe, Al)(SiAl)_2O_6$?
(b) Write the formula for the mica, $KAl_3Si_3O_{10}(OH)_2$, in two different ways.
(c) What is meant by isomorphous mixtures?

23.12. What is the reason gems such as emerald, $Be_3Al_2(SiO_3)_6$, are so valuable?

23.13. Using the following examples as a model, write formulas for five other metal silicates:

$$(K_2O \cdot SiO_2, K_2SiO_3; MgO \cdot SiO_2, MgSiO_3)$$

23.14. (a) Write the formula of orthosilicic acid and of metasilicic acid as an acid and as a hydrate of SiO_2. (b) Give the names of 5 "rock hound" minerals that are essentially silica, and which may be in small part hydrated or impurity striated, or more or less crystalline?

23.15. (a) The 20.1% average of silicon in lunar rock has united with it what weight percent of oxygen? (b) Give formulas for moon rock minerals of the following chemical names: calcium metasilicate, iron(II) orthosilicate, titania, magnesium orthosilicate, a compound formed from bonding of sodium oxide with one alumina and six silica molecules.

23.16. Geochemical weathering is in considerable part the action of carbonic acid on rock minerals. Write equations to depict the slow action of H_2CO_3 on $Na_2O \cdot Al_2O_3 \cdot 6\ SiO_2$; on $MgO \cdot SiO_2$.

23.17. Among the more abundant rock minerals are the feldspars, which are considered to be derivatives of the polysilicic acid $H_8Si_6O_{16}$. Write the formula for a sodium aluminum feldspar.

23.18. What is one important use for:
(a) quartz, (b) silica gel, (c) diatomaceous earth, (d) sodium silicates, (e) lead glass, (f) soda-lime glass.

23.19. Explain the operation of photochromic sunglasses.

23.20. Obtain the melting points of SiO_2, Na_2SiO_3, $CaSiO_3$, $PbSiO_3$, and B_2O_3 from a handbook. On the basis of these melting points and the fact that coal- or gas-fired furnaces give maximum temperatures of about 1100°C, why are mixed silicates rather than pure SiO_2 or B_2O_3 used for most glass production?

23.21. To obtain 100 kilograms of glass composed of equimolecular proportions of Na_2SiO_3 and $CaSiO_3$, what weights of Na_2CO_3, $CaCO_3$, and SiO_2 should be used?

23.22. Germanium (at. no. 32) is in the same family with carbon and silicon.
(a) Write its electronic notation to show occupancy of orbitals.
(b) Write formulas for orthogermanic acid, sodium metagermanate, germanium sulfide.

Supplementary Readings

23.1. "The Earth's Mantle," P. J. Wyllie, *Sci. American*, March 1975; p. 50.

23.2. "The Apollo Missions and the Chemistry of the Moon," R. A. Pacer and W. D. Ehmann, *J. Chem. Educ.*, **52**, 350 (1975).

23.3. "Opals," P. J. Darragh, A. J. Gaskin and J. V. Sanders, *Sci. American*, April 1976; p. 84.

23.4. *Glass in the Modern World*, F. J. T. Maloney, Doubleday, New York, 1968 (paper).

Chapter 24

A 1-cm cube has a surface area of 6 cm^2. If it is cut to 8 smaller cubes, the total surface area is now 12 cm^2. If each of the 8 little cubes is again cut three ways, the 64 cubes now obtained have a total facial area of 24 cm^2. The 6-cm^2 surface area of a 1-cm cube calculates to 6 $\times$ 10^6-cm^2 surface area for the total cubes resulting from breaking the initial cube to cubes 0.000,001 cm on a side. A gram of material subdivided to colloid-sized particles has an extremely large surface area.

The Colloidal State of Matter

24.1 CHARACTERISTICS OF COLLOIDS

When sugar is dissolved in water, individual molecules from the crystalline sugar are dispersed or scattered throughout the water. Such a mixture of two or more substances molecularly or ionically dispersed is termed a **true solution.**

If finely ground clay is mixed with water, the result is a **suspension,** in which the clay will slowly settle out. The larger the particles of clay, the faster the settling. It is impossible to reduce the size of the clay particles to molecular size, and we say that the clay is insoluble. The mixture of sugar and water is said to be **homogeneous,** while the mixture of clay and water is said to be **heterogeneous.**

Between true solutions and ordinary suspensions, there may exist mixtures of two substances, the particles of which are so finely divided that they do not settle and cannot be filtered by ordinary means. The particles in such a mixture are smaller than those in a suspension but larger than those in a true solution. These mixtures are called **colloidal dispersions** or **colloidal systems.**

Size of Colloidal Particles

In discussing the size of colloidal particles we shall use the unit of linear measurement, the **nanometer,** designated nm and defined as 10^{-9} m. If we assume that molecules are spherical, the diameter of an average-size molecule will be about $\frac{1}{10}$ nm. Particles of clay dispersed in water will have an average size of 1000 nm or more in diameter. Dispersed particles of a diameter ranging between 1 nm and 100 nm have properties characteristic of the colloidal state. Although true solutions, colloidal dispersions, and suspensions are differentiated largely on the basis of size of the dispersed particles, there is no sharp dividing line between the three kinds of mixtures. Molecules of some compounds are large enough to come within the size range of colloidal particles, and indeed these large molecules exhibit properties characteristic of the colloidal state.

The colloidal state is not peculiar to any class of substances; any material may be brought into a colloidal state under the proper conditions. For the formation of a colloid, the **dispersed phase** (the substance which is suspended) and the **dispersion medium** (the substance in which another substance is suspended) *must be mutually insoluble.* A colloidal system of sugar in water is impossible, since when sugar and water are mixed, the sugar is dissolved— that is, the particles are reduced to molecular size. Sugar may, however, be brought into colloidal dispersion in some liquid in which the sugar does not dissolve.

The particles in a colloidal dispersion are made up of aggregates or clusters of molecules or minute crystalline ionic particulates. A single particle may be made up of hundreds or thousands of units of a substance. Protein molecules are so large that a dispersion of them acts as a colloid.

Kinds of Colloids

In Chapter 13 the various kinds of true solutions were listed. Similarly, we may list the possible types of colloids:

1. GAS DISPERSED IN GAS.   Gases mix in all proportions, forming a homogeneous mixture or solution; consequently, there are *no* colloids of this type.

2. GAS DISPERSED IN A LIQUID.   Foams and whipped cream, beaten egg white are examples. (Minute bubbles of gas throughout liquid.)

3. GAS DISPERSED IN A SOLID.   Gases absorbed by charcoal are colloidally dispersed. White flowers and white hair contain air which is colloidally dispersed. Floating soap and pumice are further examples.

4. LIQUID DISPERSED IN A GAS.   Fog is composed of minute liquid-water particles dispersed in air; also clouds.

5. LIQUID DISPERSED IN A LIQUID.   This type of dispersion is called an **emulsion.** Milk contains butterfat in water; mayonnaise consists of oil dispersed in vinegar.

6. LIQUID DISPERSED IN A SOLID.   Cheese contains a dispersion of butterfat in casein; pearl consists of calcium carbonate in which sea water is colloidally dispersed.

7. SOLID DISPERSED IN A GAS.   Smoke and dust particles dispersed in the air may be of colloidal size.

8. SOLID DISPERSED IN A LIQUID.   This is the most common type. Coffee, tea, and paints are colloidal systems of a solid dispersed in a liquid.

9. SOLID DISPERSED IN A SOLID.   Carbon in glass gives brown of beer bottles; gold or selenium in glass give prized ruby-colored glasses. Colored material in porcelain is usually colloidal.

Preparation of Colloids

Two general methods are available for the preparation of colloids: (1) **dispersion** and (2) **condensation.** Dispersion methods involve the breaking up of large particles into particles of colloidal size. Condensation is the condensing of molecules (which are smaller than colloidal particles) to particles of colloidal size.

increasing particle
size (condensation)
————————→

Molecules → Colloids ← Precipitates and large particles

←————————
decreasing particle
size (dispersion)

DISPERSION   Certain devices, called **colloid mills,** are available for re-ducing the size of coarse particles to colloidal size. These mills are essentially pulverizing devices. Certain chemicals, called **peptizing agents,** augment agitation in attaining dispersion of the particles to the colloidal state. Clay particles may be peptized with an alkaline solution.

CONDENSATION   Condensation methods may employ chemical reac-tions in which the starting materials are in true solution. The chemical reac-tion forms an insoluble substance in which product molecules cluster together to form a particle of colloidal size. If a solution of iron(III) chloride is allowed to drip slowly into boiling water, a reddish-colored dispersion of iron hy-droxide is obtained.

$$FeCl_3 + 3\ H_2O \rightarrow Fe(OH)_3(s) + 3\ HCl$$

Arsenic(III) sulfide is insoluble in water and may be produced in a colloidal condition by passing hydrogen sulfide gas into a saturated solution of arsenic(III) oxide in water.

$$As_4O_6 + 6\ H_2S \rightarrow 2\ As_2S_3(s) + 6\ H_2O$$

Gold and silver colloids may be prepared by the action of certain reducing agents on solutions of salts of these metals.

$AuCl_3 + FeSO_4$ or Tannic acid solutions →
Colloidal Au (red or blue according to concentrations)

The Importance of Colloids

A few typical examples will illustrate the importance of colloids in everyday life. The protoplasm of living cells and tissue, most of the body fluids, glan-dular secretions, blood, and many foods are colloidal substances. Thus the essential vital processes—nutrition, digestion, and secretion—are concerned with colloidal systems.

Jellies of all kinds are colloidal dispersions. The color of the sky is due to a colloidal dispersion of dust. The action of soap in removing dirt and grease is due to a colloidal phenomenon.

Cleansing Action of Detergents

The term detergents applies to soap and synthetic soap substitutes (syndets). A typical soap constituent is $C_{17}H_{35}COONa$, sodium stearate. Formulas for

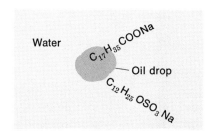

Figure 24.1
Detergent molecules link oil and water as end radicals selectively dissolve.

two syndets are $C_{12}H_{25}OSO_3Na$, sodium dodecylsulfate, and

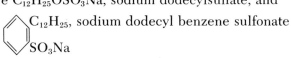

$C_{12}H_{25}$, sodium dodecyl benzene sulfonate
SO_3Na

See also Chapter 33. Note that these compounds have a water-soluble type group as a sodium compound and a water-insoluble long hydrocarbon kerosene-like group. The hydrocarbon radical is soluble in oils and fats.

Dirt or dust on fabric with no oil or grease is quite easily suspended from the fabric by agitation of water, but dirt with oil or grease on it tends to stick to fabric. Detergent molecules form links between oil or grease particles, with hydrocarbon radical penetration into surface solution in the oil and the sodium radical into the water. (See Figure 24.1.)

The oil or grease particles with dirt are more readily suspended as colloidal-sized particles in water when a detergent is used, and thus more easily washed away. The interfacial tension is lowered between oil and water by the detergent molecules.

Dispersions of Solids in Liquids

If a small amount of gelatin is dissolved in hot water and the solution cooled, the whole changes to a semisolid jellylike mass; pectin added to sugared fruit juices causes the solution to jell. The dispersed particles in these two instances have an attraction for the dispersion medium. A shell of the dispersion medium may be formed about each colloidal particle, thus preventing the free movement of these particles. In some cases the dispersed phase may surround minute pockets of the liquid. Much as beeswax encases honey, a colloidal substance in film form may enclose small droplets of the liquid dispersion medium. As a result, the colloidal system assumes a semisolid state. Colloids in which the dispersed phase has an attraction for the dispersion medium are called **lyophilic** colloids. If the liquid is water, the colloid is said to be **hydrophilic**. Lyophilic colloids are usually organic — for example, starches, proteins, soaps, glues, and gums. A semisolid colloid of this nature is called a **gel**.

Substances of an inorganic character usually form **lyophobic** colloidal solutions—dispersions in which there is little or no attraction between the two phases. If arsenic(III) sulfide is dispersed in water, the water has little attraction for the arsenic(III) sulfide, and the solution has almost exactly the same properties as water itself. A liquid colloid is called a **sol.**

Lyophilic colloids are reversible—that is, if the colloid is precipitated, the solid may again be dispersed in a colloidal condition when the dispersion medium is added. If albumen is precipitated without heat coagulation by a salt solution and is then washed free of the salt and water is added, the albumen becomes dispersed again in the colloidal state. Lyophobic colloids are not reversible; once the colloidal material aggregates to precipitation size it is difficult to again disperse it in colloidal size.

Characteristics of Colloidal Systems

Colloidal systems exhibit certain properties which enable us to differentiate them from true solutions.

THE TYNDALL EFFECT   If a beam of light is passed through a colloid sol, a very definite path of the light is visible through the material (Figure 24.2). The light ray is made visible by the scattering of light by the suspended particles, a phenomenon not observed in true solutions because the molecules present are too small to scatter the light. This property of colloidal systems was first noted by Tyndall, an English physicist. If the beam of light is observed at right angles with a high-powered microscope, the ultramicroscope, the diffraction or scattering of light by the individual particles in suspension can be seen. Although the particles in suspension cannot actually be seen, a study of the tiny specks of light leads to many interesting conclusions as to the size and properties of the particles. The Tyndall effect is analogous to the scattering of light entering a darkened room in which dust particles are suspended in the air. A well-defined path of light is visible because of the scat-

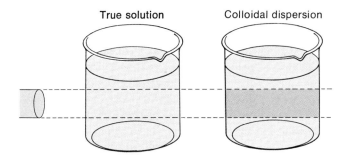

Figure 24.2

Tyndall effect.

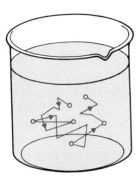

Figure 24.3
Brownian movement is random movement of a particle through the dispersion medium.

tering of light by the suspended dust particles. The observation of light flashes from colloidal particles in the ultramicroscope is analogous to the transmission of signals from a considerable distance by sunlight flashes from a pocket mirror, even though the mirror itself is too small to be seen.

THE BROWNIAN MOVEMENT (Figure 24.3) When a colloidal solution is viewed under an ultramicroscope, the suspended particles are seen as tiny specks of light dancing about in the solution. They appear to move at random, first one way, then another. This zigzag motion was first observed in a suspension of pollen dust by the botanist Robert Brown. It is believed that this motion is due to a bombardment of the suspended particles by molecules in the solution. The molecules in a liquid move at relatively high speeds, and when they collide with a larger particle (of colloidal size), the larger particle is moved in one direction or another. Since, on the average, a particle will not receive the same number of hits on all sides at the same time, the net result is a zigzag motion. This so-called Brownian movement appears to be a stability factor of colloidal dispersions; the particles are kept from settling out because of this ceaseless molecular bombardment.

ELECTRIC CHARGE, ADSORPTION   Colloidal dispersions act as poor conductors of an electric current. If a current is passed through a solution between a pair of electrodes (as in electrolysis, for example), the colloidal particles all move toward either the positive or the negative pole. This indicates that the particles in suspension carry an electric charge. The particles are discharged at one of the electrodes, and as a result the colloid is precipitated. This movement of the colloidal particles toward one of the electrodes is termed **electrophoresis**. The charge carried by the colloidal particles depends on the nature of the colloid itself. Most of the metals and their sulfides form negatively charged colloids; metallic hydroxides in general are positive. **Cataphoresis** is movement of negatively charged colloids to a cathode.

The charge on a colloidal particle is probably due to the adsorption of ions from the dispersion medium. **Adsorption** means that the adsorbed substance forms a layer of one-molecule or one-ion thickness on the surface of the adsorbing medium. Adsorption differs from absorption. When a sponge absorbs water by capillary action, the columns of water in the pores of the sponge are many molecular diameters in size.

A single colloidal particle probably adsorbs many ions. Certain dispersed substances adsorb positive ions and as a result become positively charged; others adsorb negative ions and become negatively charged. When a colloid is subjected to the potential difference between electrodes, the colloid particles move toward the anode or cathode, depending on the specific nature of the adsorbed charge on the colloid particle.

The charge on the colloidal particle explains why the solutions are stable and do not settle out. Since like charges repel, the force of repulsion prevents the particles from coalescing to form larger particles and precipitate.

Hydrogen ions (protons), small in size and positively charged, are presumed adsorbed by colloidal-sized suspended particles such as $[Fe(OH)_3]_y$ where y = a large number. Some Fe^{3+} is probably also adsorbed from solution. Thus, a hydrated iron oxide colloid selectively adsorbing positive ions acts as a large positively charged ion. $[As_2S_3]_y$ colloid-sized suspended particles, when prepared in solution containing HS^- or S^{2-} ions, seem to adsorb some of these ions to acquire a negative charge. They thus typify colloidal particles acting as large negatively charged ions.

The Precipitation of Colloidal Particles

It is reasonable to assume that if the charge on the colloidal particle could be removed or neutralized, the particles would come together and form larger particles. Eventually the particles would be large enough to settle out in the form of a precipitate; neutralization of the charge should result in coagulation of the colloid. This appears to be the case, since the addition of electrolytes (substances which ionize) causes the colloid to precipitate. Ions of charge opposite to those borne by the colloidal particle neutralize the charge on the particle. The higher the charge of the ions, the more effective the precipitating electrolyte.

Coagulation may also be effected by the addition of a colloid of opposite charge. Thus iron hydroxide colloid, which is positive, added to arsenic sulfide colloid, which is negative, results in coagulation of both colloidal solutions. This process is termed **mutual coagulation.**

Dialysis

To prepare a stable colloidal system, it is necessary to remove substances in the ionic state which would tend to precipitate the colloid. This separation may be effected by a process called **dialysis.** It is based on the fact that colloidal particles are much larger than ions. If an animal membrane bag contain-

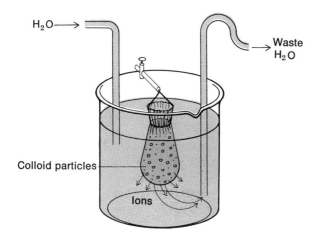

H₂O⟶

Waste
H₂O

Colloid particles

Ions

Figure 24.4
Purification of a colloid by dialysis.

ing the colloidal dispersion contaminated with electrolytes is immersed in a bath of distilled water, the ions will diffuse through the membrane until the concentration of the ions is the same inside and outside the membrane; colloidal particles are retained within the bag (Figure 24.4). By frequently renewing the water on the outside, practically all the electrolyte may be removed from the colloidal system. Dialysis is really a kind of filtration, since one kind of substance is retained on one side of the membrane, while the other passes through. Membranes used for this purpose include animal membranes, parchment paper, and collodion film.

Protective Colloids

Often a colloidal system may be stabilized by the addition of a second colloid. The second colloid probably forms a protective film around the particles and adsorbed ions of the first colloid and prevents the particles from coming together and coalescing. These protective colloids are of the lyophilic type. Gelatin may be added to an ice cream mixture to prevent the formation of large crystals. Gelatin is also employed in the photographic industry to form a film about the silver bromide particles which are dispersed on the photographic plates. Protective colloids also find application in the preparation of inks, emulsions, and paints.

Emulsions

If two mutually insoluble liquids are vigorously shaken together, small droplets of one will be dispersed in the other. Thus when kerosene and water are shaken together, the kerosene is dispersed as small droplets throughout

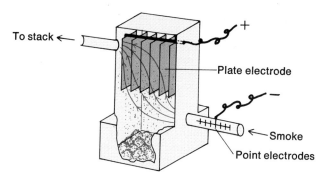

Figure 24.5

Cottrell precipitator.

the water medium. The kerosene particles soon coalesce, and finally separation of the oil and water into two distinct layers takes place. Such a temporary emulsion may be made more permanent by the addition of an **emulsifying agent,** such as soap. An emulsifying agent is a protective colloid which forms a film about each oil particle, preventing coalescence. Milk is an emulsion of drops of butterfat in water, with casein acting as an emulsifying agent. The yolk of eggs acts as an emulsifying agent in an emulsion such as olive oil in dilute vinegar (mayonnaise).

Cottrell Precipitator

Smoke, which is a colloidal dispersion of solid particles in air, is a nuisance around manufacturing centers. The·smoke from smelters and certain other industrial plants is particularly undesirable. Such smoke may be treated by a Cottrell precipitator, which operates on the principle that colloidal smoke particles in general have like adsorbed electrical charges and may be precipitated as a powder when the smoke is passed between highly charged plates or electrodes (Figure 24.5). In addition, valuable materials may be recovered as by-products from the precipitated smoke powder. For example, one of our main sources of industrial arsenic oxide is the dust recovered from smelter smoke by a Cottrell precipitator. The principle of operation will find increasing use in the worldwide drive to lessen air pollution.

Exercises

24.1. Of the three systems described below, indicate which would be classified as:
    (a) a true solution
    (b) a suspension
    (c) a colloidal system

1. Clear; no settling out of particles, light beam visible when passed through the system.
2. System contains particles visible to the eye — turbid when particles are dispersed — particles will settle out on standing — turbid dispersion fairly opaque to the light beam.
3. Clear — stable, no particles settling on standing — light beam not visible when passed through the system.

24.2. Describe four ways to distinguish between a yellow arsenic sulfide colloid and a yellow K_2CrO_4 true solution. Tell how both the colloid and K_2CrO_4 solutions will act in each case.

24.3. Below are listed several colloidal systems. Classify each as to type (solid in liquid, and so on): (a) cigarette smoke, (b) hot cocoa drink, (c) beaten raw egg white, (d) spongy platinum used as hydrogenation catalyst, (e) atomized perfume, (f) coffee cream, (g) black ink, (h) an exceedingly fine-grained alloy.

24.4. Colloids are classified by the nature of the dispersed phase (g, l, or s) and the dispersion medium (g, l, or s). Keeping this fact in mind complete the following table:

	Dispersed Phase	Dispersion Medium	Type of colloid	Example
(a)	_____	_____	foam	whipped cream
(b)	_____	_____	_____	milk, mayonnaise,
(c)	_____	_____	sol	As_2S_3 colloid
(d)	l	g	X	_____
(e)	_____	_____	smoke	X

24.5. Match the numbered items in the left column with the appropriate items in the right column.

1. Above 100 nm in diameter	hydrophilic gel
2. White hair	emulsifying agent
3. About $\frac{1}{10}$ nm in diameter	molecular size
4. Dispersed phase	air
5. Fog	dispersed phase size
6. Dispersing phase of smog	liquid in gas colloid
7. Grape jelly	gold in a red glass
8. Soap	gas in a solid
9. $(As_2S_3)_y$ dispersed in water	can absorb S^{2-} or OH^-
10. Gas in gas colloid	not possible
11. Lyophobic	suspension material
12. 1 nm to 100 nm in size	lacking in attraction

24.6. Classify the following colloids as being prepared by either dispersion or condensation methods: (a) mayonnaise, (b) smoke from airplane tank ejecting $SiCl_4$, (c) formation of gelatinous $Al(OH)_3$ from $Al_2(SO_4)_3$ in water purification, (d) milk of magnesia, (e) percolated coffee.

24.7. Define and illustrate each of the following: dispersed phase, true solution, dialysis, gel, sol, emulsifying agent, protective colloid, lyophilic, emulsion, hydrophobic, electrophoresis.

24.8. Select a term from the list on the right to match each of the following:
(a) As_2S_3 in a colloid.
(b) A dispersion in which the particles are less than 0.1 nm.
(c) Apparatus used to pulverize materials to colloid size.
(d) A colloid where the dispersed phase has an attraction for the dispersion medium.
(e) A colloid where the dispersed phase has little attraction to the dispersion medium.
(f) A light beam made visible by colloid particles.
(g) Zigzag motion of colloid particles.
(h) Process of removing ions from colloids.
(i) Stabilization of a colloid by adding another colloid; for example, gelatin added to AgBr colloid.
(j) Apparatus which precipitates smoke and dusts by removing their colloidal electric charge.

1. True solution
2. Tyndall effect
3. Brownian movement
4. Dispersed phase
5. Protective colloid
6. Dialysis
7. Colloid mill
8. Cottrell precipitator
9. A lyophilic colloid
10. A lyophobic colloid
11. Electrophoresis
12. Precipitate
13. Dispersion medium
14. Flotation

24.9. Use a word or phrase to define or describe each of the following phenomena: (a) haphazard movement of colloidal particles, (b) scattering of light beam by colloids, (c) ions adhering to the surface of particles, (d) movement of colloidal particles toward an electrode, (e) precipitation of smelter smoke electrically.

24.10. On numerous occasions it is advantageous to be able to precipitate a colloidal system. Which of the following procedures would you predict might precipitate a colloidal suspension of sulfur?
(a) Spinning in a low speed laboratory centrifuge.
(b) Filtering through a filter paper.
(c) Adding soap.
(d) Adding small quantities of salts with +3 or −3 charged ions.

24.11. Another example of such a situation is the need to remove colloidal particles from the effluent of a smoke stack. Which of the precedures below would you predict might be effective?
(a) Passing through a filter.
(b) Passing through highly charged plates.
(c) Increasing the height of the smoke stack.
(d) Increasing the air-flow in the smoke stack.

24.12. The substance "canned heat" which a camper can use as a readily ignited semisolid is made by adding saturated water solution of $Ca(C_2H_3O_2)_2$ to 95%

C_2H_5OH. Calcium acetate is water soluble but forms a gel of long flat crystals in alcohol. Since it is a gel, would it be classed as lyophilic or lyophobic? Would it be hydrophilic?

24.13. A colloid may form by a condensation method, as by growth of molecular clusters after chemical change. What reasons are given for such clusters remaining in colloidal size rather than growing further to precipitation size?

24.14. An emulsion such as an oil dispersed in water can be stabilized by the addition of an emulsifying agent. Which statement would apply to this type of compound?
(a) An emulsifying agent is a long-chain hydrocarbon.
(b) The molecule contains a water-soluble functional group and a water-insoluble region, such as soap.
(c) Sugars such as glucose are good emulsifying agents.
(d) An emulsifying agent is a strong oxidizing agent.

24.15. (a) A cube of quartz (SiO_2) exactly 1 in. along each edge weighs about 60 g (its gram molecular weight). How many SiO_2 molecules does this cube contain? If it is cut into 8 smaller cubes, what is the total surface area of the 8 cubes? If each of the 8 smaller cubes is again cut in 3 dimensions to form even smaller cubes, what will be the total area of surface? How many SiO_2 molecules will there be in each 0.25-in. cube? (b) Following the progression in (a), assume that each 0.25-in. cube is cut to 0.125-in., then to 0.0625-in., and then to 0.03125-in. cubes. What is the total number of small cubes? What is the total surface area? What is the number of SiO_2 molecules in each 0.03125-in. cube?

Supplementary Readings

24.1. "The Top Millimeter of the Ocean," F. MacIntyre, *Sci. American*, May 1974; p. 62.
24.2. "Clay Colloids," W. H. Slabaugh, *Chemistry*, **43**, 8 (1970).
24.3. A group of papers on surface chemistry, colloids, etc., H. H. G. Jellinek et al., *J. Chem. Educ.*, **49**, 148 (1972).

Chapter 25

Metals and Metallurgical Chemistry

25.1 METALS

Metals are those elements with few electrons in their outer energy level, and these electrons are comparatively readily given up to form ions. For example,

$$Na \rightarrow Na^+ + e^-$$
atom   ion
$$Zn \rightarrow Zn^{2+} + 2\ e^-$$
$$Al \rightarrow Al^{3+} + 3\ e^-$$

The electromotive series (Chapter 17) lists the metals in order of their comparative tendencies to give up electrons to form ions. As nonmetallic or negative radical ions associate themselves with these positive metallic ions, salts, oxides, and bases result.

In this chapter we shall concern ourselves particularly with the physical and chemical properties of metals as a class; the properties of specific metallic elements will be covered in following chapters.

Physical Properties of Metals

Metals are good **electrical conductors** compared with nonmetals. An electrical current in a metal wire is presumed to be a movement of the outer electrons down the wire. Such elements as silver, sodium, and copper, which have only one or two valence electrons, are among the best conductors. On the other hand, sulfur, which is a nonmetal with six outer electrons, tends to hold its outer electrons and even to capture electrons, and consequently acts as an **insulator** material.

Heat conductivity of elements closely parallels electrical conductivity; metals are good heat conductors – perhaps because of the capacity of the outside electrons to absorb energy and readily move from atom to atom in the solid metal.

Malleability, or the ease with which metals may be deformed or pounded into shape, is related to the weakness or mobility of bonding between metallic atoms. Gold, for example, is a very malleable element; it can be pounded into gold leaf as thin as $\frac{1}{300,000}$ inch. Lead is also very malleable. A **ductile** element is one that can be easily pulled into wire; copper and aluminum are very ductile. The opposite of malleability and ductility is **brittleness**. Diamond, which is typical of nonmetals in that the atoms are held together by covalent linkages, is the **hardest** element and is very brittle.

Another useful property of metals is their **elasticity** – that is, their ability to spring back to their original state. Of course, they may be deformed to such an extent that the **elastic limit** or **yield point** is exceeded, and the piece will no longer spring back to its original dimensions but undergo permanent deformation, though it remains as one piece.

Most of the metallic bonding forces between metal atoms in a solid remain, or reform, after deformation of the solid. There is a sort of plastic flow as metals undergo stretching or pounding. In contrast, a crystal of a nonmetal or of a salt fractures when slightly distorted — that is, the elastic limit is soon reached. Plastic flow of metals accounts for the fact that metals can be fashioned into various shapes by pounding, rolling, pulling or drawing through a die, extruding or squirting under high pressure through an orifice, pressing, or bending. If the force pulling on a piece of metal is increased sufficiently, it will rupture the metal. The magnitude of the force necessary to pull apart a piece of known cross-sectional area is a measure of the **tensile strength.** The yield points and tensile strengths of a number of metals are shown in Table 25.1. In general, heat treatment and mechanical working, such as rolling or drawing, increase the strength of a metal.

Table 25.1
Yield point and tensile strength of metals

Material	Yield Point lb/in^2	Ultimate Tensile Strength lb/in^2
Aluminum (cold rolled)	18,700	20,900
Lead (cast)		1,780
Nickel (cast)	25,000	60,000
Nickel (cold rolled)	120,000	160,000
Iron (wrought)	25,000	48,000
Steel	36,000–170,000	50,000–200,000
Alloy steel	39,000–276,000	60,000–340,000
Zinc	12,800	17,000–42,000

Metals are capable of giving a polished surface which reflects a high percentage of impinging light. Since smooth metal surfaces reflect light, metals are said to have **metallic luster;** nonmetals are not lustrous and at times are transparent.

Metals are, of course, selectively used according to such desired physical properties as tensile strength, malleability, ductility, electrical and heat conductivity, melting temperature, hardness, resistance to corrosion, and density. There are marked differences in properties of the various metals; for example, the metal lithium has a specific gravity (0.534) little more than one-half that of water, while osmium has a specific gravity of 22.48. Table 25.2 lists some common physical properties of some of the important metals. A more comprehensive listing of the various physical properties of metals is given in handbooks of chemistry and physics.

Note from Table 25.2 that the melting and boiling points of family IA (Li, Na, K, Cs) and family IB (Be, Mg, Ca, Sr, Ba, Ra) in general decrease with increasing atomic number and increasing radii of metal ions in crystal. Transition metals are all of fairly high melting and boiling points. The transition

Table 25.2

Physical properties of some important metals

Metal	Melting Point °C	Boiling Point °C	Specific Gravity at 20°C	Radius° of Metal Ion in Crystals cm × 10⁻⁸	Electrical Resistance at 20° ohm-cm × 10⁻⁶	Specific Heat cal/g
Lithium	179	1317	0.534	0.6	9.3	0.80
Sodium	97.8	892	0.97	0.95	4.6	0.25
Potassium	63.6	774	0.86	1.33	7.0	0.18
Cesium	28.4	690	1.87	1.69	20.0	0.05
Beryllium	1280	2970	1.85	0.31	18.5	0.42
Magnesium	651	1107	1.74	0.65	4.46	0.25
Calcium	845	1487	1.55	0.99	4.6	0.145
Strontium	769	1384	2.54	1.13		
Barium	725	1140	3.50	1.35		0.07
Radium	700	1700	5.0			
Copper	1083	2595	8.92	0.96	1.69	0.09
Silver	960	2212	10.5	1.26	1.62	0.055
Gold	1063	2966	19.3	1.37	2.4	0.031
Zinc	420	907	7.14	0.74	6.0	0.092
Mercury	−38.9	357	13.6	1.1	95.8	0.033
Aluminum	660	2467	2.71		2.62	0.22
Tin	231.8	2270	7.3	0.71	11.4	0.054
Lead	327.5	1744	11.3	0.84	21.9	0.032
Bismuth	271	1560	9.8	0.74	119.0	0.03
Chromium	1890	2482	7.1	0.52	2.6	0.11
Manganese	1244	2097	7.2		5.0	0.11
Iron	1535	3000	7.8	0.75	10.0	0.11
Nickel	1453	2782	8.9	0.69	6.9	0.10
Platinum	1769	3800	21.45		10.5	0.032

° The radius of a metal ion in a crystal will in general be considerably smaller than the radius of the metal atom.

metals of family IB (Cu, Ag, Au) have low electrical resistance and thus are good electrical conductors. Thermal conductivity of metals usually parallels electrical conductivity. The fact that the product of the specific heat of a metal and its atomic weight is nearly a constant value (Law of Dulong and Petit) indicates that atoms of all metals are nearly alike in heat absorption.

Bonding Forces of Metals and Nonmetals

Nonmetals have several electrons in the outer levels of their atoms. As a result, nonmetal atoms are probably held together in the solid state by covalent linkages — that is, by a sharing of electrons. Even in the vapor state, nonmetal atoms are frequently associated, as in O_2, S_2, S_4, S_8, C_x. On the other hand, vapors of metals are monoatomic. In the solid state, it seems that metal atoms, which have relatively few outer electrons, are not held together by covalent linkages — and certainly not by electrovalent forces; yet there is some binding force which holds metal atoms together in crystals.

The terms **metallic bond** or **metallic valency** are used in considering intermetallic bonding tendency. A marked feature of metallic bonding is that bonding remains or is instantaneously renewed on mechanical deformation of the metal. For example, a crystal of copper may be drawn into a long wire. This may be explained in part by the tendency of metals to assume a very close-packed cubic or hexagonal crystalline structure, in which a large number of planes of flow or slippage allows for deformation, but with instantaneous renewal of metallic bonding without detracting from the strength of the interatomic attraction.

Metallic bonding between atoms in a metal is a subject of conjecture, since there are inadequate facts to lead to an exact understanding. Current thinking pictures a metal as composed of spacially positioned metallic ions surrounded by a shared electron cloud. The electron cloud moves because an electric current can flow in a given metal. The proportion of metal ions to mobile electrons is such as to make a metal mass electrically neutral.

The mobile electron cloud acts as a binding force holding the positive metal ions together. This attraction is essentially nondirectional (the covalent or ionic bond is directional). The metallic bond is, however, fairly strong as factually pointed out by the close-packed crystal structure and high melting points of most metals. Ionic salt crystals are commonly brittle, but metal ion positions in an electron cloud can be altered and the crystal nature remains because of the uniform proportioning of metal ions and electron mobility in metals.

Alloys

The product obtained by melting together two or more metals and allowing the mixture to solidify is called an **alloy.** Much as the fundamental colors may be blended to many tints, a great diversity of properties of metal products is obtained by alloying. For example, copper has a melting point of 1083°C and costs about three times as much as zinc per pound. **Brass,** an alloy of about 66% Cu and 34% Zn, melts at 900°C. This alloy, then, is easier to remelt than copper and has good mechanical working qualities. An alloy of approximately 15% silicon and 85% iron, known as **ferrosilicon** or **duriron,** is very resistant to corrosion and to the action of acids. Accordingly, it is used in laboratory drain pipes. Several of the more common alloys and their compositions are shown in Table 25.3.

Binary alloys (two metals) may be of several types depending on the following factors: complete solubility, partial solubility or insolubility of the given metals in the solid state, and possible compound formation between the metals. Intermetallic compounds in alloys are fairly numerous — for example, Al_2Mg_3, Ag_3Sb, $CuMg$, $CuMg_2$.

Table 25.3

Common alloys

Trade Name	Composition by Percent
Aluminum bronze	90 Cu, 10 Al
Babbitt	90 Sn, 7 Sb, 3 Cu
Bearing bronze	82 Cu, 16 Sn, 2 Zn
German silver	60 Cu, 25 Zn, 15 Ni
Gold coinage	90 Au, 10 Cu
Magnalium	90 Al, 10 Mg
Manganese bronze	95 Cu, 5 Mn
Nickel coinage	75 Cu, 25 Ni
Pewter	85 Sn, 6 Bi, 7 Cu, 2 Sb
Red brass	90 Cu, 10 Zn
Silver coinage	90 Ag, 10 Cu
Solder	50 Pb, 50 Sn
Type metal	82 Pb, 15 Sb, 3 Sn
Yellow brass	67 Cu, 33 Zn
Wood's metal	50 Bi, 25 Pb, 12.5 Sn, 12.5 Cd

If two metals completely soluble in the melted state are also completely soluble in the solid state, a homogeneous alloy is obtained. A homogeneous alloy may also be prepared if the melt of the two metals has the exact composition of a compound of the two. For some purposes a homogeneous alloy is best. But most alloys are heterogeneous, and for most uses this is an advantage. The desired surface hardness and wearing quality of some steel is due to hard, needlelike crystals of Fe_3C throughout an iron matrix. The bearing metal, called babbitt (90 Sn, 7 Sb, 3 Cu), has a hard, antimony-containing, needlelike component embedded in a softer tin component. On wear, the soft component of babbitt smears the surface, so that there is little bearing friction, while the harder component gives rigidity. Babbitt is a ternary alloy.

Eutectics

In the same manner that a dissolved substance lowers the freezing point of water, one metal dissolved in the liquid of a second metal lowers the freezing point of the latter. This is evident from Figure 25.1.

As silver is dissolved in liquid copper, the freezing point or melting temperature of the alloy formed is progressively lower as the percentage of silver increases; as copper is dissolved in liquid silver, the freezing point or melting temperature of the alloy progressively decreases. Necessarily, a mixture exists which is the alloy of lowest melting point. It is called the **eutectic** alloy and is represented in Figure 25.1 by E, which shows a composition of 28% copper and 72% silver. If a solution of composition A is cooled, copper begins to crystalize out, and the composition of the solution alters along the curve toward E. At E the entire mass, which consists of the copper that has crystallized out

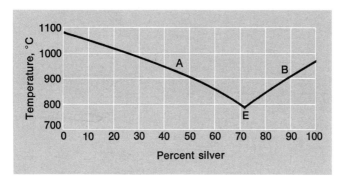

Figure 25.1
Freezing-point curve for solutions of copper and silver.

and the eutectic mixture (28% Cu, 72% Ag), solidifies. Similarly, if a solution of composition B is cooled, silver crystallizes out, with a concomitant change in composition of the liquid until E is reached and the entire mass solidifies.

Alloys of three and four metals are susceptible of forming ternary and quaternary eutectics.

25.2   METALLURGICAL CHEMISTRY

Metallurgy is the science of extracting metals from their ores. An **ore** is a naturally occurring material which consists of a mixture of the valuable metal-containing part called the **mineral** and unwanted silicate material called the **gangue**. For example, **galena** is an ore consisting of the mineral, PbS, mixed with rocky silicate material, the gangue.

Metallurgical processes may be divided into four basic operations, (1) concentration, (2) roasting of sulfide ores, (3) reduction (smelting), and (4) refining (purification), although for a number of metals the roasting step is not necessary or may be combined with reduction.

Mineral Concentration

Ore dressing and **ore benificiation** are terms used to denote the technology of treating a mined ore to obtain a product higher in concentration of a wanted mineral from the unwanted rock mass in which it occurs. Processes of breaking, crushing, grinding, pulverizing, and sizing ore particles are names applied to physical treatment steps. Gravity concentration or flotation concentration procedures follow. Suppose that a company has an ore that is 5% galena (PbS) and 95% quartz (SiO_2 or unwanted siliceous rock). Taking advantage of the difference in specific gravity between the heavy PbS particles

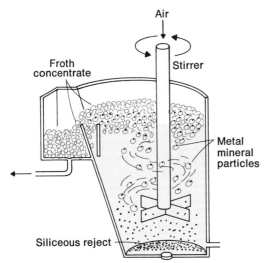

Figure 25.2
Apparatus for flotation concentration.

Figure 25.3

Flotation froth in D-R Denver machines.

and the lighter SiO_2 particles, running water in suitable gravity concentration machines can effect a considerable separation of the PbS from silica. A product of such concentration machines from the raw ore may assay 50% PbS and 50% SiO_2. Most of the silica in the original rock can be discarded with only a small loss of PbS. This is evident from the following:

500 lb PbS		475 lb PbS	+	25 lb PbS
9,500 lb SiO_2	concentration	475 lb SiO_2	+	9,025 lb SiO_2
10,000 lb ore	machines →	950 lb concentrate		9,050 lb reject

It will be noted that most of the PbS is now concentrated in less than one-tenth the original amount of rock material. If this ore concentration is effected in a mill at the mine, much hauling and transportation of waste material is saved. At the smelter to which the concentrated ore is shipped, less heat will be required in the subsequent pyrometallurgy.

Other metal minerals of comparatively high specific gravity to that of gangue siliceous material may also be effectively gravity-concentration beneficiated. Many modern ore-dressing mills use a so-called *flotation* process for the elimination of gangue. The process is based on the selective wetting of surfaces by various liquids or selective attraction or repulsion between liquids and solids. Consider the liquids water and mercury in contact with a solid such as glass. Water spreads out on a clean surface and wets the glass. Mercury, on the other hand, does not spread out but forms a globule which does not wet the glass. Certain organic liquids, such as pine oil or various organic sulfur compounds, have a marked tendency to wet and cling to metal or metal sulfide surfaces, but do not cling to siliceous matter. Powdered ore suspended in water is fed into a vat in which is placed a mechanism to beat or blow air bubbles into the suspension. A chemical such as cresylic acid is added to produce a froth and to stabilize the bubbles. Next an organic chemical such as ethyl xanthate is added in amounts as small as 0.1 lb per ton of ore. The xanthate adheres to metal and metal sulfide mineral, as well as to the air bubbles, and the powdery, metal-containing particles are floated to the surface as the air bubbles rise. They are then scraped off in a froth which, when dried, gives a metal mineral concentrate. The siliceous particles do not collect against the air bubbles but drop to the bottom of the vat (Figures 25.2 and 25.3).

Sulfide Mineral Treatment

Because of the marked insolubility of most metal sulfides, they have been geochemically precipitated in underground cracks (veins) in practical mineable concentration. Thus, metal sulfide minerals, admixed with siliceous gangue material, constitute ores for many metals. The desired metal as sulfide mineral may compose a small percentage of the ore. The ore, or ore mineral concentration product, is usually given a first chemical treatment called **roasting.** This is a treatment in an appropriate furnace to partially or completely convert sulfide minerals to oxides. Examples of roasting equations for typical sulfide minerals are:

$$2\ ZnS\ (sphalerite) + 3\ O_2 \rightarrow 2\ ZnO + 2\ SO_2$$
$$Cu_2S\ (chalcocite) + 2\ O_2 \rightarrow 2\ CuO + SO_2$$
$$CuFeS_2\ (chalcopyrite) + 3\ O_2 \rightarrow FeO + CuO + 2\ SO_2$$
$$2\ PbS\ (galena) + 3\ O_2 \rightarrow 2\ PbO + 2\ SO_2$$

In some metallurgical plants each of these respective metals occurring as sulfides is converted to oxide, thence by carbon reduction to the desired metal. SO_2 gas, the chief product of roasting other than metal oxide, is almost all converted to H_2SO_4 in an acid manufacturing plant that is commonly associated with sulfide ore roasting facilities.

Carbon Reduction

Since the practical source minerals of many metals are oxides, and since carbon (as coke or charcoal) is a cheap reducing agent, the greatest bulk of our metal is attained by chemistry such as that shown in the following equations.

$$SnO_2\ (cassiterite\ mineral) + C \xrightarrow{\Delta} Sn + CO_2$$
$$2\ C + O_2 \rightarrow 2\ CO + \Delta$$
$$Fe_2O_3\ (hematite) + 3\ CO \rightarrow 2\ Fe + 3\ CO_2$$
$$Fe_3O_4\ (magnetite) + 4\ CO \rightarrow 3\ Fe + 4\ CO_2$$
$$MnO_2\ (pyrolusite) + C \rightarrow Mn + CO_2$$

Tin is quite easily recovered from an SnO_2, siliceous mixture, by heating in a furnace with carbon. The low-melting tin (232°C) flows from the unwanted solid siliceous gangue material, and can be cast into desired blocks or ingots.

Contact between a hot gas and a solid is more conducive to chemical change in a furnace than contact between two lumpy solids. Thus, in the equations above, it is indicated that much reduction of metal oxide to metal in high-temperature furnaces is by CO rather than by carbon itself. The initial carbon entry into the furnace has two purposes — 1) for heat needed in its partial or complete oxidation, and 2) as a precursor of CO for metal oxide mineral reduction.

A greater bulk of iron is produced than of any other metal, mainly as per the reactions above as carried out in an iron blast furnace (page 539).

Many metal sulfide minerals in ores are in first treatment oxidized to metal oxides, which in secondary processing are subject to carbon reduction to yield metals.

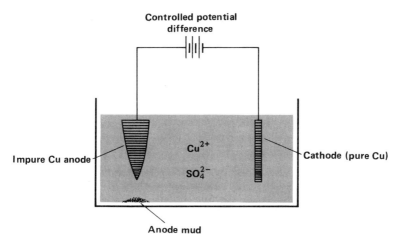

Controlled potential
difference

Impure Cu anode

Cu^{2+}

SO_4^{2-}

Cathode (pure Cu)

Anode mud

Figure 25.4

Electrolytical refining of copper.

Refining

Generally metal obtained from the reduction processes of smelters is somewhat impure and needs to be purified or refined for most uses. Copper, for
example, before refining is 97 to 99% pure. Common impurities in the copper
are iron, nickel, arsenic, antimony, bismuth, and other metals higher than
copper in the activity series as well as metals lower in the series, such as
silver, gold and the platinum metals.

Copper is refined electrolytically as shown in Figure 25.4. The impure
copper is made at the anode in an electrolysis cell, and thin sheets of pure
copper are made at the cathode. As current goes through the cell, copper and
the more active metals in the anode go into solution. Solid silver and the
less active metals drop off of the anode and settle to the bottom of the cell as a
valuable anode mud from which they are later recovered. At the cathode only
the Cu^{2+} ions are reduced to free metal. Fe^{2+} and the more active metal ions
remain in solution. Eventually essentially all of the copper in the impure
anode is transferred to the cathode which ends up nearly 100% pure copper.
The electrolysis reactions are

$$\text{Anode: Cu (impure)} \rightarrow Cu^{2+} + 2\ e^-$$
$$\underline{\text{Cathode: } Cu^{2+} + 2\ e^- \rightarrow \text{Cu (pure)}}$$
$$\text{Overall: Cu (impure)} \rightarrow \text{Cu (pure)}$$

Other metals are purified by special methods such as distillation, melting-
point differences, solubilities, or conversion to compounds which may be
readily separated from impurities and then reconverted to the free metals.

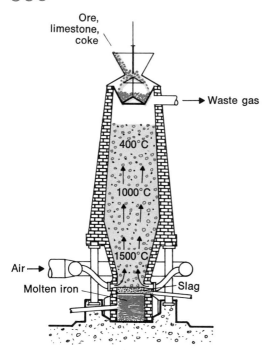

Figure 25.5
Blast furnace for the manufacture of pig iron.

25.3 METALLURGY OF IRON AND STEEL

Perhaps only the cement industry is larger in produced tonnage by applied chemistry than the iron and steel industry. There is evidence that iron was known in prehistoric times, though it seems likely that copper and bronze were known and used prior to its discovery. Oxide minerals of iron are easily reduced with carbon, and its discovery was probably made by accidentally heating the ore in a wood fire.

Occurrence

Iron stands fourth in abundance among the chemical elements present in the earth's crust (see page 499). The element is rarely found in the free state except in certain meteorites which are made of iron with a small percentage of nickel. Compounds of iron are widely distributed and are constituents of many rocks and soils.

Iron(II) compounds are greenish; iron(III) compounds are yellow to brown or red. An aesthetic chemist gazing at the magnificent coloring of the Grand Canyon of the Colorado might speculate that this would be a drab world without iron compounds. Iron(II) silicates give green to rock, the carbonates a yellow-brown, and the iron(III) oxides and hydroxides give brown to red in rock or soil. Compounds of the other abundant elements are largely colorless. The natural pink of cheeks and lips is due to the iron compound in the hemoglobin of the blood.

When traveling through a deep road cut that has been made through igneous rock, note that geochemical oxidation may be observed. Lower rock is apt to be unaltered and green tinted from iron(II) silicate (FeO · SiO$_2$). As ground water containing dissolved O$_2$ has weathered the upper part of this rock to a soil, the soil will be brown, tan, or reddish as the iron(II) minerals are converted to iron(III) oxides, hydrated oxides, and SiO$_2$. The colors of the Grand Canyon rocks, and in fact of most color in bulk rock around the world is due to iron compounds. Concentrations of naturally occurring iron oxides are the ores which are used for the production of iron and its products. Among these ores are **hematite** (Fe$_2$O$_3$) and **magnetite** (Fe$_3$O$_4$), a mineral with magnetic properties commonly called **lodestone.** Most of the iron ore mined in the United States comes from the Mesabi Range in the Lake Superior region of Minnesota. The open-pit mining method is employed in this area. Huge steam shovels scoop up the ore. Other iron-producing areas of the United States are Alabama and Michigan.

Iron pyrites (FeS$_2$), often called "fool's gold" because of its yellow, flaky appearance, is another abundant, naturally occurring compound of iron.

Metallurgy

Iron may be obtained from its oxide ores by reduction with carbon. This operation is carried out in a huge cylindrical furnace called a **blast furnace,** a cross section of which is shown diagrammatically in Figure 25.5. These furnaces vary in height from 75 to 100 ft and are 20 to 25 ft in diameter near the base. They are constructed from heavy steel plates and lined with firebrick. Ore, limestone, and coke are introduced at the top of the furnace through a series of hoppers, which are manipulated so that gases produced by the reactions are not lost. The operation of a blast furnace is continuous. The iron is tapped and run off at intervals. Since silica is the impurity usually present in iron ore, limestone is added as a flux to form a slag which may be drawn off. If the impurity in the ore is a base, such as carbonates of calcium or magnesium, silica is added as a flux to form the slag. The coke serves as a fuel to melt the iron, and as a reducing agent to convert the oxide ore to metal. Near the bottom of the furnace a hot blast of air is introduced through nozzles, called **tuyeres,** to burn the coke. An average charge for a furnace at 15-min intervals is 9 tons of ore, 3 tons of limestone, and 5 tons of coke. In addition, about 12 tons of air are necessary to burn the coke.

The chemical changes of this huge operation are essentially those written on page 536 of this chapter as relates to carbon and iron. Molten iron and slag collect in the **crucible** at the bottom of the furnace, where they may be drawn off. The temperatures in the various parts of the furnace are shown in the diagram. The iron which is tapped from the furnace is an impure variety called **pig iron** and contains a considerable amount of dissolved carbon, as well as sulfur and phosphorus, as impurities. A large blast furnace produces approximately 600 tons of pig iron every 24 hr.

A third product of the blast furnace is a mixture of gases which is sometimes called **blast furnace** or **producer gas.** Besides carbon dioxide and nitrogen (from the air), the furnace gas contains considerable carbon monoxide and some hydrogen, both of which are combustible. The mixture has fuel value, and much of it is employed in preheating the air to be used in the blast furnace. The preheated air enters the blast furnace at a high temperature, and combustion of the coke ensues rapidly, producing a higher temperature than if cold air were used. Blast furnace gas is removed near the top of the furnace. The products of the furnace, with the approximate composition of each, are shown in Table 25.4. Slag is used for road ballast and in the manufacture of certain types of cement. For the most part, it is poured and accumulated in slag dumps.

Table 25.4
Products of the blast furnace

Pig Iron		Gases		Slag
Fe	93–95%	CO	25%	Chiefly $CaSiO_3$
C	3–5%	CO_2	13%	Some $FeSiO_3$
Si	1%	N_2	57%	$Ca_3(PO_4)_2$
S	less than 1%	H_2	3%	$MgSiO_3$
P	0.1–0.3%			

In recent years 7% natural gas and 93% preheated air has been introduced into blast furnaces. The furnaces are adjusted so that there is some pressure above atmospheric build-up, and gas-solid reactions proceed a little more rapidly.

Because of the impurities in pig iron, it cannot be welded or forged, and it possesses little tensile strength. Pig iron melted with scrap iron produces **cast iron,** which is useful for objects not subject to sudden jarring or shock, such as radiators, parts for stoves, and certain types of heavy machinery. Cast iron is comparatively easily melted and expands slightly when cooled; thus it fills every part of the mold into which it is poured.

Steel

The manufacture of steel consists essentially of removing carbon, silicon, phosphorus, and sulfur from pig iron. Since the latter three impurities are objectionable, it is desirable to remove as much of them as possible. A certain amount of carbon is desirable, however, because the properties of iron are considerably modified by even small percentages of carbon. Steel is a product of almost pure iron, containing relatively small and varying amounts of carbon. The more common types of steel and their uses are shown in Table 25.5. In recent years small amounts of many other metals have been added to steel to impart certain characteristic properties. The addition of manganese produces a very tough steel; molybdenum and chromium yield a steel resistant to corrosion; tungsten forms an alloy which retains its temper when heated.

Table 25.5
Common types of steel

Type	Carbon Content	Uses
Low carbon steel	less than 0.3%	horseshoes, wire, and nails
Medium carbon steel	0.3 to 0.8%	railroad rails and axles
High carbon, or tool steel	0.8 to 2%	tools, springs, and files

The removal of the impurities from pig iron is effected by oxidation, and the process depends on the fact that the *impurities in the iron are more easily oxidized than the iron itself.* There are two principal processes for the conversion of pig iron to steel: (1) the **basic oxygen** process and (2) the **open-hearth** process.

The **basic oxygen process** is now used to produce most of the 140,000,000 tons of steel made in the United States. It is a modification of the old Bessemer Process which was used for many years. In the modern process, jets of nearly pure oxygen are introduced into a large furnace containing pig iron, scrap iron, special alloying metals, and limestone (for slag formation) which rapidly oxidizes the impurities and gives nearly smokeless gaseous products. The whole process takes 30 min to 1 hr to complete and, under high speed computer control, it gives an excellent and uniform quality steel.

Basic open-hearth process The basic open-hearth furnace consists of a shallow crucible lined with calcite and magnesia. Below the crucible is a checkerwork of brick arranged so that the heat from the spent gases is used to preheat the entering gases. Fuel for the furnace is either producer gas from the blast furnace or oil in the form of a spray. The fuel gas and air are passed over the hot brick checkerwork and allowed to come together just above the crucible of the furnace. Combustion takes place rapidly because the gases have been preheated to a high temperature and pure O_2 is introduced. The spent gases pass through the checkerwork on the other side of the furnace, where their heat is imparted to the brickwork. Every few minutes the direction of the gases is reversed, so that the incoming gases are always preheated to a high temperature. In the meantime, spent gases pass over the checkerwork, which has been cooled by the incoming gases.

The capacity of an open-hearth furnace varies between 50 and 150 tons. The furnace is charged with molten pig iron, limestone to act as a flux, and rusty scrap or iron oxide, which acts as an oxidizing agent to oxidize the carbon and other impurities. Sulfur and phosphorus are particularly objectionable in steel because they give a brittle product, and these elements must be removed by oxidation. The slag floats on top of the molten iron and prevents its rapid oxidation by the incoming air fortified with oxygen. Carbon in the pig iron is converted to its oxides, and thus removed until the desired low percentage is reached. In recent years the time required per batch-charge has been markedly reduced by the use of pure O_2 rather than air. When the carbon content has been reduced to the required percentage and the sulfur and phosphorus have been removed, the iron is tapped and run into ingot molds. After the ingots are removed, they are taken to the rolling mill and processed. If an alloy steel is wanted, the same process is followed, but just before tapping, a certain amount of the alloying element is added to the furnace.

Composition of various irons and steels

	C	Graphite	Si	S	P	Mn
Gray pig iron	2.45	1.46	2.23	.09	.21	.35
White pig iron	3.37	.35	.42	.11	.17	.61
Wrought iron	.10	.06	.16	.15	.17	.22
Bessemer steel for wire	.11	—	.01	.08	.07	.43
Boiler plates	.18	—	.03	.02	.03	.25
Railroad steel	.55	—	.12	.07	.09	1.00
Tool steel	1.05	—	.15	.05	.06	.42

Alloys of Fe

Cr 4%, Tungsten 17–20%	auto valves and high speed tools
Mn 11–14%	tough and durable; steam shovels
Cr 1.5–2%	ball bearings, safes, files, rock crushers
Ni 2–4%	great tensile strength; propeller shafts, bridge trusses
Cr 1%, V 0.15%	tough and ductile; auto springs, axles, frames, differentials
Invar 36% Ni, 5% Mn	low coefficient of expansion; pendulums, surveyors' tapes
Duriron Fe-Si	acid resistant; drains
Platinite 46% Ni, 54% Fe	same coefficient of expansion as glass; light bulbs

Crucible steel   Very high grades of steel, such as those used for watch springs, razor blades, files, and so on, are usually prepared by heating small batches of wrought iron or better grades of open-hearth steel in small crucibles. An electric furnace allows a very careful control of temperature. Carbon and alloying elements may be added in required amounts. Such high-grade steels are called **crucible steels.**

Allotropic forms of iron   The properties of steel may be modified greatly by conditioning the crystalline form of the iron. At ordinary temperatures a body-centered cubic lattice form of iron exists which is comparatively soft and magnetic and in which carbon is insoluble—it is termed **alpha iron.** When alpha iron is heated above 900°C, allotropic modification known as **gamma iron** forms as a face-centered cubic lattice. Gamma iron is very hard and nonmagnetic and possesses the important property of forming a solid solution with carbon. At 1400°C a further allotrope, called **delta iron,** is formed. The properties of this form are not well known. The equilibrium relations between the three allotropes are shown by the equilibria

$$\text{Alpha} \underset{}{\overset{900°}{\rightleftharpoons}} \text{Gamma} \underset{}{\overset{1400°}{\rightleftharpoons}} \text{Delta}$$

Tempering of steel   If a steel at a temperature above 900°C is quickly cooled by "quenching" in water or oil, the gamma iron, which is hard and dissolves carbon, does not have time to alter to alpha iron, which is soft and does not dissolve carbon. At room temperatures the unstable gamma iron would require centuries to transform completely to the alpha form. Gamma iron is much too hard to be mechanically worked or fashioned, but a steel containing some gamma iron for strength may be heated to a temperature between 200°C and 600°C for a time to enable some **tempering** of its hardness—that is, to enable a desired amount of gamma to convert to alpha iron according to the particular strength and hardness wanted in the product. During this reheating, called **annealing,** cooling stresses are also removed.

In the high-carbon steels a very hard compound of iron and carbon called **cementite** (Fe_3C) appears to be formed. Crystals of cementite, which may be viewed through a microscope, have marked effects on the properties of the steel. Heat treatment will cause some of the cementite to decompose to iron and graphite flakes, resulting in a softer product. The size of crystals formed is also important in determining properties —a steel of minute crystals or grains has more strength than a coarse-grained product. Grain size is somewhat regulated by heat treatment and by mechanical pounding or pulling of the metal. In general, a mechanically worked product has smaller grain size and greater tensile strength.

Casehardening   A low-carbon steel makes a good structural beam; it has toughness and strength and yet enough "bend" so that it does not readily break. Since a low-carbon steel is relatively soft, however, its surface is easily marred. Therefore it is often heated in a carbon or carbon-compound (often cyanide) bath, in which some carbon dissolves into the outer $\frac{1}{16}$ inch or so of the iron to yield a high-carbon, hard, wear-resisting surface. This process is called **casehardening.**

Exercises

25.1. What properties distinguish metals from nonmetals? Give 5 physical properties of Cu which illustrate its metallic character.

25.2. Explain the meaning of the terms: ore dressing, leaching, eutectic, elastic limit, ductile, malleable, metallic bond, alloy, roasting, refining.

25.3. Identify the following: (a) ore, (b) gangue, (c) mineral, (d) hydrometallurgy, (e) flux, (f) slag, (g) smelting.

25.4. Choose the item from the list on the right which matches most closely:
(a) Pig iron.   1. Metals
(b) Brass.   2. Fe_2O_3 ore
(c) Elements with 1, 2, or 3 valence electrons.   3. P

(d) Ductility.
(e) An intermetallic compound.
(f) Gangue of an ore.
(g) Hematite.
(h) One of the 4 raw materials going into a blast furnace.
(i) One of the main impurities in pig iron.
(j) Fe containing a small percent of C.
(k) Main product obtained when α-Fe is heated above the transition temperature, 900°C.
(l) A process for concentrating the mineral of an ore from the gangue.
(m) Conversion of sulfide minerals to oxides.
(n) Changing a metal oxide to the free metal.
(o) Purification of the free metal.
(p) Name of the main method used in steel making.

4. γ-Fe
5. Impure Fe from blast furnace
6. Flotation
7. Roasting
8. P-I
9. Alloy of Cu-Zn
10. Ability to be pulled into a wire
11. Ni
12. Steel
13. Al_2Mg_3
14. Refining
15. The silicate part of an ore
16. Smelting or C reduction
17. Basic oxygen process

25.5. Plot the freezing-point curve for alloys of metals A (f.p. 850°C) and B (f.p. 1320°C) which form a eutectic of 35% A at 700°C.

25.6. A freezing-point curve is given for solutions of lead and tin. From the diagram give:

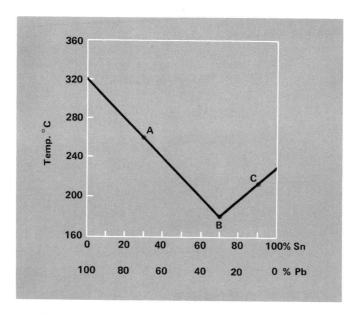

(a) The melting point of the eutectic mixture.
(b) The percent of tin in the eutectic mixture.

(c) The melting point of pure lead.

(d) The melting point of pure tin.

(e) If a mixture of Sn + Pb of composition A were cooled to 200°C, the mixture will contain both solid and liquid.

(1) What is the solid?

(2) What is the percent lead in the remaining liquid?

(f) If a mixture of tin and lead of composition B is cooled what will crystallize out first?

25.7. The following equations define chemical reactions which are utilized in the recovery of metallic copper from chalcopyrite, a sulfide ore:

1. $2 \, CuS + 3 \, O_2 \rightarrow 2 \, CuO + 2 \, SO_2$
2. $FeO + SiO_2 \rightarrow FeSiO_3$
3. $Cu^{++} + 2 \, e^- \rightarrow Cu$ (cathode)
4. $2 \, FeS + 3 \, O_2 \rightarrow 2 \, FeO + 2 \, SO_2$
5. $CuO + CO \rightarrow Cu + CO_2$

Answer the following questions concerning these equations:

(a) The roasting process would be described by this reaction.

(b) Metallic copper is initially recovered from the ore by this reduction process.

(c) This reaction removes iron impurities when the compound floats to the top of the reaction mixture.

(d) Sulfur impurities are removed in this reaction.

(e) Very pure copper is produced by this reaction in a refining step.

25.8. Complete the following table, summarizing three basic processes used in the recovery of a metal from its ore and illustrating how these steps are accomplished with copper. Insert the number of the statements in the spaces provided

Process	Definition	Process used in Cu production
Beneficiation	( )	( )
Reduction	( )	( )
Refining	( )	( )

1. Electrolysis
2. Conversion of cation to element
3. Flotation
4. Preparation of highly purified metal
5. Increasing content of desired compound in the ore
6. Carbon monoxide or chemical reduction

25.9. Explain why it is not surprising that the following ores are found in nature: (a) native ores of platinum, gold, and silver; (b) sulfate ores of barium, strontium, and lead; (c) Na^+ and Mg^{2+} in sea water; (d) sulfide ores of lead, bismuth, and nickel.

25.10. Since metal minerals are usually insoluble, write a formula for a probable naturally occurring metal (a) sulfide, (b) sulfate, (c) silicate, (d) phosphate, (e) carbonate. Write formulas of four minerals apt to occur in the dried-up residue of a large desert lake.

25.11. Describe how galena (PbS) ore, containing 2% PbS and 98% siliceous matter, can be concentrated. How can metallic lead be obtained from the concentrate?

25.12. What types of ores are roasted? Write a chemical equation for the roasting of a typical ore. Why is this procedure used?

25.13. Select the equation below which represents each of the following processes — roasting, reduction, nonaqueous electrometallurgy, use of a flux.
(a) $CaCO_3 + SiO_2 \rightarrow CaSiO_3 + CO_2$
(b) $Al^{3+} + 3\ e^- \rightarrow Al°$ (from Al_2O_3 in molten Na_3AlF_6)
(c) $CuCO_3 + H_2SO_4$ (dil) $\rightarrow CuSO_4 + CO_2 + H_2O$
(d) $4\ CuFeS_2 + 13\ O_2 \rightarrow 4\ CuO + 2\ Fe_2O_3 + 8\ SO_2$
(e) $Fe_2O_3 + 3\ CO \rightarrow 2\ Fe + 3\ CO_2$

25.14. Elements classed as metals may be part of an anion and be a unit in oxidation-reduction equations. Complete and balance:

$$CrO_4^{2-} + Fe^{2+} + H^+ \rightarrow$$
$$MnO_2 + BiO_3^- + H^+ \rightarrow HMnO_4 + Bi^{3+} + H_2O$$

25.15. (a) Write the equations for probable blast furnace chemical changes (some of them step changes) as the following mix is subjected to high temperature: C (coke), ore ($\frac{2}{3} Fe_2O_3$, $\frac{1}{3} SiO_2$, a little $FePO_4$, FeS_2), $CaCO_3$ (limestone), hot air. (b) If some MnO_2 or Cr_2O_3 were in the iron ore, what chemical changes would they undergo in the blast furnace?

25.16. Discuss the allotropic forms of iron and how they relate to various processes of quick cooling (quenching), annealing, control of dissolved or undissolved carbon.

25.17. Give equations for reactions that occur in the blast furnace for the production of pig iron.

25.18. Suggest the nature of an alloy of iron as regards amount of carbon and alloying element for each of these uses: stainless steel, prongs of a road building bulldozer, pliable iron wire, laboratory drain pipes, electrical resistance wire.

25.19. Fill in the words *highest, middle,* or *lowest:*

	Percent C	Percent S	Degree of Strength	Cost Per Pound	Degree of Hardness (and Malleability)	Melting Temp.
Pig iron						
Pure iron						
Steel						

25.20. A siliceous rock analyzes 4% Zn, 3% Cu, and 0.2 oz Au per ton. A 100-ton quantity of this ore was concentrated to 10 tons which analyzed 35% Zn, 28% Cu, and contained 18 oz of Au. What percent of each metal was recovered in the 10 tons of concentrate, 90 tons being discarded as waste?

25.21. A tool steel, after etching in acid and microscopic examination, appears to be 18% cementite (Fe_3C) crystals to 82% iron with some carbon content. The Fe_3C content would account for what percent carbon?

Supplementary Readings

25.1. *Metals in the Modern World*, E. Slade, Doubleday, New York, 1968 (paper).
25.2. "The Deformation of Metals at High Temperatures," H. J. McQueen and W. J. M. Tagart, *Sci. American*, April 1975; p. 116.
25.3. "The Direct Reduction of Iron Ore," J. R. Miller, *Sci. American*, July 1976; p. 68.

Chapter 26

Main Energy Levels	1	2	3	4	5	6	7
$^{6.941}_{3}$Li	K	s^1					
$^{22.99}_{11}$Na	K	L	s^1				
$^{39.102}_{19}$K	K	L	s^2p^6	s^1			
$^{85.47}_{37}$Rb	K	L	M	s^2p^6	s^1		
$^{132.91}_{55}$Cs	K	L	M	$s^2p^6d^{10}$	s^2p^6	s^1	
$^{9.012}_{4}$Be	K	s^2					
$^{24.305}_{12}$Mg	K	L	s^2				
$^{40.08}_{20}$Ca	K	L	s^2p^6	s^2			
$^{87.62}_{38}$Sr	K	L	M	s^2p^6	s^2		
$^{137.34}_{56}$Ba	K	L	M	$s^2p^6d^{10}$	s^2p^6	s^2	
$^{226.025}_{88}$Ra	K	L	M	N	$s^2p^6d^{10}$	s^2p^6	s^2

Group IA Metals

Group IIA Metals

The Alkali and Alkaline Earth Families

The elements of these two periodic table groups are very chemically active and thus do not occur uncombined in nature. Group IA metals exhibit an electrovalence of +1 and Group IIA an electrovalence of +2 in compounds. Since the hydroxides of these elements (except $(Be(OH)_2)$ give basic (alkaline) solutions the elements were historically called the alkali and alkaline earth metals, respectively. The metals themselves are relatively unimportant, but their compounds are exceedingly important. The salts of the sea are almost entirely compounds of some of these metals; man's bones are made of a calcium compound; a good soil for plant growth must have ions of K^+, Ca^{2+}, and Mg^{2+}

26.1   PREPARATION OF METALS

A marked feat in experimental science was the preparation of the element potassium by Humphrey Davy in October, 1907 in England. He used battery-produced electric current, platinum electrodes, and melted potash (KOH with some K_2CO_3). At the negative wire he saw globules of a silvery metal, which spontaneously took fire in air. In the following years he isolated both alkali and alkaline earth metals from compounds by the use of electrolysis. Davy was knighted for his accomplishment of the preparation of several metals never before seen by man.

Electrolysis, in the absence of air, is the current industrial method for production of alkali and alkaline earth metals. Melted metal chlorides are used (Davy largely used melted hydroxides). For example, electrolysis of melted NaCl, to which some $CaCl_2$ has been added to lower the melting point, yields industrial sodium as a cathode product and chlorine at the anode.

Most of the sodium prepared commercially is used in the preparation of certain salts, which are more easily prepared from the free metal than from its compounds. Sodium peroxide is prepared by passing oxygen over aluminum trays containing metallic sodium. Tetraethyl lead is made from a lead-sodium alloy as starting material. Sodium is an effective reducing agent in the synthesis of certain organic compounds. Recently sodium metal has been used as a heat transfer agent in power applications of nuclear reactors. Liquid sodium in pipes is used as a coolant for the nuclear reactor; the heat removed is later used to run a steam generator.

Lithium is the lightest of all the metals ($d = 0.53$ g/cm^3). It also has the highest specific heat (0.80 cal/g) of any solid element (page 530). It is a promising battery material and if proper batteries can be developed, it is estimated that some 25 million electric vehicles could be travelling the U.S. highways by the year 2000. Lithium-water and lithium-water-air batteries already are approaching the performance level of gasoline-powered engines. Thermonuclear fusion reactions involving lithium may also be a reality in the future.

Lithium compounds impart a brilliant red color when introduced into a flame. There is also considerable evidence that certain lithium salts show promise in the prevention and treatment of various mental disorders.

Physical Properties

Examine Table 26.1 for a trend in a few of the physical properties of the alkali metals from the lowest to highest in atomic number; similarly for the alkaline earth metals. In general the specific gravity and atomic radii increase with atomic number as expected.

Table 26.1

Physical properties

Metal	Symbol	Sp. Gr.	Melting point °C	Boiling point °C	Atomic radius (Å)	Ionic radius (Å)	Reduction potential, volts
			The alkali metals				
Lithium	Li	0.53	179.0	1317	1.23	0.60	−3.05
Sodium	Na	0.97	97.8	892	1.57	0.95	−2.71
Potassium	K	0.86	63.6	774	2.03	1.33	−2.95
Rubidium	Rb	1.53	38.9	688	2.16	1.48	−2.93
Cesium	Cs	1.87	28.4	690	2.35	1.69	−2.92
			The alkaline earth metals				
Beryllium	Be	1.85	1280	2970	0.89	0.31	−1.85
Magnesium	Mg	1.74	651	1107	1.36	0.65	−2.37
Calcium	Ca	1.55	845	1487	1.74	0.99	−2.87
Strontium	Sr	2.54	769	1384	1.91	1.13	−2.89
Barium	Ba	3.50	725	1140	1.98	1.35	−2.90
Radium	Ra	5.00	700	1700			−2.92

The alkali metals as a group are decidedly low in specific gravity and quite large in atomic size. Lithium, sodium, and potassium are lighter than water. The alkali metals decrease in melting and boiling temperature with increase in atomic number. All have quite low melting and boiling temperatures. Cesium will melt not far above room temperature; sodium melts below 100°C.

The alkaline earth metals are a little heavier and higher in specific gravity than the alkalies because of smaller atomic radii. They are, however, light metals as compared to the transition group metals. Boiling and freezing temperatures of alkaline earth metals decrease progressively with increase in atomic number. The oxidation potentials of the alkaline earth group toward divalent ions are almost as high as for the alkali metals.

All the Group IA and IIA metals are good electrical and heat conductors, and are relatively soft and workable. Beryllium is an exception in that it is hard and brittle; beryllium also differs from all these other metals in that its hydrated oxide acts amphoterically whereas hydrated oxides of the true alkali and alkaline earth metals are strong bases.

Occurrence and Abundance

Two alkali metals and two alkaline earth metals are among the top nine elements in abundance by weight percent in the earth's crust. They are: Na, 2.83%; K, 2.59%; Mg, 2.09%; Ca, 3.63%.

A few common silicate minerals of these elements are:

$CaO \cdot SiO_2$, $CaSiO_3$	Wollastonite	$MgO \cdot SiO_2$, $MgSiO_3$	A pyroxene
$CaO \cdot Al_2O_3 \cdot 2\ SiO_2$	Anorthite	$(MgO)_2 \cdot SiO_2$, Mg_2SiO_4	Forsterite
$Na_2O \cdot Al_2O_3 \cdot 6\ SiO_2$	Albite	$K_2O \cdot Al_2O_3 \cdot 6\ SiO_2$	Orthoclase

Similar silicate minerals of very small percentage in earth rock constitute the main existence of the other Group IA or IIA metals. At times some crystalline concentrations of the less abundant elements are found and mined. Such is the case of beryl, $(BeO)_3 \cdot Al_2O_3 \cdot 6\ SiO_2$, and spodumene, $Li_2O \cdot Al_2O_3 \cdot 4\ SiO_2$. Hundreds of silicate minerals formed from the combination of metal oxides with silica are present in the earth's crust.

Weathering of these rock minerals has produced ions of them in ground water. Since alkali metal salts are practically all water-soluble there has been little precipitation of alkali ions after they are dissolved in ground water. Thus, Na^+ has built up in sea water over geological time, and is counterpart to much Cl^- (Na^+, 1.06%, Cl^-, 1.90%; adds to 2.96%). The total salt content of the sea is 3.45%; sodium chloride is predominant.

Why is there by far more Na^+ than K^+, Mg^{2+}, or Ca^{2+} in sea water inasmuch as they are all nearly equal in earth rock? Two reasons for the mere 0.038% K^+ in sea water are (a) plant material selectively takes K^+ from soil water,[*] (b) potassium feldspar ($K_2O \cdot Al_2O_3 \cdot 6\ SiO_2$) and other potassium silicates are more insoluble than say, sodium feldspar ($Na_2O \cdot Al_2O_3 \cdot 6\ SiO_2$).

Mg^{2+} ion is 0.13% and Ca^{2+} 0.04% in sea water. Rock deposits of $MgCO_3$, **magnesite,** and $CaCO_3$, **limestone** or **marble** or **calcite** according to crystallinity, attest to precipitation of these slightly water-soluble compounds from metal ion and carbonate ion over ages of time. There are also deposits of $CaSO_4 \cdot 2\ H_2O$, **gypsum;** $BaSO_4$, **barite;** $SrSO_4$, **celestite,** as sulfates of these Group IIA metals are only very slightly soluble.

Large deposits of NaCl have been formed as a result of evaporation of inland bodies of water in desert or arid regions as, for example, the high salt content of the Great Salt Lake in Utah. One of the largest deposits in the United States is a layer several hundred feet deep beneath an area of some thousand square miles in Kansas, Oklahoma, and Texas.

---

[*] Ashes of plant material contain some K_2CO_3, and almost no Na_2CO_3. K_2CO_3 has long been called "potash," the probable source of the metal's name.

Solar evaporation of sea water is practical in certain areas where the evaporation exceeds the total rainfall. Sodium chloride obtained by this method is usually impure and contains other salts in small percentages. In general the impurities are more soluble than sodium chloride and remain in the mother liquor during the crystallization of the common salt. One of the most troublesome impurities in salt is magnesium chloride, which is hygroscopic and causes the salt to cake in a moist atmosphere. The caking can be lessened by the addition of starch, which coats the salt particles and prevents them from adhering.

Sodium chloride serves as the source for sodium and chlorine, and is the starting material in the preparation of most of the compounds of sodium. It is believed to be an essential constituent of animal food, and its use as a seasoning agent dates to ancient times. It has been estimated that the human body requires about 29 lb of salt a year. "Salt torture," in which the victim was given plenty of food but deprived of salt, was an ancient method of torture. Sodium chloride is important in the proper functioning of the human body, since it is a source material from which the body prepares hydrochloric acid in the gastric juices of the stomach for the digestion of food, and a source of some sodium hydroxide in basic intestinal digestion. It is also present in the blood, and in cases of severe shock and loss of blood, a physiological salt solution (0.8% aqueous solution of sodium chloride) may be introduced into the blood stream.

To conclude the topic of the occurrence of Group IA and IIA metals, below is a list of a few of their other important minerals.

KCl (**sylvite**) and $KCl \cdot MgCl_2 \cdot 6 H_2O$
  (**carnallite**)
$Ca_5F(PO_4)_3 \cdot Ca_5Cl(PO_4)_3$ (**apatite**)
$Ca_3(PO_4)_2$ (**phosphate rock**)
$Mg_3Ca(SiO_3)_4$ (**asbestos**)
$H_2Mg_3(SiO_3)_4$ (**talc**)

$Na_2B_4O_7 \cdot 10 H_2O$ (**borax**)
CaF_2 (**fluorite**)
$CaCO_3 \cdot MgCO_3$ (**dolomite**)
$Mg_2Si_3O_8 \cdot 2 H_2O$ (**meerschaum**)
$Na_2CO_3 \cdot NaHCO_3 \cdot 2 H_2O$ (**trona**)

26.2 CHEMICAL PROPERTIES OF GROUP IA

The atoms of Li, Na, K, Rb, Cs, and Fr are comparatively large; their electronic configuration is that of the preceding noble gas plus a single s^1 electron in the outer shell. Thus, the outer valence electron is easily given up as the metals through chemical reactions become Li^+, Na^+, K^+, Rb^+, Cs^+, Fr^+. The element francium, nearly nonexistant in nature, now prepared in trace quantities of very short-lived radioactive isotopes, will not be further considered in the practical chemistry of the alkali elements.

The atom of cesium, the largest in volume of these large Group IA atoms, has its outer s^1 electron farthest from the nucleus. Thus, cesium should be, and is, the most chemically active metal of this group; it has the lowest ionization potential among the elements (Figure 4.4, page 85). A secondary factor that alters activity is the added energy of hydration of the metal ions if the chemical change is in or related to water. The two smallest ions, Li^+ and Na^+, form hydrates with coordinated water around the cations. Li metal in particular, as a result of the small size of the Li^+ ion and its large hydration energy, is most readily changed to the hydrated 1^+ ion, as shown by its $E°$ value of -3.05 volts — highest of the alkali metals.

Since sodium compounds are used in greater bulk than compounds of the other alkali metals, and since chemical behavior of the others is about identical to that of sodium, we will now concentrate on sodium chemistry. It is a soft metal which may be easily molded into any desired form. When cut with a knife, it reveals a silver-white surface. A small piece of sodium dropped on water reacts vigorously and skitters around on the surface, since the metal's specific gravity is just 0.97; potassium metal with specific gravity of 0.83 reacts similarly.

$$2 \ Na + 2 \ H_2O \rightarrow 2 \ Na^+ + 2 \ OH^- + H_2$$

The metal combines readily with oxygen and, if heated, will burn in air. With the halogens and other nonmetals it combines readily to form salts. The metal amalgamates quickly with mercury to form a liquid alloy, which is useful as a reducing agent, since the action with water is less vigorous than that of pure sodium. Representative chemical equations follow:

$2 \ Na + Cl_2 \rightarrow 2 \ NaCl$ also with F_2, Br_2, I_2
$2 \ Na + O_2 \rightarrow Na_2O_2$ some Na_2O with the peroxide
$2 \ Na + S \rightarrow Na_2S$ also with Se, Te
$2 \ Na + H_2 \rightarrow 2 \ NaH$ considered as $Na^+ \ H^-$ (note negative hydride ion H^-)

Sodium Carbonate and Sodium Bicarbonate

So much Na_2CO_3, commercially called "soda ash" and $NaHCO_3$, called "bicarbonate of soda" or "baking soda," are used in homes and industries that the chemistry of their preparation is briefed here.

Solvay process:

$$CaCO_3 \ (\text{limestone}) \xrightarrow{\Delta} CaO + CO_2(g)$$
$$NH_3 + CO_2 + H_2O \rightarrow NH_4HCO_3 \ (\text{ammonium bicarbonate})$$
$$NH_4HCO_3 + NaCl \rightarrow NaHCO_3(s) + NH_4Cl$$

$$\text{dry } 2 \ NaHCO_3 \xrightarrow{\Delta} Na_2CO_3 + H_2O + CO_2(g)$$
$$2 \ NH_4Cl + CaO \rightarrow CaCl_2 + 2 \ NH_3$$

The materials used are cheap: limestone, salt, and water. Note that the NH_3 used is regenerated. Details of industrial plant procedure as to temperatures, concentration of water–salt solutions, use of vats and furnaces are given in industrial chemistry textbooks.

Electrolytic Process

Treatment of "caustic soda," NaOH, which is a product of the electrolysis of NaCl (page 385) with CO_2 results in the formation of Na_2CO_3 or $NaHCO_3$, depending on the amount of CO_2 added.

$$2\ NaOH + CO_2 \xrightarrow{\ H_2O\ } Na_2CO_3 + H_2O$$

$$NaOH + CO_2 \xrightarrow{\ H_2O\ } NaHCO_3$$

Uses of Sodium Salts

As a result of the hydrolysis of **sodium carbonate**, hydroxyl ions are formed, and consequently sodium carbonate exhibits a distinctly basic reaction in aqueous solution.

$$CO_3^{2-} + H_2O \rightleftarrows OH^- + HCO_3^-$$

The salt neutralizes strong acids, liberating carbon dioxide. When crystallized from aqueous solution, the decahydrate is obtained, $Na_2CO_3 \cdot 10\ H_2O$. This compound is commonly known as **washing soda**. It is an efflorescent salt which decomposes slowly into the monohydrate, $Na_2CO_3 \cdot H_2O$.

The estimated consumption of soda ash (anhydrous sodium carbonate) in the United States in a recent year by industries is shown in Table 26.2. It ranks 10th among the top chemicals.

Table 26.2

Estimated U.S. consumption of soda ash

Consuming Industry	Short Tons	Percent of Total
Glass	2,000,000	29.0
Soap	110,000	1.6
Caustic and bicarbonate	950,000	13.8
Other chemicals	1,600,000	23.5
Cleansers and modified sodas	200,000	3.0
Pulp and paper	420,000	6.1
Water softeners	150,000	2.2
Petroleum refining	50.000	0.7
Textiles	40,000	0.6
Nonferrous metallurgy	650,000	9.6
Exports	210,000	3.0
Miscellaneous	500,000	6.9
Total	6,880,000	100.0

Sodium bicarbonate or **baking soda** reaction with acids is similar to that of soda ash, liberating carbon dioxide and forming a sodium salt of the acid added. An aqueous solution is mildly alkaline, because the bicarbonate ion reacts with water (hydrolysis) to form weak carbonic acid and hydroxyl ions:

$$HCO_3^- + H_2O \rightleftharpoons H_2CO_3 + OH^-$$

Baking soda is often used, as the name implies, in cooking. Its function is to furnish carbon dioxide, which acts as a leavening agent in a moist, warm dough. A weak acid, such as lemon juice, vinegar, or sour milk, added to sodium bicarbonate liberates carbon dioxide.

Baking powders are mixtures of three essential ingredients:

1. Sodium bicarbonate which furnishes carbon dioxide.
2. Some substance which will furnish hydrogen ions to react with the sodium bicarbonate and liberate carbon dioxide when water is added to the baking powder.

$$HCO_3^- + H^+ \rightarrow CO_2(g) + H_2O$$

3. A substance, such as starch, added to prevent intimate contact and reaction between ingredients 1 and 2.

All baking powders are alike in ingredients 1 and 3, but the second ingredient is variable. **Cream of tartar** ($KHC_4H_4O_6$), an acid salt, is a constituent of certain baking powders. In water solution the reaction is

$$NaHCO_3 + KHC_4H_4O_6 \rightarrow NaKC_4H_4O_6 + H_2O + CO_2(g)$$

Phosphate baking powders contain primary calcium acid phosphate, which acts on sodium bicarbonate.

$$2\ NaHCO_3 + Ca(H_2PO_4)_2 \rightarrow CaHPO_4 + Na_2HPO_4 + 2\ CO_2(g) + 2\ H_2O$$

Alum baking powders contain potash alum. Because of the hydrolysis of aluminum sulfate contained in the alum, hydrogen ions are produced which react with the bicarbonate to liberate carbon dioxide. Certain baking powders contain mixtures of alum and calcium acid phosphate.

Sodium hydroxide (NaOH), 6th in the list of 50 top chemicals, is commonly called **caustic soda** or **lye.** Its strongly basic properties are familiar. It is manufactured along with Cl_2 by electrolysis of aqueous solutions of NaCl (page 385).

Solid sodium hydroxide is easily melted and cast in sticks, the form in which it is usually marketed. Major uses include the manufacture of artificial silk, soap, textiles, pulp and paper, petroleum refining, metal processing, and the making of other chemicals.

Sodium sulfate (Na_2SO_4) ranks 32nd in importance among the chemicals. About 70% of its use is in making paper. It is also used in the manufacture of glass and ceramics, as a builder or extender in household detergents, and in the manufacture of mineral-feed supplements, sponges, dyes, bleaches, and photography. It is produced primarily from natural brines, as a by-product of rayon manufacture, and from the reaction of sodium chloride with sulfuric acid.

$$2 \text{ NaCl} + \text{H}_2\text{SO}_4 \xrightarrow{\Delta} \text{Na}_2\text{SO}_4 + 2 \text{ HCl}$$

Other important compounds of sodium with their sources and uses are shown in Table 26.3.

Table 26.3
Compounds of sodium

Compound	Source	Use
NaNO$_3$ (Chile saltpeter)	Natural deposits in Chile	Preparation of HNO$_3$ and KNO$_3$
NaCN	Synthetically produced from soda, C, and N$_2$	Extraction of gold and silver from ores; electroplating
Na$_2$B$_4$O$_7$ (borax)	Occurs native	Washing powder and water softener
Na$_2$S$_2$O$_3$ (thiosulfate, "hypo")	Made from Na$_2$SO$_3$	Photography (fixing agent)
Na$_2$O$_2$	Na + O$_2$	Oxidizing agent

Uses of Other Alkali Compounds

K$_2$CO$_3$ called potash, in wood ashes, source of potassium.
KCl   forerunner for other salts or KOH preparation.
KOH   NaOH used much more; used in preparation of soft soaps.
KBr, KI   used in photographic labs and as medicine.
KNO$_3$ called saltpeter, good oxidizing agent.

Until the beginning of the twentieth century, large amounts of saltpeter were used in the preparation of black gunpowder. This powder was prepared by thoroughly mixing powdered saltpeter, carbon, and sulfur. When ignited, sulfur and carbon are very rapidly oxidized by the potassium nitrate, producing several gaseous products.

$$4 \text{ KNO}_3 + \text{S} + 4 \text{ C} \rightarrow 2 \text{ K}_2\text{O} + 2 \text{ N}_2 + \text{SO}_2 + 4 \text{ CO}_2 + \Delta$$

The rapid change in volume (as gases are formed and expand with heat) results in an explosion. In recent years, black gunpowder has been replaced largely by smokeless powders prepared from organic nitrogen compounds. Some black powder is still employed as a blasting agent in soft-rock mining operations.

Potassium nitrate is a particularly effective fertilizer because it contains both potassium and nitrogen. It is also used in preserving meats.

$KHC_4H_4O_6$	potassium acid tartrate, **cream of tartar,** used in some baking powders; main source is crystalline deposits from grape juice.
$KSbOC_4H_4O_6$	**tartar emetic,** potassium antimonyl tartrate, is used medicinally.
$LiK(Al_2(OH \cdot F))Al(SiO_3)_3$	lepidolite, a complex compound yet a crystalline mined mineral.
Li_2CO_3, $LiOH$	scarlet color to flames, some in ashes of tobacco or sugar beet, are relatively insoluble, as is Li_3PO_4.
$RbCl$, $RbNO_3$	dark red color to burner flame.
$CsCl$, $CsNO_3$	blue color to burner flame.

26.3   CHEMICAL PROPERTIES OF GROUP IIA (alkaline earths)

The size of the alkaline earth metal atoms increases with atomic number. Ra, with the largest atomic volume and lowest ionization potential, gives its outermost two electrons, s^2, most easily in chemical union with electron-receiving atoms. Be, on the other hand, tends somewhat to share its outer two electrons in compounds (the outer s^2 electrons of Be are comparatively near the very small beryllium positive nucleus). As expected the $E°$ values of the metal ions decrease as the atomic numbers increase.

Many compounds of the alkaline earth metals are not very soluble in water, whereas alkali metal compounds are almost all soluble. Water-soluble alkaline earth metal salts tend to form hydrated crystals.

Natural Deposits of Mineable Material

The percentage of the element calcium, 3.63% in earth crust rocks, makes it the fifth most abundant element. Magnesium, 2.09%, is eighth in abundance. The main occurrence of each is in igneous silicate rock as $CaO \cdot SiO_2$ and $MgO \cdot SiO_2$. Geochemical weathering of near-surface portions of earth rock has resulted in Ca^{2+} and Mg^{2+} in ground water. Plants have use of these ions from soil water. Vast tonnage of these ions has reached the sea. Marine life selectively picked Ca^{2+} from sea water which was united to CO_3^{2-} to fashion their $CaCO_3$ skeletal bodies; and with some PO_4^{3-} to fashion $Ca_3(PO_4)_2$. These compounds are very slightly water-soluble. As life on land developed in evolution from marine life, the pattern of a calcium compound bone persisted, tending more to the $Ca_3(PO_4)_2$ composition. As marine life from small organisms to big fishes of bony skeleton died, the $CaCO_3$ dropped to the ocean bottom and by pressure and chemical factors became a limestone rock. $Ca_3(PO_4)_2$, phosphate rock or phosphorite deposits, are existant for mining;

they are of varied purity. The limestone ($CaCO_3$) deposits may have admixed $MgCO_3$ and be called a dolomitic limestone. Some fairly pure $MgCO_3$ (magnesite) deposits are being mined.

In addition to massive bedded limestone deposits some $CaCO_3$ occurs as rhombohedral crystals and is called **calcite.** A **marble** is a limestone that by heat and pressure has been caused to become a more crystalline mass; it may be pure white or striated by colored impurities. **Chalk** is a fine grained $CaCO_3$ material. The **stalactites** and **stalagmites** of various caves and various hot springs deposits are $CaCO_3$ in composition.

Industrial Chemistry of Calcium Compounds

Large quantities of limestone are mined and form raw material for many chemical conversions to desired products. A brief of several processes follows:

Lime, calcium oxide (CaO), also called quick lime, ranks 2nd in importance among the 50 top chemicals, over 36 billion pounds being produced annually in the U.S. Some of its uses are for medicinal purposes, insecticides, soil conditioning, sulfite paper processes, dehairing hides, water softening; manufacturing of soap, rubber, varnish, and bricks; and in making mortar, plastic, and cement. Directly or indirectly lime and limestone ($CaCO_3$) are used in more industries than any other natural material. Some 70% of all stone that is mined or quarried is limestone, from which lime is made by heating.

$$CaCO_3 \overset{\Delta}{\to} CaO + CO_2$$
$$\text{limestone} \qquad \text{lime}$$

This reaction is an equilibrium system and proceeds to the right only as the carbon dioxide is removed. The heating is usually carried out in large furnaces, or **kilns,** at about 900°C. The product, if pure, is white and extremely porous. It is very stable toward heat (melting point 2572°C). Its principal use is in the manufacture of slaked lime (see below), and it is also used in the preparation of bleaching powder, and as a base in the neutralization of acid soils.

Calcium hydroxide forms as calcium oxide is added to water; a vigorous reaction ensues, with the liberation of considerable heat—enough to convert part of the water to steam. The white product is calcium hydroxide (**slaked lime** or **hydrated lime**).

$$CaO + H_2O \to Ca(OH)_2 \quad \text{or} \quad CaO \cdot x\ H_2O$$

Calcium hydroxide is only slightly soluble in water. Because of its basic characteristics it absorbs carbon dioxide from the atmosphere and is in part converted into calcium carbonate. Slaked lime has many uses, chiefly in the preparation of mortar and plaster. **Mortar** consists of a mixture of sand and slaked lime with enough water to form a paste. As the mortar sets some water evaporates, and the $Ca(OH)_2 \cdot x\ H_2O$ solidifies to a firm gel which includes sand particles. After years of time, the outer one-eighth inch or so of mortar is converted to a calcium carbonate-sand mixture because CO_2 from the air is slowly absorbed. Most of the set material seen between bricks is the $Ca(OH)_2 \cdot x\ H_2O$ gel enclosing siliceous granular materials. Plaster is identical with mortar except that hair may be added to help hold the mass together. Plaster of Paris $(CaSO_4 \cdot \frac{1}{2} H_2O)$ may be added to give a smoother plaster in the finishing coat.

Other uses of slaked lime are for milk of lime (a suspension of calcium hydroxide in water), preparation of bleaching powder, and whitewash.

Cement was produced over a century ago by Joseph Aspdin, who applied high heat to some rock near his home in Portland, England, and discovered that if he pulverized the product he had a powder which, with water, would set to a strong mass. The rock became known as Portland stone and eventually as "Portland cement." Portland cement is now prepared from a mixture of limestone and clay. In some cases natural rock formations contain these ingredients in suitable proportions and may be used directly for cement manufacture. About six $CaCO_3$ units to one clay unit is required, and cement plants mix their raw materials in this proportion. The mixture, after being ground to a fine powder, is heated in rotary kilns until the mass just begins to fuse. During the heating, CO_2 is lost from the limestone, and the CaO combines with silica and alumina from the clay to form calcium silicates and calcium aluminates.

$$6\ CaCO_3 \rightarrow 6\ CaO + 6\ CO_2$$
$$Al_2O_3 \cdot 2\ SiO_2 \cdot 2\ H_2O + 6\ CaO \rightarrow 2\ CaO \cdot Al_2O_3 + 2\ (2\ CaO \cdot SiO_2) + 2\ H_2O$$

The resulting mass, known as "clinker," is ground finely, and a small amount of gypsum is added to retard the rate of setting. When a paste of cement and water is allowed to stand, the material sets to a hard, durable, rocklike mass. Because of the insolubility of the constituents, this process occurs even if the material is under water. The mechanism of the setting process is somewhat obscure, but it is known that hydration of the constituents takes place, so that the mass is securely bound together by interlacing crystals of hydrates of CaO and Al_2O_3.

Cement is never used alone; it is mixed with sand and gravel, and the resulting set mixture is called **concrete.** Concrete which is to be subjected to considerable strain can be reinforced with steel rods or wire netting.

When the mineral gypsum $(CaSO_4 \cdot 2\ H_2O)$ is heated carefully below 125°C, it loses three-fourths of its water of hydration and forms the hemihydrate of calcium sulfate, often written $CaSO_4 \cdot \frac{1}{2} H_2O$ and commonly known as plaster of Paris:

$$2\ CaSO_4 \cdot 2\ H_2O \rightleftharpoons (CaSO_4)_2 \cdot H_2O + 3\ H_2O$$

When the hemihydrate is mixed with water to form a paste, this reaction is reversed and, on setting, a hard mass of interlacing crystals of gypsum is formed. Plaster of Paris is used in making plaster casts for holding broken bones and in making statuary casts. If the gypsum is heated too strongly, the anhydrous salt, which hydrates only slowly, is formed. This property makes the anhydrous salt useless as plaster.

Gypsum is used as fireproofing for building materials, as a soil fertilizer, and for many other purposes. About 2% gypsum is added to Portland cement to retard and regulate the setting process.

In an important process of manufacturing pulp from wood, a water solution of calcium acid sulfite [$Ca(HSO_3)_2$] is used. Wood chips are composed largely of cellulose, lignin, and resins, and a $Ca(HSO_3)_2$ solution dissolves lignin and resin from the cellulose pulp. The solution of lignin and resins issuing from pulp mills is called "waste sulfite liquor." The calcium bisulfite solution is made as follows:

$$S + O_2 \rightarrow SO_2$$
$$SO_2 + H_2O \rightarrow H_2SO_3$$
$$CaCO_3 + H_2SO_3 \rightarrow CaSO_3 + CO_2 + H_2O$$
$$CaSO_3 + H_2SO_3 \rightarrow Ca(HSO_3)_2$$

In some mills $Mg(HSO_3)_2$ or NH_4HSO_3 is used instead of $Ca(HSO_3)_2$.
Other important compounds of calcium are given in Table 26.4.

Table 26.4
Other compounds of calcium

Salt	Formula	Source	Uses
Calcium chloride	$CaCl_2$	By-product of Solvay soda process	Drying agent; lay dust on roads
Calcium cyanamide	$CaNCN$	Calcium carbide + nitrogen	Prep. of ammonia; fertilizer
Bleaching powder	$CaOCl_2$	Lime and chlorine	Bleaching
Calcium carbide	CaC_2	Lime and carbon	Source of acetylene
Calcium phosphate	$Ca_3(PO_4)_2$	Occurs as phosphate rock	Making fertilizer
Superphosphate	$Ca(H_2PO_4)_2$ + 2 $CaSO_4$	Rock phosphate + 2 H_2SO_4	Fertilizer
Calcium silicate	$CaSiO_3$	Limestone + sand + Δ	In glass along with Na_2SiO_3
Calcium carbonate	$CaCO_3$	Limestone	Flux in metallurgy

Hard Waters

Most natural waters, with the exception of rain water, contain dissolved mineral matter. This mineral matter is derived from the rocks and soil and contains calcium, magnesium, or iron salts as chlorides, sulfates, or bicarbonates. Waters containing any of these salts are objectionable for certain household and industrial uses and are termed **hard** waters. If soap is added to hard water, a precipitate forms before a lather can be obtained. This residue may stain cloth with which it comes in contact and, of course, the soap is wasted. Soap is a sodium or potassium salt of stearic, palmitic, or oleic acids (page 703). If calcium ions are present in water, the reaction with soap is

$$Ca^{2+} + C_{17}H_{35}COONa \rightarrow (C_{17}H_{35}COO)_2Ca(s) + 2\ Na^+$$

Waters containing bicarbonates of calcium, magnesium, or iron are said to possess **temporary hardness** or **carbonate hardness,** since the hardness may be removed simply by boiling the water. Thus when calcium bicarbonate is heated, the following reaction takes place:

$$Ca(HCO_3)_2 \rightarrow CaCO_3(s) + H_2O + CO_2$$

Calcium carbonate precipitates and, since the calcium ions are removed, the water will no longer react with soap. A water from which the ions responsible for the precipitation of soap are removed is termed **soft** water. Decomposition of bicarbonates is the cause of scale formation in tea kettles in which considerable water has been boiled. Water containing bicarbonates is also objectionable for industrial uses because it will form boiler-scale deposits (carbonates of calcium or magnesium) on the inside of boiler pipes. Since the scale is a poor conductor of heat, boiler efficiency is reduced.

Temporary hardness may be removed by adding the proper amount of slaked lime [$Ca(OH)_2$], household ammonia (NH_3), or soda (Na_2CO_3).

$$Ca(HCO_3)_2 + Ca(OH)_2 \rightarrow 2\ CaCO_3(s) + 2\ H_2O$$
$$Ca(HCO_3)_2 + 2\ NH_3 \rightarrow CaCO_3(s) + (NH_4)_2CO_3$$
$$Ca(HCO_3)_2 + Na_2CO_3 \rightarrow CaCO_3(s) + 2\ NaHCO_3$$

This last equation explains the use of washing soda for laundry purposes. Na_2CO_3 is present in a high percentage in most detergent powders.

Permanent hardness or **noncarbonate hardness** in water results from chlorides or sulfates of metals, such as calcium or magnesium. Boiling does not remove these salts; a substance called a **water softener** is often added to the water to form a precipitate with the salts. Common water softeners are soda (Na_2CO_3) and borax ($Na_2B_4O_7$). Either of these will precipitate calcium or magnesium carbonate or borate from solution as an insoluble substance. Typical reactions of these water softeners are

$$Ca^{2+} + CO_3^{2-} \rightarrow CaCO_3(s)$$
$$Mg^{2+} + B_4O_7^{2-} \rightarrow MgB_4O_7(s)$$

In some cases both soda and slaked lime are used to soften water—the slaked lime to remove temporary hardness and the soda to remove permanent hardness.

Water may be softened by adding a complexiny agent to it, such as sodium tripolyphosphate ($Na_5P_3O_{10}$, number 42 in the list of top chemicals), used in many detergents. The tripolyphosphate ion

$$\left[\begin{array}{ccc} O & O & O \\ \| & \| & \| \\ O-P-O-P-O-P-O \\ | & | & | \\ O & O & O \end{array} \right]^{5-}$$

reacts with Ca^{2+}, Mg^{2+} or other polyvalent cations in hard water to tie them up in soluble phosphate complexes. Sodium hexametaphosphate $[(NaPO_3)_6]$, called Calgon, behaves in a similar manner.

Another method of softening water utilizes natural or artificial **zeolites**, minerals that contain sodium aluminum silicate. For example, the mineral **natrolite** is $Na_2[Si_3Al_2O_{10}] \cdot 2\,H_2O$. The zeolite crystal structures show rather porous open three-dimensional frameworks of aluminosilicates. Na^+ (and other cations) are free to move within the lattice framework and exchange with other cations. Figure 26.1 shows the crystal structure of a highly ordered synthetic zeolite. If water is allowed to filter through this coarse-grained material, sodium ions are displaced by calcium and magnesium ions (this is called **ion exchange**).

$$Na_2 Zeolite + Ca^{2+} \rightleftarrows Ca\,Zeolite + 2\,Na^+$$

Sodium ions left in solution are not objectionable, since they do not make the water hard. The sodium zeolite may be regenerated in a water softener with a concentrated sodium chloride solution. By mass action the above reaction is reversed and the original zeolite re-formed.

$$Ca\,Zeolite + 2\,NaCl \rightleftarrows CaCl_2 + Na_2 Zeolite$$

An artificial zeolite called Permutit is also effective as a water softener. Figure 26.2 shows the arrangement of a typical system using Permutit.

Organic Ion Exchange Resins

Water of high purity which has both the objectional cations and anions removed is called **deionized** or demineralized water and is often used in place of distilled water in chemical laboratories and for small electrical appliances such as steam irons.

The **cation exchanger** is an organic polymer, such as Dowex 50, which consists of beads of styrene-divinyl benzene resins which have been sulfonated. Its formula may be represented by RSO_3H. It is actually a strong acid in solid form and reacts readily with cations, exchanging the hydrogen ions for them. For example,

$$2\,RSO_3H + CaCl_2 \rightleftarrows (RSO_3)_2Ca + 2\,HCl$$

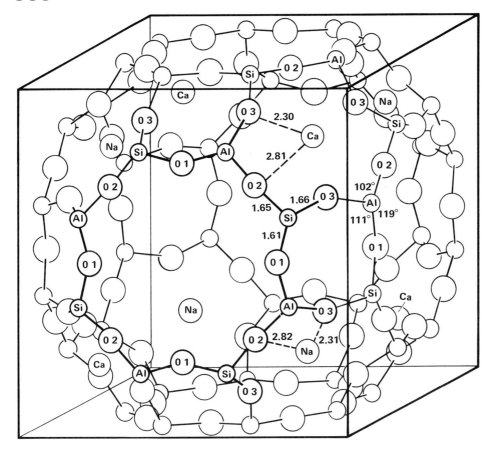

Figure 26.1

Crystal structure of a synthetic zeolite. One cubic unit cell is shown. [From K. Seff and D. P. Shoemaker, *Acta Cryst.*, **22**, 162 (1967).]

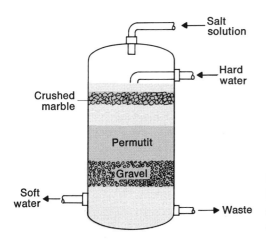

Figure 26.2

Permutit water softener. Salt solution enters and waste flows out only during regeneration.

The resin is regenerated by reversing the reaction with a strong acid, usually H_2SO_4. **Anion exchangers** are generally quaternary ammonium salts incor-

porated into resins. Dowex 1, for example, is R—⟨○⟩—$CH_2N(CH_3)_3$ OH

and is an example of a strong-base resin. Anions will react with such resins, exchanging with the OH^- ions. For example,

$$RN(CH_3)_3OH + HCl \rightleftarrows RN(CH_3)_3Cl + H_2O$$

Regeneration is accomplished with NaOH. Practically any water, even sea water, that is run first through the cation exchanger, and then through the anion exchanger will emerge demineralized (anions and cations removed, except for H^+ and OH^-), and essentially pure water will result.

26.4 MAGNESIUM, STRONTIUM, AND BARIUM

Little discussion of these three elements is necessary, since they resemble calcium very closely in both chemical and physical properties. Table 26.5 lists important compounds of them and their uses.

Magnesium is silvery white, tough, strong, and the lightest of the rigid metals. Its specific gravity is 1.74. It is produced by electrolysis of its melted salt, $MgCl_2$. Malleable and ductile, it may be rolled into sheets or drawn into wire. Magnesium is high in the activity series, though it is somewhat less active than calcium, which appears just above it in the series. It reacts slowly with hot water to liberate hydrogen and reacts readily with acids (except hydrofluoric) to produce hydrogen. It is stable in dry air but corrodes slowly in moist air to form a basic carbonate. In a finely divided state or in the form of a thin ribbon, the metal burns vigorously in air to form both magnesium oxide and magnesium nitride. During combustion a bright white light rich in short wavelengths of light is produced; thus the metal, mixed with an oxidizing agent such as potassium chlorate, is used in photographic flash powders. It is also used in military pyrotechnics for signals, flares, and incendiaries. In large pieces the metal does not ignite readily and may be forged and worked at high temperatures.

Alloyed with aluminum, magnesium forms a light, strong, durable metal for aircraft construction. The alloys with aluminum are easily machined and worked, and are resistant to corrosion. Because it has many advantages for aircraft construction, the production of magnesium has increased tremendously in the past few years. **Magnalium,** an alloy of aluminum containing 5 to 30% magnesium, and **duralumin** (Mg 5%, Cu 3 to 5%, and Al) are the more important alloys.

There is little practical use for metallic strontium, barium, or radium. Radium will be discussed separately in Chapter 35 on radioactivity.

Table 26.5

Compounds of magnesium, strontium, barium

Compound	Formula	Source	Uses
Magnesium chloride hexahydrate	$MgCl_2 \cdot 6\ H_2O$	Salt brines in Michigan; sea water; $MgCO_3 + HCl$	With MgO and H_2O makes a dental cement; production of electrolytic Mg. Objectionable in table salt, since it deliquesces.
Magnesite mineral	$MgCO_3$	Mined, particularly in Washington	Solution in acids gives salts; Mg production
Basic magnesium carbonate (magnesia alba)	$MgCO_3 \cdot Mg(OH)_2$	$Mg^{2+} + CO_3^{2-} + OH^-$	Tooth powders and pastes; polishing powders; filler
Magnesium bicarbonate	$Mg(HCO_3)_2$	Minor impurity in hard water; $MgCO_3 + H_2CO_3$	Causes soap to precipitate, and boiler scale if in water
Magnesia	MgO	$MgCO_3 + heat$	High refractory brick; suspended in H_2O as milk of magnesia for hyperacidity or mild laxative
Epsom salts	$MgSO_4 \cdot 7\ H_2O$	Certain mineral waters; $MgCO_3 + H_2SO_4$	Medicinal purgative; fireproofing agent; weighting cotton
Magnesium ammonium phosphate	$MgNH_4PO_4$	$Mg^{2+} + NH_3 + Na_2HPO_4$	Analysis for Mg; when ignited is converted to magnesium pyrophosphate, $Mg_2P_2O_7$
Strontium nitrate	$Sr(NO_3)_2$	From $SrCO_3$ or $SrSO_4$ mineral	Gives red color to flame, flares, fireworks, signal lights
Barium nitrate	$Ba(NO_3)_2$	From $BaSO_4$ barite mineral	Gives green color to flame
Barium sulfate	$BaSO_4$	Mined as such	Heavy filler for some rubber, oil well 'drill mud', in 'lithopone' paint

26.5 BERYLLIUM

The element beryllium is present in many natural silicates, the most important of which is **beryl**, an aluminum silicate with the composition $(BeO)_3 \cdot Al_2O_3 \cdot 6 SiO_2$. **Emerald** is a green, transparent form of beryl containing traces of chromium. Beryllium may be obtained by electrolytic methods, but this is difficult because of the high melting point (1350°C). Thus the price of beryllium has remained high—several dollars a pound. It alloys readily with aluminum, to which it adds strength and resistance to corrosion. Beryllium is only slightly denser than magnesium, so that if it could be prepared cheaply it might find extensive use as a light alloy for structural purposes.

Beryllium was at one time called "glucinium" because of the sweet taste of $BeCl_2$. However, care must be exercised in handling beryllium compounds because of their poisonous nature.

Exercises

26.1. Examine the physical properties of the alkali metals and alkaline earths as given in Table 26.1. (a) What is outstanding about (1) Li, (2) Ba? (b) Explain the outstanding properties of Li.

26.2. For the following important industrial chemicals, over a billion pounds per year of each of which is produced in the U.S., give (1) the formula or name, (2) a method of preparation, (3) one of the most important uses: (a) soda ash, (b) caustic soda (lye), (c) Na_2SO_4, (d) lime, (e) $CaCl_2$.

26.3. Write formulas for:
(a) borax, (b) fluorite, (c) sodium peroxide, (d) calcium hydride, (e) limestone, (f) baking soda, (g) saltpeter, (h) slaked lime, (i) magnesia, (j) barium sulfate, (k) calcium carbide, (l) lithium carbonate, (m) gypsum.

26.4. Name the following compounds (either chemical or common name): (a) $Na_2CO_3 \cdot 10H_2O$, (b) $KHC_4H_4O_6$, (c) $Ca(H_2PO_4)_2$, (d) $Na_2S_2O_3$, (e) $NaCN$, (f) $CaSO_4 \cdot \frac{1}{2}H_2O$, (g) $Ca(HSO_3)_2$, (h) $C_{17}H_{35}COONa$, (i) $Na_5P_3O_{10}$, (j) $Na_2[Si_3Al_2O_{10}] \cdot 2H_2O$, (k) $Sr(NO_3)_2$.

26.5. What is the importance of each of the above compounds listed in question 26.4?

26.6. Use your knowledge of chemistry to explain simply how you could distinguish between: (a) LiCl and NaCl, (b) CaH_2 and $CaCl_2$, (c) NaOH and $Ba(OH)_2$, (d) $CaCO_3$ and $Ca(HSO_3)_2$, (e) $NaNO_3$ and $Na_2S_2O_3$, (f) Li_2CO_3 and CsCl, (g) $RaSO_4$ and $BeSO_4$, (h) $CaCO_3$ and $Ca(HCO_3)_2$.

26.7. Write chemical equations for the following preparations:
(a) NaOH from the electrolysis of NaCl solution.

(b) $NaHCO_3$ from $NaOH$ and CO_2.
(c) Slaked lime from quick lime.
(d) Li metal from $LiCl$.

26.8. Complete and balance the following equations:
(a) $NaHCO_3 + KHC_4H_4O_6 \rightarrow NaKC_4H_4O_6 +$ _____
(b) $CaCO_3 + H_2O + CO_2 \rightarrow$ _____
(c) $CaO + H_2O \rightarrow$ _____
(d) $CaCl_2 + Na_2CO_3 \rightarrow$ _____ + _____
(e) $K + H_2O \rightarrow$ _____ + _____
(f) $Ca(HCO_3)_2$ heat _____ + _____ + _____
(g) $C_{17}H_{35}COONa + Ca^{2+} \rightarrow$ _____ + _____

26.9. Indicate which of the above reactions:
(a) is involved in the action of baking powder.
(b) is the reason why soap does not lather in hard water.
(c) will soften temporary hard water.
(d) accounts for the presence of Ca^{2+} in water supplies.
(e) could be used in softening permanently hard water.
(f) is used in "slaking lime."

26.10. Write an equation for each reaction that occurs between Na and the following:
(a) H_2, (b) H^+, (c) O_2, (d) Cl_2, (e) S, (f) P_4, (g) H_2O.

26.11. Give the formula for a calcium compound used for each of the following:
(a) production of window glass, (b) preparation of C_2H_2, (c) superphosphate fertilizer, (d) bleaching powder.

26.12. From the alkaline earth family of metals, $_{38}Sr$ is typical.
(a) Write equations for its action with warm water; its union with sulfur.
(b) Write balanced equations for strontium hydroxide + HNO_3; strontium sulfide + dilute H_2SO_4.

26.13. Write equations to show how $Ca(HCO_3)_2$ is involved in (a) stalagmite formation in caves, (b) formation of scale in tubes or pipes carrying hot water, (c) the reaction with soap $(C_{17}H_{35}COONa)$, (d) reaction with Na_2CO_3 in water softening.

26.14. NaCl obtained from sea water contains small amounts of $MgCl_2$ and $CaCl_2$, which tend to deliquesce and cause the salt to "cake." What would be the effect of bubbling CO_2 into a concentrated brine of the three salts before crystallization?

26.15. Explain briefly the approximate composition of: (a) baking powders, (b) mortar, (c) Portland cement, (d) hard water, (e) zeolites, (f) an organic cation exchange resin.

26.16. An equation is given in the text for a cream of tartar baking powder in action in hot water. From the use of 10.0 g of $NaHCO_3$ in such baking powder, what volume of CO_2 should be liberated through the dough at $273°C$?

26.17. Outline briefly:
  (a) the action of a baking powder.
  (b) the hardening of cement in concrete.
  (c) the Solvay process.
  (d) the softening of a hard water containing $CaSO_4$ by (1) chemical precipitation, (2) use of sodium tripolyphosphate, (3) use of a zeolite, (4) use of organic ion exchange resins.
  (e) how a zeolite ion exchange resin is regenerated.
  (f) why hard water is objectionable.

26.18. A red signal flare for use on railroads is made by mixing strontium nitrate, carbon, and sulfur in the proportions shown by the equation

$$2 \ Sr(NO_3)_2 + 3 \ C + 2 \ S \rightarrow$$

(a) What would you predict the products to be? (b) Calculate the weight of $N_2(g)$ which would be produced from 1.00 kg of $Sr(NO_3)_2$.

26.19. Lithopone white paint is made by heating a mixture of BaS and $ZnSO_4$ according to the equation

$$BaS + ZnSO_4 \rightarrow BaSO_4 + ZnS$$

What is the maximum weight of ZnS which could be produced by reacting together 1.00 kg each of BaS and $ZnSO_4$?

26.20. Calculate the quantity of soda ash necessary to soften 1000 l of water (sp. gr. 1.00) containing 90.0 mg of $Ca(HCO_3)_2$ and 25.0 mg of $MgSO_4$ per l.

26.21. A seller of superphosphate fertilizer [$Ca(H_2PO_4)_2 + 2 \ CaSO_4$] was suspected of diluting his product with fine sand. Analysis of a one-gram sample of the fertilizer yielded 0.35 g of $Mg_2P_2O_7$. (a) If the superphosphate was 100% pure, what weight of $Mg_2P_2O_7$ would be obtained? (b) What is the apparent percent purity of the sample?

26.22. One gram of a mixture of NaCl and LiCl is analyzed for chloride by precipitation of AgCl. A 2.815-g quantity of AgCl is obtained. Calculate the percentage of LiCl in the original mixture of chlorides.

26.23. Six cubic meters of sea water of density about 1.0 g/ml weigh how much in grams? At 3.5% dissolved salt, 90% of which is NaCl, what weight of NaCl in grams does the sea water contain?

26.24. A bath salt is a mixture of NaCl and $Na_2CO_3 \cdot 10 \ H_2O$. A one-gram sample is dissolved in water and neutralized with 19.7 ml of 0.156 N HCl. What is the weight of $Na_2CO_3 \cdot 10 \ H_2O$ in the sample? (Assume that the HCl reacts with all of the Na_2CO_3 in the sample.)

26.25. A cleaner is known to be a mixture of NaOH and SiO_2. A 0.500-g sample of the cleaner requires 36.7 ml of 0.105 N H_2SO_4 to neutralize it. What is the percentage of NaOH present?

Supplementary Readings

26.1. "Lithium and Mental Health," M. T. Doig III, M. G. Heyl, and D. R. Martin, *J. Chem. Educ.*, **50**, 343 (1973).

26.2. "How an Eggshell is Made," T. G. Taylor, *Sci. American*, March 1970; p. 89.

26.3. "How is Muscle Turned On and Off?," G. Hoyle, *Sci. American*, April 1970; p. 85.

26.4. "The Chemistry of Concrete," S. Brunauer and L. E. Copeland, *Sci. American*, April 1964; p. 80.

Chapter 27

Main Energy Levels	1	2	3	4	5	6
$^{10.81}_{5}\text{B}$	K	s^2p^1				
$^{26.98}_{13}\text{Al}$	K	L	s^2p^1			
$^{72.59}_{32}\text{Ge}$	K	L	M	s^2p^2		
$^{118.69}_{50}\text{Sn}$	K	L	M	$s^2p^6d^{10}$	s^2p^2	
$^{207.2}_{82}\text{Pb}$	K	L	M	N	$s^2p^6d^{10}$	s^2p^2

Group IIIA and Group IVA Metals

The guide to the chemistry of aluminum and boron is, of course, the common s^2p^1 outer electron configuration. Although only one of the electrons is unpaired, the three electrons behave similarly to each other for an invariable oxidation state of three for these elements. Boron bonding tends to be more covalent than ionic. Aluminum may form both covalent and electrovalent bonds. Gallium, indium and thallium, also in Group IIIA, are very rare, and extraction from earth rocks is very costly; thus they are omitted from further discussion. All the elements in this group are similar to aluminum in many respects—for example, all exhibit an oxidation state of +3. Boron, the first element of the group, is distinctly nonmetallic. Aluminum is amphoteric, and the rest exhibit metallic characteristics.

Germanium, tin (stannum), and lead (plumbum) may exhibit +2 or +4 oxidation states, in keeping with their s^2p^2 valence electrons. Both lead and tin may give up their two unpaired p electrons to form Pb^{2+} and Sn^{2+}. However, in the +4 oxidation state the elements tend toward a more covalent bond character, but of a polar nature. The hydroxides of both elements are amphoteric and form the complex ions PbO_2^{2-} (plumbite), PbO_3^{2-} (plumbate), SnO_2^{2-} (stannite), and SnO_3^{2-} (stannate). Table 27.1 shows some properties of B, Al, Ge, Sn, and Pb.

Carbon and silicon are Group IVA elements, but they do not act as metals and have been covered in previous chapters.

Table 27.1
Some properties of the IIIA, IVA elements

Element	Symbol	Sp. Gr.	Melting Point °C	Boiling Point °C	Atomic Radius (Å)	E°
Boron	B	2.35	2300	2550	0.8	−0.87 (from H_3BO_3)
Aluminum	Al	2.71	660	2467	1.25	−1.66 (from Al^{3+})
Germanium	Ge	5.3	937	2830	1.22	−0.1 (from GeO_2)
Tin	Sn	7.3	232	2270	1.41	−0.14 (from Sn^{2+})
Lead	Pb	11.3	328	1744	1.54	−0.13 (from Pb^{2+})

Occurrence

Aluminum is the third most abundant element in the earth's crust (averages 8.13%). Thus, there is tremendous tonnage of the element as Al_2O_3 (average 15.4%) in earth rock. Most of this oxide occurrence, however, is in covalent union with other metal oxides and SiO_2. For example, the common rock granite is more than one-third feldspar, a mixture of $K_2O \cdot Al_2O_3 \cdot 6\ SiO_2$, and $Na_2O \cdot Al_2O_3 \cdot 6\ SiO_2$. Many other metal alumina silicates constitute a vast bulk of rock. Weathering of these rocks over ages of time have altered parts of them to now existant deposits of clay, $(H_2O)_2 \cdot Al_2O_3 \cdot 2\ SiO_2$, also called **kaolin**; and on to **bauxite**, $Al_2O_3 \cdot 2\ H_2O$. When the available fairly pure deposits of bauxite are nearly mined out for aluminum production, industry will begin to use ferrugenous bauxite or clay for aluminum production. The unusual mineral, **cryolite**, Na_3AlF_6, is needed in current electrometallurgy of aluminum; the mining of deposits of it in Greenland is important to that country.

Boron is a relatively rare element (about 0.001% of the earth's crust). It does not occur in the free state, but its compounds are found in many areas. Boric acid (H_3BO_3) occurs naturally in certain spring waters. Borax ($Na_2B_4O_7 \cdot$ 10 H_2O) is found in certain dry lake areas of California. The most abundant boron compound is colemanite ($Ca_2B_6O_{11} \cdot 5$ H_2O), deposits of which are found in California and Nevada.

Phoenician ships sailed through the Mediterranean a few centuries ago to obtain tin, which was then melted with copper to give the alloy **bronze**. This alloy is of lower melting temperature than copper and more easily casts into molds. The production of bronze statuary required considerable use of tin during the bronze age. As the English tin ore miners had to dig deeper and even tunnel under the ocean, a water pumping problem led to Watt's invention of the steam engine.

Various estimates place the percentage of lead in the earth's crust at about 0.0005, much smaller than the percentage of such elements as zirconium, cerium, and vanadium, which we usually think of as being rare elements. There has been a fortunate concentration of the limited earth rock occurrence of lead as geochemical precipitation of PbS, **galena**, in earth cracks.

Metallurgy of Aluminum

The availability of very active potassium led Wöhler in 1827 to the first successful production of aluminum.

$$AlCl_3 + 3 \text{ K} \rightarrow Al + 3 \text{ KCl}$$

The cost of the method prohibited its production in any quantity, however, and for many years the metal was only a curiosity. It was recognized that the element would be extremely useful if it could be produced economically, and many attempts were made to develop cheaper processes of manufacture. The substitution of metallic sodium for potassium reduced the cost from $160 to $10/lb. The development of electrolytic methods for the preparation of sodium reduced the cost further, to $4/lb. In 1886 an American, Charles Martin Hall, when a student at Oberlin College, found that electrolysis of aluminum oxide could be effected if the oxide were dissolved in melted cryolite (Na_3AlF_6), a double fluoride salt found in Greenland. When Hall's process was industrialized, the price of aluminum dropped to about thirty cents a pound.

The process for the preparation of aluminum today is essentially the same as that developed by Hall in 1886. Purified dehydrated bauxite is dissolved in molten Na_3AlF_6, which acts as a solvent for the Al_2O_3. In modern practice, AlF_3 and CaF_2 are added to lower the melting point and increase fluidity. The mixture, when melted, is a good conductor of electric current, and electrolysis results in the decomposition of the Al_2O_3 to form metallic aluminum and oxygen.

The electrolysis is carried out in carbon-lined vats, which act as cathodes. The anodes are carbon rods (Figure 27.1) which dip into the electrolyte. The principal costs of this method are the electric energy required and the treatment of the bauxite ore to prepare pure Al_2O_3.

The very high melting temperature of boron attests to covalent bonding between atoms of very small atomic radii. The high reduction potential of aluminum has been referred to; aluminum metal cannot be electrodeposited from water solution; a scum of $Al_2O_3 \cdot H_2O$ forms on the cathode. (If aluminum ions are discharged at a cathode, the aluminum atoms act with water to form a hydrated oxide.) A piece of aluminum metal in water is said to have its surface atoms converted to a tenacious $Al_2O_3 \cdot x\ H_2O$ film, which protects internal aluminum from water interaction. One is always looking at a piece of aluminum, then, through an oxide film. When an aluminum-mercury amalgam is put in water, aluminum atoms migrate to the surface of the liquid mercury and act with water to form a puffy hydrated oxide mass and hydrogen gas.

For wires of equal cross section, copper is a better conductor than aluminum, but on a pound-for-pound basis, aluminum is the better conductor. Aluminum foil is used as a wrapping for soaps, food products, and tobacco. Aluminum powder is a constituent of "silver" paint and is used also in the reduction of certain metal oxides (see thermite process). Much aluminum metal is used in the manufacture of alloys, since these possess greater tensile

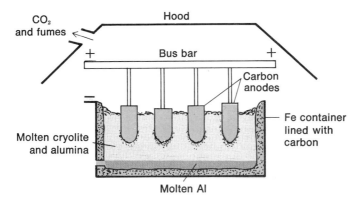

Figure 27.1
Preparation of aluminum by electrolysis (Hall process).

strength and are more easily machined than the pure metal. Among the more important alloys of aluminum are **duralumin** (Al with small amounts of Cu, Mg, and Mn) and **magnalium** (Al with a small amount of Mg), both of which are used in the construction of crankcases for automobiles and other automotive parts.

27.1 CHEMISTRY OF BORON

Boron exhibits a characteristic oxidation state of $+3$, but, because it is much more nonmetallic than aluminum, its acid-forming properties are more pronounced. Borax, $Na_2B_4O_7$, is probably the most important compound of boron. Large amounts are used in water softeners and washing powders; smaller amounts are used in making certain kinds of glass and in forming glazes or enamels for both metal and pottery. Borax in a molten condition will dissolve many metallic oxides, with the formation of glass of a characteristic color. This is the basis for borax tests for such metals as cobalt, nickel, manganese, chromium, and iron, as well as the use of borax for cleaning metallic surfaces coated with oxide deposits.

Boric acid, H_3BO_3, because of its weakly acidic character and its mild antiseptic action, is used as an eyewash. It is sparingly soluble in cold water, and is precipitated on the addition of an acid to a solution of borax. Equations representative of the above reactions are as follows:

$$Ca^{2+} + B_4O_7{}^{2-} \rightarrow CaB_4O_7(s) \text{ (in water softening)}$$
$$Na_2O \cdot 2\ B_2O_3 + CuO \rightarrow CuO \cdot B_2O_3 + Na_2O \cdot B_2O_3$$
$$\text{(borax bead test)}$$

$$Na_2B_4O_7 + H_2SO_4 \rightarrow Na_2SO_4 + H_2B_4O_7$$
$$H_2B_4O_7 + 5\ H_2O \rightarrow 4\ H_3BO_3 \text{ (boric acid)}$$

A borate may readily be tested for by the addition of sulfuric acid and ethyl alcohol. A volatile compound, **ethyl borate** $[(C_2H_5)_3BO_3]$, is produced which burns with a characteristic green-colored flame when ignited.

Boron Fuels

Boron is like the element carbon in its capacity to form such compounds with hydrogen as B_2H_6, diborane; B_4H_{10}, tetraborane; B_6H_{10}, hexaborane; and $B_{10}H_{14}$, decaborane. Diborane is a gas and decaborane is a white solid; intermediate boranes are liquids at standard conditions.

These compounds are potential fuels for rockets and missiles, since they burn with evolution of a greater quantity of heat than gasoline. Boron hydride fuels must be kept free of water, with which they react to form boric acid.

27.2 CHEMISTRY OF ALUMINUM

At high temperatures aluminum metal burns vigorously; the aluminum sheeting of airplanes can burn completely after being ignited. Hydrochloric and sulfuric acids readily dissolve aluminum to form hydrogen, but nitric acid attacks the metal only slightly. The oxidizing nature of nitric acid helps maintain a tenacious oxide film on the aluminum, and action of this acid on the metal ceases. Salesmen can safely put concentrated HNO_3 in an aluminum-ware pot, but H_2SO_4 and HCl can make an aluminumware pot a sieve. Aluminum dissolves in sodium or potassium hydroxides, with the formation of an aluminate salt and hydrogen, a behavior characteristic of amphoteric metals.

If aluminum powder is mixed with certain powdered metallic oxides and the mixture ignited, aluminum takes the oxygen from the metallic oxide and leaves the metal in the free state. At the same time, a great quantity of heat and light energy is released. Advantage is taken of the affinity of aluminum for oxygen in a mixture called **thermite**, developed by Goldschmidt. Thermite is a mixture of iron oxide and aluminum, both in granular form. When the mixture is ignited by means of magnesium ribbon, a very vigorous reaction takes place.

$$Fe_2O_3 + 2\ Al \rightarrow Al_2O_3 + 2\ Fe \qquad \Delta H = -18.4\ kcal$$

The heat energy produced in the reaction raises the temperature of the mixture to approximately 3000°C, sufficient to melt both the aluminum oxide and the iron. The thermite mixture is a convenient source of liquid iron for welding broken rails or steel shafts. It is placed in a magnesia crucible, as shown in Figure 27.2, directly above the object to be welded. The molten iron produced in the reaction is allowed to flow over the broken part and welds the pieces together.

Aluminum oxide is found in nature in several forms. **Corundum** and

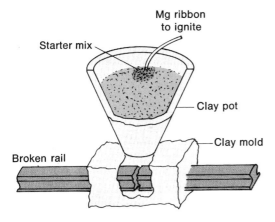

Figure 27.2
Thermite welding.

emery, naturally occurring crystalline forms of the oxide, are used extensively as abrasives in grinding and polishing.

Certain transparent crystalline forms of the oxide are prized gems. **Sapphire** is aluminum oxide containing small amounts of iron oxide and titanium oxide, while **ruby** contains a small amount of chromium oxide. Synthetic rubies and sapphires are now produced by mixing aluminum oxide with the proper ingredients to give the desired color, and passing the mixture through an oxyhydrogen flame. After crystallization, the solid product is cut and polished. These synthetic gems possess the exact composition of the natural product.

Aluminum hydroxide [$Al(OH)_3$] is formed as a white flocculent precipitate when ammonium hydroxide is added to the solution of an aluminum salt. The hydroxide is amphoteric, since it dissolves in both strong bases and acids.

If aluminum hydroxide, $Al(OH)_3$, is precipitated in a solution containing an organic dye, the hydroxide adsorbs the color and removes it from solution. The adsorptive property of aluminum hydroxide is used in the dye industry. Certain fabrics and materials will not adsorb a dye directly, and the addition of a **mordant** — something that will fix the dye to the cloth — is necessary. Aluminum hydroxide acts in this capacity if precipitated within the fibers of the cloth.

Aluminum sulfate, the cheapest of the soluble salts of aluminum, is prepared by the action of sulfuric acid on clay or bauxite. The salt crystallizes from its solutions as the hydrate $Al_2(SO_4)_3 \cdot 18\ H_2O$. It is used in the dyeing industry as a mordant, in sizing paper, in the purification of water, and in the manufacture of other aluminum compounds. It is 34th in the list of the most important chemicals.

If equimolecular quantities of potassium sulfate and aluminum sulfate are dissolved in water and the solution allowed to evaporate, crystals of $K_2SO_4 \cdot Al_2(SO_4)_3 \cdot 24\ H_2O$, which is called ordinary alum or **potash alum,** are deposited. The term **alum,** however, does not necessarily refer to a specific substance but is applied to all double sulfate salts which have the general formula

$$M_2'SO_4 \cdot M_2'''(SO_4)_3 \cdot 24\ H_2O$$

where M' may be Na^+, K^+, or NH_4^+, and M''' may be Fe^{3+}, Cr^{3+}, or Al^{3+}. Examples are [$(NH_4)_2SO_4 \cdot Cr_2(SO_4)_3 \cdot 24\ H_2O$], **ammonium chrome alum;** and [$K_2SO_4 \cdot Fe_2(SO_4)_3 \cdot 24\ H_2O$], **potassium ferric alum.** Frequently the formula of alums is contracted to the form $M'M'''(SO_4)_2 \cdot 12\ H_2O$ — for example, $KAl(SO_4)_2 \cdot 12\ H_2O$, and so on. Alums usually produce octahedral crystals and are said to be isomorphous because they crystallize in the same crystal system. In aqueous solution the double salt behaves like two single salts — a monovalent sulfate and a trivalent sulfate — giving ions of all the constituents. Alums are used in water purification because they hydrolyze with the formation of aluminum hydroxide, which adsorbs and carries down suspended matter in a precipitated form. Potash alum is a constituent of certain baking powders; the addition of water to the baking powder brings about hydrolysis of the aluminum sulfate to form hydrogen ions, which then act on the bicarbonate in the mixture to release carbon dioxide.

Ceramics

Clay is a product of the weathering of complex silicates, particularly of feldspars ($KAlSi_3O_8$, $NaAlSi_3O_8$), which are abundant minerals. Although rock silicates are in general quite insoluble, ground waters (dilute solutions of carbonic acid) have chemically altered them in geological time. A feldspar is progressively altered to the mineral **kaolin** according to such chemical changes as

$$(3 \ NaAlSi_3O_8 \ or) \ 3 \ KAlSi_3O_8 + H_2CO_3 \rightarrow KH_2Al_3(SiO_4)_3 + 6 \ SiO_2 + K_2CO_3$$
$$2 \ KH_2Al_3(SiO_4)_3 + H_2CO_3 + 3 \ H_2O \rightarrow 3 \ H_4Al_2Si_2O_9 + K_2CO_3$$
$$\text{kaolin}$$

The potassium and sodium carbonates are dissolved by water, and the clay may be carried in colloidal suspension in streams and finally deposited in beds. Ordinary clays are quite impure, usually containing considerable quantities of iron. Kaolin, also called "white clay" or "china clay," is a clay of high purity. When clay is mixed with water, a very plastic mass is formed which may be molded into any desired shape. This property of clay is the basis of the **ceramic** industries, which produce bricks, tile, pottery, stoneware, and porcelain. After the molding and drying of the clay, the product is baked. Most of the residual molecular water is lost, leaving a very hard, resistant, highly porous product. If the porosity is undesirable, the product may be glazed by producing a very thin film of fused matter on the surface. A cheap glaze can be produced by throwing salt on the object while it is still hot; the salt forms a thin liquid coating of sodium aluminum silicate, which becomes compact and impervious when cooled. Bricks and tiles are made from the cheaper and impure forms of clay.

27.3   CHEMISTRY OF TIN AND LEAD

The metal tin appears just above hydrogen in the activity series. It does not corrode or tarnish in the air, and dilute acids act on the metal only slowly. Hydrochloric and sulfuric acids act slowly on the metal to produce tin(II) (stannous) salts.

$$Sn + 2 \ H^+ \rightarrow Sn^{2+} + H_2$$

Tin forms two series of compounds, tin(II) (**stannous**) and tin(IV) (**stannic**). Both stannous and stannic hydroxides are amphoteric; the addition of sodium hydroxide to the hydroxides of tin produces stannites and stannates, respectively:

$$Sn(OH)_2 + 2\ OH^- \rightarrow Sn(OH)_4{}^{2-} \quad or \quad SnO_2{}^{2-} \cdot 2\ H_2O$$
$$Sn(OH)_4 + 2\ OH^- \rightarrow Sn(OH)_6{}^{2-} \quad or \quad SnO_3{}^{2-} \cdot 3\ H_2O$$

Brown **tin(II) sulfide** (SnS) is precipitated from solutions containing tin(II) ions (Sn^{2+}) by the action of H_2S. Since the sulfide is soluble in ammonium or sodium polysulfides, it can be separated from many other metallic sulfides in qualitative analysis.

$$SnS + S_2{}^{2-} \rightarrow SnS_3{}^{2-}$$
thiostannate ion

Tin(IV) sulfide (SnS$_2$) is precipitated as a yellow solid from solutions of stannic salts or stannates by the action of hydrogen sulfide. It is soluble in both ammonium and sodium sulfide solutions to form the thiostannate ion, $SnS_3{}^{2-}$

$$SnS_2 + S^{2-} \rightarrow SnS_3{}^{2-}$$

The low oxidation potential of tin relates to its property of resisting corrosion. Tin cans have been familiar items of use, but these are iron cans with a thin inner coating of expensive tin (variable price but near $4.00 per pound). As iron sheeting that has been cleaned of oxide is dipped into melted tin (m.p. 232°C), a thin layer of tin coats the iron surface. If the surface of tin plate is scratched so that the iron is exposed, corrosion takes place rapidly, probably because of an electric couple set up between the two metals. Tin pipes are used for conveying distilled water, and tin foil is a common wrapping material. Considerable tin is used in the production of alloys: **bronze** (Cu, Sn); **pewter** (Cu, Pb, Sb, Sn); **solder** (Pb, Sn); and **type metal** (Pb, Sn, Sb). Much zinc, aluminum, or lead foil is now substituted for the almost ten times as expensive tin foil. Tin of desired malleability, softness, and strength such that it does not crack on folding is still best for toothpaste and other container tubes.

Lead is considered a heavy metal with its specific gravity of 11.3, but not nearly as heavy as gold or platinum of specific gravity 19.3 and 21.5, respectively. In the activity series lead appears above hydrogen, but the action of acids on it is very slow. With hydrochloric and sulfuric acids, a coating of lead chloride or lead sulfate quickly covers the surface, preventing further action. It is readily soluble in hot nitric acid, with the formation of the nitrate salt and oxides of nitrogen.

$$3\ Pb + 8\ HNO_3 \rightarrow 3\ Pb(NO_3)_2 + 2\ NO + 4\ H_2O$$

The metal reacts very slowly with water which contains air, forming a loose deposit of lead hydroxide.

$$2\ Pb + 2\ H_2O + O_2 \rightarrow 2\ Pb(OH)_2$$

Because of this action, lead pipes should not be used for carrying drinking water—lead compounds are poisonous.

Because of its relative inactivity and the ease with which it may be worked, lead is used extensively in the manufacture of a wide variety of articles. One of its principal uses is in the manufacture of plates for storage batteries. Sheet lead is used for lining sinks and vats, and for conduits for corrosive chemicals. Other uses include lead shot, cable coverings, lead foil, alloys, lead pipes, and the production of white lead for paints.

Several oxides of lead are known, the more important being the **monoxide** (PbO), **red lead** (Pb_3O_4), and **lead dioxide** (PbO_2). Lead monoxide is a by-product of the refining of lead in which the crude metal is heated in contact with the air to oxidize other metals which may be present. During this process a considerable amount of lead is oxidized. The oxide is yellow, and is known commercially as **litharge.** It is used in the rubber industry and in the manufacture of optical glass; it also is mixed with glycerin for use as a plumber's cement. Pb_3O_4 may be obtained from lead or litharge by heating in air at temperatures below 500°C. PbO_2 is used as a pigment in paints, in storage battery plates, and in the manufacture of flint glass.

Lead forms many salts, only a few of which will be considered here. Lead chloride is precipitated when chloride ions are added to a solution containing lead ions.

$$Pb^{2+} + 2\ Cl^- \rightarrow PbCl_2(s)$$

Because the compound is soluble in hot water, it can be separated from mercury(I) and silver chlorides, which are insoluble in both cold and hot water. **Lead sulfide,** a black salt, is precipitated when H_2S is passed into alkaline or slightly acid solutions containing lead ions.

$$Pb^{2+} + H_2S \rightarrow PbS(s) + 2\ H^+$$

Lead chromate ($PbCrO_4$) is precipitated as a yellow salt by the addition of a soluble chromate to a solution of lead ions.

$$Pb^{2+} + CrO_4^{2-} \rightarrow PbCrO_4(s)$$

The reaction is frequently used as a test for the presence of lead ions. Commercially the chromate salt finds use as a pigment in paints. **Lead tetraethyl,** used in the manufacture of antiknock motor fuels, is prepared by the reaction of ethyl chloride and a sodium-lead alloy according to the equation

$$4\ C_2H_5Cl + Na_4Pb \rightarrow Pb(C_2H_5)_4 + 4\ NaCl$$

Paints

Ordinary paints contain three essential ingredients: (1) a **vehicle,** which is essentially a quick-drying oil; (2) a **body,** or opaque substance, of high covering power; and (3) a **pigment,** if a color other than white is desired.

The vehicle usually consists of a resin-forming compound called a **binder** and a volatile solvent or **thinner.** Typical binders, in addition to oils, are alkyds, acrylics, vinyls, silicones, latex, etc. Solvents used are such compounds as aliphatic hydrocarbons, aromatics such as toluene, alcohols, ketones, esters, and water (for latex). Linseed oil from flax plants is the major binder used in oil house paints. While other fatty acids are present in linseed oil, the major ingredient is linolenic acid, $C_{17}H_{29}COOH$. The oil oxidizes in the air to form a hard, tough film that adheres to the surface of wood or other material and protects it. Linseed oil is usually boiled with lead and manganese oxides, which act as catalysts and hasten the time of drying.

Tung, soybean, castor, and menhaden fish oils are substitutes for linseed oil. All are essentially oils of various numbers of double (unsaturated) bonds between some carbon atoms in long chains (Chapter 32).

When an oil which contains such unsaturated bonds is exposed to the air, oxygen is absorbed with the formation of an organic peroxide, for example,

$$CH_3-(CH_2)_4-\underset{\underset{\displaystyle O-O}{|}}{CH}-\underset{\underset{\displaystyle }{|}}{CH}-CH_2CH-\underset{\underset{\displaystyle O-O}{|}}{CH}-\underset{\underset{\displaystyle }{|}}{CH}-(CH_2)_7-COOR$$

Turpentine is often used to reduce viscosity and increase penetration of the vehicle oil. Lead, vanadium, or cobalt oxides may be used to catalyze drying (lessen drying time).

White lead $[(PbCO_3)_2 \cdot Pb(OH)_2]$ has been much used as a paint body because of its high covering power. The main objection to its use is the fact that it is poisonous and also is darkened by hydrogen sulfide, which produces black lead sulfide. TiO_2 is now largely replacing white lead as a body material in paints.

Interior paints contain lithopone (a $BaSO_4$-ZnS mixture) as a body, since lithopone is not darkened by hydrogen sulfide. If a colored paint is desired, a pigment is added to the lithopone. Common pigments are red lead, lead chromate, Prussian blue, iron oxide, chromium oxide, and carbon. In addition, a filler is frequently added to prevent the body from being spread too thin.

In recent years so-called "water soluble" or "rubber latex" paints for interior use have become very popular. In these preparations pigment and body are incorporated into an oil in water type emulsion. The vehicle is a water suspension containing (a) protein, such as soy bean protein, (b) peptizing agents for proteins, (c) thickeners, (d) preservatives, (e) emulsifiers and (f) pigment. A bulk of such paints contain a buna S type synthetic rubber in emulsion; a polyvinyl acetate or an acrylate synthetic is also much used.

Latex paints harden by the evaporation of the water and the aggregation of the polymer particles until a hard insoluble film is formed. They have little odor and dry rapidly. Much of their polularity is due to ease of cleaning paint brushes and wiping up spills with ordinary water.

High-gloss paints, or enamels, may contain resins (from polymerization of glycerine and phthalic anhydride) in a wide variety of oil combinations. For example, enamel used for automobile finishes contains soybean oil, alkyd resins, and resins produced from polymerization of urea or melamine with formaldehyde.

Exercises

27.1. Identify the following common substances: (a) borax, (b) bauxite, (c) cryolite, (d) bronze, (e) duralumin, (f) thermite mixture, (g) emery, (h) sapphire, (i) alum, (j) red lead, (k) white lead.

27.2. What is one use for each of the substances listed in the previous question?

27.3. Calculate the percentage aluminum in the common, igneous, feldspar rock mineral, $Na_2O \cdot Al_2O_3 \cdot 6 SiO_2$; in dry kaolin clay, $H_4Al_2Si_2O_9$.

27.4. Consider the union of the metallic oxide Na_2O with somewhat nonmetallic B_2O_3 to form borax, $Na_2O \cdot 2 B_2O_3$ in a borax bead test. If a little CoO is added to melted borax, what blue compound probably forms?

27.5. Write equations for (a) preparation of boric acid from borax, (b) burning of a possible gasoline additive of formula $(C_2H_5)_3BO_3$, (c) burning of B_4H_{10} (tetraborane), (d) hydrolysis of B_4H_{10}.

27.6. Name: (a) H_3BO_3; (b) CaB_4O_7; (c) B_2H_6; (d) $Al_2(SO_4)_3 \cdot 18 H_2O$; (e) $(NH_4)_2SO_4 \cdot Cr_2(SO_4)_3 \cdot 24 H_2O$; (f) Na_2SnO_3; (g) SnS_2; (h) PbO_2; (i) $Pb(C_2H_5)_4$; (j) TiO_2.

27.7. Write formulas for gallium nitrate, indium sulfide, thallium(III) sulfate, thallium(I) hydroxide.

27.8. Write formulas for tin(II) chloride, tin(IV) sulfide, stannite ion, stannate ion, thiostannate ion, stannous fluoride.

27.9. Write the following equations:
(a) Softening hard water (Mg^{2+}) with borax.
(b) Thermite reaction for welding rails.
(c) Formation of $Al(OH)_3$ mordant in fixing dyes to cloth:

$$Al_2(SO_4)_3 + NH_3 + H_2O \rightarrow$$

(d) Amphoterism of $Al(OH)_3$:

$$Al(OH)_3 (s) + HCl \rightarrow$$
$$Al(OH)_3 (s) + NaOH \rightarrow$$

(e) Hydrolysis of potassium alum in water purification by the formation of $Al(OH)_3 (s)$:

$$KAl(SO_4)_2 \cdot 12 H_2O + H_2O \rightarrow$$

(f) Slow dissolving of tin in concentrated HCl.
(g) Precipitation of Sn^{2+} by H_2S.
(h) Dissolving of tin(IV) sulfide by sodium sulfide:

$$SnS_2 + S^{2-} \rightarrow$$

(i) Reaction of metallic lead with dilute nitric acid.

27.10. Write an equation for the reaction that occurs between Al and each of the following: (a) Cl_2, (b) O_2, (c) S, (d) N_2, (e) H_2O, (f) H^+ (aq.).

27.11. Write an equation for the reaction that occurs between Sn and each of the following: (a) Cl_2, (b) O_2, (c) S, (d) H^+ (aq.), (e) OH^- (aq.), (f) HNO_3.

27.12. Complete the following equations:
(a) $Ge + Br_2 \rightarrow$

(b) $PbO_2(s) \xrightarrow[\text{heating}]{\text{mild}}$ red lead

(c) $PbO_2(s) \xrightarrow[\text{heating}]{\text{strong}}$ litharge

(d) $PbO(s) + H_2O + OH^-(aq) \rightarrow$ plumbite^{2-} ion dihydrate
(e) $SnS(s) + S^{2-}(aq) \rightarrow$
(f) $PbCl_2(s) + Cl^-(aq) \rightarrow$
(g) $SnF_4 + F^-(aq) \rightarrow$

27.13. A solder is a lead-tin alloy. A 1.00-g sample of the alloy treated with warm 30% HNO_3 gave a lead nitrate solution and some insoluble white residue. After drying, the white residue weighed 0.650 g and was considered SnO_2. What was the percentage tin in the sample?

$$Sn + 4\ HNO_3 \xrightarrow{\Delta} SnO_2 + 4\ NO_2 + 2\ H_2O$$

(Sn and Sb are the only metals in nonferrous alloys that are converted to insoluble oxides. $Sb + HNO_3$ yields Sb_2O_5.)

27.14. A 2.50-g sample of lead silicate optical glass is ground to a powder and digested with HF acid to eliminate the silicon as $SiF_4(g)$. The residue is converted with another acid to obtain the lead as Pb^{2+}; it is then titrated with standard K_2CrO_4 to an end-point precipitation of all Pb^{2+} as $PbCrO_4$. It requires 48.6 ml of 0.100 N K_2CrO_4 to act with the Pb^{2+}. What weight of lead is in the sample of glass?

27.15. In a Parkes process for removing silver from lead (lead from a PbS, Ag_2S-containing ore) about 20 lb of zinc is stirred into a ton of melted, impure lead. The zinc is insoluble in the lead and soon rises to a surface scum on the lead with 90% of the silver in it, silver being much more soluble in zinc than lead. If a ton of lead contains 6 oz Ag/ton, what amount of Ag in ounces per ton remains in the lead after three treatments with zinc? (Assume 90% recovery of Ag each time.)

27.16. Since an alum is an $M_2'SO_4 \cdot M_2'''(SO_4)_3 \cdot 24\,H_2O$ type compound, write the formula of an alum containing the elements potassium, gallium, sulfur and, of course, hydrogen and oxygen.

Supplementary Readings

27.1. "Superhard Materials," F. P. Bundy, *Sci. American*, Aug. 1974; p. 62.

27.2 "Lead Poisoning," J. J. Chisolm, Jr., *Sci. American*, Feb. 1971; p. 15.

27.3. "The Story of Hall and Aluminum," H. N. Holmes, *J. Chem. Educ.*, **7**, 232 (1930).

27.4. "Cage Compounds of Boron," M. S. Gaunt and A. G. Massey, *Educ. in Chemistry*, **11**, 118 (1974).

Chapter 28

IIIB	IVB	VB	VIB	VIIB	VIIIB			IB	IIB
2 8 9 2 **21** Sc 44.96	2 8 10 2 **22** Ti 47.90	2 8 11 2 **23** V 50.94	2 8 13 1 **24** Cr 52.00	2 8 13 2 **25** Mn 54.94	2 8 14 2 **26** Fe 55.85	2 8 15 2 **27** Co 58.93	2 8 16 2 **28** Ni 58.71	2 8 18 1 **29** Cu 63.546	2 8 18 2 **30** Zn 65.37
2 8 18 9 2 **39** Y 88.91	2 8 18 10 2 **40** Zr 91.22	2 8 18 12 1 **41** Nb 92.91	2 8 18 13 1 **42** Mo 95.94	2 8 18 14 1 **43** Tc 98.91	2 8 18 15 1 **44** Ru 101.07	2 8 18 16 1 **45** Rh 102.91	2 8 18 18 0 **46** Pd 106.4	2 8 18 18 1 **47** Ag 107.868	2 8 18 18 2 **48** Cd 112.40
57–71 Lantha- nide series	2 8 18 32 10 2 **72** Hf 178.49	2 8 18 32 11 2 **73** Ta 180.95	2 8 18 32 12 2 **74** W 183.85	2 8 18 32 13 2 **75** Re 186.2	2 8 18 32 14 2 **76** Os 190.2	2 8 18 32 15 2 **77** Ir 192.2	2 8 18 32 17 1 **78** Pt 195.09	2 8 18 32 18 1 **79** Au 196.97	2 8 18 32 18 2 **80** Hg 200.59
89–103 Actinide series									

Lanthanide series

2 8 18 18 9 2 **57** La 138.91	2 8 18 20 8 2 **58** Ce 140.12	2 8 18 21 8 2 **59** Pr 140.1	2 8 18 22 8 2 **60** Nd 144.24	2 8 18 23 8 2 **61** Pm (145)	2 8 18 24 8 2 **62** Sm 150.4	2 8 18 25 8 2 **63** Eu 151.96	2 8 18 25 9 2 **64** Gd 157.25	2 8 18 27 8 2 **65** Tb 158.92	2 8 18 28 8 2 **66** Dy 162.50	2 8 18 29 8 2 **67** Ho 164.93	2 8 18 30 8 2 **68** Er 167.26	2 8 18 31 8 2 **69** Tm 168.93	2 8 18 32 8 2 **70** Yb 173.04	2 8 18 32 9 2 **71** Lu 174.97

Actinide series

2 8 18 32 18 9 2 **89** Ac (227)	2 8 18 32 18 10 2 **90** Th 232.04	2 8 18 32 20 9 2 **91** Pa 231.04	2 8 18 32 21 9 2 **92** U 238.03	2 8 18 32 22 9 2 **93** Np 237.05	2 8 18 32 23 9 2 **94** Pu (242)	2 8 18 32 25 8 2 **95** Am (243)	2 8 18 32 25 9 2 **96** Cm (247)	**97** Bk (247)	**98** Cf (251)	**99** Es (254)	**100** Fm (253)	**101** Md (256)	**102** No (254)	**103** Lr (257)

Transition Elements and Coordination Compounds

Electron Structure of Representative versus Transition Elements

In discussing the periodic table in Chapter 4, elements were classified into A and B groups. The A group elements are sometimes termed the **main** or **representative** group elements, and the B group elements are termed **transition** elements. The basis for this classification is the electronic configuration of the atoms of the elements. For the A group elements the last added electrons are fitted into s and p orbitals—in other words, into the outermost or valence level, which has a capacity of eight electrons (Figure 28.1). Elements in Groups IA and IIA have only s electrons in the outer level. Elements in Groups IIIA to VIIIA have two s electrons and p electrons to a maximum of six.

For the B group elements—the transition elements—added electrons do not fill in the outer level, but rather go into an incomplete inner level, either d or f orbitals of a lower principal quantum number.

For the first horizontal period (period 4) of transition elements, $_{21}Sc$ through $_{30}Zn$, added electrons are fitted into the 3d orbitals. For the second period, $_{39}Y$ through $_{48}Cd$, electrons fill the 4d orbitals.

Following $_{57}La$, elements of atomic number 57 to 71 constitute a series in which inner 4f orbitals ($4f^{1-14}$) are filled. These fourteen elements plus

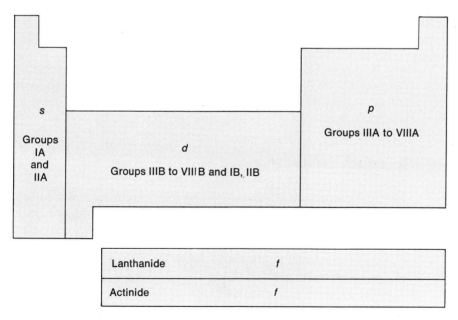

Figure 28.1
Block form of periodic table showing types of electrons most related to position of elements.

lanthanum are called the **lanthanide series, or rare earths.** Following them in increasing atomic number are $5d$-filling transition elements. Lanthanum, with $5d^1$, starts the series, and mercury, with $5d^{10}$, ends the series.

The element actinum ($_{89}$Ac) and those higher in atomic number make up a further transition metal series. The post-actinium elements have added electrons filling the $5f^{1-14}$ positions and constitute a second **rare earth** or **actinide series.**

Note that 58 elements are members of the transition series. These elements comprise more than half of the periodic table.

Beginning and End Elements of Transition Series

The starting members of the transition periods (left-hand end) — Sc, Y, La — have only one d and two outer s electrons (d^1s^2). They exhibit a positive valence of three only. The limitations in valence variability of these Group IIIB elements give them some of the characteristics of the main, or Group IIIA, elements.

The outer orbital structure of Group IIB elements — zinc, $3d^{10}4s^2$; cadmium, $4d^{10}5s^2$; and mercury, $5d^{10}6s^2$ — gives these elements, that end a transition series, a valence or oxidation number of $+2$. No electrons from the complete d orbitals are available for chemical change, and so these elements do not exhibit variable oxidation numbers, as do the other transition elements in which both d and s electrons are available. Mercury does have an oxidation number of $+1$ in mercury (I) compounds because of the formation of the unusual ion $(Hg-Hg)^{2+}$, which is usually written Hg_2^{2+}.

In many respects Group IIB elements resemble the main group elements. This may be attributed to the completeness of the d orbitals, a characteristic of main group elements. Zn, Cd, and Hg are usually considered terminal elements in the $3d^{1-10}$, $4d^{1-10}$, and $5d^{1-10}$ series, however.

Properties of the Transition Elements

From a chemical standpoint, elements which lose electrons to form positive ions (cations) are metals. Inasmuch as all of the transition elements have only one or two s electrons in the outer level and these are readily given up to form positive ions, all the transition elements are classed as metallic elements.

The transition elements exhibit both vertical and horizontal similarities in the periodic table. Vertical similarities are expected, of course, because of electron structural likeness, but striking horizontal similarities in successive members of a transition series are also observed. This is attributed to the fact that electrons are entering an incomplete inner level, while the outer or valence level remains unchanged at two electrons.

The greatest similarities are found in the **transition triads** of Group VIIIB. Fe, Co, and Ni, the first triad, are quite similar. The second and third series,

Ru, Rh, Pd and Os, Ir, Pt, are so much alike that they are usually grouped together as the **platinum** metals and their chemistry studied as a unit.

Variable Oxidation Numbers of Transition Elements

Transition elements generally exhibit variable oxidation numbers. Exceptions to this general rule are found in Groups IIB and IIIB. As pointed out above, zinc and cadmium in Group IIB show an oxidation number of only +2. Scandium, yttrium, and lanthanum in Group IIIB appear to exhibit a single oxidation number of +3 in compounds.

Group IVB elements — titanium, zirconium, and hafnium — exhibit oxidation numbers of +2, +3, and +4. The outer two s electrons, and in some cases one or two additional d electrons, are available for chemical union.

Groups VB, VIB, VIIB, and VIIIB may lose electrons to form ionic compounds, but may also form many covalent compounds. The latter fact accounts for the higher oxidation numbers they frequently exhibit.

Transition elements in their various oxidation states differ in color, which is in some way related to outer electron interchange. Iron, chromium, and vanadium compounds in particular show color. The name chromium comes from the Greek word for color, the name vanadium from Vanadis, a Scandinavian goddess of beauty. Nontransitional elements such as aluminum, calcium, sodium, and silicon are not responsible for color present in their compounds.

In the chapters which follow various transition metals will be studied, with emphasis on those of greatest industrial importance.

Use of small quantities of some of the uncommon, scarce transition metal compounds is indicated in a quote from an industrial journal. "As a harmonic generator, lithium niobate performs especially well when the laser light was produced by a crystal of neodymium doped yttrium aluminum silicate. Lithium tantalate also looks good as an optical modulator. Barium sodium niobate (sometimes called "bananas" by Bell Telephone researchers) in crystal form is a good harmonic generator material."

A new use of the less common metals is indicated in the manufacture of a pencil-sized nuclear radiation powered battery. Heart pacers are one use of this battery. The metal case is of a close tolerance machined tungsten — 10% tantalum alloy, which absorbs radiation rays that could be harmful; the essential energy source is a mixture of plutonium and promethium oxides. The energy from the radioactive metal nuclei is practically persistent for about ten years.

28.1   COMPLEX IONS AND COORDINATION COMPOUNDS

One of the characteristic properties of the transition elements is their ability to form complex ions. This involves an ion of the transition metal combining either with neutral molecules or with other ions. For example, H_2O or NH_3

molecules, or CN^- or Cl^- ions, all of which have at least one pair of unshared electrons, form coordinate bonds with many of the transition metal ions. Examples are $Cu(H_2O)_4^{2+}$; $Co(NH_3)_6^{3+}$; $Ag(CN)_2^-$; $FeCl_6^{3-}$; $Fe(CN)_6^{3-}$. Complex ions, in combination with ions of opposite charge, result in neutral **coordination** compounds. For example,

$$Cu(NH_3)_4^{2+} + SO_4^{2-} \rightarrow Cu(NH_3)_4SO_4 \text{ (ionic crystals)}$$

In the formation of complex ions, coordinating groups called **ligands** furnish electron pairs which are accommodated in the outer d and s orbitals (and perhaps in outer p orbitals as well) of the transition metal ion.

The compound NH_3 is one of the best examples of a ligand. It acts as a Lewis-base; two electrons that are not shared with the hydrogens are available to be shared with some other element.

$$H : \overset{\cdot\cdot}{\underset{\overset{\cdot\cdot}{H}}{N}} : H$$

The number of electron pairs accepted (coordinated bonds formed) by a metal ion depends on its suborbital structure. For example, the $Zn(NH_3)_4^{2+}$ complex ion stability is related to orbital structures as follows: $^{65}_{30}Zn$ has the electronic configuration: $1s^2$; $2s^2$, $2p^6$; $3s^2$, $3p^6$, $3d^{10}$; $4s^2$. The outer structure may be represented as

Zn^{2+} may be represented as

Each of four NH_3 molecules has an unused or unshared pair to donate to the unfilled one $4s$ and three $4p$ orbitals. Consequently, 4 molecules of NH_3 attach to the Zn^{2+}, to give it a coordination number of 4.

The tendency to form coordination compounds is much greater for transition metal ions than for alkali or alkaline earth ions. The latter have large ionic diameters, small charge, and no incomplete d orbitals—factors which no doubt detract from their ability to form complex ions.

Ligand field theory   A recent theory which has proven useful in connection with coordination compounds is the crystal field or ligand field theory, which states that ligands are held to the central ion of a coordination compound by electrostatic attractive forces due primarily to the charge of the central ion and the polar nature of the ligands. The extent of the attractive forces determines the stability of the complex, and is dependent on the charge and size of the central ion and ligands.

Using this theory, it is possible to calculate the energy effects for many of the charge and size factors and finally to obtain values for bond energies. The latter are useful in predicting properties and types of configurations for a large number of compounds. (A detailed treatment of this theory is beyond the scope of this book.)

Example 1: Formation of the complex ion, $Fe(CN)_6^{4-}$.
$_{26}^{56}Fe$ atom has the configuration $1s^2; 2s^2, 2p^6; 3s^2, 3p^6, 3d^6; 4s^2$

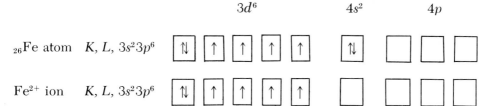

Each of six CN^- $(:C:::N:)^-$ donates an electron pair (12 electrons in total) to complete the $3d$, $4s$, and $4p$ orbitals; thus Fe^{2+} has a coordination number of 6 (is an acceptor of six electron pairs).

$Fe(CN)_6^{4-}$ ion $K, L, 3s^2 3p^6$

The acceptance of pairs of electrons by Fe^{2+} makes it an example of a Lewis acid.

Example 2: Formation of $PtCl_6^{2-}$

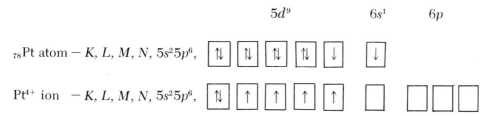

The 6 Cl^- ions furnish 12 electrons to the incomplete $5d$, $6s$, and $6p$ orbitals to form $PtCl_6^{2-}$ (hexachloroplatinate(IV)) ion. The crystalline coordinate compound $2 K^+ (PtCl_6)^{2-}$, K_2PtCl_6, is sometimes used as an electrolyte in platinum plating.

Example 3: Different groups may coordinate to a given metallic ion. Neutral molecules may be substituted for ions, and vice versa. The charge of the complex ion will be the algebraic sum of the cationic and anionic charges. For example,

$$\begin{bmatrix} NH_3 & Cl^- \\ & \diagdown \quad \diagup & \\ NH_3-Co^{3+}-Cl^- \\ & \diagup \quad \diagdown & \\ Cl^- & Cl^- \end{bmatrix}^- \qquad \begin{bmatrix} NH_3 & NH_3 \\ & \diagdown \quad \diagup & \\ NH_3-Co^{3+}-NH_3 \\ & \diagup \quad \diagdown & \\ Cl^- & Cl^- \end{bmatrix}^+ \qquad \begin{bmatrix} NH_3 & NH_3 \\ & \diagdown \quad \diagup & \\ NH_3-Co^{3+}-NH_3 \\ & \diagup \quad \diagdown & \\ NH_3 & Br^- \end{bmatrix}^{2+}$$

$(3+) + (4-) = 1-$ $\qquad\qquad$ $(3+) + (2-) = 1+$ $\qquad\qquad$ $(3+) + (1-) = 2+$

Table 28.1

Some coordinating units (ligands)

Neutral Molecules		Negative Ions	
H_2O	PH_3	CN^-	$C_2H_3O_2^-$
NH_3	CH_3NH_2	NO_2^-	Cl^-
CO	$H_2NCH_2CH_2NH_2$	NO_3^-	Br^-
		OH^-	I^-
		CNS^-	F^-
		CO_3^{2-}	SO_3^{2-}
NO		O^{2-}	
		$CH_2CO_2^-$	
		$N{-}CH_2CO_2^-$	
		$CH_2CO_2^-$	
NH_2OH			

Table 28.1 lists a few typical coordinating ligands – units which yield electron pairs in coordination bonding. Complexity increases when a given ligand may occupy more than one coordinating space.

The patterning of ligands which furnish electron pairs in outer orbitals of many metal ions becomes complicated beyond the scope of this text. Usual coordination numbers of 2, 4, and 6 (rarely 8) represent the number of accepted electron pairs in complex ion formation between a metal ion and ligands.

Ligands which occupy one coordinating space are referred to as **monodentate,**[*] (one point of attachment). Some more complex ligands may possess two or more electron donor centers and thus occupy two, three, or even four coordinating positions. The latter are termed **bidentate, tridentate,** and **quadridentate.** A polydentate unit is called a **chelate** group. Ethylenediamine ($H_2NCH_2CH_2NH_2$) behaves as.if it were two ammonia molecules, and is a good example of a bidentate and chelating unit. Other examples of bidentate ions are CO_3^{2-} and SO_3^{2-} Tridentate and quadridentate ligands are uncommon.

The anion of nitrilo tri acetic acid, $N(CH_2CO_2)_3^{3-}$, is a good example of a quadridentate ligand. The nitrogen and three oxygens contribute to coordination of four. It is currently being used by detergent manufacturers for complexing Ca^{2+} and Fe^{2+} in order to reduce hardness of water.

Werner Theory of Coordination Compounds

Much of our fundamental knowledge of coordination compounds comes from the research of Alfred Werner, who received a Nobel prize for his work. In 1920 Werner postulated the existence of an inner coordination sphere

[*] The word dentate comes from tooth; one tooth, two, three, or four teeth to bite with or hold with.

around the central atom in a complex ion or compound in which the coordinating units were tightly attached to the central atom or ion. These units are not released on ionization or dissociation of the substance; for example, a solution of hydrated copper sulfate contains $Cu(H_2O)_4^{2+}$ ions in which 4 molecules of water are attached firmly to each Cu^{2+}. Werner assigned coordination numbers of 2, 3, 4, 6, and 8 for spatial symmetry of units around the central element, with 4 and 6 being the most common.

The ligands in coordination position are added units beyond the units that are added on the basis of electrovalent or covalent bonding.

Werner's proposal of coordinated units in spatial symmetry about a central unit (often a transition element) has not been discredited by time. His ideas have been extended and nicely related to modern theory.

Naming and Examples of Coordination Compounds

1. Cations are named first, anions last.
2. The names of negative ion ligands end in *o*. Examples from Table 28.1 are **cyano-, hydroxo-, chloro-, carbonato-**. The order of ligands is alphabetic.
3. Neutral units have historical endings, such as **ammine, aqua-** (formerly **aquo-), carbonyl, nitroso-**.
4. The oxidation number of the central atom in the complex is specified by Roman numerals in parentheses following the name of the element. When a complex ion is negative, the name of the central element has -*ate* appended.
5. Neutral complexes are named as if they were cations.
6. If a complex is a positive ion, the names of the acid radicals not in the complex complete the name for the compound.

These rules will be clarified by the examples below.

$$K^+[BF_4]^-$$

Potassium tetrafluoroborate(III)

$$4\ K^+[Fe(CN)_6]^{4-}$$

Potassium hexacyanoferrate(II)
(potassium ferrocyanide)

Diamminediaquacarbonatocobalt(III) nitrate

Tetraamminedinitrocobalt(III) nitrate

$$[Ag(NH_3)_2]^+Cl^-$$

Diamminesilver(I) chloride

$$[Cu(NH_3)_4]^{2+}SO_4^{2-}$$

Tetraamminecopper(II) sulfate

$$3\ Na^+[AlF_6]^{3-}$$

Sodium hexafluoroaluminate(III) (cryolite)

Triamminetrinitrocobalt(III)

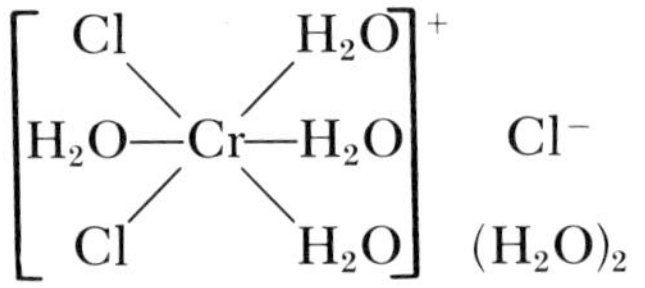

$$\begin{bmatrix} Cl & H_2O \\ & \diagdown \diagup \\ H_2O-Cr-H_2O \\ & \diagup \diagdown \\ Cl & H_2O \end{bmatrix}^+ \quad Cl^- \quad (H_2O)_2$$

Tetraaquadichlorochromium(III)
chloride dihydrate
green salt

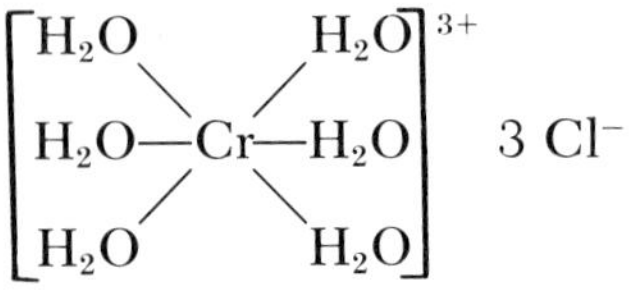

$$\begin{bmatrix} H_2O & H_2O \\ & \diagdown \diagup \\ H_2O-Cr-H_2O \\ & \diagup \diagdown \\ H_2O & H_2O \end{bmatrix}^{3+} \quad 3\ Cl^-$$

Hexaaquachromium(III) chloride
violet salt

$CrCl_3 \cdot 6\ H_2O$

$$\begin{bmatrix} UO_2 \\ (H_2O)_2 \end{bmatrix}^{2+} \quad SO_4^{2-}$$

Diaquadioxouranium(VI) sulfate

$$4\ K^+[Mo(CN)_8]^{4-}$$

Potassium octocyanomolybdate(IV)

Cis-trans Isomerism

The geometric patterning of ligands about a central element in coordination compounds leads to many examples of isomerism. The compound $Pt(NH_3)_2Cl_2$ may have two structures, **cis** and **trans.**

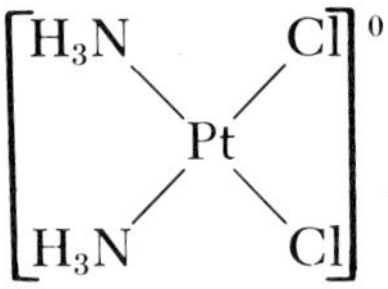

$$\begin{bmatrix} H_3N & Cl \\ & \diagdown \diagup \\ & Pt \\ & \diagup \diagdown \\ H_3N & Cl \end{bmatrix}^0 \qquad\qquad \begin{bmatrix} H_3N & Cl \\ & \diagdown \diagup \\ & Pt \\ & \diagup \diagdown \\ Cl & NH_3 \end{bmatrix}^0$$

cis   Diamminedichloroplatinum(II)   *trans*

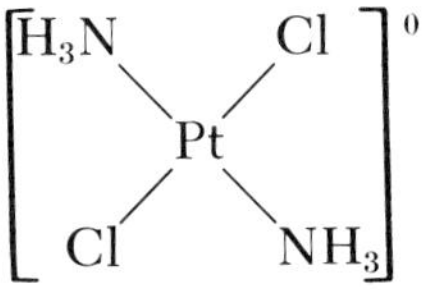

The trans compound, with the same ligands across from each other, has more symmetry than the cis. Both isomers of this compound, whose coordination number is four, are planar.

An example of isomerism for a compound of coordination number six, of octahedral structure, is $[Co(NH_3)_4Cl_2]^+Cl^-$, tetraamminedichlorocobalt(III) chloride.

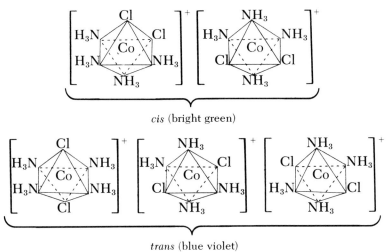

cis (bright green)

trans (blue violet)

Isomers differ from one another not only in color but also in properties such as melting temperature, solubility, etc.

Optical Isomers

Mirror image isomers exist in many coordination compounds, particularly when chelate (2 coordinate position)* groups are involved.

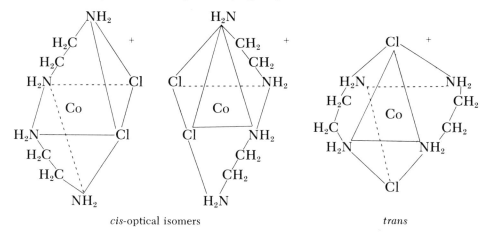

cis-*optical isomers*   *trans*

In the example of the three compounds of formula $[Co(en)_2Cl_2]$ Cl the metal-containing ion may embody two cis-optical arrangements and a trans arrangement. Many octahedral or tetrahedral complexes may exhibit spacial isomerism.

Isomers resulting from differences in spacial orientation of ligand groups are termed *stereoisomers*. Other types of isomerism are called **polymerization**, **coordination**, **hydration**, and **ionization** isomerism. Organic chemistry presents examples of stereoisomerism when different patterns of diverse groups share four bonds with carbon.

Applied Coordination Chemistry

Rust stains on a towel may be removed by treatment with a warm oxalate solution (Figure 28.2).

$$Fe^{3+} + 3 C_2O_4^{2-} \rightarrow Fe(C_2O_4)_3^{3-}$$

It may be said that thousands of complex ions and coordination compounds are possible, and that usually a transition metal is the central atom. Many of these compounds are useful in inorganic chemistry as precipitating

---

* Ethylenediamine, $H_2N—CH_2—CH_2—NH_2$, is a ligand that occupies two positions. The expression **bis** is sometimes used for **di**, denoting two.

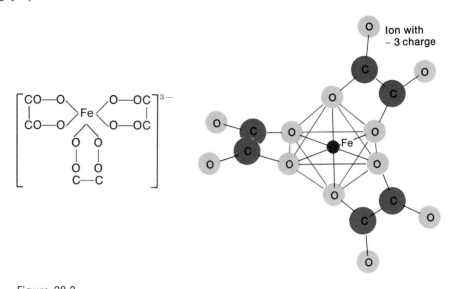

Figure 28.2

Structure of Fe(C$_2$O$_4$)$_3^{3-}$.

agents, separational agents, aids in dyeing, and so on. They assume a most important role in the make-up of chlorophyll, hemoglobin, and similar materials in living matter. The exact role of iron, magnesium, copper, and other trace metals in plant and animal life is not fully known, but it is certain that coordination compounds play key parts.

The following are a few complex ions which find application in analytical chemistry:

$Cu(NH_3)_4^{2+}$, $Ni(NH_3)_4^{2+}$ are intense blue; $Ag(NH_3)_2^+$ (NH_3 dissolves $AgCl$) $KFe(Fe(CN)_6)$, intense blue.
$Fe(OH)_3$ is insoluble; $Al(OH)_4^-$ or $Al(OH)_6^{3-}$ is soluble.
As_2S_5, HgS are insoluble; AsS_4^{3-} and HgS_2^{2-} are soluble.
$H_3C—C—C—CH_3$, dimethylglyoxime, is an example of a chelating agent and
$\quad\quad\;\; ‖\;\; ‖$
$\;\;$ HO—N$\quad$N—OH
used in obtaining an orange-red precipitate from $Ni(NH_3)_4^{2+}$. The unshared electron pairs of the two nitrogens give it a bidentate nature as it coordinates with various metal ions.

Ethylenediaminetetraacetic acid, abbreviated to EDTA, is a hexadentate ligand, which complexes numerous metal ions; may even be used in titrimetric procedures. $K_3[Co(NO_2)_6]$, insoluble yellow, named potassium hexanitritrocobaltate(III), or potassium cobaltinitrite.

As the various transitional metals are studied in the ensuing chapters, further examples of complex ions and coordination compounds will be included.

28.2 COMPLEX ION EQUILIBRIA

Having now considered complex ions it is proper to detail a few examples of complex ion equilibria.

In silver plating, a solution of $Na[Ag(CN)_2]$ is used. A low concentration of Ag^+ is maintained for good electroplating by the equilibrium with the $Ag(CN)_2^-$ complex ion. Although the dissociation of complex ions probably takes place in steps, an overall equilibrium can be written

$$Ag(CN)_2^- \rightleftarrows Ag^+ + 2CN^-$$

for which the equilibrium expression is

$$\frac{[Ag^+][CN^-]^2}{[Ag(CN)_2^-]} = K_d = 1.9 \times 10^{-19}$$

where K_d is the dissociation constant.

We can deduce the fact that $Ag(CN)^-$ is more stable than $Ag(NH_3)_2^+$ or $AgCl_2^-$ from the following.

$$K_d = \frac{[Ag^+][NH_3]^2}{[Ag(NH_3)_2^+]} = 6.0 \times 10^{-8} \qquad K_d = \frac{[Ag^+][Cl^-]^2}{[AgCl_2^-]} = 1 \times 10^{-6}$$

K_d for $AgCl_2^-$ relates to the solubility of AgCl in concentrated HCl to form the $AgCl_2^-$, which, however, still yields considerable Ag^+ in solution as seen from the comparatively high value of K_d.

Consider the sample problem.

EXAMPLE 28.1   For the complex ion equilibrium for the dissociation of $Cu(NH_3)_4^{2+}$, calculate (1) the ratio of Cu^{2+} ion to $Cu(NH_3)_4^{2+}$ in a solution 1 M in NH_3; (2) the equilibrium concentration of NH_3 needed to attain a 50% conversion of Cu^{2+} to $Cu(NH_3)_4^{2+}$. K_d for $Cu(NH_3)_4^{2+} = 2 \times 10^{-13}$.

SOLUTION:

(1)
$$K_d = 2 \times 10^{-13} = \frac{[Cu^{2+}][NH_3]^4}{[Cu(NH_3)_4^{2+}]} = \frac{[Cu^{2+}](1.0\ M)^4}{[Cu(NH_3)_4^{2+}]}$$

$$\frac{[Cu^{2+}]}{[Cu(NH_3)_4^{2+}]} = \frac{2 \times 10^{-13}}{1} = 2 \times 10^{-13}$$

An exceedingly high conversion to complex ion is indicated.

(2)
$$K_d = 2 \times 10^{-13} = \frac{[Cu^{2+}][NH_3]^4}{[Cu(NH_3)_4^{2+}]}$$

If the concentrations of Cu^{2+} and $Cu(NH_3)_4^{2+}$ are equal (50% of each in this problem), they cancel each other. It follows that $[(NH_3)]^4 = 2 \times 10^{-13} = 0.2 \times 10^{-12}$. Converting to the 4th root of each side the NH_3 concentration becomes 6.7×10^{-4}. A very small concentration or $6.7 \times 10^{-4} M$ NH_3 results in 50% conversion.

Trace Elements of Life

At the present time 24 of the known chemical elements have been found essential for living matter (Table 28.2). All but 3 of these elements have atomic numbers less than 34. Over half of the essential elements are used by living materials only in trace amounts. Many of these function in metal-organic coordination compounds such as hemoglobin (Fe), vitamin B_{12} (Co), invertebrate animal blood (Cu), chlorophyll (Mg), nitrate utilization enzyme (Mo), protein digestion enzyme (Zn), and so on.

Hemoglobin, for example, is a protein with a molecular weight of 68,000. When hydrolyzed it yields the protein globin, and four **heme** groups which are chelates of Fe^{2+} (Figure 28.3).

Chlorophyll, the green pigment in plants, is an organic catalyst in the photosynthesis of carbohydrates from CO_2 and H_2O in the presence of sun-

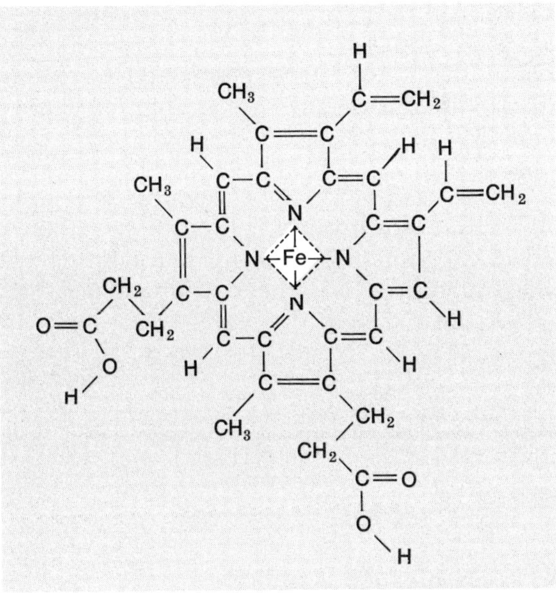

Figure 28.3
Heme.

Table 28.2

The Essential Elements for Life. Some two-thirds of the lightest elements, or 21 out of the first 34 elements in the periodic table, are now known to be essential for animal life. These 21 plus molybdenum (No. 42), tin (No. 50) and iodine (No. 53) constitute the total list of the 24 essential elements, which are here enclosed in colored boxes. It is possible that still other light elements will turn out to be essential. The most likely candidates are aluminum, nickel and germanium. The element boron already appears to be essential for some plants.

Element	Symbol	Atomic Number	Comments
HYDROGEN	H	1	Required for water and organic compounds.
HELIUM	He	2	Inert and unused.
LITHIUM	Li	3	Probably unused.
BERYLLIUM	Be	4	Probably unused; toxic.
BORON	B	5	Essential in some plants; function unknown.
CARBON	C	6	Required for organic compounds.
NITROGEN	N	7	Required for many organic compounds.
OXYGEN	O	8	Required for water and organic compounds.
FLUORINE	F	9	Growth factor in rats; possible constituent of teeth and bone.
NEON	Ne	10	Inert and unused.
SODIUM	Na	11	Principal extracellular cation.
MAGNESIUM	Mg	12	Required for activity of many enzymes; in chlorophyll.
ALUMINUM	Al	13	Essentiality under study.
SILICON	Si	14	Possible structural unit of diatoms; recently shown to be essential in chicks.
PHOSPHORUS	P	15	Essential for biochemical synthesis and energy transfer.
SULFUR	S	16	Required for proteins and other biological compounds.
CHLORINE	Cl	17	Principal cellular and extracellular anion.
ARGON	A	18	Inert and unused.
POTASSIUM	K	19	Principal cellular cation.
CALCIUM	Ca	20	Major component of bone; required for some enzymes.
SCANDIUM	Sc	21	Probably unused.
TITANIUM	Ti	22	Probably unused.
VANADIUM	V	23	Essential in lower plants, certain marine animals and rats.
CHROMIUM	Cr	24	Essential in higher animals; related to action of insulin.
MANGANESE	Mn	25	Required for activity of several enzymes.
IRON	Fe	26	Most important transition metal ion; essential for hemoglobin and many enzymes.
COBALT	Co	27	Required for activity of several enzymes; in vitamin B_{12}.
NICKEL	Ni	28	Essentiality under study.
COPPER	Cu	29	Essential in oxidative and other enzymes and hemocyanin.
ZINC	Zn	30	Required for activity of many enzymes.
GALLIUM	Ga	31	Probably unused.
GERMANIUM	Ge	32	Probably unused.
ARSENIC	As	33	Probably unused; toxic.
SELENIUM	Se	34	Essential for liver function.
MOLYBDENUM	Mo	42	Required for activity of several enzymes.
TIN	Sn	50	Essential in rats; function unknown.
IODINE	I	53	Essential constituent of the thyroid hormones.

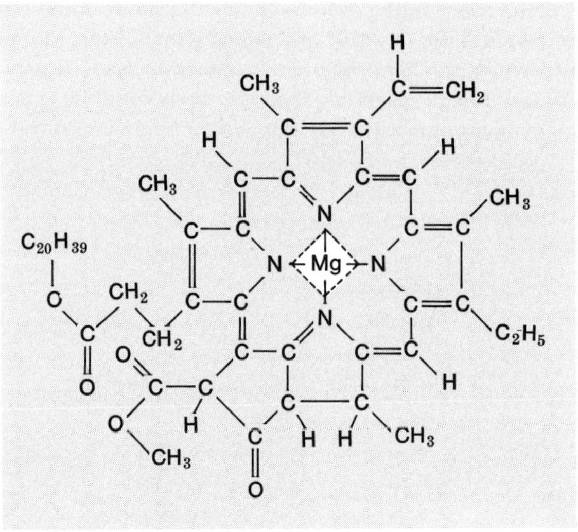

Figure 28.4

Chlorophyll A.

light. The structure of chlorophyll is a chelate similar to a heme group (Figure 28.4).

One of the most recently discovered essential trace elements is tin. It has been shown that rats fed on diets without 1 or 2 parts per million of Sn grow at only about two-thirds the normal rate. Similar results are obtained for the new trace elements vanadium, silicon, and fluorine.

Exercises

28.1. Write the electronic configuration for the 3 d and 4 s electrons for the Period 4 transition elements $_{21}$Sc through $_{30}$Zn. Check your results with Table 3.1.

28.2. On the basis of the atomic structures of the transition elements, such as those obtained in exercise 28.1, explain why the transition elements:
(a) are all metals.
(b) have variable oxidation numbers in their compounds, such as Mn varying from +2 to +7.
(c) generally form colored compounds.
(d) form complex ions readily.
(e) usually form compounds which are paramagnetic.

28.3. Consider the coordination compound, $[Pt(NH_3)_4Cl_2]Cl_2$.
(a) When dissolved in water, how many ions would result per formula unit?
(b) Write the formula and charge of the complex ion present.
(c) What is the central ion?

(d) Point out the ligands.

(e) What is the meaning of the brackets?

(f) What is the coordination number of Pt?

(g) What is the name of the compound?

(h) How many moles of AgCl would precipitate per mole of complex compound when an excess of $AgNO_3$ solution is added?

28.4. If the compound in above exercise 28.3 were replaced with (a) [Pt(NH$_3$)$_3$-Cl$_3$]Cl, (b) K[Pt(NH$_3$)Cl$_5$], repeat the exercise for any factors that would be different.

28.5. Refer to Examples 1 and 2 on page 589 of this text and then account for the formation of (a) Co(NH$_3$)$_6$$^{3+}$, (b) Zn(NH$_3$)$_4$$^{2+}$, (c) Cd(H$_2$O)$_4$$^{2+}$. (Unpaired electrons are usually considered as remaining in the d orbitals.)

28.6. Define the following: (a) lanthanide series, (b) coordination number, (c) ligand, (d) bidentate, (e) chelate, (f) cis-trans isomers.

28.7. Name the following complex compounds:

(a) K$_3$[Fe(CN)$_6$]

(b) [Pt(NH$_3$)$_4$Cl$_2$]SO$_4$

(c) [Cr(H$_2$O)$_5$Cl]Cl$_2$

(d) Na[Au(CN)$_2$]

(e) [Ni(CO)$_4$]

(f) [Co(NH$_3$)$_4$(H$_2$O)Br](NO$_3$)$_2$

(g) [Co(NH$_3$)$_5$Cl]Cl$_2$

(h) [Al(OH)(H$_2$O)$_5$]$^{2+}$

(i) [Ru(NH$_3$)$_5$(N$_2$)]Cl$_2$

28.8. Write formulas for:

(a) tetraaquadichloroplatinum(IV) chloride

(b) ammonium diamminetetrachlorochromate(III)

(c) tetraamminesulfatoplantinum(IV) nitrite

(d) tetraaquadihydroxoaluminum(III) ion

(e) sodium hexafluorosilicate(IV)

28.9. Write formulas for:

(a) hexacarbonylvanadium(0)

(b) zinc hexachloroplatinate(IV)

(c) tetraamminechloronitrocobalt(III) sulfate

(d) sodium dithiosulfatoargentate(I)

(e) tetraammineplatinum(II) amminetrichloroplatinate(II)

(f) diamminetertachloroplatinum(IV)

(g) potassium hexabromoaurate(III)

(h) hexaamminenickel(II) hexanitrocobaltate(III)

28.10. Name the following compounds:

(a) K$_4$[Ni(CN)$_4$]

(b) K$_2$[Ni(CN)$_4$]

(c) (NH$_4$)$_2$[Fe(H$_2$O)Cl$_5$]

(d) [Cu(NH$_3$)$_4$][PtCl$_4$]

(e) [Ir(NH$_3$)$_5$(ONO)]Cl$_2$

(f) [Co(NH$_3$)$_6$]$_2$[Ni(CN)$_4$]$_3$

28.11. Write formulas for (a) iron(II) hexacyanoferrate(II); (b) iron(II) hexacyano-ferrate(III); (c) iron(III) hexacyanoferrate(II); (d) iron(III) hexacyanoferrate (III).

28.12. A student prepared one sample of $CoCl_3 \cdot 4\,NH_3$ which was violet colored and another sample of $CoCl_3 \cdot 4\,NH_3$ which was green. Explain, using structural diagrams, how this could be.

28.13. One mole of $CoCl_3 \cdot 5\,NH_3 \cdot H_2O$ when reacted completely with $AgNO_3$ precipitated 3 moles of AgCl. However when one mole of $CoCl_3 \cdot 5\,NH_3$ was reacted completely with $AgNO_3$, only 2 moles of AgCl were formed. Explain.

28.14. Write balanced equations for the following reactions which occur because of the formation of complex compounds with coordination numbers 2, 4, or 6.
(a) Solid AgCl is dissolved in aqueous ammonia.
(b) Formation of a deep blue color when Cu^{2+} is reacted with ammonia.
(c) Silver bromide is dissolved off a photographic film by reaction with sodium thiosulfate (hypo) to form the trithiosulfate complex ion.
(d) Aluminum hydroxide is dissolved in excess sodium hydroxide.
(e) Mercuric iodide is dissolved in a potassium iodide solution.
(f) An electroplating bath is made from silver nitrate and sodium cyanide.
(g) Formation of a dark red-colored solution when ammonium thiocyanate ($NH_4\,CNS$) is added to a solution containing ferric ion.
(h) A lead chloride precipitate partially dissolves in concentrated hydrochloric acid.
(i) Iron stains ($Fe_2O_3 \cdot 3\,H_2O$) are removed with oxalic acid, $H_2C_2O_4$.
(j) Nitric acid is added to a clear solution prepared from AgCl and NH_3.

28.15. Write the structures for the cis and trans isomers of: (a) $[Pt(py)_2Cl_2]$, a square planar structure where py represents the unidentate ligand pyridine; (b) $[Cr(H_2O)_4(NO_2)_2]^+$, an octahedral complex ion.

28.16. In qualitative analysis, Cu^{2+} may be separated from Cd^{2+} in a mixture of the two ions by treating the mixture with cyanide ion and then precipitating the Cd^{2+} as cadmium sulfide. Give a logical explanation of this separation.

28.17. Calculate the Ag^+ concentration remaining at equilibrium when 0.001 M $AgNO_3$ is made 0.1 M in NH_3. K_d for $Ag(NH_3)_2^+ = 6.0 \times 10^{-8}$. Assume, from the value of K_d, that nearly all of the Ag^+ is converted to the complex ion.

28.18. What would be the (a) Ag^+ and (b) CN^- concentration in an electroplating bath made by dissolving 0.050 mole of $K[Ag(CN)_2]$ per l of solution? K_d for $Ag(CN)_2^- = 1.9 \times 10^{-19}$.

Supplementary Readings

28.1. **Coordination Compounds**, D. F. Martin and B. B. Martin, McGraw-Hill, New York, 1964 (paper).
28.2. "The Chemical Elements of Life," E. Frieden, *Sci. American*, July 1972; p. 52.
28.3. "Volatile Metal Chelate Complexes," C. Kutal, *J. Chem. Educ.*, **52**, 319 (1975).
28.4. **Transition Elements**, E. M. Larsen, Benjamin, New York, 1965.

Chapter 29

Main Energy Levels	1	2	3	4	5	6
$^{63.546}_{29}$Cu	K	L	M	s^1		
GROUP IB $^{107.868}_{47}$Ag	K	L	M	$s^2p^6d^{10}$	s^1	
$^{196.97}_{79}$Au	K	L	M	N	$s^2p^6d^{10}$	s^1
$^{65.37}_{30}$Zn	K	L	M	s^2		
GROUP IIB $^{112.40}_{48}$Cd	K	L	M	$s^2p^6d^{10}$	s^2	
$^{200.59}_{80}$Hg	K	L	M	N	$s^2p^6d^{10}$	s^2

Group IB and
Group IIB Metals

Copper, silver, and gold have an outer s^1 electron, and all exhibit a characteristic valence of $+1$. By transfer of one or more electrons from the d level, a higher oxidation state is attained. Copper commonly is divalent in compounds, gold is frequently trivalent, but silver normally exhibits an electrovalence of one.

In contrast to alkali metals these elements are low in the activity series, with standard reduction potentials more positive than that of hydrogen.

Group IIB elements are terminal elements of three transition series. The oxidation number of $+2$ is characteristic. These three elements, Zn, Cd, and Hg, display a greater than usual difference in properties among themselves. Zinc and cadmium are quite similar; both are slightly above the middle in the activity series of metals, but mercury is near the bottom. In compounds, the elements exist as Zn^{2+}, Cd^{2+}, Hg^{2+} and the unusual Hg_2^{2+}. This last ion is unique among cations and is considered a result of two Hg^+ ions sharing a pair of s electrons.

Typical of transition elements, Group IB and IIB metals readily form complex ions and coordination compounds.

29.1   COPPER, SILVER, AND GOLD

Copper ores are widely distributed. Some native copper is found in the Lake Superior region of Michigan, on the island of Cyprus, and in Bolivia, though the metal usually occurs in the combined state, as sulfides, oxides, or carbonates. Most of the copper produced in this country comes from sulfide minerals, the chief of which is **chalcopyrite** ($CuFeS_2$). The principal copper-producing localities in the United States are Montana, Arizona, and Utah. South America and central Africa also have large copper mines.

Copper in fairly high percentages has been found in the feathers of certain birds of brilliant plumage. The touraco bird of Africa has a quite high copper content in its feathers, which have the red color of cuprite and the green color of malachite. Traces of copper are found in several sea plants and marine animals. It seems that a copper compound is substituted for an iron compound as an oxygen carrier in the hemoglobin of the blood of oysters. There is some evidence to indicate that copper in minute amounts plays a vital role in the health and well-being of man and animals. For example, a healthy cow's liver contains between 40 and 200 parts per million of copper.

Metallic silver has been known since prehistoric times. The symbol for silver (Ag) comes from the Latin *argentum*, which means white. The alchemists associated silver with the moon and used a crescent as the symbol for the element.

In nature, silver occurs both free and combined. In the native state it is sometimes found as nuggets but usually occurs in veins in rock masses as the sulfide, Ag_2S, of the mineral named **argentite.** The element is nearly always present in copper and lead sulfide ores and follows these metals through the various metallurgical processes. Silver is recovered in the refining of copper and lead, and this constitutes one of the principal sources of silver produced; in the United States approximately 80% of the silver is obtained in this way. About 70% of the world's silver comes from the western hemisphere, with Mexico the leading producer, followed by the United States, Canada, and Peru.

Gold was one of the first metals known to man. History records the use of the metal for jewelry and ornamental purposes in very early times. The fact that the metal occurs in the free state, is pleasing in appearance, and is easily worked accounts for its early use.

Gold nearly always occurs in the native state — either in placer deposits in river sands or in veins disseminated in quartz. Native gold is occasionally found in the form of nuggets. One nugget found in Australia weighed more than a hundred pounds and was worth $50,000. The only combined source of gold of any consequence is the telluride mineral $(AgAuTe_2)$, small deposits of which are found in Colorado. Some gold is usually associated with copper and lead ores. While gold is widely distributed over the earth, relatively few deposits are worth working. A very small amount of gold is present in sea water — less than 0.00005 g per ton of water. More than 50% of the world's gold comes from mines in South Africa.

Metallurgy of Copper, Silver, Gold

In general, the metallurgy of copper is quite complex, and the method of treatment depends on several factors, including the type of ore and the facilities for treatment. The cyanide solution leaching process for gold recovery is discussed in Chapter 22. Most of the silver produced in this country is a by-product of the refining of copper and lead. In the electrolytic refining of copper, the silver finds its way into the anode sludge, or mud, along with gold, platinum, bismuth, and arsenic.

Gold production   The era of gold production that followed the discovery of America was in all probability the greatest the world had witnessed up to that time. The exploitation of mines and the looting of palaces, temples, and graves in Central and South America resulted in an influx of gold that unbalanced the economic structure of Europe and disturbed its political structure. The history and romance of Spanish galleons carrying Inca gold as treasure lead us to wonder about the bulk and value of this gold. From 1492 to 1600, more than 8,000,000 ounces of gold came from South America — 35% of the world production during this time. Eight million ounces of gold would have a bulk of an eight-foot cube and a value of $280,000,000. Contrary to common belief, it did not take many galleons to carry this bulk of gold — there were no large rooms filled with golden treasure. Actually, this amount of gold, which seemed of fabulous value to people of the fourteenth century, is small as compared with money values involved in twentieth-century world finance.

Since most mined gold goes through counting houses and mints, a fairly accurate account of all gold mined since 1493 is recorded. The world production from 1493 to 1970 closely approximates 2,000,000,000 ounces. This gold would have the bulk of a forty-seven-foot cube, which is indeed a small volume of matter to have so influenced the toil and destiny of so many people.

Placer mining is based on the high specific gravity of gold. Usually gold occurs as small particles mixed with sand and gravel. The sand is washed in pans or long troughs, called sluices, and since the particles of gold are heavier than the sand, they settle to the bottom and are retained behind cleats or riffles, while the lighter particles are washed away.

Gold ore obtained from hard-rock vein deposits is crushed to a powdery condition, and a water suspension of the ore is passed over copper plates covered with a coating of mercury. Mercury amalgamates with the gold, and at intervals the copper plates are scraped. The amalgam (Hg–Au) is heated and the mercury distilled from the gold, which remains as a residue. The recovered mercury is used over and over again. The tailings from this treatment may be leached with cyanide solution, which removes more of the gold.

Reference: Gold, *Encyclopedia Britannica*, W. E. Caldwell.

29.2 ZINC, CADMIUM, AND MERCURY

The main ore mineral of zinc is ZnS, sphalerite. This mineral after concentration from ore is given a typical roasting to oxide, followed by carbon reduction —a topic covered in the metallurgy chapter.

Cadmium is a constituent of zinc ores in the proportion of about 1 part in 200. The mineral CdS (called **greenockite**) is also known but is relatively unimportant as a source of cadmium. Cadmium is obtained as a by-product in the smelting of zinc ores. After zinc sulfide ore is roasted, both cadmium and zinc are present as oxides.

Mercury was probably known in very early times, though it seems likely that it was discovered after silver, gold, copper, tin, and lead. Mercury played an important role in the early work of the alchemists, since it was regarded by many as one of the elements of which all matter is composed. It was given the name "quicksilver" because of its liquid character and its silver-white appearance. Mercury was not recognized as a true metal or as an element until the middle of the eighteenth century. Its Latin name, *hydrargyrum*, meaning liquid silver, is the source of the symbol Hg.

Ore containing mercury(II) sulfide, **cinnabar,** is heated or roasted in small furnaces or retorts. If oxide of mercury is formed in the roasting process, it quickly decomposes because of its instability toward heat, and metallic mercury is formed.

$$HgS + O_2 \rightarrow Hg + SO_2$$

Because of its volatile nature, the mercury is distilled from the impurities present and collected in iron flasks. It may be further purified by being filtered through chamois skin or by being dropped slowly through dilute nitric acid, which oxidizes impurities more easily than it oxidizes the metal itself. Redistillation provides another means of purification. This relatively simple metallurgy of mercury accounts for its recovery from red cinnabar mineral even in biblical times.

29.3   PHYSICAL PROPERTIES OF GROUP IB AND IIB METALS

These elements have a greater bulk use as metals than in compounds, and their physical properties vary widely as shown in Table 36.1.

Table 29.1

Metal	Symbol	Sp. Gr.	Melting Point °C	Boiling Point °C	Atomic Radius (Å)	Ionic Radius (Å)	Reduction Potential (volts)
Copper	Cu	8.9	1083	2595	1.17	0.96	+0.34
Silver	Ag	10.5	960	2212	1.34	1.26	+0.80
Gold	Au	19.3	1063	2966	1.34	1.37	+1.50
Zinc	Zn	7.14	420	907	1.25	0.74	−0.76
Cadmium	Cd	8.65	321	765	1.41	0.96	−0.40
Mercury	Hg	13.60	−38.87	357	1.44	1.1	+0.85

Gold is the most malleable and ductile of the metals. It can be hammered into sheets of 0.00001 inch in thickness; one gram of the metal can be drawn into a wire 1.8 mi in length. Copper and silver are also metals that are mechanically easy to work. Zinc is a little brittle at ordinary temperatures, but may be rolled into sheets at between 120° to 150°C; it becomes brittle again about 200°C. The low-melting temperatures of zinc contribute to the preparation of zinc-coated iron, **galvanized iron;** clean iron sheet may be dipped into vats of liquid zinc in its preparation. A different procedure is to sprinkle or air blast zinc dust onto hot iron sheeting for a zinc melt and then coating; this process is called **sherardizing.**

Cadmium has specific uses because of its low-melting temperature in a number of alloys. Cadmium rods are used in nuclear reactors because the metal is a good neutron absorber.

Mercury vapor and its salts are poisonous, though the free metal may be taken internally under certain conditions. Because of its relatively low boiling point and hence volatile nature, free mercury should never be allowed to stand in an open container in the laboratory. Evidence shows that inhalation of its vapors is injurious.

The metal alloys readily with most of the metals (except iron and platinum) to form **amalgams,** the name given to any alloy of mercury.

Uses of Metallic Cu, Ag, Au; Zn, Cd, Hg

Large quantities of copper are used in the manufacture of wire, cables, and bars for transmission of electric current. Smaller amounts are used in the construction of motors, generators, and other electrical devices. Because of its resistance to corrosion, copper may be used as a roofing material and for covering ships' bottoms. Cooking vessels used for cooking food in large quantities may be made of copper; however, since copper salts are poisonous, it is necessary to clean these utensils frequently to prevent contamination of foods.

One of the principal uses of the metal is in the preparation of alloys. The two most common and important alloys are **brass,** which contains copper and zinc, and **bronze,** which contains copper and tin. "Bronze" or "brass" does not refer to alloys of a definite composition but rather to all alloys of copper with zinc and tin in varying percentages. A small amount of copper is frequently alloyed with steel because the resistance of steel to corrosion is greatly increased by as little as 0.25% copper. Copper is also alloyed with gold, silver, and nickel in coinage.

Silver is used principally for ornaments, coins, jewelry, and various articles of silverware. Pure silver is too soft for these purposes, and it is usually alloyed with copper to increase its hardness and durability. **Sterling** is 92.5% silver and 7.5% copper; it is said to be 925 fine. Silver is often plated onto cheaper metals. Copper may be readily plated on iron, and this in turn may be plated with silver—silver adheres much more readily to copper than to iron. The plating bath is usually a solution of a complex silver cyanide. Silver may be precipitated on mirrors by reducing a silver salt with an organic reducing agent; the surface is then highly polished.

Gold is used principally for coinage and jewelry. Since pure gold is too soft for these purposes, it is usually alloyed with copper to increase its hardness. The purity (1000 fine) of gold is measured in **carats**—24 carat gold is 100% gold. Thus 18 carat gold is $\frac{18}{24}$ or 75% gold, 12 carat gold is 50% and so on. American gold coinage is 21.6 carats, or 90%; British coinage is 22 carats, or 91.7%. Many alloys containing varying percentages of gold are on the market—for example, white gold (gold and silver) and green gold (gold and cadmium or silver). Various articles are electrolytically plated with a thin coating of gold.

The chief use of zinc is in the manufacture of **galvanized iron.** Galvanized iron is much more resistant to corrosion than plain iron, since the zinc forms a protective, basic carbonate coating. Its resistance to corrosion makes galvanized iron useful for roofing, gutters, and cornices. Galvanized iron has the advantage over tin plate (sheet iron coated with tin) of resisting corrosion longer after the surface is broken and the iron exposed. Since zinc is more active than iron, it corrodes first, thus protecting the iron. Once the surface of the tin plate is broken, the iron corrodes first, since it is more active than tin.

Important alloys of zinc are **brass** (Cu, Zn), **bearing metal** (Zn, Cu, Sn, Sb), and **German silver** (Cu, Zn, Ni).

Cadmium is increasing in economic importance. Its principal uses are in low-melting alloys for fuses and automatic fire sprinklers, and in the production of bearing metal. It offers promise as a plating material instead of zinc and tin. Alloyed with copper, it produces a product of durability, hardness, and tensile strength which is used for telegraphic and power transmission lines. An alloy of cadmium is used in making jewelry. Cadmium rods are used in atomic energy piles because the metal is a good neutron absorber as mentioned above.

Salts of mercury have important medicinal uses. Its convenient, wide range between freezing and boiling points and its uniform expansion with rise in temperature make mercury widely used in the manufacture of thermometers. Its high density, 13.6, makes it useful as a liquid in barometers. Some mercury is used in the preparation of dental amalgams of gold and silver for filling teeth. Because of its thermal properties, it is used in some modern boilers instead of water to attain higher heat efficiency.

Compounds of Copper

Copper lies low in the activity series, and thus is quite resistant to corrosion. Dry air, water, and nonoxidizing acids do not attack the metal. However, long exposure to air and moisture gradually forms the tight coating of black oxide or green basic carbonate on the surface which can be seen on copper drain pipe or roofing. Corrosion then ceases, in direct contrast with the rusting of iron, which in time changes the metal over completely to iron oxide. Hydrochloric acid does not act on copper unless oxygen is present, in which case it gradually dissolves.

Nitric acid acts readily on the metal to form copper nitrate, but hydrogen is *not* liberated; instead, the other products are oxides of nitrogen and water. Hot, concentrated sulfuric acid dissolves the metal to form copper sulfate, sulfur dioxide, and water.

$$Cu + 2\ H_2SO_4 \rightarrow CuSO_4 + SO_2 + 2\ H_2O$$

Copper forms two series of compounds: copper(I) (**cuprous**) and copper(II) (**cupric**) compounds. Copper(II) compounds, because of their greater stability, are more commonly used.

Copper(I) oxide (Cu_2O) occurs naturally as the mineral **cuprite**. The red oxide may also be obtained by careful oxidation of copper in air or by precipitation of Cu^+ ions from solutions with sodium or potassium hydroxide. Copper(I) hydroxide is not known.

Black **copper(II) oxide** may be obtained by heating copper in air to a fairly high temperature, or by heating $Cu(OH)_2$.

Copper sulfate, or **blue vitriol** ($CuSO_4 \cdot 5\ H_2O$) is the most important and widely used salt of copper. On heating, the salt slowly loses water to form first the trihydrate ($CuSO_4 \cdot 3\ H_2O$), then the monohydrate ($CuSO_4 \cdot H_2O$), and finally the white anhydrous salt. The anhydrous salt is often used to test for the presence of water in organic liquids. For example, some of the anhydrous copper salt added to alcohol (which contains water) will turn blue because of the hydration of the salt.

Copper sulfate is used in electroplating, and occasionally in lakes or other bodies of water to kill algae and other vegetative growth. Fishermen dip their nets in copper sulfate solution to inhibit the growth of organisms that would rot the fabric. Paints specifically formulated for use on the bottoms of marine craft contain copper compounds to inhibit the growth of barnacles and other organisms.

When dilute ammonium hydroxide is added to a solution of copper(II) ions, a greenish precipitate of $Cu(OH)_2$ or a basic copper(II) salt is formed. This dissolves as more ammonium hydroxide is added. The excess ammonia forms an ammoniated complex with the copper(II) ion of the composition, $Cu(NH_3)_4^{2+}$.

$$Cu^{2+} + 4\ NH_3 \rightleftarrows Cu(NH_3)_4^{2+}$$

This ion is only slightly dissociated; hence in an ammoniacal solution very few copper(II) ions are present. Insoluble copper compounds, except copper sulfide, are dissolved by ammonium hydroxide. The formation of the copper(II) ammonia ion is often used as a test for Cu^{2+} because of its deep, intense blue color.

Copper(II) ferrocyanide $[Cu_2Fe(CN)_6]$ is obtained as a reddish-brown precipitate on the addition of a soluble ferrocyanide to a solution of copper(II) ions. The formation of this salt is also used as a test for the presence of copper(II) ions.

Compounds of Silver and Gold

Silver nitrate, sometimes called **lunar caustic,** is the most important salt of silver. It melts readily and may be cast into sticks for use in cauterizing wounds. The salt is prepared by dissolving silver in nitric acid and evaporating the solution.

$$3\ Ag + 4\ HNO_3 \rightarrow 3\ AgNO_3 + NO + 2\ H_2O$$

The salt is the starting material for most of the compounds of silver, including the halides used in photography. It is readily reduced by organic reducing agents, with the formation of a black deposit of finely divided silver; this action is responsible for black spots left on the fingers from the handling of the salt. Indelible marking inks and pencils take advantage of this property of silver nitrate.

The **halides** of silver, except the fluoride, are very insoluble compounds and may be precipitated by the addition of a solution of silver salt to a solution containing chloride, bromide, or iodide ions. Of the three halides, silver iodide is the most insoluble. This compound is yellow; silver bromide is pale yellow; silver chloride is white. Since all three salts are darkened on exposure to light, they are used in photography.

The addition of a strong base to a solution of a silver salt precipitates brown **silver oxide** (Ag_2O). One might expect the hydroxide of silver to precipitate, but it seems likely that silver hydroxide is very unstable and breaks down into the oxide and water—if, indeed, it is ever formed at all. However, since a solution of silver oxide is definitely basic, there must be hydroxide ions present in solution.

$$Ag_2O + H_2O \rightleftharpoons 2\ Ag^+ + 2\ OH^-$$

Because of its inactivity, gold forms relatively few compounds. Two series of compounds are known—monovalent and trivalent. Monovalent (aurous) compounds resemble silver compounds (aurous chloride is water insoluble and light sensitive), while the higher valence (auric) compounds tend to form complexes. Gold is resistant to the action of most chemicals—air, oxygen, and water have no effect. The common acids do not attack the metal, but a mixture of hydrochloric and nitric acids (aqua regia) dissolves it to form gold(III) chloride or chloroauric acid. The action is probably due to free chlorine present in the aqua regia.

$$3\ HCl + HNO_3 \rightarrow NOCl + Cl_2 + 2\ H_2O$$
$$2\ Au + 3\ Cl_2 \rightarrow 2\ AuCl_3$$
$$AuCl_3 + HCl \rightarrow HAuCl_4$$
chloroauric acid ($HAuCl_4 \cdot H_2O$ crystallizes from solution)

Photography

Photography is based on the sensitivity of the silver halides to light. Silver bromide is the most commonly used of the halides. The salt is precipitated from silver nitrate and potassium bromide in the presence of gelatin, which acts as a protective colloid (page 522) to prevent the particles of silver bromide from becoming too coarse. The gelatin solution is spread as a thin coating on film made from cellulose acetate.

The film is **exposed** by opening the shutter of a camera for a moment (Figure 29.1). The image of the object is projected onto the film through the lens. The light reflected from the object causes a change, directly proportional to the intensity of the light, in the silver bromide emulsion. Light parts of the object will reflect more light than darker portions, and thus effect a greater change on the film. The film is **developed** by washing it in a solution of a mild reducing agent, such as hydroquinone or ferrous oxalate. The ease of reduc-

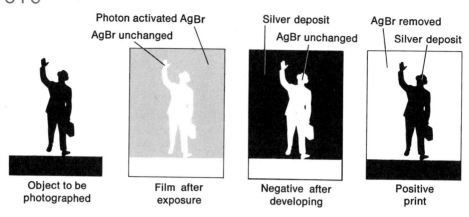

Figure 29.1

Steps in making a photograph.

tion is a direct function of the intensity of light falling on the film during exposure. The light parts of the object appear as dark patches on the plate because of the reduction of the silver bromide, probably to finely divided silver. The film is next **fixed** by washing it in a solution of sodium thiosulfate (hypo), which dissolves the unchanged silver bromide that has not been affected in the exposed stage. After being washed thoroughly with water, this **negative** is dried. **Printing** consists of allowing light to pass through the negative onto paper coated with silver bromide. The light and dark portions of the negative are again reversed, and the object now appears as a **positive.** The process of developing and fixing the positive is essentially a repetition of the treatment of the negative.

Compounds of Zinc

Zinc is fairly high in the activity series. It reacts readily with acids to produce hydrogen and displaces less active metals from their salts. The action of acids on impure zinc is much more rapid than on pure zinc, since bubbles of hydrogen gas collect on the surface of pure zinc and slow down the action. If another metal is present as an impurity, the hydrogen is liberated from the surface of the contaminating metal rather than from the zinc. An electric couple to facilitate the action is probably set up between the two metals.

$$Zn + 2 H^+ \rightarrow Zn^{2+} + H_2$$

Zinc oxide (ZnO), the most widely used zinc compound, is a white powder at ordinary temperatures, but changes to yellow on heating. When cooled, it again becomes white. Zinc oxide is obtained by burning zinc in air, by heating the basic carbonate, or by roasting the sulfide. The principal use of ZnO is as a filler in rubber manufacture, particularly in automobile tires. As a body for paints it has the advantage over white lead of not darkening on exposure to an atmosphere containing hydrogen sulfide. Its covering power, however, is inferior to that of white lead. Thus it is used primarily for interior paints. Because of its mild alkaline and antiseptic properties, zinc oxide is also used in ointments and medicinally as an antiseptic powder.

Zinc hydroxide, $Zn(OH)_2$, forms as a white precipitate when sodium or potassium hydroxide is added to a solution of a zinc salt. The hydroxide is amphoteric. An excess of OH^- will redissolve the precipitated zinc hydroxide, with the formation of zincate ions:

$$Zn^{2+} + 2\ OH^- \rightleftharpoons Zn(OH)_2(s)$$
$$Zn(OH)_2(s) + 2\ OH^- \rightleftharpoons Zn(OH)_4^{2-} \text{ or } ZnO_2^{2-} \cdot 2\ H_2O$$

Acids reverse the arrangement back to Zn^{2+}.

$$Zn(OH)_4^{2-} + 2\ H^+ \rightleftharpoons Zn(OH)_2(s) + 2\ H_2O$$
$$Zn(OH)_2(s) + 2\ H^+ \rightarrow Zn^{2+} + 2\ H_2O$$

Zinc sulfide (ZnS), a white salt, is precipitated in neutral or alkaline solution with hydrogen sulfide.

$$Zn^{2+} + H_2S \rightarrow ZnS(s)$$

Zinc sulfide is an ingredient of **lithopone,** a mixture of barium sulfate and zinc sulfide used as a paint body. It is produced by the reaction

$$BaS + ZnSO_4 \rightarrow BaSO_4(s) + ZnS(s)$$

Compounds of Cadmium and Mercury

Cadmium sulfide, a yellow salt, is precipitated from solutions of cadmium salts by hydrogen sulfide It is less soluble in acid than zinc sulfide. Cadmium sulfide is used as a yellow pigment in paint manufacture. **Cadmium sulfate hydrate** ($3\ CdSO_4 \cdot 8\ H_2O$) and metallic cadmium are used in making standard electric cells for scientific work.

Exposure control in automatic cameras depends upon the functioning of a CdS cell. The brighter the scene the greater the flow of current as the light activates the CdS cell. The magnitude of the current controls the mechanism which adjusts the lens opening.

Mercury is one of the **noble** metals; it appears near the bottom of the activity series and is very inactive. It is not affected by oxygen of the air at ordinary temperatures, but if heated to about 300°C it slowly combines with oxygen to form mercury(II) oxide. It does not dissolve in the nonoxidizing acids, but dissolves readily in nitric acid to form mercury(II) nitrate.

$$3\ Hg + 8\ HNO_3 \rightarrow 3\ Hg(NO_3)_2 + 2\ NO + 4\ H_2O \text{ (excess } HNO_3\text{)}$$

Mercury forms two series of compounds: mercury(I) (mercurous) and mercury(II) (mercuric) compounds. The most important of the mercury(I) compounds is the chloride, Hg_2Cl_2, commonly called **calomel,** which is used medicinally as a liver stimulant and cathartic. Mercury(I) chloride is insoluble, and therefore is not readily absorbed by the body when taken internally.

Mercury(I) or **mercurous nitrate** $[Hg_2(NO_3)_2]$ is prepared from nitric acid and excess mercury and is one of the few soluble mercurous salts.

Mercury(II) or **mercuric chloride** ($HgCl_2$), commonly called **corrosive sublimate,** is a white crystalline salt soluble in water. The chloride is very poisonous and is widely used in dilute solutions (about 1 percent) as a germicide. If taken internally, its action on the kidneys is destructive and causes death. The albumen combines with the mercury(II) ions to form an insoluble white solid. The action of Hg^{2+} on the protein albumen is like the action of heat on the white of an egg. Swallowing egg white is an antidote for Hg(II) salt poisoning.

The action of ammonia on mercury(I) chloride gives the amidochloride, a white solid

$$Hg_2Cl_2 + 2\ NH_3 \rightarrow Hg(l) + HgNH_2Cl(s) + NH_4Cl$$

However, the finely divided mercury also produced gives a black precipitate, and the reaction is often used as a confirmatory test for the presence of mercury(I) ions.

Exercises

29.1. The following are some common minerals found in ores containing Group IB and IIB metals. Attempt to identify the valuable metal which is present in each of these minerals.
(a) argentite, (b) cuprite, (c) aurum, (d) cinnabar, (e) chalcopyrite

29.2. (a) Consulting the appropriate tables in this text, compare the elements cesium and gold on the basis of atomic radii, specific gravity, outer electron structure, and standard electrode potentials. (b) Do the same for potassium and copper.

29.3. List an important use for each of the Group IB and IIB free metals.

29.4. (a) If each of the above metals was placed in concentrated HCl, which ones would react? (b) For those which react, write the chemical equation. (c) Which metals would react with dilute nitric acid?

29.5. Using your knowledge of chemistry, explain briefly how you would distinguish between:
(a) Au and "fool's gold" (FeS_2)
(b) Ag and Zn
(c) Hg and Cd
(d) brass and bronze

29.6. An open pit mine delivers ore of 1% copper in a siliceous gangue. The copper minerals are red Cu_2O and green $CuCO_3 \cdot Cu(OH)_2$. Write equations for reactions that occur in a dilute H_2SO_4 leach of this crushed ore. How does one get Cu from the leach solution?

29.7. A sulfide copper ore concentrate (product of gravity and flotation concentration) contains 10% Cu_2S, 10% $CuFeS_2$, and 25% FeS_2. Assume that it is given a partial roast to convert FeS_2 to FeS and CuS to Cu_2S. A matte (Cu_2S, FeS mix) remains. Explain how to get copper industrially from the matte.

29.8. Why are fruits and vegetables canned in tin-plated iron rather than cheaper galvanized iron?

29.9. Write formulas or names for:
(a) copper(I) sulfide, (b) blue vitriol, (c) silver iodide, (d) hypo, (e) sodium zincate, (f) calomel, mercurous chloride, (g) $HgNH_2Cl$, (h) $AuCl_3$, (i) $Hg(NO_3)_2$, (j) $CdSO_4 \cdot \frac{8}{3} H_2O$.

29.10. Identify the principle materials in the following common materials:
(a) active ingredient in a photographic film
(b) galvanized iron
(c) 12 carat gold
(d) sterling silver
(e) amalgams
(f) "quicksilver"
(g) brass
(h) German silver
(i) lithopone in white paints

29.11. Write the following equations:
(a) amphoteric nature of $Zn(OH)_2$:
$$Zn(OH)_2 + HCl \rightarrow$$
$$Zn(OH)_2 + NaOH \rightarrow$$
(b) precipitation of Ag^+ and Hg_2^{2+} with HCl.
(c) precipitation of Cu^{2+}, Zn^{2+}, Cd^{2+}, and Hg^{2+} with H_2S.
(d) reaction of silver metal with dilute HNO_3.
(e) reaction of gold with aqua regia:
$$Au + HCl + HNO_3 \rightarrow HAuCl_4 + NO_2 + H_2O$$
(f) dissolving AgBr in aqueous NH_3 to form a complex ion:
$$AgBr(s) + NH_3(aq) \rightarrow$$

29.12. The verdigris coating ($Cu(OH)_2$, Cu_2O, $CuCO_3$) on the outside of long-weathered Cu sheeting dissolves off with HCl. Write the equations.

29.13. Given pure silver, write equation(s) for the method of preparation of pure AgBr.

29.14. A hydrate of zinc chloride is a much-used ingredient of a soldering paste (to clean metal surface). Show by equations how zinc chloride can hydrolyze in its own water of hydration to form an acid which then dissolves metal oxide from the metal surface.

29.15. A 0.2500-g sample of cuttings from a silver coin (containing Cu) was completely dissolved in HNO_3. Then an NaCl solution was added to precipitate AgCl, which when collected and dried weighed 0.2178 g. What is the indicated percent Ag in the coin?

29.16. Mercury is sold by the "flask of mercury," which is an iron cylinder weighing 20 kg (44 lb) and containing $2\frac{2}{3}$ l of liquid mercury. What is the total approximate weight of the cylinder plus the mercury in kilograms? In pounds?

Supplementary Readings

29.1. "The Gold Content of the Sea," G. L. Putnam, *J. Chem. Educ.*, 30, 576 (1953).
29.2. "Photographic Development," T. H. James, *Sci. American*, Nov., 1952; p. 30.
29.3. "Mercury in the Environment," L. J. Goldwater, *Sci. American*, May, 1971; p. 15.

Chapter 30

Main Energy Levels	1	2	3	4	5	6	7
$^{44.96}_{21}$Sc	K	L	$s^2p^6d^1$	s^2			
$^{88.9}_{39}$Y	K	L	M	$s^2p^6d^1$	s^2		
$^{138.9}_{57}$La	K	L	M	$s^2p^6d^{10}$	$s^2p^6d^1$	s^2	
$^{227}_{89}$Ac	K	L	M	N	$s^2p^6d^{10}$	$s^2p^6d^1$	s^2

Group IIIB

Groups IIIB–VIIIB Metals

Fifty-three metals are to be briefly dealt with in this chapter. This is over half of the periodic table to be considered in a few pages of the text, and only a few highlights concerning these elements will be presented.

30.1   GROUP IIIB ELEMENTS

Group IIIB includes the elements scandium, yttrium, lanthanum, and actinium, and the two rare-earth series of fourteen elements each—the lanthanide and actinide series. The principal source of these elements is the high gravity river and beach sands built up by a water-sorting process during long periods of geologic time. **Monazite** sand, which contains a mixture of rare earth phosphates, and an yttrium silicate in a heavy sand are now commercial sources of a number of these scarce elements.

Separation of the elements is a difficult chemical operation. The solubilities of their compounds are so nearly alike that a separation by fractional crystallization is laborious and time-consuming. In recent years, ion exchange resins in high columns have proved effective. When certain acids are allowed to flow down slowly through a column containing a resin to which ions of Group IIIB metals are adsorbed, ions are successively released from the resin. The resulting solution is removed from the bottom of the column or tower in **bands** or **sections**. Successive sections will contain specific ions in the order of release by the resin. For example, lanthanum ion (La^{3+}) is most tightly held to the resin and is the last to be extracted; lutetium ion (Lu^{3+}) is less tightly held and appears in one of the first sections removed. If the solutions are recycled and the acid concentrations carefully controlled, very effective separations can be accomplished. Quantities of all the lanthanide series (except promethium, Pr, which does not exist in nature as a stable isotope) are produced for the chemical market.

Scandium, Yttrium, Lanthanum, and Actinium

These elements may be regarded as the regular members of Group IIIB and not as "rare earths," a term usually reserved for elements of the lanthanide and actinide series. Scandium and yttrium are quite rare, and so far there has developed very little use for them, either as free metals or as compounds. Lanthanum is somewhat more abundant, but still comprises only 0.00007% of the earth's crust. These metals are very active, reacting vigorously with water to yield hydrogen. Their compounds are ionic, and the metals exhibit an oxidation number of +3. Typical compounds are $Sc(NO_3)_3$, $Y_2(SO_4)_3$, $La_2(C_2O_4)_3$, Na_3ScO_3, and $AcCl_3$.

The Lanthanide Series (Rare Earths; $4f^1$ to $4f^{14}$ Fillers)

The fourteen elements of this series are

Cerium, $_{58}$Ce	Europium, $_{63}$Eu	Erbium, $_{68}$Er
Praseodymium, $_{59}$Pr	Gadolinium, $_{64}$Gd	Thulium, $_{69}$Tm
Neodymium, $_{60}$Nd	Terbium, $_{65}$Tb	Ytterbium, $_{70}$Yb
Promethium, $_{61}$Pm	Dysprosium, $_{66}$Dy	Lutetium, $_{71}$Lu
Samarium, $_{62}$Sm	Holmium, $_{67}$Ho	

The predominant group oxidation number is $+3$, but some of the elements exhibit variable oxidation states. Cerium forms cerium(III) and cerium(IV) sulfates, $Ce_2(SO_4)_3$ and $Ce(SO_4)_2$, which are employed in certain oxidation-reduction titrations. Many rare earth compounds are colored and are paramagnetic, presumably as a result of unpaired electrons in the $4f$ orbitals.

Current uses for the rare earth metals and their compounds are limited; more will be found. An alloy of cerium and iron, if scraped by a rough, hard object, produces hot sparks and may be used in laboratory burners or cigarette lighters. Gasoline vapor or propane lamps emit light by the glow of a heated ThO_2-CeO_2 filament (mantle).

The Actinide Series (Sometimes Called Rare-Rare Earths; $5f^1$ to $5f^{14}$ Fillers)

All these elements have unstable nuclei and exhibit radioactivity (Chapter 35). Those with higher atomic numbers have been obtained only in trace amounts. Actinium ($_{89}$Ac), like lanthanum, is a regular Group IIIB element and gives its name to the series of fourteen elements listed below.

Thorium, $_{90}$Th	Americium, $_{95}$Am	Fermium, $_{100}$Fm
Protactinium, $_{91}$Pa	Curium, $_{96}$Cm	Mendelevium, $_{101}$Md
Uranium, $_{92}$U	Berkelium, $_{97}$Bk	Nobelium, $_{102}$No
Neptunium, $_{93}$Np	Californium, $_{98}$Cf	Lawrencium, $_{103}$Lr
Plutonium, $_{94}$Pu	Einsteinium, $_{99}$Es	

The elements exhibit variable oxidation states.

30.2 GROUP IVB ELEMENTS

Element	Symbol	Sp. Gr.	Melting Point °C	Boiling Point °C	Oxidation States
Titanium	Ti	4.5	1675	3260	$+3, +4$
Zirconium	Zr	6.5	1825	3578	$+4$
Hafnium	Hf	13.3	2150	5400	$+4$

In chemical properties these elements resemble silicon, but they become increasingly more metallic from titanium to hafnium. The predominant oxidation state is +4 and, as with silica (SiO_2), the oxides of these elements occur naturally in small amounts. The formulas and mineral names of the oxides are TiO_2, **rutile;** ZrO_2, **zirconia;** HfO_2, **hafnia.**

Titanium is more abundant than is usually realized. It comprises about 0.44% of the earth's crust. It is over 5.0% in average composition of first-analyzed moon rock. Zirconium and titanium oxides occur in small percentages in beach sands.

Titanium and zirconium metals are prepared by heating their chlorides with magnesium metal. Both are particularly resistant to corrosion and have high melting points.

Pure TiO_2 is a very white substance which is taking the place of white lead in many paints. It is 49th in the list of top chemicals. Three-fourths of the TiO_2 is used in white paints, varnishes, and lacquers. It has the highest index of refraction (2.76) and the greatest hiding power of all the common white paint materials. The outside house paints containing TiO_2 have controlled chalking and consequently stay white longer because of the new surfaces exposed. TiO_2 also is used in the paper, rubber, linoleum, leather, and textile industries.

The compound zirconia (ZrO_2) has a high melting point and low electrical conductivity and is used as a refractory and for electrical insulators.

30.3   GROUP VB ELEMENTS

Element	Symbol	Sp. Gr.	Melting Point °C	Common Oxidation States
Vanadium	V	6.1	1890	+2, +3, +4, +5
Niobium	Nb	8.6	2468	+3, +5
Tantalum	Ta	16.7	2850	+5

Vanadium, Niobium,* and Tantalum

These are transition elements of Group VB, with a predominant oxidation number of +5. Their occurrence is comparatively rare.

These metals combine directly with oxygen, chlorine, and nitrogen to form oxides, chlorides, and nitrides, respectively. A small percentage of vanadium alloyed with steel gives a high tensile strength product which is very tough and resistant to shock and vibration. For this reason vanadium alloy steels are used in the manufacture of high-speed tools and heavy ma-

---

° **Niobium** is the accepted international name, but **Columbium** is commonly used in the United States, with symbol Cb.

chinery. Vanadium oxide is employed as a catalyst in the contact process of manufacturing sulfuric acid. Niobium is a very rare element, with limited use as an alloying element in stainless steel. Tantalum is a steel-gray metal which may be rolled or hammered with difficulty. It has a very high melting point (2850°C) and is resistant to corrosion by most acids and alkalies; thus it may be used as a platinum substitute for certain laboratory utensils. Tantalum carbide is almost as hard as diamond and is a component of some abrasive and cutting mixtures. A marked improvement in rock drilling results from the use of tantalum carbide bits on the end of drill steel. Some rotary saw blades are tipped with tantalum carbide.

Vanadium Compounds

The variable oxidation states of vanadium are represented in the following salts:

$VSO_4 \cdot 7\ H_2O$	vanadium(II) sulfate	violet
VF_3	vanadium(III) fluoride	green
$VOSO_4$	oxyvanadium(IV) sulfate	blue
$VOCl_3$	oxyvanadium(V) chloride	yellow
NH_4VO_3	ammonium vanadate(V)	colorless
KVO_4	potassium peroxyvanadate(VII)	red

If powdered zinc is added to acid solutions of vanadates, a series of successive reduction products is formed: VO_3^- (vanadate ion); VO^{2+} (vanadyl ion); V^{3+} (vanadic ion), and finally V^{2+} (vanadous ion), with concomitant color changes.

30.4   GROUPS VIB AND VIIB ELEMENTS

Element	Symbol	Sp. Gr.	Melting Point °C	Boiling Point °C	Oxidation States
Chromium	Cr	7.1	1890	2482	+2, +3, +6
Molybdenum	Mo	10.2	2620	5560	+2, +3, +4, +6
Tungsten	W	19.3	3410	5900	+2, +4, +5, +6
Manganese	Mn	7.2	1244	2097	+2, +3, +4, +6, +7
Technetium	Tc	11.5	2200	—	—
Rhenium	Re	21.0	3180	5600	+2, +3, +4, +5, +6, +7

Chromium, molybdenum, and tungsten° are Group VI*B* elements. Manganese is the only chemically important element of Group VII*B*. All these elements exhibit several oxidation states, acting as metallic elements in lower

---

° The symbol for tungsten (W) is taken from its international name, **wolfram.**

oxidation states and as nonmetallic elements in higher oxidation states. Both chromium and manganese are widely used in alloys, particularly in alloy steels.

30.5 CHROMIUM

Occurrence and Preparation

The principal ore of chromium is $FeO \cdot Cr_2O_3$ or $Fe(CrO_2)_2$, called **chromite** or **chrome iron ore.**

The pure metal is most readily obtained by the reduction of chromium(III) oxide with metallic aluminum (by the Goldschmidt process, page 575).

$$Cr_2O_3 + 2\ Al \rightarrow 2\ Cr + Al_2O_3$$

Properties and Uses

Chromium is a very hard, silvery white metal which is very resistant to corrosion. It dissolves in hydrochloric acid, with the liberation of hydrogen.

Alloyed with iron, chromium produces steels of high tensile strength, toughness, hardness, and resistance to corrosion. **Stainless steel** contains 12 to 14% chromium. Chromium steels are valuable in making cutting tools because of the hardness of the alloy; they are also used in armor plate, safes, and vaults. It has been found that engraved plates coated with a thin layer of chromium last twice as long as casehardened steel plates.

Chromium is popular as a plating for automobile parts, such as radiators or grills and bumpers, for tableware, and for various fixtures. Such plated articles are resistant to corrosion and possess a bright, lasting luster.

Other alloys of chromium include **nichrome** (Ni, Cr, Fe), which is used as electrical resistance wire; and **stellite** (Cr, W, Co), which is used for cutting tools.

The name of the element chromium is derived from the Greek word *chroma*, meaning color: the compounds of the element are highly colored. Chromium exhibits three oxidation states $(+2, +3, +6)$ and forms three series of compounds.

Chromium(II) and Chromium(III) Compounds

Chromium(III) compounds are the most stable and important. Chromium(II) salts may be prepared by reduction of chromium(III) salts with a reducing agent, such as Zn, in the presence of an acid. The chromium(II) ion is readily oxidized in the air to the chromium(III) ion; hence chromium(II) compounds are very good reducing agents.

If NaOH is added to a solution containing Cr^{3+} ions, green gelatinous $Cr(OH)_3$ is precipitated. The hydroxide is amphoteric and will dissolve in an acid to form a chromium(III) salt, or will dissolve in an excess of alkali to form a chromite.

$$Cr^{3+} + 3\ OH^- \rightarrow Cr(OH)_3(s)$$
$$Cr(OH)_3 + OH^- \rightarrow Cr(OH)_4^- \qquad\qquad Cr(OH)_4^- + 2\ OH^- \rightarrow Cr(OH)_6^{3-}$$
$$CrO_2^- \cdot 2\ H_2O \qquad\qquad\qquad\qquad CrO_3^{3-} \cdot 3\ H_2O$$

Metachromite ion   Orthochromite ion

Chromium(III) oxide (Cr_2O_3) may be produced by heating the hydroxide or by heating the metal in air at an elevated temperature. This compound, green in color, is used as a paint pigment and for coloring certain ceramic products.

Chromium(III) sulfate $[Cr_2(SO_4)_3]$, used in the tanning of leather, may be produced from chromium(III) hydroxide and sulfuric acid.

Beautiful reddish violet crystals of **chrome alum**, $[K_2SO_4 \cdot Cr_2(SO_4)_3 \cdot 24\ H_2O]$ may be obtained by crystallizing a solution containing equimolecular quantities of potassium and chromium(III) sulfates.

Chromates and Dichromates

A diagramatic representation of some chromium compound changes follows. A, B, C, etc. refer to the equations below.

$$Cr_2O_7^{2-} \underset{OH^-\ (B)}{\overset{H^+\ (A)}{\rightleftarrows}} CrO_4^{2-}$$

orange   yellow

(C) | H_2O_2, H^+ (F) | H_2O_2, OH^-

$$Cr^{3+} \underset{H^+}{\overset{OH^-\ (D)}{\rightleftarrows}} Cr(OH)_3(s) \underset{H^+}{\overset{OH^-\ (E)}{\rightleftarrows}} CrO_3^{3-}$$

green   green ppt

(A) $2\ CrO_4^{2-} + 2\ H^+ \rightleftarrows Cr_2O_7^{2-} + H_2O$
(B) $Cr_2O_7^{2-} + 2\ OH^- \rightleftarrows 2\ CrO_4^{2-} + H_2O$
(C) $Cr_2O_7^{2-} + 3\ H_2O_2 + 8\ H^+ \rightarrow 2\ Cr^{3+} + 3\ O_2 + 7\ H_2O$
(D) $Cr^{3+} + 3\ OH^- \rightarrow Cr(OH)_3(s)$
(E) $Cr(OH)_3(s) + 3\ OH^- \rightarrow CrO_3^{3-} + 3\ H_2O$
(F) $2\ CrO_3^{3-} + 3\ O_2^{2-} + 4\ H_2O \rightarrow 2\ CrO_4^{2-} + 8\ OH^-$

Both chromates and dichromates are effective oxidizing agents. Reduction products include chromium(III) ions. In the tanning of leather, a dichromate is reduced to a chromium(III) salt, which undergoes hydrolysis and precipitates $Cr(OH)_3$ within the leather, so that decay is prevented. A mixture of a dichromate with concentrated sulfuric acid is used in cleaning laboratory glassware. $PbCrO_4$ and $BaCrO_4$ are used as yellow paint pigments, and basic lead chromate, $PbO \cdot PbCrO_4$, called **chrome red,** is also used as a paint pigment.

Molybdenum and Tungsten

These elements closely resemble chromium in properties. Like chromium they exhibit variable oxidation states and form several series of compounds.

The principal source of molybdenum is **molybdenite,** MoS_2. The largest mine for this mineral is in Colorado. Molybdenite is a soft, shiny mineral which looks very much like graphite. The sulfide may be roasted to yield the oxide, which can then be reduced with carbon to produce metallic molybdenum.

Most of the molybdenum is alloyed with iron to form **ferromolybdenum,** used in the production of special alloy steels. A steel containing molybdenum is heat resistant and is used in making highspeed tools.

Tungsten is also important for alloy steels, particularly for armor plate and steels which are to be subjected to high temperatures. Tungsten steel may be heated to high temperatures without losing its temper. Tungsten also finds use as filaments in incandescent lamps and radio tubes and as contact points in spark plugs.

Tungsten is a very hard, heavy metal and has the highest melting point of all the elements except carbon. It is very resistant to corrosion and to chemicals. It is finding special use in the exhaust gas nozzles of space missiles, and a plate of it is often the target of X-ray equipment.

Manganese

Manganese is found in nature principally as the mineral MnO_2, called **pyrolusite.** It is soft and black and usually contains impurities of iron oxide and silica. Russia produces about half the world's supply; smaller amounts come from Africa, India, and Brazil. Small deposits of the ore have been found in the United States.

Properties and Uses

The pure metal resembles iron in many properties. It has a somewhat reddish color and is relatively soft. As with iron, the addition of even a small percentage of carbon produces a very hard, brittle product. If the metal is heated to redness, it reacts with steam readily to form hydrogen. In the activity series, manganese appears between aluminum and zinc; hence it dissolves readily in acids.

About 90% of the manganese consumed in this country is used in the manufacture of steel because in small amounts its presence aids in the removal of sulfur and oxygen. In larger amounts the manganese alloys with the iron to produce a hard, tough product, which is more readily forged and machined than steels free from manganese. Manganese steel is particularly well adapted to the construction of steel rails and parts of machinery subject to extreme wear.

Manganese Compounds

Manganese forms a number of compounds with oxidation numbers ranging from +2 to +7. The following are examples:

$MnSO_4$	manganese(II) sulfate or manganous sulfate	pink
$Mn(OH)_3$	manganese(III) hydroxide or manganic hydroxide	brown
MnO_2	manganese(IV) oxide or manganese dioxide	black
K_2MnO_4	potassium manganate	green
$KMnO_4$	potassium permanganate	purple

The latter salt is an important compound widely used in analytical chemistry in oxidation-reduction titrations. For example, to determine the percentage of iron in a sample of ore, the iron mineral can be dissolved in an acid and the iron in solution reduced to Fe^{2+} followed by titration with a standard solution of $KMnO_4$. The equation for the titration is

$$2\ MnO_4^- + 10\ Fe^{2+} + 16\ H^+ \rightarrow 2\ Mn^{2+} + 10\ Fe^{3+} + 8\ H_2O$$

Since the MnO_4^- ion is purple and the Mn^{2+} ion practically colorless, the endpoint of the reaction is readily evident by the disappearance of color. In neutral or alkaline solutions the permanganates are reduced to MnO_2.

30.6 GROUP VIIIB METALS

Metal	Sp. Gr.	Melting Point °C	Boiling Point °C	Oxidation States	Atomic Radius (Å)	Ionic (M^{2+}) Radius (Å)
Iron	7.86	1535	2800	+2, +3, +6	1.17	0.75
Cobalt	8.71	1495	3000	+2, +3	1.17	0.72
Nickel	8.9	1453	2800	+2, +3	1.15	0.70
Ruthenium	12.4	2250	>3900	+2, +3, +4, +6, +7, +8	1.24	−°
Rhodium	12.5	1966	>3700	+2, +3, +4	1.25	−
Palladium	12.0	1552	2927	+2, +4	1.25	−
Osmium	22.5°	3000	>5000	+2, +3, +4, +8	1.26	−
Iridium	22.4	2410	>4500	+3, +4	1.26	−
Platinum	21.5	1769	3800	+2, +3, +4, +6	1.29	−

° Densest material on earth.   ° Variable radius in different oxidation states.

Group VIII*B* contains the three triads of elements listed above. These triads appear at the middle of long periods of elements in the periodic table, and are members of the transition series. The elements of any given horizontal triad have many similar properties, but there are marked differences between the properties of the triads, particularly between the first triad and the other two. Iron, cobalt, and nickel are much more active than members of the other two triads, and are also much more abundant in the earth's crust. Metals of the second and third triads, with many common properties, are usually grouped together and called the **platinum metals.**

These elements all exhibit variable oxidation states and form numerous coordination compounds.

Iron and Steel

Without question iron is the most important and useful of the metals. The period in which we now live is often designated the age of iron and steel, and it has been said that we probably could get along without all the other metals if we had iron. Ores of the metal are easily reduced with carbon, and its discovery was probably made by accidentally heating the ore in a wood fire. The occurrence of Fe and the metallurgy of iron and steel were covered in Chapter 25.

Properties of Pure Iron

Pure iron is a soft, silver-white metal with a melting point of 1535°C. The metal is fairly high in the activity series. It does not react with cold water but decomposes steam to form hydrogen. It readily displaces hydrogen from hydrochloric and sulfuric acids. When iron is placed in concentrated nitric acid, the metal assumes a **passive** condition which, according to one explanation, is due to the formation of an adherent oxide film in which iron exhibits oxidation state six or eight.° If the surface is scratched, the metal acts in a normal manner. Iron is magnetic, which makes it useful in the construction of motors, generators, and other electrical devices.

Corrosion

Iron exposed to the action of moist air rusts rapidly, with the formation of a loose, crumbly deposit of the oxide. The oxide does not adhere to the surface of the metal, as does aluminum oxide and certain other metal oxides, but peels

---

° Chromium and nickel also become passive in highly oxidizing media.

off, exposing a fresh surface of iron to the action of the air. As a result, a piece of iron will rust away completely in a relatively short time unless steps are taken to prevent the corrosion. The chemical steps in rusting are rather obscure, but it has been established that the rust is a hydrated oxide of iron, formed by the action of both oxygen and moisture, and is markedly speeded up by the presence of minute amounts of carbon dioxide.

$$Fe + 2\ H_2CO_3 \rightarrow Fe^{2+} + 2\ H + 2\ HCO_3^-$$
$$4\ H + O_2 \rightarrow 2\ H_2O$$
$$4\ Fe^{2+} + O_2 + (4 + 2x)\ H_2O \rightarrow 2\ Fe_2O_3 \cdot xH_2O + 8\ H^+$$

Corrosion of iron is inhibited by coating it with numerous substances, such as paint, an aluminum powder gilt, tin, or organic tarry substances or by galvanizing iron with zinc. Alloying iron with metals such as nickel or chromium yields a less corrosive steel. "Cathodic protection" of iron for lessened corrosion is also practiced. For some pipelines and standpipes zinc or magnesium rods in the ground with a wire connecting them to an iron object have the following effect: With soil moisture acting as an electrolyte for an Fe-Zn couple the Fe is lessened in its tendency to become Fe^{2+}. It acts as a cathode rather than an anode.

Compounds of Iron

Iron forms two series of compounds — iron(II) (**ferrous**), and iron(III) (**ferric**). Iron(III) compounds are the more stable, since in general iron(II) compounds may be oxidized to the higher state of valence quite easily, often by the oxygen of the air.

The addition of a soluble hydroxide to a solution containing iron(II) ions results in the formation of a pale green precipitate of iron(II) hydroxide. The compound is easily oxidized by oxygen of the air, and soon after precipitation (unless oxygen is excluded), oxidation to a hydrated iron(III) oxide occurs. A hydrated iron(III) hydroxide is precipitated as a brownish-red solid by the addition of a soluble base to a solution containing iron(III) ions. Neither iron(II) nor iron(III) hydroxides are amphoteric, and thus they are insoluble in an excess of base.

Iron compounds with alkali cyanides form two series of complex salts — **ferrocyanides**, which provide $Fe(CN)_6^{4-}$ (hexacyanoferrate(II)) ions in solution; and ferricyanides, which provide $Fe(CN)_6^{3-}$ (hexacyanoferrate(III)) ions. These two types of compounds are important in testing for the presence of iron(II) and iron(III) ions. Iron(III) ions react with a soluble ferrocyanide such as $K_4Fe(CN)_6$ to form a characteristic deep blue precipitate of $KFe[Fe(CN)_6]$ called **Prussian blue**.

$$Fe^{3+} + 4\ K^+ + Fe(CN)_6^{4-} \rightarrow 3\ K^+ + KFe[\,Fe(CN)_6\,]$$
$$\text{Prussian blue}$$

Iron(II) ions react with ferricyanides such as $K_3Fe(CN)_6$ to form the same blue compound:

$$Fe^{2+} + 3\ K^+ + Fe(CN)_6{}^{3-} \rightarrow KFe[\ Fe(CN)_6] + 2\ K^+$$

If iron is dissolved in dilute sulfuric acid and the solution evaporated to dryness, light-green crystals of the heptahydrate of iron(II) sulfate ($FeSO_4 \cdot 7\ H_2O$), sometimes called **green vitriol** or **copperas**, are obtained. This salt is one of the more widely used salts of iron, and much is obtained by the evaporation of "pickling" solutions, obtained as by-products of the cleaning of iron by dipping it in sulfuric acid before galvanization. If equimolecular quantities of iron(II) sulfate and ammonium sulfate are dissolved in water and the solution allowed to evaporate, bluish-green crystals of **Mohr's salt** [$FeSO_4 \cdot (NH_4)_2SO_4 \cdot 6\ H_2O$] are obtained. This compound is ordinarily employed in quantitative procedures as a reducing agent because it is more stable than green vitriol.

Cobalt and Nickel

The largest source of nickel and cobalt ores is a deposit in the Sudbury district of Ontario, Canada. The minerals are largely sulfides and arsenides of cobalt and nickel associated with some copper, silver, and iron sulfides. Because of the mixed nature of the ore deposit, the metallurgy of obtaining cobalt and nickel metals is quite complex. Rock masses containing 2 to 3% of nickel silicate have been found in Oregon and Washington, and an iron-nickel alloy is now produced from this material.

Both cobalt and nickel depend for their major uses on their ability to resist corrosion. The metals may be plated electrolytically over iron to give a cobalt or nickel plate, or may be alloyed with other metals to give corrosion-resistant products. Nickel alloys are used in coinage, in electrical resistance wire (nichrome), and in corrosion-resistant **monel metal,** which is 72% nickel, 26.5% copper, and 1.5% iron. A very hard, corrosion-resistant alloy called **stellite** contains cobalt, chromium, and tungsten.

Properties

Like iron, cobalt and nickel form two series of compounds with characteristic oxidation states of +2 and +3. The divalent compounds are the more stable and more important. Nickel salts are usually green. A dilute solution of $CoCl_2 \cdot 6\ H_2O$, light pink in color, may be used as an invisible ink. After drying, heat will bring out a visible blue color because of the dehydration of the salt. Cobalt oxide may be incorporated into glass or glazes to give a beautiful blue color. Both cobalt and nickel in compounds may be detected by a borax bead test, in which cobalt gives a dark blue color and nickel a brown color.

The Platinum Metals

Platinum and the platinum-like elements — ruthenium, rhodium, palladium, osmium, and iridium — are somewhat like gold and silver in that they do not displace hydrogen from mineral acids. As transition elements below iron, cobalt, and nickel in Group VIIIB, they resemble these metals in electronic structure, general catalytic activity, and tendency to form complex compounds.

Predominant oxidation states for the elements are $+2$ and $+4$. **Platinum(II) chloride, ($PtCl_2$), platinum(IV) chloride ($PtCl_4$), palladium(II) chloride** ($PdCl_2$), and **palladium(IV) chloride** ($PdCl_4$) are well-known compounds.

Ruthenium and osmium have the unusual property of exhibiting an oxidation number of $+8$ in the compounds OsO_4 and RuO_4. These two compounds are produced when the finely divided metals are heated in oxygen. OsO_4 melts at 40°C and boils at 130°C; its vapor is poisonous, irritating to the eyes, and has the odor of bromine.

Platinum is placer mined as nuggets or specks from alluvial gravel in the Ural Mountains and concentrated from the rock material with which it is associated by virtue of its high specific gravity (21.5). Only osmium and iridium have higher densities than platinum. Small amounts of platinum are associated with gold in Colombia, South Africa, and the western United States. Other platinum metals are found associated with, or alloyed with, platinum.

Platinum is used in jewelry, in dentistry, and in the making of laboratory crucibles and articles which may have to withstand corrosion or high temperature. It has a specific use in finely divided form as a catalyst, usually mixed with a carrier such as asbestos fibers. Because it has about the same coefficient of expansion as glass, it may be sealed through glass for the conduction of electricity. The other platinum-like metals are used as substitutes for platinum or may be alloyed with it.

Exercises

30.1. Consider the following transition elements which are in Groups III B-VIII B:
(1) Sc, (2) Nb, (3) Os, (4) Ru, (5) W, (6) Hf, (7) Gd, (8) Pu.
(a) Attempt to name each of these metals. Check your answers with the textbook.
(b) Note the position of each of these metals in the periodic table and predict the maximum + oxidation it generally shows in compounds. Write corresponding oxides.

30.2. Name: (a) $Ac_2(SO_4)_3$, (b) $Sc(NO_3)_3$, (c) ThF_4, (d) $TiCl_4$, (e) NH_4VO_3, (f) K_2NbF_7, (g) $CaWO_4$, (h) $KMnO_4$, (i) Na_2CrO_4.

30.3. (a) Use arrows representing paired or unpaired electrons in the orbitals ($\square$) for $_{58}$Ce-K,L,M,

N	O	P
$\square$, $\square$ $\square$ $\square$, $\square$ $\square$ $\square$ $\square$ $\square$, $\square$	$\square$, $\square$ $\square$ $\square$, $\square$	$\square$

From this electron distribution pattern, account for the different valences of cerium. (b) for $_{92}$U (confer with an electron configuration table, Chapter 3) $_{92}$U-K,L,M,N,

O	P	Q
$\square$, $\square$ $\square$ $\square$, $\square$ $\square$ $\square$ $\square$ $\square$, $\square$ $\square$ $\square$	$\square$, $\square$ $\square$ $\square$, $\square$,	$\square$

How many unpaired electrons are there?
What oxidation numbers seem possible for uranium?

30.4. Write formulas for thorium(IV) sulfate, neodymium(III) carbonate, potassium uranate, uranyl acetate, gadolinium(III) bromide.

30.5. Name and give the oxidation number for the transition elements in these compounds: $La(C_2H_3O_2)_3$, $Y_2(SO_4)_3$, K_2TaF_7, NH_4VO_3.

30.6. Give the chemical name or formula of $PdCl_4$; hexaamminecobalt(III) chloride, tetraamminenickel(II) ion; nickel(III) hydroxide; $[Co(NH_3)_4(H_2O)_2]$ Br_3; $KFe[Fe(CN)_6]$; and $NiSiO_3$.

30.7. Give formulas for the following mineral or chemical names: scheelite, sodium chromite, sodium chromate, sodium dichromate, hexahydroxochromium(III) ion, dichlorodioxotungsten (wolfram)(VI).

30.8. The name of a mineral sometimes indicates the valuable element present. For the following minerals write the symbol of the indicated metal: (a) wolframite, (b) hafnia, (c) tantalite, (d) molybdenite, (e) zirconia, (f) vanadinite, (g) columbite.

30.9. Give names or formulas for the following and indicate the oxidation number of the metal in each: (a) manganous chloride, (b) manganic sulfate, (c) manganese dioxide, (d) manganic acid, (e) lithium permanganate, (f) $K_2Cr_2O_7$, (g) CrI_3, (h) H_2CrO_4, (i) RuO_4, (j) $FeSO_4 \cdot (NH_4)_2SO_4 \cdot 6H_2O$.

30.10. Carbon dioxide is known to play a part in accelerating the rusting of iron. Complete and balance the very slow chemical change equations indicated as Fe is acted on by water containing dissolved CO_2 and O_2.

$$Fe + H_2CO_3 \rightarrow Fe(HCO_3)_2 +$$
$$Fe(HCO_3)_2 + O_2 \rightarrow Fe_2O_3 \cdot H_2O + CO_2 + H_2O$$

30.11. Test your ability in equation writing with the following:
(a) yttrium + hot water → $H_2(g)$ +
(b) $LaCl_3$ + NaOH →
(c) Ce^{4+} + Fe^{2+} →
(d) americium hydroxide + nitric acid →
(e) lutetium(II) chloride + conc. H_2SO_4 →
(f) zirconium(IV) chloride + NH_4CNS →
(g) heating titanium(IV) hydroxide → titania +
(h) hafnium(IV) chloride highly heated with Mg metal out of contact with air →

30 12. Write molecular followed by ionic equations for the following:
(a) K_3CrO_3 + K_2O_2 + H_2O → KOH +
(b) $K_2Cr_2O_7$ + $SnCl_2$ + HCl → $SnCl_4$ +
(c) $KMnO_4$ + $FeSO_4$ + H_2SO_4 →
(d) $Mn(NO_3)_2$ + HNO_3 + PbO_2 → $HMnO_4$ +

30.13. Write balanced equations for
(a) chromium(VI) oxide + water.
(b) manganese dioxide + conc. HCl.
(c) calcium tungstate ore highly heated with soda ash (Na_2CO_3); then the product water leached; followed by the addition of calcium hydroxide to the leach solution.

30.14. Write equations for the following processes:
(a) roasting of NiS.
(b) iron granules added to melted $NiSiO_3$ (basis of a metallurgical process).
(c) heating of $[Co(H_2O)_6]^{2+}$ 2 Cl^- (changes pink to blue).
(d) cobalt(II) oxide heated with silica (forms blue glaze).
(e) discharge of nickel(II) ion at cathode on electrolysis.

30.15. Complete and balance the equations (coordination complexes formed, coordination number = 6):
(a) $Fe(CNS)_3$ + CNS^- →
(b) $Co(NO_2)_3$ + NO_2^- →
(c) $RhCl_3$ + Cl^- →
(d) $OsCl_4$ + Cl^- →

30.16. Name the complex ions formed in question 30.15.

30.17. Write balanced equations for
(a) Cl_2 (from Cl_2 water or aqua regia) plus Pt.
(b) Platinum(IV) chloride plus HCl forming a compound in which Pt exhibits coordination number of 6, called chloroplatinic acid.
(c) Chloroplatinic acid plus KNO_3 to form slightly soluble potassium hexachloroplatinate(IV).
(d) Platinum(IV) or palladium(IV) chloride + Mg ribbon.
(e) High heating of platinum(II) oxide.

30.18. Zirconium metal and chlorine gas react to form zirconium tetrachloride. At a temperature of 546°C and a pressure of 2.0 atm, 8500 l of $ZrCl_4$ gas were obtained. What volume of chlorine gas was necessary if measured at STP?

30.19. If chromite were pure $FeO \cdot Cr_2O_3$, what composition alloy would result on its complete reduction with carbon? If 2.00 kg of this high chromium alloy were added to 5.00 kg of pure iron, a product of what percent Cr would result?

30.20. A standard solution of $KMnO_4$ is used in determining the amount of iron in a sample. The iron is first converted to the iron(II) state, and the resulting solution may be titrated in acid solution with the standard $KMnO_4$ solution. The reaction is

$$5 \text{ Fe}^{2+} + KMnO_4 + 8 \text{ H}^+ \rightarrow K^+ + Mn^{2+} + 5 \text{ Fe}^{3+} + 4 \text{ H}_2O$$

Suppose that 35.0 ml of a solution containing Fe^{2+} react with 38.5 ml of 0.120 N $KMnO_4$ solution. (a) What is the normality of the Fe^{2+} solution? (b) What weight of Fe^{2+} is contained in the 35.0 ml of solution? (c) If this amount of Fe^{2+} were obtained from a 1.00-g sample of iron ore, what would be the percentage of iron in the sample?

30.21. A standard solution of $K_2Cr_2O_7$ is used in determining Fe^{2+} in a sample by titration in acid solution. The equation for the reaction is

$$6 \text{ Fe}^{2+} + K_2Cr_2O_7 + 14 \text{ H}^+ \rightarrow 2 \text{ K}^+ + 2 \text{ Cr}^{3+} + 6 \text{ Fe}^{3+} + 7 \text{ H}_2O$$

The molecular weight of $K_2Cr_2O_7$ is 294.2. (a) What is the equivalent weight of $K_2Cr_2O_7$ in this reaction? (b) A 1.00-g sample of iron ore is treated to obtain iron as Fe^{2+}. The insoluble residue was removed so that a clear solution with all iron from the 1-g sample as Fe^{2+} was available for titration. A 45.0-ml portion of 0.150 N $K_2Cr_2O_7$ was used to titrate the Fe^{2+} solution. What is the percentage of iron in the ore sample?

30.22. Sodium oxalate ($Na_2C_2O_4$, MW 134) may be used to standardize a solution of $KMnO_4$. The reaction is

$$5 \text{ Na}_2C_2O_4 + 2 \text{ KMnO}_4 + 16 \text{ H}^+ \rightarrow 10 \text{ Na}^+ + 2 \text{ K}^+ + 2 \text{ Mn}^{2+} + 10 \text{ CO}_2 + 8 \text{ H}_2O$$

If 36.8 ml of $KMnO_4$ solution is used to titrate 1.56 g of pure $Na_2C_2O_4$, what is the normality of the permanganate solution?

30.23. Tell how in the laboratory you would proceed to distinguish chemically between (a) Cr metal and Mn metal; (b) CuO and MnO_2; (c) $MnCO_3$ and $MnSiO_3$; (d) the yellow compounds $PbCrO_4$ and CdS.

Supplementary Readings

30.1. "Vanadium in the Living World," N. M. Senozan, *J. Chem. Educ.*, **51**, 503 (1974).
30.2. "Periodicity and the Lanthanides and Actinides," T. Moeller, *J. Chem. Educ.*, **47**, 417 (1970).
30.3. "The Oxidation States of Molybdenum," J. G. Stark, *J. Chem. Educ.*, **46**, 505 (1969).

Chapter 31

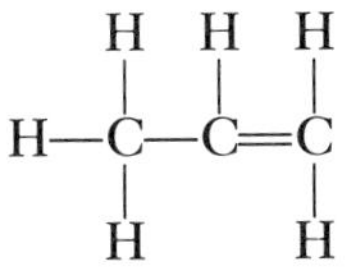

H H H
| | |
H—C—C=C
| |
H H

Propene

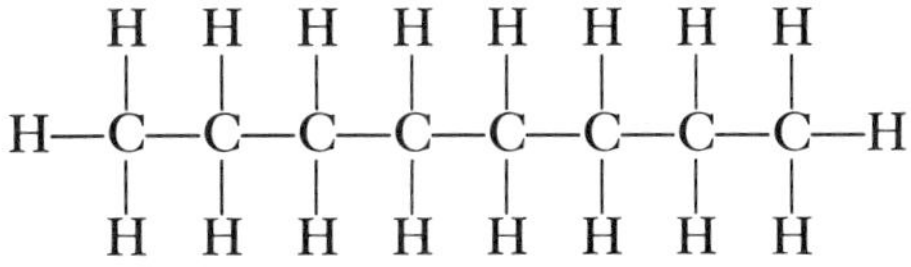

H H H H H H H H
| | | | | | | |
H—C—C—C—C—C—C—C—C—H
| | | | | | | |
H H H H H H H H

Octane

n molecules
of

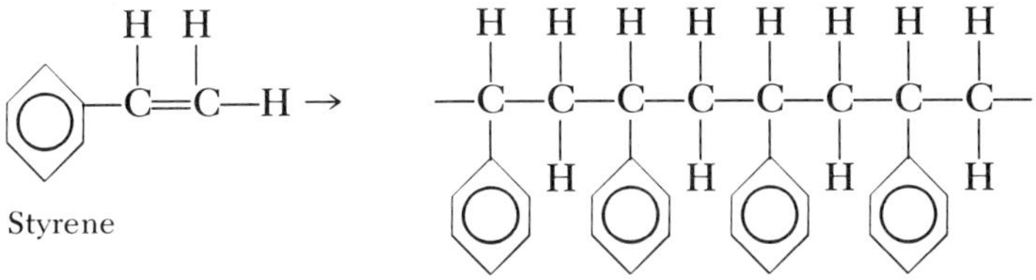

H H
| |
⬡—C=C—H →
|
Styrene

H H H H H H H H
| | | | | | | |
—C—C—C—C—C—C—C—C—
| | | | | | | |
⬡ H ⬡ H ⬡ H ⬡ H

Polystyrene
(synthetic polymer)

Introduction to Organic Chemistry I

31.1 THE NATURE OF ORGANIC CHEMISTRY

Most of the compounds thus far considered have been classified as acids, bases, salts, and oxides of metals or nonmetals—substances derived from minerals in the earth's crust. Frequently these are referred to as inorganic compounds and their chemistry as **inorganic chemistry,** one of the major divisions of the field of chemistry.

Organic chemistry is concerned with the compounds of carbon. It may seem curious that an entire division of chemistry is concerned with one element; however, this one element forms a tremendous number of compounds—several million are known. The existence of organic chemistry as a separate study is further justified by the importance of carbon compounds or mixtures in our lives. For example, carbon is common in the molecular composition of such substances as gasoline, vitamins, hormones, plastics, explosives, fats, proteins, carbohydrates, perfumes, medicines, alcohols, and wood products. Of the 50 top chemicals produced in the U.S., 32 are organic compounds. All plant and animal tissues are composed largely of carbon compounds, water, and a small mineral content.

The multiplicity of carbon compounds is explained by the exceptional ability of carbon atoms to combine with one another to form long, continuous, chainlike structures containing many carbon atoms. In inorganic compounds the number of like atoms in a molecule rarely exceeds five, but organic molecules containing a dozen or more carbon atoms are common. Carbon, with four electrons in its valence level, forms compounds by sharing its electrons with atoms of other elements or *with itself.* Two given carbon atoms may share one, two, or three pairs of electrons to give rise to three fundamental classes of compounds. Different arrangements of the same atoms within a molecule produce different compounds, and thus the number of carbon compounds is further increased. Although the number of organic compounds currently known seems very large, we can say with certainty that the field is still in its infancy; the number of possible organic substances is exceedingly great.

Many of the materials listed above as examples of organic compounds were once believed to be products of living organisms—hence the term **organic.** Following much experimentation, however, chemists were able to demonstrate that certain products which previously had been considered the results of vital, or living, processes could be produced in the laboratory. One of the classical examples of this type of work was Wöhler's synthesis of urea. In 1828, Wöhler, a German chemist, produced urea (a decomposition product of proteins) by heating ammonium cyanate, a compound of inorganic origin.

$$NH_4CNO \rightarrow CO(NH_2)_2$$

This was the beginning of a new era in the synthesis of carbon compounds which scientists had previously thought impossible to prepare in the laboratory. The vital-process idea was discarded. In a sense this was the beginning of two different branches of chemistry, organic chemistry and biochemistry.

Every day new synthetic compounds are produced in the laboratory of the organic chemist, and many of these compounds are important in everyday life.

In the discussion of the conductivities of aqueous solutions of acids, bases, and salts, it was pointed out that, as a group, they are relatively good conductors of electric current. In contrast with this the aqueous solutions of alcohol, sugar, glycerol, and the like are very poor conductors. These facts suggest that there is little or no ionization in organic compounds. Thus although ionic inorganic reactions proceed almost instantaneously, organic reactions frequently require, even under the most favorable conditions, considerable time before they approach completion.

Some Sources of Organic Compounds

Table 31.1 shows the sources of some of the more familiar organic substances.

Of the substances listed, some — such as benzene, phenol, wood alcohol, and acetic acid — are pure compounds; others — such as fats, proteins, coal, tar, and petroleum — are complex mixtures. Starch and cellulose are large molecules made up of repeating units of a sugar, glucose.

Petroleum is an extremely valuable raw material of nature. It provides such important commodities as gasoline, kerosene, fuel oils, lubricating oils, paraffin, wax, and asphalt.

A vast variety of products are obtained from coal tar, a product of the destructive distillation of coal. Benzene, toluene, phenol, naphthalene, and many other compounds obtained from coal tar are the starting materials for a great number of dyes, perfumes, medicines, plastics, and explosives.

Table 31.1
Sources of organic substances

Source	Organic Compound or Mixture
Animals	fats, proteins
Plants	starch, cellulose
Coal (destructive distillation)	coal tar, benzene, phenol, naphthalene
Wood (destructive distillation)	wood alcohol, acetone, acetic acid
Petroleum (fractional distillation)	cleaner's naphtha, gasoline, kerosene, mineral oil, vaseline
Fermentation processes	ethyl alcohol, acetone, butyl alcohol
Laboratory synthesis	almost unlimited numbers and types of compounds

Structural Formulas

While the inorganic chemist uses such molecular formulas as H_2SO_4, H_2O, and $NaCl$ to show the composition of compounds, molecular formulas are

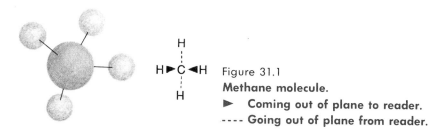

$$
\begin{array}{c}
\text{H} \\
| \\
\text{H} \blacktriangleright \text{C} \blacktriangleleft \text{H} \\
| \\
\text{H}
\end{array}
$$

Figure 31.1
Methane molecule.
▶   Coming out of plane to reader.
- - - -   Going out of plane from reader.

inadequate in representing many organic compounds. Different arrangements of the same atoms in molecules give rise to different compounds. This property is called **isomerism.**

The molecular formula C_2H_6O may represent either ethyl alcohol (CH_3CH_2OH) or dimethyl ether (CH_3OCH_3). Thus in organic chemistry it is necessary to show *how* the atoms are joined together in the molecule—in other words, to show the arrangement of the atoms in the molecule—in order to get an accurate idea of the substance. About a century ago Friedrich Kekulé instituted the use of **structural** or **graphic** formulas to clarify the molecular structure of organic compounds. Kekulé's contributions brought order out of chaos and enabled the organic chemist to make great strides in the synthesis of organic substances on the basis of this so-called "picture chemistry."

The basis for structural formulas is the covalence of 4 of the carbon atom. The carbon atom with four electrons in its outer level neither gives up nor takes up electrons, but rather shares these electrons with other atoms of carbon or with atoms of other elements; thus the compounds of carbon are covalent and nonionic. A structural formula shows how these electrons are shared with other atoms. For example, methane (CH_4) may be represented

$$
\begin{array}{ccc}
& \text{H} & \\
& \overset{\cdot\cdot}{\underset{\cdot\cdot}{\text{H:C:H}}} & \text{or} \\
& \text{H} &
\end{array}
\qquad
\begin{array}{c}
\text{H} \\
| \\
\text{H}-\text{C}-\text{H} \\
| \\
\text{H}
\end{array}
$$

Four electrons from the carbon atom and four electrons from the hydrogen atoms are shared; the formula shows four pairs of shared electrons. Each pair of electrons represents a covalence of one. In writing structural formulas, a pair of electrons is usually replaced with a dash to indicate a single bond. (See also Figure 31.1.)

Tetrahedral Bonds in Certain Carbon Compounds

Please refer to page 105, where carbon is presented as an example of sp^3 electron availability for bonding.

$$_6C - 1s^2; \quad \boxed{\downarrow} \qquad \boxed{\downarrow} \qquad \boxed{\downarrow} \qquad \boxed{\downarrow}$$
$$\qquad\qquad 2s^1 \qquad 2p_x^1 \qquad 2p_y^1 \qquad 2p_z^1$$

The four unfilled orbitals each accept an electron from another atom in form-ing four electron-pair bonds. The equality of the four bonds leads to a tetra-hedral structure for compounds of carbon such as CH_4 and CCl_4. Bond dis-tance and bond energies will vary with the different elements with which carbon unites. In all cases the carbon as the central atom will have its four bonds directed to the corners of a tetrahedron. Carbon may bond to carbon in electron sharing, but this does not change the nature of the directional bonding.

Carbon-to-Carbon Bonding

Carbon atoms can share a pair of electrons between themselves – thus ethane (C_2H_6) (Figure 31.2):

$$\begin{array}{cc} H & H \\ H\!:\!\ddot{C}\!:\!\ddot{C}\!:\!H \\ \ddot{H} & \ddot{H} \end{array} \qquad or \qquad \begin{array}{ccc} H & H \\ | & | \\ H-C-C-H \\ | & | \\ H & H \end{array}$$

and propane (C_3H_8)

$$\begin{array}{ccc} H & H & H \\ | & | & | \\ H-C-C-C-H \\ | & | & | \\ H & H & H \end{array}$$

Where there can be no question of the arrangement of the atoms in a com-pound, a condensed structural formula may be used. For example, the above hydrocarbons may be represented as CH_3-CH_3 and $CH_3-CH_2-CH_3$.

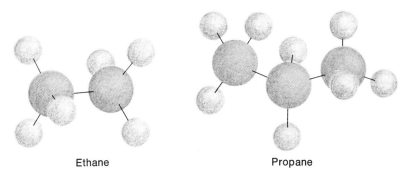

Ethane   Propane

Figure 31.2

Structural models of ethane and propane.

Chains and Rings of C Atoms

C atoms in some compounds may be arranged one after another to give what is referred to as **chain** compounds, while other compounds are known as **ring** or cyclic compounds, because the carbon atoms are joined together as a closed system or ring. For example, four C atoms may be joined as follows:

(1) a **straight chain** or **normal** compound

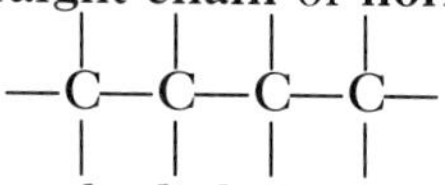

$$-\overset{|}{\underset{|}{C}}-\overset{|}{\underset{|}{C}}-\overset{|}{\underset{|}{C}}-\overset{|}{\underset{|}{C}}-$$

(2) a **branched chain** compound

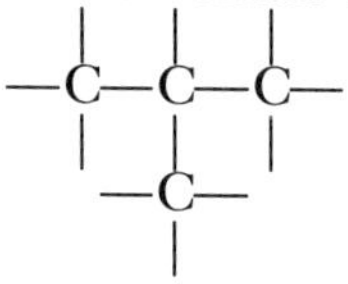

$$-\overset{|}{\underset{|}{C}}-\overset{|}{\underset{|}{C}}-\overset{|}{\underset{|}{C}}-$$
$$-\overset{|}{\underset{|}{C}}-$$

(3) **ring compounds**

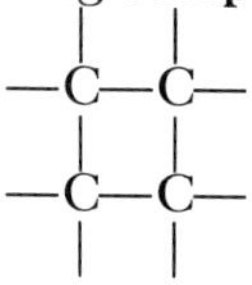

$$-\overset{|}{\underset{|}{C}}-\overset{|}{\underset{|}{C}}-$$
$$-\overset{|}{\underset{|}{C}}-\overset{|}{\underset{|}{C}}-$$ and   structure

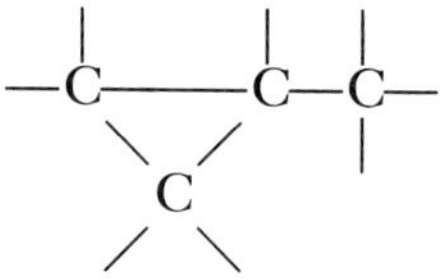

In the above structures only the carbon atoms are shown. These are called the **carbon skeletons** of the molecules. Hydrogen atoms or other elements are attached to the other bonds to give the completed structural formulas. Note that in these structures, as in nearly all organic compounds, four bonds go to each C atom (covalence of 4).

31.2   HYDROCARBONS

Compounds composed only of the elements carbon and hydrogen are termed **hydrocarbons.** They are among the simplest of organic compounds and will be considered first. Although hundreds of hydrocarbons are known, we shall limit our discussion to a few of the simpler ones.

The following are examples of hydrocarbons:

(1) $CH_3CH_2CH_2CH_3$

n-Butane
C_4H_{10}

(2)

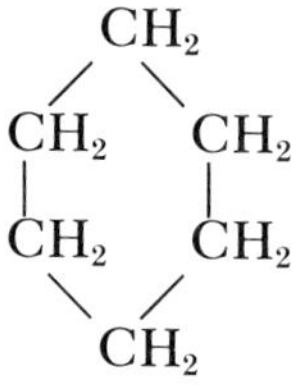

$$
\begin{array}{c}
CH_2 \\
CH_2 \quad CH_2 \\
| \qquad | \\
CH_2 \quad CH_2 \\
CH_2 \\
\end{array}
$$

Cyclohexane
C_6H_{12}

(3)

$$
\begin{array}{c}
CH_2 \\
CH \quad CH_2 \\
\| \qquad | \\
CH \quad CH_2 \\
CH_2 \\
\end{array}
$$

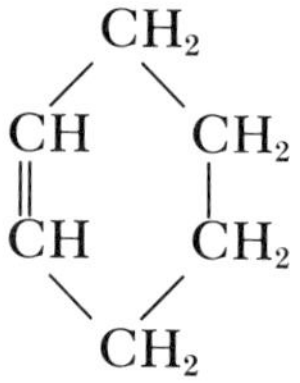

Cyclohexene
C_6H_{10}

(4) $CH_3CH_2CH{=}CH_2$

1-Butene
C_4H_8

(5) $CH{\equiv}CH$

Acetylene
C_2H_2

(6)

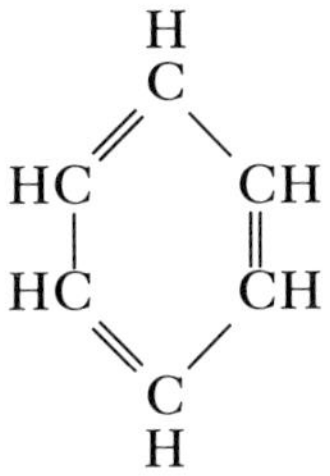

$$
\begin{array}{c}
H \\
C \\
HC \quad\quad CH \\
HC \quad\quad CH \\
C \\
H
\end{array}
$$

Benzene
C_6H_6

Compounds in which the carbons are bonded to each other by only *single* bonds are known as **saturated** compounds. Examples 1 and 2 above are saturated hydrocarbons. Examples 3, 4, 5, and 6 contain one or more double or triple bonds between the carbon atoms and are called **unsaturated** compounds. When the organic compound has only a chain structure (no rings), such as 1, 4, and 5 above, it is called an **acyclic** compound; whereas ring structures such as 2 and 3 are **alicyclic** compounds. Benzene (6) and compounds containing one or more **benzene rings** belong to a vast class of compounds known as **aromatic** compounds. The fourth main class of organic compounds, **heterocyclic** compounds, consist of ring structures which contain other elements besides carbon in the ring. Examples of heterocyclic compounds are shown below:

Pyridine   Furan   Thiophene

Heterocyclic compounds, of course, are not hydrocarbons since elements other than carbon and hydrogen are present.

The Alkane Series

Saturated hydrocarbons are of two types, acyclic compounds (alkanes) and cyclic compounds (cycloalkanes). We will consider the **alkane series** first in our brief study of the hydrocarbons.

Such a series is called a **homologous** series, since any member of the series differs from the preceding or following member by one carbon atom and two hydrogen atoms (CH_2). For example, butane (C_4H_{10}) in the alkane series differs from propane (C_3H_8), immediately preceding it, by CH_2; and differs from pentane (C_5H_{12}), the member following, by CH_2. In any given series the members possess similar chemical properties. The physical properties, however, show a gradual change as the compounds of few carbon atoms progress to those containing a large number.

The alkane series is sometimes referred to as the **methane** series, after the first member, or the **paraffin** series (paraffin means inactive). The general formula of the series is C_nH_{2n+2}, where n is an integral number—1, 2, 3, 4, and so on. The names of the compounds all end in *ane*. The student should learn the names and molecular formulas for the first 10 members of the alkane series (Table 31.2) because of their great involvement in the formulas and names of many other compounds to be considered later. The first four members of the series are gases at ordinary temperatures; members containing from five to fifteen carbon atoms are liquid, and the higher members of the series are solids. Thus it is evident that a gradual increase in the boiling points of these compounds takes place as the number of carbon atoms in the molecule increases.

Table 31.2

The first ten members of the alkane series

Name	Molecular Formula (C_nH_{2n+2})
Methane	CH_4
Ethane	C_2H_6
Propane	C_3H_8
Butane	C_4H_{10}
Pentane	C_5H_{12}
Hexane	C_6H_{14}
Heptane	C_7H_{16}
Octane	C_8H_{18}
Nonane	C_9H_{20}
Decane	$C_{10}H_{22}$

Methane is the principal constituent of natural gas, which is now used extensively in the heating of homes and as an industrial fuel. It is also formed during the decay of organic matter under water in marshes and swamps in the absence of air, and thus has the common name **marsh gas.** In coal mines, where it is known as **fire damp,** it is the principal cause of explosions. Bacteria in the rumen of cattle are able to produce methane.

Chemical Properties of the Alkanes

These hydrocarbons are characterized by their inactivity; they show little tendency to react with other substances. Strong oxidizing agents, however, may react with them. For example, if methane and chlorine gases are mixed and allowed to stand in the sunlight, chlorine atoms are substituted successively for the hydrogen atoms in the molecule according to the following equations:

$$CH_4 + Cl_2 \rightarrow CH_3Cl \text{ (methyl chloride)} + HCl$$
$$CH_3Cl + Cl_2 \rightarrow CH_2Cl_2 \text{ (methylene chloride)} + HCl$$
$$CH_2Cl_2 + Cl_2 \rightarrow CHCl_3 \text{ (chloroform)} + HCl$$
$$CHCl_3 + Cl_2 \rightarrow CCl_4 \text{ (carbon tetrachloride)} + HCl$$

Chloroform has been used as a general anesthetic in surgery; carbon tetrachloride is used for solvent and cleaning purposes. Other members of the alkane series react similarly with chlorine and the other halogens.

All hydrocarbons are combustible and, if complete burning is allowed to take place, carbon dioxide and water are formed as oxidation products.

$$CH_4 + 2 O_2 \rightarrow CO_2 + 2 H_2O$$
$$C_3H_8 + 5 O_2 \rightarrow 3 CO_2 + 4 H_2O$$
$$2 C_8H_{18} + 25 O_2 \rightarrow 16 CO_2 + 18 H_2O$$
octane

Incomplete combustion results in the formation of some carbon monoxide or free carbon.

$$2 C_8H_{18} + 23 O_2 \rightarrow 13 CO_2 + 2 CO + C + 18 H_2O$$

Because of incomplete combustion, all internal combustion engines produce some carbon monoxide.

Petroleum

This black, oily liquid—so important that nations have waged war for it—is the principal source of the paraffin hydrocarbons. Petroleum is a mixture of gaseous, liquid, and solid hydrocarbons, most of which belong to the methane series. The United States annually produces more than a billion barrels of petroleum.

The hydrocarbons contained in petroleum are separated by fractional distillation (page 297), a process which takes advantages of differences in the boiling points of the compounds. When petroleum is heated, those constituents with the lowest boiling points distil over first, followed by the higher boiling components. By changing the receivers of the distillate at intervals, several fractions are obtained. By repeated fractional distillation, the hydrocarbons may be separated in relatively pure form. The principal products of petroleum refining are shown in Table 31.3 and Figure 31.3.

One of the most important products of the refining of petroleum is gasoline. The great demand for gasoline has led to efforts to increase the yield of this important fuel. It is evident from Table 31.3 that gasoline is a mixture of the hydrocarbons which have an average near seven or eight carbon atoms. By subjecting petroleum fractions containing a higher number of carbon atoms to high heat and pressure, it is possible to break down fuel oils and

Table 31.3

Principal products of the refining of petroleum

Hydrocarbons	Products	Boiling Point Range °C	Uses
CH_4 C_2H_6 C_3H_8 C_4H_{10}	natural gas	−160 to 0	fuel gas for heat and power
C_5H_{12} C_6H_{14}	petroleum ether or ligroin	35 to 70	solvent for fats and oils
C_7H_{16} C_8H_{18}	gasoline	70 to 150	motor fuel, solvent
C_9H_{20} to $C_{14}H_{30}$	kerosene fuel oil	150 to 250	fuel, illuminating oil
$C_{15}H_{32}$ to $C_{20}H_{42}$	middle fraction or cracking oil, lubricating oil	250 and up	lubrication; cracked to gasoline
above $C_{20}H_{42}$	vaseline paraffin tar residue petroleum coke		medicines candles road surfacing fuel

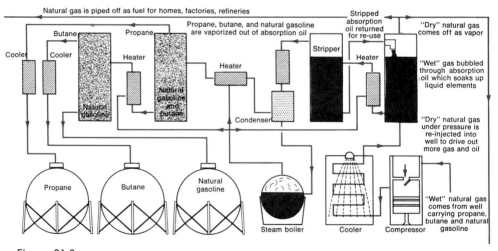

Figure 31.3

Natural gasoline plant flow chart. Light hydrocarbons are recovered from the "wet" gas that flows from an oil well. Varying from a fraction of a gallon to more than eight quarts per thousand feet of gas, natural gasoline is an important product. Blended with straight-run and cracked gasolines from the refinery, natural gasoline imparts qualities of quick starting and anti-knock to motor fuel.

higher-boiling fractions into gasoline and lower hydrocarbons. This process is referred to as **cracking** petroleum. The following equations are typical of the cracking operation:

$$C_{14}H_{30} \rightarrow C_8H_{18} + C_5H_{12} + C \qquad C_{16}H_{34} \rightarrow C_8H_{16} + C_8H_{18}$$

By means of the cracking process it is possible to utilize less desirable petroleum-distillation fractions and to increase greatly the yield of gasoline; also it results in unsaturates such as C_8H_{16}, (octene).

Isomers

If we study the structural formulas for butane and higher hydrocarbons of the methane series, we note that it is possible to arrange the atoms in the molecule in more than one way, thus giving rise to two or more different arrangements for the same formula. For example, we may write the structural formula for butane (C_4H_{10}) as

in which the carbon atoms are arranged in a continuous chain. In this formula there are never more than two carbon atoms attached to any single carbon atom. But it is possible to write a formula for butane as

in which three carbon atoms are attached to a single carbon atom. The result is called a branched chain compound. Both these butanes are known and have been prepared; the latter is called **isobutane** (or 2 methyl propane) to differentiate it from the continuous chain butane, which is called **normal** butane. Compounds with the same molecular formula but with a different arrangement of the constituent atoms in the molecule are called **isomers**, and the name applied to this phenomenon is **isomerism**. Careful study of the formulas for butane will show that these two arrangements are the only ones which will satisfy all valence requirements. As the number of carbon atoms increases, the number of isomers also increases. A study of the formulas for pentane will reveal the following three possible isomers (hydrogen atoms are not shown):

$$-\overset{|}{\underset{|}{C}}-\overset{|}{\underset{|}{C}}-\overset{|}{\underset{|}{C}}-\overset{|}{\underset{|}{C}}-\overset{|}{\underset{|}{C}}-$$

$$-\overset{|}{\underset{|}{C}}-\overset{|}{\underset{|}{C}}-\overset{|}{\underset{|}{C}}-\overset{|}{\underset{|}{C}}-$$

$$-\overset{|}{\underset{|}{C}}-$$

$$-\overset{|}{\underset{|}{C}}-\overset{|}{\underset{|}{C}}-\overset{|}{\underset{|}{C}}-$$

$$-\overset{|}{\underset{|}{C}}-$$

The student will find it interesting to draw structural formulas for the five possible isomers of hexane (C_6H_{14}).

The number of possible isomers increases rapidly as the number of carbon atoms increases. For example, decane, $C_{10}H_{22}$, has 75 isomers and calculations show that there may be 366,319 isomers for $C_{20}H_{42}$. $C_{40}H_{82}$ has 62,491,178,805,831 possible isomers.

Naming Alkanes

As the number of carbon atoms increases, the complexity of possible structures becomes greater. To show the presence of branched chains and their position, a system of nomenclature called the IUPAC° system has been devised which is used universally by organic chemists. In this system the longest chain is taken as the basic structure, and the carbon atoms are numbered from the end of this chain closest to a branch chain or other modification of simple alkane structure. The positions of substituents in the chain are indicated by the number of the carbon atom or atoms to which they are attached. Study the following examples:

$$\overset{1}{C}H_3-\overset{2}{C}H-\overset{3}{C}H_2-\overset{4}{C}H_3 \quad \text{or} \quad \overset{4}{C}H_3-\overset{3}{C}H_2-\overset{2}{C}H-\overset{1}{C}H_3$$
$$\underset{CH_3}{|} \qquad\qquad\qquad\qquad \underset{CH_3}{|}$$

2-Methylbutane

$$\overset{1}{C}H_3-\overset{2}{C}H_2-\overset{3}{C}H-\overset{4}{C}H_2-\overset{5}{C}H_3$$
$$\underset{CH_3}{|}$$

3-Methylpentane

$$\overset{5}{C}H_3-\overset{4}{C}H_2-\overset{3}{C}H-\overset{2}{C}H-\overset{1}{C}H_3$$
$$\underset{CH_3}{|}\ \underset{CH_3}{|}$$

2,3-Dimethylpentane

The CH_3- groups are called **methyl** groups (from CH_4, methane). Likewise, C_2H_5- groups are **ethyl** groups, C_3H_7- **propyl,** and so on (page 659).

The abbreviated rules for naming alkanes by the IUPAC system are as follows:

° International Union of Pure and Applied Chemistry.

1. Select the **longest continuous carbon chain** and number the carbon atoms from the end which will give any substituents as low a number as possible.
2. Name this longest chain according to the number of carbons it contains using the characteristic ending *-ane*.
3. Substituents (atoms of groups) are indicated by a prefix and a number showing their position on the carbon chain.

Cycloalkanes are saturated ring hydrocarbons with the formula C_nH_{2n}. They are analogous to the corresponding acyclic compounds except the formulas have two less H atoms and the names have a prefix of **cyclo**. Note the following examples:

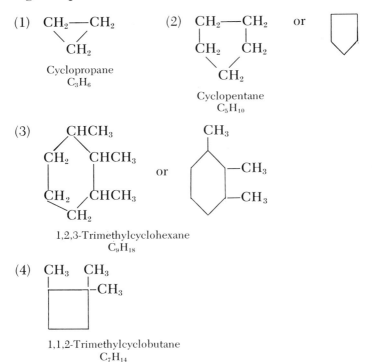

(1) CH_2——CH_2
          CH_2
   Cyclopropane
          C_3H_6

(2) CH_2——CH_2 or
       CH_2 CH_2
             CH_2
   Cyclopentane
          C_5H_{10}

(3) $CHCH_3$
    CH_2 $CHCH_3$ or
    CH_2 $CHCH_3$
          CH_2

          CH_3
                —CH_3
                —CH_3

   1,2,3-Trimethylcyclohexane
          C_9H_{18}

(4) CH_3 CH_3
                —CH_3

   1,1,2-Trimethylcyclobutane
          C_7H_{14}

In formulas 2, 3, and 4 the geometric shapes of the molecules are given and stand for the ring carbons and their attached hydrogens. Only the substituent groups need to be shown specially.

Cyclohexane, C_6H_{12}, is number 38 in the list of 50 top chemicals produced in the U.S. It is made by extracting the compound from petroleum and by the hydrogenation of benzene using a catalyst.

$$C_6H_6 + H_2 \xrightarrow[\text{110°C, 3 atm}]{\text{Ni catalyst}} C_6H_{12}$$

Most of the cyclohexane is used in manufacturing nylon (page 692).

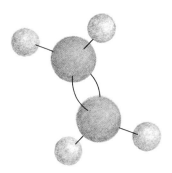

Figure 31.4
Structural model of ethene (ethylene).

The Alkene Series

Alkenes have the general formula C_nH_{2n} and the cycloalkenes, C_nH_{2n-2}. The alkenes are unsaturated acyclic hydrocarbons and are sometimes referred to as the **olefin** series, or as the **ethylene** series, after ethylene, the first member. (Figure 31.4). Names for the members of the alkene series are derived from names of compounds of the alkane series with the same number of carbon atoms, by changing the -*ane* ending to -*ene*; thus C_4H_8 is **butene,** from butane, C_4H_{10}; and so on.

Members of the series are characterized by carbon-to-carbon linkages which involve **one double bond** $\left(\begin{array}{c} \diagdown \\ \diagup \end{array} C{=}C \begin{array}{c} \diagup \\ \diagdown \end{array} \right)$. The position of the double bond is indicated by the lowest-numbered carbon atom to which it is attached (except for ethene or propene for which the number is not necessary). The carbon atoms are numbered in the longest chain so that the carbons of the double bond have the lowest possible numbers (Table 31.4).

Other examples of alkenes and cycloalkenes are:

(1) $\overset{4}{C}H_3{-}\overset{3}{C}H\overset{2}{C}H{=}\overset{1}{C}H_2$
$\qquad\quad |$
$\qquad\quad CH_3$

3-Methyl-1-butene

(2) $\overset{5}{C}H_3{-}\overset{4}{\underset{|}{C}}{-}\overset{3}{C}{=}\overset{2}{\underset{|}{C}}{-}\overset{1}{C}H_3$

with CH$_3$ and H above carbons 4 and 3, and H and CH$_3$ below

3,4-Dimethyl-2-pentene

(3)

Cyclopentene

(4)

H_3C

4-Methyl cyclohexene

The structural change of joining two carbon atoms by a double bond

Table 31.4
Names and formulas of several alkenes

Name	Molecular Formula	Structural Formula
Ethene (ethylene)	C_2H_4	$CH_2{=}CH_2$
Propene	C_3H_6	$CH_3CH{=}CH_2$
1-Butene	C_4H_8	$\overset{4}{C}H_3\overset{3}{C}H_2\overset{2}{C}H{=}\overset{1}{C}H_2$
2-Butene	C_4H_8	$\overset{1}{C}H_3\overset{2}{C}H{=}\overset{3}{C}H\overset{4}{C}H_3$

(two pairs of electrons) shortens the interatomic distance from the 1.54 Å in ethane to 1.34 Å in ethylene. The ethylene molecule also seems to have all atoms in one plane, and rotation between the respective halves of the molecule is limited.

Ethene (C_2H_4), commonly called ethylene, ranks number 5 among the top chemicals and is the simplest member of the series. It is obtained industrially by the cracking of petroleum. In the laboratory it may be prepared by the dehydration of ethyl alcohol with concentrated sulfuric acid at a temperature of about 160°C.

$$C_2H_5OH \xrightarrow[\text{heat}]{H_2SO_4} C_2H_4 + H_2O$$

Ethylene gas, either alone or mixed with ether, is now used as an anesthetic, since it causes less nausea and discomfort than ether. It is also employed to hasten the ripening of citrus fruits — it causes the development of a uniform orange or yellow color. In addition, ethylene is used as a starting material in the preparation of many organic compounds — for example, polyethylene.

The justification for the formula for ethene is that ethene has the property of adding two hydrogen atoms to form ethane, and we represent the change as

$$\underset{\underset{H}{|}\ \underset{H}{|}}{H-C{=}C-H} + H_2 \rightarrow \underset{\underset{H}{|}\ \underset{H}{|}}{\overset{\overset{H}{|}\ \overset{H}{|}}{H-C-C-H}}$$

Other elements may also add to the compound; for example, ethene readily absorbs bromine to form ethylene dibromide, $C_2H_4Br_2$.

$$\underset{\underset{H}{|}\ \underset{H}{|}}{H-C{=}C-H} + Br_2 \rightarrow \underset{\underset{H}{|}\ \underset{H}{|}}{\overset{\overset{Br}{|}\ \overset{Br}{|}}{H-C-C-H}}$$

Polymerization

Similar molecules of many organic substances may be made to combine to form complex molecules with new properties. This phenomenon is known as **polymerization**, and the product is called a **polymer**. For example, two ethylene molecules may add together.

$$
\begin{array}{ccccc}
\text{H} \; \text{H} & \text{H} \; \text{H} & & \text{H} \; \text{H} \; \text{H} \; \text{H} \\
| \; \; | & | \; \; | & & | \; \; | \; \; | \; \; | \\
\text{C}=\text{C} & + \; \text{C}=\text{C} & \rightarrow & -\text{C}-\text{C}-\text{C}-\text{C}- \\
| \; \; | & | \; \; | & & | \; \; | \; \; | \; \; | \\
\text{H} \; \text{H} & \text{H} \; \text{H} & & \text{H} \; \text{H} \; \text{H} \; \text{H}
\end{array}
$$

Since the product has free bonds (unsaturated valency), additional molecules may attach, so that the very large molecules of **polyethylenes** of the general formula $(CH_2=CH_2)_n$ may be built up, and n may have a value of many thousands. These polyethylene polymers are hard, tough, and durable products, which find application in the manufacture of many commercial items, such as kitchen utensils and household gadgets of many kinds. Synthetic plastics of many kinds are polymers. Artificial rubbers appear to be polymers of simple hydrocarbons.

Propene

Commonly called **propylene** $(CH_3CH=CH_2)$, is number 14 among the top chemicals and is obtained as a product of petroleum distillation or from petroleum cracking. It is used to make polypropylene plastics, isopropyl alcohol, propylene oxide, glycerine, gasoline, and other chemicals.

Styrene

$(C_6H_5CH=CH_2)$, number 23 among the chemicals, although not a member of the ethene series is an important hydrocarbon used in the production of polymers, the most important being polystyrene, used for cheap foam heat-insulating cups and insulation. Other uses of polystyrene are packaging and furniture. It is one of the cheapest of polymer materials.

Dienes

Compounds which contain two double bonds are called **dienes**; three double bonds, **trienes**, etc. These compounds are not members of the alkene series, but belong to other series. Some examples are:

(1) $CH_2{=}CH{-}CH{=}CH_2$ (2) $CH_2{=}CH{-}CH{=}CH{-}CH{=}CH_2$

       1,3-Butadiene   1,3,5-Hexatriene

(3)

1,4-Cyclohexadiene

Butadiene is number 31 in the list of 50 top chemicals and is used in making artificial rubber.

The Alkyne Series

Members of this series have the general formula C_nH_{2n-2} and contain one triple bond $(-C{\equiv}C-)$. The names are derived from the hydrocarbons of the methane series by replacing the *-ane* ending of the methane series with *-yne,* thus C_3H_4 is propyne, from propane, C_3H_8, and so on. This series possesses a higher degree of unsaturation than the alkene series. **Ethyne,** more commonly called **acetylene** (Figure 31.5), is written

$$H\!:\!C\!:\!:\!:\!C\!:\!H \qquad or \qquad H{-}C{\equiv}C{-}H$$

Members of the alkyne series are named similarly to the alkenes. For example, **1-butyne** has a structural formula of

$$
\begin{array}{c}
\quad\; H \;\; H \\
\quad\; | \;\;\; | \\
H{-}C{-}C{-}C{\equiv}C{-}H \\
\quad\; | \;\;\; | \\
\quad\; H \;\; H
\end{array}
$$

Alkynes, as would be expected, add more bromine or hydrogen than do corresponding members of the alkene series.

$$
H{-}C{\equiv}C{-}H + 2\ Br_2 \rightarrow
\begin{array}{c}
Br \;\; Br \\
| \;\;\;\; | \\
H{-}C{-}C{-}H \\
| \;\;\;\; | \\
Br \;\; Br
\end{array}
$$

Tetrabromoethane

Acetylene (ethyne) is readily produced by the action of water on calcium carbide.

Figure 31.5
Acetylene (ethyne).

$$CaC_2 + 2\,H_2O \rightarrow Ca(OH)_2 + C_2H_2$$

The gas has a high heat of combustion, which accounts for its use in the acetylene-oxygen torch for welding purposes. Acetylene also finds use as a starting material in the preparation of **neoprene**, an artificial rubber.

The Benzene Series of Hydrocarbons

The general formula for members of this series is C_nH_{2n-6}. These compounds may be obtained from coal tar. The ratio of hydrogen atoms to carbon atoms in the molecules of these compounds is smaller than in any of the preceding series and leads us to the conclusion that they are unsaturated. Although certain reactions show that these compounds possess some degree of unsaturation, they do not add the elements hydrogen and bromine with the same facility as members of the alkene and alkyne series.

Kekulé suggested that members of this series possess a ring structure with six carbon atoms joined together. For example, **benzene** has a structure which is represented

The carbon atoms are arranged in the form of a hexagon with double bonds between alternate carbon atoms and with a hydrogen atom attached to each carbon. The benzene ring structure is usually shown simply as the figure of a hexagon. The reactions of benzene indicate that it possesses a symmetrical structure — that is, the six carbon atoms all bear the same relationship to each other. A resonance between the two double-bonded structures is postulated, and it is now somewhat customary in chemical shorthand to use the hexagon with a circle within it. This symbol suggests that benzene has an average of $1\frac{1}{2}$ bonds between any two adjacent carbons. The benzene molecule is planar and the carbon-to-carbon bond length of 1.40 Å is reasonable for a bond with partial double-bond character. Recall that C—C in ethane is 1.54 Å and C=C in ethylene is 1.34 Å. The lesser chemical activity of the partial double bond in benzene as compared with ethylene is indicated.

Benzene ranks 13th in importance among the top chemicals. It is obtained mainly from petroleum and is used in gasoline manufacture to raise the octane rating of the gasoline. Large amounts are also used in the chemical industry as a starting material to produce a whole host of benzene derivatives.

This hexagon arrangement of carbon atoms seems to be the basic structure of all members of the benzene series of hydrocarbons and their many derivatives. **Toluene,** the member following benzene, has a CH_3 group substituted for one of the hydrogen atoms and is shown structurally as

Toluene also is obtained from petroleum and has uses similar to benzene. It ranks 17th among the chemicals.

Other members of the benzene series ranking in the 50 top chemicals are:

Ethyl benzene   o-Xylene   p-Xylene

Cumene
(Isopropyl benzene)

Ethyl benzene (20th in rank) is produced synthetically by reacting benzene with ethylene

$$CH_2CH_2 + C_6H_6 \xrightarrow[\text{heat, pressure}]{\text{catalyst}} C_6H_5C_2H_5$$

It is used to make styrene and as a solvent. There are three possible xylenes (dimethyl benzene) which are isomers of one another:

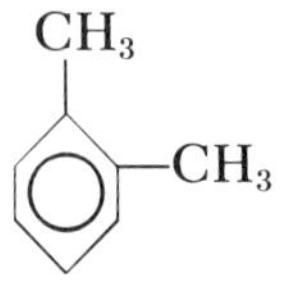

CH₃ ... o-Xylene (ortho-dimethyl benzene)

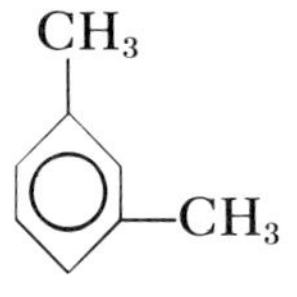

CH₃ ... m-Xylene (meta-dimethylbenzene)

CH₃ ... p-Xylene (para-dimethylbenzene)

The prefix *ortho*, abbreviated *o*-, denotes immediately adjacent positions of the two substituted groups; *meta*, *m*-, denotes replacement of two alternate hydrogens; *para*, *p*-, relates to opposite hydrogens being replaced. The functional groups replacing the hydrogens may be alike or different. There are numerous examples of ortho, meta, and para isomers of benzene derivatives.

The xylenes (particularly o-xylene, ranked 19th, and p-xylene, ranked 33rd) are used mainly in gasoline to improve octane ratings and as high-boiling solvents. They are produced from petroleum. **Cumene** (ranked 37th) is incorporated in aviation gasoline and high-octane motor fuels. It is also used to make acetone and phenol. It is made by reacting benzene with propylene using a catalyst.

Examples of aromatic hydrocarbons of other series include naphthalene:

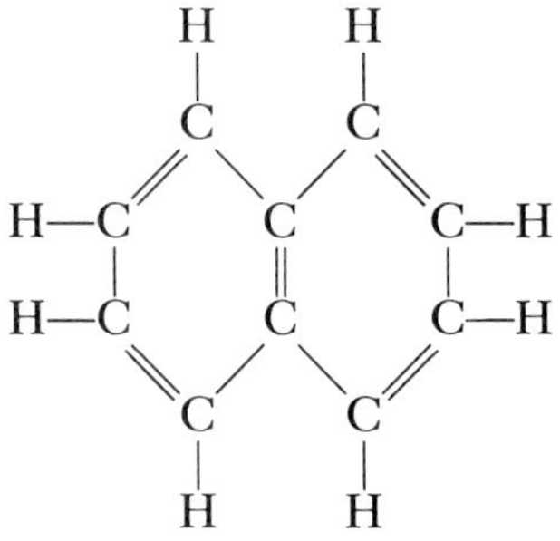

or

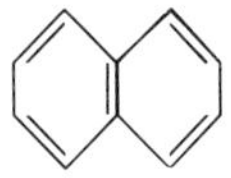

and anthracene:

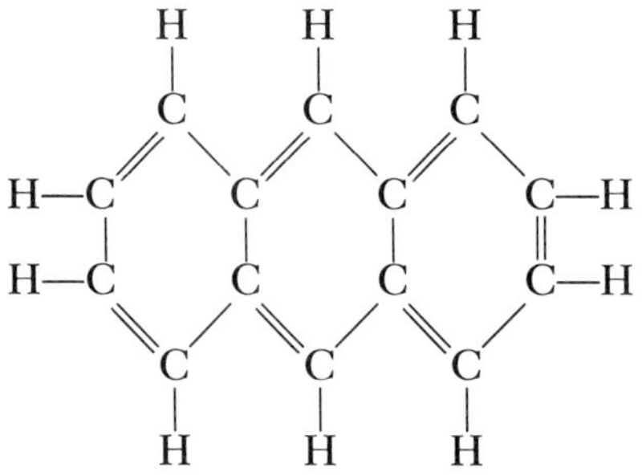

or

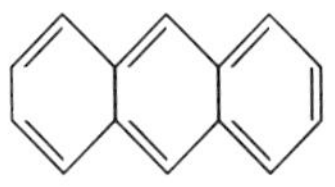

both of which are obtained as products of coal tar and serve as starting materials for the preparation of many dyes and medicines.

31.3 BONDING IN HYDROCARBONS

The valence bond theory of electron orbitals may be used to explain the bonding and structure in the various series of hydrocarbons just discussed. An example will be given of one member of each of the hydrocarbon series.

Ethane, C_2H_6

$$H-\underset{\underset{H}{|}}{\overset{\overset{H}{|}}{C}}-\underset{\underset{H}{|}}{\overset{\overset{H}{|}}{C}}-H$$

   As discussed in Chapter 5, methane, CH_4, has a tetrahedral structure as a result of the hybridization in a carbon atom of one s and three p orbitals to form four sp^3 orbitals oriented toward the corners of a regular tetrahedron. Ethane, with two carbon atoms, is pictured in Figure 31.6; two tetrahedra are joined together by an overlap of two of the sp^3 orbitals. Each of the three remaining sp^3 orbitals on each carbon atom would overlap with a $1s$ orbital from each of the three hydrogen atoms. When orbitals overlap in an endwise fashion, as is the case here with either the s-sp^3 or sp^3-sp^3 overlaps, the resulting bonds are termed **sigma** (σ) bonds. The bond between the carbon atoms is referred to as a single σ bond. The C—C distance is 1.54 Å, the C—H bond length is 1.10 Å, and the H—C—H bond angles are 109.5°. These values are characteristic of the bonds in members of the alkane series.

Ethene (ethylene), C_2H_4

$$\underset{H}{\overset{H}{\diagdown}}C=C\underset{\diagdown H}{\overset{\diagup H}{}}$$

   In the alkenes with one double bond, it is assumed that one s and two p orbitals of each carbon are hybridized to yield three sp^2 bonds, leaving one pure p orbital. The three sp^2 bonds lie in a plane with the bonds pointed toward the corners of an equilateral triangle, with an angle of 120° between bonds, as in Figure 31.7. The single p orbital left over is oriented at right angles to the plane of the sp^2 bonds.

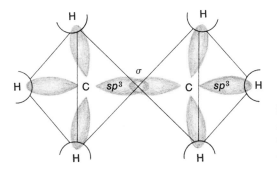

Figure 31.6
A sigma bond is formed between two carbons atoms of ethane by the overlap of sp^3 orbitals.

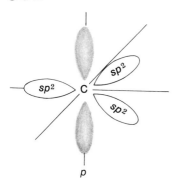

Figure 31.7

The four orbitals of each carbon atom in ethene — three sp^2 and one p orbital.

In the compound ethene, with two carbon atoms, two of the three sp^2 orbitals of each carbon atom overlap with the $1s$ orbitals of two H atoms to give two σ bonds; and one sp^2 orbital from each of the carbon atoms overlaps with one sp^2 orbital from another carbon atom to yield a σ bond between the carbon atoms. The remaining two pure p orbitals from the two carbon atoms overlap in a sidewise fashion to give a so-called **pi** (π) bond, as in Figure 31.8.

Thus the double bond between the two carbon atoms consists of one σ and one π bond. The π bond is weaker than the σ bond and is easily broken, accounting for the reactivity of the alkenes with elements such as the halogens. When Br_2 is added to ethene, $C_2H_4Br_2$ readily forms. The $C{=}C$ distance is 1.34 Å compared with 1.54 in the alkanes, $C{-}C$.

Ethyne (acetylene), C_2H_2

$$H{-}C{\equiv}C{-}H$$

The assumption is made here that one s and one p orbital from each carbon atom are hybridized to yield two sp bonds. This leaves two pure p orbitals available for bonding. One of the sp orbitals from each carbon atom forms a bond with the $1s$ orbital from a hydrogen atom; the other sp orbital forms a σ bond with an sp orbital from a second carbon atom. Meanwhile, the two pure p orbitals on each carbon atom overlap sidewise with two p orbitals from a second carbon atom to form two π bonds (Figure

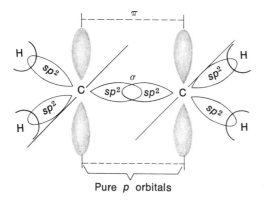

Pure p orbitals

Figure 31.8

Two pure p orbitals in ethene overlap to form a pi bond. The molecule is flat, with all atoms in a single plane.

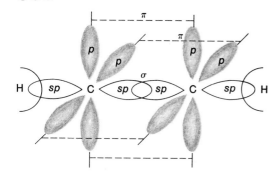

Figure 31.9
The overlap of orbitals in acetylene to yield a triple bond between carbon atoms made up of one sigma bond and two pi bonds. The molecule is linear—all four atoms lie in a straight line.

31.9). Thus the triple bond in acetylene is composed of one σ bond and two π bonds. The π bonds are easily broken or opened up with a reagent such as bromine. The C≡C distance is 1.20 Å and the C—H distance is 1.06 Å. The bond energy of the triple bond is 199 kcal/mole compared with 146 kcal/mole for C=C and 83 kcal/mole for C—C.

Benzene, C_6H_6

In benzene the carbon and hydrogen atoms lie in a plane, with the carbon atoms arranged in the form of a hexagon at 120° angles between the carbon-to-carbon bonds (Figure 31.10). The structure is usually represented as having double bonds between alternating pairs of carbon atoms; however, the actual structure, as explained on page 648, is a resonance hybrid, with an average of $1\frac{1}{2}$ bonds between each pair of carbon atoms. The bonding may be explained on the basis of each carbon atom's forming three sp^2 bonds and one pure p bond. The sp^2 orbitals for each carbon atom lie in a plane, with the p orbital at right angles and perpendicular to the plane. The p orbital of each carbon atom consists of two equal lobes—one above, the other below the plane of the three sp^2 orbitals. It is occupied by a single electron. The p orbital overlaps with the p orbitals of the two adjacent carbon atoms to form π bonds. Each carbon contributes one electron to the p orbital, so that each π bond between adjacent carbon atoms is $\frac{1}{2}$ a bond. Each of the three sp^2 orbitals on each carbon has one electron. One

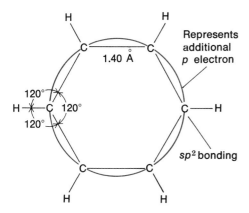

Figure 31.10
Structure of a benzene molecule.

electron is used in bonding each hydrogen atom, which leaves two electrons for the bonding between the carbon atoms. Including the one electron from the π bond, the three electrons account for the average of $1\frac{1}{2}$ bonds (one σ and $\frac{1}{2}\pi$) between adjacent carbon atoms in the benzene structure.

Exercises

31.1. For each of the following organic compounds write (1) the molecular formula, (2) a structural formula, (3) an electron-dot formula: (a) butane, (b) propene, (c) propyne, (d) benzene, (e) cyclopropane.

31.2. Using only 5 carbon atoms and all single bonds, write all the chain and ring carbon skeleton structures possible.

31.3. Write a structural formula for a 7 carbon atom member of each of the following series: (a) alkane, (b) alkene, (c) alkyne, (d) cycloalkene, (e) diene, (f) benzene.

31.4. Which of the compounds in question 3 are unsaturated?

31.5.

$$\underset{\text{A}}{CH_3-CH_2-\underset{\overset{|}{CH_3}}{CH}-CH_3} \qquad \underset{\text{B}}{CH_2\!=\!CH_2} \qquad \underset{\text{C}}{\bigcirc\!\!-CH_3}$$

$$\underset{\text{D}}{CH\!\equiv\!C-CH_2-CH_2CH_3} \qquad \underset{\text{E}}{\underset{\displaystyle O}{\overset{\displaystyle CH\!-\!\!-\!CH}{\underset{\displaystyle CH\quad CH}{\|\quad\quad\|}}}}$$

Indicate which of the compounds above would:
(a) have a characteristic formula C_nH_{2n-2}?
(b) have a molecular formula C_7H_8?
(c) be classified as an aromatic compound?
(d) be isomeric with 2,2-dimethyl propane?
(e) be classified as a heterocyclic compound?
(f) be classified as an alkane?
(g) be classified as an alkene?
(h) have a characteristic formula C_nH_{2n}?
(i) be classified as an alkyne?
(j) be saturated?

31.6. Which of the following compounds, if any, are:

(1) $CH_3-CH_2CHCH_2CH_3$
           $|$
          CH_3

(2) $\underset{\displaystyle O}{\overset{\displaystyle HC\!-\!\!-\!\!-\!CH}{\underset{\displaystyle HC\quad\;\; CH}{\|\quad\quad\quad\|}}}$

(3) (fused bicyclic ring structure)

(4) CH₂—CH₂
$$\begin{array}{c} CH_2-CH_2 \\ |\qquad| \\ CH_2-CH_2 \end{array}$$

(5)
$$\begin{array}{ccc} H & H & H \\ | & | & | \\ H-C-C=C \\ | & & | \\ H & & H \end{array}$$

(a) hydrocarbons, (b) unsaturated, (c) aromatic, (d) acyclic, (e) alicyclic, (f) heterocyclic, (g) alkanes, (h) alkenes, (i) alkynes?

31.7. In general terms, distinguish between the chemical properties of the various hydrocarbon series.

31.8. Name C_6H_{14}; C_8H_{16}; C_6H_{10}; C_4H_8; C_6H_6.

31.9. Write structural formulas for all possible isomers of C_6H_{14}. How many possibilities did you find?

31.10. Which of the following compounds are identical?

(a) $CH_3CH_2CH_2CH_2$
      |
      Br

(b) $CH_2CH_2CH_2CH_3$
    |
    Br

(c) $CH_3CH_2CHCH_3$
              |
              Br

(d) $CH_3CHCH_2CH_3$
        |
        Br

(e) CH_3CHCH_2Br
        |
        CH_3

(f) $CH_3CH_2CH_2CH_2Br$

(g)
$$\begin{array}{c} CH_3 \\ | \\ H-C-CH_2Br \\ | \\ CH_3 \end{array}$$

(h)
$$\begin{array}{c} Br \\ | \\ CH_3CHCH_2CH_3 \end{array}$$

31.11. Write structural formulas for all of the isomers of pentene, C_5H_{10}.

31.12. Using from 4 to 6 carbon atoms and any other elements necessary, write the structural formula of a logical compound in each of the following classes: (a) acyclic, (b) alicyclic, (c) aromatic, (d) heterocyclic.

31.13. Select from the list on the right the correct names for the three structures.

(a) $CH_3-CH_2-CH=CH_2$

(b)
$$\begin{array}{c} CH_3 \\ | \\ CH_3-CH-CH_2-CH-CH_2-CH_2-CH_3 \\ \qquad\qquad\qquad | \\ \qquad\qquad\qquad C_2H_5 \end{array}$$

(c)
$$\begin{array}{c} CH_3-C{\equiv}C-CH-CH_3 \\ | \\ CH_3 \end{array}$$

1. 3-butene
2. 4-ethyl-6-methyl heptane
3. 4-methyl-2-pentyne
4. 1-butene
5. hexyne
6. 2-methyl-4-ethyl heptane
7. 2-methyl-3-pentyne
8. 2-butene

31.14. Name the following compounds by the IUPAC system:
(a) $(CH_3)_2CHCH_2C(CH_3)_2CH_2CH_3$, (b) $(C_2H_5)_2CHC(CH_3)_3$, (c) △—C_3H_7,

(d)

(e) $CH_3CCH_2CHCH_3$.
$\quad\;\; CH_3\;\; CH_3$
with CH_3 on top.

31.15. Write a correct structural formula for each of the following: (a) methyl cyclo-pentane; (b) 1,1,2-trymethylcyclohexane; (c) 2-methyl butane; (d) 2,4-di-methyl-3-ethyl heptane; (e) n-decane.

31.16. Draw structural formulas for (a) 3-methyl-1-butyne; (b) cyclobutene; (c) 2-methyl 4-ethyl-2-hexene; (d) 2-5-7 decatriene.

31.17. Name: (a) CH_3—$C{\equiv}CH$; (b) $CH_2{=}CH_2$; (c) $CH_3CH{=}CHCH_2CH_2CH_3$; (d) $CH_2{=}CH$—$CH{=}CH_2$; (e) CH_3—CH—CH_2—$CH{=}CH$—CH_3;
$\qquad\qquad\qquad\qquad\qquad\quad |$
$\qquad\qquad\qquad\qquad\qquad C_2H_5$

(f)

(g)

31.18. Write structural formulas for the following compounds: (a) 3,5-dimethyl-3-ethyl-4-isopropylheptane; (b) 3-methyl-1,4-pentadiene; (c) 2,2-dibromo-propane; (d) 1,2-dichloroethane; (e) m-diethylbenzene; (f) 2,2,5,5-tetramethyl-3-hexyne; (g) 1,2,4-trichlorobenzene.

31.19. What is incorrect about each of the following names? (a) 2-ethylhexane; (b) 3-methyl-3-pentene; (c) 4-hexyne; (d) 2-methyl-2-hexyne; (e) 3,3-dimethyl-2-hexene; (f) 4,6-dimethylcyclohexene.

31.20. Write structural formulas for: (a) benzene; (b) naphthalene; (c) o,m, and p-xylenes; (d) cumene (isopropyl benzene); (e) n-propyl benzene.

31.21. Consider the compounds C_2H_2, C_4H_{10}, and CH_4. (a) Which, if any, are linear molecules? (b) Which compound has sp^3 overlaps (sigma (σ) bonds)? (c) Which compound has both σ and π bonds? (d) Which compound has tetra-hedral symmetry?

31.22. Consider ethyne (acetylene), H—$C{\equiv}C$—H.
  (a) What is the nature of the atomic orbital of C in the C—H bond in ethyne?
      (1) sp^3; (2) sp^2; (3) sp; (4) s; (5) p or π.
  (b) What is the nature of the 3 carbon-carbon bonds in the acetylene ($C{\equiv}C$) molecule? (1) all sp^3; (2) all sp^2; (3) all sp; (4) sp^3, sp^2, and sp; (5) all σ bonds; (6) σ and π bonds.
  (c) What is the shape of the ethyne molecule? (1) flat or planar, (2) tetrahedral, (3) linear, (4) a ring.
  (d) The $C{\equiv}C$ bond (1) is longer than the C—C bond, (2) is longer than the C=C bond, (3) has two bonds that are weaker and more easily broken than the other bond, (4) has less bond energy than the C=C bond. (Pick one.)

31.23. Write equations for:

(a) $CH_3CH_2CH_3 + Cl_2 \xrightarrow{\text{sunlight}}$ (first step only)

(b) cracking of a hydrocarbon, C_4H_{10} (use any one of several possible reactions producing alkanes and alkenes)

(c) complete combustion of hexane

(d) the polymerization of styrene

(e) $CH_2{=}CH_2 + \hexagon \rightarrow$

(f) $CH_3CH{=}CH_2 + Br_2 \rightarrow$

(g) $HC{\equiv}CH + H_2 \rightarrow$

31.24. What are some of the uses of the following important compounds? (a) o-xylene; (b) ethyl benzene; (c) cumene; (d) styrene; (e) C_2H_2; (f) CH_4; (g) C_6H_6; (h) $CH_2{=}CH{-}CH{=}CH_2$; (i) propene; (j) $CH_2{=}CH_2$; (k) $\hexagon$; (l) octane (and similar hydrocarbons); (m) $C_{17}H_{36}$ (and similar compounds).

31.25. Write an equation for the complete combustion of butane gas. What volume of oxygen is required to burn 100 l of the gas (consider all gases under the same conditions)? What volume of air (20% oxygen) is required?

31.26. A 4.00-l sample of octane gasoline (a little more than one gallon) weighs 3.19 kg. Calculate the volume of air at STP required for its complete combustion.

Supplementary Readings

31.1. *Modern Principles of Organic Chemistry*, J. L. Kice and E. N. Marvell, MacMillan, New York, 1974.

31.2. "Hydrocarbons in Petroleum," F. D. Rossini, *J. Chem. Educ.*, **37**, 554 (1960).

31.3. "High-Energy Reactions of Carbon," R. M. Lemmon and W. R. Erwin, *Sci. American*, Jan., 1975; p. 72.

31.4. "Oil and Gas From Coal," N. P. Cochran, *Sci. American*, May, 1976; p. 24.

Chapter 32

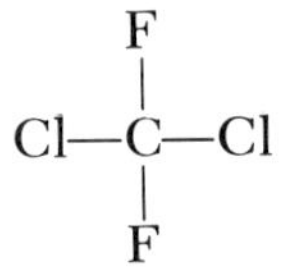

$$Cl-\underset{\underset{F}{|}}{\overset{\overset{F}{|}}{C}}-Cl$$

Dichlorodifluoromethane
(Freon – 12)

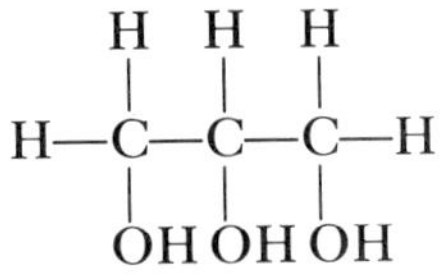

$$H-\underset{\underset{OH}{|}}{\overset{\overset{H}{|}}{C}}-\underset{\underset{OH}{|}}{\overset{\overset{H}{|}}{C}}-\underset{\underset{OH}{|}}{\overset{\overset{H}{|}}{C}}-H$$

1,2,3-Trihydroxypropane
(Glycerol)

$CH_3CH_2OCH_2CH_3$
Diethyl ether
(Ether)

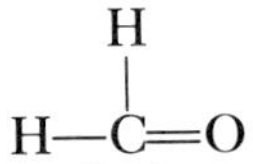

$$H-\underset{}{\overset{\overset{H}{|}}{C}}=O$$

Methanal
(Formaldehyde)

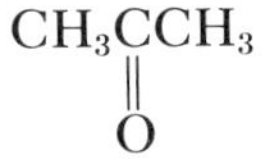

$$\underset{\underset{O}{\|}}{CH_3CCH_3}$$

Propanone
(Acetone)

$$HO-\underset{\underset{CH_2COOH}{|}}{\overset{\overset{CH_2COOH}{|}}{C}}-COOH$$

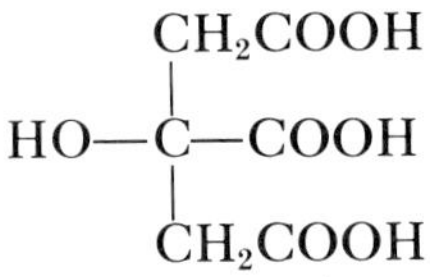

3-Hydroxy-3-carboxy-
pentanedioic acid
(Citric acid)

—OH
—COOCH₃

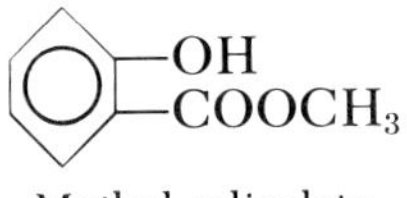

Methyl salicylate
(Oil of Wintergreen)

Introduction to Organic Chemistry II

32.1   DERIVATIVES OF THE HYDROCARBONS

The substitution of elements or groups for one or more of the hydrogen atoms in a hydrocarbon results in the formation of many types of compounds, which are referred to as **derivatives** of the hydrocarbons. For example, methyl chloride (CH_3Cl), one of the products formed by the treatment of methane with chlorine, is derived from the hydrocarbon methane.

$$
\begin{array}{ccc}
\text{H} & & \text{H} \\
| & & | \\
\text{H—C—H} & & \text{H—C—H} \\
| & & | \\
\text{H} & & \text{Cl} \\
\text{Methane} & & \text{Methyl chloride}
\end{array}
$$

The CH_3 group is termed **methyl** and derives its name from the parent hydrocarbon methane (See Chapter 31, page 642). Corresponding groups of other hydrocarbons of the methane series are similarly derived. The general term, **alkyl groups,** includes all groups derived from the methane series; they are given the general symbol of **R.** Sometimes, however, **R** may stand for more complex groups than just simple alkyl groups, and sometimes it may represent a simple hydrogen, H. The specific name of the group is obtained by replacing the ending -*ane* by -*yl.* The composition of the group will be the same as that of the parent hydrocarbon, except that it will contain one less hydrogen atom; thus, C_2H_5 is the **ethyl** group; C_3H_7 the **propyl** group, and so on. Obviously, the valence of alkyl groups must be 1.

Derivatives of the hydrocarbons may be classified into different types of compounds depending on the atom or group of atoms substituted for the hydrogen atom. Such groups are called **functional** groups and in large measure determine the properties of the compound. Compounds with the same functional group or groups exhibit very similar chemical properties. Table 32.1 is a summary of several of the more important types of derivatives with the general formula and functional group for compounds of each type.

Similar to the hydrocarbons, a homologous series of each type of compound is possible. For example, the following amines constitute such a series: CH_3NH_2, $C_2H_5NH_2$, $C_3H_7NH_2$, $C_nH_{2n+1}NH_2$.

Bond Angles and Distances in Selective Carbon Compounds

Diagrams of carbon compounds are seldom drawn to exact scale. Instrumental measurements in physical chemistry have given bonding distances between atoms and special information such as angles between three atomic centers. Molecules are seldom planar; carbon atoms in a molecule are not commonly linear with sidebonded atoms at 90°. The structure of a molecule bears on its chemical activity. In intermolecular collisions, the structure is related to those atoms which are first contacted. Close bonding between atoms denotes that greater energy is needed to disrupt the bond. Examples of some molecule types with bonding distances in angstroms and spacial angles follow.

C_4H_{10} Butane

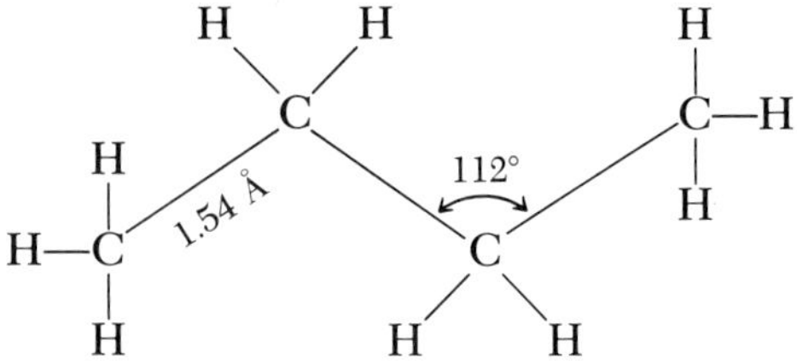

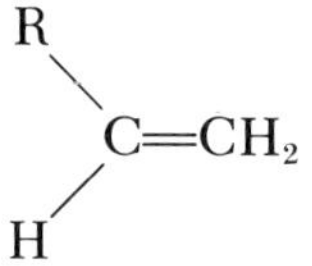

$\begin{array}{c} R \\ \diagdown \\ C{=}CH_2 \\ \diagup \\ H \end{array}$ Alkene

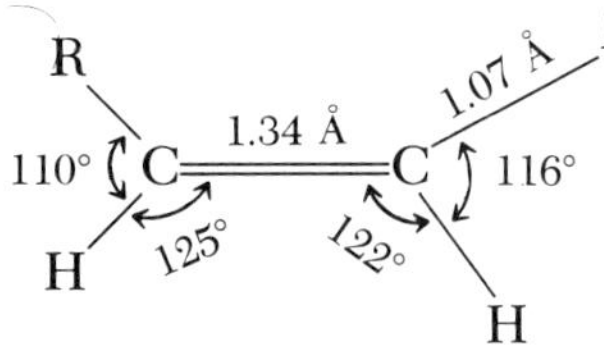

R—C≡C—H   Alkyne   R—C≡C—H with 1.20 Å and 1.06 Å

R—O—H   Alcohol

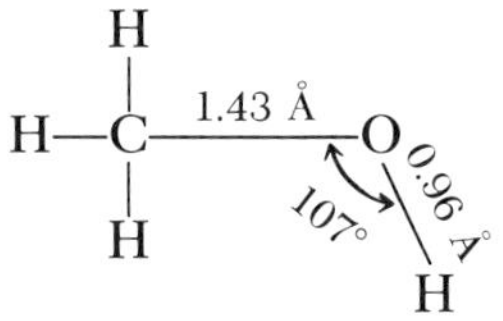

R—NH₂   Amine

$\begin{array}{c} O \\ \parallel \\ R{-}C \\ \diagdown \\ O{-}H \end{array}$ Acid

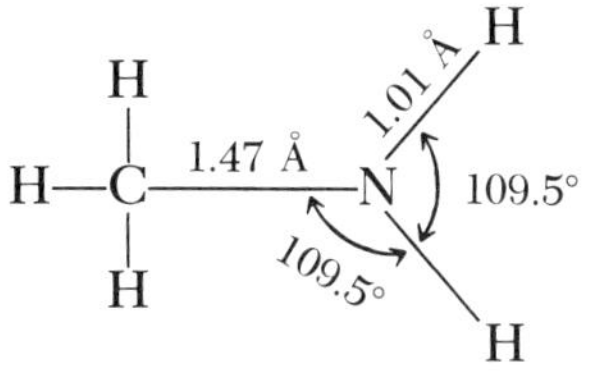

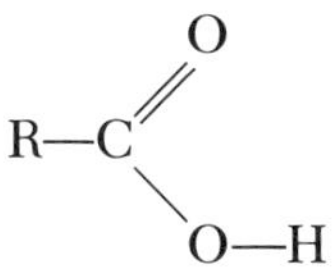

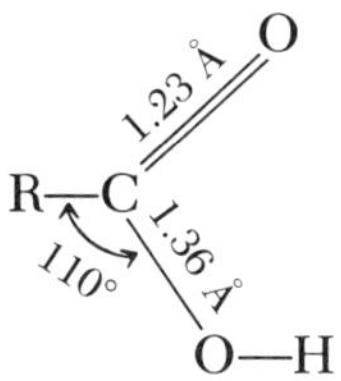

Note the bond distances C—C in an alkane hydrocarbon such as butane, 1.54 Å; of C=C in an alkene, 1.34 Å, and of C≡C in an alkyne, 1.20 Å. These diverse bond distances relate to bond energies and reactivity.

Table 32.1
Hydrocarbon derivatives and their functional groups

Type of Compound	General Formula	Functional Group	Name of Functional Group
Alkyl halide	R—X(where X=F,Cl,Br,I)	—X	halide
Alcohol	R—OH	—OH	hydroxyl
Ether	R—O—R′	—OR	alkoxy
Aldehyde	R—C—H $\overset{\|}{\underset{O}{}}$	—C—H $\overset{\|}{\underset{O}{}}$	—
Ketone	R—C—R′ $\overset{\|}{\underset{O}{}}$	—C— $\overset{\|}{\underset{O}{}}$	carbonyl
Acid	R—C—OH $\overset{\|}{\underset{O}{}}$	—C—OH $\overset{\|}{\underset{O}{}}$	carboxyl
Ester	R—C—O—R′ $\overset{\|}{\underset{O}{}}$	—C—O— $\overset{\|}{\underset{O}{}}$	—
Amine	R—NH₂	—NH₂	amino
Sulfonic acid	R—SO₃H	—SO₃H	sulfonic acid
Nitrile	R—CN	—CN	nitrile
Nitro compound	R—NO₂	—NO₂	nitro
Thiol (mercaptan)	R—SH	—SH	—

Shapes of Organic Molecules

It was pointed out in Chapter 5 that molecules may assume many shapes, such as linear, angular, planar, and tetrahedral. In such a simple molecule as CH_4, the sp^3 hybrid bonds point to the corners of a tetrahedron, and this directional feature carries over if other atoms are substituted for the hydrogens. It is also in effect in the bonding of carbon-to-carbon atoms in the higher hydrocarbons of the alkane series. The result is a staggered chain of carbon atoms. Since there is rotation about the single bonds between carbon atoms, an infinite number of arrangements or configurations is theoretically possible. For example, n-pentane, C_5H_{12}, might exhibit the following configurations (H atoms not shown).

Change from one to another could take place simply by rotation and without the rupture of any bonds. A given sample of n-pentane is actually a mixture of the possible configurations.

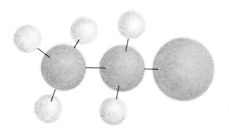

Figure 32.1
Ethyl chloride.

32.2 HALOGEN DERIVATIVES OF THE HYDROCARBONS

Examples of **alkyl halides** are shown below:

CH_3F	CH_3CH_2Cl	C_3H_7Br	C_4H_9I
Methyl fluoride	Ethyl chloride (Figure 32.1)	Propyl bromide	Butyl iodide

Substitution of more than one hydrogen of a hydrocarbon by halogen atoms is also possible. Examples are:

$$CF_2Cl_2 \qquad\qquad CFCl_3$$
Dichlorodifluoromethane (Freon-12)   Trichlorofluoromethane (Freon-11)

$$CH_2{-}CH_2$$
$$\;\;|\qquad\;\;|$$
$$Cl\quad\;\; Cl \qquad\qquad CH_2{=}CHCl \qquad\qquad CF_2{=}CF_2$$

1,2-Dichloroethane (Ethylene dichloride)   Chloroethylene (Vinyl chloride)   Tetrafluoroethylene

Vinyl chloride is one of the 50 top chemicals produced in the U.S. It ranks number 24 in the list and is used to make polymers (page 693). Ethylene dichloride is number 16 in the list. Much of this compound is converted into vinyl chloride, but it also has uses as a scavenging agent for leaded gasoline, as a fumigant, insecticide, solvent, and paint remover. Both of these industrially important compounds are made by reacting Cl_2 with ethylene. $CF_2{=}CF_2$ is the monomer from which the polymer **Teflon**, $(-CF_2-CF_2-)_n$, is made (page 693).

32.3 ALCOHOLS

Alcohols have the general formula R—OH, and in some respects are similar to metallic hydroxides or bases. They show no tendency to form hydroxide ions in solution, however, and the properties which we usually associate with bases – bitter taste, soapy feeling, and effect on litmus – are not exhibited by the alcohols. Alcohols may also be considered to be derived from water, with one of the hydrogen atoms in the water molecule replaced by an alkyl group. Actually, in most of their chemical properties alcohols resemble water more than they resemble metallic hydroxides.

While it is theoretically possible to substitute more than one OH group for hydrogen atoms in methane, such a compound is not actually known, and it seems likely that compounds in which more than one OH group is attached to the *same* carbon atom are very unstable, if they exist at all.

Structurally the oxygen atom present in the hydroxide group of an alcohol is linked to the carbon atom in the alkyl radical by a single pair of shared electrons. The oxygen atom also shares a pair of electrons with the hydrogen atom of the hydroxide group; thus methyl alcohol is shown as

$$
\begin{array}{c}
\text{H} \\
\text{H:C:O:H} \\
\text{H}
\end{array}
\quad \text{or} \quad
\begin{array}{c}
\text{H} \\
| \\
\text{H—C—O—H} \\
| \\
\text{H}
\end{array}
$$

Isomers become possible in alcohols of more than two carbon atoms; for example, propyl alcohol (C_3H_7OH) may be written structurally as

$$
\begin{array}{c}
\text{H H H} \\
|\ \ |\ \ | \\
\text{H—C—C—C—O—H} \\
|\ \ |\ \ | \\
\text{H H H}
\end{array}
\quad \text{or} \quad
\begin{array}{c}
\text{H H   H} \\
|\ \ |\ \ \ \ | \\
\text{H—C—C——C—H} \\
|\ \ |\ \ \ \ | \\
\text{H OH H}
\end{array}
$$

The first isomer is called a **primary** alcohol, since the OH group is attached to a carbon atom, which itself is linked directly to only one carbon atom. The second is called a **secondary** alcohol, since the OH group is attached to a carbon to which two other carbon atoms are directly linked. With the higher alcohols it is possible to have an OH group attached to a carbon atom which is directly linked to three other carbon atoms, and the alcohol is termed a **tertiary** alcohol. The above structural formulas for the propyl alcohols are usually written in a semistructural manner — that is,

$$CH_3—CH_2—CH_2OH \qquad CH_3—CHOH—CH_3 \ \ \text{or} \ \ (CH_3)_2CHOH$$
Normal propyl alcohol   Isopropyl alcohol

In naming alcohols (IUPAC system), the ending *-ol* is attached to the name of the hydrocarbon from which the alcohol is derived; thus methyl alcohol is **methanol,** ethyl alcohol is **ethanol,** propyl alcohol is **propanol,** and so on. Some typical examples of alcohols and their IUPAC and common names are as follows:

$$CH_3CH_2CH_2CH_2OH$$
1-Butanol
(*n*-Butyl alcohol)

$$CH_3CHCH_2OH$$
$$\ \ \ \ \ \ \ \ \ |$$
$$\ \ \ \ \ \ \ \ CH_3$$
2-Methyl-1-propanol
(Isobutyl alcohol)

$$OH$$
$$\ \ \ \ \ \ \ \ \ |$$
$$CH_3CH_2CHCH_3$$
2-Butanol
(*sec*-Butyl alcohol)

$$\begin{array}{c} \text{OH} \\ | \\ \text{CH}_3\text{---}\overset{|}{\underset{|}{\text{C}}}\text{---CH}_3 \\ | \\ \text{CH}_3 \end{array}$$

2-Methyl-2-propanol
(*tert*-Butyl alcohol)

$$\begin{array}{c} \text{CH}_3 \quad\ \text{CH}_3 \\ | \qquad\ \ | \\ \text{CH}_3\text{CHCH}_2\overset{|}{\underset{|}{\text{C}}}\text{CH}_3 \\ | \\ \text{OH} \end{array}$$

2,4-Dimethyl-2-pentanol

The first 4 of the above compounds are isomers of butanol, C_4H_9OH.

Methyl Alcohol

Methyl alcohol, or **methanol** (also called wood alcohol), is one of the products of the destructive distillation of wood. Until a few years ago this method provided almost the sole source of methanol, but today most of the methanol synthesis is accomplished by the following reaction:

$$CO + 2\ H_2 \rightarrow CH_3OH$$

The gases are passed over ZnO, Cr_2O_3 or other suitable catalysts at an elevated temperature and under pressure.

Methyl alcohol is a colorless liquid which boils at 67°C. It is extremely toxic when taken internally. Large quantities of methanol are employed commercially in solvents for varnishes, shellacs, and lacquers, which are readily soluble in the alcohol. It is used also as a denaturant for ethyl alcohol and in the preparation of formaldehyde and other chemicals. Of recent interest is the possible use of methanol in mixtures with gasoline as a motor fuel. Methanol ranks 18th in the list of top chemicals.

Ethyl Alcohol

Ethyl alcohol, or **ethanol** (C_2H_5OH, Figure 32.2), sometimes referred to as grain alcohol, is a colorless liquid which is quite similar to methanol in physical properties. It is much less poisonous than methyl alcohol, though if consumed in large amounts it may be decidedly harmful. The boiling point of

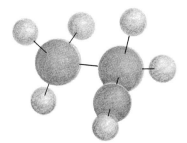

Figure 32.2
Ethanol (ethyl alcohol).

ethanol is 78°C, which is not very different from that of methanol (67°C). A mixture of the two, therefore, is not easily separated by distillation.

Two-thirds of the ethyl alcohol produced industrially is by the fermentation of glucose and fructose. Glucose or fructose may be obtained by hydrolysis of cane or malt sugar or starch (page 705). The reaction occurring during the fermentation is represented by the equation

$$C_6H_{12}O_6 \rightarrow 2\ C_2H_5OH + 2\ CO_2$$

The fermentation is catalyzed by certain **enzymes**, organic compounds produced from yeast cells. Solutions containing 10 to 15% alcohol are usually produced by these fermentation processes, since higher concentrations of alcohol kill the yeast cells. By fractional distillation of these solutions a concentration of approximately 95% alcohol is readily prepared, and industrial alcohol is usually marketed in this concentration. The other one-third of the industrial ethanol is produced synthetically from ethylene. For example, the direct hydration of ethylene is

$$C_2H_4 + H_2O \xrightarrow[\text{heat, pressure}]{\text{H}_3\text{PO}_4 \text{ catalyst}} C_2H_5OH$$

Synthetic ethanol ranks 47th in the list of top chemicals and has more different uses than almost any other organic compound. Like methanol, it is used as a solvent for varnishes and lacquers. **Tinctures** are alcoholic solutions — tincture of iodine is familiar to all. Flavoring extracts, tonics, lotions, and many medicines contain large percentages of alcohol. Ethanol is used as a starting material in the preparation of ethyl acetate, acetaldehyde, ether, and other important chemical compounds. Because of its low freezing point (−117°C), ethyl alcohol is employed as an antifreeze in automobile radiators. Many other uses of lesser importance for ethyl alcohol might be listed.

The concentration of an alcoholic solution is usually expressed as **proof** rather than percentage. A 95% alcohol solution is 190 proof, 100% alcohol is 200 proof, 45% alcohol is 90 proof, and so on. While alcohol used for beverages is taxed highly by the federal government, the alcohol used for industrial purposes is tax free. Usually some nonpalatable substance (denaturant) is added to ensure that it will not be diverted to beverage manufacture.

Other Alcohols

Ethylene glycol [$C_2H_4(OH)_2$], ranking 28th in the top chemicals list, because of its relatively high boiling point (198°C) and its noncorrosive action on metals finds use as an antifreeze in automobile radiators. **Prestone, Zerex, Dowgard,** and so forth, are trade names for antifreeze preparations containing ethylene glycol. The compound is also used in deicing fluids, in exterior latex paints to improve freeze-thaw characteristics, in automobile brake fluids, and in the manufacture of polyester and urethanes.

Figure 32.3
Glycerine.

Ethylene glycol is now produced primarily by the following reactions or modifications of these:

$$C_2H_4 + \tfrac{1}{2} O_2 \xrightarrow[\text{heat, pressure}]{\text{Ag catalyst}} CH_2OCH_2$$

Ethylene   Ethylene
                                          oxide

$$CH_2OCH_2 + H_2O \rightarrow CH_2OHCH_2OH$$

Ethylene oxide, CH_2——CH_2, is an intermediate in the above reactions.
                       $\diagdown_O\diagup$

It ranks 26 in the list of top chemicals. In addition to its use in making ethylene glycol, it is used as a sterilant for bacteria, spores, and viruses; in the control of weevils in nuts; as an aerosol propellant, and in the making of a number of insecticides, fungicides, explosives, and resins.

Glycerol or **glycerine** (CH_2CHCH_2) (Figure 32.3) is a colorless, viscous
                                  | | |
                                 OH OHOH
liquid with a sweet taste. It is produced as a by-product of soap manufacture. Besides its use in many toilet preparations, it is the starting material in the manufacture of nitroglycerin, a powerful explosive.

Isopropyl alcohol or **2-propanol** (CH_3CHOH) ranks 41st in the list of top
                                              |
                                            CH_3
chemicals and has many uses. It is a good solvent and not very poisonous, although it is not suitable for drinking. Since it is tax free, it replaces ethanol in many applications. It is used as a denaturant in ethanol; in medicine as rubbing alcohol; in deicing; as an ingredient in drugs and cosmetics; as a solvent for shellac, gums, oils, and stains; and in the synthesis of other compounds such as acetone and glycerol. It may be made industrially by the direct hydration of propene using polytungsten catalysts:

$$CH_3CH\!\!=\!\!CH_2 + H_2O \xrightarrow[\text{W catalyst}]{\text{heat, pressure}} (CH_3)_2CHOH$$

32.4   ETHERS

The general formula for ethers is R—O—R', where R and R' may be the same or different alkyl radicals. They may be considered to be derived from water by the replacement of the two hydrogen atoms in the water molecule with alkyl radicals.

In general, ethers are prepared by the dehydration of alcohols with sulfuric acid. If ethyl alcohol is mixed with concentrated sulfuric acid and the mixture warmed to 140°C, **diethyl ether** and water are formed, according to the reaction

$$\begin{array}{ccc} \text{H} \;\; \text{H} & & \text{H} \;\; \text{H} \\ | \;\; | & & | \;\; | \\ \text{H--C--C--O--} \!\! & \!\! \text{H} \;\; \text{OH} \!\! & \!\! \text{--C--C--H} \xrightarrow{\text{H}_2\text{SO}_4} \\ | \;\; | & & | \;\; | \\ \text{H} \;\; \text{H} & & \text{H} \;\; \text{H} \end{array}$$

$$\begin{array}{cc} \text{H} \;\; \text{H} & \text{H} \;\; \text{H} \\ | \;\; | & | \;\; | \\ \text{H--C--C--O--C--C--H} + \text{H}_2\text{O} \\ | \;\; | & | \;\; | \\ \text{H} \;\; \text{H} & \text{H} \;\; \text{H} \end{array}$$

If the above mixture is heated to 160°C or above, further dehydration takes place, with the formation of ethylene.

Diethyl ether, commonly called **ether,** is used as an anesthetic in surgery. It is a highly flammable and volatile liquid, with a boiling point of 35°C. Ether is also used as a solvent for waxes, gums, resins, and fats. In the analysis of certain food products for fat content, the fat is extracted with ether.

In naming simple ethers, the names of the alkyl radicals present are followed by the term **ether,** or, if the alkyl radicals are the same, the prefix *di-* may be employed; thus, $C_3H_7OC_2H_5$ is propyl ethyl ether; CH_3OCH_3 or $(CH_3)_2O$ is methyl methyl ether, or better, dimethyl ether. The latter compound is isomeric with ethyl alcohol.

Under the IUPAC system of naming ethers, the simplest R group along with the oxygen is called an **alkoxy** group. Thus $CH_3OC_2H_5$ would be called **methoxyethane** and $C_2H_5OC_2H_5$ would be **ethoxyethane.** Generally the IUPAC system is used for the more complex ethers.

Ethylene oxide, $CH_2\!\!-\!\!-\!\!CH_2$, mentioned in the previous section, and

propylene oxide, $CH_2\!\!-\!\!-\!\!CHCH_3$ (43rd among the top chemicals), are exam-

ples of cyclic ethers or epoxides. Most of the latter is used to produce polyurethane.

32.5   ALDEHYDES

The gentle oxidation of a primary alcohol containing the —CH$_2$OH group results in the formation of an **aldehyde** with the general formula

$$R—\overset{\overset{\displaystyle H}{|}}{C}{=}O$$

For example, the oxidation of ethyl alcohol produces **acetaldehyde.**

$$H—\overset{\overset{\displaystyle H}{|}}{\underset{\underset{\displaystyle H}{|}}{C}}—\overset{\overset{\displaystyle H}{|}}{\underset{\underset{\displaystyle H}{|}}{C}}—O— \boxed{H + (O)} \rightarrow H—\overset{\overset{\displaystyle H}{|}}{\underset{\underset{\displaystyle H}{|}}{C}}—\overset{\overset{\displaystyle H}{|}}{C}{=}O + H_2O$$

The name aldehyde comes from the two words al**cohol dehydr**ogenation. The symbol (O) stands for **oxidation** (addition of an oxygen atom from some oxidizing agent such as KMnO$_4$, K$_2$Cr$_2$O$_7$, CuO, O$_2$, etc.; removal of hydrogen atoms or loss of electrons). The actual reaction for the oxidation of ethanol above, using potassium dichromate and sulfuric acid is:

$$3\,CH_3CH_2OH + K_2Cr_2O_7 + 4\,H_2SO_4 \rightarrow 3\,CH_3CHO + Cr_2(SO_4)_3 + K_2SO_4 + 7\,H_2O$$

The slow oxidation of methyl alcohol yields **formaldehyde** (HCHO), which is the simplest of the aldehydes. Here the R of the general formula is a hydrogen atom rather than an alkyl group.

$$CH_3OH + (O) \rightarrow H—\overset{\overset{\displaystyle H}{|}}{C}{=}O + H_2O$$

Formaldehyde ranks 22 in the top chemicals list and is prepared principally by the oxidation of methanol using silver or iron molybdate catalysts under carefully controlled elevated temperatures.

A 40% solution of formaldehyde is marketed under the trade name of **formalin.** Formaldehyde is used as a preservative for museum specimens, as a disinfectant and insecticide, and in the preparation of synthetic resins.

Formaldehyde Phenol   Hydroxybenzyl alcohol (Bakelite)

Formaldehyde and urea (H$_2$N—CO—NH$_2$), ranked 15th among the top chemicals, heated together form a thermosetting product with the possible structure diagrammed below. (Note the cross-linking in the indicated giant molecule – typical of a formaldehyde-urea resin.)

$$-CH_2-\overset{\overset{\displaystyle H}{|}}{N}-\overset{\overset{\displaystyle O}{\|}}{C}-\overset{\overset{\displaystyle H}{|}}{N}-CH_2-\overset{\overset{\displaystyle CH_2}{|}}{N}-\overset{\overset{\displaystyle O}{\|}}{C}-\overset{\underset{\displaystyle CH_2}{|}}{N}-CH_2-\overset{\overset{\displaystyle H}{|}}{N}-\overset{\overset{\displaystyle H}{|}}{\underset{\underset{\displaystyle O}{\|}}{C}}-\overset{\overset{\displaystyle H}{|}}{N}-CH_2-\overset{\overset{\displaystyle H}{|}}{N}-\overset{\overset{\displaystyle H}{|}}{\underset{\underset{\displaystyle O}{\|}}{C}}-\overset{\overset{\displaystyle H}{|}}{N}\ldots$$

$$-\overset{\overset{\displaystyle H}{|}}{N}-\overset{\overset{\displaystyle O}{\|}}{C}-\overset{\underset{\displaystyle CH_2}{|}}{N}-CH_2-\overset{\overset{\displaystyle H}{|}}{\underset{\underset{\displaystyle O}{\|}}{C}}-N-CH_2-\overset{\overset{\displaystyle H}{|}}{\underset{\underset{\displaystyle O}{\|}}{C}}-\overset{\underset{\displaystyle CH_2}{|}}{N}-CH_2\ldots$$

$$-CH_2-\overset{\underset{\displaystyle CH_2}{|}}{N}-\overset{\overset{\displaystyle}{\underset{\underset{\displaystyle O}{\|}}{C}}}-N-CH_2-\overset{\overset{\displaystyle H}{|}}{\underset{\underset{\displaystyle O}{\|}}{C}}-\overset{\overset{\displaystyle H}{|}}{N}-CH_2-\overset{\overset{\displaystyle}{}}{N}-\overset{\overset{\displaystyle H}{|}}{\underset{\underset{\displaystyle O}{\|}}{C}}-N\ldots$$

These plastics are used in the manufacture of such items as telephone transmitters and receivers, radio equipment, other electrical equipment, combs, pencils, pipestems, and ash trays.

In naming aldehydes by the IUPAC system, the (-e) ending of the hydrocarbon name of the same number of carbon atoms is replaced by -al; i.e. HCHO is methan*al*, CH_3CHO is ethan*al*, C_4H_9CHO is pentan*al*, etc.

32.6   KETONES

If a secondary alcohol containing the —CHOH group is gently oxidized, a **ketone** of the general formula $R-\overset{\overset{\displaystyle}{}}{\underset{\underset{\displaystyle O}{\|}}{C}}-R'$ is obtained. The C=O group is termed the **carbonyl** group and is characteristic of both aldehydes and ketones. The oxidation of isopropyl alcohol yields **dimethyl** *ketone*, more commonly called **acetone.**

$$\overset{\displaystyle CH_3}{\underset{\displaystyle CH_3}{\diagdown}}\overset{\displaystyle H}{\underset{\displaystyle OH}{\diagup}}C + (O) \rightarrow CH_3-\overset{\overset{\displaystyle O}{\|}}{C}-CH_3 + H_2O$$

Acetone, ranking 40th in the list of top chemicals, is produced primarily by the above reaction using O_2 of the air and a ZnO catalyst. It is one of the more important solvents for varnishes and lacquers and for compressed acetylene. Many fingernail polish removers owe their effectiveness to the solvent acetone. It is quite volatile and readily dissolves in water as well as many organic solvents.

Most of the acetone now goes to make methyl isobutyl ketone (4-methyl-2-pentanone), $(CH_3)_2CHCH_2COCH_3$. This is an excellent solvent for many resins, adhesives, and rubber cements as well as a purification agent for pharmaceuticals, mineral oils, etc. It is an effective agent for separation of certain inorganic salts such as those of plutonium from uranium and zirconium from hafnium.

In the IUPAC system, the characteristic name ending is *-one*. Examples of some simple ketones are shown below (the common name is in parentheses):

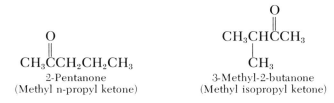

| 2-Pentanone | 3-Methyl-2-butanone |
| (Methyl n-propyl ketone) | (Methyl isopropyl ketone) |

32.7 ACIDS

Acids contain the

$$\overset{\overset{\displaystyle O}{\|}}{—C—OH},$$

carboxyl group (from the combination of **carb**onyl and hyd**roxyl** groups). This functional group is usually written on one line to save space: **—COOH**. The organic acids are generally referred to as **carboxylic acids.** The most common acids are usually called by their common names which are based on their origin. The IUPAC nomenclature uses the ending *-oic acid* added to the usual hydrocarbon name. The first four of the aliphatic carboxylic acid series are given below. The general formula of the series is **RCOOH**, where R may be a hydrogen atom or an alkyl group.

	Common Name	IUPAC Name
HCOOH	Formic acid (Latin, *formica*, ant)	Methanoic acid
CH_3COOH	Acetic acid (Latin, *acetum*, vinegar)	Ethanoic acid
or $HC_2H_3O_2$		
C_2H_5COOH	Propionic acid (Greek, *propion*, fat)	Propanoic acid
C_3H_7COOH	Butyric acid (Latin, *butyrum*, butter)	Butanoic acid

Three organic acids belong in the top 50 chemicals: **acetic acid** (36) and its anhydride $(CH_3CO)_2O$ (44), **adipic acid** (45), used to make Nylon, and **terephthalic acid** (21), used to make Dacron. Their structural formulas and those of several other common acids are shown below (the common name is shown in parentheses):

$$CH_3\overset{\overset{\displaystyle O}{\|}}{C}OH$$
Ethanoic acid
(Acetic acid)

$$CH_3\overset{\overset{\displaystyle O}{\|}}{C}—O—\overset{\overset{\displaystyle O}{\|}}{C}—CH_3$$
(Acetic anhydride)

$$HO\overset{\overset{\displaystyle O}{\|}}{C}CH_2CH_2CH_2CH_2\overset{\overset{\displaystyle O}{\|}}{C}OH$$
1,6-Hexanedioic acid
(Adipic acid)

COOH

COOH
(Terephthalic acid)

$$HO\overset{\overset{\displaystyle O}{\|}}{C}—\overset{\overset{\displaystyle O}{\|}}{C}OH$$
Ethanedioic acid
(Oxalic acid)
in rhubarb

$$\begin{array}{c} CH_2COOH \\ | \\ HO—C—COOH \\ | \\ CH_2COOH \end{array}$$
3-Hydroxy-3-carboxy-
pentanedioic acid
(Citric acid)
in citrus fruits

$$\begin{array}{c} CH_3 \\ | \\ H—C—OH \\ | \\ COOH \end{array}$$
(Lactic acid)
in sour milk

$$\begin{array}{c} COOH \\ | \\ H—C—OH \\ | \\ H—C—OH \\ | \\ COOH \end{array}$$
(Tartaric acid)
in grapes

—OH
—COOH
(Salicylic acid)

Acids may be obtained by the oxidation of alcohols and aldehydes. The oxidation of acetaldehyde, for example, yields acetic acid.

$$CH_3\overset{\overset{\displaystyle O}{\|}}{C}—H + (O) \rightarrow CH_3\overset{\overset{\displaystyle O}{\|}}{C}—OH$$

It should be pointed out that the product obtained in the oxidation of an alcohol depends on the stage of oxidation. An aldehyde may be considered the first stage of oxidation of an alcohol; acids are obtained by further oxidation of the aldehyde. Complete oxidation of an alcohol results in the formation of carbon dioxide and water. It is very often difficult to control the oxidation so that only one type of compound is formed.

The ionizable hydrogen in the —COOH group is responsible for acid properties. In water solutions these acids are weakly ionized. For example, acetic acid ionizes

$$H—\overset{\overset{\displaystyle H}{|}}{\underset{\underset{\displaystyle H}{|}}{C}}—\overset{\overset{\displaystyle O}{\|}}{C}—O—H + H_2O \rightleftharpoons \left[H—\overset{\overset{\displaystyle H}{|}}{\underset{\underset{\displaystyle H}{|}}{C}}—\overset{\overset{\displaystyle O}{\|}}{C}—O\right]^- + H_3O^+$$

Acetic acid in 0.1 N concentration is less than 2% ionized as compared with the highly ionized inorganic acids such as hydrochloric or sulfuric acids.

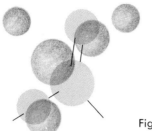

Figure 32.4
Formic acid.

The acid radical of the organic acid is the RCOO— group and derives its name from the -*ic* acid, replacing that ending with -*ate*. CH_3COO— is the acetate radical, C_3H_7COO— the butyrate radical, HCOO— the formate radical, and so on. Like the mineral acids, organic acids are neutralized with bases.

$$RCOOH + NaOH \rightarrow H_2O + RCOONa$$
sodium salt

Formic acid (HCOOH, Figure 32.4) is the simplest acid of the acid series. It is found in certain nettles and is responsible for their sting. Red ants and bees also contain the acid, which is injected when these insects bite or sting. Formic acid may be prepared by the oxidation of methyl alcohol, but actually the demand for the acid is small.

Pure acetic acid is a colorless liquid with a sharp, penetrating odor. At 18°C it freezes to a white solid, very similar in appearance to ice—hence it is often called glacial acetic acid.

If alcohol produced in the fermentation of sugars and fruit juices is allowed to come in contact with bacteria from mother of vinegar, further oxidation of the alcohol takes place, with the formation of vinegar containing about 4% acetic acid. Hard cider will in time yield a vinegar containing acetic acid.

Acetic acid is used widely in the food industry (vinegar) and in making industrial acetates such as vinyl acetate (for vinyl resins), cellulose acetate (a type of rayon), and sodium-, ethyl-, butyl-, and pentyl acetates (flavorings and solvents). Industrially, one way that acetic acid is made is by air oxidation of acetaldehyde using a manganese acetate catalyst

$$CH_3CHO + 1/2\ O_2\ (air) \xrightarrow[pressure]{catalyst} CH_3COOH$$

Acetic anhydride is a compound representative of a class of compounds called **acid anhydrides.** They may be formed by the dehydration (removing a molecule of water) of an acid. For example,

$$2\ CH_3\overset{\overset{\displaystyle O}{\|}}{C}{-}OH \xrightarrow{H_2SO_4} CH_3\overset{\overset{\displaystyle O}{\|}}{C}{-}O{-}\overset{\overset{\displaystyle O}{\|}}{C}CH_3$$

Acetic anhydride is a more active compound than acetic acid and is used for aspirin and cellulose acetate manufacture.

Stearic, palmitic, and **oleic** acids are fundamental to the formation of fats. Their formulas are, respectively, $C_{17}H_{35}COOH$, $C_{15}H_{31}COOH$, and $C_{17}H_{33}COOH$. (See page 701.)

32.8   ESTERS

Esters are formed from chemical reactions between alcohols and organic acids. For example, if ethyl alcohol is added to acetic acid in the presence of concentrated sulfuric acid as a dehydrating agent, the reaction is

$$
\underset{\substack{|\\H}}{\overset{\substack{H\;\;O\\|\;\;\|}}{H-C-C-}}\;[OH+H]\;O-\underset{\substack{|\\H\;H}}{\overset{\substack{H\;H\\|\;\;|}}{C-C}}-H \xrightarrow{\;H_2SO_4\;} \underset{\substack{|\\H}}{\overset{\substack{H\;\;O\\|\;\;\|}}{H-C-C-}}O-\underset{\substack{|\;\;|\\H\;H}}{\overset{\substack{H\;H\\|\;\;|}}{C-C}}-H + H_2O
$$

or, semistructurally,

$$
CH_3CO\;[OH+H]\;OC_2H_5 \xrightarrow{\;H_2SO_4\;} CH_3COOC_2H_5 + H_2O
$$
Ethyl acetate

The reaction is similar to the neutralization of an acid with a base in inorganic reactions. The inorganic acid-base reactions are much more rapid than those between alcohols and acids, however, because of the ionization which takes place with inorganic substances. The reaction of an alcohol with an acid is called **esterification.**

The general formula for an ester is RCOOR', where RCOO— is the acid group and R' is the group derived from the alcohol from which the ester was prepared. Esters, then, according to their formulas appear to be similar to inorganic salts; as a matter of fact, they are sometimes referred to as **ethereal salts** because of their volatile nature. One essential difference between esters and salts is the ionic nature of most salts and the nonionic character of esters.

In naming an ester, the alkyl group (R') of the alcohol is named first, followed by the name of the acid group with an *-oate* ending. The following formulas, written semistructurally with their names, will make this clear.

HCOOCH$_3$
Methyl methanoate
(Methyl formate)

$CH_3COOC_2H_5$
Ethyl ethanoate
(Ethyl acetate)

$CH_3COOC_5H_{11}$
Pentyl ethanoate
(Amyl acetate)
banana flavor

$C_3H_7COOC_2H_5$
Ethyl butanoate
(Ethyl butyrate)
strawberry flavor

$CH_3COOC_8H_{17}$
Ocyl ethanoate
(Ocyl acetate)
orange flavor

(Methyl salicylate)
wintergreen flavor

Many of the esters have a pleasant odor of fruit or flowers, and this property makes certain esters useful as perfumes and artificial flavorings. In the formulas above, **amyl acetate** has the odor of bananas and is called "banana oil"; **methyl salicylate** is "oil of wintergreen"; **ethyl butyrate** has the odor of pineapple, and **octyl acetate** smells like oranges.

Esters of inorganic acids may also be prepared. For example, glycerin reacts with nitric acid in the presence of sulfuric acid.

$$C_3H_5(OH)_3 + 3\ HNO_3 \xrightarrow{\ H_2SO_4\ } C_3H_5(NO_3)_3 + 3\ H_2O$$

Glycerin   Glyceryl trinitrate
                          (Nitroglycerine)

Esters, like inorganic salts, may undergo hydrolysis, forming an acid and an alcohol. The process is the reverse of esterification.

$$RCOOR' + H_2O \rightarrow RCOOH + R'OH$$

ester   acid   alcohol

For example, the hydrolysis of ethyl acetate in the presence of HCl or other strong inorganic acids is:

$$CH_3COOC_2H_5 + H_2O \xrightarrow{\ H^+\ } CH_3COOH + C_2H_5OH$$

Ethyl acetate   Acetic acid   Ethyl alcohol

Odor One is often amazed at how persistent or permeating an odor from a substance can be. Skunk oil, C_4H_9SH in composition, is particularly noticeable in distance and time from its point of ejection. An arithmetic analysis of the dispersion of C_4H_9SH molecules after a skunk ejects about 9 g of the odiferous material follows: One mole of C_4H_9SH is 90 g and thus contains 6.02×10^{23} molecules. The 9 g, 6.02×10^{22} molecules, is not quickly vaporized; suppose that 1% of it, 6.02×10^{20} molecules, is dispersed into a volume of air a little less than a square mile in area and 1000 ft high. The volume $5000 \times 5000 \times 1000$ equals 25×10^9 cubic feet, the area that would contain the 6.02×10^{20} molecules.

$$\frac{6.02 \times 10^{20}\ \text{molecules}}{25 \times 10^9\ \text{cubic feet}} = 2.4 \times 10^{10}\ \text{molecules per ft}^3$$

It does not take long for one to breathe in a cubic foot of air; this air, within a half mile of a dead skunk, might contain an average of some 2.4×10^{10} molecules to which one's nose is chemically sensitive!

Two esters are among the 50 top chemicals—**dimethyl terephthalate** (included with terephthalic acid, 21st), used to make polyester fibers such as dacron and mylar, and **vinyl acetate** (50th), used to make surface coatings, films and adhesives, polyvinyl acetate for latex paints, polyvinyl alcohol used in sizing of textiles and paper treatment, and other compounds.

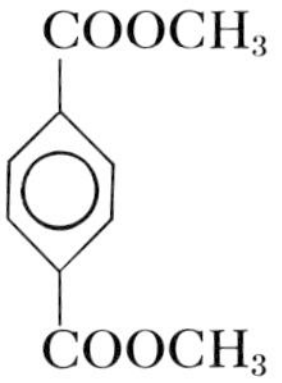

$$COOCH_3$$

$$COOCH_3$$

Dimethyl terephthalate

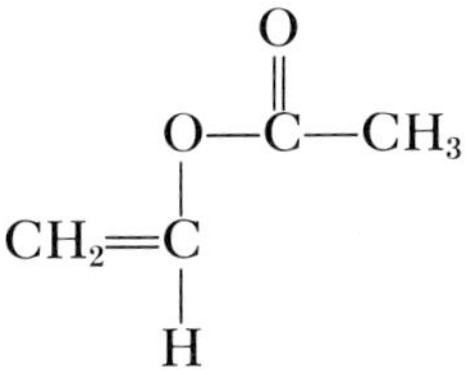

$$CH_2=C$$

$$O—C—CH_3$$

H

Vinyl acetate

32.9 AMINES

Amines may be considered derivatives of ammonia in which one or more of the hydrogen atoms of the ammonia molecule have been replaced with an organic radical. As a matter of fact, these compounds are very similar in chemical properties to ammonia. The simpler amines are gaseous at ordinary temperatures and have an odor very much like ammonia. Like ammonia, they show basic properties and react with acids to form salts. For example,

$$CH_3NH_2 + HCl \rightarrow CH_3NH_3Cl$$

Methyl ammonium chloride

One method by which amines may be prepared is by the action of ammonia on an organic halide.

$$R—X + 2\ NH_3 \rightarrow R—N\begin{smallmatrix}H\\ \\H\end{smallmatrix} + NH_4X$$

If two hydrogens in ammonia are replaced by two alkyl groups, a **secondary** amine results: $(CH_3)_2NH$ is **dimethyl amine**, $(C_6H_5)_2NH$ is **diphenyl amine**, and so on. The substitution of three alkyl groups for the three hydrogen atoms gives a **tertiary** amine: $(C_2H_5)_3N$ is **triethyl amine**, and so on.

Aniline, one of the more important aromatic amines, is shown below.

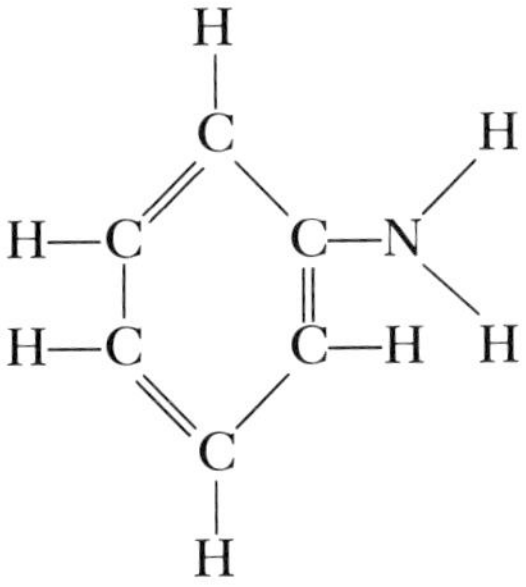

or

$$C_6H_5NH_2$$

It is a starting material in the preparation of many aniline dyes, medicines, and drugs.

Asymmetric Carbons and Enantiomers

If a given carbon atom has **four different substituents** about it, the name **asymmetric** is applied to it. The compound CCl_4 typifies a symmetrical molecule. For an asymmetric example, consider 2-butanol.

$$CH_3 - CH_2 - \overset{\displaystyle |}{\underset{\displaystyle |}{C}} \quad HO-C-H \quad CH_3$$

$$\begin{array}{cc} CH_3 & CH_3 \\ | & | \\ CH_2 & CH_2 \\ | & | \\ HO-C-H & H-C-OH \\ | & | \\ CH_3 & CH_3 \end{array}$$

The formulas represent two enantiomeric molecules, one being the mirror image of the other; each contains an asymmetric carbon and bears a right hand-left hand relationship to the other. Compounds of this kind have a special type of isomerism and are termed **stereoisomers**. Stereoisomers have nearly identical physical and chemical properties. They differ only in the spatial configuration of the atoms and they are **optically active**; that is, a beam of plane polarized light on passing through them is rotated to the right by the **dextro** form (d) and to the left by the **levo** form (l) of the compound. The d and l optical isomers of lactic acid, for example, are:

$$\begin{array}{cc} COOH & COOH \\ | & | \\ H-C-OH & HO-C-H \\ | & | \\ CH_3 & CH_3 \\ d\text{-Lactic acid} & l\text{-Lactic acid} \end{array}$$

Optical isomerism is of profound importance in biochemistry, where only one of the two isomers (generally the l isomer) is active in the body.

Another kind of isomerism is **geometric** or **cis-trans** isomerism. For example, 2-butene exists as two different compounds with different physical properties:

$$\begin{array}{cc} H \quad H & H \quad CH_3 \\ | \quad | & | \quad | \\ C = C & C = C \\ | \quad | & | \quad | \\ CH_3 \quad CH_3 & CH_3 \quad H \\ \text{cis-2-Butene} & \text{trans-2-Butene} \\ \text{mp} - 139°C & \text{mp} - 106°C \\ \text{bp } 4°C & \text{bp } 1°C \end{array}$$

The isomerism results because there is no rotation of the molecule around the double bond. The **cis** isomer (Latin, *cis*, on this side) has both methyl groups on the same side. The **trans** isomer (Latin, *trans*, across) has the methyl groups on opposite sides of the double bonds. There are many examples of cis-trans isomerism in organic and complex inorganic compounds.

Exercises

32.1. Define or explain the following: (a) alkyl group (R-), (b) secondary amine, (c) ether, (d) tertiary alcohol, (e) IUPAC system for ketones, (f) cis-trans isomerism.

32.2. The following are structural formulas of some common organic compounds:

1. CH_3SH (odor of paper mill, cheese)

2. CH_3—N—CH_3 (odor of rotten fish) with CH_3 above

3. 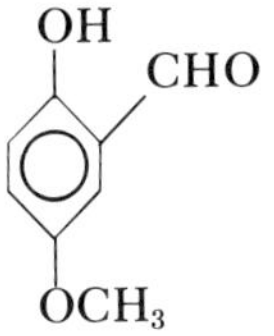 (vanilla flavor) — benzene ring with OH, CHO, OCH_3 substituents

5. (aspirin) — benzene ring with COOH and O—C—CH₃ ($\|$ O)

$$CH_3-\underset{CH_3}{N}-CH_3$$

6. (explosive, TNT) — benzene ring with CH_3, $2ON$, NO_2, NO_2

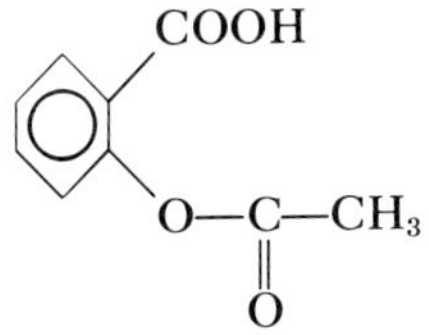

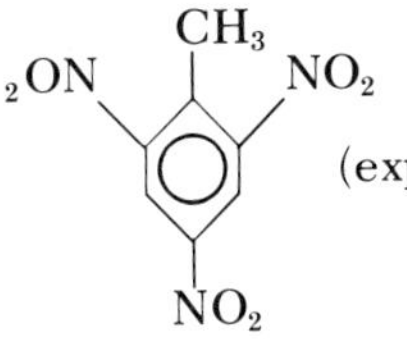

4. 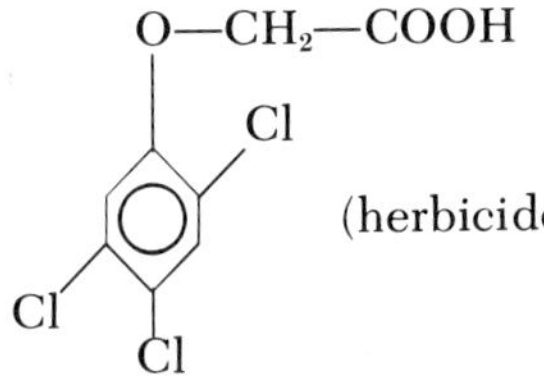 (herbicide 2,4,5-T) — benzene ring with O—CH_2—COOH, Cl, Cl, Cl

Which of these compounds is:
(a) an aldehyde
(b) a carboxylic acid
(c) a mercaptan or thiol
(d) a nitro compound
(e) an ester
(f) an amine
(g) an ether
(h) an alcohol
(i) a halide

32.3. The following are structural formulas of some organic compounds:

1. 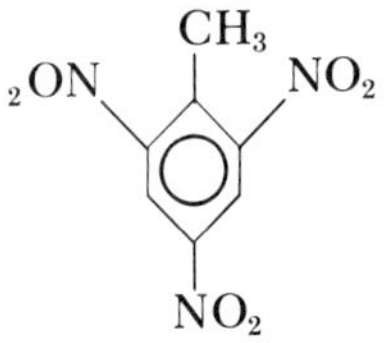 TNT, an explosive — benzene ring with CH_3, $2ON$, NO_2, NO_2

2. $CH_3-\underset{SH}{\overset{CH_3}{C}}-\underset{NH_2}{CH}-COOH$
compound derived from pencillin

3. 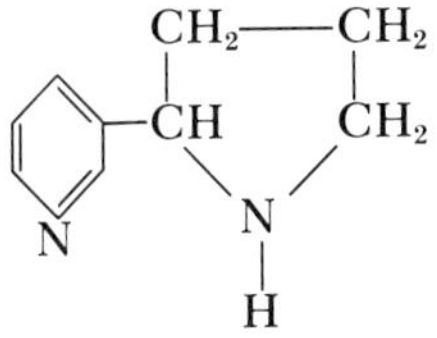 nicotine

$$CH_2\text{——}CH_2$$

4. 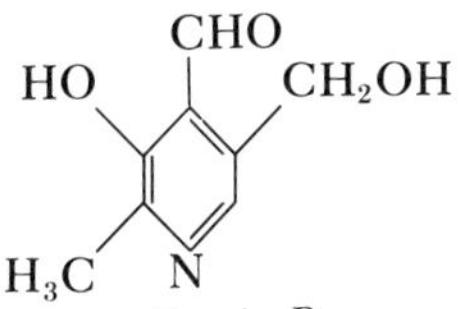 vitamin B_6 — ring with CHO, HO, CH_2OH, H_3C, N

$$O—CH_2COOC_3H_7$$

5.

Cl

Cl

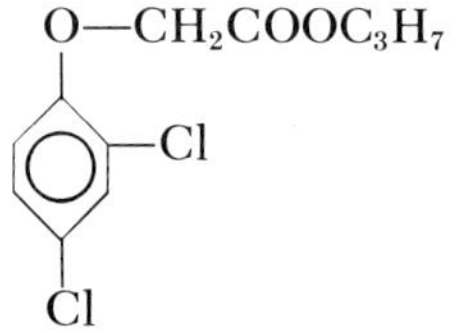

herbicide

Which of these compounds is

(a) an aldehyde
(b) an alcohol
(c) an ester
(d) an ether
(e) a primary amine

(f) a nitro compound
(g) a thiol or mercaptan
(h) a carboxylic acid
(i) a halide

32.4. Name the following halogen derivatives of the hydrocarbons: (a) CH_3CH_2I,

(b) 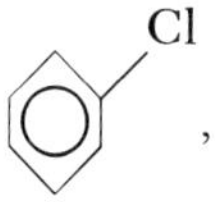 Cl , (c) $CF_2{=}CF_2$, (d) $CHCl_3$, (e) $CH_3CHBrCH_3$, (f) CF_2Cl_2.

32.5. Write structural formulas and names of four pentyl alcohols. Which are primary, secondary, and tertiary alcohols?

32.6. Write structural formulas for the following alcohols: (a) 1-pentanol, (b) 3-pentanol, (c) 3-methyl-2-pentanol, (d) 2,4,4-trimethyl-2-heptanol, (e) 3,3-dimethylcyclohexanol.

32.7. Write structural formulas for butyl amine, propyl acetate, propanoic acid, methyl butyl ether, 3-hexanone, octyl fluoride, pentanal, propyl butanoate.

32.8. Name and classify the following compounds according to type: C_3H_7Br; $HCOOH$; $CH_3COC_4H_9$; C_2H_5CHO; CH_3NH_2; $(C_2H_5)_2O$; $CH_3COOC_3H_7$; $HCOOCH_3$; C_4H_9OH; C_2H_5SH.

32.9. Name the following: (a) $C_8H_{17}OH$, (b) $C_2H_5NH_2$, (c) CH_3CHO, (d) C_3H_7COOH, (e) $C_3H_7COC_2H_5$, (f) C_2H_5COOK, (g) $CH_3COOC_2H_5$, (h) $CH_3CHOHCH_3$, (i) $(CH_3COO)_2Ca$, (j) $CH_3OC_{10}H_{21}$.

32.10. Name the following compounds: (a) CH_3CH_2COOH, (b) $CH_3COCH_2CH_2CH_3$, (c) $CH_3COOCH_2CH_3$, (d) CH_3CHO, (e) $C_4H_9OC_2H_5$.

32.11 Write formulas for: (a) trimethyl amine, (b) methanal, (c) 2-methyl-2-propanol, (d) methoxymethane, (e) n-octyl ethanoate.

32.12. From the group of formulas listed below, select an example of each of the following: (a) the product of the oxidation of acetaldehyde, (b) a primary amine,

(c) the product of the dehydration of methyl alcohol with sulfuric acid, (d) a saturated hydrocarbon, (e) a compound used as an antifreeze, (f) an ester, (g) a compound found in citrus fruits, (h) an alkyl halide.

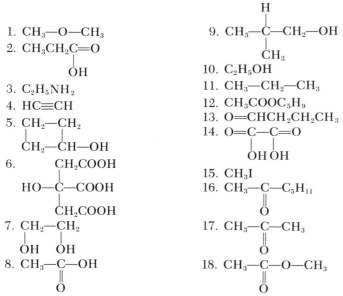

1. CH_3-O-CH_3

2. $CH_3CH_2C=O$
   $\quad\quad\quad |$
   $\quad\quad\quad OH$

3. $C_2H_5NH_2$

4. $HC\equiv CH$

5. CH_2-CH_2
   $\ |$
   $CH_2-CH-OH$

6. $\quad\quad CH_2COOH$
   $\quad\quad\quad |$
   $HO-C-COOH$
   $\quad\quad\quad |$
   $\quad\quad CH_2COOH$

7. CH_2-CH_2
   $\ |\quad\quad |$
   $\ OH\quad OH$

8. CH_3-C-OH
   $\quad\quad\quad \|$
   $\quad\quad\quad O$

9. $\quad\quad H$
   $\quad\quad |$
   CH_3-C-CH_2-OH
   $\quad\quad |$
   $\quad\quad CH_3$

10. C_2H_5OH

11. $CH_3-CH_2-CH_3$

12. $CH_3COOC_5H_9$

13. $O=CHCH_2CH_2CH_3$

14. $O=C-C=O$
    $\quad\ |\quad |$
    $\quad OH\,OH$

15. CH_3I

16. $CH_3-C-C_5H_{11}$
    $\quad\quad \|$
    $\quad\quad O$

17. CH_3-C-CH_3
    $\quad\quad \|$
    $\quad\quad O$

18. $CH_3-C-O-CH_3$
    $\quad\quad \|$
    $\quad\quad O$

32.13. Name the 18 compounds listed in question 32.12.

32.14. The following formulas are written *incorrectly*. Write a correct formula for each.
   (a) $CH_3CH\equiv CH$

   (b) 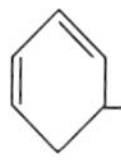 ⬡—$O-CH_2CH_3$

   (c) $CH_3C-O-O-H$

   (d) $H:\overset{H}{\underset{..}{C}}:\overset{H}{\underset{..}{C}}:H$

   (e) C_4H_{12}

32.15. The following formulas are written *incorrectly*. Write a correct formula for each.
   (a) $(CH_3)_2NH_2$

   (b) CH_3CH_2CHOH
   $\quad\quad\quad\quad |$
   $\quad\quad\quad\quad OH$

   (c) 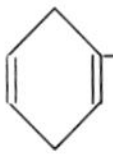 ⬡$-OH$

   (d) $CH_3CH_2-C-CH_2-C_3H_7$
   $\quad\quad\quad\quad |$
   $\quad\quad\quad\quad O$

   (e) $CH_2=CH_2=CH_2$
   $\quad |\quad\quad |\quad\quad |$
   $\quad OH\quad OH\quad OH$

32.16. Write the structural formula for *one isomer* of each of the following compounds:

(a) 2-propanol
(b) $CH_3COOC_2H_5$
(c) m-dichlorobenzene

(d) 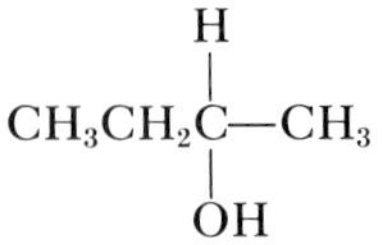 $CH_3CH_2\overset{\overset{\displaystyle H}{|}}{\underset{\underset{\displaystyle OH}{|}}{C}}-CH_3$ (d isomer)

(e) 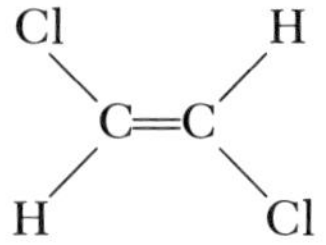 $\overset{\displaystyle Cl}{\diagdown}\underset{\displaystyle H}{\diagup}C=C\overset{\displaystyle H}{\diagup}\underset{\displaystyle Cl}{\diagdown}$ (trans isomer)

32.17. For which of the following compounds are cis- and trans-isomers possible?
(a) 2,3-dimethyl-2-butene, (b) 3-methyl-2-pentene, (c) 2-methyl-2-pentene,
(d) CH(COOH)=CH(COOH).

32.18. For which of the following compounds are optical isomers possible? (a) 2-hy-droxypropanoic acid; (b) hydroxyacetic acid; (c) 2-bromo-2-chloropropane; (d) 1-bromo-1-chloroethane; (e) the cyanohydrin prepared from butanone,

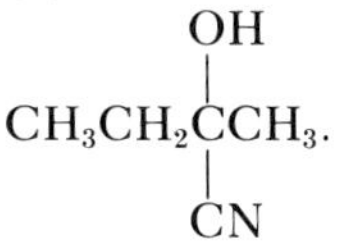

$$CH_3CH_2\overset{\overset{\displaystyle OH}{|}}{\underset{\underset{\displaystyle CN}{|}}{C}}CH_3.$$

32.19. Write equations for
(a) action of concentrated H_2SO_4 on methanol.
(b) formic acid + KOH.
(c) acetic acid + pentyl (amyl) alcohol in the presence of concentrated H_2SO_4 (gives banana oil).
(d) preparation of octyl acetate of orange odor.
(e) preparation of pentyl butyrate of apricot odor.
(f) oxidation of ethanal.

32.20. Write equations for
(a) hydrolysis of propyl ethanoate.
(b) reaction of ethyl alcohol with butanoic acid.
(c) oxidation of formaldehyde.
(d) neutralization of propanoic acid with potassium hydroxide.
(e) combustion of acetone in air.
(f) oxidation of 2-butanol.

32.21. Use equations to indicate the steps you would take to prepare
(a) butyl amine from butyl chloride.
(b) ethanol from ethylene.
(c) acetone from 2-propanol.
(d) acetic acid from ethyl alcohol.
(e) methanal from methanol.

32.22. Complete the following reactions by inserting the appropriate structural formula in the space provided.

(a) $2 \, C_2H_5OH \xrightarrow{H_2SO_4}$ _____ $+ \, H_2O$

(b) $CH_3-\underset{\underset{O}{\|}}{C}-OH +$ _____ $\rightarrow CH_3-\underset{\underset{O}{\|}}{C}-O-CH_3 + H_2O$

(c) $CH_3-CH_2-\underset{\underset{OH}{|}}{CH}-CH_3 + (O) \rightarrow$ _____

(d) $H-\underset{\underset{O}{\|}}{C}-O-CH_2CH_3 \xrightarrow{NaOH}$ _____ $+$ _____

(e) $CH_3-CH{=}CH_2 + Br_2 \rightarrow$ _____

(f) $CH_4 + Cl_2 \rightarrow$ _____ $+ \, HCl$

32.23. Which of the reactions in the previous question would be classified as:
(a) a hydrolysis or saponification   (d) a dehydration
(b) an addition   (e) a substitution
(c) an oxidation   (f) an esterification

32.24. From the list on the right, select the type of compound which would be produced by each of the reactions on the left.
(a) $CH_3CHO + (O) \rightarrow$ 1. Ester   6. Ether
(b) Secondary alcohol $+ (O) \rightarrow$ 2. Amine   7. Ketone
(c) $C_4H_9Cl + NH_3 \rightarrow$ 3. Sulfonic acid 8. Alcohol
(d) $CH_3COOH + NaOH \rightarrow$ 4. Alkyl halide   9. Aldehyde

(e) ⬡—COOH $+ CH_3OH \rightarrow$ 5. Salt   10. Acid

32.25. From the list on the right select the correct *organic* product of each of the following reactions.
(a) 2-butanol $+ (O) \rightarrow$ 1. $CH_3CH_2CH_2CHO$ 6. $HCOOC_3H_7$
(b) $CH_3CH_2CH_2CH_2OH + (O) \rightarrow$ 2. $C_2H_5OC_2H_5$ 7. $C_4H_{10} + C_4H_8$
(c) $CH_3CH_2OH \xrightarrow{H_2SO_4}$ 3. 1-butanol   8. CO_2
(d) $HCOOH + CH_3CH_2CH_2OH \rightarrow$ 4. $CH_3\underset{\underset{O}{\|}}{C}CH_2CH_3$ 9. CH_3CH_2COOH
                                                                      10. $CH_3CH_2CH_2CH_2OH$
(e) $C_8H_{18} + O_2 \xrightarrow[\text{combustion}]{\text{complete}}$ 5. $CH_3COOH + C_2H_5OH$

32.26. (a) What quantity of ethyl acetate is, in theory, obtainable from a preparation employing 30.0 g of ethanol and 30.0 g of acetic acid? (b) If 30.0 g of ethyl acetate is isolated, what percent of the theoretical yield was obtained?

32.27. (a) A 3.00-g sample of tartaric acid is burned in oxygen; 1.79 l of CO_2 (measured at STP) and 1.08 g of H_2O are collected from the combustion. Tartaric acid contains only carbon, hydrogen, and oxygen. What is the empirical formula of this compound? (b) A mixture of 0.050 g of tartaric acid and 0.50 g of camphor melts at 153°C. The melting point of pure camphor is 179°C, and the molal freezing-point depression constant is 39.7°C/m. What is the molecular formula of tartaric acid? (c) A 0.300-g sample of pure tartaric acid requires 25.0 ml of 0.160*N* NaOH for complete neutralization. What is the equivalent weight of the acid? (d) Draw a structural formula for tartaric acid consistent with the data given in this problem.

32.28. Write the structural formula of an organic compound which matches each descriptive statement concerning some common uses:
   (a) Freon — used as a refrigerant and a propellant
   (b) Acetic anhydride — used in making aspirin
   (c) Acid in vinegar
   (d) Anti-freeze (Prestone, etc.)
   (e) Wood alcohol (solvent)
   (f) Common organic solvent — very flammable
   (g) Formalin

Supplementary Readings

32.1. *Essentials of Organic and Biochemistry*, D. J. Burton and J. I. Routh, Saunders, Philadelphia, Pa, 1974.
32.2. "Food Additives," G. O. Kermode, *Sci. American*, March, 1972; p. 15.
32.3. "The Art and the Science of Perfumery," D. R. Moore, *J. Chem. Educ.*, **37**, 434 (1960).

Chapter 33

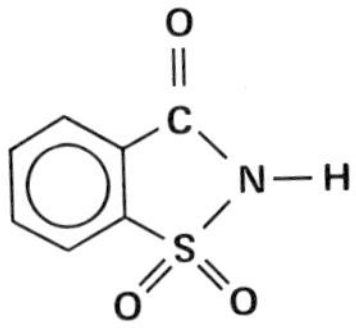

Saccharin
(Sweetener)

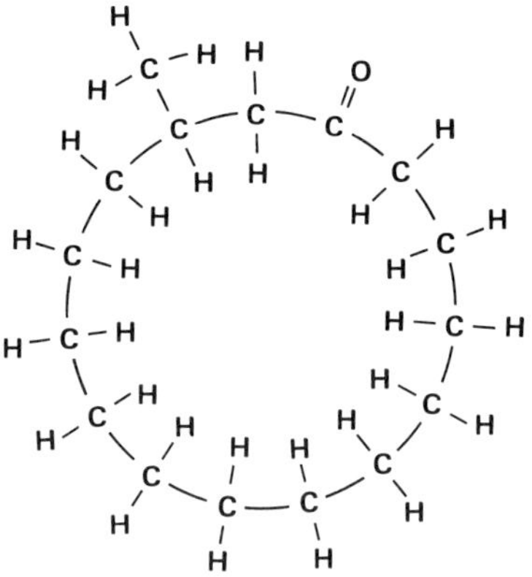

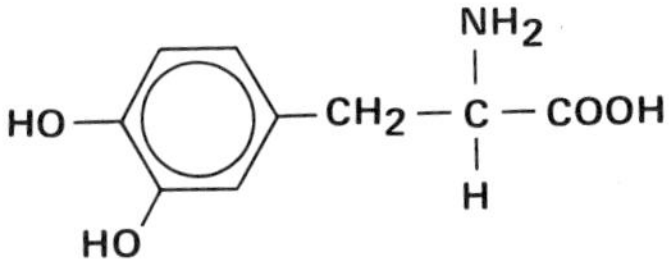

L–dopa
(Drug–Parkinson's disease)

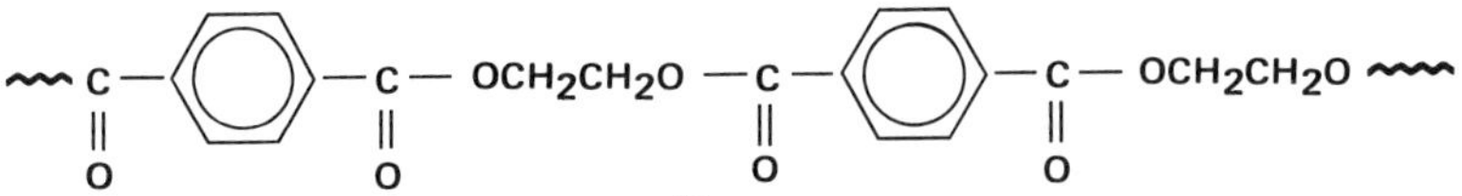

Muskone
(Musk deer pheromone)

Dacron
A polyester

Modern Organic Molecules

In 1835 Friedrich Wohler, sometimes called the "father of organic chemistry," wrote to Berzelius, "organic chemistry appears to me like a primeval forest of the tropics, full of the most remarkable things — a monstrous and boundless thicket. . . ." Considering the millions of organic compounds known today and the unlimited number possible, organic chemistry now may be more of an exciting "jungle" than ever before.

A few of the compounds resulting from modern synthesis will be examined briefly in this chapter. Most of these compounds are not used in the same tremendous amounts as the 50 top chemicals; however they are of considerable importance and often "a little bit goes a long way."

33.1   SIMPLE DERIVATIVES OF BENZENE

Several common simple derivatives of benzene are illustrated below:

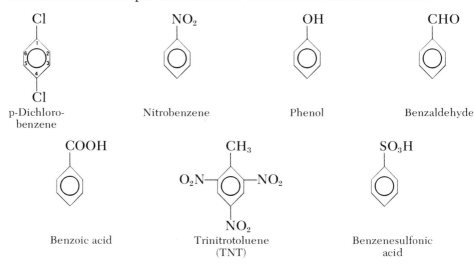

| p-Dichloro-benzene | Nitrobenzene | Phenol | Benzaldehyde |

| Benzoic acid | Trinitrotoluene (TNT) | Benzenesulfonic acid |

The first compound above, p-dichlorobenzene (also called, 1,4-dichlorobenzene), is what common moth crystals are made of.

When benzene is heated with a mixture of nitric and sulfuric acids, **nitration** — a process in which an NO_2 (nitro) group is substituted for a hydrogen atom attached to the ring — takes place to form nitrobenzene. It has a characteristic odor of shoe polish.

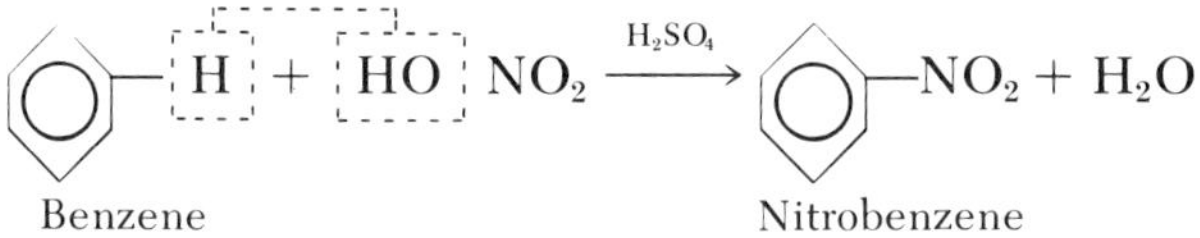

$$\text{Benzene} - H + HO - NO_2 \xrightarrow{H_2SO_4} \text{Nitrobenzene} - NO_2 + H_2O$$

Further nitration results in the substitution of two or three nitro groups on the benzene nucleus.

The highly nitrated compounds such as TNT above are powerful explosives. **Phenol,** sometimes known as **carbolic acid,** is an important industrial compound obtained from coal tar and ranks 39 in the list of top chemicals. It is a strong germicide; however it is extremely toxic and caustic. Dilute solutions of phenols (Lysol) are used in hospitals as a disinfectant. The principal uses of phenol are in phenolic resins such as Bakelite and in production of compounds — for example, pentachlorophenol (a wood preservative and fungicide). Although phenol contains the characteristic group of an alcohol (OH), it behaves chemically as a weak acid and may be neutralized with strong bases to form salts.

Benzaldehyde has an almond-like odor and is used in almond flavoring. **Benzoic acid** is only slightly soluble in water, but its sodium salt, **sodium benzoate,** is used as a food preservative.

The treatment of aromatic hydrocarbons with hot concentrated sulfuric acid results in **sulfonation** and the formation of sulfonic acids. The reaction of benzene with sulfuric acid results in the formation of benzene sulfonic acid, which is a strong acid, comparable to the inorganic acids.

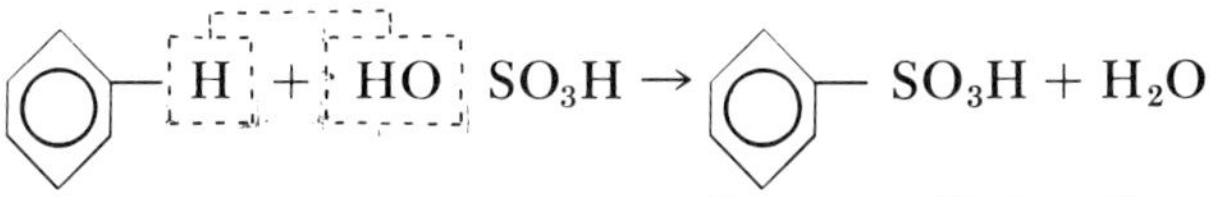

Benzene   Benzene sulfonic acid

Many dyes are produced from sulfonic acids or their derivatives. A typical formula for a dye is that of Congo red:

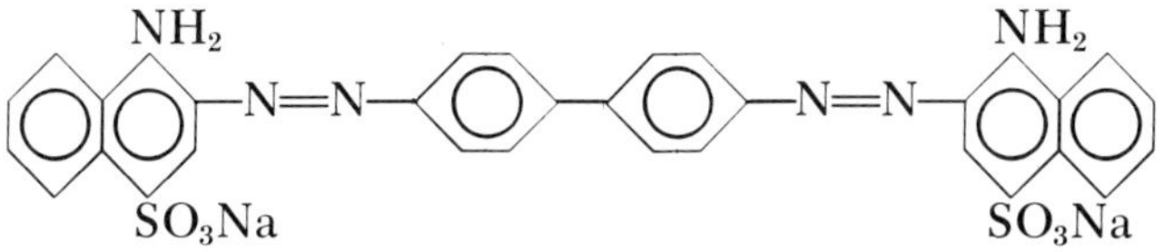

Note the great diversity of characteristics of the compounds obtained by the replacement of a hydrogen on a benzene ring with simple functional groups.

33.2 MEDICINE AND PHARMACEUTICALS

The remarkable curative power of **sulfanilamide** was discovered in 1935. Since that time, thousands of compounds have been described as researchers have tried to develop more potent germ killers with less toxicity to patients. Formulas for a few of the drugs are shown below.

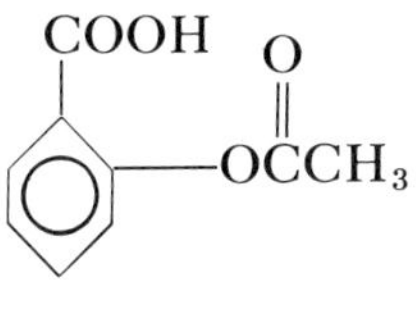

Acetylsalicylic acid
(aspirin)

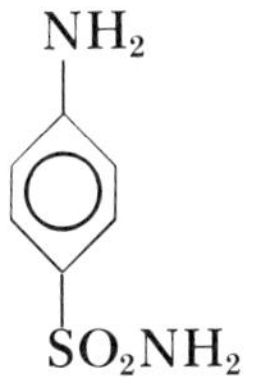

Sulfanilamide

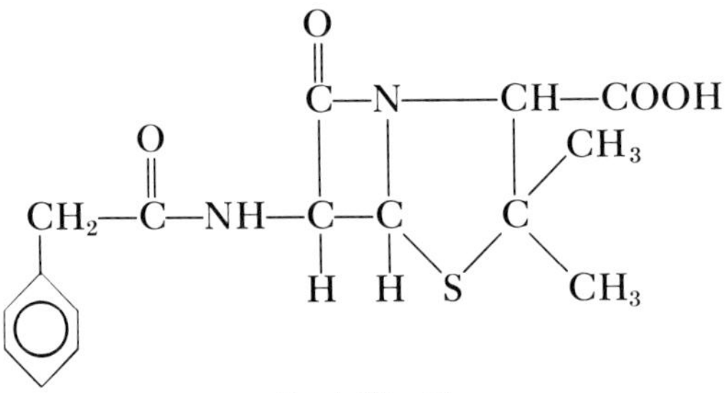

Penicillin G

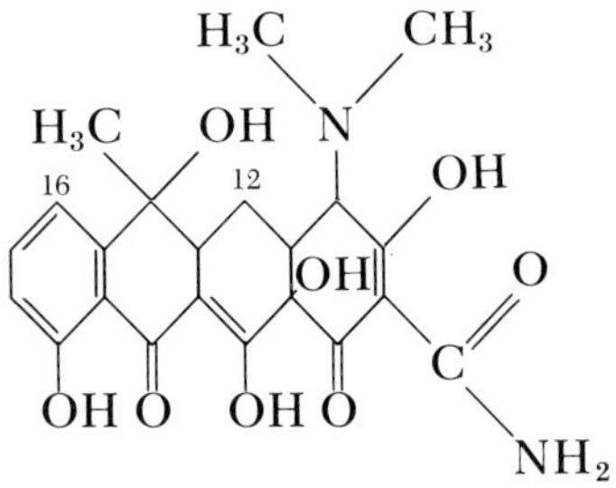

Tetracycline

Aspirin was first synthesized in 1853 and is still one of the most commonly used drugs. Some 30 million pounds are produced in the U.S. annually. It is not found in nature and is a remarkable medicine, being a pain killer, reducing fever and local inflammation, and helping eliminate uric acid (for gout).

Penicillin is a specific secretion which diffuses from a mold. In 1929 it was found to inhibit bacterial growth near the mold on bacterial test plates, and it is specifically effective against gonorrhea and certain other infections. There are many organisms for which both sulfa drugs and penicillin are effective; penicillin will act on some for which sulfa is ineffective, and sulfa is useful for treatment in some cases—such as dysentery—in which penicillin seems ineffective.

In spite of intensive research during the war, it was not until 1945 with the aid of X-ray crystallography that the riddle of the structure of the penicillin molecule was solved. It is an unusual, complex heterocyclic compound containing 4- and 5-member rings, which makes it difficult to synthesize.

Certain other antibiotics of proven effectiveness are derivatives of **tetracycline**, which kills a "broad spectrum" of bacteria. The drug **aureomycin** has a Cl atom at the carbon labeled 16, and **terramycin** has a hydroxyl at the carbon labeled 12 of the tetracycline structure.

Some other modern medicinals are shown here.

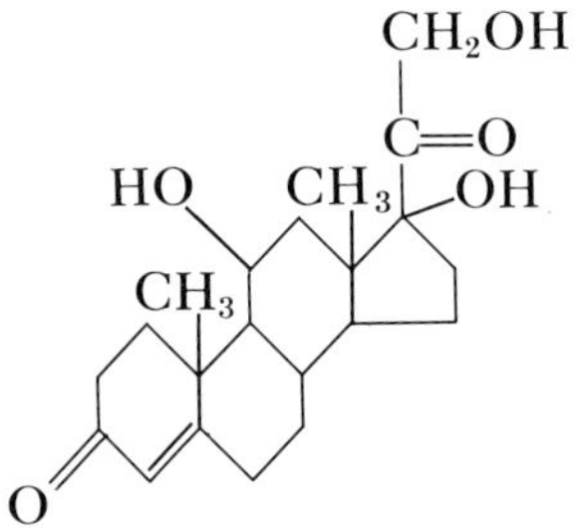

Cortisone

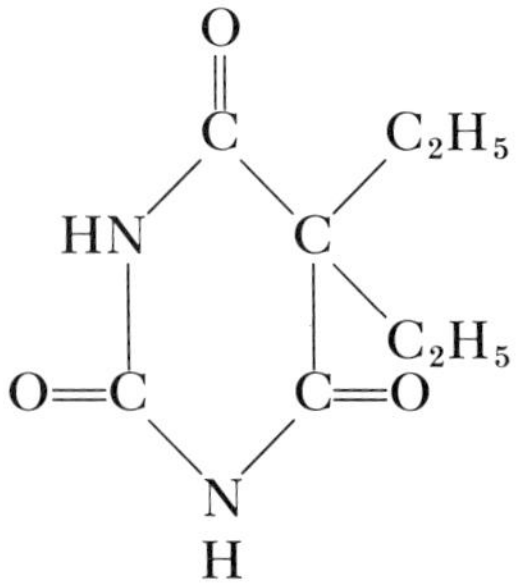

Barbital

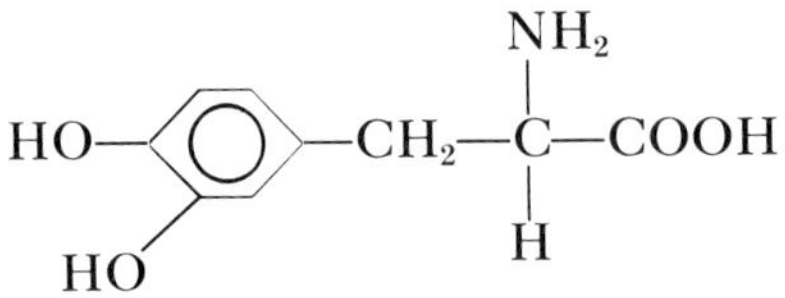

L-Dopa

The compound **cortisone** is called a sterol derivative by the chemist. It is related to cholesterol, a constituent of membranes in living organisms. **Barbital**, originally called Veronal, and its derivatives such as **phenobarbital**, **amytal**, and **seconal**, have been used for many years as sedatives and hypnotics (sleep inducers). **L-Dopa** has shown success in controlling Parkinson's disease.

Narcotics and Hallucinogens

These compounds, while having some positive medical uses owing to their action on the central nervous system, are abused by certain individuals so that they constitute a major problem in society. Four of these compounds are shown below:

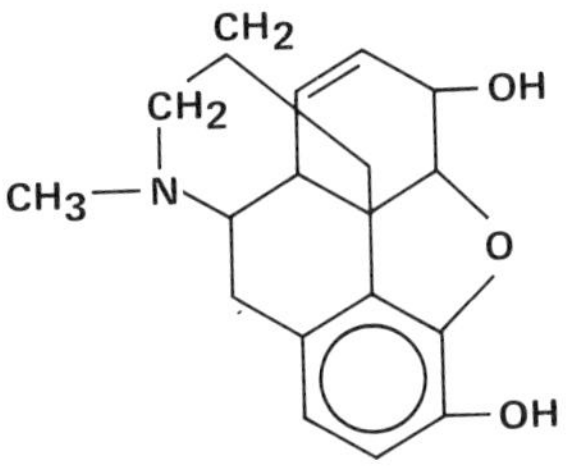

Morphine
(narcotic drug from opium)

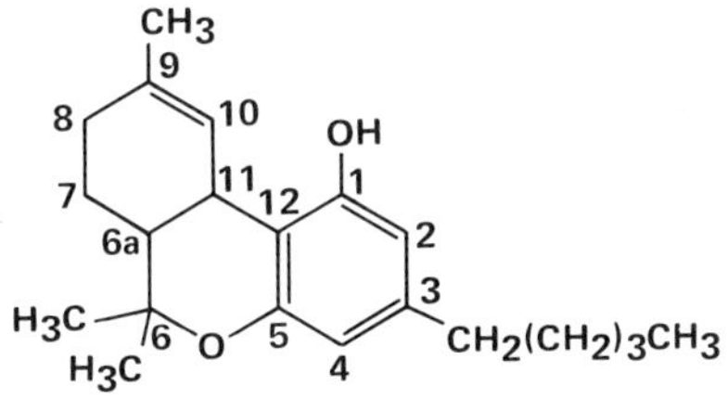

Δ^9-Tetrahydrocannabinol
(Hallucinogen in marijuana)

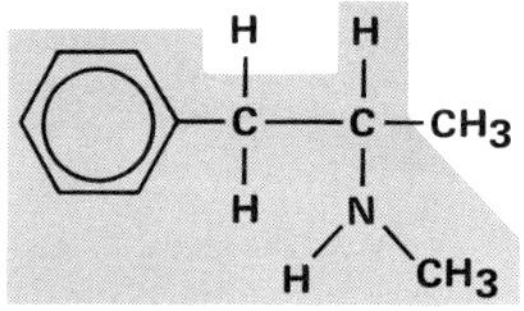

Methamphetamine
(Methedrine)

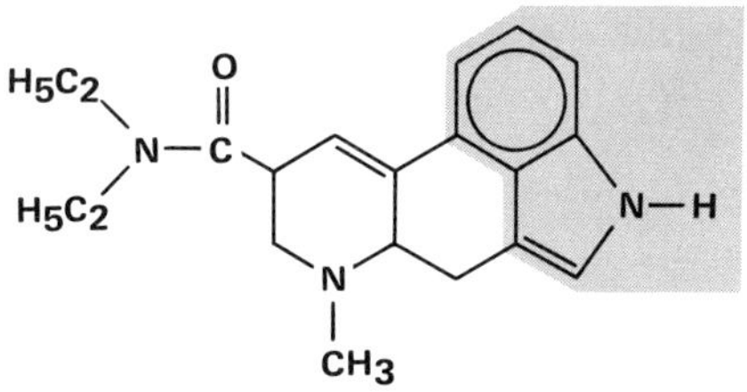

Lysergic acid diethylamide (LSD)

Note the similarity of **methedrine** and **epinephrine** (adrenalin), shown below.

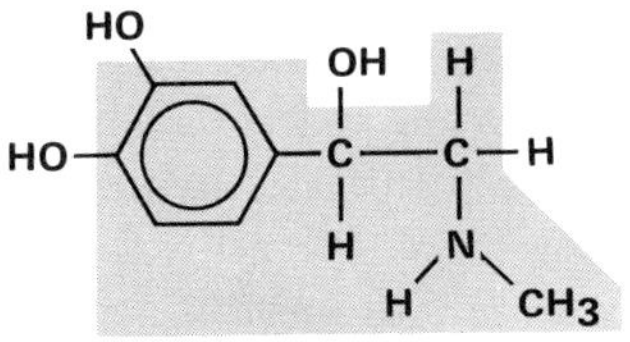

Epinephrine
(Adrenalin)

33.3 INSECTICIDES AND HERBICIDES

The structures of several common insecticides and herbicides are shown below:

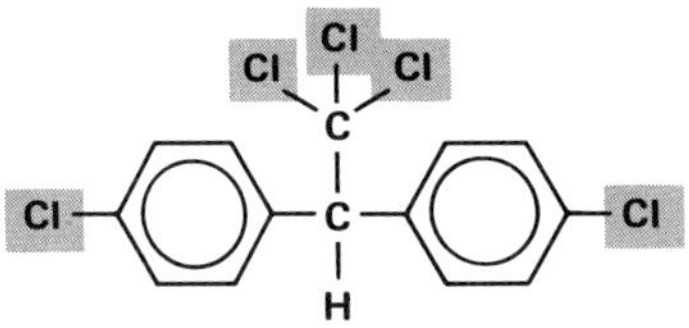

Dichlorodiphenyltrichloroethane
(DDT)

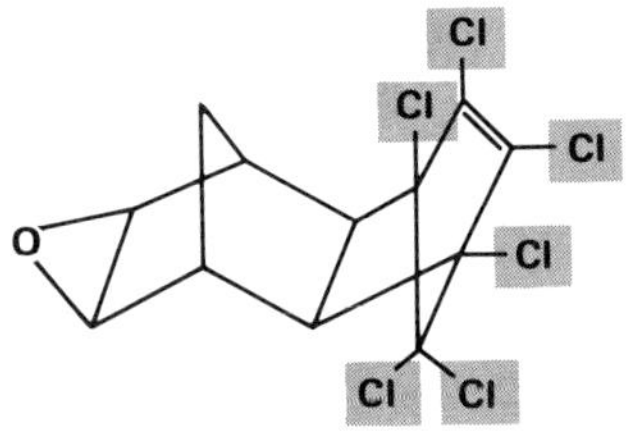

Dieldrin

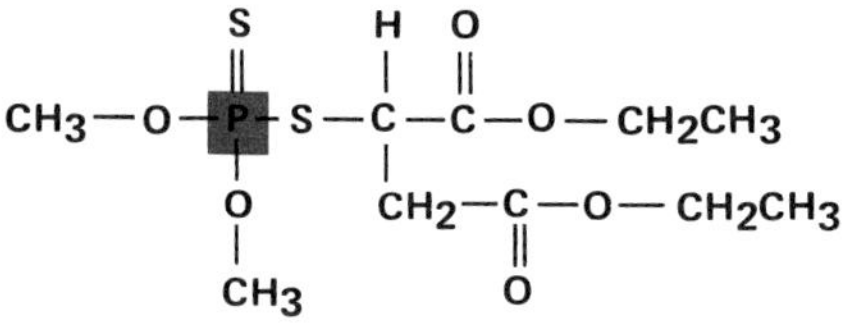

Malathion

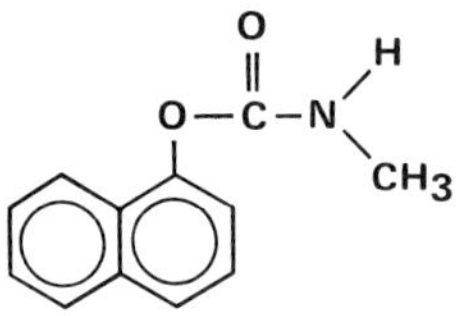

1-Naphthyl N-methyl-
carbamate
(Sevin)

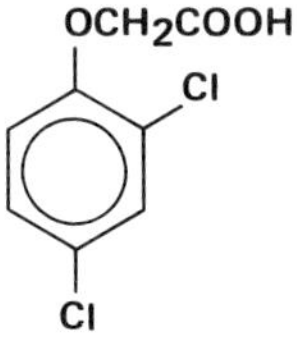

2,4-Dichlorophenoxy-
acetic acid
(2,4-D)

DDT is a powerful insecticide and was able to control successfully such things as malaria and the boll weevil. However, it was found that DDT and related compounds are long lasting and accumulate in fish and birds, threatening the survival of a number of species. In 1973, the use of DDT was banned in the U.S., except in a few approved instances. Cyclocompounds such as **dieldrin** are more toxic than DDT but break down more rapidly. The Environmental Protection Agency banned the use of many of these compounds in 1975.

Organophosphates were discovered to be potent insecticides. Some 100,000 of these compounds have been synthesized and tested. About 40 have been used commercially. **Malathion,** for example, is commonly used in

garden and household insect control. Derivatives of **carbamic acid** (HOCNH_2), such as **Sevin,** also are rapidly degraded and many are suitable for general-purpose insecticides.

The compound **2,4-D** (2,4-Dichlorophenoxyacetic acid) is one of a number of similar compounds which kill broad-leaf plants by mimicking the plants natural growth hormones, but disturbing its normal physiological functions. In the soil 2,4-D normally degrades in one to four weeks.

A new area of insect research is the **pheromones,** which are organic compounds secreted in minute quantities by insects. For example, the compound cis-7,8-epoxy-2-methyl-octadecane

$$CH_3(CH_2)_9\overset{O}{\overset{\triangle}{CHCH}}(CH_2)_4CH_2(CH_3)_2$$

acts as a sex attractant for male gypsy moths, and traps baited with traces of this pheromone catch large numbers of male moths, rendering them un-available for reproduction. As little as 30 molecules of the above compound can be detected by a male moth. Similar compounds have been isolated for the silkworm moth, black carpet beetle, and boll weevil. Among other pheromones which have been identified are trail-marking compounds used by ants, termites, bees, and other insects.

33.4   FLAVORS AND PERFUMES

There are some 1100 flavors which are approved for use as food additives, most of these being synthetic. In addition there are many rather simple com-pounds which give pleasant odors. Some examples are shown below:

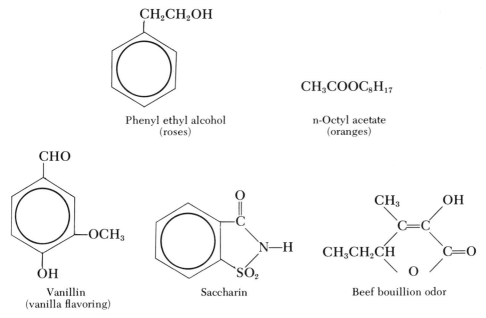

Phenyl ethyl alcohol
(roses)

n-Octyl acetate
(oranges)

Vanillin
(vanilla flavoring)

Saccharin

Beef bouillion odor

33.5   DETERGENTS

Synthetic detergents, sometimes called **syndets,** are cleansing agents which have found wide use in both home and industry. These substances lower interfacial tension and promote ease of wetting. Sale of syndets in the United States is now six times that of soaps, on a pound basis. Syndets may be used to advantage in waters in which soaps would be precipitated as a curd by ions such as Ca^{2+} and Mg^{2+}.

Formulas for three important detergent compounds are

| Sodium dodecyl benzene sulfonate | Sodium dodecylsulfate | Alkyl-dimethyl phenyl ammonium chloride |

Up until 1965 the **alkyl benzene sulfonates,** the most commonly used detergents, caused a serious problem, since they were not readily biodegradable; that is, bacteria did not act to decompose them, and large amounts of longlasting foam built up in rivers and streams. It was found, however, that by making the alkyl group a straight chain (normal) group instead of the commonly used branched-chain group, the detergent became biodegradable and the foam problem was solved. Today, by law all detergents must be biodegradable.

The term **builder** has been accepted as meaning a compound which has little or no cleansing action in itself but greatly enhances the cleansing action of a detergent, whether syndet or soap. Note the composition of a typical heavy-duty syndet.

Component	Percentage By Weight
Dodecyl benzene sulfonate	20.0
Sodium tripolyphosphate	45.0
Sodium silicate	5.0
Sodium carboxy methyl cellulose	0.5
Fatty amide	2.5
Sodium sulfate	22.0
Moisture	5.0

The phosphate compounds, which function to tie up the Ca^{2+} and Mg^{2+} ions of hard water, are themselves a problem in that rivers and lakes receiving them produce extensive algal growth which lowers the oxygen content of the water, killing much of the aquatic life. A number of phosphorus substitutes are being investigated, but at present no really satisfactory replacements have been found.

33.6 SYNTHETIC RESINS AND PLASTICS

Natural resins, such as pine resins and various tree gums, have found considerable use in varnishes. For example, **shellac** is made from the bodies of

certain insects that infest some trees in India. Modern-day demand has resulted in the construction of large synthetic resin and plastic plants. A synthetic resin is a basic material which may be made by chemical changes of
a polymerization, condensation, or combination type. The resin may be a
plastic or may be subject to plasticizing. Plastics that cannot be remelted
once they have been molded and set are called **thermosetting;** plastics that
may be remelted and reformed are called **thermoplastic.**

If phenol and formaldehyde are compounded under carefully controlled
conditions, a fairly hard product which may be ground to a powdery resin is
obtained. If polymerized to a lesser degree or with excess phenol, the product
may in itself be cast as a plastic. If the powdery resin is heated with a small
percentage of a plasticizing agent, such as dibutyl phthalate, a plastic product
results. Phenolformaldehyde polymerized products are made into plastics
with the trade name **Bakelite.**

The phenolic resins are compounded into molding compositions, such as

Material	Percent
Resin	40
Hexamethylenetetramine	5
Lime	3
Calcium stearate	1
Mixed waxes	1
Wood fiber	40
Clay	5
Dye	5

Each constituent has a specific function: the wood fiber for filler, the clay for
filler and bulk density, the waxes for a product capable of better polish, the
lime for neutralizing acidity and drying the mix, the hexamethylenetetra-
mine for conversion from thermoplastic to a thermosetting mass, and the
calcium stearate to prevent sticking to the mold.

If the resin is dissolved in organic solvents and pigment added, a coating lacquer is obtained. Urea or melamine may be used instead of phenol with formaldehyde.

Cellulose nitrate, cellulose acetate, or cellulose propionate may be plasticized by compounding with various reagents, such as triphenyl phosphate, $(C_6H_5)_3PO_4$.

It is beyond the scope of this book to describe the chemistry of the many plastic materials, but feminine interest in the plastic substance **nylon** seems to warrant a note on the chemistry of its manufacture. If a diamine and a dibasic carbon-chain acid are subjected to dehydration, long molecules constituting nylon result.

$$H-O-\overset{\overset{\displaystyle O}{\|}}{C}-(CH_2)_4-\overset{\overset{\displaystyle O}{\|}}{C}-O-H + H-\overset{\overset{\displaystyle H}{|}}{N}-(CH_2)_6-\overset{\overset{\displaystyle H}{|}}{N}-H \rightarrow$$

Adipic acid   a diamine   under controlled conditions

$$-O-\overset{\overset{\displaystyle O}{\|}}{C}-(CH_2)_4-\overset{\overset{\displaystyle O}{\|}}{C}-\overset{\overset{\displaystyle H}{|}}{N}-(CH_2)_6-\overset{\overset{\displaystyle H}{|}}{N}-\overset{\overset{\displaystyle O}{\|}}{C}-(CH_2)_4-\overset{\overset{\displaystyle O}{\|}}{C}-\overset{\overset{\displaystyle H}{|}}{N}-(CH_2)_6-\overset{\overset{\displaystyle H}{|}}{N}-\text{further similar}$$

Nylon   linkages

When nylon is melted, it can be forced through small openings in a disk to produce threads. When these threads are subjected to cold pulling or stretching, they set to a strong fiber with marked elastic and durable qualities, usable for hosiery, brush bristles, tennis racket strings, parachutes, tow lines for gliders, and so on.

Types of Polymerization

Chemical combination of identical or similar molecules to form a complex high-molecular-weight molecule may be typed as **addition** or **condensation** polymerization.

Addition polymers are formed by the union of monomeric units without the elimination of atoms. (See page 646 describing polyethylenes, which are an example.) More rigidity and less flow under pressure are qualities of polypropenes or polybutenes. Resistance to burning and other desired properties are inherent in polymers in which fluorine, chlorine, or a group such as $-COOCH_3$ replace some hydrogen atoms to give the synthetic polymers **Teflon, Saran,** and **Plexiglas,** respectively. Teflon, for instance, is polymerized $F_2C=CF_2$.

Condensation polymers result from chemical changes in which some simple molecule (such as water) is eliminated between functional groups. The phenol-formaldehyde resin and nylon are of this type. Another example is in the manufacture of **Dacron,** in which water is eliminated in the condensation polymerization of ethylene glycol, $HOCH_2CH_2OH$ and diethyl terephthalate,

$$C_2H_5OOC-\langle \bigcirc \rangle-COOC_2H_5$$

Dacron is an example of an alkyd much used in fabrics. In the preparation of surface coating materials, a drying oil such as linseed may be added to an alkyd resin.

Table 33.1 shows some common polymers and the monomers from which they are made.

Table 33.1
Some common polymers

Polymer	Monomer	Some Uses
Polyethylene	$CH_2{=}CH_2$	food bags, plastic chemical ware, cups
Polypropylene	$CH_3CH{=}CH_2$	fibers, rope, carpet, pipes
Polystyrene (Styrofoam)	$CH_2{=}CH$ (with phenyl ring)	packaging, insulation, moldings hot drink cups
Polyvinyl chloride (PVC, vinyl, Dynel Koroseal)	$CH_2{=}CHCl$	pipe, gutters, sheeting, leather, cloth, food wrap, raincoats, insulation
Saran	$CH_2{=}CCl_2$	food wrap
Orlon	$CH_2{=}CHCN$	fabrics, carpets
Teflon	$CF_2{=}CF_2$	gaskets, electrical insulation, cookware coating, stopcocks
Lucite (acrylic, Plexiglas)	$CH_2{=}\overset{CH_3}{\underset{\,}{C}}COOCH_3$	"unbreakable glass," brush handles, dental fillings, auto accessories
Polyurethane	$O{=}C{=}N{-}(CH_2)_2{-}N{=}C{=}O$ $+$ $\underset{OH\ \ \ OH}{CH_2{-}CH_2}$	upholstery foams, insulation, surface coatings, paints

The compound $CH_2{=}CHCN$ in the above table is called **acrylonitrile** and is 48 in the list of top chemicals. It is prepared by reacting hydrogen cyanide with acetylene in an addition reaction:

$$HCN + HC{\equiv}CH \rightarrow H_2C{=}CHCN$$

Polyacrylonitrile is a very hard resin and formulated with other materials produces products such as table and sink tops, shoe soles, wall panels, and luggage materials.

33.7   RUBBER AND ELASTOMERS

Natural rubber is obtained from the sap of a species of tree. The first working of natural rubber was in Brazil. Trees were selectively planted and grown in the East Indies, and a better yield of rubber tree sap (latex) was obtained. Latex when heated with acid yields crude rubber, which in turn may be compounded for desired properties by incorporation of filler materials, antioxidants, and vulcanizing agents, such as sulfur. Natural rubber is essentially a high-molecular-weight hydrocarbon with numerous unsaturated bonds.

In World War II the supply of natural rubber was cut off, and at the same time there was a great increase in the demand for rubber. Chemists solved the problem by developing rubber substitutes which go under the general name of **elastomers.** In many ways industrially produced elastomers are better than natural rubber; for example, most elastomers are more oil resistant and some are less permeable than rubber. The chemistry of production of a few synthetic rubbers will be briefly described.

Neoprene is produced from acetylene. A first operation is polymerization to form vinyl acetylene. When this compound is treated with hydrochloric acid, chloroprene is formed.

$$\text{H}-\text{C}\equiv\text{C}-\text{CH}=\text{CH}_2 + \text{HCl} \rightarrow \text{H}-\underset{\underset{\text{H}}{|}}{\text{C}}=\underset{\underset{\text{Cl}}{|}}{\text{C}}-\text{CH}=\text{CH}_2$$

Vinyl acetylene   Chloroprene

When chloroprene is treated with metallic sodium under carefully controlled conditions, carbon-to-carbon linkages are effected among many four-carbon units to give neoprene.

$$+ 2\,\text{Na} \rightarrow 2\,\text{NaCl} +$$

Chloroprene   Neoprene

Many molecules, like the last one shown, polymerize to rubberlike products.

Thiokol is obtained by controlled heating of hydrocarbon dichlorides, such as ethylene dichloride ($\text{ClCH}_2-\text{CH}_2\text{Cl}$), with sodium polysulfide (Na_2S_x). The ethylene dichloride can be made from chlorine and ethylene. Ethylene is a by-product of the cracking process of gasoline, or it may be made by dehydration of ethyl alcohol.

One of the elastomers manufactured in great bulk has been known as **buna S, GR-S,** or **butadiene-styrene** product. Butadiene,

$$H_2C=CH-CH=CH_2$$

results from dehydrogenating butane.

Styrene 1 unit wt.	Butadiene 3 units wt.	emulsified in water, catalyst for copolymerization added — temperature and time control → buna S

When the desired amount of molecular union has taken place to give a product of desired properties, a stopping agent, such as phenyl naphthalene, is added. The crude synthetic rubber latex is then separated, purified by washing, and compounded for use.

33.8   SILICONES

We have already noted the chemical similarity of carbon and silicon. Recently many compounds with chainlike structures containing both carbon and silicon have been prepared. They possess the general structure

These compounds, known as **silicones,** are remarkable for their stability and resistance to the action of chemicals. The low-molecular-weight silicones are oils which are nearly as fluid at low temperatures as they are at high. They find application as lubricants in regions of very low temperatures. High-molecular-weight silicones with elastomer properties make suitable rubber substitutes. A puttylike silicone called "bouncing putty" exhibits nearly perfect elasticity and may be bounced in the same manner as a rubber ball. Under certain conditions, silicones are plastic and can be shaped as desired. Silicones are used as moistureproofing and fireproofing agents and as insulating coatings for electrical equipment. A silicone finish for textiles makes them water repellent even after laundering or dry cleaning. Because silicone films can be made thinner than other kinds of insulation, electrical motors can now be made much smaller. The chemistry of the silicones is summarized in Figure 33.1. Note that cross-linking of chains may take place to produce three-dimensional structures of great complexity.

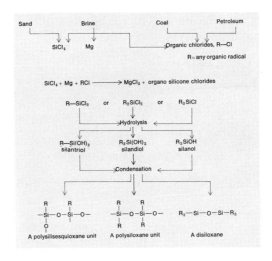

Figure 33.1
The chemistry of silicones.

Miscellaneous Compounds

Some other modern organic compounds of interest are:

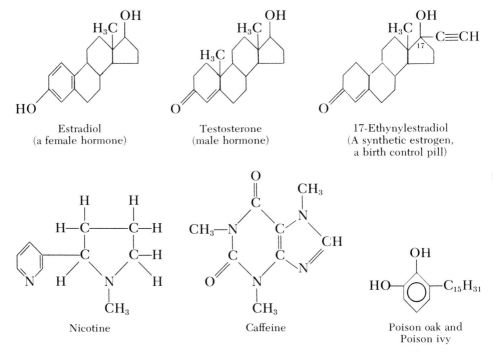

Exercises

33.1. Write structural formulas for: (a) benzene, (b) toluene, (c) m-xylene, (d) p-nitro-toluene, (e) o-bromotoluene, (f) 1,4-dichlorobenzene, (g) phenol, (h) 1,2-di-nitrobenzene.

33.2. Examine the structural formulas of the following compounds and list for each any of the functional groups included in Table 32.1: (a) aspirin, (b) sulfanila-mide, (c) penicillin, (d) cortisone, (e) morphine, (f) 2,4-D, (g) vanillin, (h) Bakelite, (i) nylon, (j) Lucite, (k) estradiol, (l) Orlon.

33.3. Attempt to write equations showing the
(a) preparation of toluene from methyl chloride and benzene.
(b) formation of chlorobenzene from benzene and chlorine.
(c) reaction of nitric acid and benzene to form nitrobenzene.
(d) reaction of nitrobenzene with bromine to form m-nitrobromobenzene.
(e) formation of benzenesulfonic acid from benzene and sulfuric acid.
(f) formation of teflon from tetrafluoroethylene.
(g) formation of styrofoam from styrene.

33.4. Write the structural formula of DDT and give its chemical name. Show how the name and the formula correspond.

33.5. A more complete chemical name for DDT is 1,1,1-trichloro-2,2-bis (p-chloro-phenyl) ethane. Correlate this name with the formula of DDT.

33.6. DDD is a compound similar to DDT. It is known as 1,1-dichloro-2,2-bis (p-chlorophenyl) ethane. Write the structure of DDD.

33.7. The compound 2,4,5-T is an organic herbicide similar to 2,4-D. Write the names and formulas for 2,4-D and 2,4,5-T.

33.8. Write structural formulas for the following flavors: (a) n-pentyl acetate (pears), (b) n-butyl acetate (raspberries), (c) ethyl butyrate (strawberries), (d) pentyl pentanoate (apples), (e) isopentyl acetate (bananas).

33.9. Barbituric acid has the same structure as barbital (veronal) except that the two ethyl groups are replaced with hydrogen atoms. Write the equation for the formation of barbituric acid by reacting malonic acid and urea, (the reaction involves the splitting off of two molecules of water).

Malonic acid   Urea

33.10. Salicylic acid has the structural formula:

Write the formulas of (a) methyl salicylate (oil of wintergreen), (b) sodium salicylate, (c) acetylsalicylic acid (aspirin), (d) sodium acetylsalicylate.

33.11. The formation of polyethylene from ethylene with the aid of a catalyst might be represented by the reaction

$$nCH_2{=}CH_2 \xrightarrow{\text{catalyst}} (CH_2{-}CH_2)_n$$

Write similar equations for the production of (a) polypropylene, (b) PVC, (c) Plexiglas, (d) Orlon.

Supplementary Readings

33.1. "The DDT Story," E. Keller, *Chemistry*, **43**, 8 (1970).
33.2. "Pheromones," E. O. Wilson, *Sci. American*, May, 1963; p. 100.
33.3. "Hallucinogenic Drugs," F. Barron, M. E. Jarvik, and S. Bunnell, Jr., *Sci. American*, April, 1964; p. 29.
33.4. "The Microstructure of Polymeric Material," D. R. Uhlmann and A. G. Kolbeck, *Sci. American*, Dec., 1975; p. 96.
33.5. *Giant Molecules*, M. Kaufman, Doubleday, New York, 1968 (paper).

Chapter 34

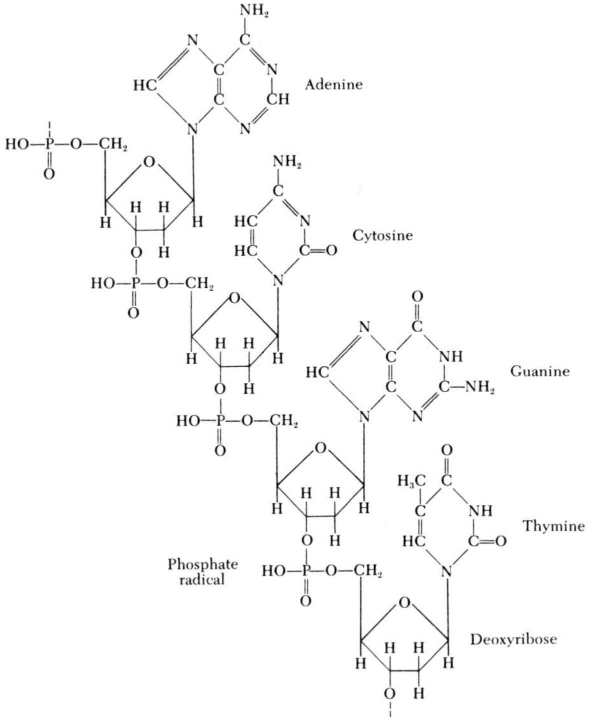

NH₂ ... Adenine

HO—P—O—CH₂ ...

Cytosine

Guanine

Thymine

Phosphate radical

Deoxyribose

A small section of a deoxyribonucleic acid, DNA

Introduction to Biochemistry

The selectivity of living cells in their choice of molecular units is a cause of wonder. Why is the five-carbon sugar, ribose, inherent in nucleic acids found in all living cells? Why are the four nitrogen bases, as shown in the structural formula on the previous page, also selective units in nucleic acids?* It is fortunate that certain basic structural patterns are common in the chemistry of living tissue. Biochemistry is the branch of chemistry that deals primarily with molecular structure and compounds from biological materials.

Biochemistry is closely aligned to biology and chemistry. The biochemist is interested in "structure-function relations." In other words, "why is a particular chemical structure specifically related to a given biological function?" Nature has been extremely ingenious in designing molecules that are involved in the various tasks such as the stretching and contraction of muscle, the transmission of an impulse in nerves, and in the ability of like to produce like. Biochemistry involves the study of the structure and function of compounds isolated from biological materials such as plants, animals, bacteria, and viruses.

Biochemical energetics relates to heat and work as the result of energy-providing reactions in a living organism. The human body operates at about 37°C and at such a relatively low temperature the word "burning" cannot be applied to body use of part of its food. Certain molecular complexes of stored bond energies, resultant from sequences of chemical change, are now known that keep the body warm or capable of movement.

A proper procedure is to first study the simpler structural compounds and then the patterns that are involved in unions to form large polymeric molecules, such as the nucleic acids and proteins. Some relatively simple molecules such as carbohydrates and fats are studied by biochemists in particular for their metabolism.

34.1 WATER

It may seem strange to begin with the simple compound water. Yet living organisms are made up of 60 to 90% water by weight. Its usefulness to living organisms derives from the high solvent ability of water, its role in chemical reactions, and as a lubricant of body tissues. Water is also of value in main-

---

* In addition to the use of non-metal organic complexes of sulfur and phosphorus in biological processes, use is made of trace amounts of certain metals, such as copper compounds in the liver. A marked oddity and further example of living-cell or cell-membrane selectivity concerns Na^+ and K^+ in animal bodies. Blood serum of all animals contains 0.32% sodium and 0.022% potassium. This relates to ocean water in which the ratio of sodium to potassium is 2.6% NaCl to only 0.07% KCl (about 37 times more sodium salt than potassium). However, in the milk of carnivora, sodium and potassium occur in approximately equivalent amounts; in that of herbivora, and in human milk, potassium predominates 3.5 to 1.0. Why and how do cell membranes select and differentiate, so as to have this higher amount of potassium in milk for the new born? These questions have not been answered.

taining a constant temperature in warm-blooded animals because of its high heat capacity. Evaporation of water from the skin is useful as a coolant.

Water occurs in living organisms in two forms—free and bound. The bulk of the water is "bound" to the chemical constituents of the cell. Free water is that found in body fluids such as blood and urine. Most molecules are carried in, through, and out of the body in an aqueous solution.

Water is markedly important in hydrolysis reactions, and also in the acidic or basic nature of hydrolytic products. In some cases H^+ is in excess in body fluid, in other cases OH^-.

34.2 FATS

Common fats and oils are biologically important compounds belonging to a more general type of material called **lipids,** which are oil-soluble substances of living matter in plants and animals. Lipids are insoluble in water but dissolve in organic solvents such as ether, benzene, chloroform, and acetone. In the human body they are present primarily in the cell membranes and in the brain and nervous tissue. Lipids include not only fats and oils, but some waxes, steroids, fat–soluble vitamins, phospholipids, and several other types of compounds.

Fats have multiple functions, but one of the more important is food storage. To the disgust of all gourmets, most excess foods end up in fat storage deposits. Of course, to the hibernating bear this is a distinct advantage. An interesting use of fats is by the camel, whose hump is composed largely of fats. During periods of lack of water the camel hydrolyzes these lipids to water, other degradation products, and some energy.

Fats and oils are **glyceryl esters** of organic acids that have in the realm of 17 carbons in a hydrocarbon chain. They are obtained from both plants and animals. Usually animal fats, such as stearin and butterfat, are solid or semi-solid, while those derived from plants—olive oil, cotton seed oil, and corn oil, for example, are liquids and are designated as oils. Oils as applied to organic compounds may be classified into three groups: (1) mineral oils, derived from petroleum or petroleum products, (2) essential oils, volatile liquids derived from plants, and (3) **glyceryl esters.** We shall limit our discussion here to fats and oils which are glyceryl esters.

The principal fats are **stearin, palmitin,** and **olein.**

$$
\begin{array}{lll}
C_{17}H_{35}COO-CH_2 & C_{15}H_{31}COO-CH_2 & C_{17}H_{33}COO-CH_2 \\
\;\;\;\;\;\;\;\;\;\;\;\;\;\;\;| & \;\;\;\;\;\;\;\;\;\;\;\;\;\;\;| & \;\;\;\;\;\;\;\;\;\;\;\;\;\;\;| \\
C_{17}H_{35}COO-CH & C_{15}H_{31}COO-CH & C_{17}H_{33}COO-CH \\
\;\;\;\;\;\;\;\;\;\;\;\;\;\;\;| & \;\;\;\;\;\;\;\;\;\;\;\;\;\;\;| & \;\;\;\;\;\;\;\;\;\;\;\;\;\;\;| \\
C_{17}H_{35}COO-CH_2 & C_{15}H_{31}COO-CH_2 & C_{17}H_{33}COO-CH_2 \\
\;\;\;\;\;\;\text{Stearin} & \;\;\;\;\;\;\text{Palmitin} & \;\;\;\;\;\;\text{Olein}
\end{array}
$$

These esters may be considered to be derived from glycerol and a fatty acid.

For example,

$$C_3H_5(OH)_3 + 3\ C_{17}H_{35}COOH \rightarrow (C_{17}H_{35}COO)_3C_3H_5 + 3\ H_2O$$

Glycerol   Stearic acid   Stearin or glyceryl
                                                                          stearate

Stearin and palmitin are saturated esters and are solids at ordinary temperatures; olein is an unsaturated ester with one double bond in each $C_{17}H_{33}$ chain
and is a liquid. The hardness of a fat depends on the relative amounts of
these three esters present. Beef tallow, a hard fat, is about 75% stearin and
25% olein. Lard is softer and is composed of about 60% olein and 40% palmitin; butterfat contains both olein and palmitin and, in addition, butyrin, the
glyceryl ester of butyric acid.

There is some evidence that the greater softness of women's skin as compared with men's skin is due to a slightly higher glyceryl oleate content.
Since all body fat is a mixture of oil and fat, all mortals are more or less fatty
and oily.

Liquid plant oils are frequently hydrogenated to convert them to saturated esters which are solid or semisolid. For example, the hydrogenation of
olein may be represented by the equation

$$(C_{17}H_{33}COO)_3C_3H_5 + 3\ H_2 \rightarrow (C_{17}H_{35}COO)_3C_3H_5$$

Olein   Stearin

Similarly, fish oil, peanut oil, soy bean oil, castor oil, and other unsaturated
liquid oils may be hydrogenated to edible solid fats. There is more than one
double bond in the hydrocarbon chain part of the acid radicals of most of the
odorous oils. They can catalytically add hydrogen to become odorless fat.

Polyunsaturates

A fat with no double bonds in the long hydrocarbon radical is said to be
harder to digest and may lead to hardening of the arteries. Thus **polyunsaturates,** a term used in food advertisements, refers to a fat-oil mixture with
a moderate degree of double bonding.

Several double bonds (unsaturation) result in a repulsive castor oil or
fish oil-like material. However, the presence of two or three double bonds in
the long hydrocarbon radical permits easier oxidation of an oil. **Linoleic
acid,** $C_{17}H_{31}COOH$, and **linolenic acid,** $C_{17}H_{29}COOH$, from the digestion of
fats, have been related to preventative or curative effects, and so indicate the
desirability of a degree of polyunsaturation in lipids.

$C_3H_5(C_{17}H_{35}COO)_3$ may be called **tri stearin,** a simple glyceride, in which
the three fatty acid radicals are the same. Mixed glycerides may contain both
saturated and unsaturated radicals and are of diverse carbon chain length.
Naturally occurring fats and oils are composed of mixed glycerides. Glycerides that are found in vegetables are usually in the form of oils instead of
fats; that is their fatty acid radicals tend toward the unsaturated linoleic and
linolenic.

Saponification

Soaps are prepared by heating a fat with sodium or potassium hydroxide solution. Sodium hydroxide (lye) produces a hard soap, while potassium hydroxide gives a soft soap (liquid). Fats, such as tallow, coconut oil, and palm oil, are usually employed for soap manufacture. These and similar reactions of an ester with a base are termed **saponification** reactions. The general reaction may be illustrated by the following equation for the preparation of soap from stearin:

$$(C_{17}H_{35}COO)_3C_3H_5 + 3\ NaOH \rightarrow 3\ C_{17}H_{35}COONa + C_3H_5(OH)_3$$

$$\text{Sodium stearate} \qquad \text{Glycerin}$$

Sodium salts of the higher fatty acids, such as sodium stearate, sodium palmitate, and sodium oleate, are the principal constituents of ordinary soaps. Glycerin is a by-product of the preparation of soap; it is removed and purified by distillation.

Sodium palmitate, $C_{15}H_{31}COONa$, will jelly gasoline, thus the name Napalm for a war-use mixture.

34.3   CARBOHYDRATES

A group of compounds that are not only building blocks in living organisms but also compounds that can be oxidized to yield energy are the carbohydrates. It is the energy yield factor that we will emphasize, since it is so needed to carry out body functions. No student needs to reflect long on the distinctive gastrointestinal features of an oyster, a chicken or a cow. Yet all of these creatures use essentially identical primary fuels. Whether mold or mosquito, mouse or man, the same molecules pass before the inquiring eye of the biochemist, and one is forced to conclude that despite the almost fantastic variety that exists in the morphology of organisms, and in individual molecules of proteins or nucleic acids, there are only a handful of compounds that can furnish energy directly to a living cell. The process by which we obtain this energy is part of metabolism and occurs in small discrete steps.

Metabolism is a term used to describe the chemical transformations that befall a given compound in an organism. The breakdown stage is referred to as **catabolism** and the synthesis as **anabolism.**

Primarily energy is derived from sunlight by a reaction carried out in plants called photosynthesis. Essentially this reaction consists in its simplest form of:

$$6\ CO_2 + 6\ H_2O \xrightarrow[\text{plants}]{\text{sunlight}} C_6H_{12}O_6 + 6\ O_2$$

For animals it is the $C_6H_{12}O_6$ (sugar) which an oxidation yields most of the energy that sustains life.

Chemistry of Carbohydrates

Prior to considering the role of these compounds in living systems, it is once again necessary to assemble a small molecular vocabulary. They are compounds of carbon, hydrogen, and oxygen. Sugar, starch, and cellulose are common examples. The term "carbohydrate" derives from the fact that the empirical formula of many sugars is $C_nH_{2n}O_n$ which technically is a hydrate of carbon. Structurally, carbohydrates are hydrated ketones or aldehydes. If a single unit is present, the compound is called a **monosaccharide;** however, if two units (two monosaccharides) are present it is a **disaccharide;** three, a **trisaccharide;** and eventually **polysaccharides** when many units are present. Complete hydrolysis of polysaccharides or di- and trisaccharides yield monosaccharides.

Monosaccharides

Monosaccharides or simple sugars have names with the characteristic ending *-ose.* The most common and important sugars contain six carbons and are called **hexoses.** Glucose, $C_6H_{12}O_6$, and one of its isomers, **fructose,** are examples.

Glucose is also called dextrose. It may be obtained by the hydrolysis of sucrose, maltose or starch. Fructose, also called levulose, is present in many fruits and is a hydrolysis product of sucrose along with glucose. Fructose is usually prepared by the hydrolysis of inulin, a polysaccharide somewhat similar to starch found in dahlia tubers and the Jerusalem artichoke. Other hexoses include mannose and galactose.

The formula of glucose may be represented as follows, and the Haworth projection is that most commonly used:

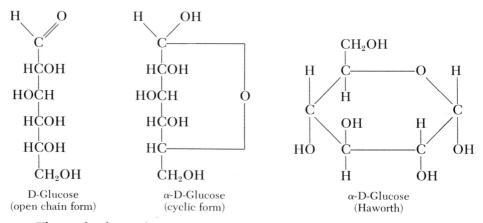

D-Glucose (open chain form)   α-D-Glucose (cyclic form)   α-D-Glucose (Haworth)

The cyclic form of glucose has an additional assymetric carbon atom, the first carbon atom, not possessed by the open chain form. Consequently, two

additional isomers of D-glucose exist and are designated **α-** and **β-D-glucose.**
The α-isomer has the OH on the right and the β-isomer on the left.

All the hexoses have the same empirical formula $C_6H_{12}O_6$, but the spatial placements of the radicals are different.

$$
\begin{array}{ccc}
H & CH_2OH & H \\
| & | & | \\
C{=}O & C{=}O & C{=}O \\
| & | & | \\
H{-}C{-}OH & HO{-}C{-}H & H{-}C{-}OH \\
| & | & | \\
HO{-}C{-}H & H{-}C{-}OH & HO{-}C{-}H \\
| & | & | \\
H{-}C{-}OH & H{-}C{-}OH & HO{-}C{-}H \\
| & | & | \\
H{-}C{-}OH & CH_2OH & H{-}C{-}OH \\
| & & | \\
CH_2OH & & CH_2OH \\
\text{D-glucose} & \text{D-fructose} & \text{D-galactose} \\
\text{(aldehyde)} & \text{(ketone)} & \text{(aldehyde)}
\end{array}
$$

Disaccharides

A common disaccharide is sucrose $(C_{12}H_{22}O_{11})$. It is obtained in a white crystalline form by evaporation of juices from the sugar beet or sugar cane and is the sugar we use on our table. Sucrose is the purest chemical compound that can be purchased in carload lots. Its formula is:

(glucose unit)   sucrose,   (fructose unit)
$C_{12}H_{22}O_{11}$

Sucrose is hydrolyzed by the enzyme, **invertase,** to one molecule each of glucose and fructose.

$$C_{12}H_{22}O_{11} + H_2O \rightarrow C_6H_{12}O_6 + C_6H_{12}O_6$$

sucrose   glucose   fructose

The term invert sugar is used to indicate this equimolecular mixture of glucose and fructose.

Lactose, or milk sugar, is a disaccharide obtained from the milk of mammals. Cow's milk contains about 5% lactose, human 8 to 10%. Hydrolysis of lactose produces one molecule each of glucose and galactose.

Trioses and Pentoses

Although hexoses are the most common sugars, mention must be made of three and five carbon-containing carbohydrates. Trioses, such as **glyceraldehyde** and **dihydroxyacetone**, $C_3H_6O_3$, are involved in muscle metabolism. Examples of pentoses, $C_5H_{10}O_5$, are **arabinose, xylose, ribose,** and **deoxyribose.** Arabinose may be obtained from gum arabic and cherry tree gum. Hydrolysis of corn cobs, straw, or wood yields xylose. Ribose and deoxyribose are the sugar constituents of the nucleic acids RNA and DNA (page 718) present in cell nuclei.

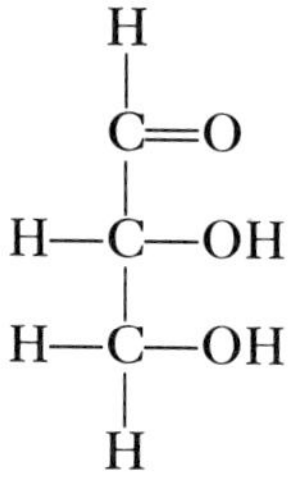

Glyceraldehyde

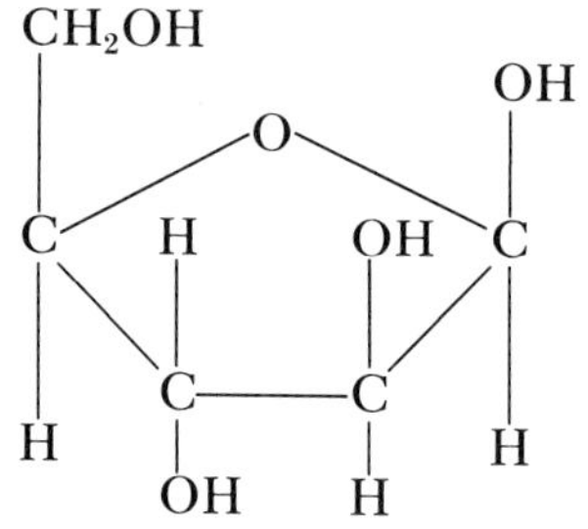

Arabinose

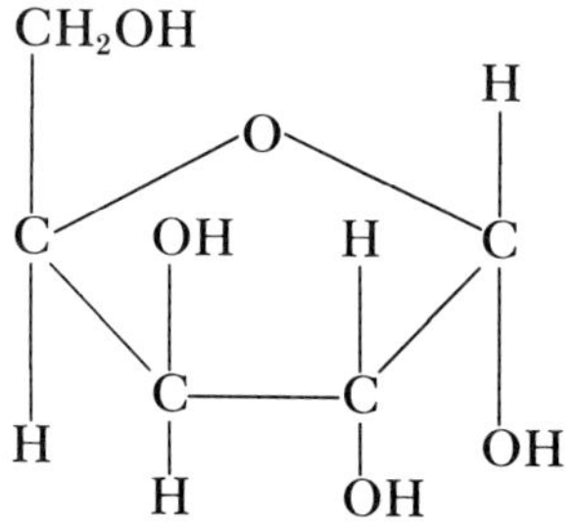

Xylose

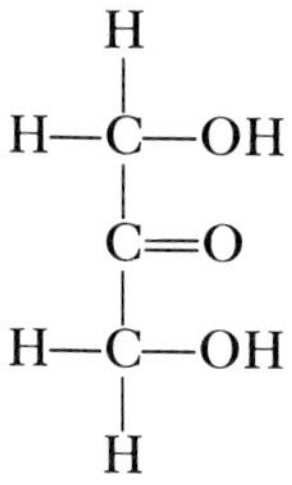

Dihydroxyacetone

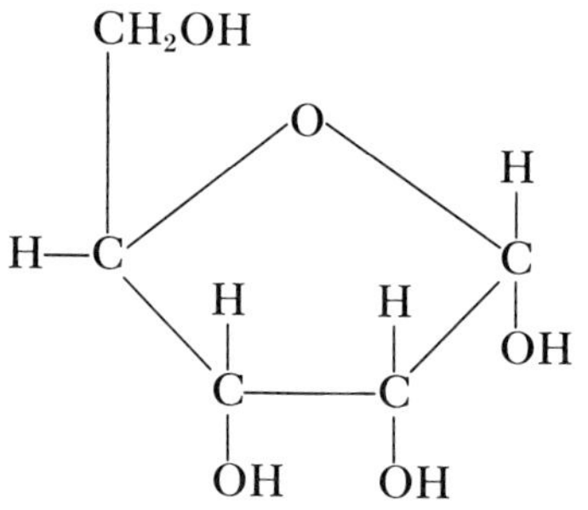

D-Ribose

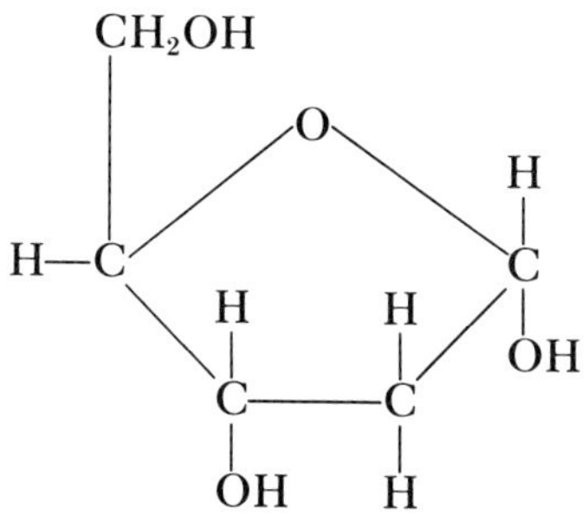

D-Deoxyribose

Starch

Starch is a polysaccharide represented by the molecular formula $(C_6H_{10}O_5)_x$. The magnitude of x is such that the molecular weight is large, from 15,000 to 300,000. As shown below, its structure consists of repeating α-D-glucose units. Some of these repeating units are straight chain and some are branched chain in the polymer.

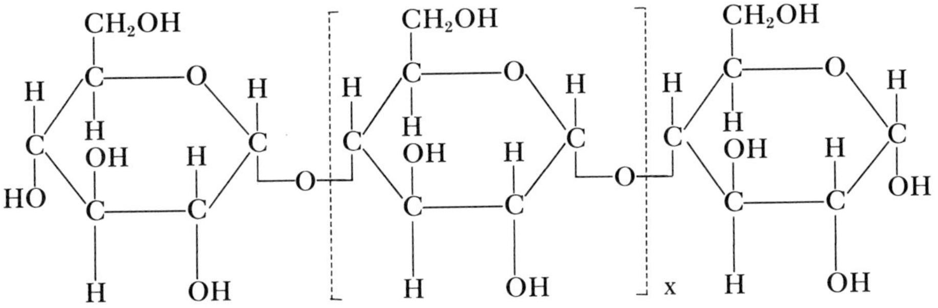

Starch
$(C_6H_{10}O_5)_x$

Starch is present in nearly all plants; raw potatoes contain about 20%, and dried rice and corn about 75 and 50%, respectively. These substances provide the source of starch for commercial production. Starch is very insoluble in cold water, but with hot water the granules swell and produce a colloidal dispersion which, if concentrated, will set to a gel when cooled. Boiling with hydrochloric acid results in hydrolysis to form glucose.

Cellulose

Cellulose $(C_6H_{10}O_5)_y$ is a polysaccharide of even higher molecular weight than starch, 150,000 to 1,000,000. Although it has the same empirical formula as starch, its properties are very different. It is insoluble in water and organic solvents and is not easily hydrolyzed, though boiling for a long time with dilute sulfuric acid gradually results in the formation of glucose.

As shown below, cellulose is a straight chain polymer of α-D-glucose molecules. While humans have the necessary enzymes to hydrolyze starch, they cannot break down cellulose. Termites and animals, however, are able to use cellulose as a source of dietary carbohydrate.

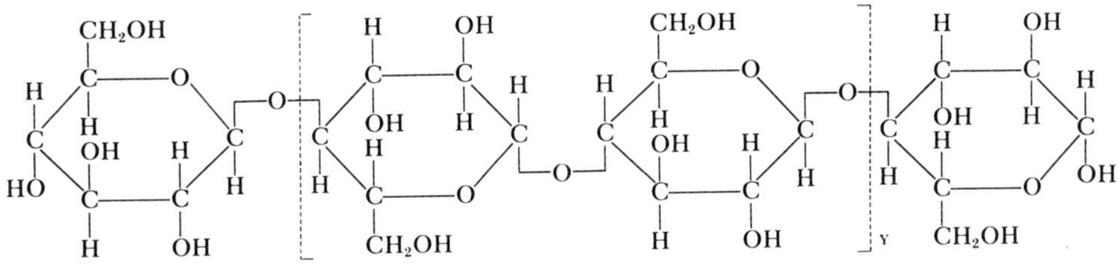

Cellulose
$(C_6H_{10}O_5)_y$

Cellulose is the principal constituent of all plant fiber and serves as a framework or supporting structure. Cotton (which is practically pure cellulose) and wood are the principal sources of cellulose. In addition, wood fibers contain lignin and resins which bind the material together. In the production of cellulose from wood, the lignin and resin are usually dissolved by means of calcium acid sulfite, or a sodium hydroxide and sodium sulfide solution to leave pure cellulose. The manufacture of paper is essentially a process of washing and bleaching the cellulose and pressing it into compact form in sheets.

Since carbohydrates contain the alcohol group (OH), we might expect them to react with acids to form esters. If cellulose is treated with nitric acid in the presence of concentrated sulfuric acid to act as a dehydrating agent, cellulose nitrates are formed. Intensive nitration results in the production of guncotton, a highly explosive compound used in the production of smokeless powders and cordite. Incomplete nitration forms **pyroxylin,** which is used in the production of artificial leathers, photographic film, lacquers, and collodion. It is soluble in mixtures of organic solvents. This permits its separation from guncotton, which is insoluble. Fabrics treated with an organic solvent solution of pyroxylin, on evaporation of the solvent, have a tough, flexible finish that resembles natural leather.

Collodion is a solution of pyroxylin in a mixture of alcohol and ether. A common lacquer is a solution of pyroxylin in amyl acetate (banana oil). **Celluloid** is produced from cellulose nitrates by forming a plastic mass with camphor; the product may be molded into desired shapes. It is highly flammable.

Rayon

The principal difference between silk and rayon is in their chemical compositions. Silk is a protein; rayon is cellulose or a cellulose derivative. Cotton or cellulose fibers are short, irregular, and dull; the fibers of silk are long, smooth, and bright. The production of rayon requires changing the structure of the cotton or wood cellulose fibers. While there are several methods employed for doing this, all methods are the same in principle. In brief, the process consists of dispersing cellulose or a cellulose derivative in a liquid in colloidal form, which can then be forced through a small orifice or nozzle into a solution that will precipitate the cellulose or its derivative in tiny filaments. These filaments are twisted together to form a larger thread, which is then used for weaving purposes. One of the more common methods of producing rayon is the viscose process, in which the cellulose is treated with sodium hydroxide and carbon disulfide. A cellulose-xanthate solution is formed and forced through small openings into a bath of sulfuric acid and sodium bisulfate, where the cellulose is precipitated in the desired fibrous form:

$$—R^{\circ}OH + NaOH \rightarrow —R^{\circ}ONa + HOH$$
Cellulose

$$—R^{\circ}ONa + CS_2 \rightarrow —R^{\circ}—O—\overset{\displaystyle S}{\overset{\displaystyle \|}{C}}—SNa$$
Cellulose° xanthate unit

$$2\text{—R}^{\circ}\text{—O—}\overset{\displaystyle S}{\overset{\|}{\text{C}}}\text{—SNa} + H_2SO_4 \rightarrow 2\ CS_2 + Na_2SO_4 + 2\text{—R}^{\circ}\text{OH}$$

regenerated cellulose

$^{\circ}$—R$^{\circ}$ represents a fragment of a large cellulose molecule.

34.4 PROTEINS

All proteins are nitrogenous substances containing, on the average, 16% nitrogen. In addition to carbon, hydrogen, and oxygen, other elements occur in small amounts and include sulfur, phosphorous, and metals as copper, iron, and zinc. The functions of proteins are variable. Included among proteins are hormones (as insulin), the contractile fiber of muscle (myosin), the oxygen carrier of blood (hemoglobin), antibodies, and enzymes. The latter are the catalysts of living systems (see p. 727). A human has more than 10,000 different enzymes each of which catalyzes a specific chemical reaction and has a different molecular structure.

Proteins (from a Greek word meaning "of prime importance") are very large, complex molecules ranging in molecular weights from 34,500 to 50,000,000. Oxyhemoglobin, for example, has a molecular weight of 68,000 and a molecular formula of $C_{2932}H_{4724}N_{828}S_8Fe_4O_{840}$. Figure 34.1 shows the structure of the cytochrome C molecule as determined by X-ray crystallography. It is one of the simpler proteins, containing 104 amino acids units, and it, or modifications of it, is found in every living cell that uses oxygen for respiration. Near the center of the molecule can be seen the heme group containing Fe.

If a protein is hydrolyzed in acid, it is degraded to its smallest chemical constituents—amino acids. An amino acid consists of an amino group (—NH$_2$), a carboxyl group (—COOH), and an additional constituent (R):

$$\begin{array}{c} H \\ | \\ R\text{—C—COOH} \\ | \\ NH_2 \end{array}$$

An amino acid has both basic and acidic properties. Depending on the acidity of the solution it may exist in a different charged form.

$$\begin{array}{c} H \\ | \\ R\text{—C—COOH} \\ | \\ NH_3{}^+ \end{array} \overset{OH^-}{\underset{H^+}{\rightleftarrows}} \begin{array}{c} H \\ | \\ R\text{—C—COO}^- \\ | \\ NH_3{}^+ \end{array} \overset{OH^-}{\underset{H^+}{\rightleftarrows}} \begin{array}{c} H \\ | \\ R\text{—C—COO}^- \\ | \\ NH_2 \end{array}$$

acid solution   neutral molecule   basic solution
protons on   (dipolar)   both protons
both ends   lost

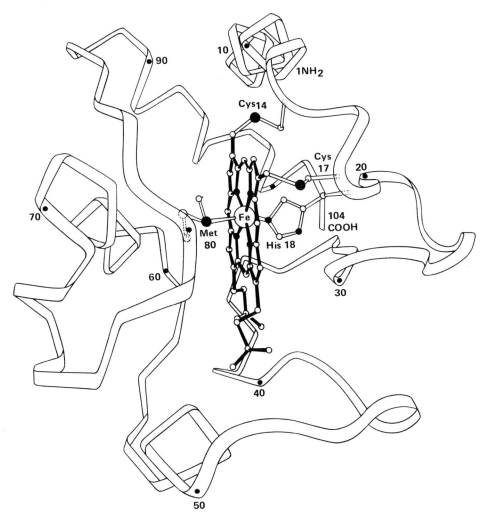

Figure 34.1
Cytochrome c molecule. The ribbon drawing shows the sequence of the 104
amino acid units comprising the molecule. The heme group is shown in black
bonds with an Fe atom in the center. NH_2 and COOH chain terminal groups, as
well as every 10th amino acid, are indicated in the chain. Sulfur atoms are
indicated as larger black balls, two of which serve as covalent attachments
between the heme and polypeptide chain (Cys 14 and 17), and one of which is a
heme iron ligand (Met 80). (Drawing by L. R. Salemme)

When an amino acid is treated with a base, protons are lost from the carboxyl group and from the nitrogen atom of the —NH$_3^+$ group. In strong acids protons are transferred to the carboxylate ion (—COO$^-$) and the NH$_2$ group. Since amino acids react with both acids and bases they are amphoteric compounds.

In general proteins consist of about twenty different amino acids that are often combined many times over in each protein molecule. A few others occasionally occur in some proteins. Different proteins contain varied amounts of amino acids, as shown in Table 34.1. Plants are able to make their own amino acids, while animals can synthesize some but require others in their diets. For example, eight amino acids are essential in human diets (**isoleucine, leucine, lysine, methionine, phenylalanine, threonine, tryptophan** and **valine**). In addition, **arginine** and **histidine** are needed by growing rats. The structures of the twenty common amino acids occurring in proteins are shown in Table 34.2.

Table 34.1

Amino acid content of some animal and plant proteins (percent)

	Muscle	Egg White	Gelatin	Whole Wheat
Arginine	6.6	5.9	7.8	4.3
Histidine	2.8	2.5	0.8	1.8
Lysine	8.5	6.4	4.8	2.5
Tyrosine	3.1	4.3	0.6	3.6
Tryptophan	1.1	1.8	0.01	1.2
Phenylalanine	4.5	6.0	2.0	4.4
Cystine	1.4	2.6	0.1	3.3
Methionine	2.5	4.0	0.9	1.2
Serine	5.1	7.3	3.4	3.8
Threonine	4.6	4.7	1.7	3.9
Leucine	8.0	9.0	3.5	6.9
Isoleucine	4.7	6.4	1.4	4.4
Valine	5.5	7.8	2.7	4.5
Glutamic acid	14.6	12.8	7.8	31.4
Aspartic acid	8.0	7.6	4.9	3.8
Glycine	5.0	3.7	15.7	3.4
Alanine	6.5	—	7.9	3.0
Proline	5.0	2.9	14.8	10.3
Hydroxyproline	4.7	—	13.3	—

The simplest amino acid is **glycine** (CH$_2$NH$_2$COOH). Proteins are built up by combination of amino acids. The basic part of one molecule may combine covalently with the acidic part of another molecule, whose basic part in turn may combine with another acidic group of still another amino acid; thus a chain containing many molecules of amino acid molecules may be built

Table 34.2
Amino acids commonly found in proteins:

$\text{R Group}-\underset{\underset{H}{|}}{\overset{\overset{NH_2}{|}}{C}}-\text{COOH}$	Name	Abbreviation
H—	Glycine	gly
CH_3—	Alanine	ala
$HOCH_2$—	Serine	ser
CH_3CH— $\quad\underset{OH}{}$	Threonine	thr
HO_2C-CH_2—	Aspartic	asp
$(CH_3)_2CH$—	Valine	val
$H_3C-SCH_2CH_2$—	Methionine	met
$HO_2C-CH_2CH_2$—	Glutamic	glu
$(CH_3)_2CHCH_2$—	Leucine	leu
CH_3CH_2CH— $\quad\underset{CH_3}{}$	Isoleucine	ile
$H_2N-CH_2CH_2CH_2CH_2$—	Lysine	lys
$HSCH_2$—	Cysteine	cys
$HO_2C-CH-CH_2S-S-CH_2$— $\quad\underset{NH_2}{}$	Cystine	cys-cys
$H_2N-\underset{\underset{NH}{\|}}{C}-NH-CH_2CH_2CH_2$—	Arginine	arg
$\underset{N\qquad NH}{\overset{HC=C-CH_2-}{}}$ $\quad C \atop H$	Histidine	his
(indole)$-CH_2$—	Tryptophan	trp
(benzene)CH_2—	Phenylalanine	phe
HO-(benzene)-CH_2—	Tyrosine	tyr
$\begin{array}{c} H_2C-CH_2 \\ \quad\quad CHCOOH \\ H_2C-N \\ \quad\quad H \end{array}$	Proline	pro
$\begin{array}{c} OH \\ HC-CH_2 \\ \quad\quad CHCOOH \\ H_2C-N \\ \quad\quad H \end{array}$	Hydroxyproline	hypro

up. We can illustrate how this combination might take place between glycine molecules:

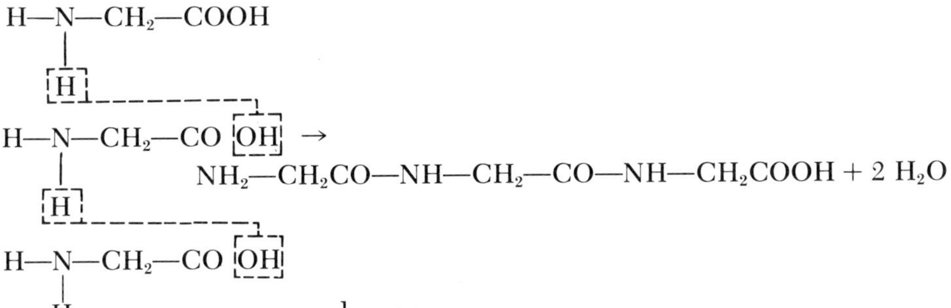

$$H—N—CH_2—COOH$$

$$\begin{matrix}|\\ [H]\end{matrix} \text{-----------}$$

$$H—N—CH_2—CO \, [OH] \rightarrow$$

$$\begin{matrix}|\\ [H]\end{matrix} \qquad NH_2—CH_2CO—NH—CH_2—CO—NH—CH_2COOH + 2\,H_2O$$

$$\text{-------------}$$

$$H—N—CH_2—CO \, [OH]$$

$$\begin{matrix}|\\ H\end{matrix} \qquad\qquad \text{and so on.}$$

Each protein is made up of a series of amino acids linked together in a unique sequence:

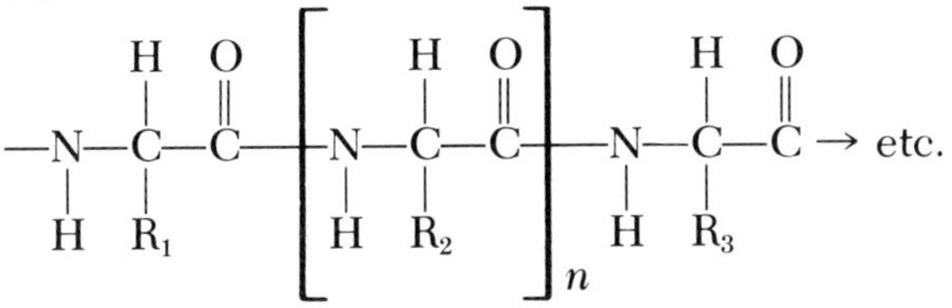

$$-N-\underset{\underset{H}{|}}{\overset{\overset{H}{|}}{C}}-\underset{}{\overset{\overset{O}{\|}}{C}}\left[-N-\underset{\underset{H}{|}}{\overset{\overset{H}{|}}{C}}-\underset{R_2}{\overset{\overset{O}{\|}}{C}}\right]_n -N-\underset{\underset{H}{|}}{\overset{\overset{H}{|}}{C}}-\underset{R_3}{\overset{\overset{O}{\|}}{C}} \rightarrow etc.$$

The R groups in the sequence may be different. Since proteins consist of hundreds of units (n) of the amino acids, a markedly high number of sequences is possible. The linkage that joins the different amino acid units together is called a **peptide bond.** (See Figure 34.2)

Each individual protein has a specific function and the specificity of this function is due to the sequence and 3-dimensional arrangement of its amino acids. By sequence we mean the order of the amino acids in this particular protein. Let us look at a simple hypothetical example to see how changes in the sequence results in different proteins. If we assume that a protein contains three different amino acids, A, B, and C, joined by two peptide bonds (—), then by changing the order, we can write six different combinations: (1)

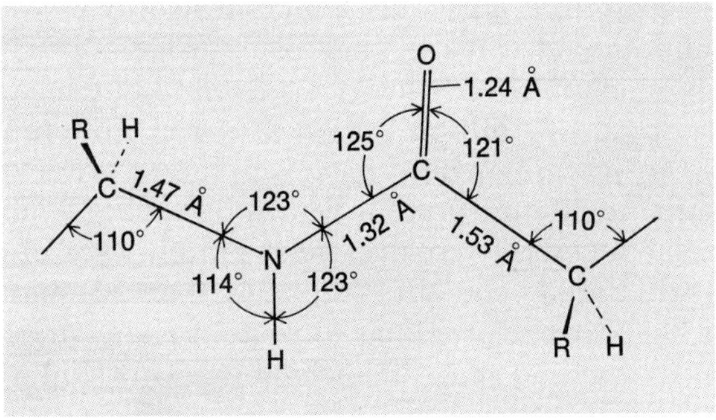

Figure 34.2
Bond distances and angles of the peptide link in proteins.

A—B—C, (2) A—C—B, (3) B—A—C, (4) B—C—A, (5) C—A—B, and (6) C—B—A.

To show that A—B—C and C—B—A are not the same it is necessary to introduce another fact. Each protein has two ends, an amino end and a carboxyl end:

$$NH_2-\overset{\overset{H}{|}}{\underset{\underset{R}{|}}{C}}-\overset{\overset{O}{||}}{C}-\left[-NH-\overset{\overset{H}{|}}{\underset{\underset{R}{|}}{C}}-\overset{\overset{O}{||}}{C}-\right]_n -NH-\overset{\overset{H}{|}}{\underset{\underset{R}{|}}{C}}-COOH$$

(amino end)   (carboxyl end)

with many amino acids in between in which the amino and carboxyl groups are tied up in peptide bonds. In our example, A is gly (NH_2CH_2COOH),

B is ala (CH_3CHNH_2—COOH), and C is ser ($HOCH_2-\overset{\overset{H}{|}}{\underset{\underset{NH_2}{|}}{C}}-COOH$); then,

A—B—C would be:

$$NH_2-\overset{\overset{H}{|}}{\underset{\underset{H}{|}}{C}}-\overset{\overset{O}{||}}{C}-NH-\overset{\overset{H}{|}}{\underset{\underset{CH_3}{|}}{C}}-\overset{\overset{O}{||}}{C}-NH-\overset{\overset{H}{|}}{\underset{\underset{CH_2OH}{|}}{C}}-COOH$$

and C—B—A:

$$NH_2-\overset{\overset{H}{|}}{\underset{\underset{CH_2OH}{|}}{C}}-\overset{\overset{O}{||}}{C}-NH-\overset{\overset{H}{|}}{\underset{\underset{CH_3}{|}}{C}}-\overset{\overset{O}{||}}{C}-NH-\overset{\overset{H}{|}}{\underset{\underset{H}{|}}{C}}-COOH$$

You can readily see that these are not the same. The sequence in A—B—C is gly-ala-ser and that in C—B—A is ser-ala-gly.

Mathematically we can represent how many different proteins are possible with only 3 different amino acids as 3 factorial (3!) ($3 \times 2 \times 1 = 6$). Consider what is possible if we have proteins with only a total of twenty amino acids and each is a different amino acid. This is analogous to having only twenty letters and calculating how many twenty letter words could be made of them. A factorial 20 (20!) or some two billion billion (2×10^{18}) different proteins are possible. Each of the proteins would have a different order or sequence of the amino acid units. A 20-unit protein would be a very small protein (approx. 10,000 molecular weight). Actually, most proteins are of a considerably greater molecular weight and contain many more than 20 amino acid units (80 to 1600). One of the current problems of protein chemistry is that of determining the sequence of amino acids in specific proteins (Figure 34.3).

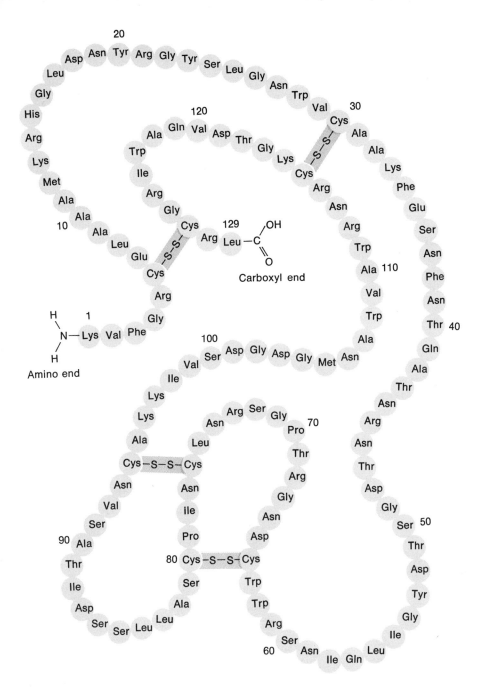

Figure 34.3
Two-dimensional model of a protein, lysozyme. This protein contains 129 amino acid residues. The illustration shows the sequence of amino acids. Note the crosslinking at four places by disulfide (—S—S—) bonds. From "The Three-dimensional Structure of an Enzyme Molecule" by David C. Phillips. Copyright © November 1966 by Scientific American, Inc. All rights reserved.

An example of how a simple change in the order of amino acids in a protein can result in disease concerns hemoglobin, the oxygen-carrying protein of blood. **Hemoglobin A** is the normal adult compound. The order of amino acids has been determined for a number of hemoglobins. In one case where one amino acid in the normal order is replaced by another amino acid a disease, sickle cell anemia, results. This disease is usually fatal and results from a genetic change. It is but one example of how a very minor change in sequence can lead to a disastrous result. The faulty hemoglobin is called **hemoglobin-S.** The amino acid glutamic acid in position 6 (i.e. starting from the amino end and counting to the amino acid in position 6) is replaced by valine. All of the other amino acids are the same and are in the same order. Thus, studies of the sequence of amino acids in proteins may have great importance in our learning more about various diseases.

The sequence of specific proteins can tell us a great deal about the course of evolution. The more similar the order the closer two organisms may be in the evolutionary scale. For example, if hemoglobins from different animals are examined it is found that the sequence of amino acids in gorilla hemoglobin is much more like human hemoglobin than horse hemoglobin.

The sequence of amino acids is called the **primary structure** of proteins. In addition proteins have **secondary, tertiary,** and **quaternary** structures. See Figure 34.4.

The **alpha helix** has a right-hand screw pattern with 3.7 amino acid residues per turn. Hydrogen bonding between $C=O$ and NH groups holds the polypeptide chains together. The alpha helix formation is referred to as the protein's secondary structure. Helices fall over themselves and intertwine in a tertiary structure. Quaternary structure arises if several polypeptide chains in some lateral fashion bond together as a unit.

Optical isomerism occurs in amino acids. All, except glycine, contain at least one asymmetric carbon atom and can exist in the D or L forms.

Of interest in biochemistry is that only the L-isomer exists in proteins except for very rare instances. This means that if an animal requires a certain amino acid for life (essential in the diet) it will require the L-amino acid. The D-amino acid would be ineffectual or, in some cases, toxic, and if the animal's diet were only D-amino acids it could not survive. Where isomers of natural products exist, living systems usually use only one of the isomers.

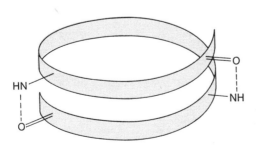

Figure 34.4
Schematic representation of the protein α-helix.

Proteins can be isolated as single compounds (macromolecules) from living organisms. If they retain their biological activity (i.e., are able to carry out the same reactions as in the animal or plant), they are called native proteins. Often they lose this activity when treated with a wide variety of reagents and conditions which do not hydrolyze the peptide bonds. For example, heating at relatively low temperatures (65°C) for short times will denature proteins. Ovalbumin is a major constituent of egg white. If a solution of ovalbumin, as in egg white, is heated or cooked, it precipitates or coagulates indicating denaturation.

34.5   NUCLEIC ACIDS

Every living organism needs to perform the essential function of reproduction. Certain **nucleic acids,** in particular DNA, are important constituents found in hereditary material responsible for this process. In terms of size the nucleic acids are much larger than proteins, having molecular weights in the millions. Nucleic acids, however, are composed of fewer constituents than the proteins. On hydrolysis, they may be broken down to two closely related sugars, inorganic phosphate ions, and, in general four nitrogenous bases. As in proteins, the individuality of each nucleic acid is expressed in the sequence of the units making up the macromolecular chain.

Nucleic acids are of two general kinds, **deoxyribonucleic acid (DNA)** and **ribonucleic acids (RNA).** These names derive from the fact that the former contains the sugar deoxyribose and the latter the sugar ribose. Hydrolysis of DNA and RNA to their smallest constituents yields as products:

DNA	*RNA*
Deoxyribose	Ribose
Phosphate	Phosphate
Adenine	Adenine
Guanine	Guanine
Cytosine	Cytosine
Thymine	Uracil

The structures of these compounds are shown in Table 34.3. The units diagrammed in the table develop into nucleic acids in the following sequence.

A purine or pyrimidine base + ribose → a nucleoside (Figure 34.5)
A nucleoside + H_3PO_4 → a nucleotide
Nucleotides on polymerization → nucleic acids

The **nucleotides** are repeating units in the very large polymeric molecules called nucleic acids. DNA is largely in the cell nucleus, whereas RNA is in the cellular fluid, the cytoplasm, outside the nucleus.

Table 34.3
Constituents of nucleic acids

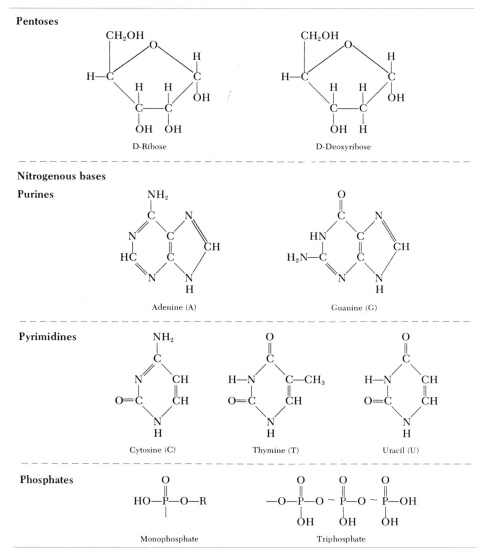

Pentoses

D-Ribose

D-Deoxyribose

Nitrogenous bases
Purines

Adenine (A)

Guanine (G)

Pyrimidines

Cytosine (C)

Thymine (T)

Uracil (U)

Phosphates

Monophosphate

Triphosphate

DNA and RNA

Both DNA and RNA consist of many nucleotides joined together by phosphate bridges between carbon atoms 3 and 5 of adjacent five-carbon sugar molecules as illustrated in Figure 34.6. RNA generally occurs as a single strand of nucleotides connected by phosphate bridges. In contrast to this, DNA consists of two single strands that are held together by hydrogen bonds be-

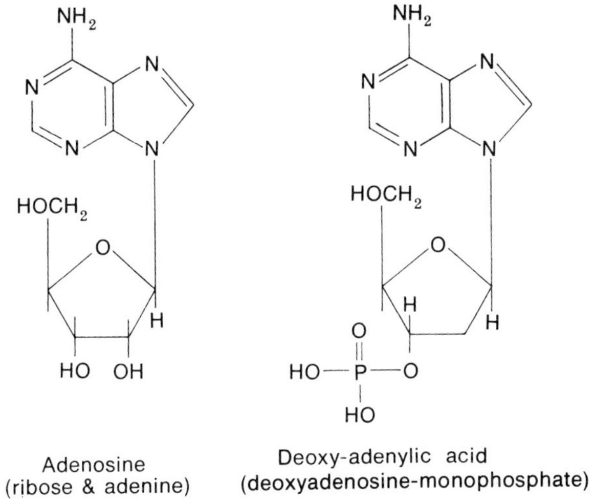

Adenosine
(ribose & adenine)

Deoxy-adenylic acid
(deoxyadenosine-monophosphate)

Fig. 34.5
A typical nucleoside and typical nucleotide.

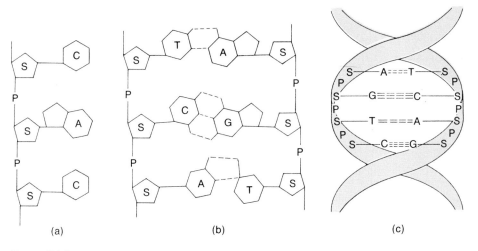

(a)   (b)   (c)

Figure 34.6
**Structure of nucleic acids: (a) RNA (a single strand); (b) DNA (a double strand);
(c) Watson-Crick helix. (A = adenine; C = cytosine; G = guanine; T = thymine;
S = sugar (ribose or deoxyribose); P = phosphate; — — — — = hydrogen bond.)**

tween the bases. As noted in the figure, adenine usually forms a hydrogen
bond with thymine, and guanine forms a hydrogen bond with cytosine. The
consequence of this is that the amount of adenine present in a particular DNA
molecule equals the amount of thymine, and the amount of cytosine equals

the amount of guanine. See page 722 for a picture of the structure of a section of a DNA molecule.

Because of the complexity of the nucleic acid molecules and the difficulty of presenting them on paper in terms of a complete chemical structure, abbreviations are used. We use A for adenine, G for guanine, C for cytosine, U for uracil, T for thymine, and S for the sugar ribose (in RNA or deoxyribose in DNA). The P indicates a phosphate between the nucleosides (base + sugar), but often the abbreviation used is only CAC, TCA, or AGT.

Genetic Material

The fact that the nucleic acid DNA is related to heredity (or genes) has been shown by two types of experiments. One consisted of isolating genetic material (nucleus of cell) and finding DNA present. Of course other compounds were also present and a role could not be definitely assigned for DNA. The answer came from some quite different experiments on types of pneumococci (bacteria causing pneumonia). Two types exist, one a smooth type and the other rough. The smooth type is virulent (that is, causes pneumonia), whereas the rough form is non-virulent and does not produce the disease. It was known that under certain conditions the non-virulent form became virulent. If the virulent form is isolated and killed by heating, and then injected into a rat, pneumonia will not result. If this heated material is combined with the non-virulent form and injected into the rat, the animal develops pneumonia. Reisolation of the bacteria showed that the living non-virulent strain had been changed to a virulent form. This occurred in spite of the fact that the heated material (virulent form) was dead and therefore could not reproduce itself. This could only mean that something in the heated material could convert the rough (non-virulent) to the smooth (virulent) form. The chemical responsible was isolated and found to be DNA. Once the non-virulent form was transformed to the virulent form it then reproduced only virulent forms. Thus the trait was inheritable and DNA was the molecule responsible.

The manner in which DNA can function as a gene (or hereditary material) has been shown by the biologist Watson and the physicist Crick. The model they developed is called the **Watson-Crick double helix.** Watson and Crick shared the Nobel prize in 1962 for this contribution and a partial representation of their model is shown in Figure 34.6. The Watson-Crick double helix consists of two single strands of DNA joined together by hydrogen bonds. In this model **adenine** (A) is *always* paired to **thymine** (T) by hydrogen bonds, and **cytosine** (C) to **guanine** (G).

Note in Figure 34.7 the hydrogen bonding between thymine and adenine and between cytosine and guanine. Each of these two hydrogen-bonded nitrogenous bases is referred to as a **base-pair,** and one base is said to be complementary to the other—that is, A is complementary to T (or U), T (or U) complementary to A, G complementary to C, and C complementary to G.

Replication of DNA may occur as represented in Figure 34.8. The two

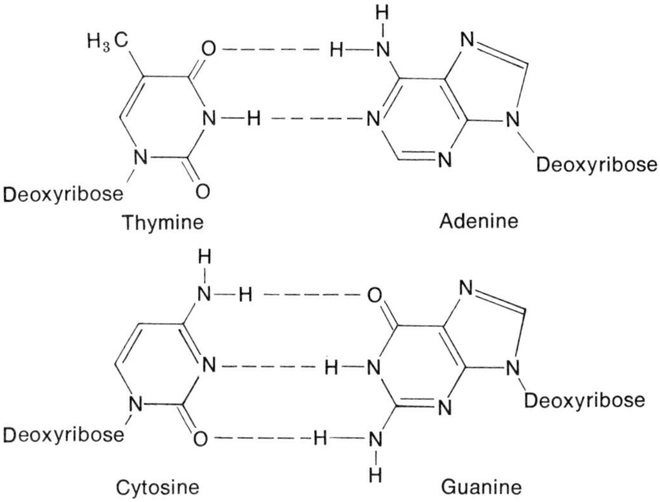

Figure 34.7
Hydrogen bonding of base pairs in DNA. (Double hydrogen bonding between thymine⚌adenine; triple bonding between cytosine☰guanine.

initial strands partially unwind or unzip right down the middle prior to cell division, the base pairs breaking apart at their hydrogen bonds. Nucleoside triphosphates occuring free in the nucleus are selectively bound so that the proper base pairing occurs (i.e., A with T and G with C). A new strand is then formed by a chemical reaction that results in a sugar phosphate ester being formed. Eventually a completely new complementary strand is made to the entire length of the parent strand, forming exact copies of the parent double helix. The result is two double-stranded DNA molecules each of which is identical to the parent DNA. Genetic material of a particular organism in such replication would result in two identical organisms.

Mutations

Suppose something happened to the gene so that one base was changed. Perhaps, for example, radiation or a chemical agent caused a deletion of a G—C base pair:

$$
\begin{array}{cc}
\text{A—T} & \text{A—T} \\
\text{T—A} & \text{T—A} \\
\text{G—C} \longrightarrow & |\quad| \\
\text{G—C} & \text{G—C} \\
\text{C—G} & \text{C—G} \\
\text{A—T} & \text{A—T} \\
\text{C—G} & \text{C—G}
\end{array}
$$

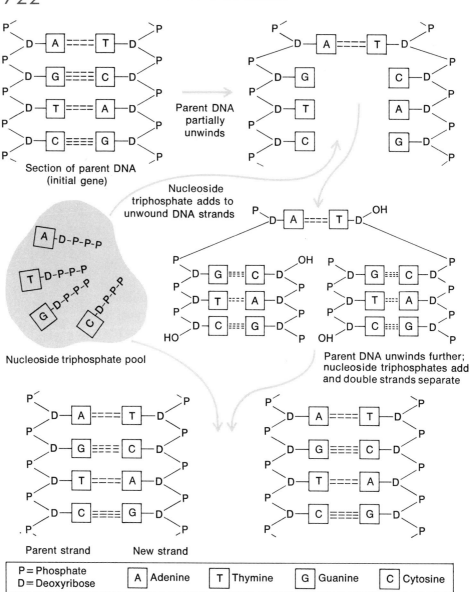

Figure 34.8

Replication of DNA.

When this altered gene is replicated, the follow-up new genes would be identical to the altered gene, except again lacking in one of the G—C factors. This is called a **mutation.** If the original gene was essential to the well-being of the organism, the result might be a malformed offspring. At least the offspring would be lacking one of the characteristics of the parent. If this gene were responsible for the synthesis of an essential part of the organism, death might result, and we would call this a lethal mutation. While it must be realized that this is an over-simplification, it points out the importance to man of our understanding of the chemical basis for heredity. If one uses his imagination, it is evident why chemists now seek to synthesize genes in a test tube. For example, if we could synthesize a beneficial gene that may be absent in a certain disease and could incorporate this into the genetic material of a cell, it might be possible to change a diseased state to a healthy one. In our earlier example with pneumococci, this is what happened in reverse, and the result is disease.

In addition to DNA, we stated that other nucleic acids exist, namely RNAs. Although RNAs are usually not associated with heredity, they play an important role in life. (An exception is the plant viruses which do not contain DNA but only RNA and are able to reproduce like kind.) This role will be discussed in the next section.

RNA occurs in at least three distinct forms, each with a different role. These are messenger RNA (**mRNA**), transfer RNA (**tRNA**), and ribosomal RNA (**rRNA**). Each "gene" in a living organism is able to synthesize a different mRNA. Thus, if we have 10,000 genes it would be possible to have 10,000 different mRNAs. In addition there are genes for rRNA (probably only a small number) and genes for tRNA (probably around 50). We shall see later why fewer genes are required for tRNA and rRNA than mRNA. In contrast to DNA, RNA does not appear to exist as two distinct strands.

Relations of DNA, RNA, and Proteins

Thus far we have discussed these important macromolecules as entities unto themselves. In reality, they are closely associated and together serve as a fundamental basis for life. Sometimes they are referred to as the "trinity" of life and in terms of reactions can be written as DNA → RNA → protein. DNA is the blueprint, RNA the carrier of information, and proteins the final product essential to carrying out living functions.

Earlier we discussed the fact that the unique characteristic of proteins is in the different sequences of amino acids that occur. Biologists in years past had shown that various characteristics were passed down by heredity, and they had come to believe that a separate gene existed for every trait. More recently biochemists have shown that genes also control the formation of specific proteins; eventually there emerged the idea of one gene being responsible for the making of one protein. We are then confronted with the problem of how DNA controls the synthesis of a specific protein molecule? What is the intermediate role of RNA?

Much experimentation has been done toward synthesizing artificial proteins in a test tube. Research with amino acids that have been prepared containing a carbon-14 isotope has been helpful toward our understanding of how proteins are made from nucleic acids. The reactions studied are as follows:

$$R—\overset{\overset{\displaystyle H}{|}}{\underset{\underset{\displaystyle NH_2}{|}}{^{14}C}}—COOH + Cells \rightarrow {}^{14}C\text{-amino acid in protein}$$

Such experiments led to the knowledge that DNA, so to speak, furnishes the blueprint for specific protein synthesis, but that each of three kinds of RNA has a role to contribute.

The DNA in every cell of an individual organism is presumably identical. It contains the information necessary to construct a whole organism. Although the DNA in such different cells as liver, skin, heart, and hair is chemically identical, only part of the DNA is used by different kinds of cells. That is, certain genes in liver may be "turned on" to guide the synthesis of specific liver proteins and not "turned on" in a heart cell even though the genes are present. This is thought to be accomplished in the following way. Basic proteins called **histones** are entwined about long stretches of DNA and prevent the DNA from making a particular messenger RNA. Some proteins may be considered "repressors." Certain proteins may unlock (remove) the repressors. The result is that a certain segment of the DNA where the histone is removed now can act as the "blueprint" for the synthesis of a particular messenger RNA. Recently discovered sigma, rho, and psi factors are suggested as playing a part in recognizing appropriate genes. Admittedly, much is yet to be learned in the complexity of cellular chemistry.

Amino Acids and Codons

DNA macromolecules, aside from phosphate, P, and deoxyribose, S, units, contain but four nitrogenous bases: adenine, A; guanine, G; cytosine, C; thymine, T. RNA along with P and ribose, R, also has adenine, guanine, and cytosine, but has uracil, U, instead of thymine. The question arises — how can DNA or RNA, containing but four diverse nitrogenous bases, act as a computer to code the 20 amino acids in the synthesis of a specific protein? The answer is relatively simple.

Sequences of three purine and pyrimidine bases on mRNA direct the amino acid sequence in protein synthesis. These triads code for particular amino acids and are called **codons.** The complementary triad — that is, the triad which contains bases which are complementary to the bases in the codon, is found as an integral part of tRNA and is called the **anticodon.** Since there are four nucleotide base residues (A, G, C, U) on the mRNA chain, there are 4^3 or 64 different codons available for coding. The scheme of codons which has been worked out for protein synthesis is shown in Table 34.4.

Table 34.4

Amino acids and respective codons of mRNA

Amino Acid	Codons	Amino Acid	Codons
alanine	GCA, GCC, GCG, GCU	leucine	CUA, CUC, CUG, CUU, UUA, UUG
arginine	CGA, CGC, CGG, CGU, AGA, AGG	lysine	AAA, AAG
asparagine	AAC, AAU	methionine	AUG
aspartic acid	GAC, GAU	phenylalanine	UUC, UUU
cystine	UGC, UGU	proline	CCA, CCC, CCG, CCU
glutamic acid	GAA, GAG	serine	UCA, UCC, UCG, UCU, AGC, AGU
glutamine	CAA, CAG	threonine	ACA, ACC, ACG, ACU
glycine	GGA, GGC, GGG, GGU	tryptophan	UGA, UGG
histidine	CAC, CAU	tyrosine	UAC, UAU
isoleucine	AUA, AUC, AUU	valine	GUA, GUC, GUG, GUU

Note: U, uracil in an RNA replaces T, thymine in a DNA. Thymine is a methyl derivative of uracil.

For example, if the amino acid sequence of a section of protein were ala-lys-phe, the messenger RNA would contain three codons. From Table 34.4 we could construct a messenger RNA for this protein and it may be G—C—A—A—A—A—U—U—U. The messenger RNA is made on a section of DNA as a complementary one-stranded RNA chain. The messenger RNA given in the example above would be made on a section of the DNA having a base sequence of —C—G—T—T—T—T—A—A—A.

Other codons have their own anticodons. The mRNA can now be the messenger to select the amino acid to be added in the synthesis of a protein.

Related to the genetic code another type of RNA is important, the transfer RNA (tRNA).

tRNA

Transfer RNA (tRNA), sometimes referred to as soluble RNA (sRNA) molecules are the smallest of the three RNA's, and have molecular weights from 25,000 to 40,000. Their function is to transport specific amino acids to an allocated place on a messenger mRNA, which in turn helps incorporate them into a growing peptide chain.

Each amino acid covalently bonds with a tRNA having a unique nucleotide (A, C, G, U) sequence, part of which is one of the anticodons mentioned previously. The reaction is catalyzed by an enzyme (discussion of enzymes, p. 727). If the transfer RNA in this reaction is for leucine the reaction proceeds with leucine but if another amino acid is used, say valine, no reaction occurs. This means that at least 20 different tRNA's occur in cells and each is specific for its own particular acid. The order of amino acids in proteins is not controlled by tRNA but by mRNA.

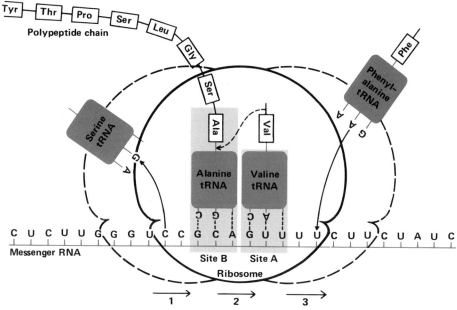

Figure 34.9

Synthesis of Protein Molecules is accomplished by intracellular particles called ribosomes. The coded instructions for making the protein molecule are carried to the ribosome by a form of ribonucleic acid (RNA) known as "messenger" RNA. The RNA code "letters" are four bases: uracil (U), cytosine (C), adenine (A), and guanine (G). A sequence of three bases, called a codon, is required to specify each of the 20 kinds of amino acid, identified here by their abbreviations. When linked end to end, these amino acids form the polypeptide chains of which proteins are composed. Each type of amino acid is transported to the ribosome by a particular form of "transfer" RNA (tRNA), which carries an anticodon that can form a temporary bond with one of the codons in messenger RNA, "reading off" the codons in sequence. It appears that the ribosome has two binding sites for molecules of tRNA: one site (A) for positioning a newly arrived tRNA molecule and another (B) for holding the growing polypeptide chain.

From "The Genetic Code: III" by F. H. C. Crick. Copyright © October 1966 by Scientific American, Inc. All rights reserved.

mRNA

The messenger RNA has a base-pair structure complementary to DNA (note again RNA has uracil, U, in place of thymine, T, of DNA). It is the more soluble tRNA that brings a specific amino acid to the proper section of the mRNA that requires it, and the mRNA is a template for the arrangement of the amino acid units in a protein chain. The mRNA is said to unite with cellular cytoplasm **ribosomes,** which are nucleoprotein particles composed of proteins and the 3rd form of RNA (ribosomal RNA, rRNA). mRNA may react with 4 to 6 ribosomes to form **polyribosomes** (polysomes).

The representative scheme for protein synthesis is shown in Figure 34.9.

Nucleic acids, the DNA and the RNA's, therefore truly constitute the thread of life. They are the chemical constituents in the body that are the determinents of the continuity of life. The expression of this thread is made by proteins. Thus, different cells in our body are unlike one another because of the protein makeup of each cell. The future holds for us the promise of learning just how the activity of a particular gene is turned on, resulting in the synthesis of a particular message whose final product is a specific protein. As an illustration of how important this is, suppose that you are diabetic. In some way your normal production of insulin is impaired. Now insulin is a protein hormone. If there were some way to turn on the gene responsible for insulin production the disease might be alleviated.

Conversely, one idea of cancer is that certain kinds are caused by viruses. Like other organisms, viruses contain nucleic acids for their reproduction. It is believed that some viruses are always present in an inactive form. Herpes simplex, the virus that causes a cold sore, is an example. When you are exposed to sunlight, a reaction occurs that activates this virus and multiplication results. A similar possibility may exist for viruses causing cancer.

34.6 ENZYMES

At the outset we can state that all enzymes are proteins, but all proteins are not enzymes.

The protein **casein** (not an enzyme) may be chemically hydrolyzed to its constituent amino acids (this occurs naturally in the digestion of milk products). In the laboratory we must add concentrated mineral acid to the casein and boil the mixture for 20 hr in order to accomplish the hydrolysis. If we use in place of mineral acids the lining of a stomach of a recently slaughtered animal, the casein will be hydrolyzed to its amino acids in a few hours at body temperature because of the enzyme reactions present. When the enzyme that hydrolyzes casein is purified, it will hydrolyze some 10,000 times its own weight of casein in a short time. This is why we call enzymes biological catalysts—a little does a great deal and goes a long way.

We may list three characteristics of an enzyme: (1) chemically they are proteins; (2) they act catalytically; (3) they are specific for particular compounds.

Vitamins

These important compounds are required in minute amounts in the diet. Chemically, they are divided into two groups depending on their solubility. The fat-soluble vitamins are those that are soluble in solvents usually used to extract fats from nature (ether, chloroform, methanol, etc.). They include vitamins A, D, E, and K.

For example, Vitamin A

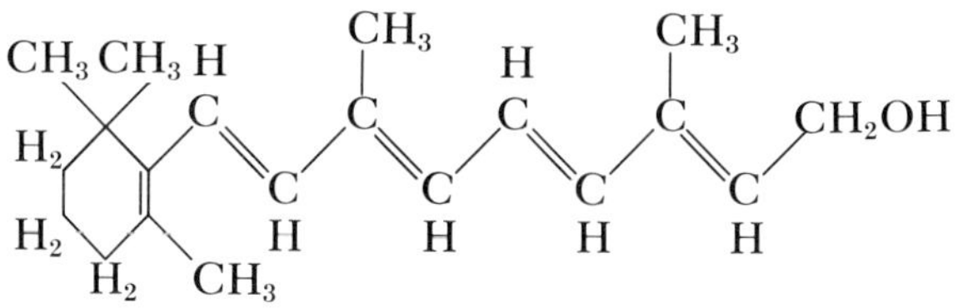

Vitamin A

is an essential part of the process of vision. Vitamin A is reversibly oxidized and reduced, being alternately bleached by light and regenerated in the dark. It is part of a visual protein called **rhodopsin.** Night blindness can result if a deficiency exists. The temporary blindness you experience at night if you stare at oncoming headlights is the result of "bleaching" of vitamin A.

The water-soluble vitamins may be extracted by water from natural materials and include the B complex plus vitamins C and choline.

Vitamin C is **ascorbic acid** ($C_6H_8O_6$). This vitamin, which helps prevent scurvy, is present in citrus fruits and in cabbage and tomatoes. Note that ascorbic acid is merely an oxidation product of a hexose.

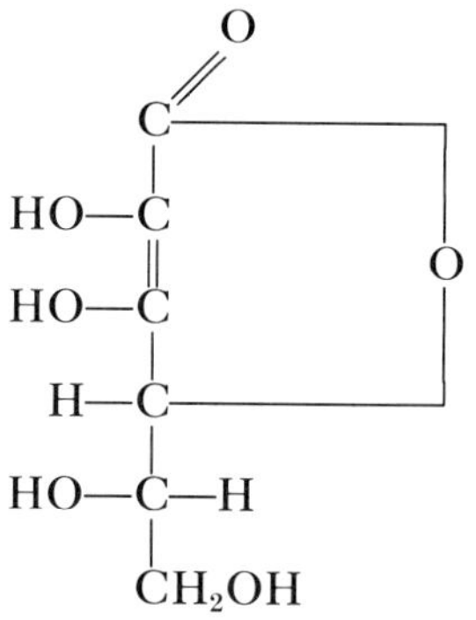

L-Ascorbic acid (water soluble)

Historically, vitamins were recognized by the fact that a lack of a certain one often resulted in a recognizable disease. Sailors of the time of Columbus or before used to carry lemons to prevent scurvy. Today it is known that the active principle was vitamin C, which occurs in lemons.

34.7   NUTRITION

Over the years biochemists have been concerned with not only the manufacture of food but also the requirements of certain nutrients by living organisms. Foods consist largely of natural products that may be treated in
various ways for preservation or improvement of their nutritional value.
Sometimes the addition of foreign chemicals to a food supply may have
undesirable effects. The establishment of the presence of such foreign compounds is the realm of the toxicologist and food scientist. The furor over the
use of chemical sprays, for example, results from a fear that they may remain
as part of the food we inject. Carefully planned and executed experiments are
essential before rational recommendations can be made.

Food produces new body tissue for growth, replaces worn-out tissues,
helps maintain body temperature, and supplies energy for work and physical
activity. The main groups of food are fats, carbohydrates, proteins, water,
oxygen, minerals, and vitamins. It is how these materials are used that has
been the subject of this introduction to biochemistry.

While water and oxygen are not frequently classified as foods, both are
necessary for the maintenance of life. Minerals are necessary for building the
bones and solid structure of the body. Calcium and phosphorus are necessary for bone structure, which is largely calcium phosphate. Iron is a constituent of hemoglobin, the oxygen-carrying protein of the red blood cells. A
small amount of iodine is necessary for proper functioning of the thyroid
gland. Although mineral matter in general is needed in only relatively small
amounts, these small amounts play an indispensable role in the vital processes. Trace amounts of the elements Co, Cr, Cu, Fe, F, I, Mn, Mo, Se, Sr,
V, Zn, Mg, Si, and Sn are known to be essential for animals. Plants also require
trace elements for their growth.

Exercises

34.1. Define or explain the following terms: (a) biochemistry, (b) lipids, (c) fats,
(d) polyunsaturates, (e) saponification, (f) glycerol esters.

34.2. Define or explain what is meant by the terms: (a) carbohydrate, (b) monosaccharide, (c) α-D-glucose, (d) hexose and pentose, (e) starch and cellulose,
(f) polysaccharide.

34.3. Explain the terms: (a) protein, (b) amino acid, (c) peptide bond, (d) α-helix,
(e) DNA and RNA, (f) nitrogenous base, (g) nucleoside and nucleotide, (h) base
pair, (i) codons and anticodons, (j) nucleic acid, (k) tRNA and mRNA.

34.4. Explain briefly the meaning of (a) enzymes and (b) vitamins.

34.5. Concerning the following carbohydrates:

(a) Which of the compounds above would be classified as a
    (1) Pentose?
    (2) Disaccharide?
    (3) Polysaccharide?
    (4) Hexose?
(b) Of the compounds whose structures are given above, which would be
    (1) A component of DNA?
    (2) A component of RNA?
    (3) The monomer from which the polysaccharide starch is formed?
(c) What are the two functional groups, such as those characteristic of alde-
    hydes, esters, etc., which these structures show? What other functional
    group does glucose form by rearrangement?

34.6. Structural formulas for a few compounds of biological significance are given
    below.

3. 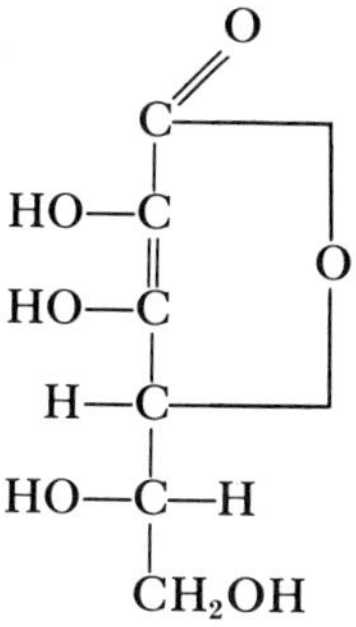
$$
\begin{array}{l}
\text{O} \\
\parallel \\
\text{C} \\
\text{HO—C} \\
\text{HO—C} \quad \text{O} \\
\text{H—C} \\
\text{HO—C—H} \\
\text{CH}_2\text{OH}
\end{array}
$$

4. 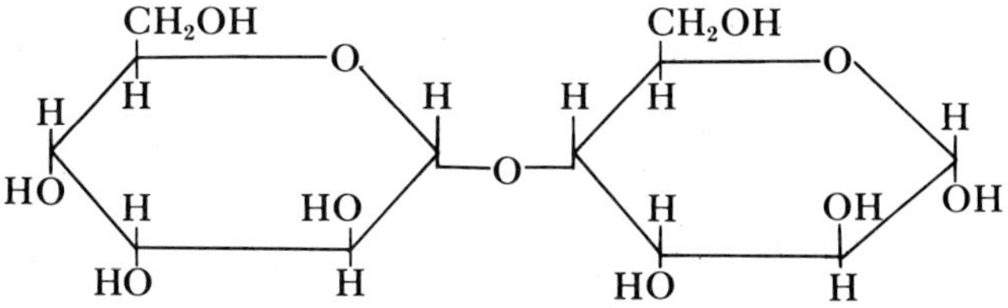 (pyranose ring structure with CH$_2$OH, H, HO, OH groups)

5. 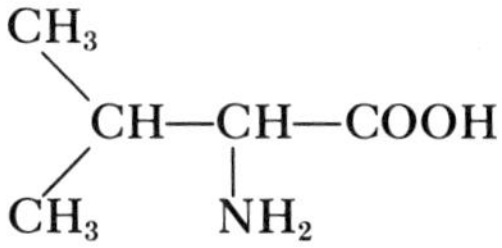
$$
\begin{array}{l}
\text{CH}_3 \\
\text{CH—CH—COOH} \\
\text{CH}_3 \quad \text{NH}_2
\end{array}
$$

6. 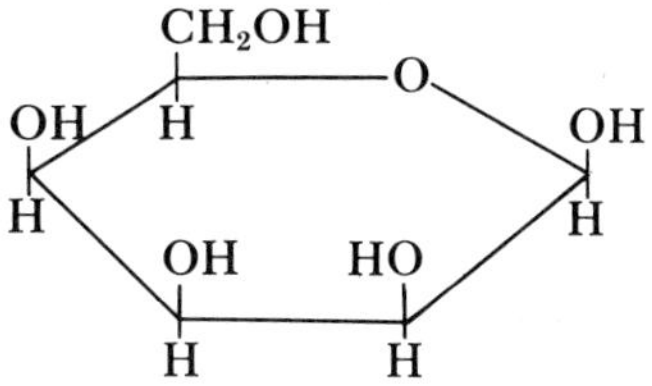 (sugar ring structure with CH$_2$OH, OH, H, HO groups)

7. 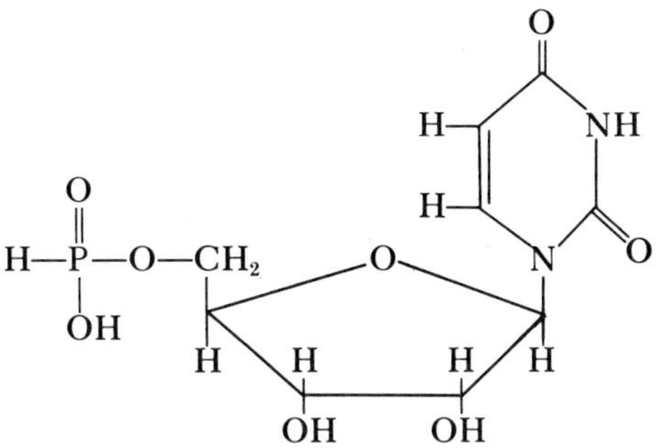
$$
\begin{array}{l}
\text{O} \\
\parallel \\
\text{H—P—O—CH}_2 \\
\text{OH}
\end{array}
$$
(nucleotide structure with ribose ring and uracil base: H, NH, O)

8. CH$_3$(CH$_2$)$_{14}$COOH

9. 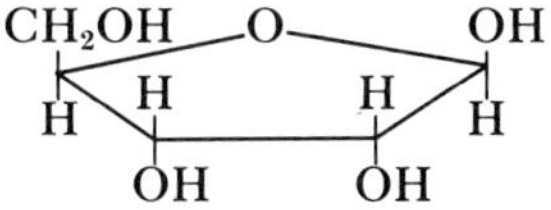 (furanose ring structure with CH$_2$OH, O, OH, H groups)

10. 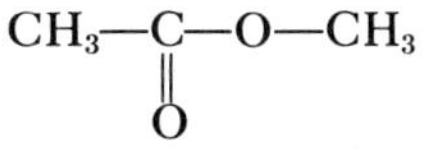
$$
\text{CH}_3\text{—C—O—CH}_3 \\
\quad \parallel \\
\quad \text{O}
$$

11. 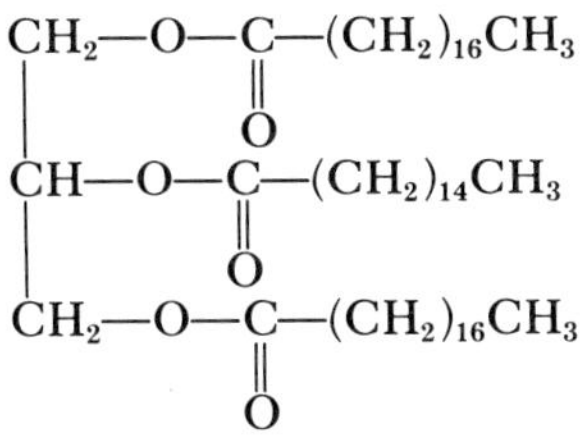
$$
\begin{array}{l}
\text{CH}_2\text{—O—C—(CH}_2\text{)}_{16}\text{CH}_3 \\
\quad\quad\quad \parallel \\
\quad\quad\quad \text{O} \\
\text{CH—O—C—(CH}_2\text{)}_{14}\text{CH}_3 \\
\quad\quad\quad \parallel \\
\quad\quad\quad \text{O} \\
\text{CH}_2\text{—O—C—(CH}_2\text{)}_{16}\text{CH}_3 \\
\quad\quad\quad \parallel \\
\quad\quad\quad \text{O}
\end{array}
$$

12. 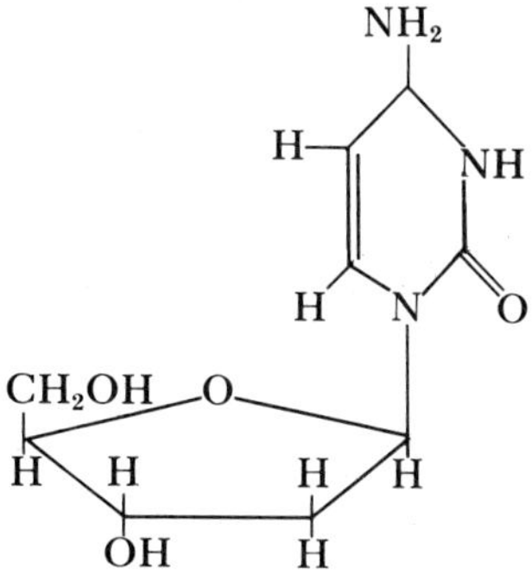 (nucleoside structure with ribose ring and cytosine base: NH$_2$, H, NH, O)

From these compounds select:
(a) an amino acid,
(b) a fat or triglyceride,
(c) a nucleotide,
(d) a protein,
(e) a nitrogen base,
(f) a nucleoside,

(g) a disaccharide,
(h) a vitamin,

(i) ribose,
(j) a fatty acid.

34.7. Match the terms with the list of statements.
   (a) A biological catalyst.
   (b) A 3-base sequence on mRNA which specifies the position of an amino acid.
   (c) RNA polymer formed from DNA which acts as the plan for protein synthesis.
   (d) Monosaccharide with 6 carbons.
   (e) RNA molecule containing both a specific binding site for an amino acid and an anticodon.
   (f) Very specific matching of adenine to thymine and cytosine to guanine.

   1. Cellulose
   2. tRNA
   3. Anticodon
   4. Hexose
   5. Enzyme
   6. β-oxidation
   7. mRNA
   8. Base-pairing (hydrogen bonding)
   9. ATP
   10. Codon

34.8. The extent to which an organism is oxidizing fat or carbohydrate is indicated by the production of CO_2 relative to the amount of O_2 used. Balanced equations are given for (1) the oxidation of glucose, (2) the oxidation of a typical fatty acid.
   (1) $C_6H_{12}O_6 + 6\ O_2 \rightarrow 6\ CO_2 + 6\ H_2O$
   (2) $CH_3(CH_2)_{14}COOH + 23\ O_2 \rightarrow 16\ CO_2 + 16\ H_2O$
   Calculate the liters of CO_2 produced per liter of O_2 used at STP for: (a) the oxidation of glucose, (b) the oxidation of a fatty acid.

34.9. Write equations for the following reactions:
   (a) Saponification of a triglyceride
   $(C_{17}H_{33}COO)_3C_3H_5 + NaOH \rightarrow$

   (b) Hydrogenation of an oil
   $(C_{17}H_{31}COO)_3C_3H_5 + H_2 \xrightarrow[\text{pressure}]{\text{heat}}$

   (c) Fermentation of a sugar
   $C_6H_{12}O_6 \xrightarrow{\text{fermentation}}$

   (d) Hydrolysis of starch
   $(C_6H_{10}O_5)_x + H_2 \xrightarrow{\text{HCl}}$

   (e) Amphoteric nature of amino acids (example, alanine)
   (f) The condensation (splitting out of a molecule of water) from a reaction of alanine with cysteine.

Supplementary Readings

34.1. "The Molecular Biology of Poliovirus," D. H. Spector and David Baltimore, *Sci. American*, May, 1975; p. 24.
34.2. "The Structure and History of an Ancient Protein," R. E. Dickerson, *Sci. American*, April, 1972; p. 58.
34.3. "The Synthesis of DNA," A. Kornberg, *Sci. American*, Oct., 1968; p. 64.
34.4. "Molecular Isomers in Vision," R. Hubbard and A. Kropf, *Sci. American*, June, 1967; p. 64.

Chapter 35

Radioactive and Stable Atomic Nuclei

Introduction

Our study of the atom so far has placed most of the emphasis on the electrons in the atom, particularly the valence electrons, since it is these that are primarily involved in chemical reactions and bonding of atoms in the molecule. It is the nuclei of atoms, however, that are involved in such important topics as radioactivity, nuclear energy, transmutation of one element to another, tracer atoms, radiocarbon dating, and energy of the sun.

35.1   RADIOACTIVITY

In 1896 H. Becquerel wrapped a photographic plate in black paper to make it light proof. He happened to place some crystals of potassium uranyl sulfate above the plate protected by the paper as he stored the plate in a cabinet drawer. Upon developing the plate later, he observed that the plate appeared to have been light struck (dark spots showed up on his negative). In various checking experiments he placed thin sheets of aluminum and copper between the plate and specimens of various uranium compounds and ores. The same results recurred—the photographic plate appeared to have been exposed to light. Furthermore, uranium salts were found to render air near them a conductor of electricity, as evidenced in experiments in the discharge of an electroscope. A type of radiation from the uranium compounds similar in action to X rays (discovered a few years earlier) seemed the answer to the evidence. This new phenomenon—the spontaneous emission of radiation by an element —was termed **radioactivity,** and those elements responsible for the radiation were termed **radioactive** elements.

Radioactive Elements

Mme. Marie Curie and her husband, Pierre, began the investigation of the new phenomenon and found that **pitchblende,** an ore of uranium consisting largely of U_3O_8, exhibited an activity greater than could be accounted for on the basis of the uranium present. They began a long series of tedious and painstaking experiments in the separation and analysis of this uranium mineral. After several months of work, Madame Curie announced the separation of an element more active than uranium, which she named **polonium,** after her native country, Poland. The work continued, and in 1898 the Curies announced the discovery of **radium,** which showed a radioactivity a million times as great as that of uranium.

Radium is similar to barium and, like $BaSO_4$ the compound $RaSO_4$ is very insoluble, even in acid solution. This fact is useful in concentrating radium from other metals with which it occurs. (Only about 0.2 g of radium is contained in a ton of high-grade uranium ore.) The element itself was isolated by Madame Curie in 1910 by the electrolysis of melted radium chloride.

During the studies with uranium minerals, Madame Curie found that all uranium minerals are radioactive, and that the activity of a sample depends on the amount of uranium present. As a result of her studies, Madame Curie concluded that radioactivity is a specific property of the atom and is not affected by the nature of the chemical combination in which the atom exists; nor is it affected by physical conditions.

Characteristics of Radioactivity

As a result of radioactive disintegrations, the following characteristics should be noted.

1. The rays emitted pass through solid objects.
2. The rays expose or blacken a photographic film.
3. The rays ionize a gas.
4. The rays cause certain compounds, such as ZnS, to glow; for example, luminous watch dials.
5. The rays leave tracks in a cloud chamber.
6. Rates of disintegration are not altered by changes in temperature, pressure, or type of chemical combination.
7. The change is a chemical change; for example,

$$Ra \rightarrow Rn + He$$

8. Enormous amounts of energy (MeV) are involved in comparison with ordinary chemical reactions (eV).

Radioactive Rays

Becquerel continued his research on the radiation properties he had discovered, and in 1899 observed that these radiations were deflected in a magnetic field, in the same manner as cathode rays. Shortly afterwards, by means of an electric field, Rutherford was able to identify three types of radiation from radioactive substances.

A small amount of a radioactive mixture was placed in a depression in a block of lead (Figure 35.1). All rays except those moving directly right were absorbed by the lead. Electrodes and a photographic plate to record the emitted radiation were placed to the right of the block of lead. This experiment shows that the radiation may be resolved into three kinds, one kind unaffected by the electrodes, the other two kinds showing opposite charges, as evidenced by the bending of the rays in opposite directions. From the extent and the direction of bending, it is possible to calculate the

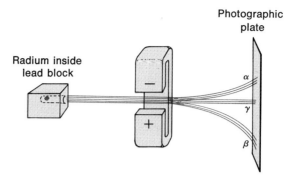

Figure 35.1
Three kinds of rays are produced by radioactive disintegrations.

masses and charges of the radiations. The three types of radiation were named alpha (α), beta (β), and gamma (γ) rays (Table 35.1).

Table 35.1
Types of emission from natural radioactive materials

Name	Symbol	Identity	Charge	Mass, amu	Velocity Miles/sec	Relative Penetrating Ability
Alpha	α, $^4_2$He	helium nuclei	$+2$	4	10,000	1
Beta	β, $^0_{-1}$e	electrons	-1	0.00055	up to 170,000	100
Gamma	γ	short wave-length x-rays	0	0	186,000	10,000

Alpha rays, positively charged particles with a mass four times that of the hydrogen atom, have been shown to be nuclei of helium atoms, each particle consisting of 2 neutrons and 2 protons. Thus their symbol is $^4_2$He, or α. These particles are expelled from radioactive substances at moderate velocities (1.5–2×10^9 cm/sec), and have a low penetrating power, since they may be stopped by aluminum foil $\frac{1}{10}$ mm in thickness. Air through which alpha particles pass becomes ionized because of electrons knocked out of molecules in the air on impact with these relatively heavy particles. If this ionized air is saturated with water vapor and suddenly cooled, the ionized particles in the air act as nuclei for the condensation of small droplets of water, and these particles are rendered visible as a fog. C. T. R. Wilson was able to follow the path of an alpha particle by photographing the fog tracks thus produced.

The expulsion of alpha particles from radioactive material may also be observed with the aid of a **spinthariscope,** a device consisting of a metal tube with a phosphorescent screen at one end and a high-grade lens at the other. Projecting from the center of the phosphorescent screen is a pin on the side of which is placed a small quantity of the radioactive salt being tested. This instrument shows the scintillations produced on the phosphorescent screen by the constantly emanating alpha particles.

Beta rays are negatively charged particles identical with electrons. Their symbol is $_{-1}^{0}e$ or β, signifying nearly zero mass number and a minus one charge. They are sometimes referred to as "high speed electrons," since their speed sometimes approaches that of light (3×10^{10} cm/sec or 186,000 mi/sec). These rays are much more penetrating than alpha particles—for example, they pass through a lead sheet of 1 mm thickness, or aluminum of 5 mm thickness. The stopping power of various metals is dependent on their density and on the speed of the radiation particles.

Gamma rays are of high frequency and are identical with X rays of very short wavelength. They possess relatively high penetrating power (up to 25 cm of lead) and, like alpha and beta rays, affect a photographic plate. They carry no charge. Gamma rays constitute the portion of radioactive emission which affected the photographic plate of Becquerel even though the plate had been wrapped in several layers of black paper. The symbol for gamma ray is γ.

Gas Filled Detectors

The simplest detector sensitive to radiation is an **electroscope.** When an electroscope is charged, metal leaves separate by the mutual repulsion of like charges. (See Figure 35.2.) The amount of displacement is a measure of the charge. When radiation passes through the air in a charged electroscope, gaseous ions are formed and collect on and neutralize the charged metal-leaf conductors, which then fall back together.

More sensitive detectors, such as the **Geiger counter,** operate on the fol-

Figure 35.2
The electroscope.

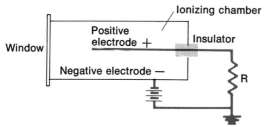

Figure 35.3
Components of a gas-filled detector.

lowing principle: (a) nuclear radiation ionizes some gas molecules in an ionization chamber; (b) an electric field pulls the charged particles to the electrodes, producing an electric current in the circuit; (c) the resulting current or current pulsations, through a resistor (R), work to produce light flashes or clicks for observation and counting. (See Figure 35.3.)

Disintegration of Atoms

In 1902 Rutherford suggested that elements may spontaneously change into other elements or break down into simpler elements. The idea was revolutionary, since elements were supposed to be permanent. Rutherford's theory has been completely confirmed by experimental evidence, and today there is no doubt that elements may be broken down into simpler elements. When radium disintegrates, an alpha particle is expelled from the nucleus of the radium atom, leaving an atom of mass number that is four mass units less than radium. The change in the nucleus is represented (p standing for proton and n for neutron) thus:

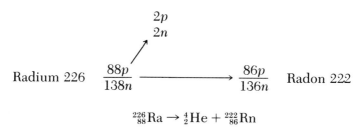

$$^{226}_{88}\text{Ra} \rightarrow {}^{4}_{2}\text{He} + {}^{222}_{86}\text{Rn}$$

Rutherford was able to collect an inert gas from the disintegration of radium. He named it **niton**; it is now called **radon**. Ramsey studied this "radium emanation" and found its atomic weight to be 222, which is what we should expect from theoretical considerations. The two outer electrons of radium drift off and may be said to be "grounded" or flow away as a micro current.

During the disintegration of radioactive substances, not all three types of rays are emitted at the same time. A pure sample of uranium emits only α rays at first, but after a time β rays are also emitted. That both α and β rays are obtained after a time is explained by the formation, during the disintegration process, of other radioactive substances which emit particles. Uranium itself emits only α particles, but **thorium**, which is formed in the first step, gives off β particles. Gamma rays are of very short wavelength, and as emitted by a specific nucleus have definite energy values which relate to energy changes within a nucleus. The emission of gamma rays alone does not cause a change in mass or atomic number of a nuclear species; gamma emission is from excited-state nuclei to nearer ground-state nuclei. An entire series of disintegration products of $^{238}_{92}\text{U}$ has been discovered; the successive changes are represented in Table 35.2.

Table 35.2
Disintegration products of uranium

Element	Isotope°	Half-Life	Element	Isotope°	Half-Life
Uranium ↓ ↘ α	$^{238}_{92}\text{U}$	4.5×10^9 years	Lead ↓ ↘ β	$^{214}_{82}\text{Pb}$	26.8 minutes
Thorium ↓ ↘ β	$^{234}_{90}\text{Th}$	24.1 days	Bismuth ↓ ↘ β	$^{214}_{83}\text{Bi}$	19.7 minutes
Protactinium ↓ ↘ β	$^{234}_{91}\text{Pa}$	1.18 minutes	Polonium ↓ ↘ α	$^{214}_{84}\text{Po}$	1.6×10^{-4} seconds
Uranium ↓ ↘ α	$^{234}_{92}\text{U}$	2.5×10^5 years	Lead ↓ ↘ β	$^{210}_{82}\text{Pb}$	19.4 years
Thorium ↓ ↘ α	$^{230}_{90}\text{Th}$	8.0×10^4 years	Bismuth ↓ ↘ β	$^{210}_{83}\text{Bi}$	5 days
Radium ↓ ↘ α	$^{226}_{88}\text{Ra}$	1622 years	Polonium ↓ ↘ α	$^{210}_{84}\text{Po}$	138 days
Radon ↓ ↘ α	$^{222}_{86}\text{Rn}$	3.8 days	Lead	$^{206}_{82}\text{Pb}$	stable
Polonium ↓ ↘ α	$^{218}_{84}\text{Po}$	2.05 minutes			

° Subscripts represent atomic numbers; superscripts represent mass numbers.

The expulsion of an alpha particle results in a decrease of 2 in atomic number (loss of 2 protons) and a decrease of 4 in mass number (2 protons and 2 neutrons). The emission of a β particle° results in no change in mass

° We should differentiate very carefully between the loss of electrons from the nucleus of an atom (radioactivity) and the loss of electrons from the valence levels of an atom (ionization). As an electron is emitted from a nucleus, some change must take place, such as conversion of a neutron into a proton and an electron. For example, $^{1}_{0}n \rightarrow {}^{1}_{1}\text{H} + {}^{0}_{-1}e$.

number (electrons have almost negligible weight), but an increase of 1 in atomic number (number of protons). A part of the uranium disintegration series is shown below.

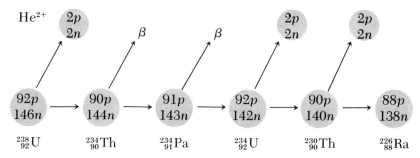

Table 35.2 shows $^{206}_{82}$Pb as being the nonradioactive end element of the series. We may note from the table that $^{234}_{92}$U and $^{230}_{90}$Th have somewhat longer half-life periods (see next section) than $^{226}_{88}$Ra; consequently, these three isotopes along with $^{206}_{82}$Pb are the principal elements found in nature associated with $^{238}_{92}$U.

The energy of the alpha particles, as well as of the beta particles or gamma rays, is very definite for each given atomic species in its decay.

35.2   HALF-LIFE PERIODS

The time required for the decomposition of one-half of any given amount of radioactive element is termed its **half-life.** The half-life of radium is 1622 yr — during this time one-half of any given sample of radium will decompose; during the next 1622 yr half of that remaining will decompose, and so on. Half-life periods vary from 0.00016 sec for $^{214}_{84}$Po to 4,500,000,000 yr for $^{238}_{92}$U.

Table 35.3 shows the concept of half-life for the radioactive element $^{210}_{83}$Bi, which has a half-life of 5 days. Figure 35.4 shows a plot of percent activity versus half-lives for $^{210}_{83}$Bi.

Table 35.3

Rate of disintegration of (30.0) thirty grams of $^{210}_{83}$Bi.

Time (Days)	Number of Half-Lives	% Activity	Grams of Bi Remaining
0	0	100	30.0
5	1	50.0	15.0
10	2	25.0	7.50
15	3	12.5	3.75
20	4	6.25	1.88
25	5	3.13	0.938
30	6	1.56	0.469

Precise calculations of radioactive decay amounts may be made from the first-order rate equation (Chapter 14):

$$2.303 \log \frac{A_0}{A} = kt \qquad (1)$$

where A_0 is the original activity or amount of radioactive material $(t = 0)$, A is the activity or amount left at time t, and k is the rate constant. The value of k may be determined by substituting the value for the half–life, $t = t_{1/2}$ and $\frac{A_0}{A} = 2.00$ in the above equation, which then becomes

$$k = \frac{0.693}{t_{1/2}} \qquad (2)$$

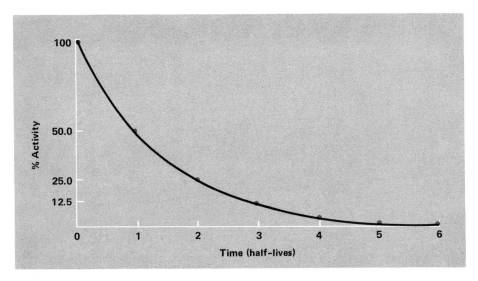

Figure 35.4
Curve of % activity versus half-lives for $^{210}_{83}$Bi.

EXAMPLE 35.1   (a) Approximately how long would it take a sample of 30.0 g of $^{210}_{83}$Bi to be reduced to an activity of 10.0%? What weight of Bi would re-remain?

SOLUTION:
(a) An approximate solution would be to construct a table such as Table 35.4. When the activity is 10% of the original activity, the time would be roughly 17 days and the weight of Bi remaining would be around 3 g. Similar results could be obtained from the curve in Figure 35.4.
(b) A precise answer is readily obtained from the rate equation. First the value of the rate constant, k, is calculated from equation (2), the half-life of $^{210}_{83}$Bi being 5 days.

$$k = \frac{0.693}{5 \text{ days}} = 0.139/\text{day}$$

Using this value of k and the values of $A_0 = 1$ (or $\%A = 100$) and $A = 0.100$ ($\%A = 10$), solve equation (1) for t.

$$2.303 \log \frac{100}{10.0} = (0.139)(t)$$

$$t = \frac{2.303 \log 10}{0.139} = 16.6 \text{ days}$$

The rate equation also is used to find the weight of Bi remaining after 16.6 days.

$$2.303 \log \frac{A_0}{A} = (0.139)(16.6 \text{ days})$$

$$\log \frac{A_0}{A} = \frac{(0.139)(16.6)}{2.303} = 1.00$$

and

$$\frac{A_0}{A} = 10.0 \text{ or } A = \frac{A_0}{10.0}$$

Since A_0 corresponds to 30.0 g. of the original $^{210}_{83}\text{Bi}$, the weight of Bi remaining (A) after 16.6 days is $\frac{30.0}{10.0} = 3.00$ g.

Isotopes of Lead

Radioactivity of ^{238}U ceases with the formation of $^{206}_{82}\text{Pb}$. In the series of changes between uranium and lead, a total of 8 α particles are lost per atom — a decrease of 32 (8×4) units of mass. Since the mass number of uranium is 238, we should expect a mass number of 206 for lead ($238 - 32$). T. W. Richard's careful determinations of the atomic weights of lead from radioactive sources yields a value of about 206. The usual atomic weight of lead is given as 207.2 – the value for lead not associated with radioactive changes. This difference between the atomic weight of lead from radioactive sources and that of usual lead is factual evidence of the radioactive disintegration series which yields ^{206}Pb rather than the usual isotopic mixture of atomic weight, 207.2.

The Age of the Earth

When ^{206}Pb is found associated with ^{238}U in nature, it is quite certain that this lead isotope has been formed as a result of the radioactive decay of uranium.

By analysis of the amount of lead in ratio to the uranium, and with application of half-life information in the radioactive series ^{238}U to ^{206}Pb, it is possible to calculate the time required to reach the uranium-to-lead ratio found in the mineral deposit. It is assumed that the original pure uranium was formed as the earth formed. Accordingly, calculation of the age of the mineral deposit gives a minimum age for the earth. This "radioactive clock" shows that some of these minerals were formed 5×10^9 years ago.

Furthermore, calculations pertaining to the low abundance of naturally radioactive potassium isotope, ^{40}K, with a mean half-life of 1.3×10^9 years, indicates this nucleus species was formed at least a few billion years ago, giving it time to radioactively decay to a large extent. Similar studies of the half-lives and scarce occurrence of ^{235}U, ^{232}Th, and ^{87}Rb, reveal similar times for formation of the earth's elements.

Radiocarbon Dating

The mysterious cosmic rays which come to this earth cause some nuclear changes within the realm of their penetration. In the upper atmosphere some neutrons resulting from disintegration of nuclei by cosmic rays produce radioactive carbon.

$$^1_0n + ^{14}_7N \rightarrow ^{14}_6C + ^1_1H$$

The ^{14}C atoms become well mixed with normal ^{12}C atoms in the CO_2 content of air. A quite constant ratio of the two isotopes of carbon exists in the CO_2 of the air, and this same ratio continues into new plant and animal life. When plant and animal life dies and are covered with soil, they leave the zone of the atmospheric CO_2 – photosynthesis process, in which the definite $^{14}_6C$ to $^{12}_6C$ ratio is known. The nonliving carbonaceous material is subject to slow decrease in the amount of $^{14}_6C$ present because these atoms, in accord with their half-life, radioactivity emit beta particles. The technique of radiocarbon dating, developed by W. F. Libby at the University of Chicago, can now be used by the geologist, archeologist, and historian to trace events which took place thousands of years ago.

$$^{14}_6C \rightarrow ^{14}_7N + _{-1}^0e$$

By measuring the radioactivity which remains in a given carbonaceous specimen, the age since its death can be determined.

The half-life of $^{14}_6C$ is 5770 years. Electronic-counting equipment now available, if carefully used, is sensitive enough to measure back through several half-life times. Fossil materials may be dated to about 30,000 yr ago.

It is interesting to note that counting the tree rings in a California giant sequoia gave it an age of 2928 ± 10 yr. Radiocarbon dating produced an age of 2900 yr.

EXAMPLE 35.2   Wood from trees buried in the volcanic eruption of Mt. Mazama when Crater Lake in the Oregon Cascades was born, shows a $^{14}_{6}C$ activity of only 45.0% of the present activity of wood. When did the volcanic eruption occur?

SOLUTION:

$$1) \ k = \frac{0.693}{5770 \ \text{yr}} = 1.20 \times 10^{-4}/\text{yr}$$

$$2) \ 2.303 \log \frac{100}{45.0} = (1.20 \times 10^{-4})(t)$$

$$\log \frac{100}{45.0} = \log 2.22 = 0.347$$

$$t = \frac{(2.303)(0.347)}{1.20 \times 10^{-4}/\text{yr}} = 6660 \ \text{yr ago}$$

Properties of Radium

Radium appears in Group IIA of the periodic table and, except for its radio-active properties, is like the other elements in this group – calcium, strontium, and barium. Although not as active as members of the alkali group, it is the most chemically active member of Group IIA and will decompose water with the formation of hydrogen. Radium has distinct metallic properties and forms the strong base radium hydroxide [$Ra(OH)_2$]. Its compounds are very similar to those of the other alkaline earth elements.

Radium atoms are spontaneously and continuously disintegrating at a constant rate, whether the element is free or combined with other elements. Furthermore, there seems to be nothing the chemist can do to accelerate or retard this rate.

Rays from radium have a powerful effect on living tissues and produce severe and painful sores if allowed to come in contact with the skin for any length of time. Fortunately, the effect is greater on diseased than on healthy tissue; if it is very carefully controlled, radium may be used in the treatment of cancerous tissue and other malignant growths. Its principal application is in the treatment of surface cancer. Since the rays of radium are very pene-trating, it is stored in heavy lead containers to prevent injury to persons working around it.

One of the principal sources of radium, the mineral pitchblende (U_3O_8), is found in the Belgian Congo and northern Canada. As we mentioned, the purest deposits of pitchblende yield about 0.2 g of radium per ton of ore.

Deposits of **carnotite,** a low-grade uranium ore, are found in Utah and Colorado.

35.3   UNITS OF RADIOACTIVITY

The uses of radioactive substances are based principally on the effect of their alpha-, beta-, and gamma-ray emissions. The intensity or quantity of the emission is measured in **curies.** A curie is **the quantity of a substance which gives 3.70×10^{10} disintegrations per second.**[*] Since the curie represents a very large amount of radioactivity, the **millicurie** (0.001 curie) and the **microcurie** (0.000001 curie) are frequently used.

Radiation dosage is measured in three common units, roentgen (r), rad, or rem. A **roentgen** is the amount of X-ray or gamma-ray radiation that creates 2.08×10^8 ion pairs in 1 cc of *air.* The **rad** is a standard amount of radiation absorbed per gram of *tissue.* It is approximately equal to a roentgen, but may differ somewhat depending on the kind of tissue and type of radiation. For example, one roentgen of gamma-ray radiation absorbed in muscle tissue will produce a dosage of 0.97 rad.

A **rem** is the product of the absorbed dosage in rads and the relative biological effectiveness (R.B.E.)

$$\text{rem} = (\text{rad})(\text{R.B.E.})$$

The R.B.E. of X-rays, γ-rays, and β-rays $= 1$; that of α-rays $= 10$; and that of neutrons ranges from 2.5 to 10.5. Table 35.4 shows typical radiation dosages received by persons in the United States.

Table 35.4
Sources of radiation exposure

	mrem[°] per Year per Person
Natural radiation, external	100
Natural radiation, internal (K-40, etc.)	25
Medical X-Rays	~72
Chest X-Ray	~45
Nuclear power plant, at the fence	1
Nuclear power plant, 30 miles away	0.02
Nuclear power plant, state average	0.005
Jet flight, per thousand miles	1
	50

[°] 1 mrem $\equiv$ 1 milliroentgen equivalent man $\approx \dfrac{0.10 \text{ erg}}{\text{g or cc tissue}}$

[*] Originally it was thought that this number of disintegrations per second was given by one gram of radium; actually 1.028 g of Ra is required.

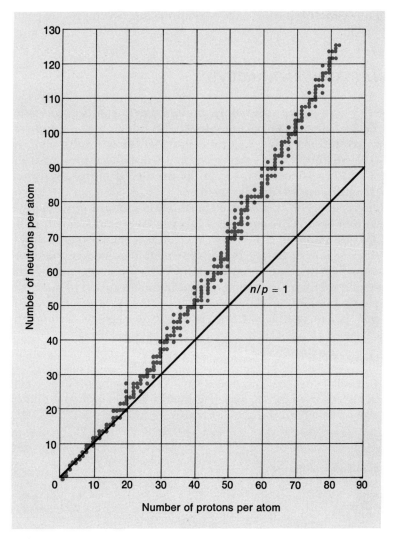

Figure 35.5
Neutron and proton content of stable nuclei. Neutron-proton
ratio of the lighter elements approaches one. Hydrogen is
unique with no neutrons; deuterium, however, has one neutron
to its proton. Heavier elements have an excess of neutrons over
protons. Elements of atomic number greater than 83 are
radioactive.

35.4 RADIOACTIVE AND STABLE ATOMIC NUCLEI

Stability

The nuclear structure of atoms was discussed in Chapter 3, in which it was pointed out that the nuclei of all atoms contain protons and neutrons. **The nuclei of the lighter stable atoms contain about an equal number of neutrons and protons;** in fact, many do contain an equal number of neutrons and protons — for example, $_2^4$He, $_6^{12}$C, $_7^{14}$N, $_8^{16}$O, $_{16}^{32}$S, and $_{20}^{40}$Ca. Atoms heavier than calcium lose the tendency to have an equal neutron-proton (n/p) ratio; **there is a tendency for neutrons to exceed protons as atomic weight increases** — for example, $_{24}^{52}$Cr, $_{42}^{96}$Mo, $_{92}^{238}$U. For each of the elements having atomic numbers 1 through 83, however, there is apparently a **narrow range or zone of stability insofar as the n/p ratio is concerned.** This range of stability for a given element is an index of the number of stable isotopes for that element (see Figure 35.5). As long as the n/p ratio is within this range for a particular element, the isotopes will be stable; but if the ratio falls outside this range, an unstable isotope subject to radioactive disintegration, results. To illustrate: three stable isotopes of oxygen are known ($_8^{16}$O, $_8^{17}$O, and $_8^{18}$O), a fact which indicates that the range of stability in the n/p ratio is from 8:8 to 10:8. Atoms of oxygen such as $_8^{13}$O or $_8^{20}$O must be too unstable to exist. The range of stability for most of the heavier elements appears to be wider than for the lighter elements; hence more stable isotopes are possible (see Table 3.2, page 64).

Both naturally occurring and man-made elements having numbers greater than 83 are characterized by radioactivity; the nuclei of these heaviest atoms are subject to alpha- or beta-particle emission. The newly formed nucleus may be stable or radioactive. A radioactive series ends with an atom which has a stable n/p ratio.

Exercises

35.1. What was the contribution of the following scientists to our understanding of radioactivity and atomic structure? (a) Becquerel, (b) Madame Curie, (c) Geiger, (d) Rutherford, (e) Roentgen, (f) Libby.

35.2. Explain each of the characteristics of radioactivity listed on page 735.

35.3. Which of the three types of radiation (α, β, or γ):
(a) is the most penetrating? (b) will produce the most ionization? (c) is a helium nucleus? (d) results from the stabilization of a nucleus excited by the emission of a particle?

35.4. The decay processes produce changes in the properties of the radioactive nucleus. Complete the following table summarizing these changes. (Use numbers $+1$, -1, $+2$, -2, etc., in your answers.)

Decay Process	Change produced in mass no. (A)	change produced in atomic no. (Z)	change produced in number of neutrons(n)
(a) alpha, α	_____	_____	_____
(b) beta, β	0	_____	_____
(c) gamma, γ	_____	0	_____

35.5. Consider the radioactive decay of $^{227}_{89}$Ac. Place mass numbers and atomic numbers in proper places for elements X, Y, and Z in the sequence below. With the aid of the periodic table, identify elements X, Y, and Z.

$$^{227}_{89}\text{Ac} \nearrow \xrightarrow{\text{alpha}} \text{X} \nearrow \xrightarrow{\text{beta}} \text{Y} \nearrow \xrightarrow{\text{beta}} \text{Z}$$

35.6. How many alpha and beta particles are emitted as the result of radioactive decay in the following overall changes? (a) $^{214}_{84}$Po $\rightarrow$ $^{206}_{82}$Pb, (b) $^{226}_{88}$Ra $\rightarrow$ $^{206}_{82}$Pb, (c) $^{238}_{92}$U $\rightarrow$ $^{222}_{86}$Rn.

35.7. An isotope has a half-life of one day. Starting with 100 g, how much will be present at the end of 10 days?

35.8. (a) If a man stores away 2.72 g of Ra in the year 1977, what weight of Ra will be left in the year 5221, assuming that the half-life of Ra is 1622 yrs? (b) $^{222}_{86}$Rn decays by alpha emission with a half-life of 3.8 days. What total weight of He gas will be produced in 15.2 days from a 1.00-g sample of Rn?

35.9. If one starts with 650 g of pure ^{238}U, what weight of ^{238}U will be left at the end of 9×10^9 yr? The half-life of ^{238}U is 4.5×10^9 yr.

35.10. A sample of a radioactive isotope of phosphorus, ^{32}P, when counted showed an activity of 4,000 d.p.s. When counted 42 days later the activity had declined to 500 d.p.s. What is the half-life of ^{32}P?

35.11. ^{24}Na is an undesirable contaminant in neutron activation analysis because its high-energy γ-rays interfere with the counting of other radioactive isotopes. To eliminate this interference, one simply allows the sample to stand until the ^{24}Na decreases to an acceptable level. If the half-life of ^{24}Na is 15 hr, how long would one hold the sample to reduce the activity of the isotope to 1.5% of that observed immediately after the neutron irradiation?

35.12. The activity of a sample of ^{121}I was found initially to be 8 microcuries. If, after 24 days, the activity of the same sample had decreased to 1 microcurie, what is the half-life in days of this radioactive isotope?

35.13. The half-life of ^{210}Bi is 5.00 days. (a) In what period of time would the activity of a given sample of this isotope be reduced to 20.0%? (b) Starting with 5 mg of ^{210}Bi, what amount will be left after 3.0 days?

35.14. The activity of a certain radioactive element is reduced to 10.0% in a period of 24 hr. What is the half-life of this element?

35.15. Radioactive iodine, ^{128}I, has a half-life of 25 min. (a) In what period of time would the radioactivity be reduced to 5.0% of the original? (b) Starting with 5.0 mg of ^{128}I, what weight will be left at the end of 5 hr?

35.16. The nuclide $^{198}_{79}Au$ has a half-life of 64.8 hr. How much of a 1.00-g sample remains at the end of 1.00 day?

35.17. The nuclide $^{22}_{11}Na$ decays by positron emission. The rate is such that 76.6% of the original quantity remains after 1.00 yr. (a) What is the rate constant? (b) What is the half-life of $^{22}_{11}Na$?

35.18. The half-life of $^{35}_{16}S$ is 86.7 days. How long will it take for 75.0% of a sample to disappear?

35.19. The half-life of $^{237}_{93}Np$ is 2.1×10^6 yr, and the age of the earth is approximately 3.5×10^9 yr. What fraction of the quantity of $^{237}_{93}Np$ that was formed when the earth was created is now present?

35.20. Use the following information to plot the curve of radioactive ray emission for ^{14}C. The vertical axis of the graph will show radioactive emission in percentages. Using suitable unit divisions on this axis, mark a scale from 0% (radioactive emission of carbonaceous materials buried for an infinitely long time) to 100% (radioactive emission from recently living carbonaceous material). The horizontal axis of the graph will show time in years. Since 5770 yr is the half-life of ^{14}C, use suitable divisions on the horizontal axis to show 5770; 11,540; 17,310; and 23,080 yr. Plot points on a curve showing that the residual radiation from ^{14}C at these periods of time will be 50%, 25%, 12.5%, and 6.25%, respectively. Use the graph to give estimates for the following:
(a) the age of charcoal from fire in a cave of prehistoric man if radioactive emission is 18.75% that of present-day new charcoal.
(b) the age of parchment paper sealed in jars in a cave and giving 75% emission compared with present-day paper.
Now calculate (a) and (b) above using the rate equation. How do the values obtained from your plot compare with the calculated values?

35.21. The carbon from the heartwood of a giant sequoia tree gives 10.8 $^{14}_{6}C$ counts per min per g of carbon, whereas the wood from the outer portion of the tree gives 15.3 $^{14}_{6}C$ counts per min per g of carbon. How old is the tree?

35.22. One curie of radioactivity gives 3.7×10^{10} d.p.s. If you purchased a sample of radioactive carbon-14 which had a specific activity of 10 microcuries per gram, how many disintegrations per second would you observe if you counted 1 mg of the sample?

35.23. With the aid of Figure 35.5, determine whether the n/p ratio for the following isotopes is above, within, or below that for a stable atom: $^{37}_{19}K$; $^{234}_{92}U$; $^{30}_{16}S$; $^{30}_{14}Si$.

35.24. From the standpoint of n/p ratio, select the member of each pair which would be expected to be stable: (a) $^{40}_{18}Ar$, $^{30}_{18}Ar$; (b) $^{24}_{12}Mg$, $^{28}_{12}Mg$; (c) $^{27}_{13}Al$, $^{23}_{13}Al$; (d) $^{200}_{80}Hg$, $^{160}_{80}Hg$; (e) $^{119}_{50}Sn$, $^{105}_{50}Sn$.

Supplementary Readings

35.1. "The Age of the Elements," D. N. Schramm, *Sci. American,* Jan. 1974; 69.

35.2. "Radiocarbon Dating," W. F. Libby, *Chem. in Britain,* **5,** 548, (1969).

35.3. "Some Recollections of Early Nuclear Age Chemistry," G. T. Seaborg, *J. Chem. Educ.,* **45,** 278 (1968).

Chapter 36

Nuclear Chemistry

36.1   NUCLEAR REACTIONS

Nuclear reactions are those reactions in which the nucleus of an atom gains or loses protons and other particles so as to cause it to change into an atom of another element. Symbols commonly used for the particles involved are shown in Table 36.1.

Table 36.1
Names and symbols of particles involved in nuclear reactions.

Type of Particle	Symbols	
Alpha particle	^4_2He	α
Proton	^1_1H	p
Deuteron	^2_1H or ^2_1D, d	
Tritium	^3_1H or ^3_1T, t	
Electron or beta particle	$^0_{-1}e$	e^- β
Positron	$^0_{+1}e$	e^+ β^+
Neutron	^1_0n	n

For each symbol the superscript number represents the approximate **mass** of the particle in amu (to the closest whole number), and the subscript represents the electric **charge**. For example, $^0_{-1}e$ designates an *electron* which has a mass closer to 0 than to 1 amu and has a charge of -1.

When writing equations for nuclear reactions, it is usually better to use the symbols which have the mass and charge designated as part of the symbol (the first symbols in Table 36.1), since the whole number sum of the subscripts on the left side of the equation always equals the sum on the right. The same is true of the superscripts. For example, the formation of $^{14}_6\text{C}$ from neutrons reacting with ordinary nitrogen gas in the upper atmosphere has been previously mentioned in connection with radiocarbon dating. The equation for the reaction is

$$^{14}_7\text{N} + {}^1_0n \rightarrow {}^{14}_6\text{C} + {}^1_1\text{H}$$

Nuclear Stability and Prediction of Emanating Particle

Alpha emission is generally observed for elements with atomic numbers greater than 82 (Pb). The n/p ratio is too high to be stable and the elements decay one after another in a series until the belt of stability is reached. An example of α decay is

$$^{226}_{88}\text{Ra} \rightarrow {}^{222}_{86}\text{Rn} + {}^4_2\text{He}$$

Beta emission also occurs with elements having a high n/p ratio. This occurs for many natural radioactive elements as well as artificial radioisotopes. The β particle results from the transformation of a neutron in the nucleus into a nuclear proton and an electron. The electron is given off as a β particle.

$$_0^1n \rightarrow {}_1^1H + {}_{-1}^0e$$

This causes the number of neutrons to decrease and the number of protons to increase, thus improving the n/p ratio. Radiocarbon decay, for example, is of this type:

$$_6^{14}C \rightarrow {}_7^{14}N + {}_{-1}^0e$$

Positron emission is one way a radioisotope with too low an n/p ratio decays. For example

$$_{16}^{30}S \rightarrow {}_{15}^{30}P + {}_{+1}^0e$$

It seems that this results from a nuclear *proton* changing into a neutron and a positron:

$$_1^1H \rightarrow {}_0^1n + {}_{+1}^0e$$

Electron capture (*K* capture) also occurs with radioisotopes of low n/p ratio. The nucleus captures an orbital electron from the atom's *K* or *L* shell and converts a nuclear proton into a neutron:

$$_{-1}^0e + {}_1^1H \rightarrow {}_0^1n$$

This results in the atom having one less proton and one more neutron, and the n/p ratio increases. An example is

$$_8^{15}O + {}_{-1}^0e \xrightarrow{\text{K capture}} {}_7^{15}N + \gamma\text{-rays}.$$

Electron capture is usually accompanied by production of gamma (γ) rays; in this case, by other electrons falling into the vacated positions resulting from the capture. Actually gamma rays often accompany the emission of α and β particles as well because of nuclear energy adjustments resulting from the emission of these particles. This does not change the mass or charge of the nucleus, however.

Artificial Transmutation of Elements

The facts of radioactivity immediately bring to mind the questions (1) Can the atoms of those elements which are not radioactive be broken down by artificial means into atoms of simpler elements? (2) Can atoms of the simpler elements be built up into more complex atoms? In either case, a transmutation of the element would result—a problem at which the alchemists worked vainly for centuries. Radioactivity proved that if we can alter the nuclei of atoms, new elements must result. Some success in transmutation has been achieved by bombarding atoms with high-speed particles, such as protons or neutrons. Since atoms somewhat resemble a solar system in that their volume is largely open space with a nucleus of exceedingly small diameter as compared with the diameter of the atom, hitting the minute, positively charged nucleus to effect nuclear change is difficult. A particle for nuclear bombardment must necessarily be smaller than the atom in order to penetrate the electron orbital atmosphere. If a properly directed positive particle is to hit the positive nucleus, it must have sufficient speed to overcome the repellent force of the positively charged nucleus. Rutherford reasoned correctly that alpha particles from naturally radioactive elements would have sufficient speed and mass to disrupt an atomic nucleus.

In 1919 Rutherford allowed a beam of alpha particles from a radioactive substance to pass through nitrogen gas. Some of the high-speed alpha particles, on collison with the nuclei of nitrogen atoms, effected a change as represented by the following equation.

$$^{14}_{7}N + ^{4}_{2}He \rightarrow ^{17}_{8}O + ^{1}_{1}H$$

In this particular instance an isotope of oxygen (mass number 17, atomic number 8) and a proton were obtained from the nitrogen and helium. Only a few alpha particles were effective in bringing about a change; atoms have so much space in their volume that only a few of the alpha particles actually collided with a nucleus. In fact, only about one alpha particle in 50,000 made an effective collision. Nevertheless, this was the first accomplishment of artificial transmutation.

Many further experiments in bombarding atoms with alpha particles resulted in small-scale transmutations. Examples are

$$^{27}_{13}Al + ^{4}_{2}He \rightarrow ^{30}_{15}P + ^{1}_{0}n$$
$$^{11}_{5}B + ^{4}_{2}He \rightarrow ^{14}_{7}N + ^{1}_{0}n$$
$$^{9}_{4}Be + ^{4}_{2}He \rightarrow ^{12}_{6}C + ^{1}_{0}n$$

This last equation denotes the way in which Chadwick discovered the neutron, $^{1}_{0}n$ (see page 37).

Particle Accelerators

To speed a beam of nuclear particles, such as helium ions, protons, or deuterons, a **linear** accelerator was invented. To further increase speed, E. O. Lawrence of the University of California invented the **cyclotron.** In this machine the particles are accelerated in a strong magnetic field between electrodes which change their polarity at intervals equal to the time of half a revolution of the particles. These accelerated particles have been effective in many transmutations. Another accelerating machine, the **betatron,** throws off high-energy electrons which are used to produce high-energy gamma rays, and these in turn initiate many nuclear changes. The **synchrotron** is a newer type of electrical machine which gives increased kinetic energy to electrons and is also applicable with positive ions. Table 36.2 gives a comparative summary of these accelerators.

Table 36.2
Acceleration of particles

Type of Accelerator	Ion Particles Accelerated	Maximum Energy (MeV)
Linear	$_1^1H$, $_1^2H$, $_2^4He$ (electrons and ions up to argon)	1–1000
Cyclic cyclotron	$_1^1H$. $_1^2H$, $_2^4He$	25
Synchrocyclotron	$_1^1H$, $_1^2H$, $_2^4He$	500
Synchrotron	electrons, $_1^1H$	8000
Betatron	electrons	340

Because of their neutral character neutrons should make good nuclear bombardment particles; but man-made electrical machines cannot accelerate neutrons. Slow neutrons, a product of certain artificial transmutations, have been effective in initiating further nuclear changes.

Typical Nuclear Changes

Further examples of artificial transmutation are[*]

$$_3^6Li + _0^1n \rightarrow _2^4He + _1^3T \qquad\qquad _{15}^{31}P + _0^1n \rightarrow _{15}^{32}P + \gamma$$
$$_7^{14}N + _0^1n \rightarrow _6^{14}C + _1^1H \qquad\qquad _{77}^{191}Ir + _2^4He \rightarrow _{79}^{194}Au + _0^1n$$
$$_{14}^{28}Si + _1^2D \rightarrow _{15}^{29}P + _0^1n \qquad\qquad _{25}^{55}Mn + _1^1H \rightarrow _{26}^{55}Fe + _0^1n$$
$$_{11}^{23}Na + _1^2D \rightarrow _{11}^{24}Na + _1^1H \qquad\qquad _{42}^{96}Mo + _1^2H \rightarrow _{43}^{97}Tc + _0^1n$$
$$_{19}^{39}K + _1^1H \rightarrow _{20}^{39}Ca + _0^1n \qquad\qquad _{83}^{209}Bi + _2^4He \rightarrow _{85}^{211}At + 2\,_0^1n$$

The last two of these nuclear changes have been used to prepare atoms of $_{43}^{97}Tc$ and $_{85}^{211}At$, which elements are so scarce that they have never been detected on earth.

The above nuclear changes may be designated, respectively, as (n,t), (n,p), (d,n), (d,p), (p,n), (n,γ), (α,n), (p,n), (d,n), $(\alpha,2n)$, according to the manner in which the bombarding and emission units are listed. These pairs represent "in" and "out" particles commonly involved, with n denoting neutron; p, a proton; α, an alpha particle; d, a deuteron; and t, tritium.

---

[*] The understood release of energy is not added on the right side of the equations.

To illustrate an abbreviated form of writing nuclear equations in which the "in" and "out" particles are shown in parentheses, the first three equations may be rewritten as

$$^{6}_{3}\text{Li} \ (n,t) \ ^{4}_{2}\text{He}$$
$$^{14}_{7}\text{N} \ (n,p) \ ^{14}_{6}\text{C}$$
$$^{28}_{14}\text{Si} \ (d,n) \ ^{29}_{15}\text{P}$$

Artificial Radioactivity

It has been found that the bombardment of nuclei of several of the lighter elements results in the formation of isotopes of elements which disintegrate spontaneously, very much in the manner of the heavier radioactive elements. For example, when aluminum is bombarded with alpha particles, an isotope of phosphorus is formed in an artificial transmutation.

$$^{27}_{13}\text{Al} + ^{4}_{2}\text{He} \rightarrow ^{30}_{15}\text{P} + ^{1}_{0}n$$

This isotope of phosphorus is radioactive and decomposes spontaneously.

$$^{30}_{15}\text{P} \rightarrow ^{30}_{14}\text{Si} + ^{0}_{+1}e$$

The above isotope of phosphorus, with a half-life of 2.5 min, is said to be artificially radioactive, since it has been produced by artificial means.

Many other artificially radioactive elements have been produced, and they promise to be effective therapeutic agents, as are the naturally radioactive elements, like radium. In general, these radioactive isotopes have short half-life periods, and this offers certain advantages in the treatment of disease. Additional examples showing the manner of transmutation preparation of the radioactive isotope and then its disintegration products are

$$^{127}_{53}\text{I} + ^{1}_{0}n \rightarrow ^{128}_{53}\text{I} + \gamma$$
$$^{128}_{53}\text{I} \ (\text{half-life 25 min}) \rightarrow ^{128}_{54}\text{Xe} + ^{0}_{-1}e$$
$$^{59}_{27}\text{Co} + ^{1}_{0}n \rightarrow ^{60}_{27}\text{Co}$$
$$^{60}_{27}\text{Co} \ (\text{half-life 5.27 years}) \rightarrow ^{60}_{28}\text{Ni} + ^{0}_{-1}e$$
$$^{35}_{17}\text{Cl} + ^{1}_{0}n \rightarrow ^{35}_{16}\text{S} + ^{1}_{1}\text{H}$$
$$^{35}_{16}\text{S} \ (\text{half-life 87 days}) \rightarrow ^{35}_{17}\text{Cl} + ^{0}_{-1}e$$

The discovery of these radioactive elements may lead to very significant new discoveries in the metabolism of plants and animals. The passage of radioactive phosphorus and sulfur through the body of an animal may be followed with a Geiger counter. $^{60}_{27}\text{Co}$ is used in cancer therapy.

As a result of research in the past thirty years, radioactive isotopes of every known element have been artificially prepared. Elements 43, 61, 85, and 97 were for years blank spaces in the periodic table. After much unsuccessful research to find them, scientists began to realize that if these elements existed in nature at all, it was in exceedingly small amounts. These elements have now been prepared in small (or trace) amounts by nuclear reactions, or have been identified as branch elements in radioactive decay series.

36.2 ATOMIC ENERGY AND THE ATOMIC BOMB

Experiments with small-scale artificial transmutations resulting in new stable or radioactive atoms gave rise to the idea that atoms heavier than uranium might be produced. Research on the small-scale production of transuranium atoms was successful, and in the process the phenomenon of nuclear fission was discovered.

In early August of 1945 came the startling announcement that a bomb of unprecedented explosive violence had been dropped on the Japanese city of Hiroshima. The war with Japan came to an end a few days later. It was then revealed that since early in the war scientists had been concentrating their efforts on the release of atomic energy, and an atomic bomb was one result. Preliminary experiments with the bomb had been carried out near Los Alamos, New Mexico. Four huge plants for the manufacture of the essential ingredients of the bomb had been erected—three in Tennessee and one on the Columbia River in Washington.

Although the details of the construction of the atomic bomb are still secret, the underlying scientific principles have been fairly well known for some time. A brief discussion of some of the principles underlying the release of atomic energy is appropriate.

Chemical Energy and Atomic Energy

In ordinary chemical changes, atoms of the elements undergo rearrangement as to bonding, but the identity of the atom is retained. Electrons in the valence level are transferred or shared, but the rest of the atom is unchanged.

$$C + O_2 \rightarrow CO_2 + heat$$

Energy associated with the electrons of the atoms is considered the basic factor of chemical energy.

Atomic energy is associated with **changes in the nuclei** of atoms. The basic nature of the atom is changed—the product atoms are not the same as the reacting atoms. An example of a nuclear change that results in release of much energy is

$$^{7}_{3}Li + ^{1}_{1}H \rightarrow 2\,^{4}_{2}He$$

In contrast with the 4.1 eV of energy produced from the combustion of one atom of carbon, 17 MeV are produced from the nuclear reaction of one atom of lithium with one atom of hydrogen. It is evident, then, that the energy produced in nuclear reactions is many million times greater than that produced as a result of ordinary chemical change. The source of this tremendous amount of energy is now known to be associated with a very small mass loss. For example,*

$$^{7.016005}_{3}Li + {}^{1.007825}_{1}H \rightarrow 2\ ^{4.002604}_{2}He$$
$$8.023830 \text{ amu} \rightarrow 8.005208 \text{ amu}$$

In this transformation, a loss of mass equal to $8.023830 - 8.005208$ or 0.018622 units of mass takes place. This disappearance of mass accounts for the resulting energy. In 1905 Einstein stated the law of equivalence of mass and energy, which says, in effect, that matter may be converted into energy, and vice versa. By using the Einstein equation

$$\text{Energy} = \text{mass} \times (\text{velocity of light})^2$$

and proper physical units, the energy evaluation associated with a given loss of mass can be calculated. Such a calculation gives the result that **1 amu of mass is equivalent to 931.43 MeV.** Therefore the energy released in the above reaction of one atom of Li combining with one atom of H to form two atoms of He would be

$$(0.018622 \text{ amu})\ (931.48 \text{ MeV/amu}) = 17.346 \text{ MeV}$$

For 1 mole of Li (6.022×10^{23} atoms or 7.016005 g) the energy released would be 1.0446×10^{25} MeV or 4.0008×10^{11} cal. This energy release of 400 billion cal from the reaction of 7 g Li with 1 g H is truly stupendous. However, if all of the 8-g mass of Li and H were converted completely to energy instead of the small fraction shown above, the energy release would be in excess of 160 trillion cal.

Chain Reactions

Why are nuclear reactions with the evolution of these tremendous amounts of energy not taking place about us in everyday life, much the same as ordinary chemical changes are taking place? The answer seems to be that, whereas most chemical changes are self-maintaining, nuclear changes are not. In a chemical reaction, once the action is started, it proceeds until the reactants are used up. A piece of wood or coal ignited in the air continues to burn until used up. In nuclear changes, however, the nuclear "fire" usually goes out soon after it is started. If some way were available to keep a nuclear "fire" burning, nuclear changes would proceed as easily as ordinary chemical

---

* The precise amu weights are used as superscripts in many instances in this chapter in place of mass numbers. $^{1.007825}_{1}H$ relates to a neutral hydrogen atom, rather than to a proton. The proton weight is $1.007825 - 0.000548$ (weight of electron) $= 1.007277$.

changes. A so-called **chain reaction,** in which the action, once started, passes from one atom to another as in the links of a chain, appears to be necessary for a self-maintaining nuclear reaction.

Atomic Fission

The transmutation of many elements has already been described. In general, in these experiments only small fragments of matter, such as protons, electrons, or helium ions, were knocked out of a nucleus or added to it. As a result, the product atom had an atomic number very near that of the bombarded atom.

Researchers in Germany in 1939 discovered a new and potentially most important type of nuclear change. By bombarding uranium with neutrons traveling at moderate speeds, they split the atom into two fragments of approximately equal size. Actually, it was only the ^{235}U isotope in natural uranium that was subject to this change. This process, now referred to as **fission,** was accompanied by the evolution of a tremendous amount of energy. Furthermore, in the process—in addition to the relatively large fragments produced—several neutrons were released in the fission of each atom. The implications of the neutron release were at once clear—these product neutrons might be used to bombard another parent atom, the fission of which in turn would produce more neutrons. Thus a self-maintaining, or chain, reaction seemed to be a definite possibility. The important point, of course, was that a chain reaction would be self-maintaining only if a neutron from each fission produced fission in at least one other atom. We may illustrate a nuclear fission by the following equations (see also Figure 36.1):

$$^{235}_{92}U + ^{1}_{0}n \rightarrow ^{94}_{38}Sr + ^{139}_{54}Xe + 3^{*}\ ^{1}_{0}n + \text{energy}$$
$$\text{(about 200 Mev)}$$

or

$$\rightarrow ^{145}_{56}Ba + ^{88}_{36}Kr + 3\ ^{1}_{0}n + \text{energy}$$

In these equations, strontium plus xenon or barium plus krypton are shown as the two large fragments in the fission. It should not be assumed, however, that these elements always constitute the major fragments of a fission. Many other elements near the middle of the periodic table have been identified as products (see Figure 36.2).

The mass of all the products of a fission is not quite equal to the mass of the reactants. The loss in mass accounts for the very large amounts of energy evolved in nuclear fissions—that is, mass is converted into energy.

No whole units of mass are lost in a fission; note that in the two fission equations given above the sum of the atomic numbers on the left is equal

---

° The average number of neutrons emitted is between two and three.

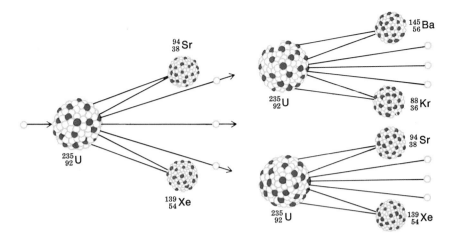

Figure 36.1
Fission of uranium and the chain reaction. The small white circles represent neutrons.

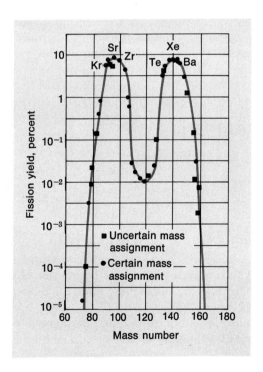

Figure 36.2
Yields of U²³⁵ fission products as a function of mass.

to that on the right, and that the sums of the mass numbers on the left and right sides of the equations are equal. The energy is derived from the disappearance of only a fractional part of one unit of mass. Note also that the $^{94}_{38}$Sr plus $^{139}_{54}$Xe, and $^{145}_{56}$Ba plus $^{88}_{36}$Kr are isotopes of these elements having high neutron-proton ratios. It is the radioactive decomposition of such atoms that accounts for the intense radiation associated with fission—radiation mainly of beta and gamma rays. For example,

$$^{145}_{56}\text{Ba} \rightarrow {}^{145}_{57}\text{La} + e^- + \gamma \rightarrow {}^{145}_{58}\text{Ce} + e^- + \gamma \text{ and so on}$$
$$^{139}_{54}\text{Xe} \rightarrow {}^{139}_{55}\text{Cs} + e^- + \gamma \rightarrow {}^{139}_{56}\text{Ba} + e^- + \gamma \text{ and so on}$$

Controlled Fissions

It is evident that if all of the neutrons liberated in the fission of each atom of uranium were utilized in producing more fissions, a rapid and violent reaction might take place. A piece of uranium, once the reaction was started, would explode. Although a violently explosive substance was the immediate objective in producing an atomic bomb, many problems had to be solved before this objective was achieved. One of the first problems to confront scientists working on the atomic energy project was to find a means of controlling these atomic fissions. We may indicate in a general way how this was accomplished.

Ordinary uranium is composed of three isotopes, which occur in the following proportions: ^{238}U, 99.276%; ^{235}U, 0.719%; and ^{234}U, 0.005%. Thus in ordinary uranium the ratio of the ^{235}U isotope (which is the isotope fissionable by slow neutrons) to ^{238}U is about 1 to 140. For a fission chain to be set up, neutrons liberated by the fission of one ^{235}U atom must hit another ^{235}U atom, and so on. But neutrons resulting from fission are moving so fast that many of them are captured by ^{238}U atoms, though not in a fission-yielding process. And if the piece of uranium is small, many or most of the fast neutrons may escape through the surface of the metal before being absorbed by other nuclei. The uranium isotope ^{238}U is particularly effective in capturing neutrons. Although this process lessens the chance for absorption of neutrons to produce fission in ^{235}U atoms, it is of utmost importance in helping control fission rate and in production of plutonium.

It is evident that at least one neutron from each fission must be effective in bringing about another fission; otherwise, the chain is broken, and reaction ceases. ^{235}U atoms seem able to capture slow neutrons in competition with the more numerous ^{238}U atoms, but ^{238}U atoms capture most of the high-energy neutrons. In order to increase the probability of capture of neutrons by ^{235}U and the subsequent fission, it is necessary to slow down the liberated neutrons. This may be accomplished with a "moderator"—graphite, for example—which absorbs a part of the energy of the neutron and slows it down. With a latticework arrangement of ^{235}U-enriched uranium lumps imbedded in graphite (called a "pile" and larger than a so-called critical size), neutrons which might ordinarily be lost from the surface will diffuse back and forth through the lattice work until they have lost sufficient energy to be captured by a ^{235}U nucleus. The capture results in fission, with liberation of more neutrons, and the chain reaction is set up. The first pile was constructed at the University of Chicago in 1942 with about six tons of ordinary uranium and many tons of graphite. Since the fission process is exorthermic, it is obvious that the pile will heat up as the fission process continues to take place more and more rapidly, and will eventually reach a temperature high enough to cause an explosion. However, the pile is controlled by providing passageways through which rods of neutron-absorbing material, such as boron steel or cadmium metal, are inserted. Neutrons are absorbed by the boron steel or cadmium rods or plates, and the chain reaction is regulated.

$$^{113}_{48}\text{Cd} + ^{1}_{0}n \rightarrow ^{114}_{48}\text{Cd} + \gamma$$

Proper adjustment of the controls allows the reaction to proceed only at a determined rate or power level. To raise the power level, the rods are partially withdrawn; to lower it, the rods are pushed in farther. Such a controlled pile produces low-temperature heat energy from atomic energy. This source of energy is being used increasingly in nuclear-powered electric generating plants.

Nuclear Power Plants

Various nuclear reactors in experimental use, with application of their released energy to power development, are classified as **burners, converters,** and **breeders.** The burners are fueled with separated ^{235}U. Converters make use of partially enriched uranium, and include the plutonium production reactors at Hanford, Washington and Oak Ridge, Tennessee. Britain, France, and Russia have similar reactors. In them, the essential fuel is ^{235}U and ^{239}Pu, the latter being a conversion by-product of fission particles from ^{235}U acting on ^{238}U. A breeder is a special type of converter in which more fissionable isotopes are produced than consumed; this is possible because an average of over two neutrons result from fission of each ^{235}U atom. The isotopes ^{232}Th and ^{233}U are involved in some converter cycles.

Figure 36.3 shows the sections necessary for a nuclear power plant, which

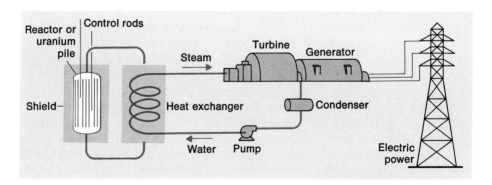

Figure 36.3
A nuclear-powered electric generating plant.

consists of a pile (as described above), a heat exchanger, and an electrical generating installation.

Plutonium

As previously stated, ordinary uranium tends to absorb neutrons and thus to slow down the chain reaction, if not actually to break it. This absorption of neutrons by ^{238}U turns out to be extremely important, since it leads to the formation of a new element, **plutonium** (Pu, atomic number 94) which, like ^{235}U, may undergo fission with the release of more neutrons. The changes taking place in the formation of Pu from ^{238}U may be represented diagrammatically.

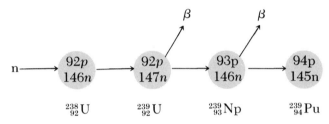

^{238}U absorbs a neutron to form ^{239}U, which then emits a beta particle, resulting in the formation of **neptunium** (Np), a new element of atomic number 93 and of short half-life. Neptunium then emits a beta particle which results in the formation of plutonium, ^{239}Pu.

With the discovery of Pu, two substances, ^{235}U and ^{239}Pu, were available for the production of atomic energy. Plutonium manufacture possesses certain definite advantages over the recovery of ^{235}U from ordinary uranium. The separation of ^{235}U from ^{238}U is expensive, since **physical means,** which take advantage of the relatively small differences in the mass of the two isotopes, must be employed. Plutonium, when produced in a uranium pile, can be separated from unconverted uranium by **chemical means,** since Pu is a different element and, as such, exhibits a different set of chemical properties.

The atomic bomb   Relatively little information about the construction of the atom bomb itself has been released. It was known early in the studies of this project, however, that a certain **critical size** of material is necessary for a chain reaction to be self-perpetuating and for an explosion to take place. In a small piece of ^{235}U or ^{239}Pu, in which the ratio of surface area to volume is large, too many neutrons escape through the surface, and consequently a chain reaction is not maintained. As the size of the piece is increased, the ratio of surface area to volume decreases, and there is more chance that neutrons will be captured before escaping through the surface. In producing an atomic bomb, it seems logical that pieces of subcritical size might be brought together rapidly, possibly by using one or more pieces as projectiles and another as a target in a firing mechanism. Just how pieces of subcritical size are brought together to form an effective fissionable mass in a bomb is still kept secret.

In the explosion of an atomic bomb, a very high temperature is attained; desert sand has been sintered at some distance from actual bomb-test sites at Los Alamos, New Mexico. Written accounts tell of vaporization of the metal of the tower holding the test bomb. Just how high a temperature is reached is difficult to estimate because man has not been able to measure temperatures above about 4000°C. Temperatures in the order of millions of degrees are believed to have been reached. This excessively high temperature consideration leads us to some understanding of the nature of possible temperature effects in an atomic bomb explosion. Probably no large mass of material in an atomic bomb blast is heated as high as the internal part of the sun, though certainly some very high temperature is reached locally. The atomic bomb explodes and causes such air expansion—and contraction on cooling—that cyclonic wind or blast effects are produced. The radioactive fragments, clinging to dust, are very dangerous.

The elements that are products of fission are commonly isotopes of quite high atomic weight for the given elements. In reverting to stable isotopes, intense beta and gamma ray activity is involved. This is the marked after-effect so dangerous to biological tissue. Radioactive strontium and cobalt isotopes have been publicized as particularly dangerous products of a fission bomb.

The **Binding Energy** of a nucleus is the energy which would be required to pull apart the protons and neutrons in the nucleus or the energy released if the nucleons were to be packed together to form the nucleus.

Exact work with the mass spectrograph shows that the **atomic masses of nuclei are always less than the additive masses of their nucleons.** The mass difference, converted to energy, is the binding energy.

For example, we may compute the mass loss when helium is formed from two hydrogens and two neutrons.

$$2\,_1^1\text{H} \quad + \quad 2\,_0^1\text{n} \quad \rightarrow \quad _2^4\text{He} \quad + \text{binding energy}$$
$$2 \times 1.007825 + 2 \times 1.008665 \rightarrow 4.002604 + 0.030376 \text{ amu}$$

Similarly, for all atoms a small fractional mass loss exists in atomic mass as compared with the total of nuclear particle masses. There is variation in the amount of binding energy per nucleon as diverse-sized atoms are built up.

$$\text{Binding energy for }_2^4\text{He} = \frac{(0.030376 \text{ amu}) (931 \text{ Mev/amu})}{4 \text{ nucleons}} = 7.07 \text{ Mev/nucleon}$$

Consider the binding-energy curve which is shown in Figure 36.4. Note that a high in the curve exists near iron of mass number 56. The $^{55.93488}_{26}$Fe isotope, that constitutes so much of the total weight of this earth, can conceivably be formed in cosmic space from the merging of 26 $^{1.007825}_1$H and 30 $^{1.008665}_0$n. The weight loss is 0.52852 amu per iron nucleus, corresponding to 492 Mev of binding energy. Iron would seem to be about the most stable atom. There is some evidence that the earth's core consists largely of iron and nickel, and meteorites have a high iron and nickel content. Perhaps this is related to the nuclear stability of these elements.

Note that the binding energy curve in Figure 36.4 gradually descends from iron

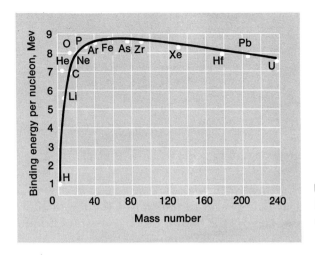

Figure 36.4
Binding energy per nucleon as a function of the mass number.

to uranium and the transuranium elements (elements above uranium in atomic number). These larger mass-number elements of lower binding energy have mass converted to energy in any fission of them into more stable elements in the middle range of mass numbers. This accounts for the very large amounts of energy accompanying atomic fission.

Atomic Fusion

Another potential source of nuclear energy exists in a **fusion** process, in which the very lightest nuclei can be made to combine to give somewhat heavier, more stable nuclei, with an attendant release of energy. Fundamentally, this is the idea behind the hydrogen bomb. Fusion of nucleons is believed to be the source of the sun's energy. Man probably will be able to control fusion processes someday as he has learned to control fission processes.

Thermonuclear Devices

Fusion of light elements requires temperatures of millions of degrees. Such high temperatures are obtained in fission bombs, and it is likely that the latter are used as detonators for fusion bombs. One possible arrangement of the ingredients in a **hydrogen** or **fusion bomb** is pictured in Figure 36.5. The core of the device contains the two hydrogen isotopes, deuterium and tritium. Around the core is a layer of fissionable matter such as ^{235}U which, when set off, "detonates" the interior. By implosion, the following reaction between the hydrogen isotopes in the core is initiated:

$$_1^2D + _1^3T \rightarrow _2^4He + _0^1n$$

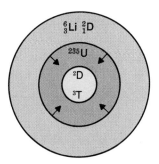

Figure 36.5
Arrangement of ingredients of an H bomb.

This reaction releases a flood of energy and further raises the temperature high enough to explode the outer layer of the bomb, which consists of lithium deuteride ($^6_3$Li $^2_1$D).

Peacetime Uses of Atomic Energy

Vast new fields which have been opened by the advent of atomic energy are only just beginning to be explored. The heat released by an atomic pile constitutes a potent source of energy in nuclear power plants. At Harwell, England, an eighty-office building is being heated by a nearby atomic pile. The atomic-powered U.S. submarines have completed many thousands of miles of travel.

In the publication *American Scientist*, Dr. Blabha is quoted concerning world power requirements. Taking the unit Q, equal to 10^{18} BTU of energy (approximately the energy given by burning 33 million tons of coal), it is stated that only $\frac{1}{2}$ Q of energy was used by man up to the year 1850. In 1950 the energy consumption by man was up to 10 Q per century; and it is increasing in rate by four percent each year. Less than 100 Q are still available as coal and oil. In estimated mineable uranium and thorium, about 1700 Q may be available. A practically unlimited supply of energy would be available in the deuterium of sea water if a means of handling a controlled fusion-thermonuclear reaction were found, but the technical difficulties are formidable.

Available Radioactive Isotopes

The use of atomic piles has made possible the preparation of artificial radioactive isotopes in amounts hitherto unimagined. These isotopes are finding application as "tracers" in medicine, scientific research, industrial research, and industrial control. The iodine isotope (^{131}I) is being used very extensively in the medical treatment of hyperthyroidism. Radioactive iodine administered to the patient tends to concentrate in the thyroid gland and provides radiation therapy by virtue of its high-energy gamma rays. **Cobalt-60,** a gamma and beta emitter, with a half-life of 5.3 yr, is currently being much used in both medicine and industry as a much cheaper substitute for radium.

Representative of the extensive use of tracers in scientific research is the application of ^{14}C, an isotope of carbon with a half-life of 5770 yr, to the study of photosynthesis. Biochemists are using ^{14}C and ^{32}P in the study of body chemistry.

Atomic piles make possible a method of analysis, called **neutron activation analysis,** which can be used for the determination of less than a billionth of a gram of an element. The method involves placing the material in an atomic reactor for a few minutes while the elements in the sample are bombarded by neutrons, changing the elements to radioactive isotopes. The γ-emission spectra of the resulting radioactivity is monitored and allows a determination of the kinds and amounts of the elements present. Each element has its own characteristic γ-ray spectrum. The method may be applied without destroying the sample as, for example, was the case when the valuable moon rocks were analyzed.

36.3   TRANSURANIUM ELEMENTS

Neptunium ($^{239}_{93}Np$) and **plutonium** ($^{239}_{94}Pu$) are artificially produced elements above uranium in atomic number. Several isotopes of each of these elements have now been prepared. These elements were discovered in connection with neutron capture by uranium and subsequent beta emissions (see diagram, page 763).

$$^{238}_{92}U + ^1_0n \rightarrow ^{239}_{92}U$$
$$\downarrow$$
$$^{239}_{93}Np + ^{\ 0}_{-1}e$$
$$\downarrow$$
$$^{239}_{94}Pu + ^{\ 0}_{-1}e$$

$^{239}_{93}Np$ has a half-life of only 2.3 days and on beta-particle ejection forms $^{239}_{94}Pu$, which is quite stable, with a half-life of 2.4×10^4 years. An isotope of neptunium of much longer half-life may be prepared.

$$^{238}_{92}U + ^2_1D \rightarrow ^{237}_{93}Np + 3\ ^1_0n$$

Neptunium shows oxidation numbers of $+3, +4, +5,$ and $+6$.

The chemical properties of plutonium were originally studied by **tracer techniques.** To understand the nature of tracer techniques, assume that a very small quantity of plutonium ions is mixed with several other metal ions in solution. A qualitative scheme of analysis involving precipitations and separations is carried out. The minute amount of plutonium may be followed through the various chemical separations because of its radioactivity, and its chemistry is then likened to that of the metal ions it follows. Solutions of plutonium ion as dilute as 10^{-8} M may be treated by such tracer techniques, and in this manner much of the chemistry of the element is worked out. Since plutonium is now available in pound quantities for current investigations, tracer techniques (**microchemistry**) are not so important for plutonium as for some other scarce elements or isotopes. Plutonium may exhibit oxidation numbers of +3, +4, +5, and +6 in various compounds.

Experimental endeavors to make still higher-atomic-number atoms has met with some success. Bombardment of ^{238}U and ^{239}Pu with very high-energy helium ions in the cyclotron has produced isotopes of elements of atomic numbers 95 (americium, symbol Am) and 96 (curium, symbol Cm).

$$^{238}_{92}\text{U} + ^4_2\text{He} \rightarrow ^{241}_{94}\text{Pu} + ^1_0 n$$

$$^{241}_{94}\text{Pu} \rightarrow e^- + ^{241}_{95}\text{Am} \quad \text{(470-yr half-life)}$$

$$^{239}_{94}\text{Pu} + ^4_2\text{He} \rightarrow ^1_0 n + ^{242}_{96}\text{Cm} \quad \text{(5-mon half-life)}$$

Researchers at the University of California, under the leadership of Professor G. T. Seaborg, have contributed markedly to the knowledge of transuranium elements. Trace quantities of the new elements berkelium, $^{247}_{97}$Bk, and californium, $^{251}_{98}$Cf, have been produced, and a few of their properties and those of their compounds established. Recent discoveries include the elements **einsteinium** ($^{254}_{99}$Es), **fermium** ($^{253}_{100}$Fm), **mendelevium** ($^{256}_{101}$Md) and **nobelium** ($^{254}_{102}$No). Element 103, whose creation was revealed at the University of California's Lawrence Radiation Laboratory, was identified on the basis of the 8.6 Mev alpha particles which the element emits. It is named **lawrencium** (Lr) and was made by bombarding 3 μg of californium with boron-10 and boron-11 nuclei. Lawrencium has a half-life of about eight seconds and an atomic weight of 257. Elements 104 and 105 have also been reported recently.

Isotopes of most of these transuranium elements have been prepared, and all of them are radioactive. Atoms of even higher atomic weight may be developed, but these larger nuclei are so unstable and have such short half-lives that progress toward heavier atoms is not expected to go on indefinitely.

The chemistry of all these transuranium elements is being studied by tracer techniques, and the following speculation may be made. In considering atomic structure and periodic-table placement of elements, it is noted that the "rare earths," which follow lanthanum, atomic number 57, are part of a transition group of elements in which the fourth level is building up from 18 to 32 electrons (see page 58). It follows, by analogy, that beginning with **actinium**, atomic number 89, a second instance may occur of a group of elements filling up (in whole or part) an inner level—that is, level 5. There is definite experimental evidence that transuranium elements have a characteristic valence of 3, and thus are a part of what may be called a "rare rare-earth" group, or **actinide series**, which terminates with lawrencium. The actinide series elements exhibit more diversity in oxidation number states than lanthanide series elements.

36.4 ENERGY OF THE SUN

The source of the energy that the sun has dissipated over many, many millions of years, and the source of the heat to maintain its estimated millions of degrees of internal temperature, have long been subjects of conjecture. Chemical changes involving mere electron interchange cannot account for solar energy. It has been speculated that the following chain of nuclear changes may be involved in hydrogen-to-helium formation in which ^{12}C acts as a catalyst.

$$^{12}_{6}C + ^{1}_{1}H \rightarrow ^{13}_{7}N + \text{energy}$$
$$^{13}_{7}N \rightarrow ^{13}_{6}C + ^{0}_{1}e + \text{neutrino}$$
$$^{13}_{6}C + ^{1}_{1}H \rightarrow ^{14}_{7}N + \text{radiation}$$
$$^{14}_{7}N + ^{1}_{1}H \rightarrow ^{15}_{8}O + \text{radiation}$$
$$^{15}_{8}O \rightarrow ^{15}_{7}N + ^{0}_{1}e + \text{neutrino}$$
$$^{15}_{7}N + ^{1}_{1}H \rightarrow ^{12}_{6}C + ^{4}_{2}He$$

Addition and canceling leaves

$$4\,^{1}_{1}H \rightarrow ^{4}_{2}He + 2\,^{0}_{+1}e + \text{energy equal to 29 MeV}$$

Astronomical evidence indicates that protons are much more abundant in stars than any other nuclear particle. A part of the energy of the sun is ascribed to proton union to form deuterons.

$$^{1}_{1}H + ^{1}_{1}H \rightarrow ^{2}_{1}D + ^{0}_{+1}e + \text{energy}$$

Formation of some heavier atoms, such as $^{16}_{8}O$, $^{28}_{14}Si$, $^{56}_{26}Fe$, with release of binding energy, is probably inherent in nuclear fusions in the solar interior.

Exercises

36.1. Complete the following equations for nuclear bombardments:

(a) $^{14}_{7}N + ^{4}_{2}He \rightarrow ^{1}_{0}n +$

(b) $^{79}_{35}Br + ^{4}_{2}He \rightarrow ^{1}_{0}n +$

(c) $^{59}_{27}Co\ (\alpha, n)$

(d) $^{16}_{8}O + ^{2}_{1}D \rightarrow ^{1}_{1}H +$

(e) $^{39}_{19}K\ (d,\alpha)$

(f) $^{37}_{17}Cl + ^{1}_{0}n \rightarrow ^{1}_{1}H +$

(g) $^{88}_{38}Sr\ (n,\alpha)$

(h) $^{106}_{46}Pd + ^{1}_{1}H \rightarrow ^{106}_{47}Ag +$

(i) $^{197}_{79}Au\ (p,n)$

(j) $^{37}_{17}Cl + ^{1}_{0}n \rightarrow \gamma +$

(k) $^{14}_{7}N + ^{1}_{0}n \rightarrow ^{14}_{6}C +$

(l) $^{59}_{27}Co + ^{1}_{0}n \rightarrow ^{60}_{27}Co +$

(m) $^{27}_{13}Al\ (\alpha,n)$

36.2. Complete the following equations as examples of radioisotopes in decay processes:

(a) $^{14}_{6}C \rightarrow e^{-} +$

(b) $^{92}_{38}Sr \rightarrow$ (2 separate emissions)

(c) $^{60}_{27}Co \rightarrow$

(d) $^{38}_{20}Ca \rightarrow$

(e) $^{20}_{8}O \rightarrow e^{-} + \_\_\_\_ \rightarrow e^{-} + \_\_\_\_$

36.3. Complete the following equations for nuclear bombardments:

(a) $^9_4\text{Be} + ^4_2\text{He} \rightarrow ^{12}_6\text{C} +$

(b) $^{27}_{13}\text{Al} + ^4_2\text{He} \rightarrow ^1_0n +$

(c) $^{12}_6\text{C} + ^1_1\text{H} \rightarrow \gamma +$

(d) $^9_4\text{Be} + ^1_1\text{H} \rightarrow ^2_1\text{D} +$

(e) $^{37}_{17}\text{Cl} + ^2_1\text{D} \rightarrow ^{37}_{18}\text{Ar} +$

(f) $^{19}_9\text{F} + ^1_0n \rightarrow ^{16}_7\text{N} +$

(g) $^{85}_{37}\text{Rb} + ^1_0n \rightarrow 2\,^1_0n +$

(h) $^{209}_{83}\text{Bi} + ^4_2\text{He} \rightarrow 2\,^1_0n + \text{At}$

(i) $^9_4\text{Be} + ^2_1\text{D} \rightarrow ^{10}_5\text{B} +$

36.4. Complete the following equations for radioactive decay. An e^- is sometimes designated $_{-1}^0e$; and e^+ is the same as $_{+1}^0e$.

(a) $^{239}_{93}\text{Np} \rightarrow _{-1}^0e +$

(b) $^{226}_{88}\text{Ra} \rightarrow ^{222}_{86}\text{Rn} +$

(c) $^{30}_{15}\text{P} \rightarrow _{+1}^0e +$

(d) $^{239}_{92}\text{U} \rightarrow _{-1}^0e +$

(e) $^{40}_{19}\text{K} \rightarrow ^{40}_{20}\text{Ca} +$

36.5. Suggest possible products of the decay of the following neutron-deficient radioisotopes.

(a) $^{25}_{13}\text{Al} \rightarrow _{+1}^0e +$

(b) $^{34}_{17}\text{Cl} \rightarrow$

(c) $^{14}_8\text{O} \rightarrow$

(d) $^{230}_{90}\text{Th} \rightarrow ^4_2\text{He} +$

(e) $^{30}_{16}\text{S} \rightarrow _{+1}^0e + \underline{\quad} \rightarrow _{+1}^0e + \underline{\quad}$

(f) $^{78}_{35}\text{Br} \rightarrow _{+1}^0e + \underline{\quad}$

36.6. Write equations for the following examples of radioactive decay: (a) alpha emission by $^{221}_{87}\text{Fr}$, (b) beta emission by $^{66}_{29}\text{Cu}$, (c) positron emission by $^{18}_9\text{F}$, (d) electron capture by $^{133}_{56}\text{Ba}$.

36.7. Write equations for the following examples of radioactive decay: (a) alpha emission by $^{227}_{91}\text{Pa}$, (b) beta emission by $^{114}_{47}\text{Ag}$, (c) positron emission by $^{39}_{20}\text{Ca}$, (d) electron capture by $^{55}_{26}\text{Fe}$.

36.8. Write equations for the following induced nuclear reactions:

(a) $^{10}_5\text{B}$ (α,n)

(b) ^6_3Li (n,α)

(c) ^7_3Li (d,n)

(d) $^{12}_6\text{C}$ (p,γ)

(e) $^{96}_{42}\text{Mo}$ (p,n)

(f) $^{75}_{33}\text{As}$ (d,p)

(g) $^{45}_{21}\text{Sc}$ (α,p)

(h) $^{51}_{23}\text{V}$ $(d,2n)$

36.9. Following are listed some transuranium nuclides and the types of induced nuclear reactions that have been used to prepare them. In each case, what isotope was used as the starting material?

(a) $(d,6n)^{231}_{93}\text{Np}$

(b) $(d,3n)^{243}_{97}\text{Bk}$

(c) $(\alpha,^3_1\text{H})^{250}_{99}\text{Es}$

(d) $(^{12}_6\text{C},4n)^{245}_{99}\text{Es}$

(e) $(^{16}_8\text{O},5n)^{249}_{100}\text{Fm}$

(f) $(\alpha,2n)^{255}_{101}\text{Md}$

(g) $(^{22}_{10}\text{Ne},4n)^{256}_{102}\text{No}$

(h) $(^{10}_5\text{B},3n)^{257}_{103}\text{Lr}$

36.10. An isotope of element 105 was recently prepared by bombarding $^{249}_{98}\text{Cf}$ target with $^{15}_7\text{N}$ nuclei. Four neutrons were emitted when a $^{249}_{98}\text{Cf}$ nucleus absorbed a $^{15}_7\text{N}$ nucleus. Write an equation for the transformation.

36.11. Complete the equations for the following nuclear bombardment and radioactive decay processes.

(a) $^{37}_{17}Cl + ^{2}_{1}D \rightarrow ^{37}_{18}Ar + \underline{\quad}$

(c) $^{239}_{92}U \quad ^{0}_{-1}e + \underline{\quad}$

(b) $^{59}_{27}Co \ (\alpha,n)$

(d) $^{197}_{79}Au \ (p,n)$

36.12. Radioactivity emitted by radioactive isotopes is comprised primarily of (1) α particles, (2) β^- particles ($_{-1}^{0}e$), (3) β^+ particles ($_{+1}^{0}e$), and (4) γ-rays. Indicate which emission produces the following changes:

(a) Increases atomic number (Z) by one.

(b) Decreases number of neutrons (n) by two.

(c) Increases n by one.

(d) Produces no change in Z

(e) Produces no change in the mass number (A).

(f) Decreases n by one.

36.13. The ionizing radiation derived from radioactive isotopes may be either (1) α-particles, (2) β^--particles, (3) γ-rays, or (4) X rays. Match the type of radiation with the following statements.

(a) An electron derived from the nucleus.

(b) High-energy electromagnetic radiation derived from orbital electrons dropping from outer shells into the K-shell.

(c) Emitted only by isotopes with Z greater than 82.

(d) A helium nucleus.

(e) Most penetrating radiation – could penetrate 20 cm of lead.

(f) Emitted in association with electron capture.

(g) Most energetic radiation – forms the greatest number of ions.

(h) Emitted by isotopes with high n/p ratios.

(i) High-energy electromagnetic radiation derived from an "excited" nucleus.

(j) Would have intermediate penetrating power and energy.

36.14. Suggest how radioisotopes might be used in the following: (a) determining the distance house flies migrate from a release point, (b) treatment of cancer of the thyroid gland which preferentially absorbs iodine, (c) measuring the wear of piston rings in an engine, (d) determining in what parts of a plant an herbicide such as 2-4-D concentrates.

36.15. What is meant by atomic energy? Where does it come from? How does it compare to ordinary chemical energy? Explain Einstein's mass-energy equation.

36.16. Define and illustrate these terms: fission, fusion, artificial radioactivity, binding energy, tracer elements, chain reaction, thermonuclear device.

36.17. Calculate in Mev the energy which is released in the following process:

$$4 \, ^{1}_{1}H \rightarrow \, ^{4}_{2}He + 2 \, ^{0}_{+1}e + \text{energy}$$

A decrease in mass of 1.0000 amu corresponds to an energy release of 931 Mev. The masses involved are $^{1}_{1}H = 1.007825$ and $^{4}_{2}He = 4.002604$. Assume that the two positrons produced and the two electrons left over (mass = 0.0005486 amu) annihilate each other, and that their masses are converted completely into energy.

36.18. For $^{12.000000}_{6}C$ calculate the binding energy. (Note: Assume $6\,^1_1H + 6\,^1_0n$ in fusion.) (Pattern after calculations on page 764 for $^{4.002604}_{2}He$.)

36.19. The nuclide $^{213}_{85}At$ decays to $^{209}_{83}Bi$ by alpha emission. The mass of $^{213}_{85}At$ is 212.9931 amu, and the mass of $^{209}_{83}Bi$ is 208.9804 amu. Calculate the energy released in this process.

36.20. Calculate the binding energy per nucleon of $^{235}_{92}U$ (mass, 235.0439 amu). The masses of the proton, neutron, and electron are 1.007277 amu, 1.008665 amu, and 0.0005486 amu, respectively.

36.21. (a) Calculate the energy released by the fission:

$$^{235}_{92}U + \,^1_0n \rightarrow \,^{94}_{38}Sr + \,^{139}_{54}Xe + 3\,^1_0n$$

The atomic masses are: $^{235}_{92}U$, 235.0439 amu; $^{94}_{38}Sr$, 93.9154 amu; $^{139}_{54}Xe$, 138.9179 amu. The mass of the neutron is 1.0087 amu. (b) What percent of the total mass of the starting materials is converted into energy?

36.22. (a) Calculate the energy released by the fusion:

$$^1_1H + \,^2_1H \rightarrow \,^3_2He$$

The atomic masses are: ^1_1H, 1.0078 amu; ^2_1H, 2.0141 amu; ^3_2He, 3.0160 amu. (b) What percent of the total mass of the starting materials is converted into energy?

36.23. Each of the following nuclides represents one of the products of a neutron-induced fission of $^{235}_{92}U$. In each case, another nuclide and three neutrons are also produced. Identify the other nuclides. (a) $^{148}_{58}Ce$, (b) $^{95}_{37}Rb$, (c) $^{134}_{52}Te$.

36.24. What are some arguments for and against nuclear power plants?

36.25. Explain briefly the principle of the operation of a nuclear power plant.

36.26. What are the functions of the moderator and the control rods in a nuclear reactor? Of what are they composed?

36.27. What is neutron activation analysis? What are some advantages of this method of quantitative analysis?

Supplementary Readings

36.1. "A Natural Fission Reactor," G. A. Cowan, *Sci. American*, July, 1976; p. 36.
36.2. "Neutron Activation Analysis," W. H. Wahl and H. H. Kramer, *Sci. American*, April, 1967; p. 68.
36.3. "The Synthetic Elements," G. T. Seaborg and J. L. Bloom, *Sci. American*, April, 1969; p. 56.
36.4. "Fusion Power," J. W. Landis, *J. Chem. Educ.*, **50**, 659 (1973).

Appendix A

Basic SI Units

Physical Quantity	Name of Unit	Symbol
length	meter	m
mass	kilogram	kg
time	second	s
electric current	ampere	A
temperature	kelvin	K or °K
amount of substance	mole	mol
luminous intensity	candela	cd

Some Physical Constants

Constant	Symbol	Value
Atomic mass unit	amu	1.6605×10^{-24} gram
		931.48 MeV
Avogadro's Number	N	6.0222×10^{23} molecules/mol
Electron mass	m	9.1096×10^{-28} gram
		5.4859×10^{-4} amu
Electron charge	e	1.6022×10^{-19} coulomb
Faraday	F	96487 coulombs/equiv
Gas constant	R	0.08206 liter atm/°K mol
		1.9872 cal/°K mol
Molar volume, ideal gas at STP		22.4136 liters
Neutron mass		1.6749×10^{-24} gram
		1.008665 amu
Proton mass		1.6726×10^{-24} gram
		1.007277 amu
Speed of light	c	2.9979×10^{10} cm/s

Some Relationships Between Units

Unit	Abbreviation	Relationship
Angstrom	Å	$1 \text{ Å} = 10^{-10}$ m $= 10^{-8}$ cm $= 0.1$ nm
Atmosphere	atm	1 atm $= 760$ torr $= 760$ mm of mercury
		$= 29.92$ in of Hg
Calories	cal	1 cal $= 10^{-3}$ kcal $= 4.1840$ joule
Electron volt	eV	1 eV $= 10^{-6}$ meV $= 2.3061 \times 10^{4}$ cal/mole
Kelvin temperature	°K	°K $=$ °C $+ 273.15$

Appendix B

Equilibrium Constants

Acids

Acetic	$HC_2H_3O_2$	$\rightleftarrows$	$H^+ + C_2H_3O_2^-$	1.8×10^{-5}
Benzoic	$HC_7H_5O_2$	$\rightleftarrows$	$H^+ + C_7H_5O_2^-$	6.4×10^{-5}
Bromoacetic	$HC_2H_2O_2Br$	$\rightleftarrows$	$H^+ + C_2H_2O_2Br^-$	2.0×10^{-3}
Butyric	$HC_4H_7O_2$	$\rightleftarrows$	$H^+ + C_4H_7O_2^-$	1.5×10^{-5}
Carbonic	H_2CO_3	$\rightleftarrows$	$H^+ + HCO_3^-$	$K_1 = 4.3 \times 10^{-7}$
	HCO_3^-	$\rightleftarrows$	$H^+ + CO_3^{2-}$	$K_2 = 5.6 \times 10^{-11}$
Chloroacetic	$HC_2H_2O_2Cl$	$\rightleftarrows$	$H^+ + C_2H_2O_2Cl^-$	1.4×10^{-3}
Chlorobenzoic	$HC_7H_4O_2Cl$	$\rightleftarrows$	$H^+ + C_7H_4O_2Cl^-$	1.5×10^{-4}
Chlorous	$HClO_2$	$\rightleftarrows$	$H^+ + ClO_2^-$	1.1×10^{-2}
Formic	$HCHO_2$	$\rightleftarrows$	$H^+ + CHO_2^-$	1.6×10^{-4}
Hydrocyanic	HCN	$\rightleftarrows$	$H^+ + CN^-$	4.0×10^{-10}
Hydrofluoric	HF	$\rightleftarrows$	$H^+ + F^-$	6.7×10^{-4}
Hydrosulfuric	H_2S	$\rightleftarrows$	$H^+ + HS^-$	$K_1 = 1.1 \times 10^{-7}$
	HS^-	$\rightleftarrows$	$H^+ + S^{2-}$	$K_2 = 1.0 \times 10^{-14}$
Hypobromous	$HBrO$	$\rightleftarrows$	$H^+ + BrO^-$	2.1×10^{-9}
Hypochlorous	$HClO$	$\rightleftarrows$	$H^+ + ClO^-$	3.2×10^{-8}
Hypoiodous	HIO	$\rightleftarrows$	$H^+ + IO^-$	2.3×10^{-11}
Iodic	HIO_3	$\rightleftarrows$	$H^+ + IO_3^-$	1.9×10^{-1}
Lactic	$HC_3H_5O_3$	$\rightleftarrows$	$H^+ + C_3H_5O_3^-$	8.4×10^{-4}
Nitrous	HNO_2	$\rightleftarrows$	$H^+ + NO_2^-$	4.5×10^{-4}
Oxalic	$H_2C_2O_4$	$\rightleftarrows$	$H^+ + HC_2O_4^-$	$K_1 = 5.9 \times 10^{-2}$
	$HC_2O_4^-$	$\rightleftarrows$	$H^+ + C_2O_4^{2-}$	$K_2 = 6.4 \times 10^{-5}$
Phosphoric	H_3PO_4	$\rightleftarrows$	$H^+ + H_2PO_4^-$	$K_1 = 7.5 \times 10^{-3}$
	$H_2PO_4^-$	$\rightleftarrows$	$H^+ + HPO_4^{2-}$	$K_2 = 6.2 \times 10^{-8}$
	HPO_4^{2-}	$\rightleftarrows$	$H^+ + PO_4^{3-}$	$K_3 = 1 \times 10^{-12}$
Trichloroacetic	$HC_2Cl_3O_2$	$\rightleftarrows$	$H^+ + C_2Cl_3O_2^-$	2×10^{-1}

Bases

Ammonia	$NH_3 + H_2O$	$\rightleftarrows$	$NH_4^+ + OH^-$	1.8×10^{-5}
Aniline	$C_6H_5NH_2 + H_2O$	$\rightleftarrows$	$C_6H_5NH_3^+ + OH^-$	4.6×10^{-10}
Dimethylamine	$(CH_3)_2NH + H_2O$	$\rightleftarrows$	$(CH_3)_2NH_2^+ + OH^-$	7.4×10^{-4}
Hydrazine	$N_2H_4 + H_2O$	$\rightleftarrows$	$N_2H_5^+ + OH^-$	9.8×10^{-7}
Hydroxylamine	$NH_2OH + H_2O$	$\rightleftarrows$	$NH_2OH_2^+ + OH^-$	1.1×10^{-8}
Methylamine	$CH_3NH_2 + H_2O$	$\rightleftarrows$	$CH_3NH_3^+ + OH^-$	4.4×10^{-4}
Pyridine	$C_5H_5N + H_2O$	$\rightleftarrows$	$C_5H_5NH^+ + OH^-$	1.5×10^{-9}
Quinoline	$C_9H_7N + H_2O$	$\rightleftarrows$	$C_9H_7NH^+ + OH^-$	6.3×10^{-10}
Trimethylamine	$(CH_3)_3N + H_2O$	$\rightleftarrows$	$(CH_3)_3NH^+ + OH^-$	7.4×10^{-5}

Solubility Products

$AgBr$	4.0×10^{-13}	$Ca_3(PO_4)_2$	1.3×10^{-32}
$AgCl$	1.1×10^{-10}	CuS	8×10^{-37}
AgI	8.5×10^{-17}	$Fe(OH)_3$	6×10^{-38}
Ag_2CrO_4	1.9×10^{-12}	FeS	4×10^{-19}
$BaCO_3$	1.6×10^{-9}	HgS	1.6×10^{-54}
$BaCrO_4$	8.5×10^{-11}	MgF_2	8×10^{-8}
$BaSO_4$	1.0×10^{-10}	NiS	3×10^{-21}
Bi_2S_3	1.6×10^{-72}	$PbCrO_4$	2×10^{-16}
$CaCO_3$	4.7×10^{-9}	$PbCl_2$	1.6×10^{-5}
CaC_2O_4	1.3×10^{-9}	PbS	7×10^{-29}
CaF_2	3.9×10^{-11}	$SrSO_4$	7.6×10^{-7}
CdS	1.0×10^{-28}	ZnS	2.5×10^{-22}

Appendix C

Standard Electrode (Reduction) Potentials

Half Reaction		$E°$, volts
$Li^+ + e^-$	$\rightleftarrows Li$	-3.05
$K^+ + e^-$	$\rightleftarrows K$	-2.95
$Ba^{2+} + 2\,e^-$	$\rightleftarrows Ba$	-2.90
$Ca^{2+} + 2\,e^-$	$\rightleftarrows Ca$	-2.87
$Na^+ + e^-$	$\rightleftarrows Na$	-2.71
$Mg^{2+} + 2\,e^-$	$\rightleftarrows Mg$	-2.37
$Al^{3+} + 3\,e^-$	$\rightleftarrows Al$	-1.66
$2\,H_2O + 2\,e^-$	$\rightleftarrows H_2 + 2\,OH^-$	-0.83
$Zn^{2+} + 2\,e^-$	$\rightleftarrows Zn$	-0.76
$Cr^{3+} + 3\,e^-$	$\rightleftarrows Cr$	-0.74
$Fe^{2+} + 2\,e^-$	$\rightleftarrows Fe$	-0.44
$Cd^{2+} + 2\,e^-$	$\rightleftarrows Cd$	-0.40
$PbSO_4 + 2\,e^-$	$\rightleftarrows Pb + SO_4^{2-}$	-0.36
$Co^{2+} + 2\,e^-$	$\rightleftarrows Co$	-0.28
$Ni^{2+} + 2\,e^-$	$\rightleftarrows Ni$	-0.25
$Sn^{2+} + 2\,e^-$	$\rightleftarrows Sn$	-0.14
$Pb^{2+} + 2\,e^-$	$\rightleftarrows Pb$	-0.13
$2\,H^+ + 2\,e^-$	$\rightleftarrows H_2$	0.00
$S + 2\,H^+ + 2\,e^-$	$\rightleftarrows H_2S$	$+0.14$
$Sn^{4+} + 2\,e^-$	$\rightleftarrows Sn^{2+}$	$+0.15$
$AgCl + e^-$	$\rightleftarrows Ag + Cl^-$	$+0.22$
$Cu^{2+} + 2\,e^-$	$\rightleftarrows Cu$	$+0.34$
$I_2 + 2\,e^-$	$\rightleftarrows 2\,I^-$	$+0.54$
$O_2 + 2\,H^+ + 2\,e^-$	$\rightleftarrows H_2O_2$	$+0.68$
$Fe^{3+} + e^-$	$\rightleftarrows Fe^{2+}$	$+0.77$
$Ag^+ + e^-$	$\rightleftarrows Ag$	$+0.80$
$Hg^{2+} + 2\,e^-$	$\rightleftarrows Hg$	$+0.85$
$NO_3^- + 4\,H^+ + 3\,e^-$	$\rightleftarrows NO + 2\,H_2O$	$+0.96$
$Br_2 + 2\,e^-$	$\rightleftarrows 2\,Br^-$	$+1.07$
$O_2 + 4\,H^+ + 4\,e^-$	$\rightleftarrows 2\,H_2O$	$+1.23$
$Cr_2O_7^{2-} + 14\,H^+ + 6\,e^-$	$\rightleftarrows 2\,Cr^{3+} + 7\,H_2O$	$+1.33$
$Cl_2 + 2\,e^-$	$\rightleftarrows 2\,Cl^-$	$+1.36$
$Au^{3+} + 3\,e^-$	$\rightleftarrows Au$	$+1.50$
$MnO_4^- + 8\,H^+ + 5\,e^-$	$\rightleftarrows Mn^{2+} + 4\,H_2O$	$+1.51$
$Ce^{4+} + e^-$	$\rightleftarrows Ce^{3+}$	$+1.61$
$PbO_2 + SO_4^{2-} + 4\,H^+ + 2\,e^-$	$\rightleftarrows PbSO_4 + 2\,H_2O$	$+1.68$
$H_2O_2 + 2\,H^+ + 2\,e^-$	$\rightleftarrows 2\,H_2O$	$+1.78$
$F_2 + 2\,e^-$	$\rightleftarrows 2\,F^-$	$+2.87$

Appendix D

Table of Logarithms

	0.0	0.1	0.2	0.3	0.4	0.5	0.6	0.7	0.8	0.9
1	000	041	079	114	146	176	204	230	255	279
2	301	322	342	362	380	398	415	431	447	462
3	477	491	505	519	532	544	556	568	580	591
4	602	613	623	634	644	653	663	672	681	690
5	699	708	716	724	732	740	748	756	763	771
6	778	785	792	799	806	813	820	826	833	839
7	845	851	857	863	869	875	881	887	892	898
8	903	909	914	919	924	929	935	940	945	949
9	954	959	964	969	973	978	982	987	991	996

Appendix E

Answers to

Numerical Problems

CHAPTER 1

1.3 (a) 160.80 g
(b) 170 g
(c) 4.410 g
(d) 2048.5 g
1.4 (a) 203 m
(b) 2.0 m
(c) 5000 m
1.5 (a) 7.23 l
(b) 9.20 l
(c) 10.68 l
1.6 0.0543 l
5.43 cl
5.43×10^{-5} kl
1.7 23 mm
23000 μ
23000000 nm
1.8 32768000 cm³

1.9 2,800,000 cm³
1.10 10Å
1.11 10^5 cm²
1.12 (a) 5
(b) 2
(c) 4
(d) 3
(e) 4
(f) 9
(g) 1, 2, or 3
1.13 (a) 169.3
(b) 0.67
1.14 85.2 cm³
1.15

	°C	°K	°F
(a)	0	273	32
(b)	−273	0	−459
(c).	−17.8	255	0

(d)	25	298	77
(e)	−23	250	−9.4
(f)	−29	244	−20
(g)	−20	253	−4.0
(h)	−98	175	−144
(i)	1927	2200	3500

1.16 37°C
1.17 −452°F
1.18 4.8×10^5 cal
1.19 9.0×10^3 kcal
1.20 22.5 g/cm³
1.21 0.68 g/cm³
1.22 480 g
1.23 35.1 ml
1.24 0.49 g/cm³
1.25 164 ml

CHAPTER 2

2.4 197.4
94.2
180.2
77.1
2.5 (a) 5.68
(b) 2.50

(c) 2.55
(d) 7.80
2.6 (a) 10 moles of C
(b) 45 moles of C
(c) 15
(d) 90

2.7 (a) 5.87 moles of N
(b) 2.01 moles of N
(c) 1.59
(d) 3.33
2.8 (a) 98 amu
(b) 31 g

(c) 98 g
(d) 16 g
(e) 98 g
(f) 5.0 moles
(g) 2.0 g

(h) 8
2.9 (a) 2000 g
(b) 5000 g
(c) 193 cm^3, 390 g
2.17 4.29 × 10^{14}/sec = ν

2.84 × 10^{-12} erg = E
7.50 × 10^{14}/sec = ν
4.97 × 10^{-12} erg = E
2.19 (a) 2.92 × 10^{15}/sec
(b) 1,030Å

CHAPTER 3

3.16 98.9% ^{12}C
1.1% ^{13}C
3.17 69.7
3.18 (a) 79.906

(b) 107.868
(c) 20.171
(d) 72.630
(e) 24.313

3.19 60.28% ^{69}Ga
39.72% ^{71}Ga
3.20 69.5% ^{63}Cu
30.5% ^{65}Cu

CHAPTER 5

5.6 (a) 3.32 Å
(b) 1.81 × 10^{-3} Å

5.25 (a) $\Delta H = -118$ kcal
or 59 kcal/mole
(b) $\Delta H = -216$ kcal
(c) $\Delta H = -16$ kcal
(d) $\Delta H = -37$ kcal

CHAPTER 6

6.2 (a) 6C, 6H, 1 oxygen
(b) 94g
(c) 6 moles C, 6 moles H,
1 mole O
(d) 5 moles
(e) 30 moles C, 30
moles H, 5 moles O
(f) 0.04 mole phenol
6.3 (a) NH$_4$Br 1.02 mol
CaCO$_3$ 1.00 mol
H$_3$PO$_4$ 1.02 mol
ZnS 1.03 mol
Br$_2$ 0.626 mol
Ba(ClO$_4$)$_2$ 0.297 mol
(b) NH$_4$Br 490 g
CaCO$_3$ 500 g
H$_3$PO$_4$ 490 g
ZnS 487 g
Br$_2$ 799 g
Ba(ClO$_4$)$_2$ 1680 g

6.4 (a) Na 27.05%
N 16.48%
O 56.57%
(b) C 76.57%
H 6.43%
O 17.00%
(c) K 26.58%
Cr 35.35%
O 38.07%
(d) C 31.97%
H 6.71%
S 42.67%
N 18.64%
6.5 (a) 30.27%
(b) 26.92%
(c) 31.35%
6.6 (a) 82.24%
(b) 46.65%
(c) 21.20%
(d) 28.18%

6.7 (a) C$_5$H$_7$N
(b) C$_{10}$H$_{14}$N$_2$
6.8 A Li$_2$CO$_3$
B C$_4$H$_7$N
C Rb$_2$SO$_4$
D K$_2$Cr$_2$O$_7$
E C$_3$H$_8$O
F K$_2$S$_2$O$_7$
6.9 B C$_4$H$_7$N
E C$_6$H$_{16}$O$_2$
6.10 (a) As$_2$O$_5$
(b) As$_4$O$_{10}$
6.11 S$_2$F$_5$O$_2$
6.12 (a) S$_6$Cl$_2$
(b) As$_4$S$_4$
(c) P$_4$S$_7$
(d) C$_3$H$_6$O$_3$
(e) B$_6$H$_{12}$
(f) B$_{20}$H$_{16}$
(g) S$_4$N$_4$F$_4$
(h) Te$_2$F$_{10}$

6.13 CHO
6.30 (a) 0.5 mol
      (b) 98 g
      (c) 10 mol
      (d) 436 g
6.31 63.0 g NH_3
6.32 58.8 mol NO

6.33 (a) 2.13 mol H_3PO_4
      (b) 108 g H_2O
      (c) 759 g H_3PO_4
6.34 (a) 1.1 mol
      (b) 0.75 mole each
      (c) 36 g
      (d) 128 g
6.35 (a) 20 mol

      (b) 640 g
      (c) 0.6 mol
      (d) 367 g
      (e) 198 g
6.36 489.9 g H_2SO_4
6.37 (a) 8.34 g NaN
      (b) 65.1%

CHAPTER 8

8.19 (a) 1.5 mol
      (b) 288 g O_2
      (c) 368 g
      (d) any amount

8.20 $KClO_3$
8.21 394 kg LiH

CHAPTER 9

9.2   (b) 1.05 atm
9.3   10 atm
9.4   2,630 ml
9.5   269 l
9.6   (a) 6.16 atm
      (b) 6060°K
           5787°C
9.7   −113°C
9.8   6.97 l
9.9   18.4 atm
9.10 179 ml
9.11 322 ml
9.12 46.1 g/mol
9.13 50 atm
9.14 79.0
9.15 130
9.16 35 atm
9.17 OsO_4

9.18 9.8 atm
9.19 (a) 0.59 He
           0.35 N_2
           0.059 Ar
      (b) 5.9 atm
9.20 (a) 0.450
      (b) 6.75 atm
9.21 P_{O_2} = 373 torr
      P_{N_2} = 427 torr
9.22 1.9 ml/sec
9.23 144
9.24 54.0
9.26 13.4 l PH_3
9.27 12.2 g $KClO_3$
9.28 24.6 l
9.29 13.7 l H_2
9.30 (a) 26.7 l NO
      (b) 2.38 mol HNO_3

9.31 (a) 50 l O_2
      (b) 75 l products
9.32 3×10^{21} molecules
9.33 232 g SiF_4
9.34 (a) 5.85×10^{-22} g
      (b) 5.6 l
9.35 (a) 0.12
      (b) 2.8 l
      (c) 7.5×10^{22}
           molecules
9.36 (a) 2.5×10^{-3} mol
      (b) 1×10^{-2} g
      (c) 1.5×10^{21}
           molecules
9.39 (c) 0.71

CHAPTER 10

10.5   12 kcal
10.6   1,440,000 cal
10.7   13 kcal

10.8   65.7 kcal
10.18 $9.44
10.19 $CuSO_4 \cdot 5 H_2O$

10.20 $ZnSO_4 \cdot 7 H_2O$
10.22 (a) 7.0 atm
       (b) 1.0 atm

CHAPTER 11

11.10 (a) $7.95 \times 10^{-23}\,cm^3$
(b) $7.64 \times 10^{-23}\,g$
(c) $0.961\,g/cm^3$
(d) $7.1\,\text{Å}$
(e) $54.2\,g/mole$
At. wt. $= 54.2$

11.11 (a) $3.54\,\text{Å}$
(b) $1.04 \times 10^{-22}\,cm^3$
(c) $3.95 \times 10^{-22}\,g$
$9.88 \times 10^{-23}\,g$
(d) 3

11.12 (a) $9.61 \times 10^{-23}\,cm^3$
(b) $28.9\,cm^3/mole$
(c) $5.25\,g/cm^3$

11.13 (a) $4.10\,\text{Å}$
(b) 7.17×10^{-22}
$g/unit\ cell$
(c) $10.4\,g/cm^3$

11.14 $5.52\,\text{Å}$

11.15 (1) $3.524\,\text{Å}$

11.16 $6.99\,g/cm^3$

11.17 2 atoms/unit cell

11.18 195

11.19 $3.57\,cm$

11.20 (a) $5.73\,\text{Å}$
(b) $1.43\,\text{Å}$

11.21 (a) $66.4\,\text{Å}^3$
(b) $49.0\,\text{Å}^3$
(c) $17.4\,\text{Å}^3$

11.22 $2.8 \times 10^{-8}\,cm$

11.23 $2.04\,\text{Å}$

11.24 $1.94\,\text{Å}$

11.25 $6°\ 44'; 13°\ 32'$

11.28 128

11.29 $0.031\,cal/g\,°C$

CHAPTER 12

12.1 $5609\,cal/g$
12.2 $1.35 \times 10^3\,kcal$
12.3 $27.30\,°C$
12.4 (a) $-369.7\,kcal/mole$ C_2H_6
(b) $-334.4\,kcal$
(c) $781\,kcal/mole$ C_2H_6
12.5 (a) $-204.9\,kcal$
(b) $171\,kcal/mole$
(c) $-337.1\,kcal$
12.6 (a) $-34.0\,kcal$
(b) $-47.3\,kcal$
(c) $17.7\,kcal$
12.7 (a) $-10.7\,kcal$
(b) $8.2\,kcal$
(c) $-4.0\,kcal$

12.8 $-6.1\,kcal$
12.9 $-94.5\,kcal$
12.10 $86.4\,kcal$ or $21.6\,kcal/mole\ NO$
12.11 $12.1\,kcal/mole\ N_2H_4$
12.12 (a) $-276.8\,kcal$
(b) $-133.1\,kcal$
12.13 $-330.8\,kcal$
12.14 (a) $-27.0\,kcal$
(b) $\Delta E° = 26.4\,kcal$
12.15 $2.42\,cal/mole\text{-}deg$
12.16 $5.27\,cal/mole\text{-}deg$
12.18 $47.0\,cal/mole\text{-}deg$
12.19 (a) $-78.0\,cal/mole\text{-}$ deg
(b) $-113.4\,kcal$

12.20 (a) $7.47\,kcal/mole$
(b) $6.80\,kcal/mole$
(c) $22.3\,kcal/°K$
(d) 0
12.21 (a) $-135.4\,kcal$
(b) $-123.0\,kcal$
(c) $7.6\,cal/°K$
12.23 $+88.0\,kcal/mole$
12.24 (a) $+48.2\,kcal$
(b) $+22.2\,kcal$
(c) $+87.2\,cal/°K$
12.25 (a) $-5500\,cal/mole$
12.26 $-9.2\,kcal$
12.27 $+36.1\,cal/°K\ mole$

CHAPTER 13

13.7 (a) NaCl, $10.2\,g$
(b) NaCl, $13\,g$
(c) $19\,g\ NaCl$, $34.5\,g\ KNO_3$
13.8 (a) $25°$
(b) $22°$
(c) below $22°$
(d) $25\,g\ NaNO_3$ $0.9\,g\ NaCl$

(g) NaCl, $4.0\,g$
KCl, $26\,g$
KNO_3, $228\,g$
13.9 $210\,g$ alcohol
13.10 (a) $10\,g\ NaCl/90\,g\ H_2O$
(b) 3 moles NaCl/liter of solution
(c) 0.1 equiv NaCl/liter of solution
(d) 2.5 moles NaCl/kg H_2O
(e) 0.3 mole NaCl/0.7 mole H_2O

13.11		Molarity	Normality
	(a)	3.00	3.00
	(b)	0.30	0.60
	(c)	6.0	6.0
	(d)	0.05	0.15
	(e)	0.25	0.50
	(f)	0.29	1.7

13.12 (a) 60 g
    (b) 3 g
    (c) 1.7 g
    (d) 22 g Na_2CO_3
    (e) 0.85 g $AgNO_3$
    (f) 0.901 g

13.13 (a) 0.025 M
    (b) 0.100 N

13.14 23.8 ml

13.15 0.30

13.16 0.1431

13.17 0.1229 equiv/l

13.18 83.0 g

13.19 (a) 0.667
    (b) 4.00%

13.20 69.6%

13.21 (a) 53.0
    (b) 4.64%

13.22 (a) 0.124
    (b) 5.59 g

13.23 0.718

13.24 (a) 30.5%

    (b) 10.0
    (c) 10.0
    (d) 11.0
    (e) 0.165

13.25 (a) 5.00
    (b) 0.0825
    (c) 4.59
    (d) 9.18

13.26 (a) 0.833
    (b) 11.1
    (c) 6.12
    (d) 12.2

13.27 22.6 torr

13.28 (a) 408 g/mole
    (b) 486 torr

13.29 (a) 102.89°C
    (b) 50 torr

13.30 −20.2°C

13.31 (a) 100.14°C = b.p.
      −0.51°C = f.p.

13.32 111 g/mole

13.33 87.4 g/mole

13.34 −98.5°C

13.35 84°C

13.36 (a) 2.29°C
    (b) 13.6°C

13.38 58.8 g/mole

13.41 3 ions/molecule

13.42 −0.359°C

13.43 3.8%

CHAPTER 14

14.4 ½ hour

14.5 (a) 13.3 hrs
    (b) 32.5 mg

14.6 (a) 9 times
    (b) 16 times

14.7 (a) 4 times
    (b) 3 times

14.8 (a) 4
    (b) 2
    (c) no effect
    (d) $3.3 \times 10^2/(mole/l)^2$ −min.

14.9 (a) 2 times
    (b) 4 times
    (c) 4 times

    (d) 9 times

14.11 (b) $0.117/M^2$ −min

14.12 (a) 1/8 of original
    (b) 1/27 of original
    (c) 2× original
    (d) 8× original

14.18 0.86 mole/l

14.19 1.0

14.20 1.7 moles/liter

14.21 0.15

14.22 0.02

14.23 (a) $[H_2] = [I_2] = 1.25$ M
      $[HI] = 7.5$ M
    (b) 0.028

14.24 (a) $[H_2] = [CO_2] = 0.509 M$
    (b) 0.930
14.25 (a) $[HCl] = 0.015 M$
        $[O_2] = 0.018 M$
        $[H_2O] = 0.030 M$
    (b) 889
14.26 1.48×10^{-5} mole/liter

14.27 0.50 mole/liter
14.28 0.43
14.29 0.028
14.30 5.00 mole/liter
14.31 0.242
14.32 0.15

CHAPTER 15

15.2 2.8×10^{-3} mole/liter
15.3 5×10^{-4}
15.4 $[H^+] = [C_7H_5O_2^-] = 5.7 \times 10^{-3} M$
    $[HC_7H_5O_2] = 0.50 M$
15.5 $1.6 \times 10^{-2} M$
15.6 1.3×10^{-5}
15.8 $7.2 \times 10^{-6} M$
15.9 $7.4 \times 10^{-7} M$
15.10 $4.0 \times 10^{-2} M$
15.11 $1.5 \times 10^{-5} M$
15.12 (a) 5.6×10^{-5}
    (b) 3.5×10^{-6} mole/liter
15.13 1.4×10^{-5} mole/liter
15.14 4.8×10^{-6} mole/liter
15.15 pH $= 9.65$
15.16 (a) pH 1.52; pOH 12.48
    (b) pH 3; pOH 11
    (c) pOH 1; pH 13
    (d) pOH 3.4; pH 10.60
15.17 (a) 0.19
    (b) 10.78
    (c) 4.48
    (d) 3.40
15.18 (a) $3.98 \times 10^{-5} M$
    (b) $2.51 \times 10^{-10} M$
    (c) $1.66 \times 10^{-2} M$
    (d) $2.82 \times 10^{-6} M$

15.19

	$[H^+]$	$[OH^-]$	pH	pOH
	10^{-5}	10^{-9}	5	9
	10^{-3}	10^{-11}	3	11
	2×10^{-6}	5×10^{-9}	5.7	8.3
	10^{-12}	10^{-2}	12	2

15.20 (a) 11
    (b) 5.0×10^{-5}

15.21

	$[H^+]$	$[OH^-]$	pH	pOH
	1.0×10^{-7}	1.0×10^{-7}	7.00	7.00
	1.0×10^{-12}	1.0×10^{-2}	12.00	2.00
	7.0×10^{-4}	1.4×10^{-11}	3.15	10.85
	2.5×10^{-2}	4×10^{-13}	1.6	12.4

15.22 (a) 3.64
    (b) 5.14
15.23 $6.8 - 7.5$
15.24 $2.6 \times 10^{-5} M$
15.25 $6.57 - 8.35$
15.26 (a) 12.70
    (b) 12.30
    (c) 10.22
    (d) 3.78
    (e) 1.85
15.27 (a) 3.89
    (b) 8.45
    (c) 12.16
15.29 (a) 3.39×10^{-6}
    (b) 1.4×10^{-15}
    (c) 3.7×10^{-11}
    (d) 1.8×10^{-18}
15.30 (a) 3×10^{-18}
    (b) 2.7×10^{-12}
    (c) 1×10^{-18}
    (d) 4.4×10^{-27}
    (e) 4×10^{-28}
    (f) 8.8×10^{-12}
    (g) 3.8×10^{-8}
    (h) 4.9×10^{-5}
    (i) 7×10^{-8}
    (j) 3×10^{-27}
    (k) 6.3×10^{-8}
    (l) 1.2×10^{-8}

15.31 (a) 6.9×10^{-5} mole/liter
(b) 3.6×10^{-5} mole/liter
(c) 7.5×10^{-2} mole/liter
(d) 2×10^{-4} mole/liter
(e) 1.6×10^{-2} mole/liter
(f) 1.7×10^{-5} mole/liter
(g) 9.2×10^{-9} mole/liter
(h) 6.3×10^{-5} mole/liter
15.32 (a) 8.7×10^{-4} mole/liter
(b) 7.6×10^{-5} mole/liter
(c) 0.087
15.33 1.6×10^{-16}
15.34 1.3×10^{-32}

15.35 1.6×10^{-5} M
15.36 1.4×10^{-7} M
15.37 (b) 4.3×10^{-5} M
15.40 5.6×10^{-10}
15.41 4.2
15.42 8.60
15.43 8.83
15.44 4.15
15.45 (c) 2.5×10^{-5}
(d) 4.5×10^{-12}
15.46 (c) 1.5×10^{-11}
(d) 8.3

CHAPTER 16

16.20	Oxidizing Agent	Reducing Agent
(a)	63	31
(b)	63	8
(c)	80	216
(d)	35.2	49.5
(e)	33.7	22.3
(f)	101	45
(g)	101	32
(h)	21	49
(i)	10.7	95
(j)	101	73
(k)	16	21
(l)	24.3	32.7
(m)	170	149
(n)	60.3	16.5
(o)	31.6	40

16.21 14.69% $CaCO_3$
16.22 (a) 17.1 g H_2S
126.9 g I_2
(b) 0.384 g H_2S
(c) 48.3% Bi_2S_3

CHAPTER 17

17.8 (f)+0.96 volt
17.9 0.056 g
17.10 5.38 g
17.11 0.499
17.12 3.48 l
17.13 3036
17.14 635
17.15 56.2
17.16 8.04
17.17 (b) +0.82 volt

17.18 (a) (2) 0.74 volt
(4) 0.89 volt
(5) 41 kcal
(6) 10^{25}
17.19 (c) (3) +2.80 volts
17.23 (a) (1) 3.17 volts
(3) −146 kcal
(b) (1) 0.473 volt
(3) −21.8 kcal

(c) (1) 2.11 volts
(3) −97.2 kcal
(d) (1) 0.440 volt
(3) −20.3 kcal
17.24 (a) 1.05 volt
17.25 −0.343 volt
17.26 (a) −0.13 volt, No
(b) +0.05 volt, Yes
17.27 (a) 0.72 volt
(b) 0.970 volt

(c) 0.91 volt
(d) 0.857 volt

17.28 (a) $E° + 0.00891$ volt
(b) $E° - 0.00891$ volt

17.29 0.92 volt

17.30 (a) 8.6×10^{-19}
(b) 2.18×10^{50}
(c) 1.65×10^{90}
(d) 1.37×10^{10}

17.31 (a) -0.652 volt
(b) 9.84×10^{-14}

CHAPTER 18

18.17 HCl, 219 g
H$_2$, 3.95 g

18.18 590 g

18.19 (a) 2.3×10^5 kg
(b) 8.1 l

CHAPTER 19

19.3 (a) 0.781 atm
(b) 209 l
(c) 4.84×10^{20}
(d) 4.7×10^{-4} g
(e) 4.5×10^{-5} moles

19.6 (a) 83%
(b) 24%

19.8 (a) 28.8 g air
(b) 108 l He

19.13 28.1 g/mole

19.14 2.8×10^{22} argon atoms

19.15 (b) 0.60
(c) 3.59 g

CHAPTER 20

20.23 (a) 46.6% N
(c) 28.2% N

20.26 23.1 l

20.27 3770 l

20.28 (a) 741 g
(b) 1059 g

20.29

Compound	Cost/ lb N
NH$_3$	$0.06
CaCN$_2$	0.08
NH$_4$NO$_3$	0.10
(NH$_4$)$_2$SO$_4$	0.12
NaNO$_3$	0.12
(NH$_4$)$_3$PO$_4$	0.30

20.30 10.8 kg Sb$_2$S$_3$
11.7 kg KClO$_3$

CHAPTER 21

21.16 (a) 640 g
(b) 2020 g

21.17 440 g

21.18 0.117% S

21.19 194 l

21.20 (a) 0.205 equiv/l
(b) 10.0 g/liter

CHAPTER 22

22.7 1.6×10^4 kg

22.8 192 kg

22.9 1.2×10^4 l

22.10 497 g

22.11 (a) 2.50 moles
(b) 62.8 l
(c) 1.51×10^{24} molecules

22.14 $87

CHAPTER 23

23.15 (a) 23% O
23.21 42 kg $CaCO_3$
45 kg Na_2CO_3
50 kg SiO_2

CHAPTER 24

24.15 (a) 6.02×10^{23}
molecules, 12 in²,
24 in², 9.41×10^{21}
molecules
(b) 32,768 cubes, 192
in², 1.84×10^{19}

CHAPTER 25

25.6 (a) 180°
(b) 70%
(c) 320°
(d) 230°
(e) (2) 40% Pb
25.20 88% Zn
93% Cu
90% Au
25.21 1.2% C

CHAPTER 26

26.16 5.33 l
26.18 (b) 132 g
26.19 577 g
26.20 81.0 g
26.21 (a) 0.440 g
(b) 80% pure
26.22 29.71% LiCl
26.23 1.9×10^5 g
26.24 0.440 g Na_2CO_3
10 H_2O
26.25 30.8% NaOH

CHAPTER 27

27.3 (a) 10.3%, 20.9%
27.13 51.2%
27.14 0.503 g
27.15 0.6 oz Ag
0.06 oz Ag
0.006 oz Ag

CHAPTER 28

28.17 6.0×10^{-9} M
28.18 (a) 1.3×10^{-7} M
(b) 2.6×10^{-7} M

CHAPTER 29

29.15 65.55% Ag
29.16 120 lbs

CHAPTER 30

30.18 11,300 l at STP
30.19 % Fe = 34.9, %
Cr = 65.1; 18.6%
30.20 (a) 0.132 equiv/liter
(b) 0.258 g
(c) 25.8% Fe
30.21 (a) 49.0
(b) 37.7% Fe
30.22 0.633 equiv/liter

CHAPTER 31

31.25 650 l O_2,
3250 l air
31.26 39200 l air

CHAPTER 32

32.26 (a) 44.0 g
(b) 68.2%
32.27 (c) 75 g

CHAPTER 34

34.18 (a) 1 liter CO_2 /liter O_2
(b) 0.7 l

CHAPTER 35

35.7 0.0977 g
35.8 (a) 0.68 g
(b) 0.0169 g
35.9 163 g
35.10 14 days
35.11 91 hrs
35.12 8 days
35.13 (a) 11.6 days
(b) 3.3 mg
35.14 7.2 hr
35.15 (a) 110 min
(b) 1.1×10^{-3} mg
35.16 0.774 g
35.17 (a) 0.267/year
(b) 2.60 years
35.18 174 days
35.19 10^{-502}
35.20 (a) 14,000 years
(b) 2,400 years
35.21 2,900 years
35.22 370

CHAPTER 36

36.17 28.8 MeV
36.18 92.1 MeV
36.19 9.40 MeV
36.20 7.59 MeV/nucleon
36.21 (a) 180 MeV
(b) 0.0818%
36.22 (a) 5.49 MeV
(b) 0.20%